PRINCIPLES OF

Environmental Science

Inquiry & Application

Seventh Edition

William P. Cunningham
University of Minnesota

Mary Ann Cunningham
Vassar College

McGraw Hill

Connect
Learn
Succeed™

PRINCIPLES OF ENVIRONMENTAL SCIENCE: INQUIRY & APPLICATIONS, SEVENTH EDITION

Published by McGraw-Hill, a business unit of The McGraw-Hill Companies, Inc., 1221 Avenue of the Americas, New York, NY 10020. Copyright © 2013 by The McGraw-Hill Companies, Inc. All rights reserved. Printed in the United States of America. Previous editions © 2011, 2009, and 2008. No part of this publication may be reproduced or distributed in any form or by any means, or stored in a database or retrieval system, without the prior written consent of The McGraw-Hill Companies, Inc., including, but not limited to, in any network or other electronic storage or transmission, or broadcast for distance learning.

Some ancillaries, including electronic and print components, may not be available to customers outside the United States.

This book is printed on acid-free paper.

2 3 4 5 6 7 8 9 0 DOW/DOW 1 0 9 8 7 6 5 4

ISBN 978–0–07–353251–6
MHID 0–07–353251–7

Senior Vice President, Products & Markets: *Kurt L. Strand*
Vice President, General Manager, Products & Markets: *Marty Lange*
Vice President, Content Production & Technology Services: *Kimberly Meriwether David*
Director of Development: *Rose Koos*
Managing Director: *Thomas Timp*
Brand Manager: *Michelle Vogler*
Development Editor: *Jodi Rhomberg*
Director of Digital Content: *Andrea M. Pellerito, Ph.D.*
Associate Marketing Manager: *Matthew Garcia*
Project Managers: *Kelly A. Heinrichs / Katie L. Fuller*
Senior Buyer: *Laura Fuller*
Design Manager: *Michelle D. Whitaker*
Interior Design: *ArtPlus Ltd.*
Cover Design: *Trevor Goodman*
Cover Image: *Tree: © plus photo/a.collectionRF/Getty Images; Grass/soil: © Jim Richardson/Getty Images; Wires: © Renold Zergat/Getty Images; Roots: © Designpics.com/PunchStock*
Content Licensing Specialist: *Lori Hancock*
Photo Research: *LouAnn Wilson*
Compositor: *ArtPlus Ltd.*
Typeface: *10/12 Times LT*
Printer: *R. R. Donnelley*

All credits appearing on page or at the end of the book are considered to be an extension of the copyright page.

Library of Congress Cataloging-in-Publication Data

Cunningham, William P.
    Principles of environmental science : inquiry & applications / William P. Cunningham, Mary Ann Cunningham. — 7th ed.
        p. cm.
    Includes index.
    ISBN 978–0–07–353251–6 — ISBN 0–07–353251–7 (hard copy : alk. paper) 1. Environmental sciences–Textbooks.
I. Cunningham, Mary Ann. II. Cunningham, Mary Ann. III. Title.
    GE105.C865 2013
    363.7–dc23   2012016055

The Internet addresses listed in the text were accurate at the time of publication. The inclusion of a website does not indicate an endorsement by the authors or McGraw-Hill, and McGraw-Hill does not guarantee the accuracy of the information presented at these sites.

www.mhhe.com

About the Authors

WILLIAM P. CUNNINGHAM

William P. Cunningham is an emeritus professor at the University of Minnesota. In his 38-year career at the university, he taught a variety of biology courses, including Environmental Science, Conservation Biology, Environmental Health, Environmental Ethics, Plant Physiology, General Biology, and Cell Biology. He is a member of the Academy of Distinguished Teachers, the highest teaching award granted at the University of Minnesota. He was a member of a number of interdisciplinary programs for international students, teachers, and nontraditional students. He also carried out research or taught in Sweden, Norway, Brazil, New Zealand, China, and Indonesia.

Professor Cunningham has participated in a number of governmental and nongovernmental organizations over the past 40 years. He was chair of the Minnesota chapter of the Sierra Club, a member of the Sierra Club national committee on energy policy, vice president of the Friends of the Boundary Waters Canoe Area, chair of the Minnesota governor's task force on energy policy, and a citizen member of the Minnesota Legislative Commission on Energy.

In addition to environmental science textbooks, Cunningham edited three editions of an *Environmental Encyclopedia* published by Thompson-Gale Press. He has also authored or co-authored about 50 scientific articles, mostly in the fields of cell biology and conservation biology as well as several invited chapters or reports in the areas of energy policy and environmental health. His Ph.D. from the University of Texas was in botany.

His hobbies include backpacking, canoe and kayak building (and paddling), birding, hiking, gardening, and traveling. He lives in St. Paul, Minnesota, with his wife, Mary. He has three children (one of whom is co-author of this book) and seven grandchildren.

MARY ANN CUNNINGHAM

Mary Ann Cunningham is an associate professor of geography at Vassar College, in New York's Hudson Valley. A biogeographer with interests in landscape ecology, geographic information systems (GIS), and remote sensing, she teaches environmental science, natural resource conservation, and land-use planning, as well as GIS and remote sensing. Field research methods, statistical methods, and scientific methods in data analysis are regular components of her teaching. As a scientist and educator, Mary Ann enjoys teaching and conducting research with both science students and non-science liberal arts students. As a geographer, she likes to engage students with the ways their physical surroundings and social context shape their world experience. In addition to teaching at a liberal arts college, she has taught at community colleges and research universities.

Mary Ann has been writing in environmental science for over a decade, and she has been co-author of this book since its first edition. She is also co-author of *Environmental Science* (now in its eleventh edition), and an editor of the *Environmental Encyclopedia* (third edition, Thompson-Gale Press). She has published work on pedagogy in cartography, as well as instructional and testing materials in environmental science. With colleagues at Vassar, she has published a GIS lab manual, *Exploring Environmental Science with GIS*, designed to provide students with an easy, inexpensive introduction to spatial and environmental analysis with GIS.

In addition to environmental science, Mary Ann's primary research activities focus on land-cover change, habitat fragmentation, and distributions of bird populations. This work allows her to conduct field studies in the grasslands of the Great Plains as well as in the woodlands of the Hudson Valley. In her spare time she loves to travel, hike, and watch birds.

Mary Ann holds a bachelor's degree from Carleton College, a master's degree from the University of Oregon, and a Ph.D. from the University of Minnesota.

Brief Contents

Contents

3

Evolution, Species Interactions, and Biological Communities   50
LEARNING OUTCOMES   50

4

Human Populations   76
LEARNING OUTCOMES   76

5

Biomes and Biodiversity   96
LEARNING OUTCOMES   96

Environmental Conservation: Forests, Grasslands, Parks, and Nature Preserves

LEARNING OUTCOMES

Food and Agriculture

LEARNING OUTCOMES

Environmental Health and Toxicology   180
LEARNING OUTCOMES   180

Climate   205
LEARNING OUTCOMES   205

Air Pollution 229

Water: Resources and Pollution 250

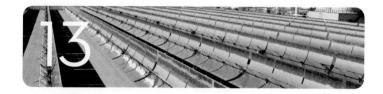

List of Case Studies

Over 200 additional Case Studies can be found online on the instructor's
resource page at www.mcgrawhillconnect.com.

Preface

THINKING TOWARD THE FUTURE

Environmental science is a study of how we use, abuse, and steward the ecological systems that support us. What are these systems, how do they function, and can they be used sustainably? Is it possible to live sustainably on this earth? Critical thinking and careful evaluation of evidence are essential for exploring these complex questions, in the face of often-contradictory evidence. Above all, students today are challenged to rethink old paradigms, assumptions, and limits, as we work to invent a new, more sustainable future for generations that follow us.

One classic parable for the possibility of sustainable living is the "ecocide" of Easter Island. This remote Pacific island, also called Rapa Nui, represents an oft-repeated morality tale of environmental destruction. The classic story is that early Polynesian colonists cleared the forest, eroded the soil, overpopulated the island, depleted the land through greed and ignorance, and finally dwindled to a war-torn, impoverished remnant of their former society. This narrative has informed environmental thought for a generation or more, but it may no longer be enough to guide us to a sustainable future.

An alternative narrative that better fits recent data is that early colonists prospered for centuries, inventing ways to farm the land sustainably, to replenish soil and conserve water, and restrain population growth. Although they felled the forests and cleared fields over the centuries, people still adapted and thrived—until the arrival of European diseases and slave traders, both of which decimated the population and reduced them to dwindling and impoverished numbers.[1]

These two narratives have contrasting morals: Do humans inevitably destroy their own homes? Or do we have the potential to invent strategies for sustaining our resources? Are the main drivers of environmental destruction overpopulation, habitat destruction, greed, injustice among societies, or some combination of these? To find better answers, today's students may need to move beyond the traditional story and embrace the idea that cooperation and creativity can lead to a sustainable future, if we put our hearts into the project.

Crises and opportunities abound

Understanding the nature of environmental destruction is necessary, but so is a commitment to progress and cooperation. We face challenges like never before—burgeoning cities, warming climates, looming water crises. We also have unprecedented opportunity to make a difference—we are seeing global expansion in access to education, healthcare, information, even political participation and human rights. Birthrates are falling almost everywhere, as women's rights gradually improve. Creative individuals are inventing new ideas for alternative energy and transportation systems that were undreamed of a generation ago. We are rethinking our assumptions about cities, food production, water use, and air quality. Local action is rewriting our expectations, and even economic and political powers feel increasingly compelled to show cooperation in improving environmental quality.

Students are leading the way in reimagining our possible futures. Student movements have led innovation in technology and science, in environmental governance, and in environmental justice around the world. The organization 350.org, highlighted in chapter 16, was started by a small group of students to address climate change. That movement has energized local communities to join the public debate on how to seek a sustainable future.

Sustainability, a central idea in this book, has grown from a fringe notion two decades ago to a widely shared framework for daily actions (recycling, reducing consumption) and civic planning (building energy-efficient buildings, investing in public transit and bicycle routes). Sustainability isn't just about the environment anymore. Increasingly we know that sustainability is also smart economics and that it is essential for social equity. Energy efficiency saves money. Alternative energy reduces our reliance on fuel sources in politically unstable regions. Healthier food options reduce medical costs. Smaller families can be happier and more secure. Accounting for the public costs and burdens of pollution and waste disposal helps us rethink the ways we dispose of our garbage and protect public health.

All of these are ideas this book explores. Our aim is to help students understand the ways in which methods and principles of environmental science apply to pressing issues around us. We also hope that this book can help each student find the ways that his or her passions can be engaged—whether they are in biology, math, journalism, politics, artistic expression, psychology, chemistry, or other subjects—in working for a smarter, more stable future.

WHAT SETS THIS BOOK APART?

Solid science: This book reflects decades of experience in the field and in the classroom, which make it up-to-date in approach, in data, and in applications of critical thinking. Emerging ideas and issues are introduced, such as ecosystem services, cooperative ecological relationships, and epigenetics, and the economics of air pollution control, in addition to basic principles such as population biology, the nature of systems, and climate processes.

---

[1] *See* Hunt, T.L., Lipo, C.P. 2009. Revisiting Rapa Nui (Easter Island) "ecocide." *Pacific Science* 63 (4): 601–616.

Demystifying science: We make science accessible by showing how and why data collection is done and by giving examples, practice, and exercises that demonstrate central principles. *Exploring Science* readings empower students by helping them understand how scientists do their work and by asking them to collect and analyze their own data. *Applications* exercises and *Data Analysis* exercises help students practice the ideas, rather than just reading about them.

Quantitative reasoning: Students need to become comfortable with graphs, data, and comparing numbers. We provide focused discussions on why scientists answer questions with numbers, the nature of statistics, of probability, and how to interpret the message in a graph. We give accessible details on population models, GIS (spatial analysis), remote sensing, and other quantitative techniques. In-text *Applications* and online, testable *Data Analysis* questions give students opportunities to gather and evaluate data on their own.

Critical thinking: We provide a focus on critical thinking, one of the most essential skills for citizens, as well as for students. We offer abundant opportunities for students to weigh contrasting evidence and evaluate assumptions and arguments, including *What Do You Think?* readings.

Synthesis: Students come to environmental science from a multitude of fields and interests. We emphasize that most of our pressing problems, from global hunger or climate change to conservation of biodiversity, draw on sciences and economics and policy. This synthesis shows students that they can be engaged in environmental science, no matter what their interests or career path.

A global perspective: Environmental science is a globally interconnected discipline. *Case studies*, data, and examples from around the world give opportunities to examine international questions. Eleven of 16 case studies examine international issues of global importance, such as forest conservation in Indonesia, soy production in Brazil, car-free cities in Germany. Half of all boxed readings and *Key Concepts* are also global. Moreover, *Google Earth* placemarks bring students virtually to locations where they can see and learn the context of the issues they read.

Key concepts: In each chapter this section draws together compelling illustrations and succinct text to create a summary "take-home" message. These key concepts draw together the major ideas, questions, and debates in the chapter but give students a central idea on which to focus. These can also serve as starting points for lectures, student projects, or discussions.

Positive perspective: All the ideas noted here can empower students to do more effective work for the issues they believe in. While we don't shy away from the bad news, we highlight positive ways in which groups and individuals are working to improve their environment. *What Can You Do?* boxes in every chapter offer practical examples of things everyone can do to make worthwhile progress towards sustainability

Thorough coverage: No other book on in the field addresses the multifaceted nature of environmental questions such as climate policy, on sustainability, or population change, with the thoroughness this book has. We cover not just climate change but also the nature of climate and weather systems that influence our climate. We explore both food shortages and the emerging causes of hunger—such as political conflict, biofuels, and global commodity trading—as well as the growing pandemic of obesity-related illness. In these and other examples, this book is a leader in in-depth coverage of key topics.

Active learning: Learning how scientists approach problems can help students develop habits of independent, orderly, and objective thought. But it takes active involvement to master these skills. This book integrates a range of learning aids—active learning exercises, *Critical Thinking and Discussion Questions*, and *Data Analysis* exercises—that push students to think for themselves. Data and interpretations aren't presented as immutable truths, but rather as evidence to be examined and tested, as they should be in the real world.

WHAT'S NEW IN THIS EDITION?

- **Gorgeous new art:** This edition introduces an entirely new art program, with beautiful new drawings that bring key concepts to life. Creative paging also gives the book an accessible, compelling appearance. Dozens of new striking, thought-provoking photographs highlight crucial topics.

- **New chapter on climate change:** Global warming may well be the most difficult and important challenge the world will face in this century. In response to suggestions from reviewers, we have split coverage of climate change from discussion of air pollution. This gives climate change the attention and emphasis it deserves without diminishing important and current topics in air pollution.

- **Up-to-date information:** Throughout the text, we provide the most current available data, as well as recent innovations in meeting environmental challenges. We introduce students to current developments such as establishment of Marine Protected Areas, REDD (reducing emissions through deforestation and degradation), renewable energy development in China, fertility declines in the developing world, and the impact of global food trade on world hunger.

- **New Case Studies:** More than two-thirds of the opening case studies are new to this edition, giving readers current and exciting examples of environmental science in action. These case studies illustrate important principles and demonstrate the importance and interconnections of these issues. Because grim stories can stop students in their tracks, a majority of our case studies offer a positive view of progress towards sustainability and environmental protection. And while we updated the case studies for this edition, all case studies from previous editions can be found on Connect at www.mcgrawhillconnect.com.

- **New Exploring Science readings:** These boxed readings illustrate how science is actually done as well as presenting information about important topics in environmental science. Many of these readings, like the case studies, give encouraging examples of progress toward sustainability.

For example, we have a current example of how technology (including GIS mapping, using mobile devices, such as iPads and smartphones) is being used to protect habitat for endangered chimpanzee populations in Gombe National Park in Tanzania. We also have an inspiring story of how inexpensive water purification systems are being made available to poor, rural villages in India.

- **New What Do Think? readings:** This popular feature invites students to think critically and creatively about current environmental dilemmas. They also serve as a springboard for class or after-class discussions. We have added challenging new topics, including the future of nuclear power following the catastrophe in Japan, and Australia's current adaptations to unprecedented, widespread drought. These important topics encourage students to examine implications of environmental science in their own lives.

- **An exciting new online learning platform:** McGraw-Hill's ConnectPlus (www.mcgrawhillconnect.com)

connect plus+ | ENVIRONMENTAL SCIENCE

is a web-based assignment and assessment platform that gives students the means to better connect with their coursework, with their instructors, and with the important concepts that they will need to know for success now and in the future. Valuable assets such as LearnSmart (an adaptive learning system), an interactive ebook, Data Analysis exercises, the extensive case study library and Google Earth exercises are all available in Connect.

DETAILED CHANGES IN THIS EDITION

- **Chapter 1** has all new art throughout the chapter. It has a new discussion of ecosystem services, including a new fig. 1.8 to illustrate these relationships. Exploring Science boxes on "How do we know the state of population and poverty?" and "Why do scientists answer questions with a number?" have been updated and expanded. Throughout the chapter, the text has been updated and enhanced.

- **Chapter 2** has a new opening case study on ecological connections in the Chesapeake Bay. The new food web diagram in fig. 2.16 now clearly shows connections between organisms in different trophic levels. All drawings of elemental cycles have been revised to a uniform appearance that makes it easier to understand sources, flows, and sinks. Estimates of total stocks and exchanges for key compartments have been added. All art has been upgraded and enhanced.

- **Chapter 3** has a new opening case study on evolution that includes new data from studies of Galápagos finches. Among the new figures are comparisons between specialist niches and generalists (fig. 3.5), speciation, resource partitioning, isolation, taxonomy, and keystone species. The chapter has a new discussion of symbiotic relationships and coevolution,

including Darwin's moth. It also has a new Data Analysis exercise on resource partitioning among Darwin's finches.

- **Chapter 4** has a new opening case study on impressive population stabilization in Brazil. Small drawings or photos are added to make graphs more interesting and understandable. Demographic data is updated throughout chapter. A new figure emphasizes the relationship between environmental impacts, population growth, affluence, and technology. A new map (fig. 4.18) shows fertility rates by country. A new Data Analysis box at the end of the chapter draws on superb demographic graphics from Gapminder.org.

- **Chapter 5** has a new opening case study on a ground-breaking field study that is helping to elucidate the effects of climate change on boreal forests. A new graph shows rates of family losses during major geologic episodes of mass extinction. A new Exploring Science box shows the effects of invasive earthworms in northern forests. New mini case studies of invasive species describe emerald ash borers and Asian carp species (species of major current concern in the U.S.)

- **Chapter 6** has a new opening case study on innovative forest protection agreements between Indonesia and Norway as part of the UN REDD program. The Active Learning box on forest area has been updated and corrected so students can calculate the amount of original forests lost. A new Exploring Science box describes the use of modern technology by native peoples in Amazonian forest protection programs. This theme continues in another new Exploring Science box on how the Jane Goodall Institute is empowering local communities around Gombe National Park in Tanzania to identify conservation priorities.

- **Chapter 7** has updated data and text throughout, including an enhanced opening case study on farming in Brazil. Our discussion of the role of global trade policy and biofuels on food supply has been refined. New data on food supply, hunger and obesity are presented. An improved discussion of global fisheries and policy options, as well as new presentation of agricultural inputs, including rising pesticide use and pesticide tolerance has been added. A new figure (7.28) shows the use of herbicide tolerant genes in genetically modified (GM) crops. And a new Data Analysis exercise on agricultural production statistics and patterns fills out the chapter.

- **Chapter 8** has a new opening case study on the dangers of Bisphenol A, a chemical found in a wide range of consumer products, including food and beverage containers. The coverage of conservation medicine now includes the dire epidemic of "white nose syndrome" that's killing millions of cave-dwelling bats in the eastern U.S. as well as Chytridiomycosis, which is wiping out thousands of amphibian species worldwide.

- **Chapter 9** has a new case study on Texas drought that illustrates the threats of weather extremes and their possible links to global climate change. New and updated figures include climate change data and greenhouse gas releases. The chapter also raises questions about why we debate climate change, and offers options for responding to climate change, including personal actions in a practical "What Can You Do?" box.

- **Chapter 10** is an entirely new chapter that treats air pollution separately from climate and climate processes. The chapter opens with a new case study on the London Smog of 1952, which is followed by details on sulfur dioxide, nitrogen oxides, and other air pollutants. We have added extensive discussion of mercury pollution, including the question of whether cap and trade is the right approach for regulating this neurotoxin. We have also added extensive discussion of halogens and their impacts on climate as well as on stratospheric ozone. A new discussion on the economic impacts of the Clean Air Act shows students why pollution control is important for economic as well as health effects. A new Data Analysis box lets students examine EPA data on air pollution data in their own area.

- **Chapter 11** has a new section "Living in an age of thirst" to expand our discussion of major droughts and to clarify the threats to freshwater supplies posed by climate change. This section includes a new map of world water scarcity, stress, and vulnerability. This theme is continued in a sobering but somewhat optimistic What Do You Think? box explaining how Australia is responding to record drought. A brief description of the megadrought in the 13th century that led to abandonment of Anasazi pueblos in America's four corners area compliments this section. A new, optimistic, Exploring Science box describes the invention and implementation by Professor Ashok Gadgil of low-cost water purification systems for developing countries.

- **Chapter 12** has a new opening case study on the environmental and social destruction caused by mountaintop removal mining. The diagrams of geological subduction and uplift have been redrawn and improved. Photos of rare earth metals are included in a box on these strategic materials along with discussion of new rare earth mines and technological advances that require less of these materials in batteries and motors. New photos have been added of historic mine reclamation, the 2011 tsunami in Japan, the 2010 eruption of Mt. Merapi in Indonesia, and of 2010 floods in Pakistan that displaced 20 million people.

- **Chapter 13** has a new opening case study about Desertech, a massive program to link high-voltage direct-current transmission lines to solar thermal energy facilities in the deserts of North Africa with off-shore wind farms, hydroelectric dams, and other renewable energy sources to provide sustainable energy in Europe. The chapter also describes how dramatic changes in unconventional fossil fuel sources, such as tar sands, shale oil and natural gas formations, and deep-sea deposits are rapidly changing our estimates of potential fossil fuel supplies. We discuss these changes along with new developments in renewable energy systems and the changing fate of nuclear power following the tsunami in Japan. Calls for a supergrid in North America are compared to developments now taking place in Europe.

- **Chapter 14** has a new case study on alternative waste management and biogas use in Kristianstad, Sweden, one of many European cities taking a new approach to handling solid waste. New figures and data have been added throughout the chapter on waste production and waste management. It also has an updated discussion of Superfund cleanup progress and costs.

- **Chapter 15** examines urban blight and emptying out of rust-belt cities along with the environmental and social impacts of urban sprawl. It describes how some of these problems can be solved through urban renewal and a new emphasis on sustainability in urban planning. The economics section of this chapter has a new section on how market mechanisms can be used to achieve environmental and social goals. New photos have been added of a walking street in Vauban, Germany, a village market as example of classical economics, communal irrigation systems in Bali, and renewable energy as example of scarcity-led technical innovation.

- **Chapter 16** has a new opening case study about how students in 350.org are working to combat climate change. A new discussion has been added to the policy section on the precautionary principle. The analysis of national environmental organizations has been streamlined in favor of more coverage of student groups and the ways in which modern electronic communications and social media are changing the world. An added emphasis on sustainability serves as a capstone for this chapter bringing us back to principles that began chapter 1.

ACKNOWLEDGMENTS

We are sincerely grateful to Jodi Rhomberg and Michelle Vogler, who oversaw the development of this edition, and to Kelly Heinrichs and Katie Fuller, who shepherded the project through production.

We would like to thank the following individuals who wrote and/or reviewed learning goal-oriented content for **LearnSmart**.
Broward College, Nilo Marin
Broward College, David Serrano
Northern Arizona University, Sylvester Allred
Palm Beach State College, Jessica Miles
Roane State Community College, Arthur C. Lee
University of North Carolina at Chapel Hill, Trent McDowell
University of Wisconsin, Milwaukee, Gina S. Szablewski

Input from instructors teaching this course is invaluable to the development of each new edition. Our thanks and gratitude go out to the following individuals who either completed detailed chapter reviews of *Cunningham, Principles of Environmental Science*, Seventh Edition, or provided market feedback for this course.
American University, Priti P. Brahma
Antelope Valley College, Zia Nisani
Arizona Western College, Alyssa Haygood
Aurora University, Carrie Milne-Zelman
Baker College, Sandi B. Gardner

Boston University, Kari L. Lavalli
Bowling Green State University, Daniel M. Pavuk
Bradley University, Sherri J. Morris
Broward College, Elena Cainas
Broward College, Nilo Marin
California Energy Commission, James W. Reede
California State University—East Bay, Gary Li
California State University, Natalie Zayas
Carthage College, Tracy B. Gartner
Central Carolina Community College, Scott Byington
Central State University, Omokere E. Odje
Clark College, Kathleen Perillo
Clemson University, Scott Brame
College of DuPage, Shamili Ajgaonkar Sandiford
College of Lake County, Kelly S. Cartwright
College of Southern Nevada, Barry Perlmutter
College of the Desert, Tracy Albrecht
Community College of Baltimore County, Katherine M. Van de Wal
Connecticut College, Jane I. Dawson
Connecticut College, Chad Jones
Connors State College, Stuart H. Woods
Cuesta College, Nancy Jean Mann
Dalton State College, David DesRochers
Dalton State College, Gina M. Kertulis-Tartar
East Tennessee State University, Alan Redmond
Eastern Oklahoma State College, Patricia C. Bolin Ratliff
Edison State College, Cheryl Black
Elgin Community College, Mary O'Sullivan
Erie Community College, Gary Poon
Estrella Mountain Community College, Rachel Smith
Farmingdale State College, Paul R. Kramer
Fashion Institute of Technology, Arthur H. Kopelman
Flagler College, Barbara Blonder
Florida State College at Jacksonville, Catherine Hurlbut
Franklin Pierce University, Susan Rolke
Galveston College, James J. Salazar
Gannon University, Amy L. Buechel
Gardner-Webb University, Emma Sandol Johnson
Gateway Community College, Ramon Esponda
Geneva College, Marjory Tobias
Georgia Perimeter College, M. Carmen Hall
Georgia Perimeter College, Michael L. Denniston
Gila Community College, Joseph Shannon
Golden West College, Tom Hersh
Gulf Coast State College, Kelley Hodges
Gulf Coast State College, Linda Mueller Fitzhugh
Heidelberg University, Susan Carty
Holy Family University, Robert E. Cordero
Houston Community College, Yiyan Bai
Hudson Valley Community College, Janet Wolkenstein
Illinois Mathematics and Science Academy, C. Robyn Fischer

Illinois State University, Christy N. Bazan
Indiana University of Pennsylvania, Holly J. Travis
Indiana Wesleyan University, Stephen D. Conrad
James Madison University, Mary Handley
James Madison University, Wayne S. Teel
John A. Logan College, Julia Schroeder
Kentucky Community & Technical College System—Big Sandy District, John G. Shiber
Lake Land College, Jeff White
Lane College, Satish Mahajan
Lansing Community College, Lu Anne Clark
Lewis University, Jerry H. Kavouras
Lindenwood University, David M. Knotts
Longwood University, Kelsey N. Scheitlin
Louisiana State University, Jill C. Trepanier
Lynchburg College, David Perault
Marshall University, Terry R. Shank
Menlo College, Neil Marshall
Millersville University of Pennsylvania, Angela Cuthbert
Minneapolis Community and Technical College, Robert R. Ruliffson
Minnesota State College—Southeast Technical, Roger Skugrud
Minnesota West Community and Technical College, Ann M. Mills
Mt. San Jacinto College, Shauni Calhoun
Mt. San Jacinto College, Jason Hlebakos
New Jersey City University, Deborah Freile
New Jersey Institute of Technology, Michael P. Bonchonsky
Niagara University, William J. Edwards
North Carolina State University, Robert I. Bruck
North Georgia College & State University, Kelly West
North Greenville University, Jeffrey O. French
Northeast Lakeview College, Diane B. Beechinor
Northeastern University, Jennifer Rivers Cole
Northern Virginia Community College, Jill Caporale
Northwestern College, Dale Gentry
Northwestern Connecticut Community College, Tara Jo Holmberg
Northwood University Midland, Stelian Grigoras
Notre Dame College, Judy Santmire
Oakton Community College, David Arieti
Parkland College, Heidi K. Leuszler
Penn State Beaver, Matthew Grunstra
Philadelphia University, Anne Bower
Pierce College, Thomas Broxson
Purdue University Calumet, Diane Trgovcich-Zacok
Queens University of Charlotte, Greg D. Pillar
Raritan Valley Community College, Jay F. Kelly
Reading Area Community College, Kathy McCann Evans
Rutgers University, Craig Phelps
Saddleback College, Morgan Barrows
Santa Monica College, Dorna S. Sakurai
Shasta College, Morgan Akin
Shasta College, Allison Lee Breedveld

Southeast Kentucky Community and Technical College, Sheila Miracle
Southern Connecticut State University, Scott M. Graves
Southern New Hampshire University, Sue Cooke
Southern New Hampshire University, Michele L. Goldsmith
Southwest Minnesota State University, Emily Deaver
Spartanburg Community College, Jeffrey N. Crisp
Spelman College, Victor Ibeanusi
St. Johns River State College, Christopher J. Farrell
Stonehill College, Susan M. Mooney
Tabor College, Andrew T. Sensenig
Temple College, John McClain
Terra State Community College, Andrew J. Shella
Texas A&M University-Corpus Christi, Alberto M. Mestas-Nuñez
Tusculum College, Kimberly Carter
Univeristy of Nebraska, James R. Brandle
University of Akron, Nicholas D. Frankovits
University of Denver, Shamim Ahsan
University of Kansas, Kathleen R. Nuckolls
University of Miami, Kathleen Sullivan Sealey
University of Missouri at Columbia, Douglas C. Gayou
University of Missouri-Kansas City, James B. Murowchick
University of North Carolina Wilmington, Jack C. Hall
University of North Texas, Samuel Atkinson
University of Tampa, Yasoma Hulathduwa
University of Tennessee, Michael McKinney
University of Utah, Lindsey Christensen Nesbitt
University of Wisconsin-Stevens Point, Holly A Petrillo
University of Wisconsin-Stout, Charles R. Bomar
Valencia College, Patricia Smith
Vance Granville Community College, Joshua Eckenrode
Villanova University, Lisa J. Rodrigues
Virginia Tech, Matthew Eick
Viterbo University, Christopher Iremonger
Waubonsee Community College, Dani DuCharme
Wayne County Community College District, Nina Abubakari
West Chester University of Pennsylvania, Robin C. Leonard
Westminster College, Christine Stracey
Worcester Polytechnic Institute, Theodore C. Crusberg
Wright State University, Sarah Harris

DIGITAL RESOURCES

McGraw-Hill offers various tools and technology products to support *Principles of Environmental Science,* 7th edition.

McGraw-Hill's ConnectPlus™

McGraw-Hill's Connect Plus (www.mcgrawhillconnect.com/environmentalscience) is a web-based assignment and assessment platform that gives students the means to better connect with their coursework, with their instructors, and with the important concepts that they will need to know for success now and in the future. The following resources are available in Connect:

- Auto-graded assessments
- LearnSmart, an adaptive diagnostic tool
- Powerful reporting against learning outcomes and level of difficulty
- McGraw-Hill Tegrity Campus, which digitally records and distributes your lectures with a click of a button
- The full textbook as an integrated, dynamic eBook which you can also assign.
- Data Analysis exercises to improve critical thinking skills
- An extensive Case Studies Library
- Instructor Resources such as an Instructor's Manual, PowerPoints, and Test Banks.
- Image Bank that includes all images available for presentation tools

With ConnectPlus, instructors can deliver assignments, quizzes, and tests online. Instructors can edit existing questions and author entirely new problems; track individual student performance—by question, assignment; or in relation to the class overall—with detailed grade reports; integrate grade reports easily with Learning Management Systems (LMS), such as WebCT and Blackboard; and much more.

By choosing Connect, instructors are providing their students with a powerful tool for improving academic performance and truly mastering course material.

Connect allows students to practice important skills at their own pace and on their own schedule. Importantly, students' assessment results and instructors' feedback are all saved online—so students can continually review their progress and plot their course to success.

LearnSmart™

LearnSmart™

Built around metacognition learning theory, LearnSmart provides your students with a GPS (Guided Path to Success) for your geology course. Using artificial intelligence, LearnSmart intelligently assesses a student's knowledge of course content through a series of adaptive questions. It pinpoints concepts the student does not understand and maps out a *personalized study plan* for success. Available as an integrated feature of McGraw-Hill's Connect, you can incorporate LearnSmart into your course in a number of ways to

- Gauge student knowledge before a lecture
- Reinforce learning after lecture
- Prepare students for assignments and exams

Visit www.mhlearnsmart.com to discover for yourself how the LearnSmart diagnostic ensures students will connect with the content, learn more effectively, and succeed in your course.

Do More

McGraw-Hill Higher Education and Blackboard

McGraw-Hill Higher Education and Blackboard have teamed up! What does this mean for you?

1. **Your life, simplified.** Now you and your students can access McGraw-Hill's Connect and Create™ right from within your Blackboard course—all with one single sign-on. Say goodbye to the days of logging in to multiple applications.

2. **Deep integration of content and tools.** Not only do you get single sign-on with Connect and Create, you also get deep integration of McGraw-Hill content and content engines right in Blackboard. Are you tired of keeping multiple gradebooks and manually synchronizing grades into Blackboard? We thought so. When a student completes an integrated Connect assignment, the grade for that assignment automatically (and instantly) feeds your Blackboard grade center.

3. **A solution for everyone.** Whether your institution is already using Blackboard or you just want to try Blackboard on your own, we have a solution for you. McGraw-Hill and Blackboard can now offer you easy access to industry leading technology and content, whether your campus hosts it, or we do. Be sure to ask your local McGraw-Hill representative for details.

www.domorenow.com

TEGRITY™

Tegrity Campus is a service that makes class time available all the time by automatically capturing every lecture in a searchable format for students to review when they study and complete assignments. With a simple one-click start and stop process, you capture all computer screens and corresponding audio. Students replay any part of any class with easy-to-use browser-based viewing on a PC or Mac.

Educators know that the more students can see, hear, and experience class resources, the better they learn. With Tegrity Campus, students quickly recall key moments by using Tegrity Campus's unique search feature. This search helps students efficiently find what they need, when they need it across an entire semester of class recordings. Help turn all your students' study time into learning moments immediately supported by your lecture.

To learn more about Tegrity, watch a 2-minute Flash demo at http://tegritycampus.mhhe.com.

Customizable Textbooks: Create™

Create what you've only imagined. Introducing McGraw-Hill Create—a new, self-service website that allows you to create custom course materials—print and eBooks—by drawing upon McGraw-Hill's comprehensive, cross-disciplinary content. Add your own content quickly and easily. Tap into other rights-secured third-party sources as well. Then, arrange the content in a way that makes the most sense for your course. Even personalize your book with your course name and information. Choose the best format for your course: color print, black and white print, or eBook. The eBook is now viewable on an iPad! And when you are finished customizing, you will receive a free PDF review copy in just minutes! Visit McGraw-Hill Create—www.mcgrawhillcreate.com—today and begin building your perfect book.

CourseSmart eBook

CourseSmart is a new way for faculty to find and review eBooks. It's also a great option for students who are interested in accessing their course materials digitally and saving money. *CourseSmart* offers thousands of the most commonly adopted textbooks across hundreds of courses. It is the only place for faculty to review and compare the full text of a textbook online, providing immediate access without the environmental impact of requesting a print exam copy. At *CourseSmart*, students can save up to 50% off the cost of a print book, reduce their impact on the environment, and gain access to powerful Web tools for learning including full text search, notes and highlighting, and email tools for sharing notes between classmates.

To review comp copies or to purchase an eBook, go to www.coursesmart.com.

ADDITIONAL MATERIALS IN ENVIRONMENTAL SCIENCE

Annual Editions: *Environment* by Richard Eathorne

ISBN: 9780073515698

Revised annually for more than 32 years, this text provides convenient inexpensive access to current articles selected from some of the most respected magazines, newspapers, and journals published today. Instructional features include: an annotated table of contents, a correlation guide to main textbooks, a topic guide for all articles, Internet references by unit for additional research, learning outcomes, and critical thinking questions. An Instructor Resource Guide with test materials is available for download as well as a practical guide for *Using Annual Editions in the Classroom.*

Taking Sides: *Clashing Views on Environmental Issues* by Tom Easton

ISBN: 9780073514512

Revised bi-annually for more than 30 years, this text is a debate-style reader designed to introduce students to controversies in environmental policy and science. The readings present arguments by leaders in the field and have been selected for their liveliness and substance. Instructional features include: an annotated table of contents, a correlation guide to main textbooks, a topic guide for all articles, internet references by unit, learning outcomes, critical thinking questions, and "Is There Common Ground?" questions to guide further research. An Instructor Resource Guide with test materials is available for download as well as a practical guide for *Using Taking Sides in the Classroom.*

Classic Edition Sources: *Environmental Studies* by Tom Easton

ISBN: 9780073527642

This collection brings together more than 40 selections of enduring intellectual value—classic articles, reviews, book excerpts, and research studies—that help define the study of the environment and our current understanding of it. These readings represent almost 150 years of ecological thought and application with dates of publication ranging from 1864 to the present. Instructional features include: an annotated table of contents, a correlation guide to main textbooks, a topic guide for all articles, Internet references by unit to facilitate further research. An Instructor Resource Guide with test materials is available for download.

Annual Editions: *Sustainability* by Nicholas Smith-Sebasto

ISBN: 9780073528694

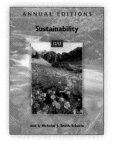

This new addition to the Annual Editions series provides carefully selected articles from the most respected magazines, newspapers, and journals published today. This volume contains interesting, well-illustrated readings by environmentalists, educators, researchers, scientists, and writers that provide perspective on the emerging field of sustainability. Instructional features include: an annotated table of contents, a correlation guide to main textbooks, a topic guide for all articles, Internet references by unit for additional research, learning outcomes, and critical thinking questions. An Instructor Resource Guide with test materials is available for download, as is the practical guide *Using Annual Editions in the Classroom.*

Taking Sides: *Sustainability* by Robert Taylor

ISBN: 9780073514505

This new addition to the Taking Sides series introduces students to controversies in the emerging field of sustainability. The text presents arguments by policy analysts, scientists, economists, and environmentalists that have been selected for their liveliness and substance. Instructional features include: an annotated table of contents, a correlation guide to main textbooks, a topic guide for all articles, Internet references by unit, learning outcomes, critical thinking questions, and "Is There Common Ground?" questions to guide further research. An Instructor Resource Guide with test materials is available for download as well as a practical guide for *Using Taking Sides in the Classroom.*

The Dictionary of Global Sustainability by Tracy Green

ISBN: 9780073514529

This textbook serves as a quick reference guide to students and professionals seeking a better understanding of sustainability concepts. The volume provides nearly 2800 key terms in this emerging field, as well as a listing of organizations and scholarly and trade journals—domestic and international—that will lead the reader to valuable research materials. It includes case studies that examine sustainability projects from around the world designed to illustrate the theory and practice of environmental, economic, technological, and social aspects of sustainability.

Guided Tour

Application-based learning contributes to engaged scientific investigation.

Key Concepts

Key concepts from each chapter are presented in a beautifully arranged layout to guide the student through the often complex network issues.

Case Studies

All chapters open with a real-world case study to help students appreciate and understand how environmental science impacts lives and how scientists study complex issues.

CASE STUDY

Protecting Forests to Prevent Climate Change

In 2010 Norway signed an agreement to support Indonesia's efforts to reduce greenhouse gas emissions from deforestation and forest degradation. Based on Indonesia's performance over the next eight years, Norway will provide up to (U.S.) $1 billion to support this partnership. Indonesia has the third largest area of tropical rainforest in the world (after Brazil and the Democratic Republic of Congo), and because it's an archipelago of more than 17,000 islands, many of which have unique assemblages of plants and animals, Indonesia has some of the highest biological diversity in the world.

Indonesia is an excellent example of the benefits of forest protection. Deforestation, land-use change, and the drying, decomposition, and burning of peatlands cause about 80 percent of the country's current greenhouse gas emissions. This means that Indonesia can make deeper cuts in CO_2 emissions and do it more quickly than most other countries. Reducing deforestation will help preserve biodiversity and protect indigenous forest people. And according to government estimates, up to 80 percent of Indonesia's logging

▲ **FIGURE 6.1** Logging valuable hardwoods is generally the first step in tropical forest destruction. Although loggers may take only one or two large trees per hectare, the damage caused by extracting logs exposes the forest to invasive species, poachers, and fires.

how much carbon is stored in a particular forest as well as how much carbon could be saved by halting or slowing deforestation. Historical forest data, on which these predictions often are based, is often unreliable or nonexistent in tropical countries. Satellite imaging and computer modeling can give answers to these questions, but the technology can be expensive. In the first phase of funding, Norway will support political and institutional reform along with building infrastructure and increasing capacity.

Like other donor nations, Norway is also concerned about how permanent the protections will be. What happens if they pay to protect a forest but a future administration decides to log it? Furthermore, loggers are notoriously mobile and adept at circumventing rules by bribing local authorities, if necessary. What's to prevent them from simply moving to new areas to cut trees? If you avoid deforestation in one place but then cut an equal number of trees somewhere else (sometimes known as "leakage"), carbon emissions won't have gone down at all.

Google Earth™ interactive satellite imagery gives students a geographic context for global places and topics discussed in the text. Google Earth™ icons indicate a corresponding exercise in Connect. In these exercises students will find links to locations mentioned in the text, and corresponding assessments that will help them understand environmental topics.

Active Learning

Students will be encouraged to practice critical thinking skills and apply their understanding of newly learned concepts and to propose possible solutions.

Active LEARNING

Examining Climate Graphs

Among the nine types of terrestrial biomes you've just read about, one of the important factors is the number of months when the average temperature is below freezing (0°C). This is because most plants photosynthesize most actively when daytime temperatures are well above freezing—and when water is fluid, not frozen (chapter 2). Among the biome examples shown, how many sites have fewer than three months when the average temperature is above 0°? How many sites have all months above freezing? Look at figure 5.3: Do all deserts have average yearly temperatures above freezing? Now look at figure 5.4: Which biome do you live in? Which biome do most Americans live in?

ANSWERS: Only the tundra site has less than three months above freezing. Three sites have all months above freezing. No. Answers will vary. Most Americans live in temperate coniferous or broadleaf forest biomes.

What Can **YOU DO?**

Working Locally for Ecological Diversity

You might think that diversity and complexity of ecological systems are too large or too abstract for you to have any influence. But you can contribute to a complex, resilient, and interesting ecosystem, whether you live in the inner city, a suburb, or a rural area.

- Take walks. The best way to learn about ecological systems in your area is to take walks and practice observing your environment. Go with friends and try to identify some of the species and trophic relationships in your area.
- Keep your cat indoors. Our lovable domestic cats are also very successful predators. Migratory birds, especially those nesting on the ground, have not evolved defenses against these predators.
- Plant a butterfly garden. Use native plants that support a diverse insect population. Native trees with berries or fruit also support birds. (Be sure to avoid non-native invasive species.) Allow structural diversity (open areas, shrubs, and trees) to support a range of species.
- Join a local environmental organization. Often, the best way to be effective is to concentrate your efforts close to home. City parks and neighborhoods support ecological communities, as do farming and rural areas. Join an organization

What Can You Do?

Students can employ these practical ideas to make a positive difference in our environment.

EXPLORING Science — Remote Sensing, Photosynthesis, and Material Cycles

Measuring primary productivity is important for understanding individual plants and local environments. Understanding the rates of primary productivity is also key to understanding global processes, such as material cycling, and biological activity:

- In global carbon cycles, how much carbon is stored by plants, how quickly is it stored, and how does carbon storage compare in contrasting environments, such as the Arctic and the tropics?
- How does this carbon storage affect global climates (chapter 9)?
- In global nutrient cycles, how much nitrogen and phosphorus wash offshore, and where?

How can environmental scientists measure primary production (photosynthesis) at a global scale? In a small, relatively closed ecosystem, such as a pond, ecologists can collect and analyze samples of all trophic levels. But that method is impossible for large ecosystems, especially for oceans, which cover 70 percent of the earth's surface. One of the newest methods of quantifying biological productivity involves remote sensing, or using data collected from satellite sensors that observe the energy reflected from the earth's surface.

As you have read in this chapter, chlorophyll in green plants *absorbs* red and blue

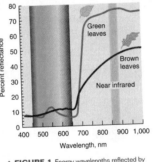

▲ **FIGURE 1** Energy wavelengths reflected by green and brown leaves.

Exploring Science

Current environmental issues exemplify the principles of scientific observation and data-gathering techniques to promote scientific literacy.

What Do **YOU THINK?**

Shade-Grown Coffee and Cocoa

Do your purchases of coffee and chocolate help to protect or destroy tropical forests? Coffee and cocoa are two of the many products grown exclusively in developing countries but consumed almost entirely in the wealthier, developed nations. Coffee grows in cool, mountain areas of the tropics, while cocoa is native to the warm, moist lowlands. What sets these two apart is that both come from small trees adapted to grow in low light, in the shady understory of a mature forest. **Shade-grown** coffee and cocoa (grown beneath an understory of taller trees) allow farmers to produce a crop at the same time as forest habitat remains for birds, butterflies, and other wild species.

Until a few decades ago, most of the world's coffee and cocoa were shade-grown. But new varieties of both crops have been developed that can be grown in full sun. Growing in full sun, trees can be crowded together more closely. With more sunshine, photosynthesis and yields increase.

◄ Cocoa pods grow directly on the trunk and large branches of cocoa trees.

coffee and cocoa plantations in these areas are converted to monocultures, an incalculable number of species will be lost.

The Brazilian state of Bahia demonstrates both the ecological importance of these crops and how they might help preserve forest species. At one time, Brazil produced much of the world's cocoa, but in the early 1900s, the crop was introduced into West Africa. Now Côte d'Ivoire alone grows more than 40 percent of the world total. Rapid increases in global supplies have made prices plummet, and the value of Brazil's harvest has dropped by 90 percent. Côte d'Ivoire is aided in this competition by a labor system that reportedly includes widespread child slavery. Even adult workers in Côte d'Ivoire get only about $165 (U.S.) per year (if they get paid at all), compared with a minimum wage of $850 (U.S.) per year in Brazil.

What Do You Think?

Students are presented with challenging environmental studies that offer an opportunity to consider contradictory data, special interest topics, and conflicting interpretations within a real scenario.

Pedagogical Features Facilitate Student Understanding of Environmental Science

CHAPTER

3 Evolution, Species Interactions, and Biological Communities

The Galápagos Islands have provided an accidental laboratory for examining biological diversity and species interactions.

LEARNING OUTCOMES

After studying this chapter, you should be able to answer the following questions:

▶ How does species diversity arise?
▶ What do we mean by tolerance limits? Give examples.
▶ How do interactions both aid and hinder species?
▶ Why don't species always reproduce up to their biotic potential?

▶ What is the relationship between species diversity and community stability?
▶ What is disturbance, and how does it affect communities?
▶ Explain ecological succession and give examples of its stages.

Learning Outcomes

Questions at the beginning of each chapter challenge students to find their own answers.

Practice Quiz

Short-answer questions allow students to check their knowledge of chapter concepts.

PRACTICE QUIZ

1. What are the two most important nutrients causing eutrophication in the Chesapeake Bay?
2. What are systems and how do feedback loops regulate them?
3. Your body contains vast numbers of carbon atoms. How is it possible that some of these carbons may have been part of the body of a prehistoric creature?
4. List six unique properties of water. Describe, briefly, how each of these properties makes water essential to life as we know it.
5. What is DNA, and why is it important?
6. The oceans store a vast amount of heat, but this huge reservoir of energy is of little use to humans. Explain the difference between high-quality and low-quality energy.
7. In the biosphere, matter follows circular pathways, while energy flows in a linear fashion. Explain.

8. Which wavelengths do our eyes respond to, and why? (Refer to fig. 2.13.) About how long are short ultraviolet wavelengths compared to microwave lengths?
9. Where do extremophiles live? How do they get the energy they need for survival?
10. Ecosystems require energy to function. From where does most of this energy come? Where does it go?
11. How do green plants capture energy, and what do they do with it?
12. Define the terms *species*, *population*, and *biological community*.
13. Why are big, fierce animals rare?
14. Most ecosystems can be visualized as a pyramid with many organisms in the lowest trophic levels and only a few individuals at the top. Give an example of an inverted numbers pyramid.

CRITICAL THINKING AND DISCUSSION

Apply the principles you have learned in this chapter to discuss these questions with other students.

1. Ecosystems are often defined as a matter of convenience because we can't study everything at once. How would you describe the characteristics and boundaries of the ecosystem in which you live? In what respects is your ecosystem an open one?
2. Think of some practical examples of increasing entropy in everyday life. Is a messy room really evidence of thermodynamics at work, or merely personal preference?
3. Some chemical bonds are weak and have a very short half-life (fractions of a second, in some cases); others are strong and stable,

lasting for years or even centuries. What would our world be like if all chemical bonds were either very weak or extremely strong?
4. If you had to design a research project to evaluate the relative biomass of producers and consumers in an ecosystem, what would you measure? (*Note:* This could be a natural system or a human-made one.)
5. Understanding storage compartments is essential to understanding material cycles, such as the carbon cycle. If you look around your backyard, how many carbon storage compartments are there? Which ones are the biggest? Which ones are the longest lasting?

Critical Thinking and Discussion Questions

Brief scenarios of everyday occurrences or ideas challenge students to apply what they have learned to their lives.

DATA ANALYSIS   Examining Nutrients in a Wetland System

As you have read, movements of nitrogen and phosphorus are among the most important considerations in many wetland systems, because high levels of these nutrients can cause excessive algae and bacteria growth. This is a topic of great interest, and many studies have examined how nutrients move in a wetland, and in other ecosystems. Taking a little time to examine these nutrient cycles in detail will draw on your knowledge of atoms, compounds, systems, cycles, and other ideas in

this chapter. Understanding nutrient cycling will also help you in later chapters of this book.
One excellent overview was produced by the Environmental Protection Agency. Go to Connect to find a description of the figure shown here, and to further explore the movement of our dominant nutrient, nitrogen, through environmental systems.

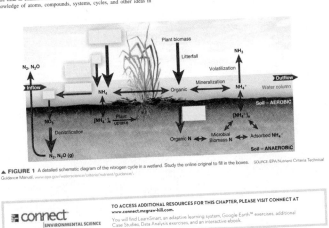

▲ **FIGURE 1** A detailed schematic diagram of the nitrogen cycle in a wetland. Study the online original to fill in the boxes. SOURCE: EPA Nutrient Criteria Technical Guidance Manual, www.epa.gov/waterscience/criteria/nutrient/guidance/.

Data Analysis

At the end of each chapter, these exercises give students further opportunities to apply critical thinking skills and analyze data. These are assigned through Connect in an interactive online environment. Students are asked to analyze data in the form of documents, videos, and animations.

connect | ENVIRONMENTAL SCIENCE

TO ACCESS ADDITIONAL RESOURCES FOR THIS CHAPTER, PLEASE VISIT CONNECT AT www.connect.mcgraw-hill.com.

You will find LearnSmart, an adaptive learning system, Google Earth™ exercises, additional Case Studies, Data Analysis exercises, and an interactive ebook.

Relevant Photos and Instructional Art Support Learning

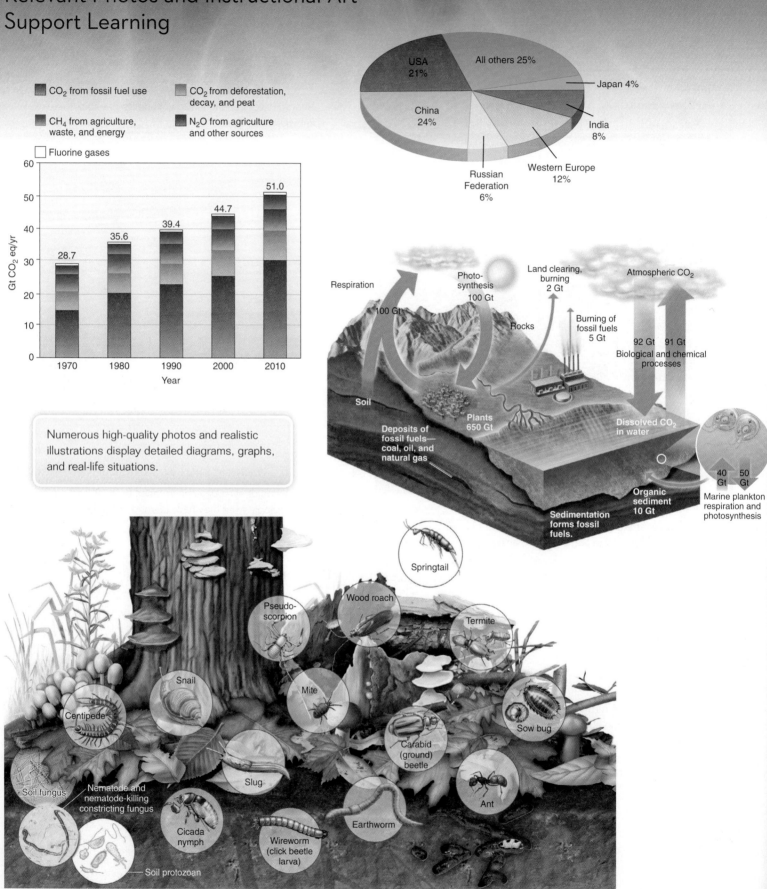

CO₂ from fossil fuel use

CO₂ from deforestation, decay, and peat

CH₄ from agriculture, waste, and energy

N₂O from agriculture and other sources

Fluorine gases

USA 21%

All others 25%

Japan 4%

China 24%

India 8%

Russian Federation 6%

Western Europe 12%

Numerous high-quality photos and realistic illustrations display detailed diagrams, graphs, and real-life situations.

Respiration

100 Gt

Photo-synthesis 100 Gt

Land clearing, burning 2 Gt

Atmospheric CO₂

Rocks

Burning of fossil fuels 5 Gt

92 Gt   91 Gt
Biological and chemical processes

Soil

Deposits of fossil fuels—coal, oil, and natural gas

Plants 650 Gt

Dissolved CO₂ in water

Sedimentation forms fossil fuels.

Organic sediment 10 Gt

40 Gt   50 Gt

Marine plankton respiration and photosynthesis

Springtail

Pseudo-scorpion

Wood roach

Termite

Snail

Mite

Centipede

Sow bug

Carabid (ground) beetle

Soil fungus

Nematode and nematode-killing constricting fungus

Slug

Earthworm

Ant

Cicada nymph

Wireworm (click beetle larva)

Soil protozoan

Understanding Our Environment

Fish and the reefs that support them are essential to the livelihoods
of people in the Philippines and most other tropical island nations.

LEARNING OUTCOMES

After studying this chapter, you should be able to answer the following questions:

▶ Describe several important environmental problems facing
the world.

▶ List several examples of progress in environmental quality.

▶ Explain the idea of sustainability and some of its aims.

▶ Why are scientists cautious about claiming absolute proof
of particular theories?

▶ What is critical thinking, and why is it important in
environmental science?

▶ Why do we use graphs and data to answer questions in science?

▶ Identify several people who helped shape our ideas of
resource conservation and preservation—why did they
promote these ideas when they did?

CASE STUDY

Saving the Reefs of Apo Island

As their outrigger canoes glide gracefully onto Apo Island's beach after an early morning fishing expedition, villagers call to each other to ask how the fishing was. "Tunay mabuti!" (very good!) is the cheerful reply. Nearly every canoe has a basketful of fish, enough to feed a family, with a surplus to send to the market. But life hasn't always been so good on the island. Forty years ago this island, like many others in the Philippines, was experiencing a catastrophic decline in fish stocks, the mainstay of the islanders' diet and livelihood. Rapid population growth coupled with destructive fishing methods had damaged critical habitat and exhausted fish stocks. Dynamite and cyanide were used to stun fish, making them easy to capture but destroying whole ecosystems at once. Fishing boats dragged heavy trawl nets across the sea floor, bulldozing the substrate on which fish depended. Fine-mesh nets were used to capture smaller and smaller fish. The island's fringe of coral reef, the breeding ground and nursery for nearly all the island's fish, was being steadily degraded and torn apart.

In 1979, scientists from Silliman University on nearby Negros Island visited Apo and proposed that residents could improve their livelihoods by managing their resources differently. Simply by protecting their coral reef, islanders could rebuild and preserve a sustainable fishery. The scientists showed villagers from Apo a no-take marine reserve at the uninhabited Sumilon Island, which was teeming with fish (fig. 1.1).

After much discussion, several families decided to establish a marine sanctuary along a short section of Apo Island's shoreline. The area still had high-quality coral, but there were few fish. Participating families took turns watching to make sure no one trespassed in the no-fishing zone. Within a few years, fish in the sanctuary became both dramatically larger and more numerous. Fishing improved outside the sanctuary, as increasingly abundant fish spread to surrounding areas. In 1985, Apo villagers voted to establish a 500 m (0.3 mi) wide marine sanctuary around the entire island.

Villagers now manage their preserve carefully. They allow fishing, but only with low-impact methods such as handheld lines, bamboo traps, large-mesh nets, spearfishing without scuba gear, and hand netting. Coral-destroying techniques, such as dynamite, cyanide, and trawling, are prohibited. By protecting the reef, villagers are guarding the nursery of an entire marine ecosystem. Young fish growing up in the shelter of the coral disperse to neighboring waters and yield abundant harvests. Fishermen report that they spend much less time traveling to distant fishing areas now that fish around the island are so much more abundant.

Apo's sanctuary has become the inspiration for more than 400 similar marine preserves in the Philippines and many others around the world.

The idea of marine protected areas (MPAs) has also spread to larger fisheries. Protecting just a small amount of critical reproductive area can ensure stability of populations and fisheries. Regional MPAs have now been established from California to New Zealand. The idea is a technologically simple and inexpensive approach to protecting the world's increasingly imperiled fisheries. The United Nations Environment Programme (UNEP) reports that 75 percent of the world's fisheries are at or beyond maximum supportable fishing levels; about a quarter have collapsed completely, with fish populations no longer able to produce enough young to support an economic fishery. MPAs are one of the best ways to reverse these declines, if the rules can be agreed upon and enforced.

Enforcement is relatively easy in a small community like Apo Island, where the benefits of the marine sanctuary are obvious and widely shared. The island's rich marine life and spectacular coral reefs have made Apo an international destination for scuba diving and snorkeling. Hotels and dive shops provide jobs for island residents. Villagers sell food and souvenirs to visitors. Fees charged to tourists for diving and snorkeling have been used to build schools, improve island water supplies, provide electricity to the island's households, and pay local people to patrol the sanctuary. Fishing is still a primary occupation, and fish are the mainstay of local diets, but it's no longer necessary to travel as far or work as hard to catch the fish. Now fishers have time for other activities, such as guiding diving tours.

Higher family incomes now allow most island children to attend high school on nearby Negros Island, and many continue on to college or technical schools. Education has greatly expanded the islanders' economic opportunities. Just as important, villagers are empowered by seeing that they can take action to improve their environment and their living conditions at the same time.

Environmental scientists often point out that crisis and opportunity go hand in hand. Faced with loss of their livelihoods, Apo Island's residents discovered new ways to cooperate and to manage their resources better than they had done before. Often the best solutions to our problems involve rethinking our accustomed approaches and actions. The idea of MPAs draws on basic ideas of ecology and population biology. Ideas of governance and economics are also essential. In this book we'll examine these and other concepts that can help us resolve the environmental, economic, and social challenges we face. That is the challenge that faces all of us. With creativity and commitment we may find strategies to rebuild the living systems that sustain us. ∎

▶ **FIGURE 1.1** Coral reefs are among the most beautiful, species-rich, and productive biological communities on the planet. They serve as the nurseries for many open-water species. At least half the world's reefs are threatened by pollution, global climate change, destructive fishing methods, and other human activities, but they can be protected and restored if we care for them.

1.1 UNDERSTANDING OUR ENVIRONMENT

Reversing the depletion of Apo Island's fishery required a scientific understanding of many aspects of the environment. Knowledge about population biology, reef ecology, the cultural history of fishing, and even economics of fishing all contributed to explaining this question of resource conservation. Similarly, the field of environmental science draws on many disciplines to help us understand pressing problems of resource supply, ecosystem stability, and sustainable living.

In this chapter we examine some of the serious challenges in environmental science and some promising signs of progress. We also explore the nature of science. What makes the scientific method different from other forms of inquiry? Why do scientists like to answer questions with numbers? Finally we review some of the fundamental ideas that have shaped the ways we think about the environment and our place in it.

What is environmental science?

We inhabit two worlds. One is the natural world of plants, animals, soils, air, and water that preceded us by billions of years and of which we are a part. The other is the world of social institutions and artifacts that we create for ourselves using science, technology, culture, and political organization. Both the natural world and the "built" or technological, social, and cultural world make up our environment (from the French *environner*, "to encircle or surround").

Environmental science is the systematic study of our environment and our place in it. Because environmental problems involve complex, interacting systems, environmental science draws on many fields of knowledge (fig. 1.2). Sciences such as biology, chemistry, earth science, and geography are central. Social sciences and humanities, from political science and economics to art and literature, help us understand how society responds to environmental challenges and opportunities. Solutions to these problems increasingly involve both social systems and natural science. One of your tasks in this class may be to discover where your knowledge and interests contribute to understanding questions in environmental science (see Active Learning, p. 4). Finding your particular interest will help you do better in this class, because you'll have more reason to explore the ideas you encounter.

Is environmental science the same thing as environmentalism? Not necessarily. Environmental science is the use of scientific methods to study processes and systems in the environment in which we live. Environmentalism is working to influence attitudes and policies that affect our environment. The two are often separate goals. Hydrologists, for example, are environmental scientists who study water, often in order to ensure ready supplies to cities

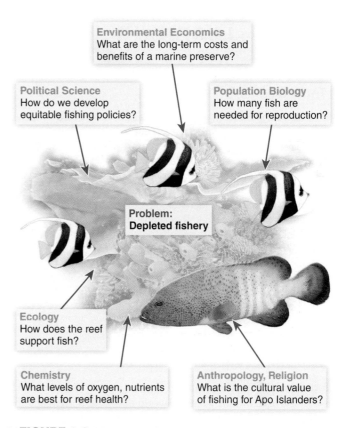

▲ **FIGURE 1.2** Many types of knowledge are needed in environmental science. A few examples are shown here.

and farms. They may work to increase access to water, but their work may or may not involve broader attention to environmental quality. On the other hand ecologists might actively work to defend the ecosystems they study. Many environmental scientists work in the public interest, to promote public health by reducing water contamination, for example, without necessarily being interested in nature or biodiversity. Whether we use science to pursue public health, economic success, environmental quality, or other goals depends on issues outside of science, such as worldviews and ethics.

Environmental science is a global subject

You are already aware of our global dependence on resources and people in faraway places, from computers built in China to oil extracted in Iraq or Venezuela. These interdependencies become clearer as we learn more about global and regional environmental systems. Often the best way to learn environmental science is to see how principles play out in real places. Familiarity with the world around us will help you understand the problems and their context. Throughout this book we've provided links to places you can see in Google Earth, a free online mapping program that you can download from googleearth.com. When you see a blue globe in the margin of this text, like the one at left, you can go to Connect and find place marks that let you virtually visit places discussed. In Google Earth you can also save your own placemarks and share them with your class.

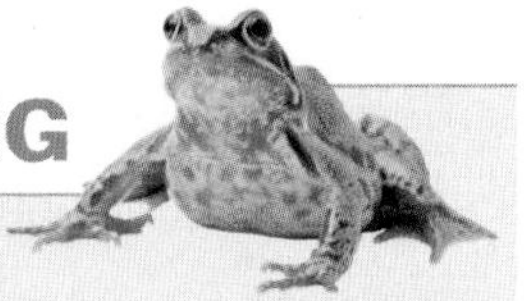

Active LEARNING

Finding Your Strengths in This Class

A key strategy for doing well in this class is to figure out where your strengths and interests intersect with the subjects you will be reading about. As you have read, environmental science draws on many kinds of knowledge (fig. 1.2). Nobody is good at all of these, but everyone is good at some of them. Form a small group of students, then select one of the questions in section 1.2. Explain how each of the following might contribute to understanding or solving that problem:

artist, writer, politician, negotiator, chemist, mathematician, hunter, angler, truck driver, cook, parent, builder, planner, economist, speaker of multiple languages, musician, business person.

ANSWERS: All of these provide multiple insights; answers will vary.

We inhabit a remarkable planet

Before turning to focus on current challenges, we should pause to consider the extraordinary natural world that we inherited and that we hope to pass on to future generations in a condition as good as—or perhaps even better than—we found it in.

Imagine that you are an astronaut returning to the earth after a trip to the moon or Mars. What a relief it would be, after the silent void of outer space, to return to this beautiful, bountiful planet (fig. 1.3). We live in an incredibly prolific and colorful world that is, as far as we know, unique in the universe. Compared with other planets in our solar system, temperatures on the earth are mild and relatively constant. Plentiful supplies of clean air, fresh water, and fertile soil are regenerated endlessly and spontaneously by biogeochemical cycles and biological communities (discussed in chapters 2 and 3). The value of these ecological services is almost incalculable, although economists estimate that they account for a substantial proportion of global economic activity (chapter 16).

Perhaps the most amazing feature of our planet is its rich diversity of life. Millions of beautiful and intriguing species populate the earth and help sustain a habitable environment (fig. 1.4). This vast multitude of life creates complex, interrelated communities where towering trees and huge animals live together with, and depend upon, such tiny life-forms as viruses, bacteria, and fungi. Together, all these organisms make up delightfully diverse, self-sustaining ecosystems, including dense, moist forests; vast, sunny savannas; and richly colorful coral reefs.

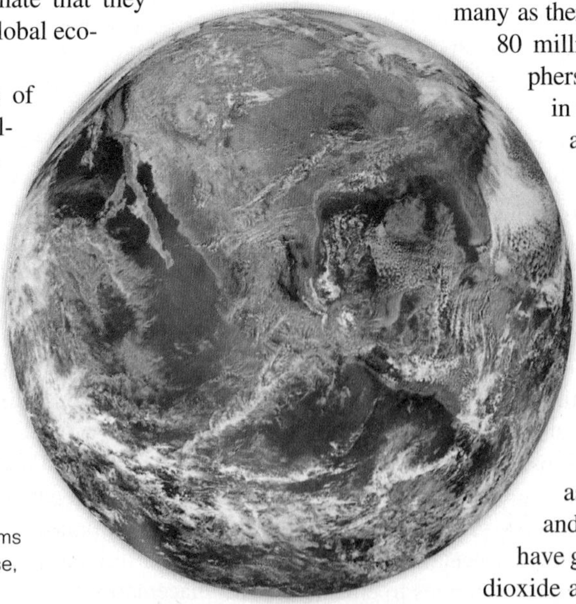

FIGURE 1.3 The life-sustaining ecosystems on which we all depend are unique in the universe, as far as we know.

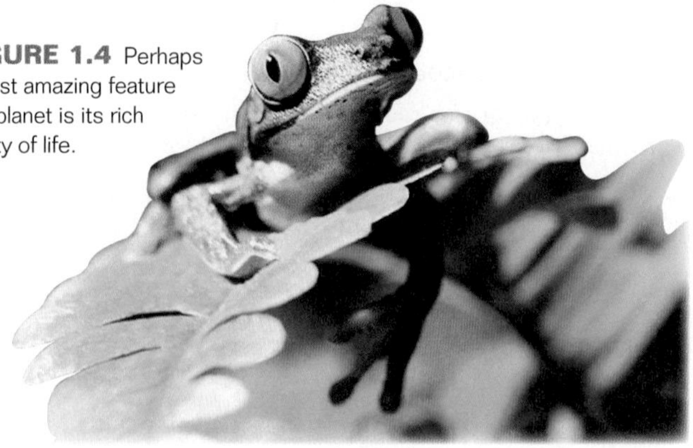

FIGURE 1.4 Perhaps the most amazing feature of our planet is its rich diversity of life.

From time to time we should pause to remember that, in spite of the challenges of life on earth, we are incredibly lucky to be here. We should ask ourselves what we can do, and what we *ought* to do, to ensure that future generations have the same opportunities to enjoy this bounty.

1.2 CRISES AND OPPORTUNITIES

As we have noted, crisis and opportunity often go hand in hand, because serious problems can drive us to seek better solutions. A first step in understanding environmental science is to understand some of the principal challenges we face and some of the recent changes in environmental quality and environmental health. We give a brief overview here, and we explore each topic in detail in later chapters. Most of these issues are influenced by multiple factors. As you read, consider what those factors are and what steps might be taken to resolve some of these problems.

What persistent challenges do we face?

There are over 7 billion people on earth, about twice as many as there were 40 years ago. We are adding about 80 million more each year. Although demographers report a transition to slower growth rates in most countries, with improved education and health care, present trends project a population between 8 and 10 billion by 2050 (fig. 1.5). The impact of that many people on our natural resources and ecological systems strongly influences many of the other problems we face.

Climate Change The atmosphere normally captures heat near the earth's surface, which is why it is warmer here than in space. But human activities such as burning fossil fuels, clearing forests and farmlands, and raising ruminant animals have greatly increased concentrations of carbon dioxide and other "greenhouse gases." In the past

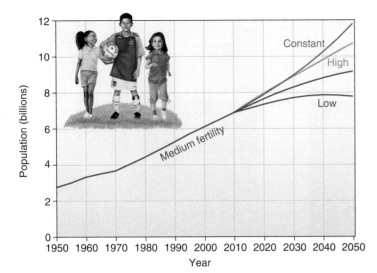

(a) Possible population trends

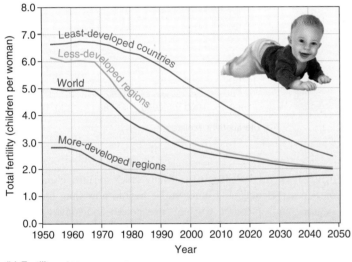

(b) Fertility rates

▲ **FIGURE 1.5** Bad news and good news: globally, populations continue to rise (a), but our rate of growth has plummeted (b). Some countries are below the replacement rate of about two children per woman. SOURCE: United Nations Population Program, 2011.

200 years, atmospheric CO_2 concentrations have increased about 30 percent. Climate models indicate that by 2100, if current trends continue, global mean temperatures will probably increase by 2° to 6°C compared to 1990 temperatures (3.6° and 12.8°F: fig. 1.6a), far warmer than the earth has been since the beginning of human civilization. For comparison, the last ice age was about 4°C cooler than now. Increasingly severe droughts and heat waves are expected in many areas. Greater storm intensity and flooding are expected in other places. Disappearing mountaintop glaciers and snowfields threaten the water supplies on which cities such as Los Angeles and Delhi depend. Canadian Environment Minister David Anderson said that global climate change is a greater threat than terrorism because it could force hundreds of millions of people from their homes and trigger an economic and social catastrophe.

Hunger Over the past century, global food production has increased faster than human population growth, but hunger remains a chronic problem because food resources are unevenly distributed. At the same time, soil scientists report that about two-thirds of all agricultural lands show signs of degradation. The biotechnology and intensive farming techniques responsible for much of our recent production gains are too expensive for many poor farmers. Can we find ways to produce the food we need without further environmental degradation? And can we distribute food more equitably? In a world of food surpluses, currently more than 850 million people are chronically undernourished, and at least 60 million people face acute food shortages due to weather, politics, or war (fig. 1.6b).

Clean Water Water may be the most critical resource in the twenty-first century. Already at least 1.1 billion people lack access to safe drinking water, and twice that many don't have adequate sanitation. Polluted water contributes to the death of more than 15 million people every year, most of them children under age 5. About 40 percent of the world population lives in countries where water demands now exceed supplies, and the UN projects that by 2025 as many as three-fourths of us could live under similar conditions (fig. 1.6c).

Energy Resources How we obtain and use energy will greatly affect our environmental future. Fossil fuels (oil, coal, and natural gas) presently provide around 80 percent of the energy used in industrialized countries. Supplies of these fuels are diminishing, however, and the costs of extracting and using these fuels are high in terms of air and water pollution, mining damage, political conflicts, and of course climate change. Energy conservation and cleaner, renewable energy resources—solar, wind, geothermal, and biomass power—could give us cleaner, less destructive options if we decide to invest in them.

Air Quality Air quality has worsened dramatically in newly industrializing areas, including much of China and India. An "Asian brown cloud," a 3-km (2-mile)-thick toxic haze of ash, acids, aerosols, dust, and photochemical smog, regularly covers the entire Indian subcontinent for much of the year. Nobel laureate Paul Crutzen estimates that at least 3 million people die each year from diseases triggered by air pollution. Worldwide, the United Nations estimates, more than 2 billion metric tons of air pollutants (not including carbon dioxide or wind-blown soil) are released each year. These air pollutants travel easily around the globe. On some days 75 percent of the smog and airborne particulates in California originate in Asia; mercury, polychlorinated biphenyls (PCBs), and other industrial pollutants accumulate in arctic ecosystems and in the tissues of native peoples in the far north.

Biodiversity Loss Biologists report that habitat destruction, overexploitation, pollution, and the introduction of exotic organisms are eliminating species as quickly as the great extinction that marked the end of the age of dinosaurs. The UN Environment Programme reports that over the past century more than 800 species have disappeared, and at least 10,000 species are now considered threatened. This includes about half of all primates and freshwater fish, together with around 10 percent of all plant species. Top

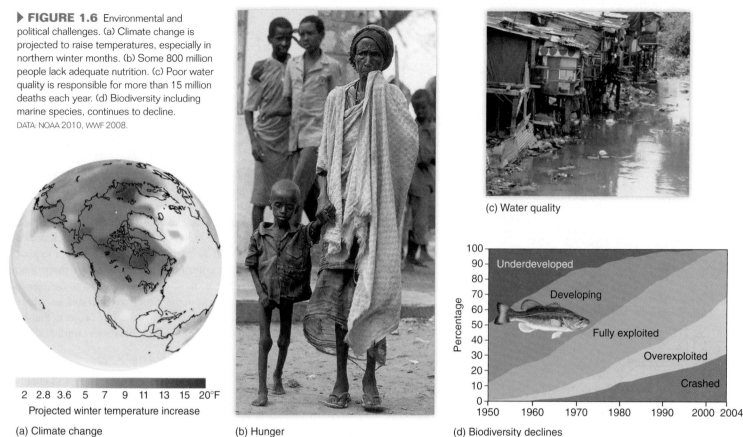

▶ **FIGURE 1.6** Environmental and political challenges. (a) Climate change is projected to raise temperatures, especially in northern winter months. (b) Some 800 million people lack adequate nutrition. (c) Poor water quality is responsible for more than 15 million deaths each year. (d) Biodiversity including marine species, continues to decline.
DATA: NOAA 2010, WWF 2008.

(a) Climate change

(b) Hunger

(c) Water quality

(d) Biodiversity declines

predators, including nearly all the big cats in the world, are particularly rare and endangered. A nationwide survey of the United Kingdom in 2004 found that most bird and butterfly populations had declined by 50 to 75 percent over the previous 20 years. At least half of the forests existing before the introduction of agriculture have been cleared, and many of the ancient forests, which harbor some of the greatest biodiversity, are rapidly being cut for timber, for oil extraction, or for agricultural production of globally traded commodities such as palm oil or soybeans.

Marine Resources As noted in the opening case study, the ocean is an irreplaceable food resource. More than a billion people in developing countries depend on seafood for their main source of animal protein, but most commercial fisheries around the world are in steep decline. (fig. 1.6d). According to the World Resources Institute, more than three-quarters of the 441 fish stocks for which information is available are severely depleted or in urgent need of better management. Canadian researchers estimate that 90 percent of all the large predators, including bluefin tuna, marlin, swordfish, sharks, cod, and halibut, have been removed from the ocean.

There are many signs of progress

Despite this dire list of problems, increasing awareness is leading to progress in many areas. As on Apo Island, both environmental scientists and ordinary people are inventing new strategies for protecting resources and improving people's lives.

Population and Pollution Many cities in Europe and North America are cleaner and much more livable now than they were a century ago, and as a consequence health and longevity have improved sharply. Population has stabilized in most industrialized countries and even in some very poor countries where social security, education, and democracy have been established. Since 1960 the average number of children born per woman worldwide has decreased from 5 to 2.45 (see fig. 1.5). By 2050 the UN Population Division predicts, most countries will have fertility rates below the replacement rate of 2.1 children per woman. If this happens, the world population will stabilize at about 8.9 billion rather than the 9.3 billion previously expected.

Health The incidence of life-threatening infectious diseases has been reduced sharply in most countries during the past century, while life expectancies have nearly doubled, on average (fig. 1.7a). Smallpox has been completely eradicated and polio has been vanquished except in a few countries. Since 1990 more than 800 million people have gained access to improved water supplies and modern sanitation. In spite of population growth that added nearly a billion people to the world during the 1990s, the number of people facing food insecurity and chronic hunger during this period actually declined by about 40 million.

Renewable Energy There has been dramatic and surprising progress in the transition to renewable energy sources. Growth in wind energy, solar, and biomass power and improvements in efficiency are beginning to reduce reliance on fossil fuels. The cost of solar power has plummeted, and in many areas solar costs the same as conventional electricity over time. Solar and wind power are now far cheaper, easier, and faster to install than nuclear power or new coal plants.

Information and Education   Because so many environmental issues can be fixed by new ideas, technologies, and strategies, expanding access to knowledge is essential to progress. The increased speed at which information now moves around the world offers unprecedented opportunities for sharing ideas. At the same time, literacy and access to education are expanding in most regions of the world (fig. 1.7b). Rapid exchange of information on the Internet also makes it easier to quickly raise global awareness of environmental problems, such as deforestation or pollution, that historically would have proceeded unobserved and unhindered. Expanding education for girls is a primary driver for declining birth rates worldwide.

Conservation of Forests and Nature Preserves   The rate of deforestation has slowed in Asia, from over 8 percent during the 1980s to less than 1 percent in the 1990s. Brazil, which has led global deforestation rates for decades, is working to protect forests. Nature preserves and protected areas have increased dramatically over the past few decades. In 2010 there were more than 100,000 parks and nature preserves in the world, representing more than 20 million km^2 (about 7.7 million mi^2), or about 13.5 percent of the world's land area (fig. 1.7c). Ecoregion and habitat protection remains uneven, and some areas are protected only on paper. Still, this is dramatic progress in biodiversity protection.

Protection of Marine Resources   Although marine resources continue to be badly overexploited, countries are gradually beginning to acknowledge the problem and find solutions. Marine protected areas and improved monitoring of fisheries provide opportunities for sustainable management (fig. 1.7d). The strategy of protecting fish nurseries is an altogether new approach to sustaining ocean systems and the people who depend on them. Marine reserves have been established in California, Hawaii, New Zealand, Great Britain, and many other areas.

1.3 HUMAN DIMENSIONS OF ENVIRONMENTAL SCIENCE

Aldo Leopold, one of the greatest thinkers on conservation, observed that the great challenges in conservation have less to do with managing resources than with managing people and our demands on resources. Foresters have learned much about how to grow trees, but still we struggle to establish conditions under which villagers in developing countries can manage plantations for themselves. Engineers know how to control pollution but not how to persuade factories to install the necessary equipment. City planners know how to design urban areas, but not how to make them affordable for everyone. In this section we'll review some key ideas that guide our understanding of human dimensions of environmental science and resource use. These ideas will be useful throughout the rest of this book.

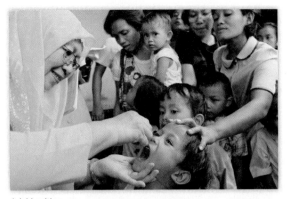
(a) Health care

(b) Education

(c) Sustainable resource use

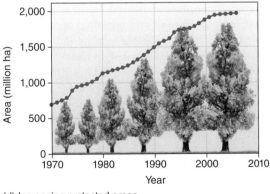
(d) Increasing protected areas

▲ **FIGURE 1.7** Conditions are improving in many areas, including access to (a) health care and (b) education. In many areas, (c) sustainable resource use is being improved by expanding (d) networks of protected areas. DATA: IUCN and UNEP 2010.

How do we describe resource use and conservation?

The natural world supplies the water, food, metals, energy, and other resources we use. Some of these resources are finite; some are constantly renewed (chapter 14). Often, renewable resources can be destroyed by excessive exploitation, as in the fisheries and reefs of Apo Island. When we consider resource consumption, an important idea is **throughput**, the amount of resources we use and dispose of. A household that consumes abundant consumer goods, foods, and energy brings in a great deal of natural resource-based materials; that household also disposes of a great deal of materials. Conversely a household that consumes very little also produces little waste (chapter 2).

Ecosystem services, another key idea, refers to services or resources provided by environmental systems (fig. 1.8). *Provisioning* of resources, such as the fuels we burn, may be the most obvious service we require. *Supporting* services are less obvious until you start listing them: these include water purification, production of food and atmospheric oxygen by plants, or decomposition of waste by fungi and bacteria. *Regulating* services include maintenance of temperatures suitable for life by the earth's atmosphere, or carbon capture by green plants, which maintains a stable atmospheric composition. Cultural services include a diverse range of recreation, aesthetic, and other nonmaterial benefits. Usually we rely on these resources without thinking about them. They support all our economic activities in some way, but we don't put a price on them because nature doesn't force us to pay for them.

Are there enough resources for all of us? One of the answers to this basic question was given in an essay entitled "**Tragedy of the Commons**," published in 1968 in the journal *Science* by ecologist Garret Hardin. In this classic framing of the problem, Hardin argues that population growth leads inevitably to overuse and then destruction of common resources—such as shared pastures, unregulated fisheries, fresh water, land, and clean air. This classic essay has challenged many to explore alternative ideas about resource management. In Apo Island's degraded fishery, the answer was agreed-upon regulation and monitoring. Another strategy is to assign prices to ecological services, to force businesses and economies to account for damages to life-supporting systems. This approach is discussed in chapter 14. The idea of sustainable development is yet another answer.

Sustainability means environmental and social progress

Sustainability is a search for ecological stability and human progress that can last over the long term. Of course, neither ecological systems nor human institutions can continue forever. We can work, however, to protect the best aspects of both realms and to encourage resiliency and adaptability in both of them. World Health Organization director Gro Harlem Brundtland has defined **sustainable development** as "meeting the needs of the present without compromising the ability of future generations to meet their own needs." In these terms, development means bettering people's lives. Sustainable development, then, means progress in human well-being that we can extend or prolong over many generations, rather than just a few years.

This idea became widely publicized after the 1992 Earth Summit, a United Nations meeting held in Rio de Janeiro, Brazil. The Rio meeting was a pivotal event. It brought together many diverse groups—environmentalists and politicians from wealthy countries, indigenous people and workers struggling for rights and land, and government representatives from developing countries. The meeting helped these diverse groups better understand their common needs, and it forced wealthy nations to admit that poorer populations also had a right to a healthy and comfortable life.

Addressing uneven distribution of resources is one of the first tasks of sustainable development. While a few of us live in increasing luxury, the poorest populations suffer from inadequate diet, housing, basic sanitation, clean water, education, and medical care, while the wealthiest consume far more resources than we can readily understand. Policy makers now recognize that eliminating poverty and protecting our common environment are inextricably interlinked. The world's poorest people are both the victims and the agents of environmental degradation (fig. 1.9). The poorest people are often forced to meet short-term survival needs at the cost of long-term sustainability. Desperate for croplands to feed

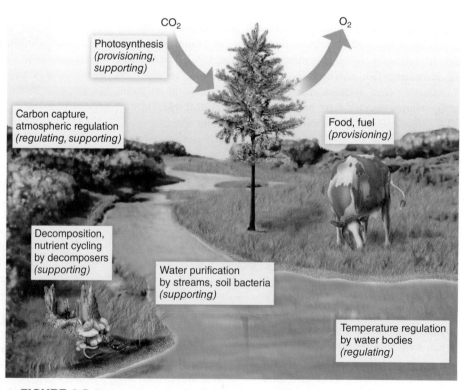

CO$_2$

O$_2$

Photosynthesis
(*provisioning,
supporting*)

Carbon capture,
atmospheric regulation
(*regulating, supporting*)

Food, fuel
(*provisioning*)

Decomposition,
nutrient cycling
by decomposers
(*supporting*)

Water purification
by streams, soil bacteria
(*supporting*)

Temperature regulation
by water bodies
(*regulating*)

▲ **FIGURE 1.8** Ecosystem services we depend on are countless and often invisible.

▲ **FIGURE 1.9** In impoverished areas, survival can mean degrading resources that are already overstressed. Helping the poorest populations is not only humane, it is essential for protecting our shared environment.

▲ **FIGURE 1.10** "And may we continue to be worthy of consuming a disproportionate share of this planet's resources." © Lee Lorenz/ The New Yorker Collection/www.cartoonbank.com

themselves and their families, many move into virgin forests or cultivate steep, erosion-prone hillsides, where soils are depleted after only a few years. Others migrate to the grimy, crowded slums and ramshackle shantytowns that now surround most major cities in the developing world. With no way to dispose of wastes, the residents have no choice but to foul their environment further and contaminate the air they breathe and the water they use for washing and drinking. Children raised in poverty and illness, with few economic opportunities, often are condemned to perpetuate a cycle of poverty.

Affluence is a goal and a liability

Economic growth offers a better life, more conveniences, and more material goods to the billions of people currently living in dire poverty. But social scientists have frequently pointed out that a major reason for both poverty and environmental degradation is that the wealthy consume a disproportionate share of food, water, energy, and other resources, and we produce a majority of the world's waste and pollutants. The United States, for instance, with less than 5 percent of the world's total population, consumes about one-quarter of most commercially traded commodities, such as oil, and produces a quarter to half of most industrial wastes, such as greenhouse gases, pesticides, and other persistent pollutants.

To get an average American through the day takes about 450 kg (nearly 1,000 lb) of raw materials, including 18 kg (40 lb) of fossil fuels, 13 kg (29 lb) of other minerals, 12 kg (26 lb) of farm products, 10 kg (22 lb) of wood and paper, and 450 liters (119 gal) of water. Every year Americans throw away some 160 million tons of garbage, including 50 million tons of paper, 67 billion cans and

bottles, 25 billion styrofoam cups, 18 billion disposable diapers, and 2 billion disposable razors (fig. 1.10).

As the rest of the world seeks to achieve a similar standard of living, with higher consumption of conveniences and consumer goods, what will the effects be on the planet? What should we do about this? Can we reduce our consumption rates? Can we find alternative methods to maintain conveniences and a consumption-based economy with lower environmental costs? These are critical questions as we seek to ensure a reasonable future for our grandchildren.

What is the state of poverty and wealth today?

In 2011 the student-led Occupy Wall Street movement used the statistic "99 percent" to draw attention to growing economic disparities in the United States. The 99 percent statistic was notable because the other 1 percent of the U.S. population controls a disproportionate share—over 35 percent—of the nation's wealth. This imbalance has not been seen since the years leading up to the Great Depression. Students leading the Occupy movement argued that such imbalance destabilizes both democracy and the economy, because a small but powerful elite can easily make shortsighted policy decisions that undermine the rest of society.

Wealth is also unevenly divided at the global scale. The world's richest 200 people have a combined wealth greater than that of the 3.5 billion people who make up the poorest half of the world's population. Countries with the highest per capita income, more than $40,000 (U.S.) per year, make up only 10 percent of the world's population. These countries are all in Europe or North America (the average U.S. income in 2010 was about $48,000), plus Japan, Singapore, Australia, and the United Arab Emirates (fig. 1.11).

More than 70 percent of the world's population—some 5 billion people—live in countries where the average per capita income is less than $5,000, roughly one-tenth of the U.S. average. These countries include China and India, the world's most populous countries, with a combined population of over 2.5 billion people. Of the

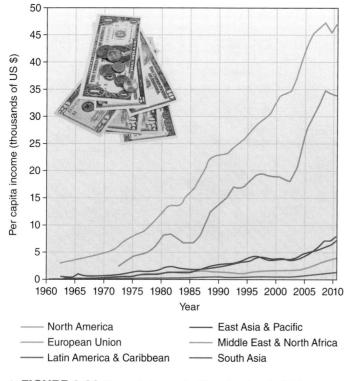

Legend:
—— North America
—— European Union
—— Latin America & Caribbean
—— East Asia & Pacific
—— Middle East & North Africa
—— South Asia

▲ **FIGURE 1.11** Per capita income in different regions (in 2010 U.S. dollars). Overall income has climbed, but the gap between rich and poor countries has grown faster. DATA SOURCE: World Bank 2010.

TABLE 1.1 | Quality-of-Life Indicators

	LEAST-DEVELOPED COUNTRIES	MOST-DEVELOPED COUNTRIES
GDP/Person[1]	(U.S.)$1,671	(U.S.)$35,768
Poverty Index[2]	78.1%	~0
Life Expectancy	58 years	80 years
Adult Literacy	58%	99%
Female Secondary Education	11%	95%
Total Fertility[3]	4.8	1.8
Infant Mortality[4]	120	5
Improved Sanitation	23%	100%
Improved Water	61%	100%
CO_2/capita[5]	0.2 tons	13 tons

[1]ANNUAL gross domestic product.
[2]PERCENT living on less than (U.S.)$2/day.
[3]AVERAGE births/woman.
[4]PER 1,000 live births.
[5]METRIC tons/yr/person.
SOURCE: UNDP Human Development Index, 2011, http://hdr.undp.org/en/statistics/

50 poorest countries, where income is less than $2.50 per day, 33 are in sub-Saharan Africa. There the destabilizing and impoverishing effects of colonialism continue to influence ongoing conflict and underdevelopment.

The gulf between the richest and the poorest nations affects many quality-of-life indicators (table 1.1). Where poverty is widespread and health care is not, life spans are shorter and illness is common. Because of high infant mortality rates and low access to education, the average family in the poorest countries has more than four times as many children as those in richer countries—although that number is dropping rapidly in much of the world. In most wealthy countries, total fertility is slightly less than the replacement rate of 2 children per woman. The poorest countries continues to grow at 2.6 percent per year (see Exploring Science, p. 11).

Indigenous peoples safeguard biodiversity

In both rich and poor countries, native or **indigenous peoples** are generally the least powerful, most neglected groups. Typically descendants of the original inhabitants of an area taken over by more powerful outsiders, they are distinct from their country's dominant language, culture, religion, and racial

communities. Of the world's nearly 6,000 recognized cultures, 5,000 are indigenous; and these account for only about 10 percent of the total world population. In many countries, traditional caste systems, discriminatory laws, economics, or prejudice repress indigenous people. At least half of the world's 6,000 distinct languages are dying because they are no longer taught to children. When the last elders who still speak the language die, so will the culture that was its origin. Lost with those cultures will be a rich repertoire of knowledge about nature and a keen understanding of a particular environment and way of life (fig. 1.12).

Nonetheless, the 500 million indigenous people who remain in traditional homelands still possess valuable ecological wisdom and remain the guardians of little-disturbed habitats that are refuges for rare and endangered species and undamaged ecosystems. The eminent ecologist E. O. Wilson argues that the cheapest and most effective way to preserve species is to protect the natural ecosystems in which they now live.

Recognizing native land rights and promoting political pluralism can be among the best ways to safeguard ecological processes and endangered species. A few countries, such as Papua New Guinea, Fiji, Ecuador, Canada, and Australia, acknowledge indigenous title to extensive land areas. As the Kuna Indians of Panama say, "Where there are forests, there are native people, and where there are native people, there are forests."

◀ **FIGURE 1.12** Indigenous cultures may have unique and important traditional knowledge about their environment.

How do we know about changes in global problems such as hunger, food production, or health, which are much too large to observe directly? We use data sets, usually collected by governments, such as the United States Census (www.census.gov) or the U.S. Census of Agriculture (www.agcensus.usda.gov/), or by organizations such as the United Nations Food and Agriculture Organization (www.faostat.fao.org/) or the World Bank (www.worldbank.org/). If you have a question and some time, you can use these data sets to examine trends, too.

In general, a census agency contacts as many individuals in a country as they can reach. They ask a standard list of questions (for example, how old are you, how many people are in your household?). You may have answered some of these questions in the 2010 United States census, a count that happens every ten years. The census agency enters all the answers into an enormous set of data tables that anybody can access, with a little practice. International organizations such as the United Nations can't contact all persons in the world, but they can survey governments and attempt to gather answers to a standard set of questions (how many citizens are there, how many children died this year, how much clean water is available per person?). Not all countries are able—or willing—to answer all questions, so sometimes there are "no data" values in global data sets. In the map here, for example, Somalia and North Korea are among the countries with "no data."

From these tables, we can calculate averages, high and low values, changes from previous surveys, or comparisons among regions. The graphs and maps you see in this book originate from these types of data.

Newspapers and news magazines rely on these large data sources, too. Take a look at a some maps and graphs in your favorite newspapers, and see what data sources were used.

You can access these databases yourself. Some are very easy to use; others require some patience and persistence. Most data-distributing agencies also provide summaries of important findings in their data. Many educational and business agencies also compile and reorganize data from public sources. For example, the Gapminder project (www.gapminder.org) has entertaining animation to help you visualize global trends. Your school or university library may also keep data, and your reference librarians may be trained to help you use them.

Once you have a graph or map, how do you interpret it? Take a look at the map shown here, which is a representation of Human Development Index (HDI) statistics—here are a few steps to follow in interpreting it:

1. Look for areas of especially high or low values (for example, identify some blue countries and some red countries).

2. In the legend, find the difference in values between those high and low areas.

3. Find a region you know, and think about reasons for the values. Does the blue color in the United States fit your expectations? The blue in China?

4. Look for areas with contrasting values. For example, why does HDI in the United Stages differ from that of India, Brazil, or Congo? Think of several possible explanations for those differences.

Data sources like these provide a large-scale view of issues such as hunger, poverty, education, or health across space or time. This view complements more subtle and complex, but less global, insights from case studies. The Apo Island case study, for example, provides a detailed perspective on the interaction of different factors, or the ways that individuals can influence progress. Both local and global views are often necessary for describing trends in environmental science.

Try exploring the websites noted above. They provide rich and valuable information and entirely new insights on the issues that interest you.

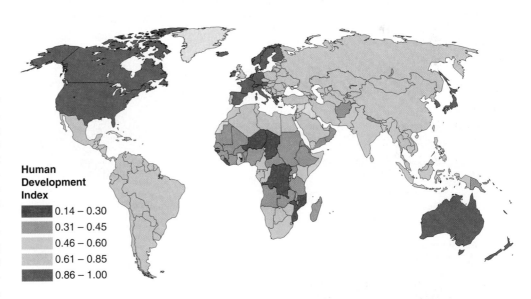

Human Development Index

- 0.14 – 0.30
- 0.31 – 0.45
- 0.46 – 0.60
- 0.61 – 0.85
- 0.86 – 1.00

▲ Statistics such as the Human Development Index (HDI) help us compare quality of life in different places. The HDI combines life expectancy, literacy, and income measures.   DATA SOURCE: World Bank 2011.

Sustainable development

What does it mean? What does it have to do with environmental science?

Sustainable development is a goal. The aim is to meet the needs of people today without compromising resources and environmental systems for future generations. In this context, the term "development" refers to improving access to health care, education, and other conditions necessary for a healthy and productive life, especially in regions of extreme poverty. Meeting the needs of people now, while also guarding those resources for their great-great grandchildren, is both a steep challenge and a good idea.

What parts of it are achievable, and how? In general, development means equitable economic growth, which supports better education, housing, and health care. Often development involves accelerated extraction of natural resources, such as more mining, forestry, or conversion of forests and wetlands to farmlands. Sometimes development involves more efficient use of resources or growth in parts of the economy that don't depend on resource extraction, such as education, health care, or knowledge-based economic activities.

Some resources can be enhanced, for example, through reforestation, maintaining fish nurseries, or careful management of soil resources, to use them without depletion for future generations.

Here are ten key factors necessary for sustainable development, according to the United Nations agreement on development, Agenda 21.

2. **Reducing resource consumption** is a global consideration, but wealthy regions are responsible for most of the world's consumption. For example, the United States and Europe have less than 15 percent of the world's population, but these regions consume about half of the world's metals, food, energy, and other resources.

3. **Population growth** leads to ever-greater resource demands, because all people need some resources. Better family planning, ensuring that all children are wanted, is a matter of justice, resource supply, and economic and social stability for states as well as for families.

1. **Combating poverty** is a central goal because poverty reduces access to health care, education, and other essential components of development.

KC 1.1

KC 1.2

4. **Health care**, especially for children and mothers, is essential for a productive life. Underdeveloped areas such as that shown above can lead to disease, accidents, respiratory and digestive impairments, and other conditions. Without health, economic security is at risk, and poverty can persist through generations.

KC 1.3

5. **Sustainable cities** are key because over half of humanity now lives in cities. Sustainable development involves ensuring that cities are healthy places to live and that they cause minimal environmental impact.

KC 1.4

Environmental science is essential to sustainable development because it helps us understand how environmental systems work, how they are degraded, and what factors can help restore them. Studying environmental science can prepare you to aid human development and environmental quality, both at home and abroad, through better policies, resource protection, and planning.

KC 1.7

6. **Environmental policy** needs to guide decision making in local and national governments, to ensure that environmental quality is protected before it gets damaged, and to set agreed-upon rules for resource use.

7. **Protection of the atmosphere** is essential for minimizing the rate of climate change and for reducing impacts of air pollution on people, plants, and infrastructure.

KC 1.8

8. **Combating deforestation and protecting biodiversity** go together because much of the world's biodiversity is in forests. We also depend on forests for water resources, climate regulation, and resources including food, wood, medicines, and building materials. Other key zones of biodiversity include coral reefs, wetlands, and coastal areas.

9. **Combating desertification and drought** through better management of water resources can save farms, ecosystems, and lives. Often removal of vegetation and soil loss make drought worse, and a few bad rainfall years can convert a landscape to desert-like conditions.

10. **Agriculture and rural development** affect the lives of the nearly half of humanity who don't live in cities. Improving conditions for billions of rural people, including more sustainable farming systems, soil stewardship to help stabilize yields, and access to land, can help reduce populations in urban slums.

KC 1.5

KC 1.6

These ten ideas and others were described in Agenda 21 of the United Nations Conference on Environment and Development (the "Earth Summit") in Rio de Janeiro, Brazil, in 1992. Laying out priorities for stewardship of resources and equity in development, the document known as Agenda 21 was a statement of principles for guiding development policies. This document has no legal power, but it does represent an agreement in principle by the more than 200 countries participating in that 1992 conference.

CAN YOU **EXPLAIN?**

1. What is the relationship between environmental quality and health?

2. Why is sustainable development an issue for people in wealthy countries to consider?

3. Examine the central photo carefully. What health risks might affect the people you see? What do you suppose the rate of material consumption is here, compared to your neighborhood? Why?

1.4 SCIENCE HELPS US UNDERSTAND OUR WORLD

Because environmental questions are complex, we need orderly methods of examining and understanding them. Environmental science provides such an approach. In this section, we'll investigate what science is, what the scientific method is, and why that method is important.

What is science? **Science** (from *scire*, "to know" in Latin) is a process for producing knowledge based on observations (fig. 1.13). We develop or test theories (proposed explanations of how a process works) using these observations. "Science" also refers to the cumulative body of knowledge produced by many scientists. Science is valuable because it helps us understand the world and meet practical needs, such as finding new medicines, new energy sources, or new foods. In this section, we'll investigate how and why science follows standard methods.

Science rests on the assumption that the world is knowable and that we can learn about it by careful observation and logical reasoning (table 1.2). For early philosophers of science, this assumption was a radical departure from religious and philosophical approaches. In the Middle Ages the ultimate sources of knowledge about matters such as how crops grow, how diseases spread, or how the stars move were religious authorities or cultural traditions. Although these sources provided many useful insights, there was no way to test their explanations independently and objectively. The benefit of scientific thinking is that it searches for testable evidence. As evidence improves, we can seek better answers to important questions.

Science depends on skepticism and reproducibility

Ideally scientists are skeptical. They are cautious about accepting a proposed explanation until there is substantial evidence to support

▲ **FIGURE 1.13** Scientific studies rely on repeated, careful observations to establish confidence in their findings.

TABLE 1.2 | Basic Principles of Science

1. *Empiricism:* We can learn about the world by careful observation of empirical (real, observable) phenomena; we can expect to understand fundamental processes and natural laws by observation.

2. *Uniformitarianism:* Basic patterns and processes are uniform across time and space; the forces at work today are the same as those that shaped the world in the past, and they will continue to do so in the future.

3. *Parsimony:* When two plausible explanations are reasonable, the simpler (more parsimonious) one is preferable. This rule is also known as Ockham's razor, after the English philosopher who proposed it.

4. *Uncertainty:* Knowledge changes as new evidence appears, and explanations (theories) change with new evidence. Theories based on current evidence should be tested on additional evidence, with the understanding that new data may disprove the best theories.

5. *Repeatability:* Tests and experiments should be repeatable; if the same results cannot be reproduced, then the conclusions are probably incorrect.

6. *Proof is elusive:* We rarely expect science to provide absolute proof that a theory is correct, because new evidence may always improve on our current explanations. Even evolution, the cornerstone of modern biology, ecology, and other sciences, is referred to as a "theory" because of this principle.

7. *Testable questions:* To find out whether a theory is correct, it must be tested; we formulate testable statements (hypotheses) to test theories.

it. Even then, every explanation is considered only provisionally true, because there is always a possibility that some additional evidence may appear to disprove it. Scientists also aim to be methodical and unbiased. Because bias and methodical errors are hard to avoid, scientific tests are subject to review by informed peers, who can evaluate results and conclusions (fig. 1.14). The peer review process is an essential part of ensuring that scientists maintain good standards in study design, data collection, and interpretation of results.

Scientists demand **reproducibility** because they are cautious about accepting conclusions. Making an observation or obtaining a result just once doesn't count for much. You have to produce the same result consistently to be sure that your first outcome wasn't a fluke. Even more important, you must be able to describe the conditions of your study so that someone else can reproduce your findings. Repeating studies or tests is known as **replication**.

We use both deductive and inductive reasoning

Ideally scientists deduce conclusions from general laws that they know to be true. For example, if we know that massive objects attract each other (because of gravity), then it follows that an apple will fall to the ground when it releases from the tree. This logical reasoning from general to specific is known as **deductive reasoning**. Often, however, we do not know general laws that guide natural systems. Then we must rely on observations to find general rules. We observe, for example, that birds appear and disappear as a year goes by. Through many repeated observations in different places,

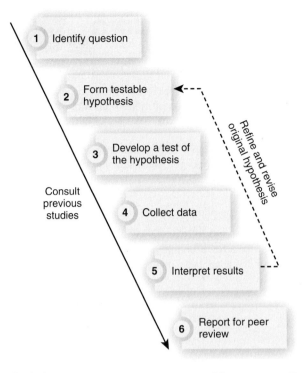

1 Identify question

2 Form testable hypothesis

3 Develop a test of the hypothesis

Consult previous studies

4 Collect data

Refine and revise original hypothesis

5 Interpret results

6 Report for peer review

▲ **FIGURE 1.14** Ideally, scientific investigation follows a series of logical, orderly steps to formulate and test hypotheses.

we can infer that the birds move from place to place in the spring and fall. We can develop a general rule that birds migrate seasonally. Reasoning from many observations to produce a general rule is **inductive reasoning**. Although deductive reasoning is more logically sound than inductive reasoning, it only works when our general laws are correct. We often rely on inductive reasoning to understand the world because we have few absolute laws.

Insight, creativity, and experience can also be essential in science. Often discoveries are made by investigators who are passionately interested in their subjects and who pursue hunches that appear unreasonable to other scientists. For example, some of our most basic understanding of plant genetics comes from the intuitive guesses of Barbara McClintock, a geneticist who discovered that genes in corn can move and recombine spontaneously. Where other corn geneticists saw random patterns of color and kernel size, McClintock's years of experience in corn breeding and her uncanny ability to recognize patterns led her to guess that genes could recombine in ways that no one had previously imagined. This intuition helped to transform our understanding of genetics.

The scientific method is an orderly way to examine problems

You may use the scientific method even if you don't think about it. Suppose you have a flashlight that doesn't work. The flashlight has several components (switch, bulb, batteries) that could be faulty. If you change all the components at once, your flashlight might work, but a more methodical series of tests will tell

you more about what was wrong with the system—knowledge that may be useful next time you have a faulty flashlight. So you decide to follow the standard scientific steps:

1. *Observe* that your flashlight doesn't light, and that there are three main components of the lighting system (batteries, bulb, and switch).
2. Propose a **hypothesis**, a testable explanation: "The flashlight doesn't work because the batteries are dead."
3. Develop a *test* of the hypothesis and *predict* the result that would indicate your hypothesis was correct: "I will replace the batteries; the light should then turn on."
4. Gather *data* from your test: After you replaced the batteries, did the light turn on?
5. *Interpret* your results: If the light works now, then your hypothesis was right; if not, then you should formulate a new hypothesis—perhaps that the bulb is faulty—and develop a new test for that hypothesis.

In systems more complex than a flashlight, it is almost always easier to prove a hypothesis wrong than to prove it unquestionably true. This is because we usually test our hypotheses with observations but there is no way to make every possible observation. The philosopher Ludwig Wittgenstein illustrated this problem as follows: Suppose you saw hundreds of swans, and all were white. These observations might lead you to hypothesize that all swans were white. You could test your hypothesis by viewing thousands of swans, and each observation might support your hypothesis, but you could never be entirely sure that it was correct. On the other hand, if you saw just one black swan, you would know with certainty that your hypothesis was wrong.

As you'll read in later chapters, the elusiveness of absolute proof is a persistent problem in environmental policy and law. Rarely can you absolutely prove that the toxic waste dump up the street is making you sick. You could, however, collect evidence to show that it is very probable that the waste has made you and your neighbors sick. The elusiveness of proof often decides environmental liability lawsuits (fig. 1.15).

When an explanation has been supported by a large number of tests, and when a majority of experts have reached a general consensus that it is a reliable description or explanation, we call it a **scientific theory**. Note that scientists' use of this term is very different from the way the public uses it. To many people, a theory is speculative and unsupported by facts. To a scientist, it means just the opposite: While all explanations are tentative and open to revision and correction, an explanation that counts as a scientific theory is supported by an overwhelming body of data and experience, and it is generally accepted by the scientific community, at least for the present.

Understanding probability reduces uncertainty

One strategy to improve confidence in the face of uncertainty is to focus on probability. **Probability** is a measure of how likely something is to occur. Usually probability estimates are based on a set of previous observations or on standard statistical measures.

▲ **FIGURE 1.15** How can you prove who is responsible for environmental contamination, such as the orange ooze in this stream? Careful, repeated measurements and well-formed hypotheses are essential.

Probability does not tell you what *will* happen, but it tells you what *is likely* to happen. If you hear on the news that you have a 20 percent chance of catching a cold this winter, that means that 20 of every 100 people are likely to catch a cold. This doesn't mean that *you* will catch one. In fact, it's more likely, an 80 percent chance, that you *won't* catch a cold. If you hear that 80 out of every 100 people will catch a cold, you still don't know whether you'll get sick, but there's a much higher chance that you will.

Science often involves probability, so it is important to be familiar with the idea. Sometimes probability has to do with random chance: If you flip a coin, you have a random chance of getting heads or tails. Every time you flip, you have the same 50 percent probability of getting heads. The chance of getting ten heads in a row is small (in fact, the chance is 1 in 2^{10}, or 1 in 1,024), but on any individual flip, you have exactly the same 50 percent chance, since this is a random test. Sometimes probability is weighted by circumstances: Suppose that about 10 percent of the students in your class earn an A each semester. Your likelihood of being in that 10 percent depends a great deal on how much time you spend studying, how many questions you ask in class, and other factors. Sometimes there is a combination of chance and circumstances: The probability that you will catch a cold this winter depends partly on whether you encounter someone who is sick (largely random chance) and also on whether you take steps to stay healthy (get enough rest, wash your hands frequently, eat a healthy diet, and so on).

Probability is often a more useful idea than proof. This is because absolute proof is hard to achieve, but we can frequently demonstrate a strong trend or relationship, one that is unlikely to be achieved by chance. For example, suppose you flipped a coin and got heads 20 times in a row. That could happen by chance, but it would be pretty unlikely. You might consider it very likely that there was a causal explanation, such as that the coin was weighted toward heads. Often we consider a causal explanation reliable (or "significant") if there is less than 5 percent probability that it happened by random chance (see Exploring Science, p. 15).

Experimental design can reduce bias

The study of fishing success (opening case study) is an example of an observational experiment, one in which you observe natural events and interpret a causal relationship between the variables. This kind of study is also called a **natural experiment**, one that involves observation of events that have already happened. Many scientists depend on natural experiments: A geologist, for instance, might want to study mountain building, or an ecologist might want to learn about how species evolve, but neither scientist can spend millions of years watching the process happen. Similarly, a toxicologist cannot give people a disease just to see how lethal it is.

Other scientists can use **manipulative experiments**, in which conditions are deliberately altered and all other variables are held constant. Most manipulative experiments are done in the laboratory, where conditions can be carefully controlled. Suppose you were interested in studying whether lawn chemicals contribute to deformities in tadpoles. You might keep two groups of tadpoles in fish tanks, and expose one to chemicals. In the lab you could ensure that both tanks had identical temperatures, light, food, and oxygen. By comparing a treatment (exposed) group and a control (unexposed) group, you have also made this a **controlled study**.

Often there is a risk of experimenter bias. Suppose the researcher sees a tadpole with a small nub that looks like it might become an extra leg. Whether she calls this nub a deformity might depend on whether she knows that the tadpole is in the treatment group or the control group. To avoid this bias, **blind experiments** are often used, in which the researcher doesn't know which group is treated until after the data have been analyzed. In health studies, such as tests of new drugs, **double-blind experiments** are used, in which neither the subject (who receives a drug or a placebo) nor the researcher knows who is in the treatment group and who is in the control group.

F or many people, a little bit of evidence is convincing. Suppose you lived on Apo Island (opening story) and you had a good day fishing. You might conclude that fisheries were doing really well. But how could you be sure your observation was right? Or that it represented something larger than one person's lucky day of fishing? Scientists try to be cautious and avoid jumping to conclusions, so they usually rely on *trends in data*, rather than *anecdotal observations*, when they evaluate the evidence.

Finding a trend in data requires collecting many observations, and usually we can see trends better by plotting those observations on graphs. Graphs are one of our easiest and most useful ways to see patterns and changes in something we care about. Here are a few ways we use graphs and data to increase our confidence in our conclusions.

1. ***A mean describes the middle of the group.*** To understand what's really going on with fishing rates, you can examine the number of fish caught per hour. Of course, some days are unlucky, and fishers vary in their fishing abilities. One way to describe the overall fishing rate per hour is by calculating the **mean** (average) number of fish caught in the hours a number of people spent fishing: Count all the fish caught in a day, and divide that count by the number of person-hours spent fishing. Try this for the four hypothetical fishing boats listed in the following table.

Fishing Rate

BOAT	NUMBER OF FISH	HOURS
1	10	1
2	15	4
3	5	2
4	20	3
Sum:	_____	_____
Number of fish per hour = _____		

Is this rate high or low? Suppose your records show that 10 years ago the mean was 12 fish per hour spent fishing. If you calculated a mean of 5 fish per hour earlier, you know the fishing rate has fallen.

2. ***Histograms show a group at a glance.*** You would know still more about the general situation if you examined more than just one village. After all, your village might be an *outlier*, or an unusual case. Records from 10 villages, or 50, would tell a much more complete story. In science, more observations always increase confidence in our conclusions.

You can't survey each of the thousands of villages in the Philippines, but you could examine a random subset, or sample, of them. The sample should be random to avoid biasing your results: if you just pick villages nearest a big city, for example, they might have especially poor fisheries. This might make your village fishery look great, but it might not tell you much that's useful. Once you have your sample, you can sort the observations and create a **histogram** to

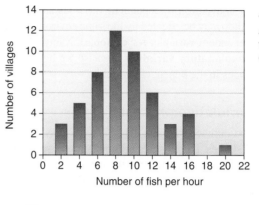

◀ **FIGURE 1**
A bar graph shows values for classes or groups.

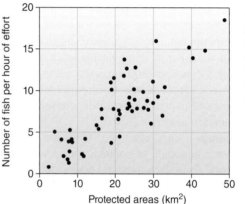

◀ **FIGURE 2**
A scatter plot shows the relationship of *x* and *y* for many observations. Here each observation is a village.

show the number of villages with small catches and large catches (fig. 1). The histogram here has many villages with about 6–11 fish per hour and just a few with 2 or 20. A bell-shaped distribution like this is known as a **normal distribution** because large sets of randomly selected observations tend to have this distribution.

3. ***Graphs show relationships.*** Suppose you suspect that the size of marine reserves tends to improve fishing rates. You can test this hypothesis by plotting your two variables against each other. Here the explanatory variable is the size of marine reserves (plotted on the horizontal axis, fig. 2) and fishing rates are the dependent variable (plotted on the vertical axis). Here the pattern of dots shows that fishing does increase as preserve size increases, for this set of villages. Now you can say with good confidence that there is a positive relationship between preserve size and fishing success.

Answering a question with numbers can give you confidence that you're drawing reasonable conclusions about your data. As a critical thinker, you can also examine the numbers in news reports and in public policies.

Are policy positions based on anecdotal evidence or on trends among many observations? Are samples selective and biased, or are they random? Asking these sorts of questions as you hear the news is a good practice for being a well-informed voter. Asking these questions may even help you do well in this course.

In each of these studies there is one **dependent variable** and one, or perhaps more, **independent variables**. The dependent variable, also known as a response variable, is affected by the independent variables. In a graph, the dependent variable is on the vertical (Y) axis, by convention. Independent variables are rarely really independent (they may be affected by the same environmental conditions as the dependent variable, for example). Many people prefer to call them **explanatory variables**, because we hope they will explain differences in the dependent variable.

Science is a cumulative process

The scientific method outlined in figure 1.14 is the process used to carry out individual studies. Larger-scale accumulation of scientific knowledge involves cooperation and contributions from countless people. Good science is rarely carried out by a single individual working in isolation. Instead, a community of scientists collaborates in a cumulative, self-correcting process. You often hear about big breakthroughs and dramatic discoveries that change our understanding overnight, but in reality these changes are usually the culmination of the labor of many people, each working on different aspects of a common problem, each adding small insights to solve a problem. Ideas and information are exchanged, debated, tested, and retested to arrive at **scientific consensus**, or general agreement among informed scholars.

The idea of consensus is important. For those not deeply involved in a subject, the multitude of contradictory results can be bewildering: Are coral reefs declining, and does it matter? Is climate changing, and how much? Among those who have done many studies and read many reports, there tends to emerge a general agreement about the state of a problem. Scientific consensus now holds that many coral reefs are in danger, as a result of pollution, physical damage, and warming seas. Consensus is that global climate conditions are changing, though models differ somewhat on how rapidly they will change in different regions.

Sometimes new ideas emerge that cause major shifts in scientific consensus. These great changes in explanatory frameworks were termed **paradigm shifts** by Thomas Kuhn (1967), who studied revolutions in scientific thought. According to Kuhn, paradigm shifts occur when a majority of scientists accept that the old explanation no longer describes new observations very well. For example, two centuries ago geologists explained many of the earth's features in terms of Noah's flood. The best scientists held that the flood created beaches well above modern sea level, scattered boulders erratically across the landscape, and gouged enormous valleys where there is no water now (fig. 1.16). Then the Swiss glaciologist Louis Agassiz and others suggested that the earth had once been much colder and that glaciers had covered large areas. Periodic ice ages proved to be a more durable explanation for geologic features than did a flood, and this new idea completely altered the way geologists explained their subject. Similarly, the idea of tectonic plate movement, in which continents shift slowly around the earth's surface (chapter 11), revolutionized the ways geologists, biogeographers, ecologists, and others explained the development of the earth and its life-forms.

▲ **FIGURE 1.16** Paradigm shifts change the ways we explain our world. Geologists now attribute Yosemite's valleys to glaciers, where once they believed catastrophes like Noah's flood carved its walls.

What is sound science?

Environmental science often deals with questions that are emotionally or politically charged. Scientific studies of climate change may be threatening to companies that sell coal and oil; studies of the health costs of pesticides worry companies that use or sell these chemicals. When controversy surrounds science, claims about **sound science** and accusations of "junk science" often arise. What do these terms mean, and how can you evaluate who is right?

When you hear arguments about whose science is valid, you need to remember the basic principles of science: Are the disputed studies reproducible? Are conclusions drawn with caution and skepticism? Are samples large and random? Are conclusions supported by a majority of scholars who have studied the problem? Do any of the experts have an economic interest in the outcome?

Often media figures on television or radio will take a position contrary to the scientific majority. A contrarian position gains them publicity and political allies (and sometimes money). This strategy has been especially popular around large issues such as climate change. For decades now, almost all climate scientists have agreed that human activities, such as fossil fuel burning and land clearing, are contributing to climate change. But it is always possible to find a contrarian scientist who is happy to contradict the majority of evidence. Especially when political favors, publicity, or money are involved, there is always "expert" witnesses who will testify on opposite sides of a case.

If you see claims of sound science and junk science, how can you evaluate them? How can you identify bogus analysis that is dressed up in quasi-scientific jargon but that has no objectivity? This is such an important question that astronomer Carl Sagan has proposed a "Baloney Detection Kit" (table 1.3) to help you out.

Uncertainty, proof, and group identity

Scientific uncertainty is frequently invoked as a reason to postpone actions that a vast majority of informed scientists consider to be prudent. In questions of chemical safety, energy conservation, climate change, or air pollution control, opponents of change may charge that the evidence doesn't constitute absolute proof,

TABLE 1.3 | Questions for Baloney Detection

1. How reliable are the sources of this claim? Is there reason to believe that they might have an agenda to pursue in this case?

2. Have the claims been verified by other sources? What data are presented in support of this opinion?

3. What position does the majority of the scientific community hold in this issue?

4. How does this claim fit with what we know about how the world works? Is this a reasonable assertion or does it contradict established theories?

5. Are the arguments balanced and logical? Have proponents of a particular position considered alternate points of view or only selected supportive evidence for their particular beliefs?

6. What do you know about the sources of funding for a particular position? Are studies financed by groups with partisan goals?

7. Where was evidence for competing theories published? Has it undergone impartial peer review or is it only in proprietary publication?

SOURCE: Carl Sagan, *The Demon Haunted World: Science as a Candle in the Dark* (1997).

so that no action needs to be taken. You will see examples of this in later chapters on environmental health, climate, air and water pollution, and other topics.

Similarly, disputes over evolution often hinge on the concept of uncertainty and proof in science. Opponents of teaching evolution in public schools often charge that because scientists call evolution a "theory," evolution is just a matter of conjecture. This is a confused use of terminology. The theory of evolution is supported by overwhelming evidence, but we still call it a theory because scientists prefer to be cautious about proof.

In recent years, sociologists have pointed out that our decisions to accept or dispute scientific evidence often depend on group identity. We like to associate with like-minded people, so we tend to adhere to a group viewpoint. Subconsciously we may ask, "Does the community I belong to agree with evolution? Does it accept the evidence for climate change?" Our urge to be agreeable to our group can be surprisingly strong, compared to our interest in critically analyzing evidence. Expectations of group behavior can shift over time, though. In decades past, you might have asked, "Am I the kind of person who recycles?" Today recycling is normal for most people, and few people probably decline to recycle just because their friends don't. Resolving differences on environmental policy sometimes requires recognition of group identity in our attitudes toward science, as well as our attitudes toward policies and issues beyond science.

1.5 CRITICAL THINKING

In science we frequently ask, "How do I know that what you just said is true?" Part of the way we evaluate arguments in science has to do with observable evidence, or data. Logical reasoning from evidence is also essential. And part of the answer lies in critical evaluation of evidence.

An ability to think critically, clearly, and analytically about a problem may be the most valuable skill you can learn in any of your classes. As you know by now, many issues in environmental science are hotly disputed, with firm opinions and plenty of evidence on both sides. How do you evaluate contradictory evidence and viewpoints? **Critical thinking** is a term we use to describe logical, orderly, analytical assessment of ideas, evidence, and arguments. Developing this skill is essential for the course you are taking now. Critical thinking is also an extremely important skill for your life in general. You can use it when you evaluate the claims of a car salesman, a credit card offer, or the campaign rhetoric of a political candidate.

Critical thinking helps us understand why prominent authorities can vehemently disagree about a topic. Disagreements may be based on contradictory data, on different interpretations of the same data, or on different priorities. One expert might consider economic health the overriding priority; another might prioritize environmental quality. A third might worry only about company stock prices, which might depend on the outcome of an environmental policy debate. You can examine the validity of contradictory claims by practicing critical thinking.

Critical thinking helps us analyze information

There are many different aspects of critical thinking. Much of the process is about carefully examining what you are trying to accomplish, and how you will know when you have the answer. Critical thinking also involves examining the source of information and how much that source influences the kind of information you receive. **Analytical thinking** helps you break a problem down into its constituent parts. **Creative thinking** asks, "How might I approach this problem in new and inventive ways?" **Logical thinking** evaluates whether the structure of your argument make sense. **Reflective thinking** asks, "What does it all mean?"

All these strategies will help you in this class. They can help you discover hidden ideas and meanings in claims, evaluate arguments, recognize the differences between facts and values, and avoid jumping to conclusions (table 1.4).

Notice that many critical thinking processes are self-reflective and self-correcting. Critical thinking isn't critical in the sense of finding fault; rather it is about identifying unspoken assumptions, beliefs, priorities, or motives (fig. 1.17). Uncovering these unspoken

TABLE 1.4 | Steps in Critical Thinking

1. What is the purpose of my thinking?

2. What precise question am I trying to answer?

3. Within what point of view am I thinking?

4. What information am I using?

5. How am I interpreting that information?

6. What concepts or ideas are central to my thinking?

7. What conclusions am I aiming toward?

8. What am I taking for granted; what assumptions am I making?

9. If I accept the conclusions, what are the implications?

10. What would the consequences be if I put my thoughts into action?

SOURCE: R. Paul, National Council for Critical Thinking.

▲ **FIGURE 1.17** Critical thinking evaluates premises, contradictions, and assumptions. Was this sign, in the middle of a popular beach near Chicago, the only way to reduce human exposure to bacteria? What other strategies might there be? Why was this one chosen? Who might be affected?

factors can contribute to honesty and humility in yourself, too: If you ask, "How do I know that what I just said is true?," then you are practicing critical thinking, and that question is likely to lead you to new and interesting insights.

We all use critical thinking to examine arguments

We all use critical or reflective thinking at times. Suppose a television commercial tells you that a new breakfast cereal is tasty and good for you. You might be skeptical and ask, What do they mean by *good*? Good for whom or what? Does *tasty* simply mean more sugar and salt? Might the sources of this information have other motives besides your health and happiness? You probably practice this kind of critical analysis regularly.

Here are some steps you can use in critical thinking:

1. *Identify and evaluate premises and conclusions in an argument.* What is the basis for the claims made? What evidence is presented to support these claims, and what conclusions are drawn from this evidence? If premises and evidence are reasonable, do the conclusions truly follow from them?

2. *Acknowledge and clarify uncertainties, vagueness, equivocation, and contradictions.* Are terms used in vague or ambiguous ways? Are all participants in an argument using the same meanings? Is ambiguity or equivocation deliberate?

3. *Distinguish between facts and values.* Can claims be tested, or are statements based on untestable assumptions and beliefs? Are claims made about the worth or lack of worth of something? (If so, these are value statements or opinions and probably cannot be verified objectively.)

4. *Recognize and assess assumptions.* Consider the backgrounds and views behind an argument: what underlying reasons might there be for the premises, evidence, or conclusions presented? Does anyone have an "axe to grind" or a personal agenda in this issue? What does s/he think you know, need, want, or believe? Do hidden biases based on race, gender, ethnicity, economics, or belief systems distort arguments?

5. *Distinguish source reliability or unreliability.* What qualifies the experts on this issue? Is that qualification sufficient for you to believe them? Why or why not?

6. *Recognize and understand conceptual frameworks.* What basic beliefs, attitudes, and values underlie an argument or action? How do these beliefs and values affect the way people view themselves and the world around them?

Critical thinking helps you learn environmental science

In this book you will have many opportunities to practice critical thinking skills. Every chapter includes facts, figures, opinions, and theories. Are all of them true? Probably not. They were the best information available when this text was written, but new evidence is always emerging. Data change constantly, as does our interpretation of data.

When reading this text, hearing the news, or watching television, try to distinguish between statements of fact and opinion. Ask yourself if the premises support the conclusions drawn from them. Although most of us try to be fair and even-handed in presenting controversies, our personal biases and values—some of which we may not even recognize—affect how we see issues and present arguments. Watch for cases in which you need to think for yourself, and use your critical and reflective thinking skills to uncover the truth.

You'll find more on critical thinking, as well as some useful tips on how to study effectively, on our website at www.mhhe.com/cunningham7e.

1.6 WHERE DO OUR IDEAS ABOUT THE ENVIRONMENT COME FROM?

Historically, many societies have degraded the resources on which they depended, while others have lived in relative harmony with their surroundings. Today our burgeoning population and our technologies that accelerate resource exploitation have given the problems of environmental degradation increased urgency.

Many of our current responses to these changes are rooted in the writings of relatively recent environmental thinkers. For simplicity, their work can be grouped into about four distinct stages: (1) resource conservation for optimal use; (2) nature preservation for moral and aesthetic reasons; (3) concern over health and ecological consequences of pollution; (4) global environmental citizenship. These stages are not mutually exclusive. You might embrace them all simultaneously. As you read this section, consider why you agree with those you find most appealing.

Environmental protection has historic roots

Recognizing human misuse of nature is not unique to modern times. Plato complained in the fourth century B.C. that Greece once was blessed with fertile soil and clothed with abundant forests of fine trees. After the trees were cut to build houses and ships, however, heavy rains washed the soil into the sea, leaving only a rocky "skeleton of a body wasted by disease." Springs and rivers dried up, and farming became all but impossible. Despite these early observations, most modern environmental ideas developed in response to resource depletion associated with more recent agricultural and industrial revolutions.

Some of the earliest recorded scientific studies of environmental damage were carried out in the eighteenth century by French or British colonial administrators, many of whom were trained scientists and who observed rapid soil loss and drying wells that resulted from intensive colonial production of sugar and other commodities. Some of these colonial administrators considered responsible environmental stewardship as an aesthetic and moral priority, as well as an economic necessity. These early conservationists observed and understood the connections between deforestation, soil erosion, and local climate change. The pioneering British plant physiologist Stephen Hales, for instance, suggested that conserving green plants preserves rainfall. His ideas were put into practice in 1764 on the Caribbean island of Tobago, where about 20 percent of the land was marked as "reserved in wood for rains."

Pierre Poivre, an early French governor of Mauritius, an island in the Indian Ocean, was appalled at the environmental and social devastation caused by destruction of wildlife (such as the flightless dodo) and the felling of ebony forests on the island by early European settlers. In 1769 Poivre ordered that one-quarter of the island be preserved in forests, particularly on steep mountain slopes and along waterways. Mauritius remains a model for balancing nature and human needs. Its forest reserves shelter more original species than are found on most other populated islands.

Resource waste triggered pragmatic resource conservation (stage 1)

Many historians consider the publication of *Man and Nature* in 1864 by geographer George Perkins Marsh as the wellspring of environmental protection in North America. Marsh, who also was a lawyer, politician, and diplomat, traveled widely around the Mediterranean as part of his diplomatic duties in Turkey and Italy. He read widely in the classics (including Plato) and personally observed the damage caused by excessive grazing by goats and sheep and by the deforestation of steep hillsides. Alarmed by the wanton destruction and profligate waste of resources still occurring on the American frontier in his lifetime, he warned of its ecological consequences. Largely because of his book, national forest reserves were established in the United States in 1873 to protect dwindling timber supplies and endangered watersheds.

Among those influenced by Marsh's warnings were U.S. President Theodore Roosevelt and his chief conservation adviser, Gifford Pinchot (fig. 1.18a and b). In 1905 Roosevelt, who was the leader of the populist Progressive movement, moved forest management out of the corruption-filled Interior Department into the Department of Agriculture. Pinchot, who was the first American-born professional forester, became the first chief of the new Forest Service. He put resource management on an honest, rational, and scientific basis for the first time in American history. Together with naturalists and activists such as John Muir, Roosevelt and Pinchot established the framework of the national forest, park, and wildlife refuge system. They passed game protection laws and tried to stop some of the most flagrant abuses of the public domain. In 1908 Pinchot organized and chaired the White House Conference on Natural Resources, perhaps the most prestigious and influential environmental meeting ever held in the United States. Pinchot also

(a) President Teddy Roosevelt

(b) Gifford Pinchot

(c) John Muir

(d) Aldo Leopold

▲ **FIGURE 1.18** Some early pioneers of the American conservation movement. (a) President Teddy Roosevelt and his main advisor (b) Gifford Pinchot emphasized pragmatic resource conservation, whereas (c) John Muir and (d) Aldo Leopold focused on ethical and aesthetic relationships.

was governor of Pennsylvania and founding head of the Tennessee Valley Authority, which provided inexpensive power to the southeastern United States.

The basis of Roosevelt's and Pinchot's policies was pragmatic **utilitarian conservation**. They argued that the forests should be saved, "not because they are beautiful or because they shelter wild creatures of the wilderness, but only to provide homes and jobs for people." Resources should be used "for the greatest good, for the greatest number, for the longest time." "There has been a fundamental misconception," Pinchot wrote, "that conservation means nothing but husbanding of resources for future generations. Nothing could be further from the truth. The first principle of conservation is development and use of the natural resources now existing on this continent for the benefit of the people who live here now. There may be just as much waste in neglecting the development and use of certain natural resources as there is in their destruction." This pragmatic approach still can be seen in the multiple-use policies of the U.S. Forest Service.

Ethical and aesthetic concerns inspired the preservation movement (stage 2)

John Muir (fig. 1.18c), amateur geologist, popular author, and first president of the Sierra Club, strenuously opposed Pinchot's utilitarian policies. Muir argued that nature deserves to exist for

▲ **FIGURE 1.19** A conservationist might say this forest was valuable as a supplier of useful resources, including timber and fresh water. A preservationist might argue that this ecosystem is important for its own sake. Many people are sympathetic with both outlooks.

its own sake, regardless of its usefulness to us. Aesthetic and spiritual values formed the core of his philosophy of nature protection. This outlook prioritizes **preservation** because it emphasizes the fundamental right of other organisms—and nature as a whole—to exist and to pursue their own interests (fig. 1.19). Muir wrote, "The world, we are told, was made for man. A presumption that is totally unsupported by the facts. . . . Nature's object in making animals and plants might possibly be first of all the happiness of each one of them. . . . Why ought man to value himself as more than an infinitely small unit of the one great unit of creation?"

Muir, who was an early explorer and interpreter of California's Sierra Nevada range, fought long and hard for establishment of Yosemite and Kings Canyon national parks. The National Park Service, established in 1916, was first headed by Muir's disciple, Stephen Mather, and has always been oriented toward preservation of nature rather than consumptive uses. Muir's preservationist ideas have often been at odds with Pinchot's utilitarian approach. One of Muir and Pinchot's biggest battles was over the damming of Hetch Hetchy Valley in Yosemite. Muir regarded flooding the valley a sacrilege against nature. Pinchot, who championed publicly owned utilities, viewed the dam as a way to free San Francisco residents from the clutches of greedy water and power monopolies.

In 1935, pioneering wildlife ecologist Aldo Leopold (fig. 1.18d) bought a small, worn-out farm in central Wisconsin. A dilapidated chicken shack, the only remaining building, was remodeled into a rustic cabin. Working together with his children, Leopold planted thousands of trees in a practical experiment in restoring the health and beauty of the land. "Conservation," he wrote, "is the positive exercise of skill and insight, not merely a negative exercise of abstinence or caution." The shack became a writing refuge and became the main focus of *A Sand County Almanac*, a much beloved collection of essays about our relation with nature. In it, Leopold wrote, "We abuse land because we regard it as a commodity belonging to us. When we see land as a community to which we belong, we may begin to use it with love and respect." Together with Bob Marshall and two others, Leopold was a founder of the Wilderness Society.

Rising pollution levels led to the modern environmental movement (stage 3)

The undesirable effects of pollution probably have been recognized as long as people have been building smoky fires. In 1723 the acrid coal smoke in London was so severe that King Edward I threatened to hang anyone who burned coal in the city. In 1661 the English diarist John Evelyn complained about the noxious air pollution caused by coal fires and factories and suggested that sweet-smelling trees be planted to purify city air. Increasingly dangerous smog attacks in Britain led, in 1880, to formation of a national Fog and Smoke Committee to combat this problem. But nearly a century later, London's air (like that of many cities) was still bad. In 1952 an especially bad episode turned midday skies dark and may have caused 12,000 deaths. This event was extreme, but noxious air was common in many large cities.

The tremendous expansion of chemical industries during and after World War II added a new set of concerns to the environmental agenda. *Silent Spring*, written by Rachel Carson (fig. 1.20a) and published in 1962, awakened the public to the threats of pollution and toxic chemicals to humans as well as other species. The movement she engendered might be called **modern environmentalism** because its concerns extended to include both natural resources and environmental pollution.

Two other pioneers of this movement were activist David Brower and scientist Barry Commoner (fig. 1.20b and c). Brower,

(a) Rachel Carson

(b) David Brower

(c) Barry Commoner

(d) Wangari Maathai

▲ **FIGURE 1.20** Among many distinguished environmental leaders in modern times, (a) Rachel Carson, (b) David Brower, (c) Barry Commoner, and (d) Wangari Maathai stand out for their dedication, innovation, and bravery.

as executive director of the Sierra Club, Friends of the Earth, and Earth Island Institute, introduced many of the techniques of environmental lobbying and activism, including litigation, testifying at regulatory hearings, book and calendar publishing, and use of mass media for publicity campaigns. Commoner, who was trained as a molecular biologist, has been a leader in analyzing the links between science, technology, and society. Both activism and research remain hallmarks of the modern environmental movement.

Under the leadership of a number of other brilliant and dedicated activists and scientists, the environmental agenda was expanded in the 1970s to most of the issues addressed in this textbook, such as human population growth, atomic weapons testing and atomic power, fossil fuel extraction and use, recycling, air and water pollution, and wilderness protection. Environmentalism has become well established in the public agenda since the first national Earth Day in 1970. A majority of Americans now consider themselves environmentalists, although there is considerable variation in what that term means.

Environmental quality is tied to social progress (stage 4)

In recent years some people have argued that the roots of the environmental movement are elitist—promoting the interests of a wealthy minority who can afford to vacation in wilderness. In fact, most environmental leaders have seen social justice and environmental equity as closely intertwined. Gifford Pinchot, Teddy Roosevelt, and John Muir all strove to keep nature accessible to everyone, at a time when public lands, forests, and waterways were increasingly controlled by a few wealthy individuals and private corporations. The idea of national parks, one of our principal strategies for nature conservation, is to provide public access to natural beauty and outdoor recreation. Aldo Leopold, a founder of the Wilderness Society, promoted ideas of land stewardship among farmers, fishers, and hunters. Robert Marshall, also a founder of the Wilderness Society, campaigned all his life for social and economic justice for low-income groups. Both Rachel Carson and Barry Commoner were principally interested in environmental health—an issue that is especially urgent for low-income, minority, and inner-city residents. Many of these individuals grew up in working-class families, so their sympathy with social concerns is not surprising.

Increasingly, environmental activists are making explicit the links between environmental quality and social progress on a global scale (fig. 1.21). The idea of sustainable development, an idea that has run through this chapter and is a core concept of modern environmental thought, is that economic improvement for the world's poorest populations is possible without devastating the environment.

Some of today's leading environmental thinkers come from developing nations, where poverty and environmental degradation together plague hundreds of millions of people. Dr. Wangari Maathai of Kenya (1940–2011) was a notable example. In 1977, Dr. Maathai (fig. 1.20d) founded the Green Belt Movement in her native Kenya as a way to both organize poor rural women and restore their environment. Beginning at a small, local scale, this organization has grown to more than 600 grassroots networks

▲ **FIGURE 1.21** Environmental scientists increasingly try to address both public health and environmental quality. The poorest populations often suffer most from environmental degradation.

across Kenya. They have planted more than 30 million trees while mobilizing communities for self-determination, justice, equity, poverty reduction, and environmental conservation. Dr. Maathai was elected to the Kenyan Parliament and served as Assistant Minister for Environment and Natural Resources. Her leadership has helped bring democracy and good government to her country. In 2004 she received the Nobel Peace Prize for her work, the first time a Nobel has been awarded for environmental action. In her acceptance speech she said, "Working together, we have proven that sustainable development is possible; that reforestation of degraded land is possible; and that exemplary governance is possible when ordinary citizens are informed, sensitized, mobilized and involved in direct action for their environment."

Photographs of the earth from space (see fig. 1.3) provide powerful icons for the fourth wave of ecological concern, which might be called **global environmentalism**. Such photos remind us how small, fragile, beautiful, and rare our home planet is. We all share an environment at this global scale. As Ambassador Adlai Stevenson noted in his 1965 farewell address to the United Nations, we now need to worry about the life-support systems of the planet as a whole. He went on to say in this speech, "We cannot maintain it half fortunate, half miserable, half confident, half despairing, half slave to the ancient enemies of mankind and half free in a liberation of resources undreamed of until this day. No craft, no crew, can travel with such vast contradictions. On their resolution depends the security of us all."

CONCLUSION

Environmental science gives us useful tools and ideas for understanding environmental problems and for finding new solutions to those problems. Environmental science draws on many disciplines, and on people with diverse interests, to understand the persistent problems we face, including human population growth, contaminated water and air, climate change, and biodiversity losses. There are also encouraging examples of progress. Population growth has slowed, the extent of habitat preserves has expanded greatly in recent years, we have promising new energy options, and in many regions we have made improvements in air and water quality.

Both poverty and affluence contribute to environmental degradation. Impoverished populations often overexploit land and water supplies, while wealthy populations consume or degrade extraordinary amounts of energy, water, forest products, food, and other resources. Differences in wealth lead to real contrasts in life expectancy, infant mortality, and other measures of well-being. Resolving these multiple problems together is the challenge for sustainability.

Our ideas about conservation and environment have evolved in response to environmental conditions, from a focus on conservation of usable resources to preservation of nature for its own sake. Throughout these ideas has been a concern for social equity, for the rights of low-income people to have access to resources and to a healthy environment. In recent years these twin concerns have expanded to recognize the possibilities of change in developing countries, and the global interconnections of environmental and social concerns.

Science gives us an orderly, methodical approach to examining environmental problems. The scientific method provides an approach to doing this: it involves developing a hypothesis and collecting data to test it. Ideally, scientists are skeptical about evidence and cautious about conclusions. This is one of the reasons they rely on analysis of repeated observations, on data, and on statistics to evaluate and comparing observations. Critical thinking also provides orderly steps for analyzing the assumptions and logic of arguments. Critical thinking is an essential skill worth developing both in school and in life.

PRACTICE QUIZ

1. Describe why fishing has changed at Apo Island, and the direct and indirect effects on people's lives.
2. What is the idea of "ecological services"? Give an example.
3. Distinguish between a hypothesis and a theory.
4. Describe the steps in the scientific method.
5. What is probability? Give an example.
6. Why are scientists generally skeptical? Why do tests require replication?
7. What is the first step in critical thinking, according to table 1.4?
8. Distinguish between utilitarian conservation and biocentric preservation. Name two environmental leaders associated with each of these philosophies.
9. Why do some experts regard water as the most critical natural resource for the twenty-first century?
10. Where in figure 1.6a does the most dramatic warming occur?
11. Describe some signs of progress in overcoming global environmental problems.
12. What is the link between poverty and environmental quality?
13. Define *sustainability* and *sustainable development*.

CRITICAL THINKING AND DISCUSSION

Apply the principles you have learned in this chapter to discuss these questions with other students.

1. How do you think the example of Apo Island's marine sanctuary meets the criteria of being scientifically sound, economically sustainable, and socially acceptable? If you were studying this situation, what information would you look for to support or refute this conclusion?
2. The analytical approaches of science are suitable for answering many questions. Are there some questions that science cannot answer? Why or why not?
3. Often opinions diverge sharply in controversial topics, such as the allowable size of fish catches, or the balance of environmental and economic priorities in land management. Think of a controversial topic with which you are familiar: What steps can you take to maintain objectivity and impartiality in evaluating the issue?
4. Does the world have enough resources for 8 or 10 billion people to live decent, secure, happy, fulfilling lives? What do those terms mean to you? Try to imagine what they mean to others in our global village.
5. Suppose you wanted to study the environmental impacts of a rich versus a poor country. What factors would you examine, and how would you compare them?

To understand trends and compare values in environmental science, we need to examine a great many numbers. Most people find it hard to quickly assess large amounts of data in a table. Graphing a set of data makes it easier to see patterns, trends, and relationships. For example, scatter plots show relationships between two variables, while bar graphs show the range of values in a set (figures 1 and 2). Reading graphs takes practice, but it is an essential skill that will serve you well in this course and others.

You will encounter several common types of graphs in this book. Go to the Data Analysis exercise on Connect to practice these skills and demonstrate your knowledge of how to read and use graphs.

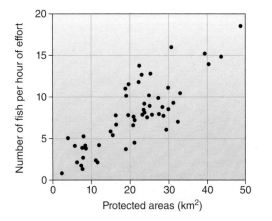
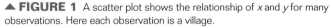

▲ **FIGURE 1** A scatter plot shows the relationship of *x* and *y* for many observations. Here each observation is a village.

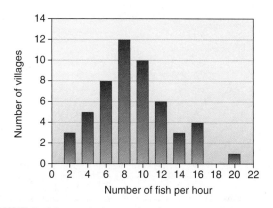

▲ **FIGURE 2** A bar graph shows values by groups or classes.

TO ACCESS ADDITIONAL RESOURCES FOR THIS CHAPTER, PLEASE VISIT CONNECT AT www.mcgrawhillconnect.com.

You will find LearnSmart, an adaptive learning system, Google Earth™ exercises, additional Case Studies, Data Analysis exercises, and an interactive ebook.

Environmental Systems: Matter, Energy, and Life

Chesapeake Bay's ecosystem supports fisheries, recreation, and communities. But the estuary is an ecosystem out of balance.

LEARNING OUTCOMES

After studying this chapter, you should be able to answer the following questions:

▶ What are systems, and how do feedback loops affect them?

▶ Explain the first and second laws of thermodynamics.

▶ Ecologists say there is no "away" to throw things to, and that everything in the universe tends to slow down and fall apart. What do they mean?

▶ Explain the processes of photosynthesis and respiration.

▶ What qualities make water so unique and essential for life as we know it?

▶ Why are big, fierce animals rare?

▶ How and why do elements such as carbon, nitrogen, phosphate, and sulfur cycle through ecosystems?

CASE STUDY

Working to Rescue an Ecosystem

Each year Chesapeake Bay, the largest estuary in the United States, gets a report card, just as you do at the end of a semester. Like your report card, this one summarizes several key performance measures. Unlike your grades, the bay's grades are based on measures such as water clarity, oxygen levels, health of sea grass beds, and the condition of the microscopic plankton community. These factors reflect the overall stability of fish and shellfish populations, critical to the region's ecosystems and economy. Since record keeping began, the bay's performance has been poor, with scores hovering between 35 and 57 out of 100, and an average grade of low C–. The main reason for the bad grades? Excessive levels of nitrogen and phosphorus, two common life-supporting elements that have destabilized the ecosystem.

Chesapeake Bay's watershed is a vast and complex system, with over 17,600 km (11,000 mi) of tidal shoreline in 6 states, and a population of 20 million people. Approximately 100,000 streams and rivers drain into the bay. All these streams carry runoff from forests, farmlands, cities, and suburbs from as far away as New York (fig. 2.1a).

The system has consistently bad grades, but it's clearly worth saving. Even in its impaired state, the bay provides 240 million kg (500 million lb) of seafood every year. It supports fishing and recreational economies worth $33 billion a year. But this is just a fraction of what it should be. The bay once provided abundant harvests of oysters, blue crabs, rockfish, white perch, shad, sturgeon, flounder, eel, menhaden, alewives, and soft-shell clams. Overharvesting, disease, and declining ecosystem productivity have decimated fisheries. Blue crabs are just above population survival levels. The oyster harvest, which was 15 to 20 million bushels per year in the 1890s, has declined to less than 1 percent of that amount. According to the Environmental Protection Agency (EPA), the bay should support more than twice the fish and shellfish populations that are there today. Human health is also at risk. After heavy rainfall, people are advised to stay out of the water for 48 hours, to avoid contamination from sewer overflows and urban and agricultural runoff.

Among the many challenges for Chesapeake Bay, the principal problem is simply excessive levels of nitrogen and phosphorus. These two elements are essential for life, but the system is overloaded by excess loads from farm fields, livestock manure, urban streets, suburban lawn fertilizers, half a million aging household septic systems, and the legal discharges of over 3,000 sewage treatment plants. Air pollution from cars, power plants, and factories also introduces nitrogen to the bay (fig. 2.1b). Sediment is also a key issue: it washes in from fields and streets, smothers eelgrass beds, and blocks sunlight, reducing photosynthesis in the bay.

Just as too many donuts are bad for you, an excessive diet of nutrients is bad for an estuary. Excess nutrients fertilize superabundant growth of algae, which further blocks sunlight and reduces photosynthesis and oxygen levels in the bay. Lifeless, oxygen-depleted areas result, leading to fish die-offs as well as poor reproduction in oysters, crabs, and fish. These algal blooms in nutrient-enriched waters are increasingly common in bays and estuaries worldwide.

Progress has been discouragingly slow for decades, but in 2010 the EPA finally addressed the problem seriously, complying with its charge from Congress (under the Clean Water Act) to protect the bay. Where piecemeal, mostly voluntary efforts by individual states had long failed to improve the Chesapeake's report cards, the EPA brought all neighboring states to the negotiating table. Total maximum daily loads (TMDLs) for nutrients and sediments were established, and states were given freedom to decide how to meet their share of nitrogen reductions. But the EPA has legal authority from the Clean Water Act to enforce reductions. The aim is to cut nitrogen levels by 25 percent, phosphorus by 24 percent, and sediment by 20 percent. The nitrogen target of 85 million kg (186 million lb) per year is still 4–5 times greater than would be released by an undisturbed watershed, but it's a huge improvement.

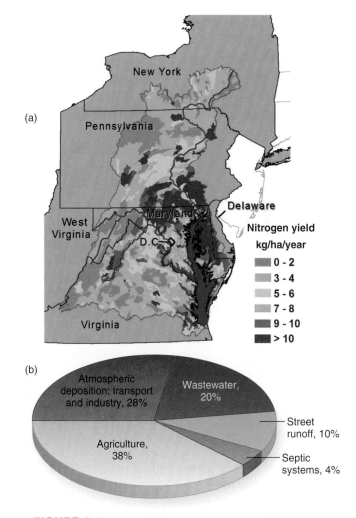

▲ **FIGURE 2.1** America's largest and richest estuary, Chesapeake Bay (shown in blue on map) suffers from pollutants from six states (a), and many sources (b).

(continued)

States from Virginia to New York have chosen their own strategies to meet limits. Maryland plans to capture and sell nitrogen and phosphorus from chicken manure. New York promises better urban wastewater treatment. Pennsylvania is strengthening soil conservation efforts to retain nutrients on farmland. These plans will be implemented gradually, but together, by addressing upstream land uses, they seem likely to turn around this magnificent estuary.

Chesapeake Bay has long been a symbol of the intractable difficulty of managing large, complex systems. Progress has required better understanding of several issues: the integrated functioning of the uplands and the waterways, the interdependence of the diverse human communities and economies that depend on the bay, and the pathways of nitrogen and phosphorus through an ecosystem. In this chapter we'll examine how these and other elements move through systems, and why they are important. Understanding these basic ideas will help you explain the functioning of many different systems, including Chesapeake Bay, your local ecosystem, even your own body.

Environmental scientists have led the way to the EPA's solution with years of ecosystem research and data collection. Through their efforts, and with EPA leadership, Chesapeake Bay could become the largest, and perhaps the most broadly beneficial, ecosystem restoration ever attempted in the United States. ∎

Most institutions demand unqualified faith; but the institution of science makes skepticism a virtue.
—ROBERT KING MERTON

2.1 SYSTEMS DESCRIBE INTERACTIONS

Managing nutrients and water in the Chesapeake Bay is an effort to restore a stable system, one with an equal amount of inputs and outputs and with balanced populations of animals and plants. This balance maintains overall stability and prevents dramatic change or collapse. In general, a **system** is a network of interdependent components and processes, with materials and energy flowing from one component of the system to another. The term "ecosystem" is probably familiar to you. This simple word represents complex assemblages of animals, plants, and their environment, through which materials and energy move. In a sense, you are a system consisting of billions of cells and thousands of species that live in or on your body, as well as the energy and matter that move through you.

The idea of systems is useful because it helps us organize our thoughts about the inconceivably complex phenomena around us. For example, an ecosystem might consist of countless animals, plants, and their physical surroundings. Keeping track of all the elements and relationships in an ecosystem would probably be an impossible task. But if we step back and think about components in terms of their roles—plants, herbivores, carnivores, and decomposers—and the relationships among them, then we can start to comprehend how the system works (fig. 2.2).

We can use some general terms to describe the components of a system. A simple system consists of compartments (also called state variables), which store resources such as energy or matter, and the flows, or the pathways, by which those resources move from one compartment to another. In figure 2.2, the plants and animals represent elements in this cycle of life. We can describe the flows in terms of

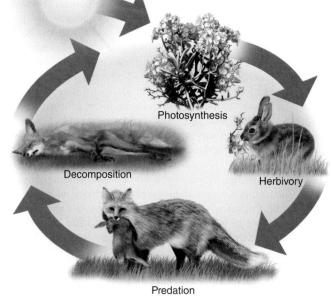

Photosynthesis

Herbivory

Decomposition

Predation

▲ **FIGURE 2.2** A system can be described in very simple terms.

photosynthesis, herbivory, predation, and decomposition. The plants use photosynthesis to create carbohydrates from carbon, water, and sunlight. The rabbit is a herbivore that consumes plants and uses the energy and chemicals stored there for its own life functions. The fox, in turn, is a carnivore that eats other animals, while the decomposers recycle wastes from all previous compartments.

It may seem cold and analytical to describe a rabbit or a flower as a state variable, but it is also helpful to do so. The energy and matter in the flower or rabbit are really the same, they just change their physical location (or state) when moving from one organism to another.

Understanding the characteristics of ecological systems can help us diagnose disturbances or changes in those systems: for example, if rabbits become too numerous, herbivory can become too rapid for plants to sustain. This could lead to collapse of the system. Similarly, in the Chesapeake Bay, excess nitrogen and phosphorus have led to excess algal growth—this results in reduced oxygen levels in the water, which kills fish and other aquatic organisms. Let's examine some of the common characteristics we can find in systems.

Systems can be described in terms of their characteristics

Open systems are those that receive inputs from their surroundings and produce outputs that leave the system. The Chesapeake Bay is an open system. Fresh water, nutrients, and some organism flow into the bay from rivers while energy comes from the sun. And other chemicals, water, and organisms (along with the energy they've captured) flow out of the bay into the ocean. In general, all natural systems are open to some extent.

In theory, a **closed system** exchanges no energy or matter with its surroundings, but these are rare. Some biological communities may appear to be closed. For example, a carefully balanced aquarium with just a few plants, fish, and decomposers or detritus-eaters may exist for a long time without any input of external materials, but plants need light to survive, and excess heat must be dissipated into the surrounding environment. So although closed to material flow, the aquarium is still open with respect to energy. We'll discuss why this is, later in this chapter.

Throughput is the flow of energy and matter into, through, and out of a system. We can measure the throughput of materials through the Chesapeake Bay, for example: If 100 million kg of nitrogen enters the bay, then ideally 100 million kg will be exported. Some will flow out to sea, some will leave the bay with fish or birds that feed on nutrient-enriched phytoplankton or aquatic plants. Ecologists have observed, though, that this level of throughput supports an overabundance of algae that blocks sunlight in the water column. To return the system to one that includes more fish and oysters, throughput should be reduced to 85 million kg of N or less.

Throughput is a useful idea in many systems. Compare the throughputs in houses, for example. An uninsulated house might consume and emit millions of calories of heat in a winter month, as the furnace struggles to keep a steady temperature inside. An efficiently insulated house might consume (and lose) a small fraction of that heat. Households also vary in throughput of material goods, water, food, and other resources.

Ideally systems tend to be stable over time. If the Chesapeake's nitrogen inputs equal outputs, then the amount of phytoplankton, algae, and other organisms should remain relatively constant. When a system is in a stable balance, we say it is in **equilibrium**. Systems can change, though, sometimes suddenly. Often there are **thresholds** or "tipping points," where rapid change suddenly occurs if you pass certain limits. Chesapeake Bay's oxygen levels may decrease gradually without much visible effect until they reach a level at which fish can't survive. Then suddenly, decaying bodies of a massive fish kill release toxins and use up even more oxygen, which throws the whole system into a downward spiral from which it's difficult to recover.

Feedback loops help stabilize systems

Systems function in cycles, with each component eventually feeding back to influence the size or rate of itself. A **positive feedback loop** tends to increase a process or component, whereas a **negative feedback loop** diminishes it. Feedbacks occur in countless familiar systems. Think of a sound system, when a microphone picks up the sound coming from speakers. The microphone amplifies the sound and sends it back to the speakers, which get louder and louder (a positive feedback) until the speakers blow out or someone cuts the power. Your body has feedback loops that regulate everything from growth and development to internal temperature. When you exercise and get hot, you sweat, and evaporation cools your skin (a negative feedback) to maintain a stable temperature. When you are cold, you shiver, and that activity helps return your temperature to normal.

Positive and negative feedbacks are a fundamental part of ecosystems. Consider the simple system in figure 2.2. A pair of rabbits produces several baby rabbits, and reproduction can lead to greater and greater numbers of rabbits (a positive feedback). If there were an unlimited food supply, the rabbit population would increase indefinitely (fig. 2.3). But if the growing rabbit population

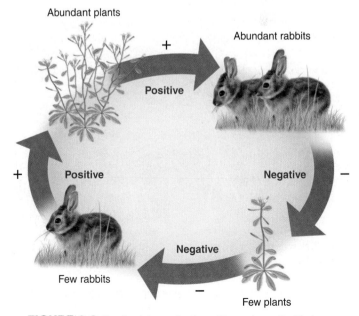

▲ **FIGURE 2.3** Feedback loops (both positive and negative) help regulate and stabilize processes in ecosystems.

depletes its food supply, starvation will slow or reverse population growth (a negative feedback). When there are fewer rabbits, the plants recover and the cycle starts again.

Multiple loops interact in ecosystems. As the number of rabbits rises, the fox in figure 2.2 has more to eat and will raise more healthy fox kits. A growing fox population reduces the number of rabbits, which in turn allows plants to increase. In general, the number of plants, rabbits, and foxes should oscillate, or vary, around an average that remains stable over time. Often we call this variation around a stable average a "dynamic equilibrium." It's not quite equilibrium because numbers vary, but the overall trend is relatively stable.

One of the more important global feedback systems now being disrupted is Arctic sea ice. As you can read in chapter 9, we have doubled the amount of heat-trapping greenhouse gases in the atmosphere, and resulting warming is melting much of the ice in the Arctic Ocean. Normally Arctic sea ice reflects most sunlight back to space, but open water absorbs sunlight. Melting thus leads to more warming, which leads to more melting, in a positive feedback loop that is affecting the state of temperatures globally.

Disturbances, events such as fire, flooding, climate change, invasion by new species, or destructive human activities, also interrupt normal feedback loops. Sometimes systems are resilient, so that they return to something like their previous states after a disturbance. Sometimes disturbances causes a system to shift to a new state, so that conditions become permanently altered.

Emergent properties are another interesting aspect of systems, in which the characteristics of a whole system are greater than the sum of its parts. Consider mangrove forests, which grow along coastlines throughout the tropical world (fig. 2.4). Because their direct economic value can be modest, thousands of kilometers of mangrove-lined coasts have been cleared to make room for shrimp ponds or beach resorts. But the mangrove forest provides countless

▲ **FIGURE 2.4** Mangrove forests, like many complex ecosystems, have emergent properties beyond being mere collection of trees. They provide many valuable ecosystem services, such as stabilizing shorelines, providing habitat, and protecting the land from tidal waves and storms.

indirect economic values. Prop-roots trap sediment and stabilize mud flats, improving water clarity and protecting offshore coral reefs from silt. Mangroves also protect small fish and other marine species. Many commercial ocean fish and shrimp species spend juvenile stages hiding among the mangrove roots. And the forest protects the shoreline from damaging storms and tsunamis. Collectively these services can be worth hundreds of times the value of a shrimp farm, but they exist only when the many parts of the system function together.

2.2 ELEMENTS OF LIFE

What exactly are the materials that flow through a system like the Chesapeake Bay? If a principal concern is the control of nutrients, such as NO_3 and PO_4, as well as maintaining O_2 levels, just what exactly are these elements, and why are they important? What does "O_2" or "NO_3" mean? In this section we will examine matter and the elements and compounds on which all life depends. In the sections that follow, you'll consider how organisms use those elements and compounds to capture and store solar energy, and how materials cycle through global systems, as well as ecosystems.

To understand how these compounds form and move, we need to begin with some of the fundamental properties of matter and energy.

Matter is recycled but doesn't disappear

Everything that takes up space and has mass is **matter**. Matter exists in three distinct states—solid, liquid, and gas—due to differences in the arrangement of its constitutive particles. Water, for example, can exist as ice (solid), as liquid water, or as water vapor (gas).

Matter also behaves according to the principle of **conservation of matter**: Under ordinary circumstances, matter is neither created nor destroyed but rather is recycled over and over again. It can be transformed or recombined, but it doesn't disappear; everything goes somewhere. Some of the molecules that make up your body probably contain atoms that once were part of the body of a dinosaur and most certainly were part of many smaller prehistoric organisms, as chemical elements have been used and reused by living organisms.

How does this principle apply to human relationships with the biosphere? Particularly in affluent societies, we use natural resources to produce an incredible amount of "disposable" consumer goods. If everything goes somewhere, where do the things we dispose of go after the garbage truck leaves? As the sheer amount of "disposed-of stuff" increases, we are having greater problems finding places to put it. Ultimately, there is no "away" where we can throw things we don't want any more.

Elements have predictable characteristics

Matter consists of **elements** such as P (phosphorus) or N (nitrogen), which are substances that cannot be broken down into simpler forms by ordinary chemical reactions. Each of the 116 accepted elements (92 natural, plus 24 created under special conditions) has distinct chemical characteristics.

Just four elements—oxygen, carbon, hydrogen, and nitrogen (symbolized as O, C, H, and N)—make up more than 96 percent of the mass of most living organisms. Water is composed of two H atoms and one O atom (written H_2O). All the elements are listed in the Periodic Table of the Elements, which you can find on the foldout map in the back of this book. Often, though, it's enough to pay attention to just a few (table 2.1).

Atoms are the smallest particles that exhibit the characteristics of an element. As difficult as it may be to imagine when you look at a solid object, all matter is composed of tiny, moving particles, separated by space and held together by energy. Atoms are composed of a nucleus, made of positively charged protons and electrically neutral neutrons, circled constantly by negatively charged electrons (fig. 2.5). Electrons, which are tiny in comparison to the other particles, orbit the nucleus at the speed of light.

Each element is listed in the periodic table according to the number of protons per atom, called its **atomic number**. The number of neutrons in the atoms of an element can vary slightly. Thus the atomic mass, which is the sum of the protons and neutrons in each nucleus, also can vary. We call forms of a single element that differ in atomic mass **isotopes**. For example, hydrogen, the lightest element, normally has only one proton (and no neutrons) in its nucleus. A small percentage of hydrogen atoms have one proton and one neutron. We call this isotope deuterium (^2H). An even smaller percentage of natural hydrogen occurs as the isotope tritium (^3H), with one proton plus two neutrons. The heavy form of nitrogen (^{15}N) has one more neutron in its nucleus than does the more common ^{14}N. Both of these nitrogen isotopes are stable, but some isotopes are unstable—that is, they may spontaneously emit electromagnetic energy, or subatomic particles, or both. Radioactive waste and nuclear energy result from unstable isotopes of elements such as uranium and plutonium.

Why should you know about isotopes? Although you might not often think of it this way, both stable and unstable types are in

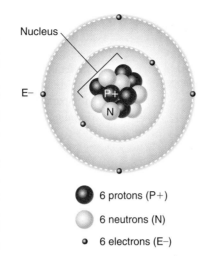

Nucleus

E−

P+
N

6 protons (P+)

6 neutrons (N)

6 electrons (E−)

◀ FIGURE 2.5 This model represents a carbon-12 atom, containing six protons and six neutrons in its nucleus, as well as six electrons orbiting in outer shells. Many physicists say electrons should be shown as a fuzzy cloud of wave energy rather than individual particles.

the news every day. Unstable, radioactive isotopes are distributed in the environment by both radioactive waste (potentially from nuclear power plants) and nuclear bombs. Every time you hear debates about nuclear power plants—which might produce the energy that powers the lights you are using right now—or about nuclear weapons in international politics, the core issue is radioactive isotopes. Understanding why they are dangerous, because emitted particles damage living cells, is important as you form your opinions about these policy issues.

Stable isotopes—those that do not change mass by losing neutrons—are important because they help us understand climate history and many other environmental processes. This is because lightweight isotopes move differently than heavier ones do. For example, oxygen occurs as both ^{16}O (a lighter isotope) and ^{18}O (a heavier isotope). Some water molecules (H_2O) contain the lightweight oxygen isotopes. These lightweight molecules evaporate, and turn into rain or snowfall, *slightly* more easily than heavier water molecules, especially in cool weather. Consequently, ice stored in glaciers during cool periods in the earth's history will contain slightly higher proportions of light-isotope water, with ^{16}O. Ice from warm periods contains relatively higher proportions of ^{18}O. By examining the proportions of isotopes in ancient ice layers, climate scientists can deduce the earth's temperature from hundreds of thousands of years ago (see chapter 9).

So although atoms, elements, and isotopes may seem a little arcane, they actually are fundamental to understanding climate change, nuclear weapons, and energy policy—all issues you can read about in the news almost any day of the week.

Electric charges keep atoms together

Atoms frequently gain or lose electrons, acquiring a negative or positive electrical charge. Charged atoms (or combinations of atoms) are called **ions**. Negatively charged ions (with one or more extra electrons) are *anions*. Positively charged ions are *cations*. A sodium (Na) atom, for example, can give up an electron to become a sodium ion (Na^+). Chlorine (Cl) readily gains electrons, forming chlorine ions (Cl^-).

Atoms often join to form **compounds**, or substances composed of different kinds of atoms (fig. 2.6). A pair or group of atoms that can exist as a single unit is known as a **molecule**. Some elements commonly occur as molecules, such as molecular oxygen (O_2) or molecular nitrogen (N_2), and some compounds can exist as molecules, such as glucose ($C_6H_{12}O_6$). In contrast to these molecules, sodium chloride (NaCl, table salt) is a compound that

TABLE 2.1 | Functions of Some Common Elements

FUNCTION	ELEMENTS	COMMENTS
Fertilizers	N nitrogen P phosphorus K potassium	Essential components of proteins, cells, other biological compounds; essential fertilizers for plants.
Organic compounds	C carbon O oxygen H hydrogen	Form the basic structure of cells and other components of living things, in combination with many other elements.
Metals	Fe iron Al aluminum Au gold	Generally malleable; most (not all) react readily with other elements
Toxic elements	Pb lead Hg mercury As arsenic	Many are metals that can interfere with processes in nervous systems

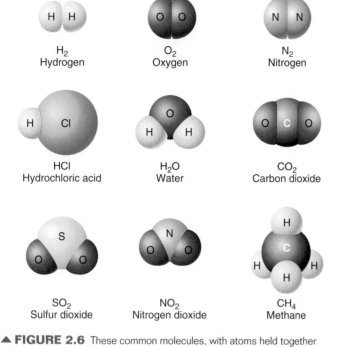

H₂
Hydrogen

O₂
Oxygen

N₂
Nitrogen

HCl
Hydrochloric acid

H₂O
Water

CO₂
Carbon dioxide

SO₂
Sulfur dioxide

NO₂
Nitrogen dioxide

CH₄
Methane

▲ **FIGURE 2.6** These common molecules, with atoms held together by covalent bonds, are important components of the atmosphere or are important pollutants.

cannot exist as a single pair of atoms. Instead it occurs in a large mass of Na and Cl atoms or as two ions, Na⁺ and Cl⁻, in solution. Most molecules consist of only a few atoms. Others, such as proteins, can include millions or even billions of atoms.

When ions with opposite charges form a compound, the electrical attraction holding them together is an *ionic bond*. Sometimes atoms form bonds by *sharing* electrons. For example, two hydrogen atoms can bond by sharing a single electron—it orbits the two hydrogen nuclei equally and holds the atoms together. Such electron-sharing bonds are known as *covalent bonds*. Carbon (C) can form covalent bonds simultaneously with four other atoms, so carbon can create complex structures such as sugars and proteins. Atoms in covalent bonds do not always share electrons evenly. An important example in environmental science is the covalent bonds in water (H_2O). The oxygen atom attracts the shared electrons more strongly than do the two hydrogen atoms. Consequently, the hydrogen portion of the molecule has a slight positive charge, while the oxygen has a slight negative charge. These charges create a mild attraction between water molecules, so that water tends to be somewhat cohesive. This fact helps explain some of the remarkable properties of water (see Exploring Science, p. 35).

When an atom gives up one or more electrons, we say it is *oxidized* (because it is very often oxygen that takes the electron, as bonds are formed with this very common and highly reactive element). When an atom gains electrons, we say it is *reduced*. Chemical reactions necessary for life involve oxidation and reduction: Oxidation of sugar and starch molecules, for example, is an important part of how you gain energy from food.

Breaking bonds requires energy, whereas forming bonds generally releases energy. Burning wood in your fireplace breaks up large molecules, such as cellulose, and forms many smaller ones, such as carbon dioxide and water. The net result is a release of energy (heat). Generally, some energy input (activation energy) is needed to initiate these reactions. In your fireplace, a match might provide the needed activation energy. In your car, a spark from the sparkplug provides activation energy to initiate the oxidation (burning) of gasoline.

Acids and bases release reactive H⁺ and OH⁻

Substances that readily give up hydrogen ions in water are known as **acids**. Hydrochloric acid, for example, dissociates in water to form H⁺ and Cl⁻ ions. In later chapters, you may read about acid rain (which has an abundance of H⁺ ions), acid mine drainage, and many other environmental problems involving acids. In general, acids cause environmental damage because the H⁺ ions react readily with living tissues (such as your skin or tissues of fish larvae) and with nonliving substances (such as the limestone on buildings, which erodes under acid rain).

Substances that readily bond with H⁺ ions are called **bases** or alkaline substances. Sodium hydroxide (NaOH), for example, releases hydroxide ions (OH⁻) that bond with H⁺ ions in water. Bases can be highly reactive, so they also cause significant environmental problems. Acids and bases can also be essential to living things: The acids in your stomach help you digest food, for example, and acids in soil help make nutrients available to growing plants.

We describe acids and bases in terms of **pH**, the negative logarithm of its concentration of H⁺ ions (fig. 2.7). Acids have a pH below 7; bases have a pH greater than 7. A solution of exactly pH 7 is "neutral." Because the pH scale is logarithmic, pH 6 represents *ten times* more hydrogen ions in solution than pH 7.

A solution can be neutralized by adding buffers, or substances that accept or release hydrogen ions. In the environment, for example, alkaline rock can buffer acidic precipitation, decreasing its acidity. Lakes with acidic bedrock, such as granite, are especially vulnerable to acid rain because they have little buffering capacity.

Organic compounds have a carbon backbone

Organisms use some elements in abundance, others in trace amounts, and others not at all. Certain vital substances are concentrated within cells, while others are actively excluded. Carbon is a particularly important element because chains and rings of carbon atoms form the skeletons of **organic compounds**, the material of which biomolecules, and therefore living organisms, are made.

The four major categories of organic compounds in living things ("bio-organic compounds") are lipids, carbohydrates, proteins, and nucleic acids. Lipids (including fats and oils) store energy for cells, and they provide the core of cell membranes and other structures. Many hormones are also lipids. Lipids do not readily dissolve in water, and their structure is a chain of carbon atoms with attached hydrogen atoms. This structure makes them part of the family of hydrocarbons (fig. 2.8a). Carbohydrates (including sugars, starches, and cellulose) also store energy and provide structure to cells. Like lipids, carbohydrates have a basic

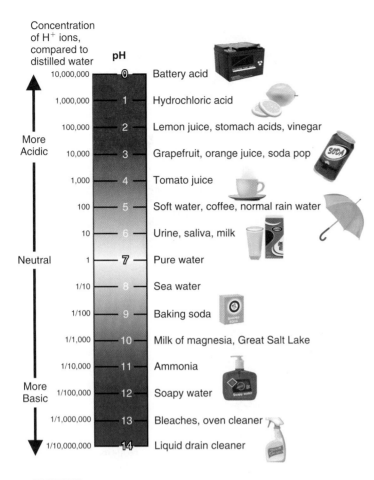

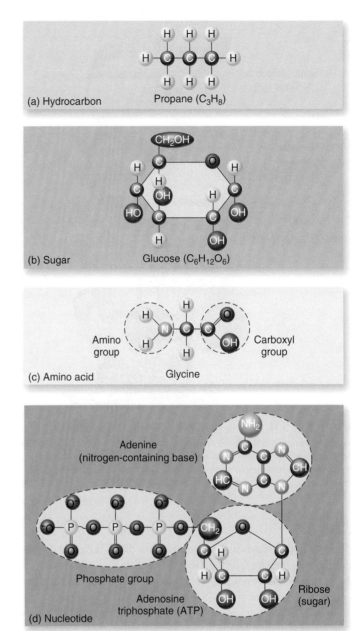

▲ FIGURE 2.7 The pH scale. The numbers represent the negative logarithm of the hydrogen ion concentration in water. Alkaline (basic) solutions have a pH greater than 7. Acids (pH less than 7) have high concentrations of reactive H^+ ions.

▲ FIGURE 2.8 The four major groups of biologically important organic molecules are based on repeating subunits of these carbon-based structures. Basic structures are shown for (a) butyric acid (a building block of lipids) and a hydrocarbon, (b) a simple carbohydrate, (c) a protein, and (d) a nucleotide (a component of nucleic acids).

structure of carbon atoms, but hydroxyl (OH) groups replace half the hydrogen atoms in their basic structure, and they usually consist of long chains of simple sugars. Glucose (fig. 2.8b) is an example of a very simple sugar.

Proteins are composed of chains of subunits called amino acids (fig. 2.8c). Folded into complex three-dimensional shapes, proteins provide structure to cells and are used for countless cell functions. Enzymes, such as those that release energy from lipids and carbohydrates, are proteins. Proteins also help identify disease-causing microbes, make muscles move, transport oxygen to cells, and regulate cell activity.

Nucleotides are complex molecules made of a five-carbon sugar (ribose or deoxyribose), one or more phosphate groups, and an organic nitrogen-containing base (fig. 2.8d). Nucleotides are extremely important as signaling molecules (they carry information between cells, tissues, and organs) and as sources of energy within cells. They also form long chains called *ribonucleic acid* (RNA) or ***deoxyribonucleic acid* (DNA)** that are essential for storing and expressing genetic information. Only four kinds of nucleotides (adenine, guanine, cytosine, and thymine) occur in DNA, but DNA contains millions of these molecules arranged in

very specific sequences. These sequences of nucleotides provide genetic information, or instructions, for cells. These instructions direct the growth and development of an organism. They also direct the formation of proteins or other compounds, such as those in melanin, a pigment that protects your skin from sunlight.

Long chains of DNA bind together to form a double helix (a two-stranded spiral, fig. 2.9). These chains replicate themselves when cells divide, so that as you grow, your DNA is reproduced in all your cells, from blood cells to hair cells. Everyone (even identical twins) has a distinctive DNA pattern, making DNA useful in

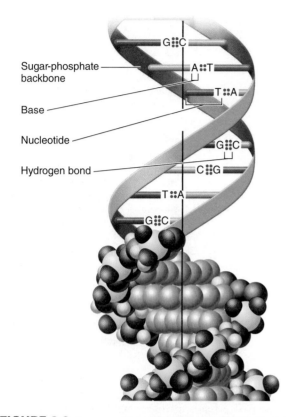

▲ **FIGURE 2.9** A composite molecular model of DNA. The lower part shows individual atoms, while the upper part has been simplified to show the strands of the double helix held together by hydrogen bonds (small dots) between matching nucleotides (A, T, G, and C). A complete DNA molecule contains millions of nucleotides and carries genetic information for many specific, inheritable traits.

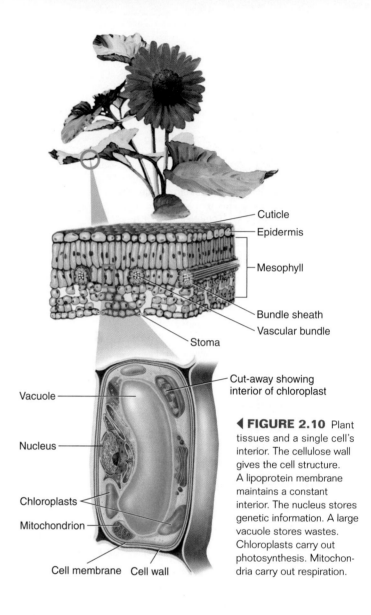

◄ **FIGURE 2.10** Plant tissues and a single cell's interior. The cellulose wall gives the cell structure. A lipoprotein membrane maintains a constant interior. The nucleus stores genetic information. A large vacuole stores wastes. Chloroplasts carry out photosynthesis. Mitochondria carry out respiration.

identifying individuals in forensics. Because DNA is passed down to us from our ancestors, it has allowed scientists to establish relationships, such as that dinosaurs were more closely related to birds than to lizards. Understanding DNA's structure has also allowed us to combine genes, for example, inserting disease-resistant traits into food crops (chapter 7).

Cells are the fundamental units of life

All living organisms are composed of **cells**, minute compartments within which the processes of life are carried out (fig. 2.10). Microscopic organisms such as bacteria, some algae, and protozoa are composed of single cells. Most higher organisms are multicellular, usually with many different cell varieties. Your body, for instance, is composed of several trillion cells of about two hundred distinct types. Every cell is surrounded by a thin but dynamic membrane of lipid and protein that receives information about the exterior world and regulates the flow of materials between the cell and its environment. Inside, cells are subdivided into tiny organelles and subcellular particles that provide the machinery for life. Some of these organelles store and release energy. Others manage and distribute information. Still others create the internal structure that gives the cell its shape and allows it to fulfill its role.

Nitrogen and phosphorus are key nutrients

As you know from the opening case study, nitrogen and phosphorus are key components of ecosystems. These are limiting elements because they are essential for plant and animal growth, but normally they are not abundant in ecosystems.

Notice that in figure 2.8 there are only a few different elements that are especially common. Carbon (C) is captured from air by green plants, and oxygen (O) and hydrogen (H) derive from air or water. The most important additional elements are nitrogen (N) and phosphorus (P), which are essential parts of the complex proteins, lipids, sugars, and nucleic acids that keep you alive. Of course, your cells use many other elements, but these are the most abundant. You derive all these elements by consuming molecules produced by green plants. Plants, however, must extract these elements from their environment. Low levels of N and P often limit growth in ecosystems where they are scarce. Abundance of N and P can cause runaway growth. In fertilizers, these elements often occur in the form of nitrate (NO_3), ammonium (NH_4), and phosphate (PO_4). Later in this chapter you will read more about how C, H_2O, N, and P circulate in our environment.

If travelers from another solar system were to visit our lovely, cool, blue planet, they might call it Aqua rather than Terra because of its outstanding feature: the abundance of streams, rivers, lakes, and oceans of liquid water. Our planet is the only place we know where water exists as a liquid in any appreciable quantity. Water covers nearly three-fourths of the earth's surface and moves around constantly through evaporation, precipitation, and runoff that distribute nutrients, replenish freshwater supplies, and shape the land. Water makes up 60 to 70 percent of the weight of most living organisms. It fills cells, giving form and support to tissues. Among water's unique, almost magical qualities, are the following:

▲ Surface tension is demonstrated by the resistance of a water surface to penetration, as when a water strider walks on it.

1. Water molecules are polar—that is, they have a slight positive charge on one side and a slight negative charge on the other side. Therefore, water readily dissolves polar or ionic substances, including sugars and nutrients, and carries materials to and from cells.

2. Water is the only inorganic liquid that occurs in nature under normal conditions at temperatures suitable for life. Most substances exist as either a solid or a gas, with only a very narrow liquid temperature range. Organisms synthesize organic compounds such as oils and alcohols that remain liquid at ambient temperatures and are therefore extremely valuable to life, but the original and predominant liquid in nature is water.

3. Water molecules are cohesive, tending to stick together tenaciously. You have experienced this property if you have ever done a belly flop off a diving board. Water has the highest surface tension of any common, natural liquid. Water also adheres to surfaces. As a result, water is subject to *capillary action*: it can be drawn into small channels. Without capillary action, movement of water and nutrients into groundwater reservoirs and through living organisms might not be possible.

4. Water is unique in that it expands when it crystallizes. Most substances shrink as they change from liquid to solid. Ice floats because it is less dense than liquid water. When temperatures fall below freezing, the surface layers of lakes, rivers, and oceans cool faster and freeze before deeper water. Floating ice then insulates underlying layers, keeping most water bodies liquid (and aquatic organisms alive) throughout the winter in most places. Without this feature, many aquatic systems would freeze solid in winter.

5. Water has a high heat of vaporization, using a great deal of heat to convert from liquid to vapor. Consequently, evaporating water is an effective way for organisms to shed excess heat. Many animals pant or sweat to moisten evaporative cooling surfaces. Why do you feel less comfortable on a hot, humid day than on a hot, dry day? Because water-vapor-laden air inhibits the rate of evaporation from your skin, thereby impairing your ability to shed heat.

6. Water also has a high specific heat; that is, a great deal of heat is absorbed before it changes temperature. The slow response of water to temperature change helps moderate global temperatures, keeping the environment warm in winter and cool in summer. This effect is especially noticeable near the ocean, but it is important globally.

All these properties make water a unique and vitally important component of the ecological cycles that move materials and energy and make life on earth possible.

2.3 ENERGY

If matter is the material of which things are made, energy provides the force to hold structures together, tear them apart, and move them from one place to another. In this section we will look at some fundamental characteristics of these components of our world.

Energy occurs in different types and qualities

Energy is the ability to do work, such as moving matter over a distance or causing a heat transfer between two objects at different temperatures. Energy can take many different forms. Heat, light, electricity, and chemical energy are examples that we all experience. The energy contained in moving objects is called **kinetic energy**. A rock rolling down a hill, the wind blowing through the trees, water flowing over a dam (fig. 2.11), or electrons speeding around the nucleus of an atom are all examples of kinetic energy. **Potential energy** is stored energy that is available for use. A rock poised at the top of a hill and water stored behind a dam are examples of potential energy. **Chemical energy** stored in the food that you eat and the gasoline that you put into your car are also examples of potential energy that can be released to do useful work. Energy is often measured in units of heat (calories) or work (joules). One joule (J) is the

▲ **FIGURE 2.11** Water stored behind this dam represents potential energy. Water flowing over the dam has kinetic energy, some of which is converted to heat.

work done when one kilogram is accelerated at one meter per second per second. One calorie is the amount of energy needed to heat one gram of pure water one degree Celsius. A calorie can also be measured as 4.184 J.

Heat describes the energy that can be transferred between objects of different temperature. When a substance absorbs heat, the kinetic energy of its molecules increases, or it may change state: a solid may become a liquid, or a liquid may become a gas. We sense change in heat content as change in temperature (unless the substance changes state).

An object can have a high heat content but a low temperature, such as a lake that freezes slowly in the fall. Other objects, like a burning match, have a high temperature but little heat content. Heat storage in lakes and oceans is essential to moderating climates and maintaining biological communities. Heat absorbed in changing states is also critical. As you will read in chapter 9, the evaporation and condensation of water in the atmosphere help distribute heat around the globe.

Energy that is diffused, dispersed, and low in temperature is considered low-quality energy because it is difficult to gather and use for productive purposes. The heat stored in the oceans, for instance, is immense but hard to capture and use, so it is low quality. Conversely, energy that is intense, concentrated, and high in temperature is high-quality energy because of its usefulness in carrying out work. The intense flames of a very hot fire or high-voltage electrical energy are examples of high-quality forms that are easy to use. Many of our alternative energy sources (such as wind) are diffuse compared to the higher-quality, more concentrated chemical energy in oil, coal, or gas.

Thermodynamics describes the conservation and degradation of energy

Atoms and molecules cycle endlessly through organisms and their environment, but energy flows in a one-way path. A constant supply of energy—nearly all of it from the sun—is needed to keep

biological processes running. Energy can be used repeatedly as it flows through the system, and it can be stored temporarily in the chemical bonds of organic molecules, but eventually it is released and dissipated.

The study of thermodynamics deals with how energy is transferred in natural processes. More specifically, it deals with the rates of flow and the transformation of energy from one form or quality to another. Thermodynamics is a complex, quantitative discipline, but you don't need a great deal of math to understand some of the broad principles that shape our world and our lives.

The **first law of thermodynamics** states that energy is *conserved;* that is, it is neither created nor destroyed under normal conditions. Energy may be transformed—for example, from the energy in a chemical bond to heat energy—but the total amount does not change.

The **second law of thermodynamics** states that with each successive energy transfer or transformation in a system, less energy is available to do work. That is, as energy is used, it is degraded to lower-quality forms, or it dissipates and is lost. When you drive a car, for example, the chemical energy of the gas is degraded to kinetic energy and heat, which dissipates, eventually, to space. The second law recognizes that disorder, or **entropy**, tends to increase in all natural systems. Consequently, there is always less *useful* energy available when you finish a process than there was before you started. Because of this loss, everything in the universe tends to fall apart, slow down, and get more disorganized.

How does the second law of thermodynamics apply to organisms and biological systems? Organisms are highly organized, both structurally and metabolically. Constant care and maintenance are required to keep up this organization, and a continual supply of energy is required to maintain these processes. Every time some energy is used by a cell to do work, some of that energy is dissipated or lost as heat. If cellular energy supplies are interrupted or depleted, the result—sooner or later—is death.

2.4 ENERGY FOR LIFE

Where does the energy needed by living organisms come from? How is it captured and transferred among organisms? For nearly all life on earth, the sun is the ultimate energy source, and the sun's energy is captured by green plants. Green plants are often called **primary producers** because they create carbohydrates and other compounds using just sunlight, air, and water.

There are organisms that get energy in other ways, and these are interesting because they are exceptions to the normal rule. Deep in the earth's crust, or deep on the ocean floor, and in hot springs such as those in Yellowstone National Park, we can find extremophiles, organisms that gain their energy from **chemosynthesis**, or extracting energy from inorganic chemical compounds such as hydrogen sulfide (H_2S). Until 30 years ago we knew almost nothing about these organisms and their ecosystems. Recent deep-sea exploration has shown that an abundance of astonishingly varied and abundant life occurs hundreds of meters deep on the ocean floor. These ecosystems cluster around thermal vents. Thermal vents are cracks where boiling-hot water, heated by magma in the earth's crust, escapes from the ocean floor. Here, microorganisms

▲ **FIGURE 2.12** A colony of tube worms and mussels clusters over a cool, deep-sea methane seep in the Gulf of Mexico.

grow by oxidizing hydrogen sulfide; bacteria support an ecosystem that includes blind shrimp, giant tube worms, hairy crabs, strange clams, and other unusual organisms (fig. 2.12).

These fascinating systems are exciting and mysterious because we have discovered them so recently. They are also interesting because of their contrast to the incredible profusion of photosynthesis-based life we enjoy here at the earth's surface.

Green plants get energy from the sun

Our sun is a star, a fiery ball of exploding hydrogen gas. Its thermonuclear reactions emit powerful forms of radiation, including potentially deadly ultraviolet and nuclear radiation (fig. 2.13), yet life here is nurtured by, and dependent upon, this searing energy source.

Solar energy is essential to life for two main reasons. First, the sun provides warmth. Most organisms survive within a relatively narrow temperature range. In fact, each species has its own range of temperatures within which it can function normally. At high temperatures (above 40°C), most biomolecules begin to break down or become distorted and nonfunctional. At low temperatures (near 0°C), some chemical reactions of metabolism occur too slowly to enable organisms to grow and reproduce. Other planets in our solar system are either too hot or too cold to support life as we know it. The earth's water and atmosphere help to moderate, maintain, and distribute the sun's heat.

Second, nearly all organisms on the earth's surface depend on solar radiation for life-sustaining energy, which is captured by green plants, algae, and some bacteria in a process called **photosynthesis**. Photosynthesis converts radiant energy into useful, high-quality chemical energy in the bonds that hold together organic molecules.

How much of the available solar energy is actually used by organisms? The amount of incoming solar radiation is enormous, about 1,372 watts/m^2 at the top of the atmosphere (imagine thirteen 100-watt light bulbs on every square meter of your ceiling). However, more than half of the incoming sunlight is reflected or absorbed by atmospheric clouds, dust, and gases. In particular, harmful, short wavelengths are filtered out by gases (such as ozone) in the upper atmosphere; thus the atmosphere is a valuable shield, protecting life-forms from harmful doses of ultraviolet and other forms of radiation. Even with these energy reductions, however, the sun provides much more energy than biological systems can harness, and more than enough for all our energy needs if technology could enable us to tap it efficiently.

Of the solar radiation that does reach the earth's surface, about 10 percent is ultraviolet, 45 percent is visible, and 45 percent is infrared. Most of that energy is absorbed by land or water or is reflected into space by water, snow, and land surfaces. (Seen from outer space, Earth shines about as brightly as Venus.)

Of the energy that reaches the earth's surface, photosynthesis uses only certain wavelengths, mainly red and blue light. Most plants reflect green wavelengths, so that is the color they appear to us.

Half of the energy plants absorb is used in evaporating water. In the end, only 1 to 2 percent of the sunlight falling on plants is available for photosynthesis. This small percentage represents the energy base for virtually all life in the biosphere!

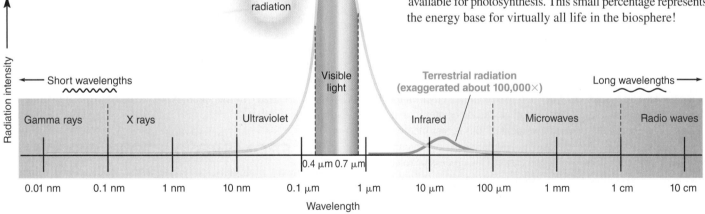

▲ **FIGURE 2.13** The electromagnetic spectrum. Our eyes are sensitive to visible-light wavelengths, which make up nearly half the energy that reaches the earth's surface (represented by the area under the "solar radiation" curve). Photosynthesizing plants use the most abundant solar wavelengths (light and infrared). The earth reemits lower-energy, longer wavelengths (shown by the "terrestrial radiation" curve), mainly the infrared part of the spectrum.

How does photosynthesis capture energy?

Photosynthesis occurs in tiny organelles called chloroplasts that reside within plant cells (see fig. 2.10). The most important key to this process is chlorophyll, a unique green molecule that can absorb light energy and use it to create high-energy chemical bonds in compounds that serve as the fuel for all subsequent cellular metabolism. Chlorophyll doesn't do this important job all alone, however. It is assisted by a large group of other lipid, sugar, protein, and nucleotide molecules. Together these components carry out two interconnected cyclic sets of reactions (fig. 2.14).

Photosynthesis begins with a series of steps called light-dependent reactions: These occur only while the chloroplast is receiving light. Enzymes split water molecules and release molecular oxygen (O_2). This is the source of nearly all the oxygen in the atmosphere on which all animals, including you, depend for life. The light-dependent reactions also create mobile, high-energy molecules (adenosine triphosphate, or ATP, and nicotinamide adenine dinucleotide phosphate, or NADPH), which provide energy for the next set of processes, the light-independent reactions. As their name implies, these reactions do not use light directly. Here, enzymes extract energy from ATP and NADPH to add carbon atoms (from carbon dioxide) to simple sugar molecules, such as glucose. These molecules provide the building blocks for larger, more complex organic molecules.

In most temperate-zone plants, photosynthesis can be summarized in the following equation:

$$6H_2O + 6CO_2 + \text{solar energy} \xrightarrow{\text{chlorophyll}} C_6H_{12}O_6 \text{ (sugar)} + 6O_2$$

We read this equation as "water plus carbon dioxide plus energy produces sugar plus oxygen." The reason the equation uses six water and six carbon dioxide molecules is that it takes six carbon atoms to make the sugar product. If you look closely, you will see that all the atoms in the reactants balance with those in the products. This is an example of conservation of matter.

You might wonder how making a simple sugar benefits the plant. The answer is that glucose is an energy-rich compound that serves as the central, primary fuel for all metabolic processes of cells. The energy in its chemical bonds—the ones created by photosynthesis—can be released by other enzymes and used to make other molecules (lipids, proteins, nucleic acids, or other carbohydrates), or it can drive kinetic processes such as movement of ions across membranes, transmission of messages, changes in cellular shape or structure, or movement of the cell itself in some cases. This process of releasing chemical energy, called **cellular respiration**, involves splitting carbon and hydrogen atoms from the sugar molecule and recombining them with oxygen to recreate carbon dioxide and water. The net chemical reaction, then, is the reverse of photosynthesis:

$$C_6H_{12}O_6 + 6O_2 \longrightarrow 6H_2O + 6CO_2 + \text{released energy}$$

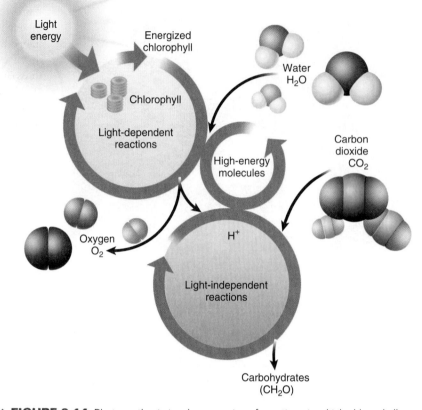

▲ **FIGURE 2.14** Photosynthesis involves a series of reactions in which chlorophyll captures light energy and forms high-energy molecules, ATP and NADPH. Light-independent reactions then use energy from ATP and NADPH to fix carbon (from air) in organic molecules.

Note that in photosynthesis, energy is *captured*, whereas in respiration, energy is *released*. Similarly, photosynthesis *uses* water and carbon dioxide to *produce* sugar and oxygen, whereas respiration does just the opposite. In both sets of reactions, energy is stored temporarily in chemical bonds, which constitute a kind of energy currency for the cell. Plants carry out both photosynthesis and respiration, but during the day, if light, water, and CO_2 are available, they have a net production of O_2 and carbohydrates.

We animals don't have chlorophyll and can't carry out photosynthetic food production. We do have the components for cellular respiration, however. In fact, this is how we get all our energy for life. We eat plants—or other animals that have eaten plants—and break down the organic molecules in our food to obtain energy (fig. 2.15). Later in this chapter we'll see how these feeding relationships work.

2.5 FROM SPECIES TO ECOSYSTEMS

While many biologists study life at the cellular and molecular level, ecologists study interactions at the species, population, biotic community, or ecosystem level. In Latin, *species* literally means "kind." In biology, **species** refers to all organisms of the same kind that are genetically similar enough to breed in nature and produce live, fertile offspring. There are several qualifications

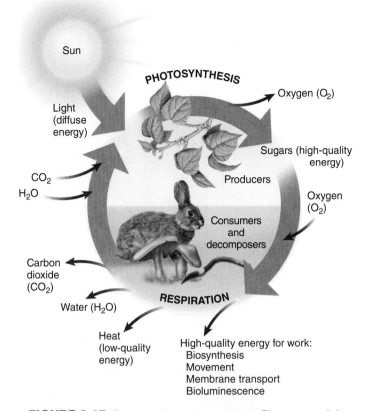

▲ **FIGURE 2.15** Energy exchange in ecosystems. Plants use sunlight, water, and carbon dioxide to produce sugars and other organic molecules. Consumers use oxygen and break down sugars during cellular respiration. Plants also carry out respiration, but during the day, if light, water, and CO_2 are available, they have a net production of O_2 and carbohydrates.

and some important exceptions to this definition of species (especially among bacteria and plants), but for our purposes this is a useful working definition.

Organisms occur in populations, communities, and ecosystems

A **population** consists of all the members of a species living in a given area at the same time. Chapter 4 deals further with population growth and dynamics. All of the populations of organisms living and interacting in a particular area make up a **biological community**. What populations make up the biological community of which you are a part? The population sign marking your city limits announces only the number of humans who live there, disregarding the other populations of animals, plants, fungi, and microorganisms that are part of the biological community within the city's boundaries. Characteristics of biological communities are discussed in more detail in chapter 3.

An ecological system, or **ecosystem**, is composed of a biological community and its physical environment. The Chesapeake ecosystem, for example, is a complex community of different species that rely on water, sunlight, and nutrients from the surrounding environment. It is useful to think about the biological

community and its environment together, because energy and matter flow through both. Understanding how those flows work is a major theme in ecology.

Food chains, food webs, and trophic levels link species

Photosynthesis (and rarely chemosynthesis) is the base of all ecosystems. Organisms that produce organic material by photosynthesis, mainly green plants and algae, are therefore known as **producers**. One of the most important properties of an ecosystem is its **productivity**, the amount of **biomass** (biological material) produced in a given area during a given period of time. Photosynthesis is described as *primary productivity* because it is the basis for almost all other growth in an ecosystem. A given ecosystem may have very high total productivity, but if decomposers consume organic material as rapidly as it is formed, the *net primary productivity* will be low. Remote sensing from aircraft or satellites allows us to measure productivity in large systems (see Exploring Science, p. 40).

In ecosystems, some consumers feed on a single species, but most consumers have multiple food sources. Similarly, some species are prey to a single kind of predator, but many species in an ecosystem are beset by several types of predators and parasites. In this way, individual food chains become interconnected to form a **food web**. Figure 2.16 shows feeding relationships among some of the larger organisms in an African savanna. If we were to add all the insects, worms, and microscopic organisms that belong in this picture, however, we would have overwhelming complexity. Perhaps you can imagine the challenge ecologists face in trying to quantify and interpret the precise matter and energy transfers that occur in a natural ecosystem!

An organism's feeding status in an ecosystem can be expressed as its **trophic level** (from the Greek *trophe*, food). In a savanna, grasses and trees are the primary producers (see bottom level of fig. 2.16). We call them autotrophes because they feed themselves using only sunlight, water, carbon dioxide, and minerals. Other organisms in the ecosystem are **consumers** of the chemical energy harnessed by the producers. An organism that eats primary producers is a primary consumer. Organisms that eat primary consumers are secondary consumers, which may, in turn, be eaten by a tertiary consumer, and so on. The highest tropic level is called the top predator. The complexity

Active **LEARNING**

Food Webs

To what food webs do you belong? Make a list of what you have eaten today, and trace the energy it contained back to its photosynthetic source. Are you at the same trophic level in all the food webs in which you participate? Are there ways that you could change your ecological role? Might that make more food available for other people? Why, or why not?

Measuring primary productivity is important for understanding individual plants and local environments. Understanding the rates of primary productivity is also key to understanding global processes, such as material cycling, and biological activity:

- In global carbon cycles, how much carbon is stored by plants, how quickly is it stored, and how does carbon storage compare in contrasting environments, such as the Arctic and the tropics?
- How does this carbon storage affect global climates (chapter 9)?
- In global nutrient cycles, how much nitrogen and phosphorus wash offshore, and where?

How can environmental scientists measure primary production (photosynthesis) at a global scale? In a small, relatively closed ecosystem, such as a pond, ecologists can collect and analyze samples of all trophic levels. But that method is impossible for large ecosystems, especially for oceans, which cover 70 percent of the earth's surface. One of the newest methods of quantifying biological productivity involves remote sensing, or using data collected from satellite sensors that observe the energy reflected from the earth's surface.

As you have read in this chapter, chlorophyll in green plants *absorbs* red and blue wavelengths of light and *reflects* green wavelengths. Your eye receives, or senses, these green wavelengths. A white-sand beach, on the other hand, reflects approximately equal amounts of all light wavelengths that reach it from the sun, so it looks white (and bright!) to your eye. In a similar way, different surfaces of the earth reflect characteristic wavelengths. Snow-covered surfaces reflect light wavelengths; dark green forests with abundant chlorophyll-rich leaves—and ocean surfaces rich in photosynthetic algae and plants—reflect greens and near-infrared wavelengths. Dry, brown forests with little active chlorophyll reflect more red and less infrared energy than do dark green forests (fig. 1).

To detect land cover patterns on the earth's surface, we can put a sensor on a satellite that orbits the earth. As the satellite travels, the sensor receives and transmits to earth a series of "snapshots." One of the best-known earth-imaging satellites, *Landsat 7*, produces images that cover an area 185 km (115 mi) wide, and each pixel represents an area of just 30 × 30 m on the ground. *Landsat* orbits approximately from pole to pole, so as the earth spins below the satellite, it captures images of the entire surface every 16 days. Another satellite, *SeaWiFS*, was designed mainly for monitoring biological activity in oceans (fig. 2). *SeaWiFS* follows a path similar to *Landsat*'s but it revisits each point on the earth every day and produces images with a pixel resolution of just over 1 km.

Because satellites detect a much greater range of wavelengths than our eyes can, they are able to monitor and map chlorophyll abundance. In oceans, this is a useful measure of ecosystem health, as well as carbon dioxide uptake. By quantifying and mapping primary production in oceans, climatologists are working to estimate the role of ocean ecosystems in moderating climate change: for example, they can estimate the extent of biomass production in the cold, oxygen-rich waters of the North Atlantic (fig. 2). Oceanographers can also detect near-shore areas where nutrients washing off the land surface fertilize marine ecosystems and stimulate high productivity, such as near the mouth of the Amazon or Mississippi River. Monitoring and mapping these patterns helps us estimate human impacts on nutrient flows from land to sea.

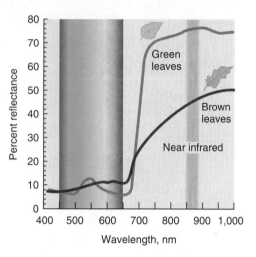

▲ Energy wavelengths reflected by green and brown leaves.

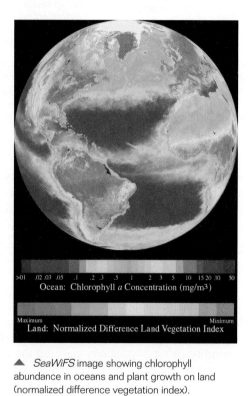

▲ *SeaWiFS* image showing chlorophyll abundance in oceans and plant growth on land (normalized difference vegetation index).

---

of a food chain depends on both the number of species available as well as the physical characteristics of a particular ecosystem. A harsh arctic landscape generally has a much simpler food chain than a temperate or tropical one.

Organisms can be identified both by the trophic level at which they feed and by the *kinds* of food they eat. **Herbivores** are plant eaters, **carnivores** are flesh eaters, and **omnivores** eat both plant and animal matter.

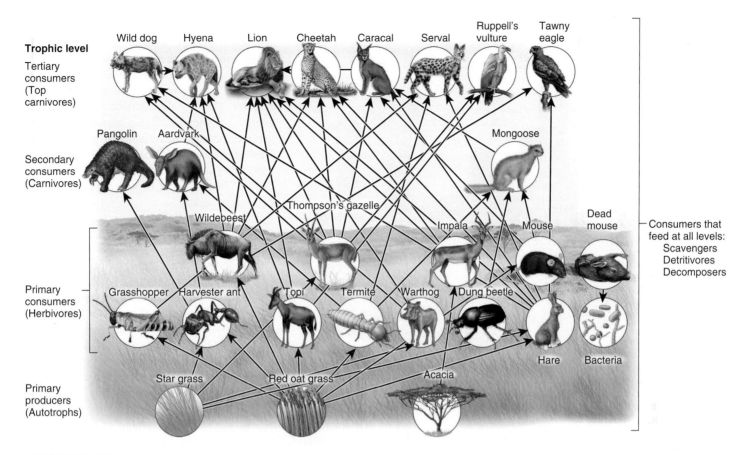

Trophic level

Tertiary consumers (Top carnivores): Wild dog, Hyena, Lion, Cheetah, Caracal, Serval, Ruppell's vulture, Tawny eagle

Secondary consumers (Carnivores): Pangolin, Aardvark, Mongoose

Primary consumers (Herbivores): Wildebeest, Thompson's gazelle, Impala, Mouse, Dead mouse, Grasshopper, Harvester ant, Topi, Termite, Warthog, Dung beetle, Hare, Bacteria

Primary producers (Autotrophs): Star grass, Red oat grass, Acacia

Consumers that feed at all levels: Scavengers, Detritivores, Decomposers

▲ **FIGURE 2.16** Each time an organism feeds, it becomes a link in a food chain. In an ecosystem, food chains become interconnected when predators feed on more than one kind of prey, thus forming a food web. The arrows in this diagram indicate the directions in which matter and energy are transferred through feeding relationships.

One of the most important feeding categories is made up of the parasites, scavengers, and decomposers that remove and recycle the dead bodies and waste products of others. Like omnivores, these recyclers feed on all the trophic levels. **Scavengers,** such as jackals and vultures, clean up dead carcasses of larger animals. **Detritivores,** such as ants and beetles, consume litter, debris, and dung, while **decomposer** organisms, such as fungi and bacteria, complete the final breakdown and recycling of organic materials. It could be argued that these cleanup organisms are second in importance only to producers, because without their activity nutrients would remain locked up in the organic compounds of dead organisms and discarded body wastes, rather than being made available to successive generations of organisms.

Ecological pyramids describe trophic levels

If we consider organisms according to trophic levels, they often form a pyramid, with a broad base of primary producers and only a few individuals in the highest trophic levels. Top predators are generally large, fierce animals, such as wolves, bears, sharks, and big cats. It usually takes a huge number of organisms at lower trophic levels (and thus a very large territory) to support a few top carnivores. While there is endless variation in the organization of ecosystems, the pyramid idea helps us describe generally how energy and matter move through ecosystems (see Key Concepts, pp. 44–45).

2.6 BIOGEOCHEMICAL CYCLES AND LIFE PROCESSES

The elements and compounds that sustain us are cycled endlessly through living things and through the environment. As the great naturalist John Muir said, "When one tugs at a single thing in nature, he finds it attached to the rest of the world." On a global scale, this movement is referred to as biogeochemical cycling. Substances can move quickly or slowly: Carbon might reside in a plant for days or weeks, in the atmosphere for days or months, in your body for hours, days, or years. The earth stores carbon (in coal or oil, for example) for millions of years. When human activities increase flow rates or reduce storage time, these materials can become pollutants. Here we will explore some of the pathways involved in cycling several important substances: water, carbon, nitrogen, sulfur, and phosphorus.

The hydrologic cycle

The path of water through our environment is perhaps the most familiar material cycle, and it is discussed in greater detail in chapter 10. Most of the earth's water is stored in the oceans, but solar energy continually evaporates this water, and winds distribute water vapor around the globe (fig. 2.17). Water that condenses over land surfaces, in the form of rain, snow, or fog, supports all

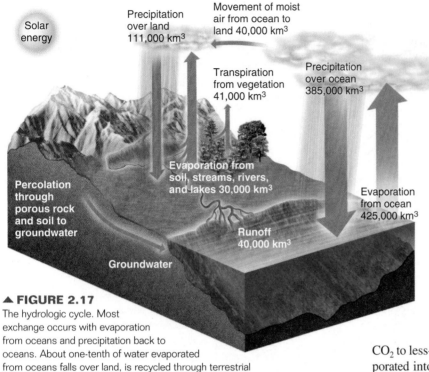

Solar energy

Precipitation over land 111,000 km³

Movement of moist air from ocean to land 40,000 km³

Transpiration from vegetation 41,000 km³

Precipitation over ocean 385,000 km³

Evaporation from soil, streams, rivers, and lakes 30,000 km³

Evaporation from ocean 425,000 km³

Percolation through porous rock and soil to groundwater

Runoff 40,000 km³

Groundwater

▲ FIGURE 2.17
The hydrologic cycle. Most exchange occurs with evaporation from oceans and precipitation back to oceans. About one-tenth of water evaporated from oceans falls over land, is recycled through terrestrial systems, and eventually drains back to oceans in rivers.

As it moves through living things and through the atmosphere, water is responsible for metabolic processes within cells, for maintaining the flows of key nutrients through ecosystems, and for global-scale distribution of heat and energy (chapter 9). Water performs countless services because of its unusual properties (see Exploring Science, p. 35). Water is so important that, when astronomers look for signs of life on distant planets, traces of water are the key evidence they seek.

The carbon cycle

Carbon serves a dual purpose for organisms: (1) it is a structural component of organic molecules, and (2) chemical bonds in carbon compounds provide metabolic energy. The **carbon cycle** begins with photosynthetic organisms taking up carbon dioxide (CO_2) (fig. 2.18). This is called carbon fixation because carbon is changed from gaseous CO_2 to less-mobile organic molecules. Once a carbon atom is incorporated into organic compounds, its path to recycling may be very quick or extremely slow. Imagine for a moment what happens to a simple sugar molecule you swallow in a glass of fruit juice. The sugar molecule is absorbed into your bloodstream, where it is made available to your cells for cellular respiration or the production of more complex biomolecules. If it is used in respiration, you may exhale the same carbon atom as CO_2 in an hour or less, and a plant could take up that exhaled CO_2 the same afternoon.

terrestrial (land-based) ecosystems. Living organisms emit the moisture they have consumed through respiration and perspiration. Eventually this moisture reenters the atmosphere or enters lakes and streams, from which it ultimately returns to the ocean again.

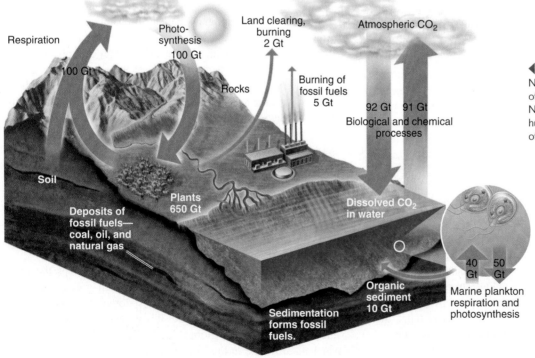

Respiration

100 Gt

Photo-synthesis 100 Gt

Rocks

Land clearing, burning 2 Gt

Atmospheric CO_2

Burning of fossil fuels 5 Gt

92 Gt   91 Gt
Biological and chemical processes

Soil

Plants 650 Gt

Deposits of fossil fuels—coal, oil, and natural gas

Dissolved CO_2 in water

Organic sediment 10 Gt

Sedimentation forms fossil fuels.

40 Gt   50 Gt

Marine plankton respiration and photosynthesis

◀ FIGURE 2.18 The carbon cycle. Numbers indicate approximate exchange of carbon in gigatons (Gt) per year. Natural exchanges are balanced, but human sources produce a net increase of CO_2 in the atmosphere.

Alternatively, your body may use that sugar molecule to make larger organic molecules that become part of your cellular structure. The carbon atoms in the sugar molecule could remain a part of your body until it decays after death. Similarly, carbon in the wood of a thousand-year-old tree will be released only when fungi and bacteria digest the wood and release carbon dioxide as a by-product of their respiration.

Sometimes recycling takes a very long time. Coal and oil are the compressed, chemically altered remains of plants and micro-organisms that lived millions of years ago. Their carbon atoms (and hydrogen, oxygen, nitrogen, sulfur, etc.) are not released until the coal and oil are burned. Enormous amounts of carbon also are locked up as calcium carbonate ($CaCO_3$) in the shells and skeletons of marine organisms from tiny protozoans to corals. The world's extensive surface limestone deposits are biologically formed calcium carbonate from ancient oceans, exposed by geological events. The carbon in limestone has been locked away for millennia, which is probably the fate of carbon currently being deposited in ocean sediments. Eventually even the deep ocean deposits are recycled as they are drawn into deep molten layers and released via volcanic activity. Geologists estimate that every carbon atom on the earth has made about 30 such round trips over the past 4 billion years.

Materials that store carbon, including geologic formations and standing forests, are known as carbon sinks. When carbon is released from these sinks, as when we burn fossil fuels and inject CO_2 into the atmosphere, or when we clear extensive forests, natural recycling systems may not be able to keep up. This is the root of the global warming problem, discussed in chapter 9. Alternatively, extra atmospheric CO_2 could support faster plant growth, speeding some of the recycling processes.

The nitrogen cycle

Organisms cannot exist without amino acids, peptides, and proteins, all of which are organic molecules that contain nitrogen. Nitrogen is therefore an extremely important nutrient for living things. This is why nitrogen is a primary component of household and agricultural fertilizers. Nitrogen makes up about 78 percent of the air around us.

Plants cannot use N_2, the stable two-atom form most common in air. But bacteria can. So plants acquire nitrogen from nitrogen-fixing bacteria (including some blue-green algae or cyanobacteria) that live in and around their roots. These bacteria can "fix" nitrogen, or combine gaseous N_2 with hydrogen to make ammonia (NH_3) and ammonium (NH_4^+). Nitrogen fixing by bacteria is a key part of the **nitrogen cycle** (fig. 2.19).

Other bacteria then combine ammonia with oxygen to form nitrite (NO_2^-). Another group of bacteria converts nitrites to nitrate (NO_3^-), which green plants can absorb and use. Plant cells reduce nitrate to ammonium (NH_4^+), which is used to build amino acids that become the building blocks for peptides and proteins.

Members of the bean family (legumes) and a few other kinds of plants are especially useful in agriculture because nitrogen-fixing bacteria actually live in their root tissues (fig. 2.20). Legumes and their associated bacteria add nitrogen to the soil, so interplanting and rotating legumes with crops, such as corn, that use but cannot replace soil nitrates are beneficial farming practices that take practical advantage of this relationship.

Nitrogen reenters the environment in several ways. The most obvious path is through the death of organisms. Fungi and bacteria decompose dead organisms, releasing ammonia and ammonium

▶ **FIGURE 2.19** The nitrogen cycle. Human sources of nitrogen fixation (conversion of molecular nitrogen to ammonia or ammonium) are now about 50 percent greater than natural sources. Bacteria convert ammonia to nitrates, which plants use to create organic nitrogen. Eventually nitrogen is stored in sediments or converted back to molecular nitrogen (1 Tg = 1,012 g).

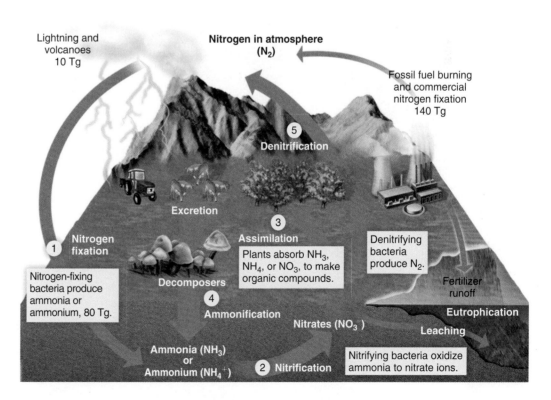

How do energy and matter move through systems?

Movement of energy and matter unites the parts of a system. In the Chesapeake Bay (opening case study), movement of water and nutrients supports photosynthesis, which supports the ecosystem. Recently the Everglades ecosystem has been destabilized by increased nutrient input, which increases photosynthesis and the accumulation of biomass (in invasive cattails).

For ecosystems in general, it is helpful to group organisms by **trophic levels** (feeding levels). In general, *primary producers* (organisms that produce organic matter, mainly green plants) are consumed by *herbivores* (plant eaters), which are consumed by *primary carnivores* (meat eaters), which are consumed by *secondary carnivores*. *Decomposers* consume at all levels and provide energy and matter to producers.

KC 2.1

Why do we find a pyramid of biomass?

Each trophic level requires a great deal of biomass at lower levels because energy is lost through growth, heat, respiration, and movement. This inefficiency is consistent with the second principle of thermodynamics, that energy dissipates and degrades to lower levels as it moves through a system.

A **general rule of thumb** is that only about 10 percent of the energy in one trophic level is represented in the next higher level. For example, it takes roughly 100 kg of clover to make 10 kg of rabbit, and 10 kg of rabbit to make 1 kg of fox.

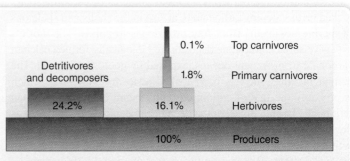

	0.1% Top carnivores
Detritivores and decomposers	1.8% Primary carnivores
24.2%	16.1% Herbivores
	100% Producers

▲ In this example, numbers show the percentage of energy that is incorporated into biomass at the next level. Here, decomposers are grouped with producers.

Why is there so much less energy in each successive trophic level?

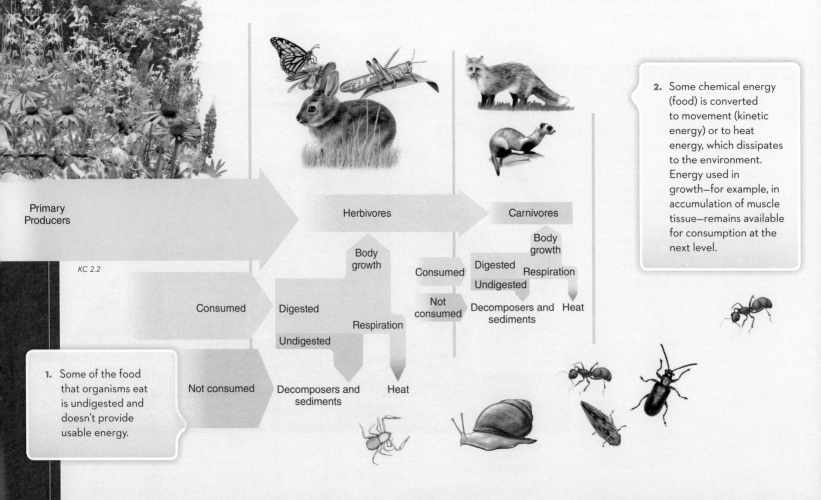

Primary Producers

KC 2.2

Consumed

Not consumed

Digested

Undigested

Decomposers and sediments

Respiration

Heat

Herbivores

Body growth

Consumed

Not consumed

Digested

Undigested

Decomposers and sediments

Respiration

Heat

Carnivores

Body growth

2. Some chemical energy (food) is converted to movement (kinetic energy) or to heat energy, which dissipates to the environment. Energy used in growth—for example, in accumulation of muscle tissue—remains available for consumption at the next level.

1. Some of the food that organisms eat is undigested and doesn't provide usable energy.

What happens if the pyramid is disrupted?

Ecosystems undergo many types of disturbances and disruptions. Often ecosystems recover in time; sometimes they shift to a new type of system structure. Forest fire is a disturbance that eliminates primary production for a short time. Fire also accelerates movement of nutrients through the system, so that nutrients once locked up in standing trees become available to support a burst of new growth.

Removal of other trophic levels also disturbs an ecosystem. If there are too many predators, prey species will decline or disappear. An overabundance of foxes, for example, may eliminate the rabbit population. With too few rabbits, the foxes may die off, or they may find alternate prey, which can further destabilize the system.

On the other hand, removal of a higher trophic level can also destabilize a system: if foxes were removed, rabbits might become overabundant and overgraze the primary producers (plants).

Sometimes a pyramid can be temporarily inverted. The biomass pyramid, for instance, can be inverted by periodic fluctuations in producer populations. For example, low plant and algal biomass are present during winter in temperate aquatic ecosystems.

KC 2.3

KC 2.4

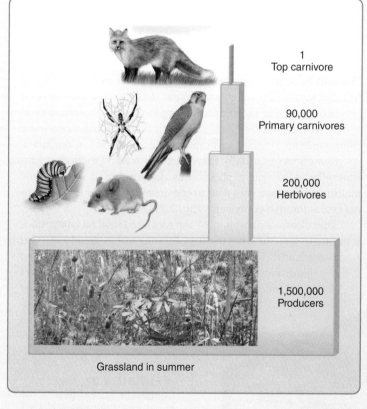

1
Top carnivore

90,000
Primary carnivores

200,000
Herbivores

1,500,000
Producers

Grassland in summer

By the numbers

▲ We often think of a pyramid in terms of the number of organisms, rather than amount of biomass in each level. The pyramid is a general model. In this pyramid, many smaller organisms support one organism at the next trophic level. So 1,000 m² of grassland might contain 1,500,000 producers (plants), which support 200,000 herbivores, which support 90,000 primary carnivores, which support one top carnivore.

Don't forget the little things.
A single gram of soil can contain hundreds of millions of bacteria, algae, fungi, and insects.

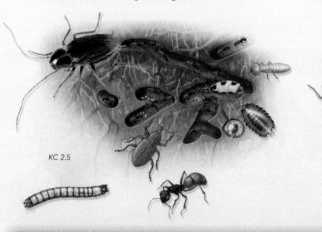

KC 2.5

CAN YOU **EXPLAIN?**

1. How many trophic levels do you eat? Is your food pyramid large or small?

2. Does your trophic level matter in terms of the structure and stability of the ecosystems you occupy?

3. Explain the food pyramid in terms of the two principles of thermodynamics.

▲ **FIGURE 2.20** Nitrogen molecules (N_2) are converted to useable forms in the bumps (nodules) on the roots of this bean plant. Each nodule is a mass of root tissue containing many bacteria that help convert nitrogen in the soil to a form that the bean plant can assimilate and use to manufacture amino acids.

ions, which then are available for nitrate formation. Organisms don't have to die to donate proteins to the environment, however. Plants shed their leaves, needles, flowers, fruits, and cones; animals shed hair, feathers, skin, exoskeletons, pupal cases, and silk. Animals also produce excrement and urinary wastes that contain nitrogenous compounds. Urine is especially high in nitrogen because it contains the detoxified wastes of protein metabolism.

All of these by-products of living organisms decompose, replenishing soil fertility (see related Case Study, "Why Trees Need Salmon," at www.mhhe.com/cunningham7e).

How does nitrogen reenter the atmosphere, completing the cycle? Denitrifying bacteria break down nitrates (NO_3^-) into N_2 and nitrous oxide (N_2O), gases that return to the atmosphere. Thus denitrifying bacteria compete with plant roots for available nitrates. Denitrification occurs mainly in waterlogged soils that have low oxygen availability and a large amount of decomposable organic matter. These are suitable growing conditions for many wild plant species in swamps and marshes, but not for most cultivated crop species, except for rice, a domesticated wetland grass.

In recent years humans have profoundly altered the nitrogen cycle. By using synthetic fertilizers, cultivating nitrogen-fixing crops, and burning fossil fuels, we now convert more nitrogen to ammonia and nitrates through industrial reactions than all natural land processes combined. This excess nitrogen input causes algal blooms and excess plant growth in water bodies, called eutrophication, which we will discuss in more detail in chapter 10. Excess nitrogen also causes serious loss of soil nutrients such as calcium and potassium; acidification of rivers and lakes; and rising atmospheric concentrations of nitrous oxide, a greenhouse gas. It also encourages the spread of weeds into areas such as prairies, where native plants are adapted to nitrogen-poor environments.

The phosphorus cycle takes millions of years

Minerals become available to organisms after they are released from rocks or salts (which are ancient sea deposits). Two mineral cycles of particular significance to organisms are phosphorus and sulfur. At the cellular level, energy-rich phosphorus-containing compounds are primary participants in energy-transfer reactions.

Phosphorus is usually transported in water. Producer organisms take in inorganic phosphorus, incorporate it into organic molecules, and then pass it on to consumers. In this way, phosphorus cycles through ecosystems (fig. 2.21).

The release of phosphorus from rocks and mineral compounds is normally very slow, but mining of fertilizers has greatly speeded the use and movement of phosphorus in the environment. Most phosphate ores used for detergents and inorganic fertilizers come from salt deposits from ancient, shallow sea beds. Most of the phosphorus used in agriculture winds up in the ocean again, from

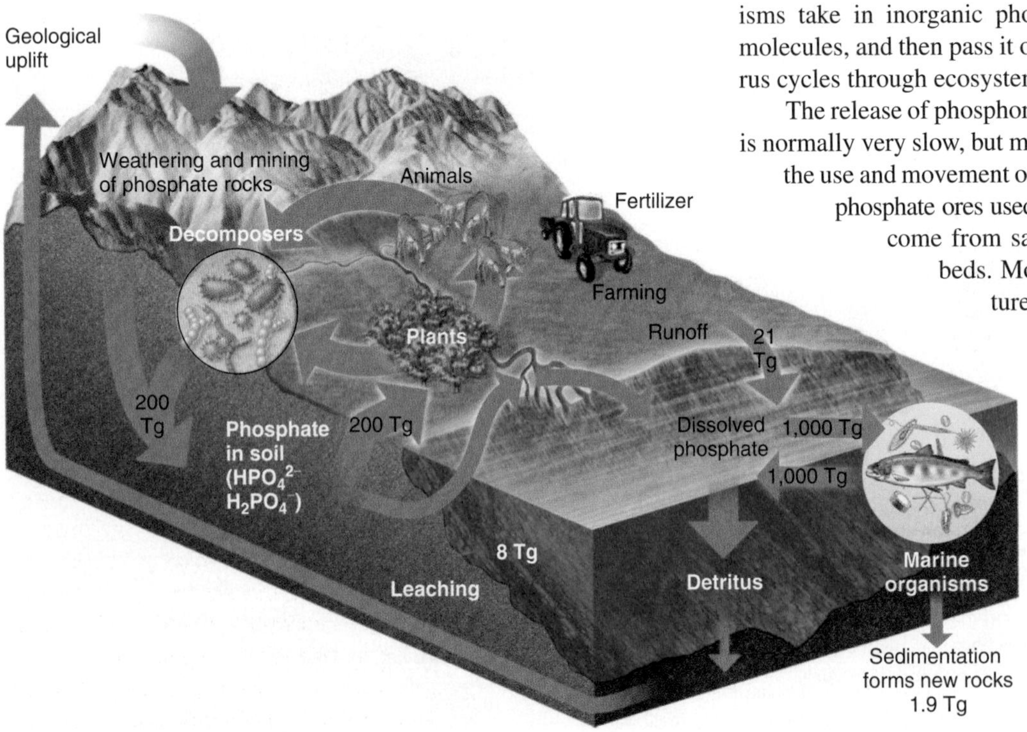

Geological uplift

Weathering and mining of phosphate rocks

Animals

Fertilizer

Decomposers

Farming

Plants

Runoff

21 Tg

200 Tg

200 Tg

Phosphate in soil (HPO_4^{2-} $H_2PO_4^-$)

Dissolved phosphate

1,000 Tg

1,000 Tg

8 Tg

Leaching

Detritus

Marine organisms

Sedimentation forms new rocks 1.9 Tg

◀ **FIGURE 2.21** The phosphorus cycle. Natural movement of phosphorus is slight, involving recycling within ecosystems and some erosion and sedimentation of phosphorus-bearing rock. Use of phosphate (PO_4^{-3}) fertilizers and cleaning agents increases phosphorus in aquatic systems, causing eutrophication. Units are teragrams (Tg) phosphorus per year.

field runoff or through human and animal waste that is released to rivers. Over millions of years, this phosphorus will become part of mineral deposits, but on shorter time scales, many earth scientists worry that we could use up our available sources of phosphorus, putting our agricultural systems at risk.

As you have read, phosphorus is an important water pollutant because excess phosphates can stimulate explosive growth of algae and photosynthetic bacteria populations (algae blooms), upsetting ecosystem stability (see related Case Study, "The Environmental Chemistry of Phosphorus," at www.mhhe.com/cunningham7e). Can you think of ways we could reduce the amount of phosphorus we put into our environment?

The sulfur cycle

Sulfur plays a vital role in organisms, especially as a minor but essential component of proteins. Sulfur compounds are important determinants of the acidity of rainfall, surface water, and soil. In addition, sulfur in particles and tiny airborne droplets may act as critical regulators of global climate. Most of the earth's sulfur is tied up underground in rocks and minerals, such as iron disulfide (pyrite) and calcium sulfate (gypsum). Weathering, emissions from deep seafloor vents, and volcanic eruptions release this inorganic sulfur into the air and water (fig. 2.22).

The sulfur cycle is complicated by the large number of oxidation states the element can assume, producing hydrogen sulfide (H_2S), sulfur dioxide (SO_2), sulfate ion (SO_4^{2-}), and others. Inorganic processes are responsible for many of these transformations, but living organisms, especially bacteria, also sequester sulfur in biogenic deposits or release it into the environment. Which of the several kinds of sulfur bacteria prevails in any given situation depends on oxygen concentrations, pH level, and light level.

Human activities also release large quantities of sulfur, primarily through burning fossil fuels. Total yearly anthropogenic sulfur emissions rival those of natural processes, and acid rain (caused by sulfuric acid produced as a result of fossil fuel use) is a serious problem in many areas (see chapter 9). Sulfur dioxide and sulfate aerosols cause human health problems, damage buildings and vegetation, and reduce visibility. They also absorb ultraviolet (UV) radiation and create cloud cover that cools cities and may be offsetting greenhouse effects of rising CO_2 concentrations.

Interestingly, the biogenic sulfur emissions of oceanic phytoplankton may play a role in global climate regulation. When ocean water is warm, tiny, single-celled organisms release dimethylsulfide (DMS), which is oxidized to SO_2 and then SO_4^{2-} in the atmosphere. Acting as cloud droplet condensation nuclei, these sulfate aerosols increase the earth's albedo (reflectivity) and cool the earth. As ocean temperatures drop because less sunlight gets through, phytoplankton activity decreases, DMS production falls, and clouds disappear. Thus DMS, which may account for half of all biogenic sulfur emissions, is one of the feedback mechanisms that keep temperature within a suitable range for all life.

CONCLUSION

The movement of matter and energy through systems, as in the Everglades, maintains the world's living environments. Matter consists of atoms, which make up molecules or compounds. Among the principal substances we consider in ecosystems are water, carbon, nitrogen, phosphorus, and sulfur. Nitrogen and phosphorus, especially, are key nutrients for living things. Energy also moves through systems. The laws of thermodynamics tell us that energy is neither created nor destroyed (first law), but it is degraded and dissipates as it moves through ecosystems (second law). For example, the chemical energy in food molecules is concentrated potential energy that degrades to less concentrated forms such as heat or kinetic energy as we use it. Matter, similarly, is neither created nor destroyed; it is reused continually.

Primary producers provide the energy and matter in an ecosystem. Nearly all ecosystems rely on green plants, which photosynthesize and create organic compounds that store energy, nutrients, and carbon. Excessive amounts of nutrients can create a positive feedback in plant or algae reproduction and growth, and positive feedbacks can destabilize a system. Negative feedbacks tend to maintain

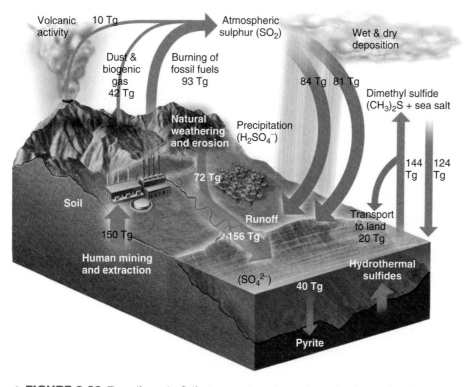

▲ **FIGURE 2.22** The sulfur cycle. Sulfur is present mainly in rocks, soil, and water. It cycles through ecosystems when it is taken in by organisms. Combustion of fossil fuels causes increased levels of atmospheric sulfur compounds, which create problems related to acid precipitation.

system stability. Cellular respiration is the reverse of photosynthesis: this is how organisms extract energy and nutrients from organic molecules.

Primary producers support smaller numbers of consumers in an ecosystem. Thus in the Chesapeake Bay saltgrass meadows support hundreds of bird, fish, and insect species. Top level predators, such as osprey, are relatively rare because large numbers of organisms are needed at each lower trophic level that supports them. We can think about this pyramid structure of trophic levels in terms of energy, biomass, or numbers of individuals. We can also understand these organisms as components of a system, through which carbon, water, and nutrients move.

PRACTICE QUIZ

1. What are the two most important nutrients causing eutrophication in the Chesapeake Bay?
2. What are systems and how do feedback loops regulate them?
3. Your body contains vast numbers of carbon atoms. How is it possible that some of these carbons may have been part of the body of a prehistoric creature?
4. List six unique properties of water. Describe, briefly, how each of these properties makes water essential to life as we know it.
5. What is DNA, and why is it important?
6. The oceans store a vast amount of heat, but this huge reservoir of energy is of little use to humans. Explain the difference between high-quality and low-quality energy.
7. In the biosphere, matter follows circular pathways, while energy flows in a linear fashion. Explain.
8. Which wavelengths do our eyes respond to, and why? (Refer to fig. 2.13.) About how long are short ultraviolet wavelengths compared to microwave lengths?
9. Where do extremophiles live? How do they get the energy they need for survival?
10. Ecosystems require energy to function. From where does most of this energy come? Where does it go?
11. How do green plants capture energy, and what do they do with it?
12. Define the terms *species*, *population*, and *biological community*.
13. Why are big, fierce animals rare?
14. Most ecosystems can be visualized as a pyramid with many organisms in the lowest trophic levels and only a few individuals at the top. Give an example of an inverted numbers pyramid.
15. What is the ratio of human-caused carbon releases into the atmosphere shown in figure 2.18 compared to the amount released by terrestrial respiration?

CRITICAL THINKING AND DISCUSSION

Apply the principles you have learned in this chapter to discuss these questions with other students.

1. Ecosystems are often defined as a matter of convenience because we can't study everything at once. How would you describe the characteristics and boundaries of the ecosystem in which you live? In what respects is your ecosystem an open one?
2. Think of some practical examples of increasing entropy in everyday life. Is a messy room really evidence of thermodynamics at work, or merely personal preference?
3. Some chemical bonds are weak and have a very short half-life (fractions of a second, in some cases); others are strong and stable, lasting for years or even centuries. What would our world be like if all chemical bonds were either very weak or extremely strong?
4. If you had to design a research project to evaluate the relative biomass of producers and consumers in an ecosystem, what would you measure? (*Note:* This could be a natural system or a human-made one.)
5. Understanding storage compartments is essential to understanding material cycles, such as the carbon cycle. If you look around your backyard, how many carbon storage compartments are there? Which ones are the biggest? Which ones are the longest lasting?

As you have read, movements of nitrogen and phosphorus are among the most important considerations in many wetland systems, because high levels of these nutrients can cause excessive algae and bacteria growth. This is a topic of great interest, and many studies have examined how nutrients move in a wetland, and in other ecosystems. Taking a little time to examine these nutrient cycles in detail will draw on your knowledge of atoms, compounds, systems, cycles, and other ideas in this chapter. Understanding nutrient cycling will also help you in later chapters of this book.

One excellent overview was produced by the Environmental Protection Agency. Go to Connect to find a description of the figure shown here, and to further explore the movement of our dominant nutrient, nitrogen, through environmental systems.

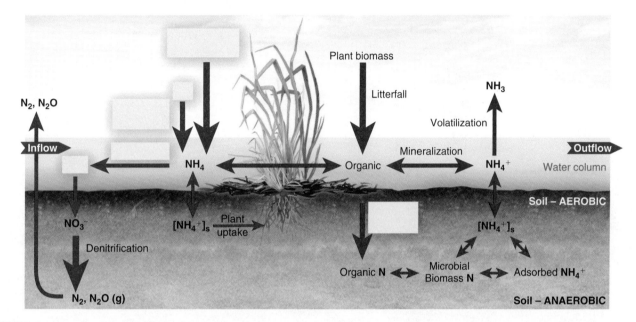

▲ **FIGURE 1** A detailed schematic diagram of the nitrogen cycle in a wetland. Study the online original to fill in the boxes.

SOURCE: EPA Nutrient Criteria Technical Guidance Manual, www.epa.gov/waterscience/criteria/nutrient/guidance/.

TO ACCESS ADDITIONAL RESOURCES FOR THIS CHAPTER, PLEASE VISIT CONNECT AT www.mcgrawhillconnect.com.

You will find LearnSmart, an adaptive learning system, Google Earth™ exercises, additional Case Studies, Data Analysis exercises, and an interactive ebook.

3 Evolution, Species Interactions, and Biological Communities

The Galápagos Islands have provided an accidental laboratory for examining biological diversity and species interactions.

LEARNING OUTCOMES

After studying this chapter, you should be able to answer the following questions:

▶ How does species diversity arise?

▶ What do we mean by tolerance limits? Give examples.

▶ How do interactions both aid and hinder species?

▶ Why don't species always reproduce up to their biotic potential?

▶ What is the relationship between species diversity and community stability?

▶ What is disturbance, and how does it affect communities?

▶ Explain ecological succession and give examples of its stages.

CASE STUDY

Natural Selection and the Galápagos Finches

The Galápagos Islands are a small archipelago of arid volcanic islands, isolated and remote—nearly 1,000 km from mainland South America. These small, rocky islands lack the profusion of life seen on many tropical islands, yet they are renowned as the place where Charles Darwin revolutionized our understanding of evolution, biodiversity, and biology in general. Why is this?

Charles Darwin (1809–1882) visited the islands in 1835 while serving as ship's naturalist on the *Beagle*. His job was to collect specimens and record observations for general interest. He found there a variety of unusual creatures, most occurring only in the Galápagos and some on just one or two islands. Giant land tortoises fed on tree-size cacti. Unique marine iguanas lived by grazing on algae scraped from rocky shoals. Sea birds were so unafraid of humans that Darwin could pick them off their nests. The islands also had a variety of small brown finches that differed markedly in appearance, food preferences, and habitat. Most finches forage for small seeds, but in the Galápagos there were fruit eaters with thick, parrot-like beaks; seed eaters with heavy, crushing beaks; and insect eaters with thin, probing beaks to catch their prey. The woodpecker finch even pecks at tree bark for hidden insects. Lacking the woodpecker's long tongue, the finch uses a cactus spine as a tool to extract insects.

Like other naturalists, Darwin was intrigued by the question of how such variety came to be. Most Europeans at the time believed that all living things had existed, unchanged, since a moment of divine creation, just a few thousand years ago. But Darwin had observed fossils of vanished creatures in South America. And he had read the new theories of geologist Charles Lyell (1797–1875), who argued that fossils showed that the world was much older than previously thought, and that species could undergo gradual but profound change over time. After his return to England, Darwin continued to ponder his Galápagos specimens. They seemed to be adapted to particular food resources on the different islands. It seemed likely that these birds were related, but somehow they had been modified to survive under different conditions.

Observing that dog breeders created new varieties of dogs, from Dachshunds to Great Danes, by selecting for certain traits, Darwin proposed that "natural selection" could explain the origins of species. Just as dog breeders favored individuals with particular characteristics, such as long legs or short noses, environmental conditions in an area could favor certain characteristics. On an island where only large seeds were available, finches with larger-than-average beaks could have more success in feeding—and reproducing—than smaller-beaked individuals. Thus competition for limited food resources could explain the prevalence of particular traits in a population. Darwin derived this idea from Thomas Malthus, a minister whose *Essay on the Principle of Population* (1798) argued that growth of human populations is always held in check by food scarcity (together with war and disease). Only the best competitors are likely to survive, according to Malthus. Darwin proposed that individuals with traits suitable for their environment are the best competitors for scarce resources. As those individuals survive and reproduce, their traits eventually become common in the population.

The explanation of evolution by natural selection has been supported by 150 years of observations and experiments, and Darwin's theory has provided explanation for countless examples of species variations. In a now-classic study of competition, for example, ecologist David Lack carefully measured the sizes and shapes of finches' beaks on several islands, to see how beaks varied with resources. Lack showed not only that beaks vary with resources, but that they specialize further in cases where multiple species compete for those resources. On the islands of Daphne Major and Los Hermanos, two finch species, *Geospiza fuliginosa* and *Geospiza fortis*, had beaks of moderate depth (thickness). But on islands where the two species coexisted, beak sizes shifted to the extremes, a shift that minimized competition for food resources (fig. 3.1). Where three species coexisted, traits again shifted to minimize competition. Thus competition among species had led to shifts in beak traits. Among Darwin's Galápagos finches, 13 modern species are now recognized, probably descended from a few seed-eating ancestors that blew to the islands from South America, where a similar species still exists.

Evolution of species through natural selection is now a cornerstone of biology and its many subfields, from ecology to medicine and health care. Subsequent discoveries have filled in many details. The discovery of DNA in the 1950s, in particular, allowed us to

(continued)

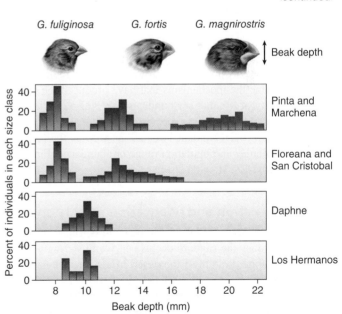

▲ **FIGURE 3.1** Two finch species have similar beaks when they occur separately (on this islands of Daphne and Los Hermanos), but beak sizes shift when the species occur together, as they specialize in different feeding strategies. When three finches coexist, feeding strategies and beak sizes are further differentiated. (After D. Lack, 1947.)

understand how random mutations (changes) in genes can account for the development of the variation in a population on which natural selection acts. In this chapter we'll look at some of the ways species and communities adapt to their environments. We'll also consider the ways populations interact, and the adaptations that make some species abundant and those that make others rare.

FURTHER READING:

Lack, D. 1947. *Darwin's Finches*. Cambridge University Press.
Stix, Gary. 2009. Darwin's living legacy. *Scientific American* 300(1): 138–43. ∎

> *When I view all beings not as special creations, but as lineal descendents of some few beings which have lived long before the first bed of the Cambrian system was deposited, they seem to me to become ennobled.*
> —CHARLES DARWIN

3.1 EVOLUTION LEADS TO DIVERSITY

Why does the earth support the astonishing biological diversity that we see around us? What determines which species will survive in one environment or another, and why are some species abundant, while others are rare? These are fundamental questions in ecology and biology, and in subfields such as population biology, the study of why populations grow and decline. In this chapter we will examine the mechanisms that have produced the extraordinary diversity of life that surrounds us. We'll also consider how certain traits can lead to some species' being rare and specialized, while others are overabundant "weeds," as well as the processes that lead to changes in population range or abundance. First we'll start with the basics: How do species arise?

Natural selection and adaptation modify species

How does a polar bear endure the long, sunless, super-cold arctic winter? How does the saguaro cactus survive blistering temperatures and extreme drought of the desert? Each species has inherited characteristics, or traits, that help it survive. The polar bear has heat-capturing fur, insulating fat layers, wide feet for swimming, and white hair that make it nearly invisible to the seals on which it preys. The saguaro has specially adapted leaves (spines), water-retaining cells, water-saving mechanisms in photosynthesis, and other traits that help it survive conditions that would kill most plants. We refer to the acquisition of these advantageous traits in a species as **adaptation**.

Adaptation involves changes in a population, with characteristics that are passed from one generation to the next. This is different from acclimation—an individual organism's changes in response to an altered environment. If you spend the summer outside, you may acclimate to the sunlight: your skin will increase its concentration of dark pigments that protect you from the sun. This is a temporary change, and you won't pass the temporary change on to future generations. However, the *capacity* to produce skin pigments is inherited. For populations living in intensely sunny environments, individuals with a good ability to produce skin pigments are more likely to thrive, or to survive, than people with a poor ability to produce pigments, and that trait becomes increasingly common in subsequent generations. If you look around, you can find countless examples of adaptation. The distinctive long neck of a giraffe, for example, developed as individuals that happened to have longer necks had an advantage in browsing on the leaves of tall trees (fig. 3.2).

▼ **FIGURE 3.2** Giraffes have long necks because in previous generations long-necked individuals happened to have an advantage in finding food, and in reproducing. This trait has become fixed in the population.

This process was explored in detail in Charles Darwin's 1859 book, *On the Origin of Species by Means of Natural Selection.* Darwin was one of many who observed and pondered the origins of natural variation, one of the great scientific questions of his time. He concluded that species change over generations because individuals compete for limited resources. Better competitors in a population are more likely to survive—giving them greater potential to produce offspring. We use the term **natural selection** to refer to the process in which individuals with useful traits pass on those traits to the next generation, while others reproduce less successfully.

We now know that these traits are encoded in genes, which are portions of an individual's DNA. Every organism has thousands of genes. Occasionally, random changes occur as DNA is replicated. If those random changes ("mutations") occur in reproductive cells, then they are passed on to offspring. (Mutations in nonreproductive cells, such as cancers, are not inherited.) Most mutations have little effect on fitness, and some can have a negative effect. But some happen to be useful in helping individuals exploit new resources or survive more successfully in new environmental conditions. We now understand the development of new species to result from many small mutations accumulating over time.

Limiting factors influence species distributions

An organism's physiology and behavior allow it to survive only in certain environments. Temperature, moisture level, nutrient supply, soil and water chemistry, living space, and other environmental factors must be at appropriate levels for organisms to persist. Generally some critical limiting factor keeps an organism from expanding everywhere. Limitations can include (1) physiological stress due to inappropriate levels of some critical environmental factor, such as moisture, light, temperature, pH, or specific nutrients; (2) competition with other species; and (3) predation, including parasitism and disease. Any one of these could constrain a species in different circumstances. For example nutrients are often a limiting factor in aquatic environments such as Chesapeake Bay (chapter 2). An infusion of nitrogen or phosphorus allows an explosion of algae, which continues until clouds of algae block sunlight in the water column. Then solar energy becomes a limiting factor, and the algae, unable to photosynthesize, die off rapidly.

In 1840 the chemist Justus von Liebig proposed that the single factor in shortest supply relative to demand is the **critical factor** determining where a species lives. The giant saguaro cactus (*Carnegiea gigantea*), which grows in the dry, hot Sonoran Desert of southern Arizona and northern Mexico, for example, is tolerant of extreme heat and drought, intense sunlight, and nutrient-poor soils, but it is extremely sensitive to freezing temperatures (fig. 3.3). A single winter night with temperatures below freezing for 12 or more hours kills growing tips on the branches, preventing further development. Because of this sensitivity, the northern edge of the saguaro's range corresponds to a zone where freezing temperatures last less than half a day at any time.

Ecologist Victor Shelford (1877–1968) later expanded Liebig's principle by stating that each environmental factor has both minimum and maximum levels, called **tolerance limits**, beyond which a particular species cannot survive or is unable to

▲ **FIGURE 3.3** The northern limit of Saguaro cactus is partly controlled by its low tolerance of freezing temperatures. In some cases, frost damage may not be visible in adult plants, but it can limit distribution by reducing reproduction.

reproduce (fig. 3.4). The single factor closest to these survival limits, Shelford postulated, is the critical factor that limits where a particular organism can live. At one time ecologists tried to identify unique factors limiting the growth of every plant and animal population. We now know that several factors working together usually determine a species' distribution. If you have ever explored the rocky coasts of New England or the Pacific Northwest, you have probably noticed that mussels and barnacles grow thickly in the intertidal zone, the place between high and low tides. Multiple factors, including temperature extremes, drying time between tides, salt concentrations, competitors, and food availability, determine the distribution of these animals.

In some species, tolerance limits affect the distribution of young differently than adults. The desert pupfish, for instance, lives in small, isolated populations in warm springs in the northern Sonoran Desert. Adult pupfish can survive temperatures between 0° and 42°C (a remarkably high temperature for a fish) and tolerate an equally wide range of salt concentrations. Eggs and juvenile fish, however, can survive only between 20° and 36°C and are killed by high salt levels. Reproduction, therefore, is limited to a small part of the range of the adult fish. Many species have greater sensitivity to salinity and other factors in young (or larvae or seedlings) than in adults.

Sometimes the requirements and tolerances of species are useful indicators of general environmental characteristics. Lichens and eastern white pine, for example, are highly sensitive to sulfur dioxide and ozone, respectively. The presence or absence of lichens or white pines, then, can indicate whether these pollutants are abundant in an area. **Indicator species** is a general term for organisms whose sensitivities can tell about environmental conditions in an area. Similarly, anglers know that trout require cool, clean, well-oxygenated water. The presence or absence of trout is used as an indicator of good water quality. Trout streams are often protected with special care, because the presence of trout indicates that the stream ecosystem as a whole is healthy.

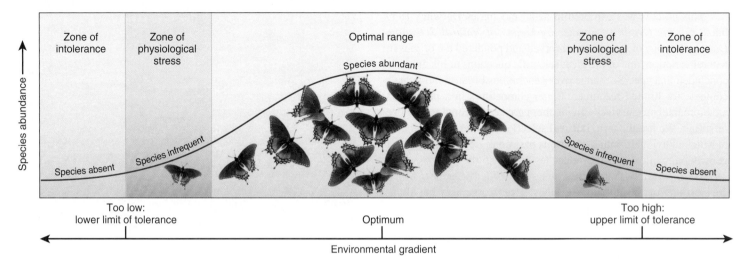

▲ FIGURE 3.4 Tolerance limits affect species distributions. For every environmental factor, there is an optimal range within which a species lives or reproduces most easily, so abundance is high. The horizontal axis here could represent a factor such as temperature, rainfall, vegetation height, or availability of some critical resource.

A niche is a species' role and environment

Habitat describes the place or set of environmental conditions in which a particular organism lives. The term **ecological niche** is more functional, describing both the role played by a species in a biological community and the set of environmental factors that determine its distribution. The concept of niche was first defined in 1927 by the British ecologist Charles Elton (1900–1991). To Elton, each species had a role in a community of species, and the niche defined its way of obtaining food, the relationships it had with other species, and the services it provided to its community.

Thirty years later, the American limnologist G. E. Hutchinson (1903–1991) proposed a more physical as well as biological definition of the niche. Every species, he pointed out, exists within a range of physical and chemical conditions, such as temperature, light levels, acidity, humidity, or salinity. It also exists within a set of biological interactions, such as the presence of predators or prey, or the availability of nutritional resources.

Some species tolerate a wide range of conditions or exploit a wide range of resources. These species are known as **generalists** (fig. 3.5). Generalists often have large geographic ranges, as in the case of the black bear (*Ursus americana*), which is omnivorous, abundant, and ranges across most of North America's forested regions. Sometimes generalists are also "weedy" species or pests, such as rats, cockroaches, or dandelions, because they thrive in a broad variety of environments.

Other species, such as the giant panda (*Ailuropoda melanoleuca*), are **specialists** and have a narrow ecological niche (fig. 3.6). Pandas have evolved from carnivorous ancestors to subsist almost entirely on bamboo, a large but low-nutrient grass. To acquire enough nutrients, pandas must spend as much as 16 hours a day eating. The panda's slow metabolism, slow movements, and low reproductive rate help it survive on this highly specialized diet. Its narrow niche now endangers the giant panda, though, as recent destruction of most of its native bamboo forest has reduced its range and its population to the margin of survival.

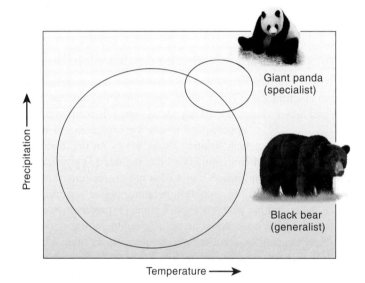

▲ FIGURE 3.5 Generalists, such as the American black bear, tolerate a wide range of environmental conditions. Specialists, in contrast, have narrower tolerance of environmental conditions.

▶ FIGURE 3.6 The giant panda has evolved from carnivorous, cat-like ancestors to live on a diet composed almost exclusively of bamboo. Adaptations include "thumbs" that help it grasp bamboo leaves, and teeth that help it chew the grass.

The saguaro is also a specialist, slow-growing and finely adapted to certain climatic conditions but unable to persist in wetter or cooler environments. Because of their exacting habitat requirements, specialists tend not to tolerate environmental change well. Often specialists are also **endemic species**—they occur only in one area (or one type of environment). The giant panda, for example, is endemic to the mountainous bamboo forests of southwestern China; the saguaro is endemic to the Sonoran Desert.

In most organisms, genetic traits and instinctive behaviors restrict the ecological niche. But some species have complex social structures that help them expand the range of resources or environments they can use. Elephants, chimpanzees, and dolphins, for example, learn from their social group how to behave and can invent new ways of doing things when presented with novel opportunities or challenges. In effect, they expand their ecological niche by transmitting cultural behavior from one generation to the next.

When two species compete for limited resources, one eventually gains the larger share, while the other finds different habitat, dies out, or experiences a change in its behavior or physiology so that competition is minimized. Consequently, as explained by the Russian biologist G. F. Gause (1910−1986), "complete competitors cannot coexist." The general term for this idea is the **principle of competitive exclusion**: no two species can occupy the same ecological niche for long. The species that is more efficient in using available resources will exclude the other. The other species disappears or develops a new niche, exploiting resources differently, a process known as **resource partitioning**. Partitioning can allow several species to utilize different parts of the same resource and coexist within a single habitat. A classic example of resource partitioning is that of woodland warblers, studied by ecologist Robert MacArthur (fig. 3.7). Although several similar warblers species foraged in the same trees, they avoided competition by specializing in different levels of the forest canopy, or in inner and outer branches.

Resources can be partitioned in time as well as space. Swallows and insect-eating bats both live by capturing flying insects, but bats hunt for night-flying insects, while swallows hunt during the day. Thus the two groups have noncompetitive feeding strategies for similar insect prey.

Speciation leads to species diversity

As a population becomes more adapted to its ecological niche, it may develop specialized or distinctive traits that eventually differentiate it entirely from its biological cousins. The development of a new species is called **speciation** (see Key Concepts, p. 56). In the case of Galápagos finches, evidence from body shape, behavior, and genetic similarity suggests that the 13 current species of finch derive from an original seed-eating finch species that probably blew to the islands from the mainland, perhaps in a storm, since finches are land birds. Accidental invasions, such as those by storms, winds, or ocean currents are probably rare. In the Galápagos, though, all land plants and animals (except those introduced by humans) derive from a few accidental colonizers. We know this because, as volcanic seamounts, the islands were never connected to a continental source of species.

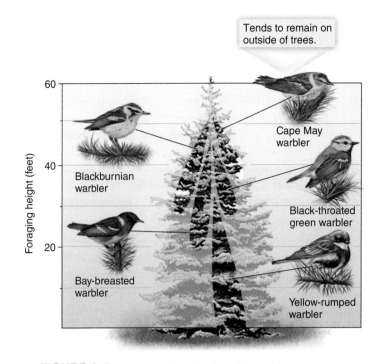

▲ **FIGURE 3.7** Several species of insect-eating wood warblers occupy the same forests in eastern North America. The competitive exclusion principle predicts that the warblers should partition the resource—insect food—in order to reduce competition. And in fact, the warblers feed in different parts of the forest. ADAPTED FROM: R. H. MacArthur (1958) *Ecology* 39:599–619.

In the Galápagos finches, speciation occurred largely because of **geographic isolation**. The islands are far enough apart that, in many cases, populations were genetically isolated: they couldn't interbreed with populations on other islands. These isolated populations gradually changed in response to their individual environments, some of which were extremely dry, with sparse vegetation, while others were relatively moist, with greater abundance and diversity of food resources. Speciation that occurs when populations are geographically separated is known as **allopatric speciation**.

The barriers that divide subpopulations aren't always physical. For example, two virtually identical tree frogs (*Hyla versicolor, H. chrysoscelis*) live in similar habitats of eastern North America but have different mating calls. This difference, which is enough to prevent interbreeding, is known as behavioral isolation. Speciation that occurs within one geographic area is known as **sympatric speciation**. Fern species and other plants sometimes undergo sympatric speciation by doubling or quadrupling the chromosome number of their ancestors, which makes them reproductively incompatible.

A general term for factors that make certain mutations advantageous is **selection pressure**. Normally we assume this selection happens slowly, but some degree of selection can happen in just a few years. A study of the finch species *Geospiza fortis* on the Galápagos island of Daphne Major, for example, showed selection after a two-year drought. As the drought reduced the availability of small seeds on the island, large-billed individuals in the finch population had better success in opening the remaining larger seeds. Within two years, the large-billed trait came to dominate the population. This shift toward one extreme of a trait is known

Where do species come from?

Evolution by natural selection

Charles Darwin is known for explaining evolution, but he was one of many in the nineteenth century who pondered the great question, **how do new species appear**? Darwin's contribution, which he developed simultaneously with biogeographer Alfred Russell Wallace, was the idea of "**natural selection**." Darwin noted that horse breeders and dog breeders selected for certain advantageous traits in their animals—speed, strength, and so on—by allowing individuals with those traits to reproduce. Breeders tended to avoid breeding individuals with less desirable traits.

Darwin suggested that natural chance might act the same way. For example, a population of birds normally has some slight variation in wing shape, size, shape of bill, and other traits. Sometimes environmental conditions make some of those traits advantageous: for example, if the main available food is hard-shelled seeds, then a slightly stronger bill might make it easier for some individuals to break open hard-shelled seeds. The stout bill gives those individuals a slight advantage in acquiring food and thus in producing offspring. Over time, the individuals with strong beaks might come to dominate the population. Those with lightweight beaks might disappear entirely.

On the other hand, if the dominant food source is small, soft grubs, then a thin and movable beak might be advantageous, and thick-billed individuals might disappear from the population.

What are key ideas of natural selection?

1. **Natural selection** occurs when circumstances make one type of trait especially advantageous, so that individuals with that trait reproduce more frequently or successfully than do others. Food resources, climate, presence of predators, or other factors can cause natural selection.

2. **Mutation** (changes) can occur randomly in a population, through reproduction. Some birds have slightly longer, or shorter bills, for example.

3. **Genetic drift**, or shifting traits in a population, is more likely in a small or isolated population. If an unusual trait (for example, red hair in humans) is relatively common in a small population, that trait has a relatively high probability of passing to later generations.

4. **Isolation** separates populations, making genetic drift more likely. Isolation of the Galápagos meant that the island ecosystems had few birds species and few food sources. Species have specialized for the types of foods in different islands. Geographic isolation also reduces interbreeding with larger population—which might mix other traits into island populations.

The Galápagos Islands, nearly 1,000 km from South America, are renowned as the place whose unusual species helped Darwin form his ideas of natural selection. Finches, mockingbirds, and other species on the isolated islands showed distinctive adaptations to different food sources and conditions. Yet they had similarities that pointed to common ancestors.

Craters show these islands are volcanic. They have been isolated since they emerged above the ocean's surface. Relatively few species have reached the Galápagos from the distant mainland. So it's relatively easy to detect divergence from a common ancestor.

KC 3.2

Large ground finch
(seeds)

Cactus ground finch
(cactus fruits and flowers)

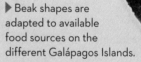

▶ Beak shapes are adapted to available food sources on the different Galápagos Islands.

GALÁPAGOS
ISLANDS

40 km (25 mi)

KC 3.3

KC 3.1

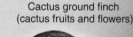

CENTRAL AMERICA

The Galápagos Islands are nearly 1,000 km (600 mi) from the mainland

SOUTH AMERICA

Vegetarian finch
(buds)

Selective pressure is a general term for factors that modify species' traits. **Competition** over resources can exert selective pressure by causing species to partition, or separate, their use of the resource. Where there is overlap in resource use, individuals that share the resource (orange shading, **a)** should be at a disadvantage, and individuals that specialize should be more abundant. Over time the traits of the populations diverge, leading to specialization, narrower niche breadth, and less competition between species. **(b)**. **Speciation**, or separation into (▶) entirely separate species, can result from competition or from isolation.

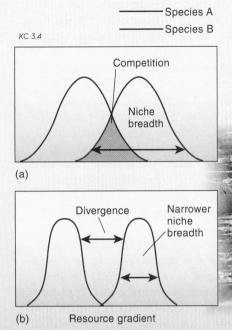

KC 3.4

— Species A
— Species B

(a)

Competition

Niche breadth

Abundance

(b)   Resource gradient

Divergence

Narrower niche breadth

▼ Two wading birds **partition** a mud flat: the Northern jacana captures insects at the surface with its short bill. The long-legged black-necked stilt probes deeper with its long bill.

KC 3.5

▶ **Intraspecific competition** (within a species) can lead to colorful and surprising traits.

KC 3.6

▼ The woodpecker finch uses cactus thorns to probe for insects under tree bark.

Woodpecker finch (insects)

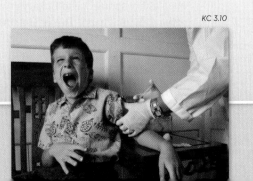

KC 3.8

▼ Galápagos tortoises have been shaped by the arid conditions of these isolated islands.

KC 3.7

Why does this matter?

Since Darwin published *On the Origin of Species* in 1859, countless studies have supported his **theory of evolution**. The idea of evolution by natural selection, or "descent with modification," as Darwin called it, has allowed us to describe evolutionary relationships among the millions of organisms around us, and to understand why they feed, breathe, reproduce, and live as they do.

We now understand most biological processes in terms of the mechanism of natural selection. Every time you get a flu shot, that vaccine is created by careful observation of adaptations in the rapidly evolving flu virus, in efforts to create a vaccine suited to the latest variety of the virus. ▼

KC 3.10

KC 3.9

CAN YOU **EXPLAIN?**

1. **What factors made natural selection relatively likely to occur in the Galápagos?**

2. **Think of several birds where you live. What kinds of food resources or feeding strategies might be reflected in their bill shapes?**

3. **Think of an example of resource partitioning between two organisms with which you are familiar.**

as directional selection (fig. 3.8). In this case the changes were not dramatic enough to result in speciation. When the drought ended, the population shifted back toward moderate-size beaks, which aided exploitation of a wider range of seeds.

Sometimes environmental conditions can reduce variation in a trait (stabilizing selection), or they can cause traits to diverge to the extremes (disruptive selection). Competition can cause disruptive selection, which allows for better partitioning of a resource (see fig. 3.1). Directional selection can be observed in the emergence of antibiotic-resistant bacteria (chapter 8) and of pesticide-tolerant insects (chapter 7). In both cases, individuals that happen to have better-than-average tolerance of these compounds tend to survive in a population, while other individuals die off. Resistant survivors produce new generations, leading to a population with the resistant traits.

New environmental conditions often lead to speciation, as new opportunities become available—as in the case of finch diversification on the previously unoccupied Galápagos Islands. Geologic time is marked by periods of tremendous diversification that have followed the sudden extinctions of species (chapter 5). The end of the age of dinosaurs, for example, was followed by dramatic diversification of mammals, which expanded to fill newly available niches. The fossil record is one of ever-increasing species diversity, despite several events that wiped out large proportions species.

We generally believe that species arise slowly, and many have existed unchanged for tens of millions of years. (The American alligator has existed unchanged for over 150 million years.) But some organisms evolve swiftly. New flu viruses, for instance, evolve every season. Fruit flies in Hawaii mutate frequently and can give rise to new species in just a few years. Reproductive isolation has led to the development of new traits in populations of red squirrels in Arizona in just the past 10,000 years or so, since the last glacial period (fig. 3.9).

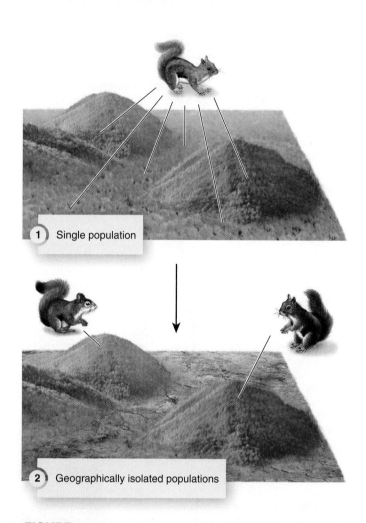

▲ **FIGURE 3.9** Geographic isolation can lead to allopatric (geographically separated) speciation. For example, in cool, moist glacial periods, Arizona was forested and red squirrels interbred freely (1). As the climate warmed and dried, desert replaced forests on the plains. Isolated in remnant forest on mountains, populations became reproductively isolated and began to develop different traits (2).

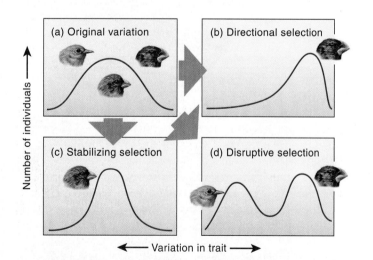

▲ **FIGURE 3.8** A trait such as beak shape can change as changing environmental conditions make one type more advantageous for survival. From an original population with wide variation in the trait (a), environmental conditions might favor one extreme (b), or neither extreme (c). Where a population occupies contrasting conditions, traits in the different areas may diverge (d), producing two distinct populations

Taxonomy describes relationships among species

Taxonomy is the study of types of organisms and their relationships. Taxonomists are biologists who use various lines of evidence, such as cell structures or genetic similarities, to trace how organisms have descended from common ancestors. Taxonomists are continually refining their understanding, but most now classify all life into three general domains according to cell structure. These domains are Bacteria (whose cells have no membrane around the nucleus), Archaea (whose DNA differs from bacteria, and whose cell functions allow them to survive in extreme environments, such as hot springs), and Eukarya (whose cells do have a membrane around the nucleus). The organisms most familiar to us are in the four kingdoms of the Eukarya (fig. 3.10). These kingdoms include animals, plants, fungi (molds and mushrooms), and protists (algae, protozoans, slime molds). Within these kingdoms are millions of different species, which we discuss further in chapter 5. The

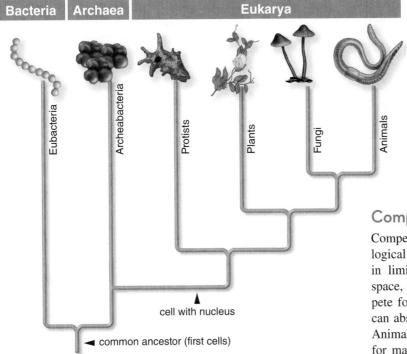

| Bacteria | Archaea | Eukarya |

Eubacteria

Archeabacteria

Protists

Plants

Fungi

Animals

▲ cell with nucleus

◄ common ancestor (first cells)

▲ **FIGURE 3.10** Taxonomists divide living organisms into three domains (Bacteria, Archaea, and Eukarya) based on fundamental cell structures. These arose from a common ancestor over the past 4 billion years. The Animal, Fungi, Plant, and Protist kingdoms are all part of the domain Eukarya.

taxonomic tree groups organisms at multiple levels of specificity. Two familiar examples—one you know and one you eat—are shown in table 3.1.

Botanists, ecologists, and other scientists often use the two most specific levels of taxonomy, genus and species, to refer to species. These two-part binomial names, also called scientific or Latin names, are more precise than common names. Many different types of yellow flowers, growing in different places, are called "buttercups," for example. This can lead to serious confusion when we try to describe these flowers. At the same time, a single species might have multiple common names. The bionomial name *Pinus resinosa,* however, always refers to the same tree, whether you call it a red pine, Norway pine, or just pine.

TABLE 3.1 | Taxonomy of Two Common Species

TAXONOMIC LEVEL	HUMANS	CORN
Kingdom	Animalia	Plantae
Phylum	Chordata	Anthophyta
Class	Mammalia	Monocotyledons
Order	Primates	Commenales
Family	Hominidae	Poaceae
Genus	*Homo*	*Zea*
Species	*Homo sapiens*	*Zea mays*
Subspecies	*H. sapiens sapiens*	*Zea mays mays*

3.2 SPECIES INTERACTIONS

Resources limitations influence a species' adaptation to its environment and its ecological niche, but interactions with other species also help shape species' traits and behaviors. Competition and predation cause species to evolve in response to each other's attributes; cooperative interactions and even interdependent relationships also confer advantages, which can lead to evolution of traits. In this section we will look at the interactions within and between species that affect their success and shape biological communities.

Competition leads to resource allocation

Competition is a type of antagonistic relationship within a biological community. Organisms compete for resources that are in limited supply: energy and matter in usable forms, living space, and specific sites to carry out life's activities. Plants compete for growing space to develop roots and leaves so that they can absorb and process sunlight, water, and nutrients (fig. 3.11). Animals compete for living, nesting, and feeding sites, and also for mates. Competition among members of the same species is called **intraspecific competition**, whereas competition between members of different species is called **interspecific competition**. Recall the competitive exclusion principle, that no two species can occupy the same ecological niche for long. Competition causes individuals and species to shift their use of a common resource: thus warblers that compete with each other for insect food in deciduous forests tend to specialize on different areas of the trees, reducing or avoiding competition. There have been hundreds of interspecific competition studies in natural populations showing a wide variety of evolutionary adaptations.

When different species use a common resource, competition often leads to resource partitioning, as in the case of some of the Galápagos finches (see fig. 3.1). Alternatively, one species can exclude another. When two species compete, the one living in the center of its tolerance limits has an advantage and, more often than not, prevails in competition with another species living at the margins of its optimal environmental conditions.

In intraspecific competition, members of the same species compete directly with each other for resources. Competition can hone a species' attributes—for example, as the fastest cheetahs prosper by catching more prey than slower ones, or as warblers with the best insect-foraging bills prosper more than poorer foragers. Direct intraspecific fighting for resources can occur, as when two bull elk battle for territory and for females. But most animals avoid fighting if possible. They posture and challenge each other, but usually the weaker individual eventually backs off. Physical injury is too high a cost to risk in most cases of intraspecific competition.

Several avenues exist to reduce competition within a species. First, the young of the year can disperse. Young animals move in search of new territories; in plants, seeds travel with wind, water, or passing animals, and at least some land in available habitat away from the parent plants. Second, by exhibiting strong territoriality, many animals force their offspring or other trespassers out of their

▲ **FIGURE 3.11** In this tangled Indonesian rainforest, plants compete for light and space. Special adaptations allow ferns, mosses, and bromeliads to find light by perching high on tree trunks and limbs; vines climb toward the canopy. These and other adaptations lead to profuse speciation.

vicinity. In this way territorial species, such as bears, songbirds, ungulates, and fish, minimize competition between individuals and generations. A third way to reduce intraspecific competition is resource partitioning between generations. The adults and juveniles of these species occupy different ecological niches. For instance, monarch caterpillars munch on milkweed leaves, while metamorphosed butterflies sip nectar. Crabs begin as swimming larvae and do not compete with bottom-dwelling adult crabs.

Predation affects species relationships

All organisms need food to live. Producers make their own food, whereas consumers eat organic matter created by other organisms. As we saw in chapter 2, photosynthetic plants and algae are the producers in nearly all communities. Consumers include herbivores, carnivores, omnivores, scavengers, detritivores, and decomposers. Often we think only of carnivores as predators, but ecologically a predator is any organism that feeds directly on another living organism, whether or not this kills the prey (fig. 3.12). Herbivores, carnivores, and omnivores, which feed on live prey, are predators, but scavengers, detritivores, and decomposers, which feed on dead things, are not.

▲ **FIGURE 3.12** Insect herbivores are predators as much as lions and tigers. Insects consume the vast majority of biomass in the world. Complex predation and defense mechanisms have evolved between insects and their plant prey.

In this sense, parasites (organisms that feed on a host organism or steal resources from it without necessarily killing it) and even pathogens (disease-causing organisms) can be considered predator organisms. Herbivory is the type of predation practiced by grazing and browsing animals on plants.

Predation is a powerful but complex influence on species populations in communities. It affects (1) all stages in the life cycles of predator and prey species; (2) many specialized food-obtaining mechanisms; and (3) the evolutionary adjustments in behavior and body characteristics that help prey escape being eaten and help predators more efficiently catch their prey. Predation also interacts with competition. In **predator-mediated competition**, a superior competitor in a habitat builds up a larger population than its competing species; predators take note and increase their hunting pressure on the superior species, reducing its abundance and allowing the weaker competitor to increase its numbers. To test this idea, scientists remove predators from communities of competing species. Often the superior competitors eliminated other species from the habitat. In a classic example, the ochre starfish (*Pisaster ochraceus*) was removed from Pacific tidal zones and its main prey, the common mussel (*Mytilus californicus*), exploded in numbers and crowded out other species.

Knowing how predators affect prey populations has direct application to human needs, such as pest control in cropland. The cyclamen mite (*Phytonemus pallidus*), for example, is a pest of California strawberry crops. Predatory mites (*Typhlodromus* and *Neoseiulus*), which arrive naturally or are introduced into fields, reduce the damage caused by the cyclamen mite. Spraying pesticides to control the cyclamen mite can actually increase the infestation because it also kills the beneficial predatory mites.

Predatory relationships may change as the life stage of an organism changes. In marine ecosystems, crustaceans, mollusks, and worms release eggs directly into the water, where they and hatchling larvae join the floating plankton community (fig. 3.13). Planktonic animals eat each other and are food for larger carnivores, including fish. As prey species mature, their predators change. Barnacle larvae are planktonic and are eaten by small fish, but as adults their hard shells protect them from fish, but not from starfish and predatory snails. Predators often switch prey in the course of their lives. Carnivorous adult frogs usually begin their lives as plant-eating tadpoles. Some predators also switch prey easily: house cats, for example, prey readily on a wide variety of small birds, mammals, lizards, and even insects. Other predators, such as the polar bear, are highly specialized in their prey preferences—a loss of access to seals can lead to polar bear starvation.

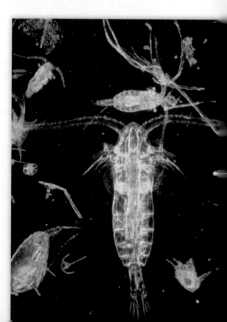

▶ **FIGURE 3.13** Microscopic plants and animals form the basic levels of many aquatic food chains and account for a large percentage of total world biomass.

Predation leads to adaptation

Predator-prey relationships exert selection pressures that favor evolutionary adaptation. Predators become more efficient at searching and feeding, and prey become more effective at escape and avoidance. Prey organisms have developed countless strategies to avoid predation, including toxic or bad-tasting compounds, body armor, extraordinary speed, and the ability to hide. Plants have evolved thick bark, spines, thorns, or distasteful and even harmful chemicals in tissues—poison ivy and stinging nettle are examples. In response, animals have found strategies for avoiding spines, eating through thick bark, or tolerating chemicals. Arthropods, amphibians, snakes, and some mammals produce noxious odors or poisonous secretions that cause other species to leave them alone. Speed is a common defense against predation. On the Serengeti Plain of East Africa, the swift Thomson's gazelle and even swifter cheetah are engaged in an arms race of speed and endurance. The cheetah has an edge in a surprise attack, because it can accelerate from 0 to 72 kph in 2 seconds. But the gazelle often escapes because the cheetah lacks stamina. A general term for this close adaptation of two species is **coevolution**.

Species with chemical defenses often display distinct coloration and patterns to warn away enemies (fig. 3.14). Species also display forms, colors, and patterns that help them hide. Insects that look exactly like dead leaves or twigs are among the most remarkable examples (fig. 3.15). Predators also use camouflage to conceal themselves as they lie in wait for their next meal. In a neat evolutionary twist, certain species that are harmless resemble poisonous or distasteful ones, gaining protection against predators who remember a bad experience with the actual toxic organism. This is called **Batesian mimicry**, after the English naturalist H. W. Bates (1825–1892). Many wasps, for example, have bold patterns of black and yellow stripes to warn off potential predators (fig. 3.16a). A harmless variety of longhorn beetle has evolved to look and act like a wasp, tricking predators into avoiding it (fig. 3.16b). Similarly, the benign viceroy butterfly has evolved to closely resemble the distasteful monarch butterfly. When two unpalatable or dangerous species who look alike, we call it **Müllerian mimicry** (after the biologist Fritz Müller). When predators learn to avoid either species, both benefit.

Symbiosis involves cooperation

In contrast to predation and competition, some interactions between organisms can be non-antagonistic, even beneficial (table 3.2). In such relationships, called **symbiosis**, two or more species live

▲ **FIGURE 3.15** This walking stick is highly camouflaged to blend in with the forest floor, a remarkable case of selection and adaptation.

(a) Wasp

(b) Beetle

▲ **FIGURE 3.16** In Batesian mimicry, a stinging wasp (a) has bold yellow and black bands, which a harmless long-horned beetle mimics (b) to avoid predators.

▶ **FIGURE 3.14** Poison arrow frogs of the family Dendrobatidae display striking patterns and brilliant colors that alert potential predators to the extremely toxic secretions on their skin. Indigenous people in Latin America use the toxin to arm blowgun darts.

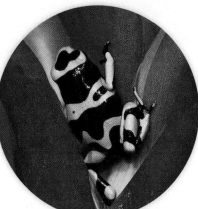

TABLE 3.2 | Types of Species Interactions

INTERACTION BETWEEN TWO SPECIES	EFFECT ON FIRST SPECIES	EFFECT ON SECOND SPECIES
Mutualism	+	+
Commensalism	+	0
Parasitism	+	−
Predation	+	−
Competition	±	±

(+beneficial; −harmful; 0 neutral; ±varies)

▲ **FIGURE 3.17** Coevolution has led to close evolutionary relationships between many species, as in this star orchid and the specially adapted hawk moth that pollinates it.

intimately together, with their fates linked. Symbiotic relationships often involve coevolution. Many plants and pollinators have forms and behaviors that benefit each other. Many moths, for example, are adapted to pollinate particular flowering plants (fig. 3.17).

Symbiotic relationships often enhance the survival of one or both partners. In lichens, a fungus and a photosynthetic partner (either an alga or a cyanobacterium) combine tissues to mutual benefit. A symbiotic relationship such as this, in which both species clearly benefit, is also called **mutualism** (fig. 3.18). Competition and predation were long thought to drive most adaptation and speciation, but ecologists have become increasingly interested in cooperative and mutualistic relationships. Survival of the fittest often means survival of organisms that can live together.

The interdependence of coral polyps and algae in coral reefs is a globally important form of mutualism, in which the polyp provides structure and safety for algae, while the photosynthetic algae provide nutrients to the coral polyp as it builds a coral reef system. Another widespread mutualistic relationship is that between ants and acacia trees in Central and South America. Colonies of ants live inside protective cover of hollow thorns on the acacia tree branches. Ants feed on nectar that is produced in glands at the leaf bases and also eat special protein-rich structures that are produced on leaflet tips. The acacias thus provide shelter and food for the ants. What do the acacias get in return? Ants aggressively defend their territories, driving away herbivorous insects that might feed on the acacias. Ants also trim away vegetation that grows around the tree, reducing competition by other plants for water and nutrients. This mutualistic relationship thus affects the biological community around an acacias, just as competition or predation shapes communities.

Commensalism is a type of symbiosis in which one member clearly benefits and the other apparently is neither benefited nor harmed. Many mosses, bromeliads, and other plants growing on trees in the moist tropics are considered commensals (fig. 3.18c). These epiphytes are watered by rain and obtain nutrients from leaf litter and falling dust, and often they neither help nor hurt the trees on which they grow. Robins and sparrows that inhabit suburban yards are commensal with humans. **Parasitism**, a form of predation, may also be considered symbiosis because of the dependency of the parasite on its host.

Keystone species play critical roles

A **keystone species** plays a critical role in a biological community that is out of proportion to its abundance. Originally, keystone species were thought to be only top predators—lions, wolves, tigers—which limited herbivore abundance and reduced the herbivory of plants. Scientists now recognize that less-conspicuous species also play keystone roles. Tropical fig trees, for example, bear fruit year-round at a low but steady rate. If figs were removed from a forest, many fruit-eating animals (frugivores) would starve in the dry season when fruit of other species is scarce. In turn, the disappearance of frugivores would affect plants that depend on them for pollination and seed dispersal. The effect of a keystone species on communities ripples across multiple trophic levels.

(a) Symbiosis   (b) Mutualism   (c) Commensalism

▲ **FIGURE 3.18** Symbiosis refers to species living together: for example, lichens (a) consist of a fungus, which gives structure, and an alga or cyanobacterium, which photosynthesizes. Mutualism is a symbiotic relationship that benefits both species, such as a lichen or a parasite-eating red-billed oxpicker and a parasite-infested impala (b). Commensalism benefits one species but has little evident effect on the other, as with a tropical tree and a free-loading bromeliad (c).

(a) Kelp shelter fish, seals, and other species

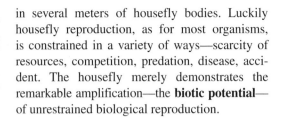

(c) Sea otters protect kelp ecosystem by preying on urchins

(b) Sea urchins graze on kelp

▲ **FIGURE 3.19** Sea otters protect kelp ecosystems on the Pacific coast by eating sea urchins, which could otherwise destroy the kelp.

Off the northern Pacific coast, a giant brown alga (*Macrocystis pyrifera*) forms dense "kelp forests," which shelter fish and shellfish species from predators, allowing them to become established in the community. Within this kelp forest are also sea urchins, which graze on the kelp on the seafloor, and sea otters, which eat the sea urchins. When sea otters have been eliminated—by trapping or by predation, for example—the urchins overgraze and diminish the kelp forests, potentially causing collapse of this complex system (fig. 3.19). Because of their critical role in supporting the entire kelp forest, otters are seen as a classic example of a keystone species.

Keystone functions have been documented for vegetation-clearing elephants, the predatory ochre sea star, and frog-eating salamanders in coastal North Carolina. Even microorganisms can play keystone roles. In many temperate forest ecosystems, groups of fungi that are associated with tree roots (mycorrhizae) facilitate the uptake of essential minerals. When fungi are absent, trees grow poorly or not at all.

3.3 POPULATION GROWTH

Apart from their interactions with other species, organisms have an inherent rate of reproduction that influences population size. Many species have potential to produce almost unbelievable numbers of offspring. Consider the a single female housefly (*Musca domestica*), which can lay 120 eggs. In 56 days those eggs become mature adults, and each female—suppose half are female—can lay another 120 eggs. At this rate, there can be seven generations of flies in a year, and that original fly would be the proud grandparent of 5.6 trillion offspring. If this rate of reproduction continued for ten years, the entire earth would be covered

in several meters of housefly bodies. Luckily housefly reproduction, as for most organisms, is constrained in a variety of ways—scarcity of resources, competition, predation, disease, accident. The housefly merely demonstrates the remarkable amplification—the **biotic potential**—of unrestrained biological reproduction.

Growth without limits is exponential

Understanding population dynamics, or the rise and fall of populations in an area, is essential for understanding how species interact and use resources. As discussed in chapter 2, a population consists of all the members of a single species living in a specific area at the same time. The growth of the housefly population just described is **exponential**, having no limit and possessing a distinctive shape when graphed over time. An exponential growth rate (increase in numbers per unit of time) is expressed as a constant fraction, or exponent, which is used as a multiplier of the existing population. The mathematical equation for exponential growth is:

$$\frac{dN}{dt} = rN$$

Here "*d*" means "change," so the change in number of individuals (*dN*) per change in time (*dt*) equals the rate of growth (*r*) times the number of individuals in the population (*N*). The *r* term (intrinsic capacity for increase) is a fraction representing the average individual contribution to population growth. If *r* is positive, the population is increasing. If *r* is negative, the population is shrinking. If *r* is zero, there is no change, and *dN/dt* = 0.

A graph of exponential population growth is described as a **J curve** (fig. 3.20) because of its shape. As you can see, the number of individuals added to a population at the beginning of an exponential growth curve can be rather small. But the numbers begin to increase quickly with a fixed growth rate. For example, when a population has just 100 individuals, a 2 percent growth rate adds just 2 individuals. For a population of 10,000, that 2 percent growth adds 200 individuals.

The exponential growth equation is a very simple model; it is an idealized description of a real process. Essentially the same. The same equation is used to calculate growth in your bank account due to compounded interest rates; achieving the maximum growth potential would require that you never withdraw any money. But in fact, some money probably will be withdrawn. Similarly, not all individuals in a population survive, so actual growth rates are something less than the full biotic potential.

Carrying capacity limits growth

In the real world there are limits to growth. Around 1970, ecologists developed the concept of **carrying capacity** to mean the number or biomass of animals that can be supported (without harvest) in a

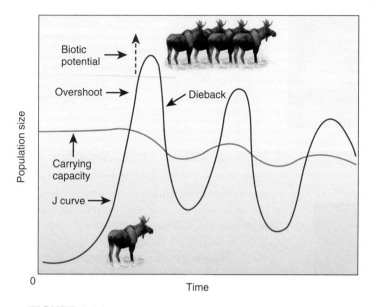

▲ **FIGURE 3.20** A J curve, or exponential growth curve, leads to repeated overshoot and dieback cycles. The environment's ability to support the species (carrying capacity) may diminish as overuse degrades habitat. Moose on Isle Royales in Lake Superior seem to have exhibited this pattern.

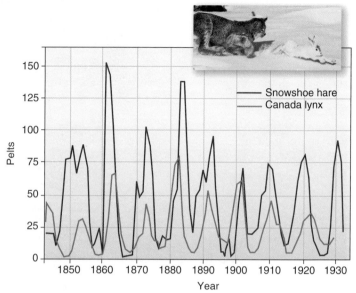

▲ **FIGURE 3.21** Ten-year oscillations in the populations of snowshoe hares and lynx in Canada suggest a close linkage of predator and prey. These data are based on the number of pelts received by the Hudson Bay Company from fur traders   SOURCE: Data from D.A. MacLulich. Fluctuations in the Numbers of the Varying Hare (*Lepus americus*). Toronto: University of Toronto Press, 1937, reprinted 1974.

certain area of habitat. The concept is now used more generally to suggest a limit of sustainability that an environment has in relation to the size of a species population. Carrying capacity is helpful in understanding the population dynamics of some species, perhaps even humans.

When a population overshoots or exceeds the carrying capacity of its environment, resources become limited and death rates rise. If deaths exceed births, the growth rate becomes negative and the population may suddenly decrease, a change called a population crash or dieback (fig. 3.20). Populations may oscillate from high to low levels around the habitat's carrying capacity, which may be lowered if the habitat is damaged. Moose and other browsers or grazers sometimes overgraze their food plants so that future populations in the same habitat find less preferred food to sustain them, at least until the habitat recovers. Some species go through predictable cycles if simple factors are involved, such as the seasonal light- and temperature-dependent bloom of algae in a lake. Cycles can be irregular if complex environmental and biotic relationships exist. Irregular cycles include outbreaks of migratory locusts in the Sahara or tent caterpillars in temperate forests—these represent irruptive population growth. Often immigration of a species into an area, or emigration from an area, also affects population growth and declines.

Sometimes predator and prey populations oscillate in synchrony with each other. One classic study (fig. 3.21) employed the 200-year record of furs sold at Hudson Bay Company trading posts in Canada (the figure shows a portion of that record). The ecologist Charles Elton showed that numbers of Canada lynx (*Lynx canadensis*) fluctuate on about a ten-year cycle that mirrors, slightly out of phase, the population peaks of snowshoe hares (*Lepus americanus*). When the hare population is high, the lynx prosper on abundant prey; they reproduce well, and their population grows. Eventually the abundant hares overgraze the vegetation, decreasing their food supplies, and the hare populations shrink. For a while the lynx benefits because starving hares are easier to catch than healthy ones. As hares become scarce, however, so do lynx. When hares are at their lowest levels, their food supply recovers and the whole cycle starts over again. This predator-prey oscillation is described mathematically in the Lotka-Volterra model, named for the scientists who developed it.

Environmental limits lead to logistic growth

Not all biological populations cycle through exponential overshoot and catastrophic dieback. Many species are regulated by both internal and external factors and come into equilibrium with their environmental resources while maintaining relatively stable population sizes. When resources are unlimited, they may even grow exponentially, but this growth slows as the carrying capacity of the environment is approached. This population dynamic is called **logistic growth** because of its changes in growth rate over time.

Mathematically, this growth pattern is described by the following equation, which adds a feedback term for carrying capacity (K) to the exponential growth equation:

$$\frac{dN}{dt} = rN\frac{(K - N)}{K}$$

The logistic growth equation says that the change in numbers over time (*dN/dt*) equals the exponential growth rate (*rN*) times the portion of the carrying capacity (K) not already taken by the current population size (N). The term ($K − N$)/K establishes the relationship between population size at any given time and the carrying capacity (K). If N is less than K, the rate of population change will be positive. If N is greater than K, then change will be negative (see Active Learning, below).

The logistic growth curve has a different shape than the exponential growth curve. It is a sigmoidal-shaped, or **S curve** (fig. 3.22). It describes a population whose growth rate decreases if its numbers approach or exceed the carrying capacity of the environment.

Population growth rates are affected by external and internal factors. External factors include habitat quality, food availability, and interactions with other organisms. As populations grow, food becomes scarcer and competition for resources more intense. With a larger population, there is an increased risk that disease or parasites will spread, or that predators will be attracted to the area. Internal factors, such as slow growth and maturity, body size, metabolism, or homonal status can reduce reproductive output. Often crowding increases these factors. Overcrowded house mice (>1,600/m^3), for instance, average 5.1 babies per litter, while uncrowded house mice (<34/m^3) produce 6.2 babies per litter. All these factors are **density-dependent**: as population size increases, the effect intensifies. With **density-independent** factors, a population is affected no matter what its size. Drought, an early killing frost, flooding, landslide, or habitat destruction by people—all increase mortality rates regardless of the population size. Density-independent limits to population are often nonbiological, capricious acts of nature.

Species respond to limits differently: r- and K-selected species

Some organisms, such as dandelions and barnacles, depend on a high rate of reproduction and growth (r) to secure a place in the environment. These organisms are called **r-selected species** because they are adapted to employ a high reproductive rate to overcome the high mortality of virtually ignored offspring. These species may even overshoot carrying capacity and experience

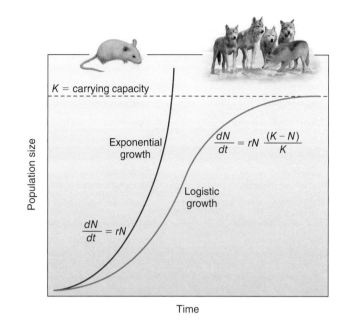

$$\frac{dN}{dt} = rN \frac{(K - N)}{K}$$

$$\frac{dN}{dt} = rN$$

▲ **FIGURE 3.22** Exponential growth rises in a J-shaped curve. In contrast, logistic growth rates from an S-shaped curve as carrying capacity slows or stops population growth.

Active **LEARNING**

Effect of K on Population Growth Rate (rN)

In logistic growth, the term ($K − N$)/K creates a fraction that is multiplied by the growth rate (rN). Suppose carrying capacity (K) is 100. If N is 150, then is the term ($K − N$)/K positive or negative? Is population change positive or negative? What if N is 50? If N were 100?

ANSWERS: negative, positive, no growth.

population crashes, but as long as vast quantities of young are produced, a few will survive. Other organisms that reproduce more conservatively—longer generation times, late sexual maturity, fewer young—are referred to as **K-selected species**, because they are adapted for slower growth conditions near the carrying capacity (K) of their environment.

Many species blend exponential (r-selected) and logistic (K-selected) growth characteristics. Still, it's useful to contrast the advantages and disadvantages of organisms at the extremes of the continuum. It also helps if we view differences in terms of "strategies" of adaptation and the "logic" of different reproductive modes (table 3.3).

So-called K-selected organisms are usually larger, live long lives, mature slowly, produce few offspring in each generation, and have few natural predators (fig. 3.23, Type I). Elephants, for example, are not reproductively mature until they are 18 to 20 years old. In youth and adolescence, a young elephant belongs to an extended family that cares for it, protects it, and teaches it how to behave. A female elephant normally conceives only once every 4 or 5 years. The gestation period is about 18 months; thus an elephant herd doesn't produce many babies in any year. Because elephants have few enemies and live a long life (60 to 70 years), this low reproductive rate produces enough elephants to keep the population stable, given good environmental conditions and no poachers.

Organisms with r-selected, or exponential, growth patterns tend to occupy low trophic levels in their ecosystems (chapter 2) or they are successional pioneers. Niche generalists occupy disturbed or new environments, grow rapidly, mature early, and produce many offspring with excellent dispersal abilities. As individual parents, they do little to care for their offspring or protect

TABLE 3.3 | Reproductive Strategies

r-SELECTED SPECIES	K-SELECTED SPECIES
1. Short life	1. Long life
2. Rapid growth	2. Slower growth
3. Early maturity	3. Late maturity
4. Many, small offspring	4. Few, large offspring
5. Little parental care and protection	5. High parental care or protection
6. Little investment in individual offspring	6. High investment in individual offspring
7. Adapted to unstable environment	7. Adapted to stable environment
8. Pioneers, colonizers	8. Later stages of succession
9. Niche generalists	9. Niche specialists
10. Prey	10. Predators
11. Regulated mainly by intrinsic factors	11. Regulated mainly by extrinsic factors
12. Low trophic level	12. High trophic level

them from predation. They invest their energy in producing huge numbers of young and count on some surviving to adulthood (fig. 3.23, Type III).

A female clam, for example, can release up to 1 million eggs in her short lifetime. The vast majority of young clams die before reaching maturity, but if even a few survive, the species will continue. Many marine invertebrates, parasites, insects, rodents, and annual plants follow this reproductive strategy. Also included in this group are most invasive and pioneer organisms, weeds, and pests.

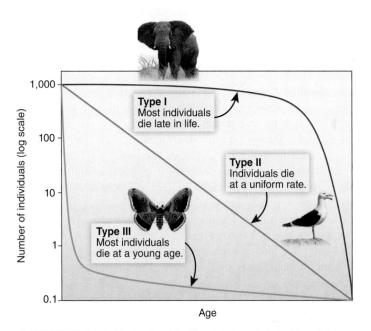

▲ **FIGURE 3.23** Idealized survivorship curves include high survival rates for juveniles (Type I), uniform mortality risk throughout life (Type II), and high mortality for juveniles, with long survival for a few individuals that reach maturity (Type III).

Many species exhibit some characteristics of both r and K selection. These might approximately follow a Type II survivorship pattern (fig. 3.23), with similar likelihood of mortality at multiple life stages.

When you consider the species you recognize from around the world, can you classify them as r- or K-selected species? What strategies seem to be operating for ants, bald eagles, cheetahs, clams, giraffes, pandas, or sharks? An important question in the back of your mind might be: Where do people fit? Are we like wolves and elephants or more similar to clams and rabbits in our population growth strategy?

3.4 COMMUNITY DIVERSITY

No species is an island. It always lives with other species in a biological community in a particular environment. You've seen how interactions among species affect biological communities. In this section, we'll consider how the many species in an area together produce the basic properties of biological communities and ecosystems. The main properties of interest to ecologists are (1) diversity and abundance; (2) community structure and patchiness; and (3) complexity, resilience, productivity, and stability.

Diversity and abundance

Diversity is the number of different species in an area, or the number per unit area. The number of herbaceous plant species per square meter of African savanna, the number of bird species in Costa Rica, all the insect species on earth (tens of millions of them!)—all describe diversity. Diversity is important because it indicates the variety of ecological niches and genetic variation in a community. **Abundance** refers to the number of individuals of a particular species (or of a group) in an area. Diversity and abundance are often related. Communities with high diversity often have few individuals of any one species, because there are many different species sharing available resources. Most communities contain a few common species and many rarer ones.

As a general rule, species diversity is greatest at the equator and drops toward the poles. Though the number of species is lower near the poles, the abundance of particular species can be very high. In the Arctic there are few insect species overall, but the abundance of one or two, especially mosquitoes, is almost incalculable. In the tropics, on the other hand, hundreds of thousands of different insect species coexist. Highly specialized forms and behaviors allow them to exploit narrow niches, but at any one location only a few individuals of a species might occur. The same pattern is seen in trees of high-latitude boreal forests versus tropical rainforests, and in bird populations. In Greenland there are 56 species of breeding birds, whereas in Colombia, with one-fifth the area, there are 1,395.

Climate explains much of this difference in diversity. Greenland has a harsh climate and a short, cool growing season that restricts the biological activity that can take place. During half the year, solar energy is abundant, but during winter incoming energy is almost absent. Only species mobile enough to take advantage of

| (a) Random | (b) Uniform | (c) Clustered |

▲ **FIGURE 3.24** Distribution of a population can be random (a), uniform (b), or clustered (c).

fleeting resources can survive. History also matters: Greenland's coast has been free of glaciers for only about 10,000 years, so that new species have had little time to develop.

Many areas in the tropics, by contrast, were never covered by glacial ice, and have abundant rainfall and warm temperatures year-round, so that ecosystems there are highly productive. The year-round availability of food, moisture, and warmth supports an exuberance of life and allows a high degree of specialization in physical shape and behavior. Many niches exist in small areas, with associated high species diversity. Coral reefs are similarly stable, productive, and conducive to proliferation of diverse and exotic life-forms. An enormous abundance of brightly colored and fantastically shaped fishes, corals, sponges, and arthropods live in the reef community. Increasingly, human activities also influence biological diversity today. The cumulative effects of our local actions can dramatically alter biodiversity (see What Can You Do?, at right). We discuss this issue in chapter 5.

Patterns produce community structure

The spatial distribution of individuals, species, and populations can influence diversity, productivity, and stability in a community. Niche diversity and species diversity can increase as the complexity increases at the landscape scale, for example. **Community structure** is a general term we use for spatial patterns. Ecologists focus on several aspects of community structure, which we discuss here.

Distribution can be random, ordered, or patchy Even in a relatively uniform environment, individuals of a species population can be distributed randomly, arranged in uniform patterns, or clustered together. In randomly distributed populations, individuals live wherever resources are available and chance events

allow them to settle (fig. 3.24a). Uniform patterns arise from the physical environment also, but more often are caused by competition and territoriality. For example, penguins or seabirds compete fiercely for nesting sites in their colonies. Each nest tends to be just out of reach of neighbors sitting on their own nests. Constant squabbling produces a highly regular pattern (fig. 3.24b). Plants also compete, producing a uniform pattern. Sagebrush releases toxins from roots and fallen leaves, which inhibit the growth of

What Can YOU DO?

Working Locally for Ecological Diversity

You might think that diversity and complexity of ecological systems are too large or too abstract for you to have any influence. But you can contribute to a complex, resilient, and interesting ecosystem, whether you live in the inner city, a suburb, or a rural area.

- Take walks. The best way to learn about ecological systems in your area is to take walks and practice observing your environment. Go with friends and try to identify some of the species and trophic relationships in your area.

- Keep your cat indoors. Our lovable domestic cats are also very successful predators. Migratory birds, especially those nesting on the ground, have not evolved defenses against these predators.

- Plant a butterfly garden. Use native plants that support a diverse insect population. Native trees with berries or fruit also support birds. (Be sure to avoid non-native invasive species.) Allow structural diversity (open areas, shrubs, and trees) to support a range of species.

- Join a local environmental organization. Often, the best way to be effective is to concentrate your efforts close to home. City parks and neighborhoods support ecological communities, as do farming and rural areas. Join an organization working to maintain ecosystem health; start by looking for environmental clubs at your school, park organizations, a local Audubon chapter, or a local Nature Conservancy branch.

- Live in town. Suburban sprawl consumes wildlife habitat and reduces ecosystem complexity by removing many specialized plants and animals. Replacing forests and grasslands with lawns and streets is the surest way to simplify, or eliminate, ecosystems.

competitors and create a circle of bare ground around each bush. Neighbors grow up to the limit of this chemical barrier, and regular spacing results.

Other species cluster together for protection, mutual assistance, reproduction, or access to an environmental resource. Ocean and freshwater fish form dense schools, increasing their chances of detecting and escaping predators (fig. 3.24c). Meanwhile many predators—whether wolves or humans—hunt in packs. When blackbirds flock in a cornfield, or baboons troop across the African savanna, their group size helps them evade predators and find food efficiently. Plants also cluster for protection in harsh environments. You often see groves of wind-sheared evergreens at mountain treelines or behind foredunes at seashores. These treelines protect the plants from wind damage, and incidentally shelter other animals and plants creating a cluster of communities.

Individuals can also be distributed vertically in a community. Forests, for instance, have many layers, each with different environmental conditions and combinations of species. Distinct communities of plants, animals, and microbes live in the treetops, at mid-canopy level, and near the ground. This layering, known as vertical stratification, is best developed in tropical rainforests (fig. 3.25). Aquatic communities also are often stratified into layers formed by species responding to varying levels of light, temperature, salinity, nutrients, and pressure.

Communities form patterns in landscapes   If you fly in an airplane, you can see that the landscapes consists of patches of different colors and shapes (fig. 3.26). Some appear long and narrow (hedgerows or rivers), while others are rectangular (pastures and cropfields), or green and lumpy in summer (forests). Each patch represents a biological community with its own set of species and environmental conditions. Most landscapes exhibit patchiness in some way. The largest patches might contain **core habitat**, a relatively uniform environment that is free of the influence of edges. Often we consider generalist species to occur on edges, while many specialists may require the more consistent conditions of core habitat. The northern spotted owl, for example, nests in the interior of large patches of mature coniferous forest in the Pacific Northwest (chapter 6). In smaller patches the owl fares poorly, possibly due to competition with the closely related barred owl. A single pair of northern spotted owls may require over 1,000 hectares of core habitat to survive.

Where communities meet, the environmental conditions blend and the species and microclimate of one community can penetrate the other. Called **edge effects**, the penetrating influences may extend hundreds of meters into an adjacent community. A forest edge adjacent to open grassland is sunnier, drier, hotter, and more

▲ **FIGURE 3.25** Vertical layering of plants and animals is an important type of community structure, and is especially evident in tropical rainforests.

▲ **FIGURE 3.26** Complex landscapes include contrasting environments, edges where they meet, and corridors connecting larger patches. Edges are biologically rich, but core areas are critical for many species.

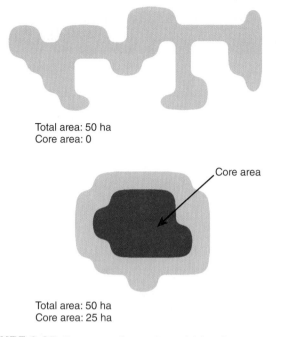

Total area: 50 ha
Core area: 0

Core area

Total area: 50 ha
Core area: 25 ha

▲ **FIGURE 3.27** Shape can influence the availability of core area in a habitat area or a preserve.

susceptible to storm damage than the center of the forest. Generalist grassland species, including weeds and predators, may move into the forest and negatively affect forest species. The shape of a patch also can affect the amount of core habitat. In a narrow, irregularly shaped patch, far-reaching edge effects would leave no core habitat. In a similar-sized square patch, however, interior species would still find core habitat (fig. 3.27). Human activities often produce fragmented habitat with increased edge effects and decreased or loss of core habitat. Suburban expansion and forest clearing, for example, tend to occur in patchy patterns with a high density of edges in the landscape.

Edges are often rich in species because individuals from both environments occupy the boundary area. Many species prefer edges and use the resources of both environments. Many game animals, such as deer and pheasants, are most plentiful on edges. To boost these populations, North American game managers in the 1930s worked to create openings in forests and to plant trees and shrubs in grasslands. More recently, wildlife biologists have recognized that these edges reduce habitat for species that need interior conditions. Habitat managers have therefore worked to preserve larger patches, and to connected smaller patches where possible (see fig. 3.27).

Resilience seems related to complexity

The relationship between complexity and resilience has long been an important question in ecology. The issue was framed most famously by ecologist Robert MacArthur (1930−1972), who proposed that the more complexity a community possesses, the more resilient it is when disturbance strikes. He reasoned that if many different species occupy each trophic level, some can fill in if others are stressed or eliminated by external forces. The

whole community has **resilience** and either resists or recovers quickly from disturbance. Some important experimental evidence for the relationship between resilience and diversity comes from the studies of ecologist David Tilman and his colleagues at Cedar Creek Natural History Area in Minnesota (see Exploring Science, p. 70).

Often we think of diversity in terms of species counts, but another aspect is community **complexity**. Complexity refers to the number of trophic levels in a community and to the number of species at each of those trophic levels. A complex community might have many trophic levels and groups of species performing the same functions. In tropical rainforests and many other communities, herbivores form guilds based on the specialized ways they feed on plants. There may be fruit-eaters, leaf-nibblers, root-borers, seed-gnawers, and sap-suckers—each guild is composed of species of different sizes, shapes, and even biological kingdoms, but they feed in similar ways.

Similarly, we can see community complexity in an Antarctic ecosystem. There again the sun powers the system, with floating algae performing photosynthesis. These feed tiny crustaceans called krill, which in turn support subsequent trophic levels (table 3.4). Most ecosystems exhibit even greater diversity of species in the multiple trophic levels.

Another factor that can contribute to resilience is productivity. **Primary productivity** is the production of biomass by photosynthesis. Plants, algae, and some bacteria produce biomass by converting solar energy into chemical energy, which they use or pass on to other organisms. We can measure primary productivity in terms of units of biomass per unit area per year—for example, in grams per m^2 per year. Because cellular respiration in producing organisms uses much of that energy, a more useful term is **net primary productivity**, or the amount of biomass stored after respiration. Productivity depends on light levels, temperature, moisture, and nutrient availability, so it varies dramatically among different ecosystem types (fig. 3.28). Tropical forests, coral reefs, and the bays and estuaries (where rivers meet the ocean) have high productivity because abundant energy and moisture are available. In deserts, a lack of water limits photosynthesis, and productivity

TABLE 3.4	Community Complexity in the Antarctic Ocean
TYPE OF FUNCTION	**MEMBERS OF FUNCTIONAL GROUP**
Top ocean predator	Sperm and killer whales, leopard and elephant seals
Aerial predator	Albatross, skuas
Other ocean predator	Weddell and Ross seals, king penguin, pelagic fish
Krill/plankton-feeder	Minke, humpback, fin, blue, and sei whales
Ocean herbivore	Krill, zooplankton (many species)
Ocean-bottom predator	Many species of octopods and bottom-feeding fish
Ocean-bottom herbivore	Many species of echinoderms, crustaceans, mollusks
Photosynthesizer	Many species of phytoplankton and algae

Does diversity matter? Ecologists long believed that a diversity of species in a community is not just beautiful and fascinating, but also functionally important. One hypothesis has been that diverse communities are more stable, able to recover more quickly after a disturbance than a species-poor commmunity. This idea, raised by ecologists in the 1950s, suggests a strong reason to preserve the earth's biodiversity, but not until 1994 did someone present data that helped answer it. In a long-term study of prairie plant communities, ecologist Dave Tilman and his colleagues planted dozens of experimental plots at Cedar Creek Natural History Area, each plot containing different numbers of species (fig. 1). In 1988, central Minnesota experienced its hottest, driest summer in 50 years. As Tilman's team watched their plots recover from the drought, they measured growth in the plots by carefully clipping, counting, and weighing every plant. By 1992 it was clear that plots with five or fewer species were recovering slowly, while higher-diversity plots had reached or exceeded their pre-drought productivity.

Because the question of diversity and recovery was controversial, the strength of Tilman's results depended on duplicated plots (replicates) and a long period of study. If he had used only one plot for each level of species diversity, chance alone might have depressed productivity. Another question in any experiment is how well experimental results represent other situations and other samples. Tilman's team approximated an answer to this question by reporting a range of values around the average recovery rates, in this case the "standard error." This standard statistical measure and its related value, confidence interval, show the range within which nearly all means should fall if someone else did the same experiment. In this way we describe how confident we are that the results can be applied to other situations (fig. 2).

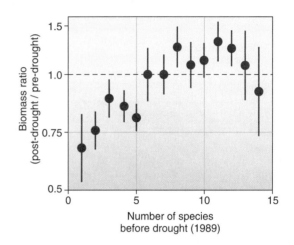

▲ FIGURE 2 Regrowth after a drought was poor for plots with five or fewer species (red). Bars slow standard error for groups of plots; dots show means.   SOURCE: Tilman and Dowling (1994) *Nature* 367:363–365.

Later experiments at Cedar Creek showed that species-rich plots recovered more fully after drought because some species were severely harmed by drought, others went dormant but didn't die, and still others continued growing but at a slow rate. When rain returned, surviving plants revived and filled the plots with green leaves and stems. The chance that a species-poor plot contained quickly recovering species was lower than for a species-rich plot. Consequently, low-diversity plots were more likely to attain their former productivity slowly, if at all. This explanation was bolstered by another experimental discovery: that nitrogen—a limiting factor in the sterile, sandy soils of the plots—was used up more completely in species-rich plots than in species-poor plots.

The experimental data from Cedar Creek prairies stimulated tests in other ecosystems. Is resilience after disturbance always evident in species-rich plant communities, such as rainforests of Central and South America? (Preliminary results show that it depends on how big and severe the disturbance is.) Do species-poor plant communities, such as dune grasslands and salt marshes, recover only slowly following disturbance? Actually, some simple plant communities appear quite stable where disturbances like wind, waves, and storms are frequent.

What are the implications of Tilman's research for systems we depend on, such as agricultural fields populated by a single crop plant? One important application may be in biofuels production. Ethanol (a biomass fuel) made from starch or cellulose may be an attractive alternative to petroleum-based fuels (chapter 12). Tilman's study suggests that, given variable weather over several years, a diverse biomass crop can achieve an economical net energy yield while providing greater environmental benefits than does a single-species crop. The general lesson is that keeping a diversity of species as a backdrop to human existence may serve us best in the long run; and if species diversity is maintained, nature may be more resilient than we imagine.

▲ FIGURE 1 Experimental grassland plots at Cedar Creek.

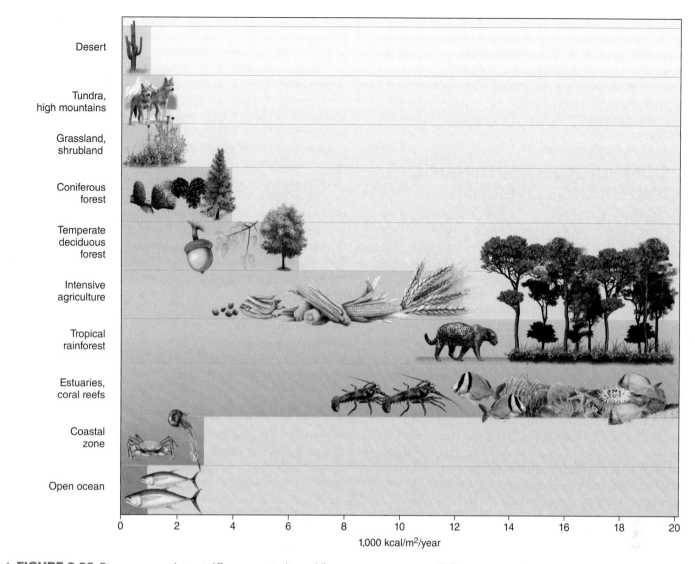

▲ FIGURE 3.28 Biomass accumulates at different rates in the world's major ecosystem types. Differences in net primary production is chiefly due to temperature, rainfall, and nutrients. Interactions among species also boost productivity.

is low. On the arctic tundra or on high mountains, low temperatures and low solar energy inputs inhibit plant growth and productivity. In the open ocean, a lack of nutrients reduces the ability of algae to make use of plentiful sunshine and water.

Even the most photosynthetically active ecosystems capture only a small percentage of the available sunlight and use it to make energy-rich compounds. In a temperate-climate oak forest, for example, leaves absorb only about half the available light on a midsummer day. Of this absorbed energy, 99 percent is used in respiration and in evaporating water to cool leaves. A large oak tree can transpire (evaporate) several thousand liters of water on a warm, dry, sunny day, while making only a few kilograms of sugars and other energy-rich organic compounds.

Stability is another property that varies among ecosystems. When we say a community or ecosystem is stable, we mean it resists changes despite disturbance, springs back resiliently after disturbance, and supports the same species in about the same

numbers as before the disturbance. We often think of tropical rainforests as stable over long periods of time. They may have annual rainy and dry seasons, but we don't expect dramatic changes in species composition. In contrast, many coniferous forests are periodically "reset" by fires. There is a turnover in dominant species that might last for decades.

Sometimes disturbance dramatically changes an ecosystem. When the Great Plains experienced devastating droughts in the 1930s and early 1940s, overall productivity declined, some species populations practically vanished, and others held their own but at lower abundance. When the rains returned, the plant communities were different, not just because of drought, but also because millions of cattle had overgrazed the range—an additional, simultaneous disturbance. Today's remnant Great Plains grasslands are very different from the grasslands of two centuries ago. Yet they remains grasslands, produce forage, and feed cattle. Are the Great Plains grasslands stable because they are still relatively productive, or are they unstable because disturbances have dramatically altered species diversity and abundance? If the range were grazed according to ecological principles, would that restore the original species diversity and abundance, and raise productivity? Asking questions about diversity, productivity, resilience, and stability helps us consider what aspects of an ecosystem we value most, and what properties help maintain those characteristics of the community.

3.5 COMMUNITIES ARE DYNAMIC AND CHANGE OVER TIME

If fire sweeps through a forest, we often say that the forest was destroyed. But usually that is an inaccurate description of what happened. Often fire is good for a community—releasing nutrients in a burned grassland, or allowing for regeneration of aging trees in a coniferous forest. The idea that dramatic, periodic change could be a part of normal ecosystems is relatively new in ecology. We used to consider stability the optimal state of ecosystems. But with more observations and more studies, we have learned that communities can be dynamic, with dramatic changes over time.

Are communities organismal or individualistic?

For several decades, starting in the early 1900s, ecologists in North America and Europe argued about the basic nature of communities. This debate doesn't make great party conversation (unless you're an ecologist), but it has long influenced the ways we study and understand communities, the ways we view changes in a community, and ultimately the ways we use them. On one side of the debate was an idea proposed by J. E. B. Warming (1841–1924) in Denmark and Henry Chandler Cowles (1869–1939) in the United States. These two proposed that communities develop in a sequence of stages, starting either from new land or after a severe disturbance. Working in sand dunes, they examined the changes as plants first took root in bare sand and, with further development, ultimately created a forest. The community that developed last and lasted the longest was called the **climax community**.

The importance of climax communities was championed by the biogeographer F. E. Clements (1874–1945). He viewed the process as a relay—species replace each other in predictable groups and in a fixed, regular order. He argued that every landscape has a characteristic climax community, determined mainly by climate. If left undisturbed, this community would mature to a characteristic set of species, each performing its optimal functions. A climax community to Clements represented the maximum possible complexity and stability in a given situation. He and others made the analogy that the development of a climax community resembled the maturation of an

organism. Like organisms, they argued, communities began simply and primitively, maturing until a highly integrated, complex, and stable condition developed.

On the other side of the debate was an individualistic view of community change, which was championed by Clements' contemporary, H. A. Gleason (1882–1975). Gleason saw community history as an unpredictable process. He argued that species are individualistic, each establishing in an environment according to its own ability to colonize, tolerate the environmental conditions, and reproduce there. This idea allows for myriad temporary associations of plants and animals to form, fall apart, and reconstitute in slightly different forms, depending on environmental conditions and the species in the neighborhood. Imagine a time-lapse movie of a busy airport terminal. Passengers come and go; groups form and dissipate. Patterns and assemblages that seem significant may not mean much a year later. General growth forms might be predictable—grasses grow in some conditions while trees grow in others—but the exact composition of a community is not necessarily predictable. Gleason suggested that we think ecosystems are uniform and stable only because our lifetimes are too short and our geographic scope too limited to understand their actual dynamic nature. Most ecologists now find that Gleason's explanation fits observed communities better than Clements' does.

Succession describes community change

Succession is a process in which organisms occupy a site and change its environmental conditions, gradually making way for another type of community. In **primary succession**, land that is bare of soil—a sandbar, rock surface, volcanic flow—is colonized by living organisms where none lived before (fig. 3.29). **Secondary succession**, meanwhile, occurs after a disturbance, when a new community develops from the biological

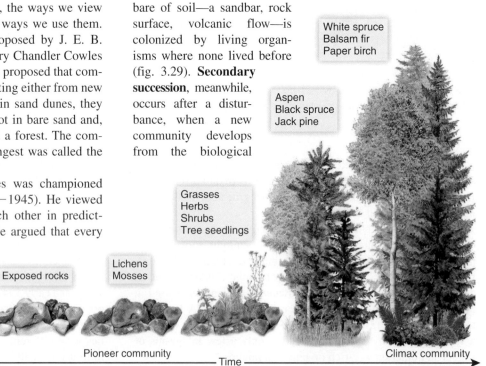

White spruce
Balsam fir
Paper birch

Aspen
Black spruce
Jack pine

Grasses
Herbs
Shrubs
Tree seedlings

Lichens
Mosses

Exposed rocks

Pioneer community

Time

Climax community

▲ **FIGURE 3.29** Primary succession in boreal forest involves five main stages. Exposed rocks are colonized by lichens and mosses, which trap moisture and build soil for grasses, shrubs, and eventually trees. Periodic fires can set back trees, initiating secondary succession. Aspens spring back from roots, pines germinate in openings, and a new forest begins.

legacy of the previous one. In both kinds of succession, organisms change the environment by modifying soil, light levels, food and water supplies, and microclimate. This change permits new species to colonize and eventually replace the previous species, a process known as ecological development or facilitation.

In primary succession on land, the first colonists are hardy **pioneer species**, often microbes, mosses, and lichens that can withstand a harsh environment with few resources. Pioneer species create patches of organic matter and debris that accumulate in pockets and crevices, retaining water and creating soil where seeds of more plants lodge and grow. As succession proceeds, the community becomes more diverse and interspecies competition arises. Larger plants grow, creating vertical structure. Pioneers disappear as the environment favors new colonizers that have competitive abilities more suited to the new environment.

You can see secondary succession all around you, in abandoned farm fields, in clear-cut forests, and in disturbed suburbs and lots. Soil, seeds, and residual plant roots may be present. Because disturbed soil lacks vegetation, plants that live one or two years (annuals and biennials) do well. Their lightweight seeds travel far on the wind, and their seedlings tolerate full sun and extreme heat. When they die, they lay down organic material that improves the soil's fertility and shelters other seedlings. Soon long-lived and deep-rooted perennial grasses, herbs, shrubs, and trees take hold, building up the soil's organic matter and increasing its ability to store moisture. Forest species that cannot survive bare, dry, sunny ground eventually find ample food, a diverse community structure, and shelter from drying winds and low humidity.

Generalists figure prominently in early succession. Over thousands of years, however, competition should decrease as niches proliferate and specialists arise. In theory, long periods of community development lead to greater community complexity, high nutrient conservation and recycling, stable productivity, and great resistance to disturbance.

Some communities depend on disturbance

Disturbances occur frequently in ecosystems. Landslides, mudslides, hailstorms, earthquakes, hurricanes, tornadoes, tidal waves, wildfires, and volcanoes are just a few obvious types. Ecologically, a **disturbance** is any force that disrupts the established patterns of species diversity and abundance, community structure, or community properties. Disturbances are not always sudden, like a hurricane. Slow events like drought can also dramatically alter ecosystems and their functions (see Exploring Science, p. 70). Animals can also cause disturbance, as when African elephants rip out small trees, trample shrubs, and tear down tree limbs as they forage and move about, opening up forest communities and creating savannas.

People also cause disturbances in many ways. Aboriginal people set fires, introduced new species, harvested resources, or changed communities in many places. Sometimes, those disturbances had major ecological effects. Archaeological evidence suggests that when humans colonized islands, such as New Zealand, Hawaii, and Madagascar, large-scale extinction of many animal species followed. Although there is debate about whether something

similar happened in the Americas at the end of the last ice age, about the same time that humans arrived, many of the large mammal species disappeared. In some cases, the landscapes created by aboriginal people may have been maintained for so long that we assume that those conditions always existed there. In eastern North America, for example, fires set by native people maintained grassy, open savannas that early explorers assumed were the natural state of the landscape.

The disturbances caused by modern technological societies are often much more obvious and irreversible. You may have seen scars from road building, mining, clear-cut logging, or other disruptive activities. Obviously, once a landscape has been turned into a highway, a parking lot, or a huge hole in the ground, it will be a long time before it returns to its former condition. But sometimes even if the disturbance seems relatively slight, it can have profound effects.

Consider the Kingston Plains in Michigan's Upper Peninsula, for example. Clear-cut logging at the end of the nineteenth century removed the white pine forest that once grew there. Repeated burning by pioneers, who hoped to establish farms, removed nutrients from the sandy soil and changed ecological conditions so that more than a century later, the forest still hasn't regenerated (fig. 3.30). Given extensive changes by either humans or nature, it may take centuries for a site to return to its predisturbance state, and if climate or other conditions change in the meantime, it may never recover.

Clements' organismal perspective would suggest that most disturbance is similarly harmful. In the early 1900s this view guided efforts to protect timber supplies in the American West from ubiquitous wildfires. Establishing stability and reducing disturbance was also an argument for building dams on rivers all across the West—to prevent disturbance from flooding and to store water for dry seasons. Fire suppression and flood control became the central policies in American natural resource management for most of the twentieth century.

▲ **FIGURE 3.30** Sometimes recovery from disturbance is extremely slow, or the system shifts to a new type of ecosystem. At Kingston Plains in Michigan's eastern Upper Peninsula, stumps remain a century after the trees were cleared by logging and fire.

Recently, new ideas about natural disturbances have entered land management discussions and brought change to some land management policies. Grasslands and some forests are now considered "fire-adapted," and fires are allowed to burn in them if weather conditions are appropriate. Floods also are seen as crucial for maintaining floodplain and river health. Policy makers and managers increasingly consider ecological information when deciding on new dams and levee construction projects.

Ecologists have found that disturbance benefits many species, much as predation does, because it sets back the most competitors and allows less-competitive species to persist. In northern deciduous forests, maples (especially sugar maple and the more recently arrived Norway maple) are more prolific seeders and more shade tolerant than most other tree species. After decades without disturbance, maples outcompete other trees for a place in the forest canopy. The dense shade of maples basically starves other species for light. Most species of oak, hickory, and other light-requiring trees diminish in abundance, as do forest understory plants. When windstorms, tornadoes, wildfires, or ice storms hit a maple-dominated forest, trees are toppled, branches broken, and light once again reaches the forest floor. This stimulates seedlings of oaks and hickories, as well as understory plants. Breaking the grip of the strongest competitor is the helpful role disturbances often play. The 1988 Yellowstone fires set back the lodgepole pine (*Pinus contorta*), which had expanded its acreage in the park. After a few years, successional processes created dense forests in some areas and open savannas in others (fig. 3.31). This resulted in a greater variety of plant species, and wildlife responded vigorously to the greater habitat diversity.

In some landscapes, periodic disturbance is a basic characteristic, and communities are made up of **disturbance-adapted species**. Some survive fires by hiding underground, others reseed quickly after fires. Grasslands, the chaparral scrubland of California and the Mediterranean region, savannas, and some kinds of coniferous forests are shaped and maintained by periodic fires that have long been a part of their history. In fact, many of the dominant plant species in these communities need fire to suppress competitors, to prepare the ground for seeds to germinate, or to pop open cones or split thick seed coats and release seeds. Without fire, community structure would be quite different.

From another view, disturbance resets the successional clock that always operates in every community. Even though all seems

▲ **FIGURE 3.31** The Yellowstone National Park fires of 1988 burned 1.4 million acres. Recovery of the region, which was 80 percent lodgepole pine before the fire, has produced greater diversity of vegetation, including savannas and more mixed forest types.

chaotic after a disturbance, it may be that preserving species diversity by allowing in natural disturbances (or judiciously applied human disturbances) actually ensures stability over the long run, just as diverse prairies managed with fire recover after drought.

CONCLUSION

Evolution is one of the key organizing principles of biology. It explains how species diversity originates, and how organisms are able to live in highly specialized ecological niches. Natural selection, in which beneficial traits are passed from survivors in one generation to their progeny, is the mechanism by which evolution occurs. Species interactions—competition, predation, symbiosis, and coevolution—are important factors in natural selection. The unique set of organisms and environmental conditions in an ecological community give rise to important properties, such as productivity, abundance, diversity, structure, complexity, connectedness, resilience, and succession. Human-caused introduction of new species as well as removal of existing ones can cause profound changes in biological communities and can compromise the life-supporting ecological services on which we all depend. These community ecology principles are useful in understanding many aspects of population change and ecosystem function.

PRACTICE QUIZ

1. Explain how tolerance limits to environmental factors determine distribution of a highly specialized species such as the saguaro cactus.

2. What is allopatric speciation? Sympatric speciation?

3. Define *selective pressure* and explain how it can alter traits in a species.

4. Explain the three types of survivorship curves shown in figure 3.23.

5. Describe several types of symbiotic relationships. Give an example of a symbiotic relationship that benefits both species.

6. What is coevolution? Give an example, either in a predatory relationship or a symbiotic relationship.

7. Competition for a limited quantity of resources occurs in all ecosystems. This competition can be interspecific or intraspecific. Explain some of the ways an organism might deal with these different types of competition.

8. Explain the idea of *K*-selected and *r*-selected species. Give an example of each.

9. Describe the process of succession that occurs after a forest fire destroys an existing biological community. Why may periodic fire be beneficial to a community?

10. Which world ecosystems are most productive in terms of biomass (fig. 3.28)? Which are least productive? What units are used in this figure to quantify biomass accumulation?

11. What do ecologists mean by the term *resilience*? In what ways might diversity contribute to resilience in an ecosystem?

CRITICAL THINKING AND DISCUSSION

Apply the principles you have learned in this chapter to discuss these questions with other students.

1. The concepts of natural selection and evolution are central to how most biologists understand and interpret the world. Why is Darwin's explanation so useful in biology? Does this explanation necessarily challenge traditional religious views? In what ways?

2. What is the difference between saying that a duck has webbed feet because it needs them to swim and saying that a duck is able to swim because it has webbed feet?

3. Given the information you've learned from this chapter, how would you explain the idea of speciation through natural selection and adaptation to a nonscientist?

4. Productivity, diversity, complexity, resilience, and structure are exhibited to some extent by all communities and ecosystems.

Describe how these characteristics apply to an ecosystem with which you are familiar.

5. Is what ways is disturbance good or bad in an ecosystem? Can it be both? Give an example from an ecosystem you know, and consider some of the disturbances that affect it. What are some negative and positive effects?

6. Ecologists debate whether biological communities have self-sustaining, self-regulating characteristics or are highly variable, accidental assemblages of individually acting species. What outlook or worldview might lead scientists to favor one or the other of these theories?

7. Many rare and endangered species are specialists. Explain what this term means. As environments change, should we worry about losing specialists? Why or why not?

DATA ANALYSIS Competitive Exclusion

The principle of competitive exclusion is one of the observations that can be explained by the process of evolution by natural selection. David Lack's classic study of finches on several Galápagos Islands is one example of this. These species are all closely related, in a genus of ground finches, *Gespiza*. (Often biologists abbreviate the genus name when it is used repeatedly: for example, *Gespiza fuliginosa* is shortened, for convenience, to *G. fuliginosa*). Go to Connect to demonstrate your understanding of these graphs and what they tell us about how Lack demonstrated competitive exclusion in these species.

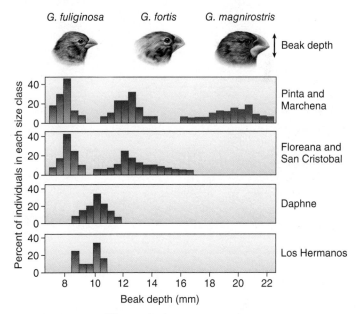

Plots of beak size on different islands (after D. Lack, 1947.)

Human Populations

Improved standards of living, women's rights, education for girls, and new visions of what life could hold for a modern family have stabilized population size in Brazil without specific government family planning policies.

After studying this chapter, you should be able to answer the following questions:

▸ Why are we concerned about human population growth?

▸ Will the world's population triple in the twenty-first century as it did in the twentieth?

▸ What is the relationship between population growth and environmental impact?

▸ Why did human populations grow so rapidly in the last century?

▸ How is human population growth changing in different parts of the world?

▸ How does population growth change as a society develops?

▸ What factors slow down or speed up human population growth?

Population Stabilization in Brazil

Can TV soap operas help control population growth? In Brazil, a mix of economic growth, female empowerment, urbanization, and the widespread popularity of television and the images it portrays of modern life have resulted in one of the most abrupt birth rate declines over the past few decades of any country in the world.

The largest country in South America, Brazil has enormous regional differences in geography, race, culture, and environmental conditions, ranging from the lush tropical forests of the Amazon basin to the immense urban agglomerations of Rio de Janeiro and Sao Paulo. With an annual GDP growth over 5 percent per year, Brazil is a member of the rapidly growing BRIC economies (Brazil, Russia, India, and China). It has the sixth largest population in the world (203 million in 2012), and the sixth largest economy.

This swift economic progress has brought a dramatically improved standard of living for most Brazilians. In 1960 the average wage in Brazil was less than $2,000 (U.S.) per year, only 19 percent of Brazilian households had electricity, TV was rare, the average education for females was 2 years, the typical woman had 6.2 children, and the annual population growth rate was 3.0 percent. Fifty years later, the average wage had risen fivefold, 95 percent of households had both electricity and TV, the average education for females was 8.6 years (a year more than for males), the fertility rate (number of children per woman in a lifetime) was 1.8, and the growth rate was 0.9 percent (fig. 4.1). Currently, 99 percent of urban residents have improved drinking water and 87 percent have improved sanitation. Infant mortality has fallen from 204 per 1,000 children in 1960 to 19 per 1,000 today. Knowing that your children are likely to survive to adulthood makes a big difference in how many you choose to have.

The industrialization that fueled this rapid economic growth also brought urbanization as people moved from rural areas to the cities in search of jobs. Currently 87 percent of Brazilians live in urban areas where smaller families are an economic advantage. Apartments are small. With both parents working, having a big family with lots of children is a burden. And with better educational opportunities for children, it makes sense to focus your resources on one or two children who can find better jobs and advance economically.

Although Brazil, like most of Latin America, still has a machismo culture, women are gaining respect and freedom. Women are more likely now to have paying jobs outside the home and have more voice in how their wages are used than ever before. In 2010, Dilma Rousseff was elected the first woman president in Brazil's history. Before her election, Rousseff was minister of energy and chief of staff for President Luiz Inacio Lula da Silva. In those positions, Rousseff gained public popularity when she pushed a program to extend electricity to remote rural areas and to the favelas (slums) that crowd the hills around major cities.

But what does this have to do with soap operas? As people moved to the city, they had both more access to TV and more leisure time. Daytime soap operas offered a view of what modern life might be like. Followed avidly by a majority of the population, the actors, plots, and situations in the telenovela were widely discussed and admired by the public. In general, the programs showed small, affluent, modern families with lots of material possessions. Women in particular are shown as powerful executives and business owners who have successful careers and considerable personal freedom. This image has changed aspirations for many women. The desire to have large families, as their mothers or grandmothers did, is no longer popular among young women.

No official government policy in Brazil has ever promoted family planning. The country is overwhelmingly Roman Catholic (with the largest catholic population in world), but most women quietly chose to ignore church teaching about birth control. Abortion is illegal (except in rare cases), but both birth control and morning-after pills are widely available over the counter. Female sterilization is one of the most widely used contraceptive techniques. A high proportion of births are cesarean, and the doctor can discreetly and cheaply perform a tubal ligation at the same time.

So, Brazil has stabilized its population without direct intervention in family planning. Economic growth, better educational opportunities for girls, women's rights, and a vision of what life could hold for a modern family have brought about this change spontaneously.

This case study introduces several important themes of this chapter. What's the best way to achieve sustainable population sizes? What are the links between poverty (or wealth), birth rates, and our common environment? Keep in mind, as you read this chapter, that resource limits aren't simply a matter of total number of people on the planet—they also depend on consumption levels and the types of technology used to produce the things we use. ∎

▲ **FIGURE 4.1** The number of children per woman and population growth have dropped sharply in Brazil over the past 50 years as per capita income and standard of living have risen.

Legend:
■ GDP per capita   —●— Population growth (%)   —●— Children per woman

> *For every complex problem there is an answer that is clear, simple, and wrong.*
>
> —H. L. MENCKEN

4.1 PAST AND CURRENT POPULATION GROWTH ARE VERY DIFFERENT

Every second, on average, four or five children are born somewhere on the earth. In that same second, one or two other people die. This difference between births and deaths means a net gain of about 2.5 more humans per second (on average) in the world's population. In 2011 the United Nations announced that we had reached 7 billion people, having added the most recent billion in only 12 years. We're now growing at 1.13 percent per year. This means we are adding about 79 million more people to the planet every year. Humans are now one of the most numerous vertebrate species on the earth. We also are more widely distributed and manifestly have a greater global environmental impact than any other species. For the families to whom these children are born, each birth may be a joyous and long-awaited event. But is a continuing increase in humans good for the planet in the long run?

Many people worry that overpopulation will cause—or perhaps already is causing—resource depletion and environmental degradation that threaten the ecological life-support systems on which we all depend. These fears often lead to demands for immediate, worldwide birth control programs to reduce fertility rates and to eventually stabilize or even shrink the total number of humans.

Others believe that human ingenuity, technology, and enterprise can expand the world's carrying capacity and allow us to overcome any problems we encounter. From this perspective, more people may be beneficial, rather than disastrous. A larger population means a larger workforce, more geniuses, and more ideas about what to do. Along with every new mouth comes a pair of hands. Proponents of this worldview argue that continued economic and technological growth can both feed the world's billions and enrich everyone enough to end the population explosion voluntarily.

Still another opinion on this subject derives from social justice concerns. According to this worldview, resources are sufficient for everyone. Current shortages are only signs of greed, waste, and oppression. The root cause of environmental degradation, in this perspective, is inequitable distribution of wealth and power rather than merely population size. Fostering democracy, rights for women and minorities, and improving the lives of the poor are essential for sustainability (fig. 4.2). A narrow focus on population growth fosters racism and blames the poor for their problems, while ignoring the high resource consumption of richer nations, according to this view.

Whether human populations will continue to grow at present rates and what that growth would imply for environmental quality and human life are among the most central and pressing questions in environmental science. Will the example of Brazil apply to other developing countries, or is it a unique situation? In this chapter, we will look at some causes of population growth, as well as at how populations are measured and described. Family planning and birth control are essential for stabilizing populations. The number of children a couple decides to have and the methods they use to regulate fertility, however, are strongly influenced by culture, religion, politics, and economics, as well as basic biological and medical considerations.

Human populations grew slowly until recently

For most of our history, humans were not very numerous, compared with many other species. Studies of hunting-and-gathering societies suggest that the total world population was probably only a few million people before the invention of agriculture and the domestication of animals around 10,000 years ago. The agricultural revolution produced a larger and more secure food supply and allowed the human population to grow, reaching perhaps 50 million people by 5000 B.C. For thousands of years, the number of humans increased very slowly. Archaeological evidence and historical descriptions suggest that only about 300 million people were living in the first century A.D. (table 4.1).

▲ **FIGURE 4.2** The number of children a family chooses to have is determined by many factors. In urban, industrialized countries, such as Brazil, most families now want only one or two children.

TABLE 4.1	World Population Growth and Doubling Times	
DATE	POPULATION	DOUBLING TIME
5000 B.C.	50 million	?
800 B.C.	100 million	4,200 years
200 B.C.	200 million	600 years
A.D. 1200	400 million	1,400 years
A.D. 1700	800 million	500 years
A.D. 1900	1,600 million	200 years
A.D. 1965	3,200 million	65 years
A.D. 2000	6,100 million	51 years
A.D. 2050 (estimate)	8,920 million	215 years

SOURCE: United Nations Population Division.

Active LEARNING

Population Doubling Time

If the world population is growing at 1.14 percent per year and continues at that rate, how long before it doubles? The "rule of 70" is a useful way to calculate the approximate doubling time in years for anything growing exponentially. For example, a savings account (or biological population) growing at a compound interest rate of 1 percent per year will double in about 70 years. Using the formula below, calculate doubling time for the world (growth rate = 1.14 percent/year), Uganda (3.2 percent), Nicaragua (2.7 percent), India (1.7 percent), United States (0.6 percent), Japan (0.1 percent), and Russia (−0.6 percent).

Example: $\dfrac{70 \text{ years}}{(\text{growth percent})}$ = doubling time in years

ANSWERS: World = 61 years; Uganda = 22 years; Nicaragua = 26 years; India = 41 years; United States = 117 years; Japan = 700 years; Russia = never

As you can see in figure 4.3, human populations began to increase rapidly after about A.D. 1600. Many factors contributed to this rapid growth. Increased sailing and navigating skills stimulated commerce and communication among nations. Agricultural developments, better sources of power, and improved health care and hygiene also played a role. We are now in an exponential, or J curve, pattern of growth, described in chapter 3.

It took all of human history to reach 1 billion people in 1800 but only 156 more years to get to 3 billion in 1960. It took us about 12 years to add the seventh billion. Another way to look at population growth is that the number of living humans tripled during the twentieth century. Will it do so again in the twenty-first century? If it does, will we overshoot our environment's carrying capacity and experience a catastrophic dieback similar to those described in chapter 3? As you will see later in this chapter, there is some evidence that population growth already is slowing, but whether we will reach equilibrium soon enough and at a size that can be sustained over the long term remains a difficult but vital question.

4.2 PERSPECTIVES ON POPULATION GROWTH

As with many topics in environmental science, people have widely differing opinions about population and resources. Some believe that population growth is the ultimate cause of poverty and environmental degradation. Others argue that poverty, environmental degradation, and overpopulation are all merely symptoms of deeper social and political factors. The worldview we choose to believe will profoundly affect our approach to population issues. In this section, we will examine some of the major figures and their arguments in this debate.

Does environment or culture control human population growth?

Since the time of the Industrial Revolution, when the world population began growing rapidly, individuals have argued about the causes and consequences of population growth. In 1798 Thomas Malthus (1766–1834) wrote *An Essay on the Principle of Population*, changing the way European leaders thought about population growth. Malthus collected data to show that populations tended to increase at an exponential, or compound, rate, whereas food production either remained stable or increased only slowly. Eventually, he argued, human populations would outstrip their food supply and collapse into starvation, crime, and misery. He converted most economists of the day from believing that high fertility increased industrial output and national wealth, to believing that per capita output actually fell with rapidly rising population.

In Malthusian terms, growing human populations are limited only by disease or famine, or social constraints that compel people to reduce birth rates—late marriage, insufficient resources, celibacy, and "moral restraint." However, the economist Karl Marx (1818–1883) presented an opposing view that population growth results from poverty, resource depletion, pollution, and other social ills. Slowing population growth, claimed Marx, requires that people be treated justly, and that exploitation and oppression be eliminated from social arrangements.

Both Marx and Malthus developed their theories about human population growth when the world, technology, and

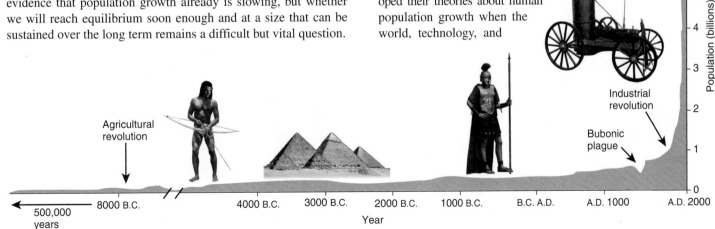

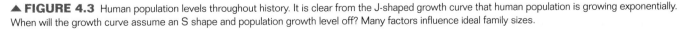

▲ **FIGURE 4.3** Human population levels throughout history. It is clear from the J-shaped growth curve that human population is growing exponentially. When will the growth curve assume an S shape and population growth level off? Many factors influence ideal family sizes.

▲ **FIGURE 4.4** Is the world overcrowded already, or are people a resource? In large part, the answer depends on the kinds of resources we use and how we use them. It also depends on democracy, equity, and justice in our social systems.

society were understood much differently from how they are today. Some believe that we are approaching, or may have surpassed, the earth's carrying capacity. Joel Cohen, a mathematical biologist at Rockefeller University, reviewed published estimates of the maximum human population size the planet can sustain. The estimates, spanning 300 years of thinking, converged on a median value of 10–12 billion. We are more than 7 billion strong today, and still growing, an alarming prospect for some (fig. 4.4). Cornell University entomologist David Pimental, for example, has said: "By 2100, if current trends continue, twelve billion miserable humans will suffer a difficult life on Earth." In this view, birth control should be our top priority.

Technology increases carrying capacity for humans

Optimists argue that Malthus was wrong in his predictions of famine and disaster 200 years ago because he failed to account for scientific and technical progress. In fact, food supplies have increased faster than population growth since Malthus' time. For example, according to the UN FAO Statistics Division, in 1970 world food supplies provided 2,435 calories of food per person per day, while in 2010 there was enough for 3,130 calories per person. Even poorer, developing countries saw a rise, from an average of 2,135 calories per day in 1970 to 2,730 in 2010. In that same period the world population grew from 3.7 to nearly 7 billion people. Certainly terrible famines have stricken various locations in the past 200 years, but many observers argue they were caused more by politics and economics than by lack of resources or population size. Whether the world can continue to feed its growing population remains to be seen, but technological advances have vastly increased human carrying capacity—so far (chapter 7).

The burst of world population growth that began 200 years ago was stimulated by scientific and industrial revolutions. Progress in agricultural productivity, engineering, information technology, commerce, medicine, sanitation, and other achievements of modern life have made it possible to support approximately 1,000 times as many people per unit area as was possible 10,000 years ago. Economist Stephen Moore of the Cato Institute in Washington, D.C., regards this achievement as "a real tribute to human ingenuity and

our ability to innovate." There is no reason, he argues, to think that our ability to find technological solutions to our problems will diminish in the future.

Much of our rising standard of living in the past two centuries, however, has been based on easily acquired natural resources, especially cheap, abundant fossil fuels. Many people are concerned about whether limited supplies of these fuels or adverse consequences of their use will result in a crisis in food production, transportation, or in some other critical factor in human society.

Moreover, technology can be a double-edged sword. Our environmental effects aren't just a matter of sheer population size; they also depend on what kinds of resources we use and how we use them. This concept is summarized as the **I = PAT** formula. It says that our environmental impacts (I) are the product of our population size (P) times affluence (A) and the technology (T) used to produce the goods and services we consume (fig. 4.5).

While increased standards of living in Brazil, for example, have helped stabilize population, they also bring about higher technological impacts. A family living an affluent lifestyle that depends on high levels of energy and material consumption, and that produces excessive amounts of pollution, could cause greater environmental damage than a whole village of hunters and gatherers or subsistence farmers.

Put another way, if the billions of people in Asia, Africa, and Latin America were to reach the levels of consumption now enjoyed by rich people in North America or Europe using the same technology that provides that lifestyle today, the environmental effects will undoubtedly be disastrous. Growing wealth in China—its middle class is now estimated at about 300 million, or nearly the entire population of the United States—is already stressing world resources, for example, and has made China the largest emitter of CO_2. There are now more millionaires in China than in all of Europe, and China has passed the United States in annual automobile production.

But China has also become the global leader in renewable energy, and is now promoting electric vehicles. Ideally, all of us will begin to use nonpolluting, renewable energy and material sources. Better yet, we'll extend the benefits of environmentally friendly technology to the poorer people of the world so everyone can enjoy the benefits of a better standard of living without degrading our shared environment.

Impact = Population × Affluence × Technology

◀ **FIGURE 4.5** Environmental impacts of population growth (I) are the product of population size (P), times affluence (A), times the technology (T) used to create wealth.

One way to estimate our environmental impacts is to express our consumption choices into the equivalent amount of land required to produce goods and services. This gives us a single number, called our **ecological footprint**, which estimates the relative amount of productive land required to support each of us. Services provided by nature make up a large proportion of our ecological footprint. For example, forests and grasslands store carbon, protect watersheds, purify air and water, and provide wildlife habitat.

Footprint calculations are imperfect, but they do give us a way to compare different lifestyle effects (see Key Concepts, pp. 82–83). The average resident of the United States, for instance, lives at a level of consumption that requires 9.7 ha of bioproductive land, whereas the average Malawian has an ecological footprint of less than 0.5 ha. Worldwide, we're currently using about one-third more resources than the planet can provide on a sustainable basis. That means we're running up an ecological debt that future generations will have to pay. Or, another way to look at it is that it would take 3.5 more earths to support the world at a current American lifestyle. If everyone lived like Malawians, on the other hand, the planet could sustainably house more than 20 billion people.

Population growth could bring benefits

Think of the gigantic economic engines that large counties, such as the United States and China, represent. More people mean larger markets, more workers, and efficiencies of scale in mass production of goods. Moreover, adding people boosts human ingenuity and intelligence that can create new resources by finding new materials and discovering new ways of doing things. Economist Julian Simon (1932–1998), a champion of this rosy view of human history, believed that people are the "ultimate resource" and that no evidence shows that pollution, crime, unemployment, crowding, the loss of species, or any other resource limitations will worsen with population growth.

In a famous wager in 1980, Simon challenged Paul Ehrlich, author of *The Population Bomb*, to pick five commodities that would become more expensive by the end of the decade. Ehrlich chose a group of metals that actually became cheaper, and he lost the bet. Leaders of many developing countries insist that, instead of being obsessed with population growth, we should focus on the inordinate consumption of the world's resources by people in richer countries. For his part, Ehrlich did not really want to bet on metals, but on renewable resources or certain critical measures of environmental health. He and Simon were negotiating a second bet just before Simon's death.

4.3 MANY FACTORS DETERMINE POPULATION GROWTH

Demography (derived from the Greek words *demos,* "people," and *graphein,* "to write" or "to measure") encompasses vital statistics about people, such as births, deaths, and where they live, as well as total population size. In this section, we will investigate ways to measure and describe human populations and discuss demographic factors that contribute to population growth.

How many of us are there?

The United Nations estimate of 7 billion people in 2011 is only an estimate. Even in this age of information technology and advanced communication, counting the number of people in the world is an inexact science. Some countries have never even taken a census, and some that have been done may not be accurate. Governments overstate or understate their populations to make their countries appear larger and more important or smaller and more stable than they really are. Some individuals, especially if they are homeless, refugees, or illegal aliens, may not want to be counted or identified.

We really live in two very different demographic worlds. One of these worlds is poor, young, and growing rapidly, while the other is rich, old, and shrinking in population size. The poorer world is occupied by the vast majority of people who live in the less-developed countries of Africa, Asia, and Latin America (fig. 4.6). These countries represent 80 percent of the world population but will contribute more than 90 percent of all projected future growth. The richer world is made up of North America, western Europe, Japan, Australia, and New Zealand. The average age in richer countries is 40, and life expectancy of their residents will exceed 90 by 2050. With many couples choosing to have either one or no children, the populations of these countries are expected to decline significantly over the next century.

The highest population growth rates occur in a few "hot spots" in the developing world, such as sub-Saharan Africa and the Middle East, where economics, politics, religion, and civil unrest keep birth rates high and contraceptive use low. In Niger, for example, annual population growth is currently 3.9 percent. Less than 10 percent of all couples use any form of birth control, women average 7.6 children each, and nearly half the population is less than 15 years old. Even faster growth is occurring in Qatar, where the population doubling time is only 7.3 years.

Some countries in the developing world are growing so fast that they will reach immense population sizes by the middle of the twenty-first century (table 4.2). China was the most populous country throughout the twentieth century; India is expected to pass

▲ **FIGURE 4.6** We live in two demographic worlds. One is rich, is technologically advanced, and has an elderly population that is growing slowly, if at all. The other is poor, crowded, underdeveloped, and growing rapidly.

How big is your footprint?

Human populations are rising, and our resource use per person is also growing. How can we assess the ways our resource consumption is changing the world? One approach is **ecological footprint analysis**—estimating the amount of territory needed to support all your consumption of food, paper, computers, energy, water, and other resources. This analysis obviously simplifies and approximates your real use, but the aggregate measure allows us to compare resource use among places or over time.

Perhaps the most comprehensive analysis has been done by the Worldwide Fund for Nature (WWF). The summary you see here gives key points. For the full story, see the original document at www.wwf.org/footprint.

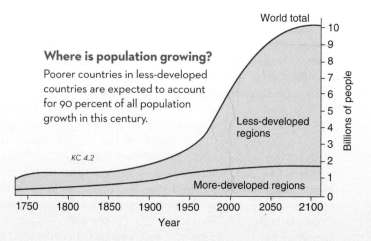

Where is population growing?
Poorer countries in less-developed countries are expected to account for 90 percent of all population growth in this century.

KC 4.2

Terms to note:

Biocapacity is the capacity of living systems to provide for our needs. Both biocapacity and global footprints can be measured in gigahectares (gha). **One ha = 2.59 acres. One gha = one billion ha**. The WWF calculated that an average hectare of land could store carbon equivalent to 1,450 liters of gasoline.

We consume more than the earth's biocapacity by mining ancient energy, soil, and other resources at a rate faster than these resources can be reproduced.

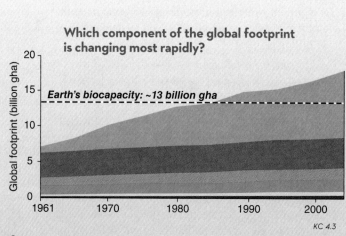

Which component of the global footprint is changing most rapidly?

Earth's biocapacity: ~13 billion gha

KC 4.3

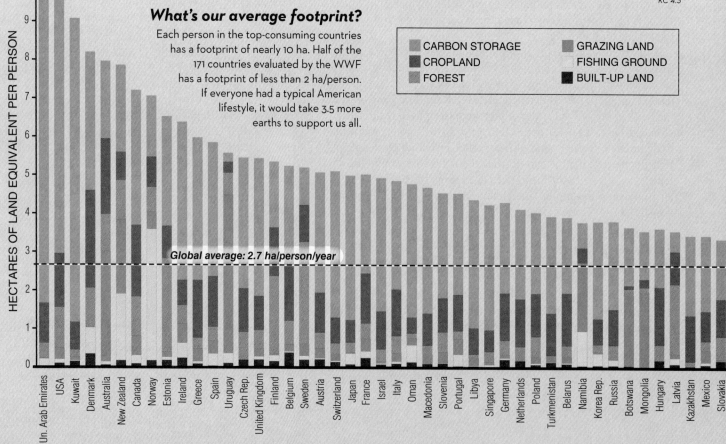

KC 4.1

What's our average footprint?

Each person in the top-consuming countries has a footprint of nearly 10 ha. Half of the 171 countries evaluated by the WWF has a footprint of less than 2 ha/person. If everyone had a typical American lifestyle, it would take 3.5 more earths to support us all.

Global average: 2.7 ha/person/year

- CARBON STORAGE
- CROPLAND
- FOREST
- GRAZING LAND
- FISHING GROUND
- BUILT-UP LAND

HECTARES OF LAND EQUIVALENT PER PERSON

Un. Arab Emirates, USA, Kuwait, Denmark, Australia, New Zealand, Canada, Norway, Estonia, Ireland, Greece, Spain, Uruguay, Czech Rep., United Kingdom, Finland, Belgium, Sweden, Austria, Switzerland, Japan, France, Israel, Italy, Oman, Macedonia, Slovenia, Portugal, Libya, Singapore, Germany, Netherlands, Poland, Turkmenistan, Belarus, Namibia, Korea Rep., Russia, Botswana, Mongolia, Hungary, Latvia, Kazakhstan, Mexico, Slovakia

1. **CARBON STORAGE Why are carbon emissions the biggest part of our footprint?** Burning fossil fuels, clearing forests, and oxidation of agricultural soils emit climate-changing gases. These gases account for nearly all of our rising carbon footprint in the past 50 years. Wealthy countries vary greatly in their carbon emissions: compare, for example, Sweden and the United States, which have similar wealth and standards of living. At nearly 10 gha our carbon footprint alone preempts a majority of the earth's biocapacity.

KC 4.4

2. **CROPLAND What are some of the resources used in farming?** Costs vary greatly. Some farming systems deplete soil and depend on fossil fuels; others build soil and require few inputs. Grain-fed beef is perhaps our most costly agricultural product.

KC 4.5

KC 4.6

3. **FORESTLAND What benefits do we get from forests?** Forestry provides our wood and paper products, and many other useful products. Forests also protect watersheds, provide wildlife habitat, and purify and store water.

6. **BUILT-UP LAND How much land is occupied by roads and buildings?** They take up less space than other uses, but preempt important ecological services.

KC 4.9

KC 4.7

4. **GRAZING LAND Can grazing deplete land?** Expansive area is needed for grazing. Overgrazing can badly degrade biodiversity and cause soil erosion. Less intense grazing can be an efficient way to convert grass to protein.

5. **FISHING GROUND How much sea do we depend on?** Fishery impact is large for some countries. Globally, 90 percent of all large marine predators are gone, and 13 of 17 major fisheries are exhausted (chapter 9).

KC 4.8

Nassau grouper

CAN YOU **EXPLAIN?**

1. Which factors are largest for the United States? Why?
2. Which countries have the greatest footprint per person in forestry? Fishing? Carbon? Grazing? Why?

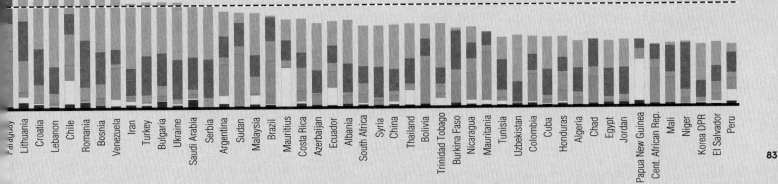

Paraguay · Lithuania · Croatia · Lebanon · Chile · Romania · Bosnia · Venezuela · Iran · Turkey · Bulgaria · Ukraine · Saudi Arabia · Serbia · Argentina · Sudan · Malaysia · Brazil · Mauritius · Costa Rica · Azerbaijan · Ecuador · Albania · South Africa · Syria · China · Thailand · Bolivia · Trinidad Tobago · Burkina Faso · Nicaragua · Mauritania · Tunisia · Uzbekistan · Colombia · Cuba · Honduras · Algeria · Chad · Egypt · Jordan · Papua New Guinea · Cent. African Rep. · Mali · Niger · Korea DPR · El Salvador · Peru

TABLE 4.2 | The World's Largest Countries

2010		2050	
COUNTRY	POPULATION (MILLIONS)	COUNTRY	POPULATION (MILLIONS)
China	1,339	India	1,628
India	1,204	China	1,437
United States	313	United States	420
Indonesia	240	Nigeria	299
Brazil	203	Pakistan	295
Pakistan	178	Indonesia	285
Bangladesh	159	Bangladesh	231
Nigeria	155	Brazil	220
Russia	142	Dem. Rep. of Congo	183

SOURCE: Data from the U.S. Census Bureau, 2012.

China in the twenty-first century. Nigeria, which had only 33 million residents in 1950, is forecast to have 299 million in 2050. Ethiopia, with about 18 million people 50 years ago, is likely to grow nearly tenfold over a century. In many of these countries, rapid population growth poses huge challenges to food supplies and stability. Bangladesh, about the size of Iowa, is already overcrowded at 153 million people. If rising sea levels flood one-third of the country by 2050, as some climatologists predict, adding another 80 million people will make their situation completely impossible.

On the other hand, some richer countries have shrinking populations. Japan, which has 128 million residents now, is expected to shrink to about 90 million by 2050. Europe, which now makes up about 12 percent of the world population, will constitute less than 7 percent in 50 years, if current trends continue. Even the United States and Canada would have stable populations if immigration were stopped. For the moment, the U.S. population continues to grow. In 2011 the U.S. population was about 313 million and growing at 0.86 percent per year.

It isn't only wealthy countries that have declining populations. Russia, for instance, is now shrinking by nearly 1 million people per year as death rates have soared and birth rates have plummeted. A collapsing economy, hyperinflation, crime, corruption, and despair have demoralized the population. Horrific pollution levels left from the Soviet era, coupled with poor nutrition and health care, have resulted in high levels of genetic abnormalities, infertility, and infant mortality. Abortions are twice as common as live births. Death rates, especially among adult men, have risen dramatically. According to some demographers, the life expectancy of Russian males fell by 10 years after 1990; it's slightly higher now, but still low by the standards of industrialized countries. It's expected that by 2050 Russia will have a smaller population than Vietnam, the Philippines, and the Democratic Republic of Congo.

The situation is even worse in many African countries, where AIDS and other communicable diseases are killing people at a terrible rate. In Zimbabwe, Botswana, Zambia, and Namibia, for example, up to 39 percent of the adult population have AIDS or are HIV positive. Health officials predict that more than two-thirds of

the 15-year-olds now living in Botswana will die of AIDS before age 50. Without AIDS, the average life expectancy would have been nearly 70 years. Now, with AIDS, Botswana's life expectancy has dropped to only 31.6 years. The populations of many African countries are now falling because of this terrible disease (fig. 4.7). Altogether, Africa's population is expected to be nearly 200 million lower in 2050 than it would have been without AIDS.

The world population density map in appendix 2 on page A-3 shows human population distribution around the world. Notice the high densities supported by fertile river valleys of the Nile, Ganges, Yellow, Yangtze, and Rhine Rivers and the well-watered coastal plains of India, China, and Europe. Historic factors, such as technology diffusion and geopolitical power, also play a role in geographic distribution.

Fertility varies among cultures and at different times

Fecundity is the physical ability to reproduce, whereas fertility is the actual production of offspring. Those without children may be fecund but not fertile. The most accessible demographic statistic of fertility is usually the **crude birth rate**, the number of births in a year per thousand persons. It is statistically "crude" in the sense that it is not adjusted for population characteristics, such as the number of women of reproductive age.

The **total fertility rate** is the number of children born to an average woman in a population during her entire reproductive life. Upper-class women in seventeenth- and eighteenth-century Europe, whose babies were given to wet nurses immediately after birth, sometimes had 25 or 30 pregnancies. The highest recorded total fertility rates for working-class people are among some

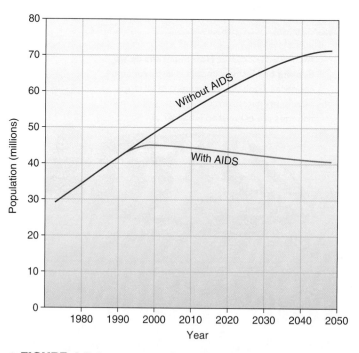

▲ **FIGURE 4.7** Projected population of South Africa with and without AIDS. DATA SOURCE: UN Population Division, 2006.

Anabaptist agricultural groups in North America, who have averaged up to 12 children per woman. In most tribal or traditional societies, food shortages, health problems, and cultural practices limit total fertility to about 6 or 7 children per woman, even without modern methods of birth control.

Zero population growth (ZPG) occurs when births plus immigration in a population just equal deaths plus emigration. It takes several generations of replacement-level fertility (in which people just replace themselves) to reach ZPG. Where infant mortality rates are high, the replacement level may be 5 or more children per couple. In the more highly developed countries, however, this rate is usually about 2.1 children per couple because some people are infertile, have children who do not survive, or choose not to have children.

As was the case in Brazil, fertility rates have declined dramatically in every region of the world except Africa over the past 50 years (fig. 4.8). In the 1960s, total fertility rates above 6 were common in many countries. The average family in Mexico in 1975, for instance, had 7 children. By 2010, however, the average Mexican woman had only 2.3 children. Similarly, in Iran, total fertility fell from 6.5 in 1975 to 2.04 in 2010. According to the World Health Organization, nearly half the world's 192 countries are now at or below a **replacement rate** of 2.1 children per couple.

For many of these countries, population growth will continue for a generation because they have such a large number of young people. Brazil, for example, now has a fertility rate of only 1.8 children per woman. But 26 percent of its population is under 14 years. Those children will mature and start to have families before their parents and grandparents die, so the population will continue to grow for a few decades. Demographers call this **population momentum**.

Even some of the poorest countries in the world have been remarkably successful in lowering growth rates. Bangladesh, for instance, reduced its fertility rate from 6.9 in 1980 to only 2.8 children per woman in 2009. China's one-child-per-family policy decreased the fertility rate from 6 in 1970 to 1.7 in 2010. This program has been remarkably successful in reducing population growth, but has raised serious questions about human rights (see What Do You Think?, p. 86).

Although the world as a whole still has an average fertility rate of 2.5, growth rates are lower now than at any time since World War II. If fertility declines like those in Brazil were to occur everywhere in the world, our total population could begin to decline by the end of the twenty-first century.

Mortality offsets births

A traveler to a foreign country once asked a local resident, "What's the death rate around here?" "Oh, the same as anywhere," was the reply, "about one per person." In demographics, however, **crude death rates** (or crude mortality rates) are expressed in terms of the number of deaths per thousand persons in any given year. Countries in Africa where health care and sanitation are limited may have mortality rates of 20 or more per 1,000 people. Wealthier countries generally have mortality rates around 10 per 1,000. The number of deaths in a population is sensitive to the population's age structure.

Rapidly growing, developing countries, such as Brazil, often have lower crude death rates (6 per 1,000 currently) than do the more-developed, slowly growing countries, such as Denmark (12 per 1,000), even though their life expectancies are considerably lower. This is because a rapidly growing country has proportionately more youths and fewer elderly than a more slowly growing country.

Life expectancy is rising worldwide

Life span is the oldest age to which a species is known to survive. Although there are many claims in ancient literature of kings living a thousand years or more, the oldest age that can be certified by written records was that of Jeanne Louise Calment of Arles, France, who was 122 years old at her death in 1997. Though modern medicine has made it possible for many of us to survive much longer than our ancestors, it doesn't appear that the maximum life span has increased much at all. Apparently, cells in our bodies have a limited ability to repair damage and produce new components. Sooner or later they simply wear out, and we fall victim to disease, degeneration, accidents, or senility.

Life expectancy is the average age that a newborn infant can be expected to attain in any given society. It is another way of expressing the average age at death. For most of human history, life expectancy in most societies probably has been 35 to 40 years. This means, not that no one lived past age 40, but instead that many people died at earlier ages (mostly early childhood), which balanced out those who managed to live longer.

The twentieth century saw a global transformation in human health unmatched in history. This revolution can be seen in the dramatic increases in life expectancy in most places (table 4.3).

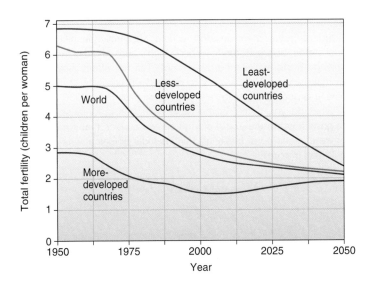

▲ **FIGURE 4.8** Average total fertility rates for less-developed countries fell by more than half over the past 50 years, mostly due to China's one-child policy. By 2050 even the least-developed countries should approach the replacement rate of 2.1 children per reproductively fertile woman.
SOURCES: Data from UN Population Division, *World Population Prospects, 1996*, and Population Reference Bureau, 2004.

China's One-Child Policy

When the People's Republic of China was founded in 1949, it had about 540 million residents, and official government policy encouraged large families. The Republic's First Chairman, Mao Zedong, proclaimed, "Of all things in the world, people are the most precious." He thought that more workers would mean greater output, increasing national wealth, and higher prestige for the country. This optimistic outlook was challenged, however, in the 1960s, when a series of disastrous government policies triggered massive famines and resulted in at least 30 million deaths.

When Deng Xiaoping became Chairman in 1978, he reversed many of Mao's policies, instead privatizing farms, encouraging private enterprise, and discouraging large families. Deng recognized that, with an annual growth rate of 2.5 percent, China's population, which had already reached 975 million, would double in only 28 years. China might have nearly 2 billion residents now if that growth had continued. Feeding, housing, educating, and employing all those people would put a severe strain on China's (and the world's) limited resources.

Deng introduced a highly successful—but also controversial—one-child-per-family policy. Rural families and ethnic minorities were supposedly exempt from this rule, but local authorities often were capricious and tyrannical in applying sanctions. Ordinary families were punished harshly for having unauthorized children, while government officials and other powerful individuals could have as many as they wanted. There were many reports of bribery, forced abortions, coerced sterilizations, and even infanticide as a result of this policy.

Critics claim that other approaches to family planning could have reduced population growth while also preserving human rights. The shift to an urban, industrialized society, they argue, might have reduced family size without such draconian intervention. Some point to the dramatic birth reductions in Brazil (see the opening case study for this chapter) as evidence that universal education, improved standards of living, democracy, and personal freedom can achieve population stabilization without coercion.

Another result of China's one-child policy is called the 4:2:1 problem. That is, there are now often four grandparents and two parents

China's one-child-per-family policy has been remarkably successful in reducing birth rates, but with a high social cost.

doting on a single child. Excessive attention seems to have spoiled many children. Social scientists often refer to this generation as "little emperors." Difficulties also occur as those parents and grandparents age. With only one adult child to support and care for elderly relatives, many seniors have to work far beyond normal retirement age because their only child can't provide for them all.

The Chinese government, also, is beginning to worry about a "birth dearth." Will there be enough workers, soldiers, farmers, scientists, inventors, and other productive individuals to keep society functioning in the future? The one-child policy has been eased recently. Couples with no siblings to help care for elderly relatives are being allowed to have two or more children as are racial minorities and others with a legitimate hardship claim.

However, the Chinese experiment in population control has been effective. China's population in 2012 was about 1.34 billion, or roughly 660 million less than it might have been given its trajectory in 1978. Its annual growth rate is now 0.51 percent, or about 40 percent less than the U.S. rate. China is already the largest contributor to global warming and is driving up world prices for many commodities with its rapidly growing middle class. Think what it would have meant to add about twice as many more people to China than there are in the United States today.

China has also been far more successful in controlling population growth than India. At about the same time that Deng introduced his one-child plan, India, under Indira Gandhi, started a program of compulsory sterilization in an effort to reduce population growth. This draconian policy caused so much public outrage and opposition that the federal government decided to delegate family planning to individual states. Some states have been highly successful in their family planning efforts, while others have not. The net effect, however, is that India is expected to grow to about 1.65 billion by 2050, while China is expected to reach zero population growth by 2030.

What do you think? Is the rapid reduction in Chinese population growth worth the social disruption and abuses that it caused? If you were in charge of family planning in India or China, what policies would you pursue?

Worldwide, the average life expectancy rose from about 40 to 67.2 years over the past 100 years. The greatest progress was in developing countries. For example, in 1900 the average Indian man or woman could expect to live about 23 years. A century later, although India had an annual per capita income of only $3,500 (U.S.), the average life expectancy for both men and women had nearly tripled and was very close to that of countries with ten times its income level. Longer lives were due primarily to better

nutrition, improved sanitation, clean water, and education, rather than to miracle drugs or high-tech medicine.

Although the gains were not as great for the already industrialized countries, residents of the United States, Sweden, and Japan, for example, now live about half-again as long as they did at the beginning of the twentieth century, and they can expect to enjoy much of that life in relatively good health. The Disability Adjusted Life Years (DALYs, a measure of disease burden that

TABLE 4.3	Life Expectancy at Birth for Selected Countries in 1900 and 2009			
	1900		2009	
COUNTRY	MALES	FEMALES	MALES	FEMALES
India	23	23	67	73
Russia	31	33	59	73
United States	46	48	76	81
Sweden	57	60	79	83
Japan	42	44	79	86

SOURCE: Data from Population Reference Bureau, 2010.

combines premature death with loss of healthy life resulting from illness or disability) that someone living in Japan can expect is now 74.5 years, compared with only 64.5 DALYs two decades ago.

As figure 4.9 shows, annual income and life expectancy are strongly correlated up to about $4,000 (U.S.) per person. Beyond that level—which is generally enough for adequate food, shelter, and sanitation for most people—life expectancies level out at about 75 years for men and 85 for women.

Large discrepancies in how the benefits of modernization and social investment are distributed within countries are revealed in differential longevities of various groups. The greatest life expectancy reported anywhere in the United States is for Asian American women in New Jersey, who live to an average age of 91. By contrast, Native American men on the Pine Ridge Indian Reservation in South Dakota live, on average, only to age 48. Two-thirds of the countries in Africa have a higher life expectancy. The Pine Ridge Reservation is the poorest area in America, with an unemployment rate near 75 percent and high rates of poverty, alcoholism, drug use, and alienation. Similarly, African-American men in Washington, D.C., live, on average, only 57.9 years, which is less than the life expectancy in Lesotho or Swaziland.

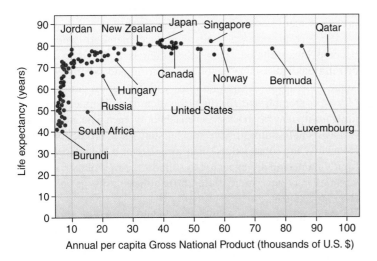

▲ FIGURE 4.9 As incomes rise, so does life expectancy up to about $4,000 (U.S.). Above that amount the curve levels off. Some countries, such as South Africa and Russia, have far lower life expectancies than their GDP would suggest. Jordan, on the other hand, which has only one-tenth the per capita GDP of the United States, actually has a higher life expectancy.
SOURCE: CIA Factbook, 2009.

Living longer has profound social implications

A population growing rapidly by natural increase has more young people than does a stationary population. One way to show these differences is to graph age classes in histograms (fig. 4.10). In Niger, which was growing at a rate of 3.9 percent per year in 2012, about half the population is in the prereproductive category (below age 15). Even if total fertility rates fell abruptly, the total number of births, and the population size, would continue to grow for some years as these young people entered reproductive age (an example of population momentum).

By contrast, a country, such as Sweden, with a relatively stable population will have nearly the same number in most cohorts. Notice that females outnumber males in Sweden's oldest group because of differences in longevity between sexes. A country that has only recently reached zero population growth, such as Singapore, can have a pronounced bulge in middle-age cohorts as fewer children are born than in their parents' generation.

Both rapidly growing countries and slowly growing countries can have a problem with their **dependency ratio**, or the number of nonworking compared with working individuals in a population. In Niger, for example, each working person supports a high number of children. In the United States, by contrast, a declining working population is now supporting an ever larger number of retired persons.

These changing age structures and shifting dependency ratios are occurring worldwide (fig. 4.11). In 1950 there were only 130 million people in the world over 65 years old. In 2012 more than 540 million had reached this age. By 2050, the UN predicts, there will be two older persons for every child in the world. Countries such as Japan, France, and Germany already are concerned that they don't have enough young people to fill jobs and support their retirement system. They are encouraging couples to have more children. Immigrants can reduce the average age of the population and create a fresh supply of workers. But nativist groups in many countries resist the integration of foreigners who may not share their culture, religion, or language. Others, however, argue that immigrants can rejuvenate and revitalize an aging society. Where do you stand on this contentious issue?

4.4 FERTILITY IS INFLUENCED BY CULTURE

A number of social and economic pressures affect decisions about family size, which in turn affects the population at large. In this section we will examine both positive and negative pressures on reproduction.

People want children for many reasons

Factors that increase people's desires to have babies are called **pronatalist pressures**. Raising a family may be the most enjoyable and rewarding part of many people's lives. Children can be a source of pleasure, pride, and comfort. They may be the only source of support for elderly parents in countries without a social security system. Where infant mortality rates are high, couples may need to have many children to ensure that at least a few will

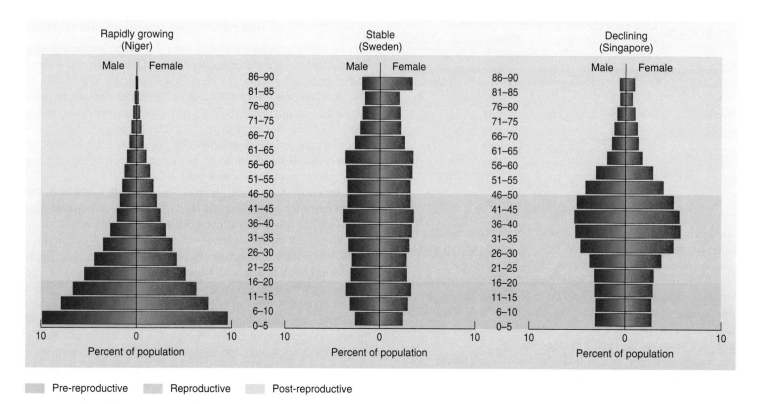

Pre-reproductive   Reproductive   Post-reproductive

▲ **FIGURE 4.10** The shape of each age-class histogram is distinctive for a population that is rapidly growing (Niger), stable (Sweden), or declining (Singapore). Horizontal bars represent the percentage of the country's population in consecutive age classes (0–5 yrs., 6–10 yrs., etc.). SOURCE: U.S. Census Bureau, 2003.

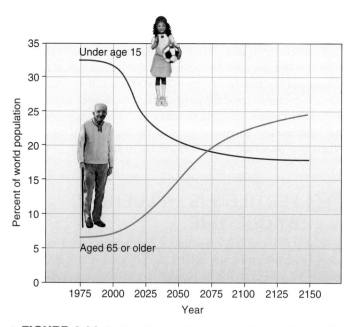

▲ **FIGURE 4.11** By the mid-twenty-first century, children under age 15 will make up a smaller percentage of world population, whereas people over age 65 will contribute an increasing share of the population.

survive to take care of them when they are old. Where there is little opportunity for upward mobility, children give status in society, express parental creativity, and provide a sense of continuity and accomplishment otherwise missing from life.

Often children are valuable to the family not only for future income but even more as a source of current income and help with household chores. In much of the developing world, small children tend domestic animals and younger siblings, fetch water, gather firewood, help grow crops, or sell things in the marketplace (fig. 4.12). Parental desire for children, rather than an unmet need for contraceptives, may be the most important factor in population growth in many cases.

Society also has a need to replace members who die or become incapacitated. This need often is codified in cultural or religious values that encourage bearing and raising children. Some societies look upon families with few or no children with pity or contempt, and for them the idea of deliberately controlling fertility may be shocking, even taboo. Women who are pregnant or have small children have special status and protection. Boys frequently are more valued than girls because they carry on the family name and are expected to support their parents in old age. Couples may have more children than they desire in an attempt to produce a son.

Male pride often is linked to having as many children as possible. In Niger and Cameroon, for example, men, on average, want 12.6 and 11.2 children, respectively. Women in these countries want only 5 or 6 on average. Even though a woman might desire fewer children, however, she may have few choices and little control over her own fertility. In many societies, a woman has no status outside of her role as wife and mother. Yet without children, she may have no source of support in her old age.

(a)   (b)

▲ **FIGURE 4.12** In rural areas with little mechanized agriculture (a), children are needed to tend livestock, care for younger children, and help parents with household chores. Where agriculture is mechanized (b), rural families view children just as urban families do—helpful, but not critical to survival. This affects the decision about how many children to have.

Education and income affect the desire for children

In more highly developed countries, many pressures tend to reduce fertility. Higher education and personal freedom for women often result in decisions to limit childbearing. A desire to spend time and money on other goods and activities offsets the desire to have children. When women have opportunities to earn a salary, they are less likely to stay home and have many children. Not only do many women find the challenge and variety of a career attractive, but the money that they earn outside the home becomes an important part of the family budget. Thus education and socioeconomic status are usually inversely related to fertility in richer countries. In some developing countries, however, fertility initially increases as educational levels and socioeconomic status rise. With higher income, families are better able to afford the children they want. More money also means that women are healthier and therefore better able to conceive and carry a child to term. It may be a generation before this unmet desire for children abates.

In less-developed countries, where feeding and clothing children can be a minimal expense, adding one more child to a family usually doesn't cost much. By contrast, raising a child in a developed country can cost hundreds of thousands of dollars by the time the child finishes school and is independent. Under these circumstances, parents are more likely to choose to have one or two children on whom they can concentrate their time, energy, and financial resources.

Figure 4.13 shows U.S. birth rates between 1910 and 2000. As you can see, birth rates fell and rose in an irregular pattern. The period between 1910 and 1930 was a time of industrialization and urbanization. Women were getting more education than ever before and entering the workforce in large numbers. The Great Depression in the 1930s made it economically difficult for families to have children, and birth rates were low. The birth rate increased at the beginning of World War II (as it often does in wartime). For reasons that are unclear, a higher percentage of boys are usually born during war years.

A "baby boom" followed World War II, as couples were reunited and new families started. During this time, the government encouraged women to leave their wartime jobs and stay home. A high birth rate persisted through the times of prosperity and optimism of the 1950s but began to fall in the 1960s. Part of this decline was caused by the small number of babies born in the 1930s, which resulted in fewer young adults to give birth in the 1960s. Part was due to changed perceptions of the ideal family size. Whereas in the 1950s women typically wanted four children or more, the norm dropped to one or two (or no) children in the 1970s. A small "echo boom" occurred in the 1980s, as baby boomers began to have babies, but changing economics and attitudes seem to have permanently altered our view of ideal family size in the United States.

4.5 A DEMOGRAPHIC TRANSITION CAN LEAD TO STABLE POPULATION SIZE

In 1945 demographer Frank Notestein pointed out that a typical pattern of falling death rates and birth rates due to improved living conditions usually accompanies economic development. He called this pattern the **demographic transition** from high birth

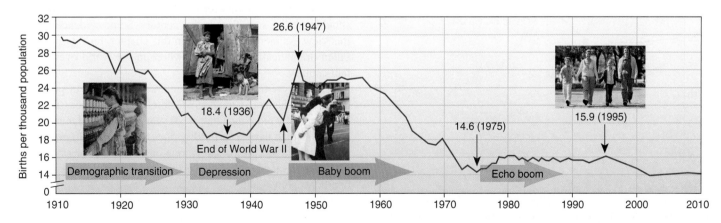

▲ **FIGURE 4.13** Birth rates in the United States, 1910–2010. The falling birth rate from 1910 to 1929 represents a demographic transition from an agricultural to an industrial society. The baby boom following World War II lasted from 1945 to 1965. A much smaller "echo boom" occurred around 1980 when the baby boomers started to reproduce. SOURCES: Data from Population Reference Bureau and U.S. Bureau of the Census.

and death rates to lower birth and death rates. Figure 4.14 shows an idealized model of a demographic transition. This model is often used to explain connections between population growth and economic development.

Economic and social conditions change mortality and births

Stage I in figure 4.14 represents the conditions in a premodern society. Food shortages, malnutrition, lack of sanitation and medicine, accidents, and other hazards generally keep death rates in such a society around 30 per 1,000 people. Birth rates are correspondingly high to keep population densities relatively constant. Economic development in Stage II brings better jobs, medical care, sanitation, and a generally improved standard of living, and death rates often fall very rapidly. Birth rates may actually rise at first as more money and better nutrition allow people to have the children they always wanted. In a couple of generations, however, birth rates fall as people see that all their children are more likely to survive and that the whole family benefits from concentrating more resources on fewer children. Note that populations grow rapidly during Stage III, when death rates have already fallen but birth rates remain high. Depending on how long it takes to complete the transition, the population may go through one or more rounds of doubling before coming into balance again.

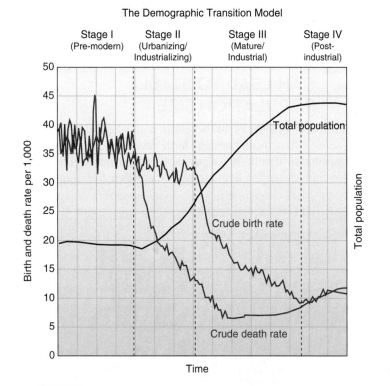

The Demographic Transition Model

▲ **FIGURE 4.14** Theoretical birth, death, and population growth rates in a demographic transition accompanying economic and social development. In a predevelopment society, birth and death rates are both high, and total population remains relatively stable. During development, death rates tend to fall first, followed in a generation or two by falling birth rates. Total population grows rapidly until both birth and death rates stabilize in a fully developed society.

Stage IV represents conditions in developed countries, where the transition is complete and both birth rates and death rates are low, often a third or less than those in the predevelopment era. The population comes into a new equilibrium in this phase, but at a much larger size than before. Most of the countries of northern and western Europe went through a demographic transition in the nineteenth or early twentieth century similar to the curves shown in figure 4.14. In countries such as Italy, where fertility levels have fallen below replacement rates, there are now fewer births than deaths, and the total population curve has started to decline.

A huge challenge facing countries in the final stage of the demographic transition is the imbalance between people in their most productive years and people who are retired or in their declining years. The continuing debate in the U.S. Congress about how to fund the social security system is due to the fact that when this program was established, the United States was in the middle of the demographic transition, with a large number of young people relative to older people. In 10 to 15 years, that situation will change, with many more elderly people living longer, and fewer younger workers to support them.

Many of the most rapidly growing countries in the world, such as Kenya, Yemen, Libya, and Jordan, now are in Stage III of this demographic transition. Their death rates have fallen close to the rates of the fully developed countries, but birth rates have not fallen correspondingly. In fact, their birth rates and total populations are higher than those in most European countries when industrialization began 300 years ago. The large disparity between birth and death rates means that many developing countries now are growing at 3 to 4 percent per year. Such high growth rates in developing countries could boost total world population to over 9 billion by the end of the twenty-first century. This raises what may be the two most important questions in this entire chapter: Why are birth rates not yet falling in these countries, and what can be done about it?

Many countries are in a demographic transition

Some demographers claim that a demographic transition already is in progress in most developing nations. They believe that problems in taking censuses and a normal lag between falling death and birth rates may hide this for a time but that the world population should stabilize sometime in this century. Some evidence supports this view. As mentioned earlier in this chapter, fertility rates have fallen dramatically nearly everywhere in the world over the past half century.

Some countries have had remarkable success in population control. In Thailand, China, and Colombia, for instance, total fertility dropped by more than half in 20 years. Morocco, Jamaica, Peru, and Mexico all have seen fertility rates fall by 30 to 40 percent in a single generation. Surprisingly, one of the most successful family planning advances in recent years has been in Iran, a predominantly Muslim country.

The following factors help stabilize populations:

• Growing prosperity, urbanization, and social reforms that accompany development reduce the need and desire for large families in most countries.

- Technology is available to bring advances to the developing world much more rapidly than was the case a century ago, and the rate of technology exchange is much faster than it was when Europe and North America were developing.

- Less-developed countries have historic patterns to follow. They can benefit from the mistakes of more-developed countries and chart a course to stability relatively quickly.

- Modern communications (especially television and the Internet) provide information about the benefits of and methods for social change.

Two ways to complete the demographic transition

The Indian states of Kerala and Andra Pradesh exemplify two very different approaches to regulating population growth. In Kerala, providing a fair share of social benefits to everyone is seen as the key to family planning. This social justice strategy assumes that the world has enough resources for everyone, but inequitable social and economic systems cause maldistributions of those resources. Hunger, poverty, violence, environmental degradation, and overpopulation are symptoms of a lack of justice, rather than a lack of resources. Although overpopulation exacerbates other problems, a focus on growth rates alone encourages racism and hatred of the poor. Proponents of this perspective argue that richer people should recognize the impacts their exorbitant consumption has on others (fig. 4.15).

The leaders of Andra Pradesh, on the other hand, have adopted a strategy of aggressively emphasizing birth control, rather than promoting social justice. This strategy depends on policies, similar to those in China, that assume providing carrots (economic rewards for reducing births) along with sticks (mandates for limiting reproduction together with punishment for exceeding limits) are the only effective ways to regulate population size.

Both states have slowed population growth significantly. And though they employ very different strategies, both aim to avoid a "demographic trap" in which rapidly growing populations exceed the sustainable yield of local forests, grasslands, croplands, and water resources. The environmental deterioration, economic decline, and political instability caused by resource shortages may prevent countries caught in this trap from ever completing modernization. Their populations may continue to grow until catastrophe intervenes.

What do you think? If you were advising developing countries on their population policies, which approach would you adopt?

Improving women's lives helps reduce birth rates

As the opening case study for this chapter shows, empowering women is key to achieving population stability. The 1994 International Conference on Population and Development in Cairo, Egypt, supported this approach to population issues. A broad consensus reached by the 180 participating countries agreed that responsible economic development, education, and women's rights, along with high-quality health care (including family planning services), must be accessible to everyone if population growth is to be slowed. Child survival is one of the most critical factors in stabilizing population. When infant and child mortality rates are high, as they are in much of the developing world, parents tend to have high numbers of children to ensure that some will survive to adulthood. There has never been a sustained drop in birth rates that was not first preceded by a sustained drop in infant and child mortality.

However, increasing family income doesn't always translate into better welfare for children, since men in many cultures control most financial assets. As the UN Conference in Cairo noted, often the best way to improve child survival is to ensure the rights of mothers. Opportunities for women's education, for instance, as well as land reform, political rights, opportunities to earn an independent income, and improved health status of women often are better indicators of family welfare than is rising gross national product (fig. 4.16).

FIGURE 4.15 Do we reduce pressure on the environment by achieving population control in developing countries, or by limiting resource use in developed countries? It depends on whom you ask.

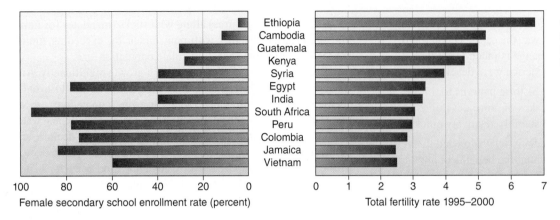

Female secondary school enrollment rate (percent)

Total fertility rate 1995–2000

Ethiopia
Cambodia
Guatemala
Kenya
Syria
Egypt
India
South Africa
Peru
Colombia
Jamaica
Vietnam

◀ **FIGURE 4.16** Total fertility declines as women's education increases. SOURCE: Data from Worldwatch Institute, 2003.

4.6 FAMILY PLANNING GIVES US CHOICES

Family planning allows couples to determine the number and spacing of their children. It doesn't necessarily mean fewer children—people could use family planning to have the maximum number of children possible—but it does imply that the parents will control their reproductive lives and make rational, conscious decisions about how many children they will have and when those children will be born, rather than leaving it to chance. As the desire for smaller families becomes more common, birth control often becomes an essential part of family planning. In this context, **birth control** usually means any method used to reduce births, including celibacy, delayed marriage, contraception, methods that prevent embryo implantation, and induced abortions.

Humans have always regulated their fertility

The high human birth rate of the last two centuries is not the norm, compared to previous millennia of human existence. Evidence suggests that people in every culture and every historic period used a variety of techniques to control population size. Studies of hunting and gathering people, such as the !Kung, or San, of the Kalahari Desert in southwest Africa, indicate that our early ancestors had stable population densities, not because they killed each other or starved to death regularly but because they controlled fertility.

For instance, San women breast-feed children for three or four years. When calories are limited, lactation depletes body fat stores and suppresses ovulation. Coupled with taboos against intercourse while breast-feeding, this is an effective way of spacing children. (However, breast-feeding among well-nourished women in modern societies doesn't necessarily suppress ovulation or prevent conception.) Other ancient techniques to control population size include celibacy, folk medicines, abortion, and infanticide. We may find some or all of these techniques unpleasant or morally unacceptable, but we shouldn't assume that other people are too ignorant or too primitive to make decisions about fertility.

Today there are many options

Modern medicine gives us many more options for controlling fertility than were available to our ancestors. More than 100 new contraceptive methods are now being studied, and some appear to have great promise. Nearly all are biologically based (e.g., hormonal) rather than mechanical (e.g., condom, IUD). Recently the U.S. Food and Drug Administration approved five new birth control products. Four of these use various methods to administer female hormones that prevent pregnancy.

Other methods are years away from use, but take a new direction entirely. Vaccines for women are being developed that will prepare the immune system to reject the hormone chorionic gonadotropin, which maintains the uterine lining and allows egg implantation, or that will cause an immune reaction against sperm. Injections for men are focused on reducing sperm production, and have proven effective in mice. Without a doubt, the contemporary couple has access to many more birth control options than their grandparents had.

4.7 WHAT KIND OF FUTURE ARE WE CREATING NOW?

Because there's often a lag between the time when a society reaches replacement birth rate and the end of population growth, we are deciding now what the world will look like in a hundred years. How many people will be in the world a century from now? Most demographers believe that world population will stabilize sometime during the twenty-first century. When we reach that equilibrium, the total number of humans is likely to be somewhere around 8 to 10 billion, depending on the success of family planning programs and the multitude of other factors affecting human populations. The United Nations Population Division projects four population scenarios (fig. 4.17). The optimistic (low) projection suggests that world population might stabilize by about 2030 and then drop back below current levels. This doesn't seem likely. The medium projection shows a population of about 9 billion in 40 years, while the high projection would reach 12 billion by midcentury.

Which of these scenarios will we follow? As you have seen in this chapter, population growth is a complex subject. Stabilizing or reducing human populations will require substantial changes from business as usual.

An encouraging sign is that worldwide contraceptive use has increased sharply in recent years. About half of the world's married couples used some family planning techniques in 2000, compared

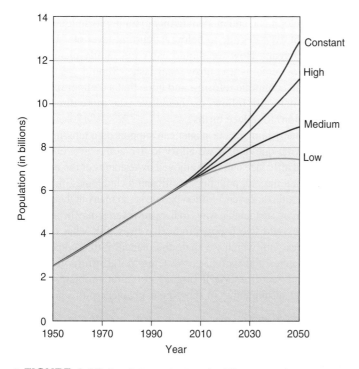

▲ **FIGURE 4.17** Population projections for different growth scenarios. Recent progress in family planning and economic development have led to significantly reduced estimates compared to a few years ago. The medium projection is 8.9 billion in 2050, compared to previous estimates of over 10 billion for that date.   SOURCE: UN Population Division, 2008.

with only 10 percent 30 years earlier, but another 100 million couples say they want, but do not have access to, family planning. If given a choice, people prefer smaller families.

Successful family planning programs often require significant societal changes. Among the most important of these are (1) improved social, educational, and economic status for women (birth control and women's rights are often linked); (2) improved status for children (fewer children are born if they are not needed as a cheap labor source); (3) acceptance of calculated choice as a valid element in life in general and in fertility in particular (the belief that we have no control over our lives discourages a sense of responsibility); (4) social security and political stability that give people the means and the confidence to plan for the future; and (5) the knowledge, availability, and use of effective and acceptable means of birth control.

The current world average fertility rate of 2.6 births per woman is less than half what it was 50 years ago. If similar progress could be sustained for the next half century, fertility rates could fall to the replacement rate of 2.1 children per woman. Whether this scenario comes true or not depends on choices that all of us make.

Already, nearly half the world population lives in countries where the total fertility rate is at or close to the replacement rate (fig. 4.18). The example of Brazil gives us hope that with rising standards of living, population growth will spontaneously slow without harsh government intervention. However, increasing wealth creates worries that consumption supported by destructive technologies will be unsustainable. The trade-off between

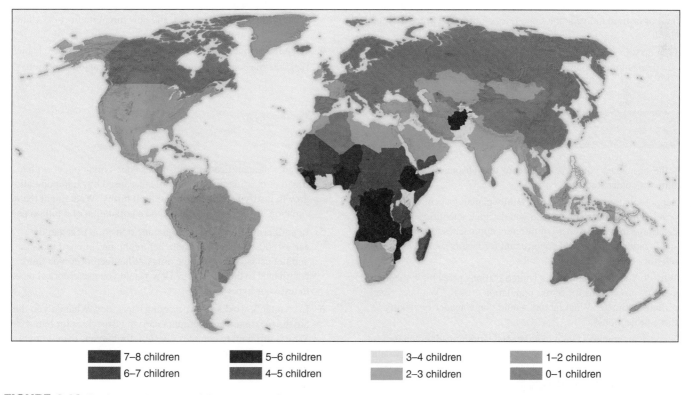

■ 7–8 children	■ 5–6 children	■ 3–4 children	■ 1–2 children
■ 6–7 children	■ 4–5 children	■ 2–3 children	■ 0–1 children

▲ **FIGURE 4.18** Fertility rates by country. Although average fertility in the United States is currenltly 2.06, it's below the replacement rate of 2.1 children per woman.

population size and affluence may still create unacceptable environmental conditions. Furthermore, as figure 4.18 shows, there are countries, especially in Africa, where wars, corruption, colonial history, religious tensions, and other factors have prevented economic and social development while perpetuating high population growth. Can we overcome all these problems and create a more humane, sustainable world?

CONCLUSION

A few decades ago, we were warned that a human population explosion was about to engulf the world. Exponential population growth was seen as a cause of nearly every important environmental problem. Some people still warn that the total number of humans might grow to 30 or 40 billion by the end of this century. Birth rates have fallen, however, almost everywhere, and most demographers now believe that we will reach an equilibrium around 9 billion people in about 2050. Some claim that if we promote equality, democracy, human development, and modern family planning techniques, population might even decline to below its current level of 7 billion in the next 50 years. How we should carry out family planning and birth control remains a controversial issue. Should we focus on political and economic reforms, and hope that a demographic transition will naturally follow; or should we take more direct action (or any action) to reduce births?

How many humans our planet can support on a long-term basis remains a vital question. If we all try to live at the level of material comfort and affluence currently enjoyed by residents of the wealthiest nations, using the old, polluting, inefficient technologies, the answer is almost certain that even 7 billion people is too many in the long run. If we find more sustainable ways to live, however, it may be that many more of us could live happy, productive lives. But what about the other species with which we share the planet? Will we leave room for wild species and natural ecosystems in our efforts to achieve comfort and security? We'll discuss pollution problems, energy sources, and sustainability in subsequent chapters of this book.

PRACTICE QUIZ

1. About how many years of human existence passed before the world population reached its first billion? What factors restricted population before that time, and what factors contributed to growth after that point?
2. Describe the pattern of human population growth over the past 200 years. What is the shape of the growth curve (recall chapter 3)?
3. Define *ecological footprint*. How many more earths would it take if all of us tried to live at the same level of affluence as the average North American?
4. Why do some economists consider human resources more important than natural resources in determining a country's future?

5. In which regions of the world will most population growth occur during the twenty-first century? What conditions contribute to rapid population growth in these locations?
6. Define *crude birth rate, total fertility rate, crude death rate,* and *zero population growth*.
7. What is the difference between life expectancy and life span? Why are they different?
8. What is the dependency ratio, and how might it affect the United States in the future?
9. What factors increase or decrease people's desire to have babies?
10. Describe the conditions that lead to a demographic transition.

CRITICAL THINKING AND DISCUSSION

Apply the principles you have learned in this chapter to discuss these questions with other students.

1. Suppose that you were head of a family planning agency in a developing country. How would you design a scientific study to determine the effectiveness of different approaches to population stabilization? How would you account for factors such as culture, religion, education, and economics?
2. Why do you suppose that the United Nations gives high, medium, and low projections for future population growth? Why not give a single estimate? What factors would you consider in making these projections?
3. Some demographers claim that the total world population has already begun to slow, while others dispute this claim. How would you recognize a true demographic transition, as opposed to mere random fluctuations in birth and death rates?

4. Discuss the ramifications of China's "one-child policy" with a friend or classmate. Do the problems caused by rapid population growth justify harsh measures to limit births? What might the world situation be like today if China had a population of 2 billion people?
5. In northern Europe, the demographic transition began in the early 1800s, a century or more before the invention of modern antibiotics and other miracle drugs. What factors do you think contributed to this transition? How would you use historical records to test your hypothesis?
6. In chapter 3, we discussed carrying capacities. What do you think are the maximum and optimum carrying capacities for humans? Why is this a more complex question for humans than it might be for other species? Why is designing experiments in human demography difficult?

Brazil's population trends have shifted dramatically in recent years. Is Brazil unusual in this change? Take a look at population size and trends in different regions, and at some of the factors that influence growth rates. Gapminder.org is a rich source of data on global population, health, and development, including animated graphs showing change over time. Go to Connect to find a link to Gapminder graphs, and answer questions about what they tell you.

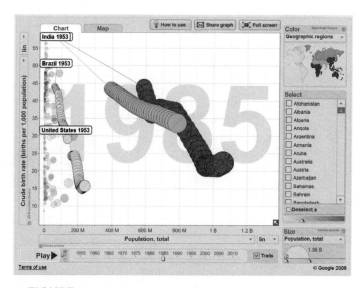

▲ **FIGURE 1** Go to Connect to examine changes in population trends over time.

Student interns measure plant growth in experimental plots in the B4Warmed study in northern Minnesota.   SOURCE: Peter Reich.

LEARNING OUTCOMES

After studying this chapter, you should be able to answer the following questions:

▶ What are nine major terrestrial biomes, and what environmental conditions control their distribution?

▶ How does vertical stratification differentiate life zones in oceans?

▶ Why are coral reefs, mangroves, estuaries, and wetlands biologically important?

▶ What do we mean by *biodiversity*? List several regions of high biodiversity.

▶ What are the major benefits of biodiversity?

▶ What are the major human-caused threats to biodiversity?

▶ How can we reduce these threats to biodiversity?

Forest Responses to Global Warming

How will biological communities respond to climate change? This is one of the great unknowns in environmental science today. Will northern regions that now support boreal (northern) forest, for example, shift to another biome—hardwood forest, open savannah, grassland, or something entirely different? With rising emissions of CO_2 and other greenhouse gases, climate models predict that boreal forests will move north by about 480 km (300 mi) within this century. But there's a great deal of uncertainty in this prediction.

How do environmental scientists approach and analyze such complex questions? One strategy is to grow plants in a greenhouse, and test plant responses to different temperature and moisture levels. By changing just one variable at a time, we can get an approximation of responses to environmental change. But this approach misses the complex species interactions that influence plant growth in a real ecosystem, so an alternative approach is to use field tests in which mixtures of plants are grown in natural settings that include competition for resources, predator/prey interactions, natural climatic variations, and other ecological factors.

Professor Peter Reich and his colleagues and student research assistants are now carrying out such a field study in a patch of boreal forest in Minnesota. Calling this experiment B4Warmed, which stands for Boreal Forest Warming at an Ecotone in Danger, they are artificially raising ambient temperatures in a series of boreal forest plots, to emulate warming climate conditions.

The group established 96 circular experimental plots, each 3 meters (9.8 ft) in diameter (fig. 5.1). Each plot was planted with a mixture of tree species and annual understory plants. The plots were then randomly assigned to one of four treatments. Half the plots are in mature forest, and half are in forest openings. Half are kept 2°C above ambient temperatures, and half are kept 4°C higher than ambient temperatures, using infrared lamps placed around the plots, as well as buried heat cables (fig. 5.2). Control plots (with no temperature manipulations) are also maintained for comparison with treatments.

It's too early to know exactly what the long-term effects of warming will be on the northern forest community. It seems likely that species, such as aspen,

spruce, and birch, that are now at the southern edge of their range in the study area won't do as well under a warmer climate as the temperate maple-oak forests now growing farther south. However, both northern and temperate species may perform poorly under warmer conditions. If so, neither our current forest trees nor their potential replacements may be well suited to our future climate. This experiment will enable us to assess the potential for climate change to alter future forest composition.

One preliminary result from this study that appears to offer good news is that the CO_2 emissions both from forest plants and from the soil are lower than expected at higher temperatures.

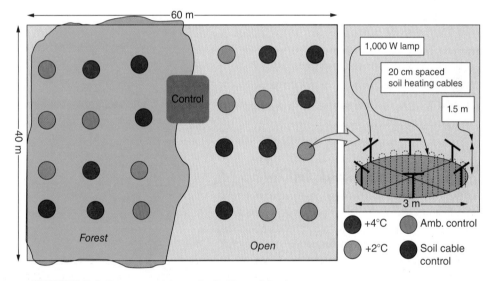

▲ **FIGURE 5.1** Experimental design for B4Warmed Study.

◀ **FIGURE 5.2**
Students and technicians adjust the electrical panel that controls heat lamps and heating cables.
SOURCE: Peter Reich.

(continued)

Apparently both standing vegetation and soil microbes alter their metabolic rates to acclimate to ambient environmental conditions. Thus the feedback cycles predicted to exacerbate global warming effects may not be as bad as we feared.

This study attempts to understand how climate changes may affect a broad biological community, or biome, and the species that make it up. To understand our environment and our impacts on it, it's important to be aware of the natural communities that existed before we arrived, as well as how the plants, animals, and physical conditions that create those communities interact. The careful, systematic approach demonstrated here is a hallmark of science and a way to reveal those connections. ■

In the end, we conserve only what we love. We will love only what we understand. We will understand only what we are taught.

—BABA DIOUM

5.1 TERRESTRIAL BIOMES

As the B4Warmed study shows, the increasing temperatures and drier climate expected if we continue our current path toward global climate change are likely to bring dramatic transformations to the boreal forest along the Canadian border from Minnesota to Maine. To evaluate the potential effects of environmental changes, it's useful to understand the patterns of plant and animal distribution that occur in various environmental conditions. We call large biological communities **biomes**. If we know the range of temperature and precipitation in a particular place, we can generally predict what kind of biome is likely to occur there, in the absence of human disturbance (fig. 5.3).

An important characteristic of each biome is its **biodiversity**, or the number and variety of different biological species that live there. Species not only create much of the structure and functions of an ecosystem, but, as we discussed in chapter 2, they also generate emergent properties, such as productivity, homeostasis, and resilience. Productivity, the rate at which plants produce biomass, varies a great deal from warm to cold climates, and from wet to dry environments. The amount of resources we can extract, such as timber or fish or crops, depends largely on a biome's biological productivity. Similarly,

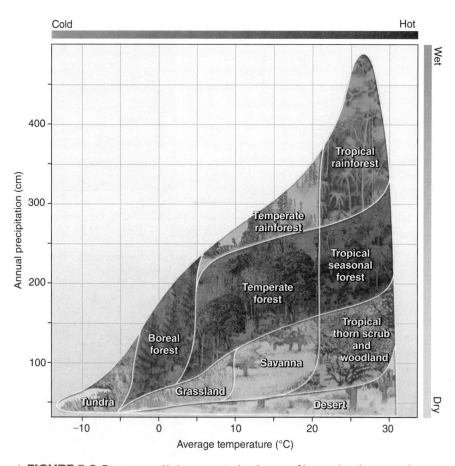

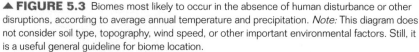

▲ **FIGURE 5.3** Biomes most likely to occur in the absence of human disturbance or other disruptions, according to average annual temperature and precipitation. *Note:* This diagram does not consider soil type, topography, wind speed, or other important environmental factors. Still, it is a useful general guideline for biome location.

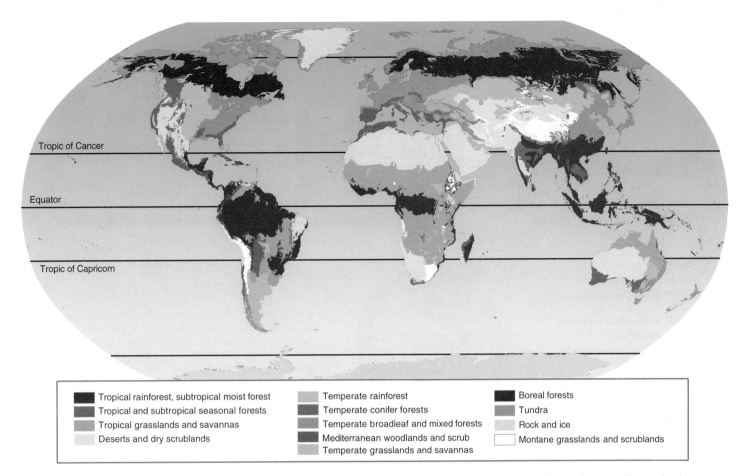

▲ FIGURE 5.4 Major world biomes. Compare this map with figure 5.3 for generalized temperature and moisture conditions that control biome distribution. Also compare it with the satellite image of biological productivity (fig. 5.15).   SOURCE: WWF Ecoregions.

Legend:
- Tropical rainforest, subtropical moist forest
- Tropical and subtropical seasonal forests
- Tropical grasslands and savannas
- Deserts and dry scrublands
- Temperate rainforest
- Temperate conifer forests
- Temperate broadleaf and mixed forests
- Mediterranean woodlands and scrub
- Temperate grasslands and savannas
- Boreal forests
- Tundra
- Rock and ice
- Montane grasslands and scrublands

homeostasis (stability) and resilience (the ability to recover from disturbance) also depend on biodiversity and productivity. Clear-cut forests, for example, regrow very quickly in the warm, moist Amazon, but slowly, if at all, in northern Canada's harsh cold, dry climate.

In the sections that follow, you will learn about nine major biome types. These nine can be further divided into smaller classes: for example, some temperate forests have mainly cone-bearing trees (conifers), such as pines or fir, while others have mainly broadleaf trees, such as maples or oaks (fig. 5.4). Many temperature controlled biomes occur in latitudinal bands. A band of boreal forest crosses Canada, Europe, and Siberia, tropical forests occur near the equator, and expansive grasslands lie near—or just beyond—the tropics. Some biomes are named for their latitudes: tropical rainforests occur between the Tropic of Cancer (23° north) and the Tropic of Capricorn (23° south); arctic tundra lies near or above the Arctic Circle (66.6° north).

Temperature and precipitation change with elevation as well as with latitude. In mountainous regions, temperatures are cooler and precipitation is usually greater at high elevations. Mountains are cooler, and often wetter, than low elevations. **Vertical zonation** is a term applied to vegetation zones defined by altitude. A 100-km transect from California's Central Valley up to Mount Whitney, for example, crosses as many vegetation zones as you would find on a journey from southern California to northern Canada (fig. 5.5).

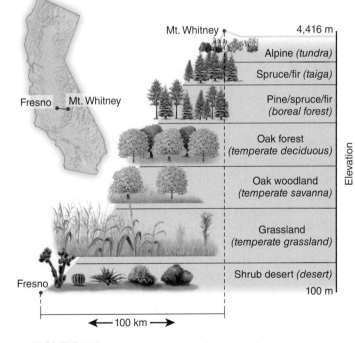

▲ FIGURE 5.5 Vegetation changes with elevation because temperatures are lower and there is more precipitation high on a mountainside. A 100-km transect from Fresno, California, to Mt. Whitney (California's highest point) crosses vegetation zones similar to about seven different biome types.

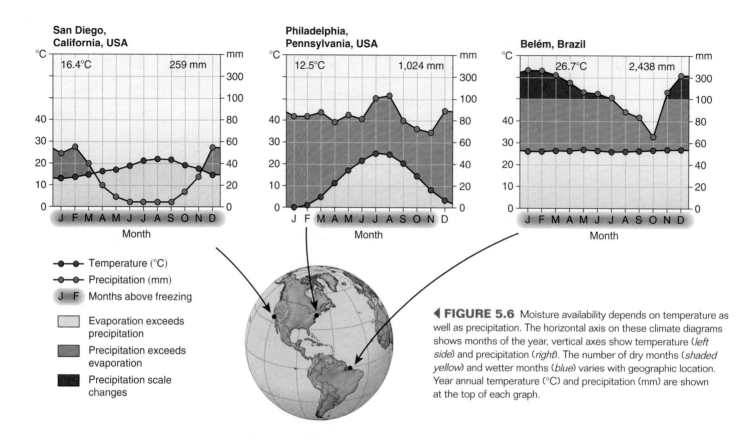

◄ FIGURE 5.6 Moisture availability depends on temperature as well as precipitation. The horizontal axis on these climate diagrams shows months of the year, vertical axes show temperature (*left side*) and precipitation (*right*). The number of dry months (*shaded yellow*) and wetter months (*blue*) varies with geographic location. Year annual temperature (°C) and precipitation (mm) are shown at the top of each graph.

As you consider the terrestrial biomes, compare the climatic conditions that help shape them. To begin, examine the three climate graphs in figure 5.6. These graphs show annual trends in temperature and precipitation (rainfall and snowfall). They also indicate the relationship between potential evaporation, which depends on temperature, and precipitation. When evaporation exceeds precipitation, dry conditions result (marked yellow). Moist climates may vary in precipitation rates, but evaporation rarely exceeds precipitation. Months above freezing temperature (marked brown) have most evaporation. Comparing these climate graphs helps us understand the different seasonal conditions that control plant and animal growth in the different biomes.

Tropical moist forests are warm and wet year-round

The humid tropical regions support one of the most complex and biologically rich biome types in the world (fig. 5.7). Although there are several kinds of moist tropical forests, all have ample rainfall and uniform temperatures. Cool **cloud forests** are found high in the mountains where fog and mist keep vegetation wet all the time. **Tropical rainforests** occur where rainfall is abundant—more than 200 cm (80 in.) per year—and temperatures are warm to hot year-round.

The soil of both these tropical moist forest types tends to be thin, acidic, and nutrient-poor, yet the number of species present can be mind-boggling. For example, the number of insect species in the canopy of tropical rainforests has been estimated to be in the millions! It is estimated that one-half to two-thirds of all species of terrestrial plants and insects live in tropical forests.

The nutrient cycles of these forests also are distinctive. About 90 percent of all the nutrients in the rainforest are contained in the bodies of the living organisms. This is a striking contrast to temperate

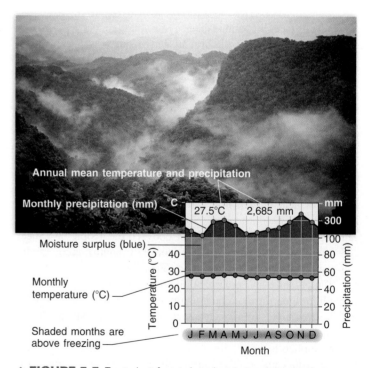

▲ FIGURE 5.7 Tropical rainforests have luxuriant and diverse plant growth. Heavy rainfall in most months, shown in the climate graph, supports this growth.

Active LEARNING

Comparing Biome Climates

Look back at the climate graphs for San Diego, California, an arid region, and Belém, Brazil, in the Amazon rainforest (fig. 5.6). How much colder is San Diego than Belém in January? In July? Which location has the greater range of temperature through the year? How much do the two locations differ in precipitation during their wettest months?

Compare the temperature and precipitation in these two places with those in the other biomes shown in the pages that follow. How wet are the wettest biomes? Which biomes have distinct dry seasons? How do rainfall and length of warm seasons explain vegetation conditions in these biomes?

ANSWERS: San Diego is about 13°C colder in January, about 6°C colder in July; San Diego has the greater range of temperature; there is about 250 mm difference in precipitation in December–February.

forests, where nutrients are held within the soil and made available for new plant growth. The luxuriant growth in tropical rainforests depends on rapid decomposition and recycling of dead organic material. Leaves and branches that fall to the forest floor decay and are incorporated almost immediately back into living biomass.

When the forest is removed for logging, agriculture, and mineral extraction, the thin soil cannot support continued cropping and cannot resist erosion from the abundant rains. And if the cleared area is too extensive, it may not be repopulated by the rainforest community.

Tropical seasonal forests have annual dry seasons

Many tropical regions are characterized by distinct wet and dry seasons, although temperatures remain hot year-round. These areas support **tropical seasonal forests**: drought-tolerant forests that look brown and dormant in the dry season but burst into vivid green during rainy months. These forests are often called dry tropical forests because they are dry much of the year; however, there must be some periodic rain to support plant growth. Many of the trees and shrubs in a seasonal forest are drought-deciduous: they lose their leaves and cease growing when no water is available. Seasonal forests are often open woodlands that grade into savannas.

Tropical dry forests are generally more attractive than wet forests for human habitation and have, therefore, suffered greater degradation from settlement. Clearing a dry forest with fire is relatively easy during the dry season. Soils of dry forests often have higher nutrient levels and are more agriculturally productive than those of a rainforest. Finally, having fewer insects, parasites, and fungal diseases than a wet forest makes a dry or seasonal forest a healthier place for humans to live. Consequently, these forests are highly endangered in many places. Less than 1 percent of the dry tropical forests of the Pacific coast of Central America or the Atlantic coast of South America, for instance, remain in an undisturbed state.

Tropical savannas and grasslands are dry most of the year

Where there is too little rainfall to support forests, we find open **grasslands** or grasslands with sparse tree cover, which we call **savannas** (fig. 5.8). Like tropical seasonal forests, most tropical savannas and grasslands have a rainy season, but generally the rains are less abundant or less dependable than in a forest. During dry seasons, fires can sweep across a grassland, killing off young trees and keeping the landscape open. Savanna and grassland plants have many adaptations to survive drought, heat, and fires. Many have deep, long-lived roots that seek groundwater and that persist when leaves and stems above the ground die back. After a fire or drought, fresh green shoots grow quickly from the roots. Migratory grazers, such as wildebeest, antelope, or bison, thrive on this new growth. Grazing pressure from domestic livestock is an important threat to both the plants and animals of tropical grasslands and savannas.

Deserts are hot or cold, but always dry

You may think of deserts as barren and biologically impoverished. Their vegetation is sparse, but it can be surprisingly diverse, and most desert plants and animals are highly adapted to survive long droughts, extreme heat, and often extreme cold. **Deserts** occur where precipitation is sporadic and low, usually with less than 30 cm of rain per year. Adaptations to these conditions include water-storing leaves and stems, thick epidermal layers to reduce water loss, and salt tolerance. As in other dry environments, many plants are drought-deciduous. Most desert plants also bloom and set seed quickly when rain does fall.

▲ **FIGURE 5.8** Tropical savannas and grasslands experience annual drought and rainy seasons and year-round warm temperatures. Thorny acacias and abundant grazers thrive in this savanna. Yellow areas show moisture deficit.

Warm, dry, high-pressure climate conditions (chapter 9) create desert regions at about 30° north and south. Extensive deserts occur in continental interiors (where rain is rare and evaporation rates are high) of North America, Central Asia, Africa, and Australia (fig. 5.9). The rain shadow of the Andes produces the world's driest desert in coastal Chile. Deserts can also be cold. Most of Antarctica is a desert; some inland valleys apparently get almost no precipitation at all.

Like plants, animals in deserts are specially adapted. Many are nocturnal, spending their days in burrows to avoid the sun's heat and desiccation. Pocket mice, kangaroo rats, and gerbils can get most of their moisture from seeds and plants. Desert rodents also have highly concentrated urine and nearly dry feces, which allow them to eliminate body waste without losing precious moisture.

Deserts are more vulnerable than you might imagine. Sparse, slow-growing vegetation is quickly damaged by off-road vehicles. Desert soils recover slowly. Tracks left by army tanks practicing in California deserts during World War II can still be seen today.

Deserts are also vulnerable to overgrazing. In Africa's vast Sahel (the southern edge of the Sahara Desert), livestock are destroying much of the plant cover. Bare, dry soil becomes drifting sand, and restabilization is extremely difficult. Without plant roots and organic matter, the soil loses its ability to retain what rain does fall, and the land becomes progressively drier and more bare. Similar degradation of dryland vegetation is happening in many desert areas, including Central Asia, India, and the American Southwest and Plains states.

Temperate grasslands have rich soils

As in tropical latitudes, temperate (midlatitude) grasslands occur where there is enough rain to support abundant grass but not enough for forests (fig. 5.10). Usually grasslands are a complex, diverse mix of grasses and flowering herbaceous plants, generally known as forbs. Myriad flowering forbs make a grassland colorful and lovely in summer. In dry grasslands, vegetation may be less than a meter tall. In more humid areas, grasses can exceed 2 m. Where scattered trees occur in a grassland, we call it a savanna.

Deep roots help plants in temperate grasslands and savannas survive drought, fire, and extreme heat and cold. These roots, together with an annual winter accumulation of dead leaves on the surface, produce thick, organic-rich soils in temperate grasslands. Because of this rich soil, many grasslands have been converted to farmland. The legendary tallgrass prairies of the central United States and Canada are almost completely replaced by corn, soybeans, wheat, and other crops. Most remaining grasslands in this region are too dry to support agriculture, and their greatest threat is overgrazing. Excessive grazing eventually kills even deep-rooted plants. As ground cover dies off, soil erosion results, and unpalatable weeds, such as cheatgrass or leafy spurge, spread.

Temperate scrublands have summer drought

Often, dry environments support drought-adapted shrubs and trees, as well as grass. These mixed environments can be highly variable. They can also be very rich biologically. Such conditions are often described as Mediterranean (where the hot season coincides with the dry season, producing hot, dry summers and cool, moist winters). Evergreen shrubs with small, leathery, sclerophyllous (hard, waxy) leaves form dense thickets. Scrub oaks, drought-resistant pines, or other small trees often cluster in sheltered valleys. Periodic fires burn fiercely in this fuel-rich plant assemblage and are a major factor in plant succession. Annual spring flowers often

▲ **FIGURE 5.9** Deserts generally receive less than 300 mm (30 cm) of precipitation per year. Hot deserts, as in the American Southwest, endure year-round drought and extreme heat in summer.

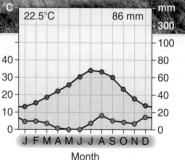

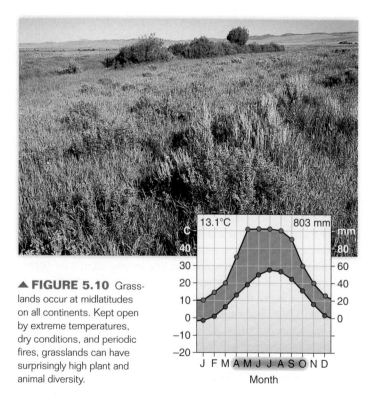

▲ **FIGURE 5.10** Grasslands occur at midlatitudes on all continents. Kept open by extreme temperatures, dry conditions, and periodic fires, grasslands can have surprisingly high plant and animal diversity.

bloom profusely, especially after fires. In California, this landscape is called **chaparral**, Spanish for "thicket." Resident animals are drought-tolerant species, such as jackrabbits, kangaroo rats, mule deer, chipmunks, lizards, and many bird species. Very similar landscapes are found along the Mediterranean coast as well as southwestern Australia, central Chile, and South Africa. Although this biome doesn't cover a very large total area, it contains a high number of unique species and is often considered a "hot-spot" for biodiversity. It also is highly desired for human habitation, often leading to conflicts with rare and endangered plant and animal species.

Temperate forests can be evergreen or deciduous

Temperate, or midlatitude, forests occupy a wide range of precipitation conditions but occur mainly between about 30° and 55° latitude (see fig. 5.4). In general we can group these forests by tree type, which can be broad-leaved **deciduous** (losing leaves seasonally) or evergreen **coniferous** (cone-bearing).

Deciduous Forests Broad-leaf forests occur throughout the world where rainfall is plentiful. In midlatitudes, these forests are deciduous and lose their leaves in winter. The loss of green chlorophyll pigments can produce brilliant colors in these forests in autumn (fig. 5.11). At lower latitudes, broad-leaf forests may be evergreen or drought-deciduous. Southern live oaks, for example, are broad-leaf evergreen trees.

Although these forests have a dense canopy in summer, they have a diverse understory that blooms in spring, before the trees leaf out. Spring ephemeral (short-lived) plants produce lovely flowers, and vernal (springtime) pools support amphibians and insects. These forests also shelter a great diversity of songbirds.

North American deciduous forests once covered most of what is now the eastern half of the United States and southern Canada. Most of western Europe was once deciduous forest but was cleared a thousand years ago. When European settlers first came to North America, they quickly settled and cut most of the eastern deciduous forests for firewood, lumber, and industrial uses, as well as to clear farmland. Many of those regions have now returned to deciduous forest, though the dominant species may have changed.

Deciduous forests can regrow quickly because they occupy moist, moderate climates. But most of these forests have been occupied so long that human impacts are extensive, and most native species are at least somewhat threatened. The greatest current threat to temperate deciduous forests is in eastern Siberia, where deforestation is proceeding rapidly. Siberia may have the highest deforestation rate in the world. As forests disappear, so do Siberian tigers, bears, cranes, and a host of other endangered species.

Coniferous Forests Coniferous forests grow in a wide range of environmental conditions. Often they occur where moisture is limited: in cold climates, moisture is unavailable (frozen) in winter; hot climates may have seasonal drought; sandy soils hold little moisture, and they are often occupied by conifers. Thin, waxy

▲ **FIGURE 5.11** Temperate deciduous forests have year-round precipitation and winters near or below freezing.

leaves (needles) help these trees reduce moisture loss. Coniferous forests provide most wood products in North America. Dominant wood production regions include the southern Atlantic and Gulf coast states, the mountain West, and the Pacific Northwest (northern California to Alaska), but coniferous forests support forestry in many regions.

The coniferous forests of the Pacific coast grow in extremely wet conditions. The wettest coastal forests are known as **temperate rainforests**, a cool, rainy forest often enshrouded in fog (fig. 5.12). Condensation in the canopy (leaf drip) is a major form of precipitation in the understory. Mild year-round temperatures and abundant rainfall, up to 250 cm (100 in.) per year, result in luxuriant plant growth and giant trees such as the California redwoods, the largest trees in the world and the largest above-ground organism ever known to have existed. Redwoods once grew along the Pacific coast from California to Oregon, but logging has reduced their range to a few small fragments of those areas.

Boreal forests lie north of the temperate zone

Because conifers can survive winter cold, they tend to dominate the **boreal forest**, or northern forests, that lie between about 50° and 60° north (fig. 5.13). Mountainous areas at lower latitudes may also have many characteristics and species of the boreal forest. Dominant trees are pines, hemlocks, spruce, cedar, and fir. Some deciduous trees are also present, such as maples, birch, aspen, and alder. These forests are slow-growing because of the cold temperatures and short frost-free growing season, but they are still an expansive resource. In Siberia, Canada, and the

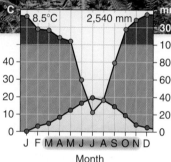

▲ **FIGURE 5.12** Temperate rainforests have abundant but often seasonal precipitation that supports magnificent trees and luxuriant understory vegetation. Often these forests experience dry summers.

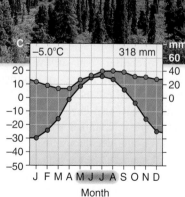

▲ **FIGURE 5.13** Boreal forests have moderate precipitation but are often moist because temperatures are cold most of the year. Cold-tolerant and drought-tolerant conifers dominate boreal forests and taiga, at the forest fringe.

western United States, large regional economies depend on boreal forests. They are favorite places for hunting, fishing, and recreation as well as extractive resource use. If those forests disappear as a result of global warming, both human uses and the unique species, such as moose, lynx, and otter, that depend on that biome will be sorely missed.

The extreme, ragged edge of the boreal forest, where forest gradually gives way to open tundra, is known by its Russian name, **taiga**. Here extreme cold and short summer limits the growth rate of trees. A 10-cm-diameter tree may be over 200 years old in the far north.

Tundra can freeze in any month

Where temperatures are below freezing most of the year, only small, hardy vegetation can survive. **Tundra**, a treeless landscape that occurs at high latitudes or on mountaintops, has a growing season of only two to three months, and it may have frost any month of the year. Some people consider tundra a variant of grasslands because it has no trees; others consider it a very cold desert because water is unavailable (frozen) most of the year.

Arctic tundra is an expansive biome that has low productivity because it has a short growing season. During midsummer, however, 24-hour sunshine supports a burst of plant growth and an explosion of insect life. Tens of millions of waterfowl, shorebirds, terns, and songbirds migrate to the Arctic every year to feast on the abundant invertebrate and plant life and to raise their young on the brief bounty. These birds then migrate to wintering grounds, where they may be eaten by local predators— effectively they carry energy and protein from high latitudes to low latitudes. Arctic tundra is essential for global biodiversity, especially for birds.

Alpine tundra, occurring on or near mountaintops, has environmental conditions and vegetation similar to arctic tundra (fig. 5.14). These areas have a short, intense growing season. Often one sees a splendid profusion of flowers in alpine tundra; everything must flower at once in order to produce seeds in a few weeks before the snow comes again. Many alpine tundra plants also have deep pigmentation and leathery leaves to protect against the strong ultraviolet light in the thin mountain atmosphere.

Compared to other biomes, tundra has relatively low diversity. Dwarf shrubs, such as willows, sedges, grasses, mosses, and lichens, tend to dominate the vegetation. Migratory musk ox, caribou, or alpine mountain sheep and mountain goats can live on the vegetation because they move frequently to new pastures.

Because these environments are too cold for most human activities, they are not as badly threatened as other biomes. There are important problems, however. Global climate change may be altering the balance of some tundra ecosystems, and air pollution from distant cities tends to accumulate at high latitudes (chapter 9). In eastern Canada, coastal tundra is being badly depleted by overabundant populations of snow geese, whose numbers have exploded due to winter grazing on the rice fields of Arkansas and Louisiana. Oil and gas drilling—and associated truck traffic— threatens tundra in Alaska and Siberia. Clearly, this remote biome is not independent of human activities at lower latitudes.

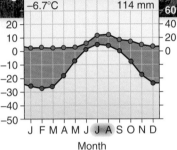

▲ **FIGURE 5.14** This landscape in Canada's Northwest Territories has both alpine and arctic tundra. Plant diversity is relatively low, and frost can occur even in summer.

Active LEARNING

Examining Climate Graphs

Among the nine types of terrestrial biomes you've just read about, one of the important factors is the number of months when the average temperature is below freezing (0°C). This is because most plants photosynthesize most actively when day-time temperatures are well above freezing—and when water is fluid, not frozen (chapter 2). Among the biome examples shown, how many sites have fewer than three months when the average temperature is above 0°? How many sites have all months above freezing? Look at figure 5.3: Do all deserts have average yearly temperatures above freezing? Now look at figure 5.4: Which biome do you live in? Which biome do most Americans live in?

ANSWERS: Only the tundra site has less than three months above freezing. Three sites have all months above freezing. No. Answers will vary. Most Americans live in temperate coniferous or broadleaf forest biomes.

5.2 MARINE ENVIRONMENTS

The biological communities in oceans and seas are poorly understood, but they are probably as diverse and complex as terrestrial biomes. In this section, we will explore a few facets of these fascinating environments. Oceans cover nearly three-fourths of the earth's surface, and they contribute in important, although often unrecognized, ways to terrestrial ecosystems. Like land-based systems, most marine communities depend on photosynthetic organisms. Often it is algae or tiny, free-floating photosynthetic plants (**phytoplankton**) that support a marine food web, rather than the trees and grasses we see on land. In oceans, photosynthetic activity tends to be greatest near coastlines, where nitrogen, phosphorus, and other nutrients wash offshore and fertilize primary producers. Ocean currents also contribute to the distribution of biological productivity, as they transport nutrients and phytoplankton far from shore (fig. 5.15).

As plankton, algae, fish, and other organisms die, they sink toward the ocean floor. Deep-ocean ecosystems, consisting of crabs, filter-feeding organisms, strange phosphorescent fish, and many other life-forms, often rely on this "marine snow" as a primary nutrient source. Surface communities also depend on this material. Upwelling currents circulate nutrients from the ocean floor back to the surface. Along the coasts of South America, Africa, and Europe, these currents support rich fisheries.

Vertical stratification is a key feature of aquatic ecosystems. Light decreases rapidly with depth, and communities below the photic zone (light zone, often reaching about 20 m deep) must rely on energy sources other than photosynthesis to persist. Temperature also decreases with depth. Deep-ocean species often grow slowly in part because metabolism is reduced in cold conditions. In contrast, warm, bright, near-surface communities, such as coral

▶ **FIGURE 5.15** Satellite measurements of chlorophyll levels in the oceans and on land. Dark green to blue land areas have high biological productivity. Dark blue oceans have little chlorophyll and are biologically impoverished. Light green to yellow ocean zones are biologically rich. SOURCE: SeaWiFS/NASA.

reefs and estuaries, are among the world's most biologically productive environments. Temperature also affects the amount of oxygen and other elements that can be absorbed in water. Cold water holds abundant oxygen, so productivity is often high in cold oceans, as in the North Atlantic, North Pacific, and Antarctic.

Open ocean communities vary from surface to hadal zone

Ocean systems can be described by depth and proximity to shore (fig. 5.16). In general, **benthic** communities occur on the bottom, and **pelagic** (from "sea" in Greek) zones are the water column. The epipelagic zone (*epi* = on top) has photosynthetic organisms. Below this are the mesopelagic (*meso* = medium) and bathypelagic (*bathos* = deep) zones. The deepest layers are the abyssal zone (to 4,000 m) and hadal zone (deeper than 6,000 m). Shorelines are known as littoral zones, and the area exposed by low tides is known as the intertidal zone. Often there is a broad, relatively shallow region along a continent's coast, which may reach a few kilometers or hundreds of kilometers from shore. This undersea area is the continental shelf.

We know relatively little about marine ecosystems and habitats, and much of what we know we have learned only recently. The open ocean has long been known as a biological desert, because it has relatively low productivity, or biomass production. Fish and plankton abound in many areas, however. Sea mounts, or undersea mountain chains and islands, support many commercial fisheries and much newly discovered biodiversity. In the equatorial Pacific and Antarctic oceans, currents carry nutrients far from shore, supporting biological productivity. The Sargasso Sea, a large region of the Atlantic near Bermuda, is known for its free-floating mats of brown algae. These algae mats support a phenomenal diversity of animals, including sea turtles, fish, and other species. Eels that hatch amid the algae eventually migrate up rivers along the Atlantic coasts of North America and Europe.

Deep-sea thermal vent communities are another remarkable type of marine system that was completely unknown until 1977, when the deep-sea submarine *Alvin* descended to the deep-ocean floor. These communities are based on microbes that capture chemical energy, mainly from sulfur compounds released from thermal vents—jets of hot water and minerals on the ocean floor (fig. 5.17). Magma below the ocean crust heats these vents. Tube worms, mussels, and microbes on the vents are adapted to survive both extreme temperatures, often above 350°C (700°F), and intense water pressure at depths of 7,000 m (20,000 ft) or more. Oceanographers have discovered thousands of different types of organisms, most of them microscopic, in these communities. Some estimate that the total mass of microbes on the seafloor represents one-third of all biomass on the planet.

Tidal shores support rich, diverse communities

As in the open ocean, shoreline communities vary with depth, light, and temperature. Some shoreline communities, such as estuaries, have high biological productivity and diversity because

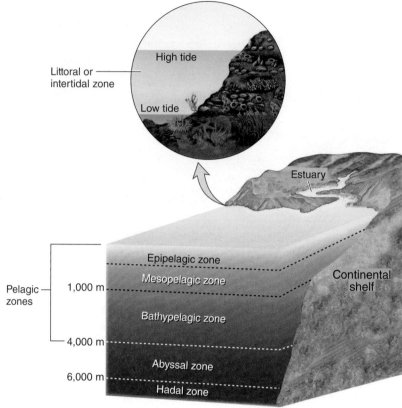

▲ **FIGURE 5.16** Light penetrates only the top 10–20 m of the ocean. Below this level, temperatures drop and pressure increases. Nearshore environments include the intertidal zone and estuaries.

▼ **FIGURE 5.17** Deep-ocean thermal vent communities have great diversity and are unusual because they rely on chemosynthesis, not photosynthesis, for energy.

they are enriched by nutrients washing from the land. Others, such as coral reefs, occur where there is little runoff from shore, but where shallow, clear warm water supports photosynthesis.

Coral reefs are among the best-known marine systems, because of their extraordinary biological productivity and their diverse and beautiful organisms (fig. 5.18a). Reefs are colonies of minute, colonial animals ("coral polyps") that live symbiotically with photosynthetic algae. Calcium-rich coral skeletons shelter the algae, and algae nourish the coral animals. The complex structure of a reef also shelters countless species of fish, worms, crustaceans, and other life-forms. Reefs occur where the water is shallow and clear enough for sunlight to reach the photosynthetic algae. They cannot tolerate abundant nutrients in the water, as nutrients support tiny floating plants and animals called plankton, which block sunlight.

Reefs are among the most endangered biological communities. Sediment from coastal development, farming, sewage, or other pollution can reduce water clarity and smother coral. Destructive fishing practices, including dynamite and cyanide poison, have destroyed many Asian reefs. Reefs can also be damaged or killed by changes in temperature, by invasive fish, and by diseases. **Coral bleaching**, the whitening of reefs due to stress, often followed by coral death, is a growing and spreading problem that worries marine biologists (chapter 1).

Sea-grass beds, or eel-grass beds, occupy shallow, warm, sandy coastlines. Like reefs, these support rich communities of grazers, from snails to turtles to Florida's manatees.

Mangroves are a diverse group of salt-tolerant trees that grow along warm, calm marine coasts around the world (fig. 5.18b). Growing in shallow, tidal mudflats, mangroves help stabilize shorelines, blunt the force of storms, and build land by trapping sediment and organic material. After the devastating Indonesian tsunami of 2004, studies showed that mangroves, where they still stood, helped reduce the speed, height, and turbulence of the tsunami waves. Detritus, including fallen leaves, collects below mangroves and provides nutrients for a diverse community of animals and plants. Both marine species (such as crabs and fish) and terrestrial species (such as birds and bats) rely on mangroves for shelter and food.

(a) Coral reefs and islands

(b) Mangroves

(c) Estuary and salt marsh

(d) Tide pool

▲ **FIGURE 5.18** Coastal environments support incredible diversity and help stabilize shorelines. Coral reefs (a), mangroves (b), and estuaries (c) also provide critical nurseries for marine ecosystems. Tide pools (d) also shelter highly specialized organisms.

Like coral reefs and sea-grass beds, mangrove forests provide sheltered nurseries for juvenile fish, crabs, shrimp, and other marine species on which human economies depend. However, like reefs and sea-grass beds, mangroves have been devastated by human activities. More than half of the world's mangroves that stood a century ago, perhaps 22 million ha, have been destroyed or degraded. They are clear-cut for timber or cleared to make room for fish and shrimp ponds. They are also poisoned by sewage and industrial waste near cities. Some parts of Southeast Asia and South America have lost 90 percent of their mangrove forests. Most have been cleared for fish and shrimp farming.

Estuaries are bays where rivers empty into the sea, mixing fresh water with salt water. **Salt marshes**, shallow wetlands flooded regularly or occasionally with seawater, occur on shallow coastlines, including estuaries (fig. 5.18c). Usually calm, warm, and nutrient-rich, estuaries and salt marshes are biologically diverse and productive. Rivers provide nutrients and sediments, and a muddy bottom supports emergent plants (whose leaves emerge above the water surface), as well as the young forms of crustaceans, such as crabs and shrimp, and mollusks, such as clams and oysters. Nearly two-thirds of all marine fish and shellfish rely on estuaries and saline wetlands for spawning and juvenile development.

Estuaries near major American cities once supported an enormous wealth of seafood. Oyster beds and clam banks in the waters adjacent to New York, Boston, and Baltimore provided free and easy food to early residents. Sewage and other contaminants long ago eliminated most of these resources, however. Recently, major efforts have been made to revive Chesapeake Bay, America's largest and most productive estuary. These efforts have shown some success, but many challenges remain (see case study in chapter 2).

In contrast to the shallow, calm conditions of estuaries, coral reefs, and mangroves, there are violent, wave-blasted shorelines that support fascinating life-forms in **tide pools**. Tide pools are depressions in a rocky shoreline that are flooded at high tide but retain some water at low tide. These areas remain rocky where wave action prevents most plant growth or sediment (mud) accumulation. Extreme conditions, with frigid flooding at high tide and hot, desiccating sunshine at low tide, make life impossible for most species. But the specialized animals and plants that do occur in this rocky intertidal zone are astonishingly diverse and beautiful (fig. 5.18d).

5.3 FRESHWATER ECOSYSTEMS

Freshwater environments are far less extensive than marine environments, but they are centers of biodiversity. Most terrestrial communities rely, to some extent, on freshwater environments. In deserts, isolated pools, streams, and even underground water systems support astonishing biodiversity as well as provide water to land animals. In Arizona, for example, many birds are found in trees and bushes surrounding the few available rivers and streams.

Lakes have extensive open water

Freshwater lakes, like marine environments, have distinct vertical zones (fig. 5.19). Near the surface a subcommunity of plankton, mainly microscopic plants, animals, and protists (single-celled

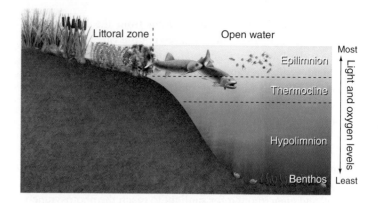

▲ **FIGURE 5.19** The layers of a deep lake are determined mainly by gradients of light, oxygen, and temperature. The epilimnion is affected by surface mixing from wind and thermal convections, while mixing between the hypolimnion and epilimnion is inhibited by a sharp temperature and density difference at the thermocline.

organisms, such as amoebae), float freely in the water column. Insects such as water striders and mosquitoes also live at the air–water interface. Fish move through the water column, sometimes near the surface and sometimes at depth.

Finally, the bottom, or *benthos*, is occupied by a variety of snails, burrowing worms, fish, and other organisms. These make up the benthic community. Oxygen levels are lowest in the benthic environment, mainly because there is little mixing to introduce oxygen to this zone. Anaerobic (not using oxygen) bacteria may live in low-oxygen sediments. In the littoral zone, emergent plants, such as cattails and rushes, grow in the bottom sediment. These plants create important functional links between layers of an aquatic ecosystem, and they may provide the greatest primary productivity to the system.

Lakes, unless they are shallow, have a warmer upper layer that is mixed by wind and warmed by the sun. This layer is the *epilimnion*. Below the epilimnion is the hypolimnion (*hypo* = below), a colder, deeper layer that is not mixed. If you have gone swimming in a moderately deep lake, you may have discovered the sharp temperature boundary, known as the **thermocline**, between these layers. Below this boundary, the water is much colder. This boundary is also called the mesolimnion.

Local conditions that affect the characteristics of an aquatic community include (1) nutrient availability (or excess), such as nitrates and phosphates; (2) suspended matter, such as silt, that affects light penetration; (3) depth; (4) temperature; (5) currents; (6) bottom characteristics, such as muddy, sandy, or rocky floor; (7) internal currents; and (8) connections to, or isolation from, other aquatic and terrestrial systems.

Wetlands are shallow and productive

Wetlands are shallow ecosystems in which the land surface is saturated or submerged at least part of the year. Wetlands have vegetation that is adapted to grow under saturated conditions. These legal definitions are important because, although wetlands make up only a small part of most countries, they are disproportionately important in conservation debates and are the focus of continual

(a) Swamp, or wooded wetland

legal disputes in North America and elsewhere around the world. Beyond these basic descriptions, defining wetlands is a matter of hot debate. How often must a wetland be saturated, and for how long? How large must it be to deserve legal protection? Answers can vary, depending on political, as well as ecological, concerns.

These relatively small systems support rich biodiversity, and they are essential for both breeding and migrating birds. Although wetlands occupy less than 5 percent of the land in the United States, the Fish and Wildlife Service estimates that one-third of all endangered species spend at least part of their lives in wetlands. Wetlands retain storm water and reduce flooding by slowing the rate at which rainfall reaches river systems. Floodwater storage is worth $3 billion to $4 billion per year in the United States. As water stands in wetlands, it also seeps into the ground, replenishing groundwater supplies. Wetlands filter, and even purify, urban and farm runoff, as bacteria and plants take up nutrients and contaminants in water. They are also in great demand for filling and development. They are often near cities or farms, where land is valuable, and, once drained, wetlands are easily converted to more lucrative uses. At least half of all the wetlands that existed in the United States when Europeans first arrived have been drained, filled, or degraded. In some major farming states, losses have been even greater. Iowa, for example, has lost 99 percent of its original wetlands.

Wetlands are described by their vegetation (fig. 5.20a–c). **Swamps**, also called forested wetlands, are wetlands with trees. **Marshes** are wetlands without trees. **Bogs** are areas of water-saturated ground, and usually the ground is composed of deep layers of accumulated, undecayed vegetation known as peat. **Fens** are similar to bogs except that they are mainly fed by groundwater, so that they have mineral-rich water and specially adapted plant species. Many bogs are fed mainly by precipitation. Swamps and marshes have high biological productivity. Bogs and fens, which are often nutrient-poor, have low biological productivity. They may have unusual and interesting species, though, such as sundews and pitcher plants, which are adapted to capture nutrients from insects rather than from soil.

The water in marshes and swamps usually is shallow enough to allow full penetration of sunlight and seasonal warming. These mild conditions favor great photosynthetic activity, resulting in high productivity at all trophic levels. In short, life is abundant and varied. Wetlands are major breeding, nesting, and migration staging areas for waterfowl and shorebirds.

Streams and rivers are open systems

Streams form wherever precipitation exceeds evaporation and surplus water drains from the land. Within small streams, ecologists distinguish areas of riffles, where water runs rapidly over a rocky substrate, and pools, which are deeper stretches of slowly moving current. Water tends to be well mixed and oxygenated in riffles; pools tend to collect silt and organic matter. If deep enough, pools can have vertical zones similar to those of lakes. As streams collect water and merge, they form rivers, although there isn't a universal definition of when one turns into the other. Ecologists consider a river system to be a continuum of constantly changing environmental conditions and community inhabitants, from the headwaters

(b) Marsh

(c) Bog

▲ **FIGURE 5.20** Wetlands provide irreplaceable ecological services, including water filtration, water storage and flood reduction, and habitat. Forested wetlands (a) are often called swamps; marshes (b) have no trees; bogs (c) are acidic and accumulate peat.

to the mouth of a drainage or watershed. The biggest distinction between stream and lake ecosystems is that, in a stream, materials, including plants, animals, and water, are continually moved downstream by flowing currents. This downstream drift is offset by active movement of animals upstream, productivity in the stream itself, and input of materials from adjacent wetlands or uplands.

5.4 BIODIVERSITY

The biomes you've just learned about shelter an astounding variety of living organisms. From the driest desert to the dripping rainforests, from the highest mountain peaks to the deepest ocean trenches, life occurs in a marvelous spectrum of sizes, colors, shapes, life cycles, and interrelationships. The varieties of organisms and complex ecological relationships give the biosphere its unique, productive characteristics. **Biodiversity**, the variety of living things, also makes the world a more beautiful and exciting place to live. Three kinds of biodiversity are essential to preserve ecological systems and functions: (1) *genetic diversity* is a measure of the variety of versions of the same genes within individual species; (2) *species diversity* describes the number of different kinds of organisms within individual communities or ecosystems; and (3) *ecological diversity* specifies the number of niches, trophic levels, and ecological processes that capture energy, sustain food webs, and recycle materials within this system. Redundancy in each of these categories enhances resiliency in a biome.

Increasingly, we identify species by genetic similarity

The concept of a species is fundamental in understanding biodiversity, but what is a species? In general, species are distinct organisms that persist because they can produce fertile offspring. But many organisms reproduce asexually; others don't reproduce in nature just because they don't normally encounter one another. Because of such ambiguities, evolutionary biologists favor the **phylogenetic species concept**, which identifies genetic similarity. Alternatively, the **evolutionary species concept** defines species according to evolutionary history and common ancestors. Both of these approaches rely on DNA analysis to define similarity among organisms.

How many species are there? Biologists have identified about 1.5 million species, but these probably represent only a small fraction of the actual number (table 5.1). Based on the rate of new discoveries by research expeditions—especially in the tropics—taxonomists estimate that somewhere between 3 million and 50 million different species may be alive today. About 70 percent of all known species are invertebrates (animals without backbones, such as insects, sponges, clams, and worms) (fig. 5.21). This group probably makes up the vast majority of organisms yet to be discovered and may constitute 90 percent of all species.

Biodiversity hot spots are rich and threatened

Most of the world's biodiversity concentrations are near the equator, especially tropical rainforests and coral reefs (fig. 5.22). Of all the world's species, only 10 to 15 percent live in North America and Europe. Many of the organisms in megadiversity countries have never been studied by scientists. The Malaysian Peninsula, for

TABLE 5.1 | Estimated Number of Species

CLASS	NUMBER DESCRIBED	NOT YET EVALUATED[1]	PERCENTAGE THREATENED[2]
Mammals	5,491	0%	21%
Birds	9,998	0%	12%
Reptiles	9,084	82%	28%
Amphibians	6,433	2%	30%
Fishes	31,300	86%	32%
Insects	1,000,000	100%	27%
Mollusks	85,000	97%	45%
Crustaceans	47,000	96%	35%
Other invertebrates	173,250	99%	30%
Mosses	16,236	99%	86%
Ferns and Allies	12,000	98%	66%
Gymnosperms	1,021	11%	35%
Flowering Plants	281,821	96%	73%
Fungi, Lichens, Protists	51,563	100%	50%

[1]Evaluated by IUCN for threatened status.

[2]Number of species as a percentage of those evaluated. Includes IUCN categories critically endangered, endangered, or vulnerable.

SOURCE: IUCN Red List, 2012

◀ **FIGURE 5.21** Insects and other invertebrates make up more than half of all known species. Many, like this blue morpho butterfly, are beautiful as well as ecologically important.

instance, has at least 8,000 species of flowering plants, while Britain, with an area twice as large, has only 1,400 species. There may be more botanists in Britain than there are species of higher plants. South America, on the other hand, has fewer than 100 botanists to study perhaps 200,000 species of plants.

Areas isolated by water, deserts, or mountains can also have high concentrations of unique species and biodiversity. Madagascar, New Zealand, South Africa, and California are all midlatitude areas isolated by barriers that prevent mixing with biological communities from other regions and produce rich, unusual collections of species.

5.5 BENEFITS OF BIODIVERSITY

We benefit from other organisms in many ways, some of which we don't appreciate until a particular species or community disappears. Even seemingly obscure and insignificant organisms can play irreplaceable roles in ecological systems or be the source of genes or drugs that someday may be indispensable.

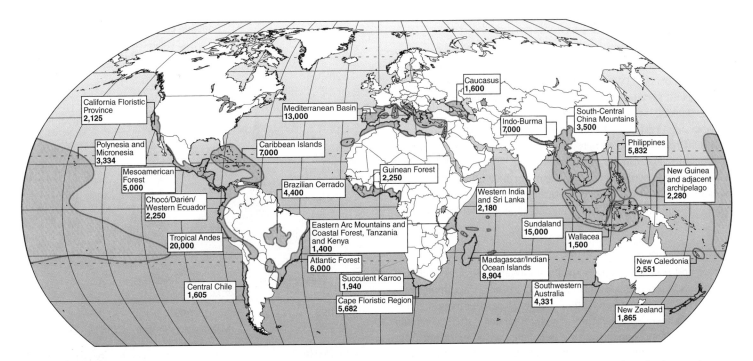

▲ **FIGURE 5.22** Biodiversity "hot spots" identified by Conservation International tend to be in tropical or Mediterranean climates and on islands, coastlines, or mountains where many habitats exist and physical barriers encourage speciation. Numbers represent estimated endemic (locally unique) species in each area.
SOURCE: Data from Conservation International.

Human life depends on ecological services provided by other organisms. Soil formation, waste disposal, air and water purification, nutrient cycling, solar energy absorption, and food production all depend on wild ecosystems and the interactions of organisms in them. Total value of these ecological services is at least $33 trillion per year, or more than double total world GNP.

Biodiversity provides food and medicines

Wild plant species make important contributions to human food supplies. Genetic material from wild plants has been used to improve domestic crops. Noted tropical ecologist Norman Myers estimates that as many as 80,000 edible wild plant species could be utilized by humans. Villagers in Indonesia, for instance, are thought to use some 4,000 native plant and animal species for food, medicine, and other products. Few of these species have been explored for possible domestication or more widespread cultivation. Wild bees, moths, bats, and other organisms provide pollination for most of the world's crops. Without these we would have little agriculture in much of the world.

Pharmaceutical products derived from developing world plants, animals, and microbes have a value of more than $30 billion per year, according to the United Nations Development Programme (table 5.2). Consider the success story of vinblastine and vincristine. These anticancer alkaloids are derived from the Madagascar periwinkle (*Catharanthus roseus*). They inhibit the growth of cancer cells and are very effective in treating certain kinds of cancer. Twenty years ago, before these drugs were introduced, childhood

TABLE 5.2 | Some Natural Medicinal Products

PRODUCT	SOURCE	USE
Penicillin	Fungus	Antibiotic
Bacitracin	Bacterium	Antibiotic
Tetracycline	Bacterium	Antibiotic
Erythromycin	Bacterium	Antibiotic
Digitalis	Foxglove	Heart stimulant
Quinine	Chincona bark	Malaria treatment
Diosgenin	Mexican yam	Birth control drug
Cortisone	Mexican yam	Anti-inflammation treatment
Cytarabine	Sponge	Leukemia cure
Vinblastine, vincristine	Periwinkle plant	Anticancer drugs
Reserpine	Rauwolfia	Hypertension drug
Bee venom	Bee	Arthritis relief
Allantoin	Blowfly larva	Wound healer
Morphine	Poppy	Analgesic

leukemias were invariably fatal. Now the remission rate for some childhood leukemias is 99 percent. Hodgkin's disease was 98 percent fatal a few years ago but is now only 40 percent fatal, thanks to these compounds. The total value of the periwinkle crop is roughly $15 million per year, although Madagascar gets little of those profits.

Biodiversity can aid ecosystem stability

High diversity may help biological communities withstand environmental stress better and recover more quickly than those with fewer species (see chapter 3). A reason may be that in a diverse community, some species survive disturbance, so ecological functions are maintained. Because we don't fully understand the complex interrelationships among organisms, we often are surprised and dismayed at the effects of removing seemingly insignificant members of biological communities. For instance, it is estimated that 95 percent of the potential pests and disease-carrying organisms in the world are controlled by natural predators and competitors. Maintaining biodiversity can be essential for pest control and other ecological functions.

Aesthetic and existence values are important

Nature appreciation is economically important. The U.S. Fish and Wildlife Service estimates that Americans spend $104 billion every year on wildlife-related recreation. This is 25 percent more than the $81 billion spent each year on new automobiles. Often recreation is worth even more than the resources that can be extracted from an area. Fishing, hunting, camping, hiking, and other nature-based activities also have cultural value. These activities provide exercise, and contact with nature can be emotionally restorative. In many cultures, nature carries spiritual connotations and observing and protecting nature has religious or moral significance.

For many people, just the idea that wildlife exists has value. This idea is termed "existence value." Even if they will never see a tiger or a blue whale, many find it gratifying to know they exist.

5.6 WHAT THREATENS BIODIVERSITY?

Extinction, the elimination of a species, is a normal process of the natural world. Species die out and are replaced by others, often their own descendants, as part of evolutionary change. In undisturbed ecosystems, the rate of extinction appears to be about one species lost every decade. Over the past century, however, human impacts on populations and ecosystems have accelerated that rate, possibly causing untold thousands of species, subspecies, and varieties to become extinct every year. Many of these are probably invertebrates, fungi, and microbes, which are unstudied but may perform critical functions for ecosystems (see table 5.1).

In geologic history, extinctions are common. Studies of the fossil record suggest that more than 99 percent of all species that ever existed are now extinct. Most of those species were gone long before humans came on the scene. Periodically, mass extinctions have wiped out vast numbers of species and even whole families (fig. 5.23). The best studied of these events occurred at the end of the Cretaceous period, when dinosaurs disappeared, along with at least 50 percent of existing species. An even greater disaster occurred at the end of the Permian period, about 250 million years ago, when 95 percent of species and perhaps half of all families died out over a period of about 10,000 years—a mere moment in geologic time. Current theories suggest that these catastrophes were caused by climate changes, perhaps triggered when large asteroids struck the earth.

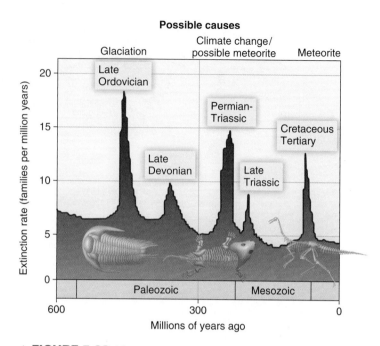

▲ **FIGURE 5.23** Major mass extinctions through history. We may be in a sixth mass extinction now, caused by human activities.

Many ecologists worry that global climate change caused by our release of greenhouse gases in the atmosphere could have similarly catastrophic effects (chapter 9).

HIPPO summarizes human impacts

The rate at which species are disappearing has increased dramatically over the past 150 years. Between A.D. 1600 and 1850, human activities appear to have eliminated two or three species per decade, about double the natural extinction rate. In the past 150 years, the extinction rate has increased to thousands per decade. Conservation biologists call this the sixth mass extinction, but note that this time it's not asteroids or volcanoes but human impacts that are responsible. E. O. Wilson summarizes human threats to biodiversity with the acronym **HIPPO**, which stands for *H*abitat destruction, *I*nvasive species, *P*ollution, *P*opulation of humans, and *O*verharvesting. Let's look in more detail at each of these issues.

Habitat destruction is usually the main threat

The most important extinction threat for most species—especially terrestrial ones—is habitat loss. Perhaps the most obvious example of habitat destruction is conversion of forests and grasslands to farm land (fig. 5.24). Over the past 10,000 years, humans have transformed billions of hectares of former forests and grasslands to croplands, cities, roads, and other uses. These human-dominated spaces aren't devoid of wild organisms, but they generally favor weedy species adapted to coexist with us.

Today, forests cover less than half the area they once did, and only around one-fifth of the original forest retains its old-growth characteristics. Species that depend on the varied structure

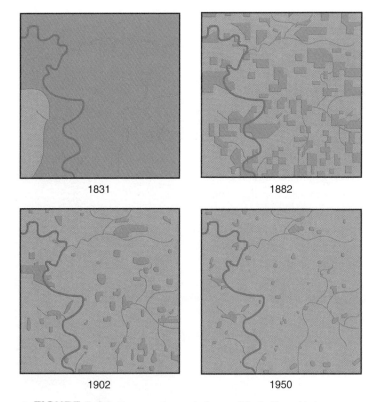

1831   1882

1902   1950

▲ **FIGURE 5.24** Decrease in wooded area of Cadiz Township in southern Wisconsin during European settlement. Green areas represent the amount of land in forest each year.

and resources of old-growth forest, such as the northern spotted owl (*Strix occidentalis caurina*), vanish as their habitat disappears (chapter 6). Grasslands currently occupy about 4 billion ha (roughly equal to the area of closed-canopy forests). Much of the most highly productive and species-rich grasslands—for example, the tallgrass prairie that once covered the U.S. corn belt—has been converted to cropland. Much more may need to be used as farmland or pasture if human populations continue to expand.

Sometimes we destroy habitat as a side effect of resource extraction, such as mining, dam-building, and indiscriminate fishing methods. Surface mining, for example, strips off the land covering along with everything growing on it. Waste from mining operations can bury valleys and poison streams with toxic material. Dam-building floods vital stream habitat under deep reservoirs and eliminates food sources and breeding habitat for some aquatic species. Our current fishing methods are highly unsustainable. One of the most destructive fishing techniques is bottom trawling, in which heavy nets are dragged across the ocean floor, scooping up every living thing and crushing the bottom structure to lifeless rubble. Marine biologist Jan Lubechenco says that trawling is "like collecting forest mushrooms with a bulldozer."

Fragmentation Reduces Habitat to Small, Isolated Areas In addition to the loss of total habitat area, the loss of large, contiguous areas is a serious problem. A general term for this is habitat **fragmentation**—the reduction of habitat into small, isolated

patches. Breaking up habitat reduces biodiversity because many species, such as bears and large cats, require large territories to subsist. Other species, such as forest interior birds, reproduce successfully only in deep forest far from edges and human settlement. Predators and invasive species often spread quickly into new regions following fragment edges.

Fragmentation also divides populations into isolated groups, making them much more vulnerable to catastrophic events, such as storms or diseases. A very small population may not have enough breeding adults to be viable even under normal circumstances. An important question in conservation biology is what is the **minimum viable population** size for a species, and when dwindling populations have grown too small to survive.

Much of our understanding of fragmentation was outlined in the theory of **island biogeography**, developed by R. H. MacArthur and E. O. Wilson in the 1960s. Noticing that small islands far from a mainland have fewer terrestrial species than larger, nearer islands, MacArthur and Wilson proposed that species diversity is a balance between colonization and extinction rates. An island far from a population source generally has a lower rate of colonization than a nearer island because it is hard for terrestrial organisms to reach. At the same time, the population of any single species is likely to be small, and therefore vulnerable to extinction, on a small island. By contrast, a large island can support more individuals of a given species and is, therefore, less vulnerable to natural disasters or genetic problems.

Large islands also tend to have more variation in habitat types than small islands do, and this also contributes to higher species counts in large islands.

The effect of island size has been observed in many places. Cuba, for instance, is 100 times as large and has about 10 times as many amphibian species as its Caribbean neighbor Montserrat. Similarly, in a study of bird species on the California Channel Islands, Jared Diamond observed that on islands with fewer than 10 breeding pairs, 39 percent of the populations went extinct over an 80-year period, whereas only 10 percent of populations with 10 to 100 pairs went extinct in the same time (fig. 5.25). Only one species numbering between 100 and 1,000 pairs went extinct, and no species with over 1,000 pairs disappeared over this time.

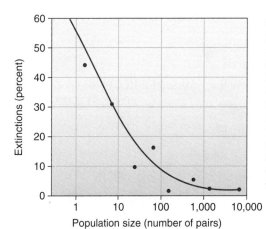

◀ **FIGURE 5.25** Extinction rates of bird species on the California Channel Islands as a function of population size over 80 years.
SOURCE: Data from H. L. Jones and J. Diamond, "Short-term-base Studies of Turnover in Breeding Bird Populations on the California Coast Island," in *Condor*, vol. 78:526–549, 1976.

What is biodiversity worth?

Often we consider biodiversity conservation a luxury: it's nice if you can afford it, but most of us need to make a living. We find ourselves weighing the pragmatic economic value of resources against ethical or aesthetic value of ecosystems. **Is conservation necessarily contradictory to good economic sense?** This question can only be answered if we can calculate the value of ecosystems and biodiversity. For example, how does the value of a standing forest compare to the value of logs taken from the forest? Assigning value to ecosystems has always been hard. We take countless ecosystem services for granted: water purification, prevention of flooding and erosion, soil formation, waste disposal, nutrient cycling, climate regulation, crop pollination, food production, and more. We depend on these services, but because nobody sells them directly, it's harder to name a price for these services than for a truckload of timber.

In 2009-2010, a series of studies called The Economics of Ecosystems and Biodiversity (TEEB) compiled available research findings on valuing ecosystem services. TEEB reports found that **the value of ecological services is more than double the total world GNP**, or at least $33 trillion per year.

The graphs below show values for two sample ecosystems: tropical forests and coral reefs. These graphs show average values among studies, because values vary widely by region. For details, find the TEEB reports at www.teebweb.org/.

Note that these graphs have different scales.

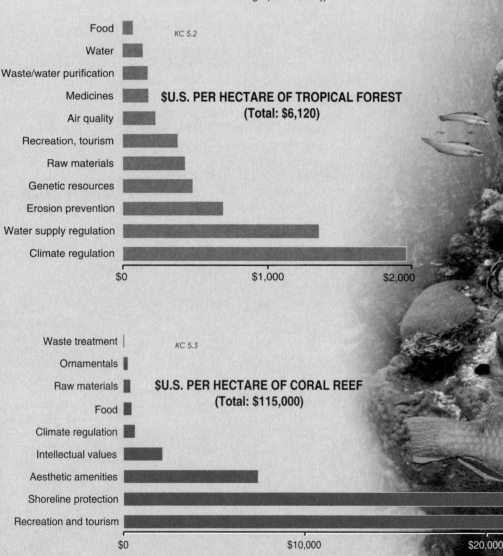

KC 5.1

$U.S. PER HECTARE OF TROPICAL FOREST
(Total: $6,120)

KC 5.2

Food
Water
Waste/water purification
Medicines
Air quality
Recreation, tourism
Raw materials
Genetic resources
Erosion prevention
Water supply regulation
Climate regulation

$0 $1,000 $2,000

$U.S. PER HECTARE OF CORAL REEF
(Total: $115,000)

KC 5.3

Waste treatment
Ornamentals
Raw materials
Food
Climate regulation
Intellectual values
Aesthetic amenities
Shoreline protection
Recreation and tourism

$0 $10,000 $20,000 $30,000

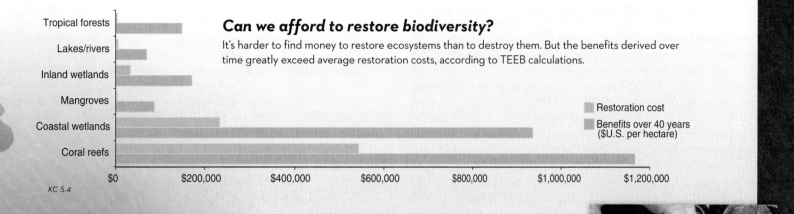

Can we afford to restore biodiversity?

It's harder to find money to restore ecosystems than to destroy them. But the benefits derived over time greatly exceed average restoration costs, according to TEEB calculations.

Tropical forests
Lakes/rivers
Inland wetlands
Mangroves
Coastal wetlands
Coral reefs

■ Restoration cost
■ Benefits over 40 years ($U.S. per hectare)

$0 $200,000 $400,000 $600,000 $800,000 $1,000,000 $1,200,000

KC 5.4

Foods and wood products These are easy to imagine but much lower in value than erosion prevention, climate controls, and water supplies provided by forested ecosystems. Still, we depend on biodiversity for foods. By one estimate, Indonesia produces 250 different edible fruits. All but 43, including this mangosteen, are little known outside the region.

KC 5.5

KC 5.6

Pollination Most of the world is completely dependent on wild insects to pollinate crops. Natural ecosystems support populations year-round, so they are available when we need them.

SOME NATURAL MEDICINE PRODUCTS

KC 5.8

Product	Source	Use
Penicillin	Fungus	Antibiotic
Bacitracin	Bacterium	Antibiotic
Tetracycline	Bacterium	Antibiotic
Erythromycin	Bacterium	Antibiotic
Digitalis	Foxglove	Heart stimulant
Quinine	Chincona bank	Malaria treatment
Diosgenin	Mexican yam	Birth control drug
Cortisone	Mexican yam	Anti-inflammation treatment
Cytarabine	Sponge	Leukemia cure
Vinblastine, vincristine	Periwinkle plant	Anticancer drugs
Reserpine	Rauwolfia	Hypertension drugs
Bee venom	Bee	Arthritis relief
Allantoin	Blowfly larva	Wound healer
Morphine	Poppy	Analgesic

Medicines More than half of all prescriptions contain some natural products. The United Nations Development Programme estimates the value of pharmaceutical products derived from developing world plants, animals, and microbes to be more than $30 billion per year.

Climate and water supplies These may be the most valuable aspects of forests. Effects of these services impact areas far beyond forests themselves.

KC 5.7

Fish nurseries As discussed in chapter 1, the biodiversity of reefs and mangroves is necessary for reproduction of the fisheries on which hundreds of millions of people depend. Marine fisheries, including most farmed fish, depend entirely on wild food sources. These fish are worth a great deal as food, but they are worth far more for their recreation and tourism value.

CAN YOU EXPLAIN?

1. Do the relative costs and benefits justify restoring a coral reef? A tropical forest?

2. Identify the primary economic benefits of tropical forest and reef systems. Can you explain how each works?

$40,000 $50,000 $60,000 $70,000 $80,000

The island idea has informed our understanding of parks and wildlife refuges, which are effectively islands of habitat surrounded by oceans of inhospitable territory. Like small, remote islands, they may be too isolated to be reached by new migrants, and they can't support large enough populations to survive catastrophic events or genetic problems. Often they are also too small for species that require large territories. Tigers and wolves, for example, need large expanses of contiguous range relatively free of human incursion to survive. Glacier National Park in Montana, for example, is excellent habitat for grizzly bears. It can support only about 100 bears, however, and if there isn't migration at least occasionally from other areas, this probably isn't a large enough population to survive in the long run.

Invasive species are a growing threat

A major threat to native biodiversity in many places is from accidentally or deliberately introduced species. Called a variety of names—alien, exotic, non-native, nonindigenous, pests—**invasive species** are organisms that thrive in new territory where they are free of predators, diseases, or resource limitations that may have controlled their population in their native habitat. Note that not all exotic species expand after introduction: for example, peacocks have been introduced to many city zoos, but they tend not to

escape and survive in the wild in most places. At the same time, not all invasive species are foreign. But most of the uncontrollable invasive species are introduced from elsewhere.

Humans have always transported organisms into new habitats, but the rate of movement has risen sharply in recent years with the huge increase in speed and volume of travel by air, water, and land. Some species are deliberately released because people believe they will be aesthetically pleasing or economically beneficial. Some of the worst are pets released to the wild when owners tire of them. Many hitch a ride in ship ballast water, in the wood of packing crates, inside suitcases or shipping containers, or in the soil of potted plants (fig. 5.26). Sometimes we introduce invasive species into new habitats thinking that we're being kind and compassionate without being aware of the ecological consequences (see Exploring Science, p. 117).

Over the past 300 years, approximately 50,000 non-native species have become established in the United States. Many of these introductions, such as corn, wheat, rice, soybeans, cattle, poultry, and honeybees, were intentional and mostly beneficial. At least 4,500 of these species have established wild populations, of which 15 percent cause environmental or economic damage (fig. 5.27). Invasive species are estimated to cost the United States $138 billion annually and are forever changing a variety of ecosystems (see What Can You Do?, p. 119).

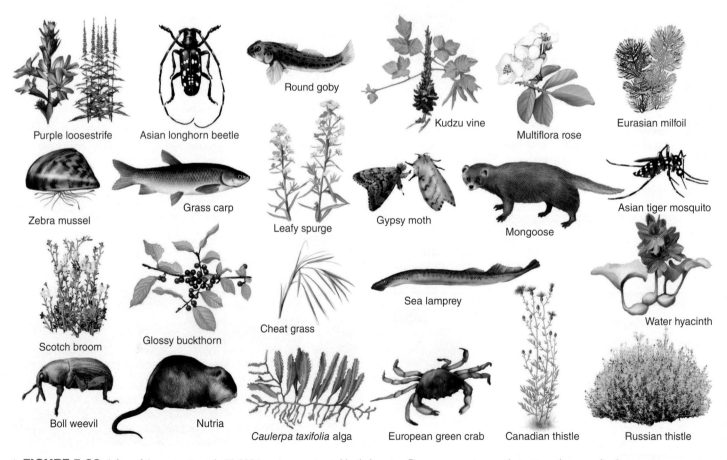

Round goby

Purple loosestrife   Asian longhorn beetle   Kudzu vine   Multiflora rose   Eurasian milfoil

Zebra mussel   Grass carp   Leafy spurge   Gypsy moth   Mongoose   Asian tiger mosquito

Sea lamprey   Water hyacinth

Scotch broom   Glossy buckthorn   Cheat grass

Boll weevil   Nutria   *Caulerpa taxifolia* alga   European green crab   Canadian thistle   Russian thistle

▲ **FIGURE 5.26** A few of the approximately 50,000 invasive species in North America. Do you recognize any that occur where you live? What others can you think of?

Thirty years ago, visitors to the boreal forests of northern Minnesota, Wisconsin, and Michigan noticed that some areas of the forest floor looked strangely denuded of leaf litter and were missing many familiar flower species. Ecologists suspected that exotic worm species might be responsible, especially because these impoverished areas seemed to be around boat landings and along shorelines where anglers discard unwanted bait. People often think they are being benevolent when they release unwanted worms. They don't realize the ecological effects they may be unleashing.

Northern forests in North America normally lack earthworms because they were removed thousands of years ago when glaciers bulldozed across the landscape. Vegetation has returned since the glaciers retreated, but worms never made it back to these forests. Over the past 10,000 years, or so, local flora and fauna have adapted to the absence of earthworms. For successful growth, seedlings depend on a thick layer of leaf mulch along with associations with fungi and invertebrates that live in the upper soil horizons. Earthworms eat up the litter layer and disrupt nutrient cycling, soil organism populations, and other aspects of the forest floor community.

But how do we analyze these changes and show they're really the work of worms? Perhaps pathogens, nutrient deficiencies, excess herbivory, or some other stressor caused this degradation. A direct experimental approach would be to set up two identical forest patches, something like the opening case study for this chapter, with the same flora, fauna, climate, and soils but with worms in one and no worms in the other. But this is difficult and expensive to do in a biome as complex as the boreal forest. Another approach is an observational study (sometimes called a natural experiment). You look for statistical correlations between worm numbers and depletion of the biotic community across a large number of similar sites.

The latter approach was taken by Andy Holdsworth, a graduate student in conservation biology at the University of Minnesota, who studied earthworm invasion in the Chequamegon and

▲ No worm invasion.

▲ Heavy worm invasion.

Chippewa National Forests in Wisconsin and Minnesota. He chose 20 areas in each forest with similar forest type, soils, and management history. Each area bordered on lakes and had no logging activity in the last 40 years. On transects across each area, all vascular plants were identified and recorded. Earthworm populations were sampled both by hand-sifting dirt samples and by pouring mustard extraction solutions on plots to drive worms out. Soil samples were taken for pH, texture, and density analysis.

Holdsworth found a mixture of European worm species in most of his sites, reflecting the diversity used as fishing bait. By plotting worm biomass against plant diversity, he showed that worm infestation rates correlated with decreased plant species richness and abundance. Among the species most likely to be missing in worm-invaded plots were wild sarsaparilla (*Aralia nudicaulis*), big-leaved aster (*Aster macrophyllus*), rose twisted stalk (*Streptopus roseus*), hairy Solomon's seal (*Polygonatum pubescens*), and princess pine (*Lycopodium obscurum*).

Perhaps most worrisome were low numbers of some tree species, especially sugar maple (*Acer saccharum*) and basswood (*Tilia americana*), which are among the defining species of these forests. Adult trees don't seem to be adversely affected by the presence of exotic worms, but their seedlings require deep leaf litter to germinate, litter that is consumed by earthworms when infestations are high.

This study and others like it suggest that invasions of the lowly worm may lead eventually to dramatic changes in the composition and structure of whole forests. Often we introduce exotic species into new biomes either without much thought or with the mistaken idea that we're doing a good thing. This example shows how careful scientific study can determine causes and effects of our actions that aren't immediately obvious to the casual observer. Can you think of examples of similar unintended consequences on biological communities in your neighborhood?

For more information, see A. Holdsworth, L. Frelich, and P. Reich. 2007. *Conservation Biology* 21(4): 997–1008.

▲ FIGURE 5.27 Invasive leafy spurge (*Euphorbia esula*) blankets a formerly diverse pasture. Introduced accidentally, and inedible for most herbivores, this plant costs hundreds of millions of dollars each year in lost grazing value and in weed control.

A few important examples of invasive species include the following:

- Eurasian milfoil (*Myriophyllum spicatum* L.) is an exotic aquatic plant native to Europe, Asia, and Africa. Scientists believe that milfoil arrived in North America during the late nineteenth century in shipping ballast. It grows rapidly and tends to form a dense canopy on the water surface, which displaces native vegetation, inhibits water flow, and obstructs boating, swimming, and fishing. Humans spread the plant between water body systems from boats and boat trailers carrying the plant fragments. Herbicides and mechanical harvesting are effective in milfoil control but can be expensive (up to $5,000 per hectare per year). There is also concern that the methods may harm nontarget organisms. A native milfoil weevil, *Euhrychiopsis lecontei*, is being studied as an agent for milfoil biocontrol.

- Water hyacinth (*Eichhornia crassipes*) is a free-floating aquatic plant that has thick, waxy, dark green leaves with bulbous, spongy stalks. It grows a tall spike of lovely blue or purple flowers. This South American native was introduced into the United States in the 1880s. Its growth rate is among the highest of any plant known: hyacinth populations can double in as little as 12 days. Many lakes and ponds are covered from shore to shore with up to 500 tons of hyacinths per hectare. Besides blocking boat traffic and preventing swimming and fishing, water hyacinth infestations also prevent sunlight and oxygen from getting into the water. Thus water hyacinth infestations reduce fisheries, shade-out submersed plants, crowd-out immersed plants, and diminish biological diversity. Water hyacinth is controlled with herbicides, machines, and biocontrol insects.

- The emerald ash borer (*Agrilus planipennis*) is an invasive wood-boring beetle from Siberia and northern China. It was first identified in North America in the summer of 2002 in southeast Michigan and in Windsor, Ontario. It's believed to have been introduced into North America in shipping pallets and wooden containers from Asia. In just eight years the beetle spread into 13 states from West Virginia to Minnesota. Adult emerald ash borers have golden or reddish-green bodies with dark metallic emerald green wing covers. More than 40 million ash trees have died or are dying from emerald ash borer attack in the United States, and more than 7.5 billion trees are at risk.

- In the 1970s several carp species, including bighead carp (*Hypophthalmichthys nobilis*), grass carp (*Ctenopharyngodonidella*), and silver carp (*Hypophthalmichthys molitrix*), were imported from China to control algae in aquaculture ponds. Unfortunately they escaped from captivity and have become established—often in very dense populations—throughout the Mississippi River Basin. Silver carp can grow to 100 pounds (45 kg). They are notorious for being easily frightened by boats and personal watercraft, which causes them to leap as much as 8–10 feet (2.5–3 m) into the air. Getting hit in the face by a large carp when you're traveling at high speed in a boat can be life threatening. Large amounts of money have been spent trying to prevent Asian carp from spreading into the Great Lakes, but carp DNA has already been detected in every Great Lake except Lake Superior.

- Zebra mussels (*Dreissena polymorpha*) probably made their way from their home in the Caspian Sea to the Great Lakes in ballast water of transatlantic cargo ships, arriving sometime around 1985. Attaching themselves to any solid surface, zebra mussels reach enormous densities—up to 70,000 animals per square meter—covering fish spawning beds, smothering native mollusks, and clogging utility intake pipes. Found in all the Great Lakes, zebra mussels have moved into the Mississippi River and its tributaries. Public and private costs for zebra mussel removal now amount to some $400 million per year. On the good side, mussels have improved water clarity in Lake Erie at least fourfold by filtering out algae and particulates.

Disease organisms, or pathogens, may also be considered predators. When a disease is introduced into a new environment, an epidemic may sweep through the area.

The American chestnut (*Castanea dentata*) was once the heart of many eastern hardwood forests. In the Appalachian Mountains, at least one of every four trees was a chestnut. Often over 45 m (150 ft) tall, 3 m (10 ft) in diameter, fast growing, and able to sprout quickly from a cut stump, it was a forester's dream. Its nutritious nuts were important for birds (such as the passenger pigeon), forest mammals, and humans. The wood was straight-grained, light, and rot-resistant, and it was used for everything from fence posts to fine furniture. In 1904 a shipment of nursery trees from China brought a fungal blight to the United States, and within 40 years the American chestnut had all but disappeared from its native range. Efforts are now under way to transfer

What Can YOU DO?

You Can Help Preserve Biodiversity

Our individual actions are some of the most important obstacles—and most important opportunities—in conserving biodiversity.

Pets and Plants

- Help control invasive species. Never release fish or vegetation from fish tanks into waterways or sewers. Pet birds, cats, dogs, snakes, lizards, and other animals, released by well-meaning owners, are widespread invasive predators.

- Keep your cat indoors. House cats are major predators of woodland birds and other animals.

- Plant native species in your garden. Exotic nursery plants often spread from gardens, compete with native species, and introduce parasites, insects, or diseases that threaten ecosystems. Local-origin species are an excellent, and educational, alternative.

- Don't buy exotic birds, fish, turtles, reptiles, or other pets. These animals are often captured, unsustainably, in the wild. The exotic pet trade harms ecosystems and animals.

- Don't buy rare or exotic house plants. Rare orchids, cacti, and other plants are often collected and sold illegally and unsustainably.

Food and Products

- When buying seafood, inquire about the source. Try to buy species from stable populations. Farm-raised catfish, tilapia, trout, Pacific pollack, Pacific salmon, mahimahi, squid, crabs, and crayfish are some of the stable or managed species that are good to buy. Avoid slow-growing top predators, such as swordfish, marlin, bluefin tuna, and albacore tuna.

- Buy shade-grown coffee and chocolate. These are also organic and often "fair trade" varieties that support workers' families as well as biodiversity in growing regions.

- Buy sustainably harvested wood products. Your local stores will start carrying sustainable wood products if you and your friends ask for them. Persistent consumers are amazingly effective forces of change!

blight-resistant genes into the few remaining American chestnuts that weren't reached by the fungus or to find biological controls for the fungus that causes the disease.

The flow of organisms happens everywhere. The Leidy's comb jelly (*Mnemiopsis leidyi*), native to North American coastal areas, has devastated the Black Sea, where it now makes up more than 90 percent of all biomass at certain times of the year. Similarly, the bristle worm from North America has invaded the coast of Poland and now is almost the only thing living on the bottom

of some of Poland's bays and lagoons. A tropical seaweed named *Caulerpa taxifolia*, originally grown for the aquarium trade, has escaped into the northern Mediterranean, where it covers the shallow seafloor with a dense, meter-deep shag carpet from Spain to Croatia. Producing more than 5,000 leafy fronds per square meter, this aggressive weed crowds out everything in its path. This type of algae grows low and sparse in its native habitat, but aquarium growers transformed it into a robust competitor that has transformed much of the Mediterranean.

Pollution poses many types of risk

We have long known that toxic pollutants can have disastrous effects on local populations of organisms. The links between pesticides and the declines of fish-eating birds were well documented in the 1970s (fig. 5.28). Population declines are especially likely in species high in the food chain, such as marine mammals, alligators, fish, and fish-eating birds. Mysterious, widespread deaths of thousands of Arctic seals are thought to be linked to an accumulation of persistent chlorinated hydrocarbons, such as DDT, PCBs, and dioxins, in the food chain. These chemicals accumulate in fat and cause weakened immune systems. Mortality of Pacific sea lions, beluga whales in the St. Lawrence estuary, and striped dolphins in the Mediterranean is similarly thought to be caused by accumulation of toxic pollutants.

▲ **FIGURE 5.28** Bald eagles, and other bird species at the top of the food chain, were decimated by DDT in the 1960s. Many such species have recovered since DDT was banned in the United States, and because of protection under the Endangered Species Act.

Lead poisoning is another major cause of mortality for many species of wildlife. Bottom-feeding waterfowl, such as ducks, swans, and cranes, ingest spent shotgun pellets that fall into lakes and marshes. They store the pellets, instead of stones, in their gizzards and the lead slowly accumulates in their blood and other tissues. The U.S. Fish and Wildlife Service (USFWS) estimates that 3,000 metric tons of lead shot are deposited annually in wetlands and that between 2 and 3 million waterfowl die each year from lead poisoning (fig. 5.29).

Population growth consumes space, resources

Even if per capita consumption patterns remain constant, more people will require more timber harvesting, fishing, farmland, and extraction of fossil fuels and minerals. In the past 40 years, the global population has doubled from about 3.5 billion to about 7 billion. In that time, according to calculations of the Worldwide Fund for Nature (WWF), our consumption of global resources has grown from 60 percent of what the earth can support over the long term to 150 percent. At the same time, global wildlife populations have declined by more than a third because of expanding agriculture, urbanization, and other human activities.

The human population growth curve is leveling off (chapter 4), but it remains unclear whether we can reduce global inequality and provide a tolerable life for all humans while also preserving healthy natural ecosystems and a high level of biodiversity.

▲ **FIGURE 5.30** A pair of stuffed passenger pigeons (*Ectopistes migratorius*). The last member of this species died in the Cincinnati Zoo in 1914.   Courtesy of Bell Museum, University of Minnesota.

Overharvesting depletes or eliminates species

Overharvesting involves taking more individuals than reproduction can replace. A classic example is the extermination of the American passenger pigeon (*Ectopistes migratorius*). Even though it inhabited only eastern North America, 200 years ago this was the world's most abundant bird, with a population of 3 to 5 billion animals (fig. 5.30). It once accounted for about one-quarter of all birds in North America. In 1830 John James Audubon saw a single flock of birds estimated to be ten miles wide and hundreds of miles long, and thought to contain perhaps a billion birds. In spite of this vast abundance, market hunting and habitat destruction caused the entire population to crash in only about 20 years between 1870 and 1890. The last known wild bird was shot in 1900 and the last existing passenger pigeon, a female named Martha, died in 1914 in the Cincinnati Zoo.

At about the same time that passenger pigeons were being extirpated, the American bison, or buffalo (*Bison bison*), was being hunted to near extinction on the Great Plains. In 1850 some 60 million bison roamed the western plains. Many were killed only for their hides or tongues, leaving millions of carcasses to rot. Much of the bison's destruction was carried out by the U.S. Army to eliminate them as a source of food, clothing, and shelter for Indians, thereby forcing Indians onto reservations. After 40 years, there were only about 150 wild bison left and another 250 in captivity.

Fish stocks have been seriously depleted by overharvesting in many parts of the world. A huge increase in fishing fleet size and efficiency in recent years has led to a crash of many oceanic

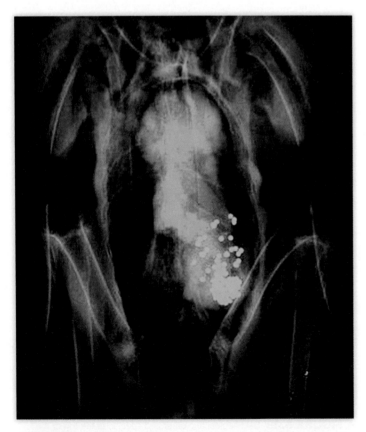

▲ **FIGURE 5.29** Lead shot shown in the stomach of a bald eagle, consumed along with its prey. Fishing weights and shot remain a major cause of lead poisoning in aquatic and fish-eating birds.

populations. Worldwide, 13 of 17 principal fishing zones are now reported to be commercially exhausted or in steep decline. At least three-quarters of all commercial oceanic species are overharvested. Canadian fisheries biologists estimate that only 10 percent of the top predators, such as swordfish, marlin, tuna, and shark, remain in the Atlantic Ocean. Groundfish, such as cod, flounder, halibut, and hake, also are severely depleted. You can avoid adding to this over-harvest by eating only abundant, sustainably harvested varieties.

Perhaps the most destructive example of harvesting terrestrial wild animal species today is the African bushmeat trade. Wild-life biologists estimate that 1 million tons of bushmeat, including antelope, elephants, primates, and other animals, are sold in African markets every year. Thousands of tons more are probably sold illegally in New York, Paris, and other global cities each year. For many poor Africans, this is the only source of animal protein in their diet. If we hope to protect the animals targeted by bushmeat hunters, we will need to help them find alternative livelihoods and replacement sources of high-quality protein. Bushmeat can endanger those who eat it, too. Outbreaks of ebola and other diseases are thought to result from capturing and eating wild monkeys and other primates. The emergence of the respiratory disease known as SARS in 2003 (chapter 8) resulted from the wild food trade in China and Southeast Asia, where millions of civets, monkeys, snakes, turtles, and other animals are consumed each year as luxury foods.

Collectors Serve Medicinal and Pet Trades   In addition to harvesting wild species for food, we also obtain a variety of valuable commercial products from nature. Much of this represents sustainable harvest, but some forms of commercial exploitation are highly destructive and a serious threat to certain rare species (fig. 5.31). Despite international bans on trade in products from endangered species, smuggling of furs, hides, horns, live specimens, and folk medicines amounts to millions of dollars each year.

Developing countries in Asia, Africa, and Latin America with the richest biodiversity in the world are the main sources of wild animals and animal products, while Europe, North America, and some of the wealthy Asian countries are the principal importers. Japan and China (including Taiwan and Hong Kong) buy three-quarters of all cat and snake skins, for instance, while European countries buy most live wild birds, such as South American parrots. The United States imports 99 percent of all live cacti and 75 percent of all orchids sold each year.

The profits to be made in wildlife smuggling are enormous. Tiger or leopard fur coats can bring $100,000 in Japan or Europe. The population of African black rhinos dropped from approximately 100,000 in the 1960s to about 3,000 in the 1980s because of a demand for their horns. In Asia, where it is prized for its supposed medicinal properties, powdered rhino horn fetches $28,000 per kilogram. The entire species is classified as critically endangered, and the western population is officially extinct.

Plants also are threatened by overharvesting. Wild ginseng has been nearly eliminated in many areas because of the Asian demand for the roots, which are used as an aphrodisiac and folk medicine. Cactus "rustlers" steal cacti by the ton from the American Southwest and Mexico. With prices as high as $1,000 for rare specimens, it's not surprising that many are now endangered.

▲ **FIGURE 5.31** Parts from rare and endangered species for sale on the street in China. Use of animal products in traditional medicine and prestige diets is a major threat to many species.

The trade in wild species for pets is a vast global business. Worldwide, some 5 million live birds are sold each year for pets, mostly in Europe and North America. Pet traders import (often illegally) into the United States some 2 million reptiles, 1 million amphibians and mammals, 500,000 birds, and 128 million tropical fish each year. About 75 percent of all saltwater tropical aquarium fish sold come from coral reefs of the Philippines and Indonesia. Some of these wild animal harvests are sustainable; many are not.

Many of these fish are caught by divers using plastic squeeze bottles of cyanide to stun their prey (fig. 5.32). Far more fish die with this technique than are caught. Worst of all, it kills the coral animals that create the reef. A single diver can destroy all of the life on 200 m² of reef in a day. Altogether, thousands of divers destroy about 50 km² of reefs each year. Net fishing would prevent this destruction, and it could be enforced if pet owners would insist on net-caught fish. More than half the world's coral reefs are potentially threatened by human activities, with up to 80 percent at risk in the most populated areas.

▲ **FIGURE 5.32** A diver uses cyanide to stun tropical fish being caught for the aquarium trade. Many fish are killed by the method itself, while others die later during shipment. Even worse is the fact that cyanide kills the coral reef itself.

Predator and Pest Control Is Expensive But Widely Practiced   Some animal populations have been greatly reduced, or even deliberately exterminated, because they are regarded as dangerous to humans or livestock or because they compete with our use of resources. Every year, U.S. government animal control agents trap, poison, or shoot thousands of coyotes, bobcats, prairie dogs, and other species considered threats to people, domestic livestock, or crops.

This animal control effort costs about $20 million in federal and state funds each year and kills some 700,000 birds and mammals, about 100,000 of which are coyotes. Defenders of wildlife regard this program as cruel, callous, and mostly ineffective in reducing livestock losses. Protecting flocks and herds with guard dogs or herders or keeping livestock out of areas that are the home range of wild species would be a better solution, they believe. Ranchers, on the other hand, argue that without predator control western livestock ranching would be impossible.

5.7   BIODIVERSITY PROTECTION

We have gradually become aware of our damage to biological resources and of reasons for conserving them. Slowly, we are adopting national legislation and international treaties to protect these irreplaceable assets. Parks, wildlife refuges, nature preserves, zoos, and restoration programs have been established to protect nature and rebuild depleted populations. There has been encouraging progress in this area, but much remains to be done.

In this section we examine legal protections for species in the United States, but keep in mind that this is only a small part of species protection measures worldwide. Most countries now have laws protecting endangered species (though many laws remain unenforced), and dozens of international treaties aim to reduce the decline of biodiversity worldwide.

Hunting and fishing laws protect useful species

In 1874 a bill was introduced in the U.S. Congress to protect the American bison, whose numbers were falling dramatically. This initiative failed, partly because most legislators could not imagine that wildlife that was so abundant and prolific could ever be depleted by human activity. By the end of the nineteenth century, bison numbers had plunged from some 60 million to only a few hundred animals.

By the 1890s, though, most states had enacted some hunting and fishing restrictions. The general idea behind these laws was to conserve the resource for future human use rather than to preserve wildlife for its own sake. The wildlife regulations and refuges established since that time have been remarkably successful for many species. A hundred years ago, there were an estimated half a million white-tailed deer in the United States; now there are some 14 million—more in some places than the environment can support. Wild turkeys and wood ducks were nearly gone 50 years ago. By restoring habitat, planting food crops, transplanting breeding stock, building shelters or houses, protecting these birds during breeding season, and using other conservation measures, we have restored populations of these beautiful and iconic birds to several million each. Snowy egrets, which were almost wiped out by plume hunters 80 years ago, are now common again.

The Endangered Species Act protects habitat and species

The Endangered Species Act is one of our most powerful tools for protecting biodiversity and environmental quality. It not only defends rare and endangered organisms, but helps protect habitat that benefits a whole biological community and safeguards valuable ecological services.

What does the ESA do? It provides (1) criteria for identifying species at risk, (2) directions for planning for their recovery, (3) assistance to landowners to help them find ways to meet both economic needs and the needs of a rare species, and (4) enforcement of measures for protecting species and their habitat.

The act identifies three degrees of risk: **endangered species** are those considered in imminent danger of extinction; **threatened species** are likely to become endangered, at least locally, within the foreseeable future. **Vulnerable species** are naturally rare or have been locally depleted by human activities to a level that puts them at risk. Vulnerable species are often candidates for future listing as endangered species. For vertebrates, a protected subspecies or a local race or ecotype can be listed, as well as an entire species.

Currently the United States has 1,372 species on the endangered and threatened species lists and about 386 candidate species waiting to be considered. The number of listed species in different taxonomic groups reflects much more about the kinds of organisms that humans consider interesting and desirable than the actual number in each group. In the United States, invertebrates make up about three-quarters of all known species but only 9 percent of those considered worthy of protection.

Worldwide, the International Union for Conservation of Nature and Natural Resources (IUCN) lists 17,741 endangered and threatened species, including nearly one-fifth of mammals, nearly one-third amphibians, reptiles, and fish, and most of the few mosses and flowering plants that have been evaluated (see table 5.1). IUCN has no direct jurisdiction for slowing the loss of those species. Within the United States, the ESA provides mechanisms for reducing species losses.

Recovery plans aim to rebuild populations

Once a species is listed, the Fish and Wildlife Service (FWS) is given the task of preparing a recovery plan. This plan details how populations will be stabilized or rebuilt to sustainable levels. A recovery plan could include many different kinds of strategies, such as buying habitat areas, restoring habitat, reintroducing a species to its historic ranges (as with Yellowstone's gray wolves), captive breeding programs, and plans for negotiating the needs of a species and the people who live in an area.

The FWS can then help landowners prepare Habitat Conservation Plans. These plans are specific management approaches that identify steps to conserve particular pieces of critical habitat. For example, the red-cockaded woodpecker is an endangered species that preys on insects in damaged pine forests from North Carolina to Texas. Few suitable forests remain on public lands, so much of the remaining population occurs on privately owned lands that are actively managed for timber production.

International Paper and other corporations have collaborated with the FWS to devise management strategies that conserve specified amounts of damaged tree stands while harvesting other areas. These plans restrict cutting of some trees, but they also ensure that the FWS will not interfere further with management of the timber, as long as the provisions of the plan continue to protect the woodpecker. This approach has helped to stabilize populations of the red-cockaded woodpecker, and timber companies have gained goodwill and sustainable forestry certification for their products. Habitat conservation plans are not always perfect, but often they can produce mutually satisfactory solutions.

Restoration can be slow and expensive, because it tries to undo decades or centuries of damage to species and ecosystems. About half of all funding is spent on a dozen charismatic species, such as the California condor, the Florida panther, and the grizzly bear, which receive around $13 million per year. By contrast, the 137 endangered invertebrates and 532 endangered plants get less than $5 million per year altogether. This disproportionate funding results from political and emotional preferences for large, charismatic species (fig. 5.33). There are also scientifically established designations that make some species merit special attention:

- *Keystone species* are those with major effects on ecological functions and whose elimination would affect many other members of the biological community; examples are prairie dogs (*Cynomys ludovicianus*) and bison (*Bison bison*).

- *Indicator species* are those tied to specific biotic communities or successional stages or environmental conditions. They can be reliably found under certain conditions but not others; an example is brook trout (*Salvelinus fontinalis*).

- *Umbrella species* require large blocks of relatively undisturbed habitat to maintain viable populations. Saving this habitat also benefits other species. Examples of umbrella species are the northern spotted owl (*Strix occidentalis caurina*), tiger (*Panthera tigris*), and gray wolf (*Canis lupus*).

- *Flagship species* are especially interesting or attractive organisms to which people react emotionally. These species can motivate the public to preserve biodiversity and contribute to conservation; an example is the giant panda (*Ailuropoda melanoleuca*).

Landowner collaboration is key

Two-thirds of listed species occur on privately owned lands, so cooperation between federal, state, and local agencies and private and tribal landowners is critical for progress. Often the ESA is controversial because protecting a species is legally enforceable, and that protection can require that landowners change their plans for their property. On the other hand, many landowners and communities appreciate the value of biodiversity on their land and like the idea of preserving species for their grandchildren to see. Others, like International Paper, have decided they can afford to allow some dying trees for woodpeckers, and they benefit from goodwill generated by preserving biodiversity.

A number of provisions protect landowners, and these serve as incentives for them to participate in developing habitat conservation

▲ **FIGURE 5.33** The Endangered Species Act seeks to restore populations of species, such as the bighorn sheep, which has been listed as endangered in much of its range. Charismatic species are easier to get listed than obscure ones.

plans. For example, permits can be issued to protect landowners from liability if a listed species is accidentally harmed during normal land-use activities. In a Candidate Conservation Agreement, the FWS helps landowners reduce threats to a species in an effort to avoid listing it at all. A Safe Harbor Agreement is a promise that, if landowners voluntarily implement conservation measures, the FWS will not require additional actions that could limit future management options. For example, suppose a landowner's efforts to improve red-cockaded woodpecker habitat lead to population increases. A Safe Harbor agreement ensures that the landowners would not be required to do further management for additional woodpeckers attracted to the improved habitat.

The ESA has seen successes and controversies

The ESA has held off the extinction of hundreds of species. Some have recovered and been delisted, including the brown pelican, the peregrine falcon, and the bald eagle, which was delisted in 2007. In 1967, before the ESA was passed, only about 800 bald eagles remained in the contiguous United States. DDT poisoning, which prevented the hatching of young eagles, was the main cause. By 1994, after the banning of DDT, the population rebounded to 8,000 birds; by 2007 the population was up to about 20,000, enough to ensure a stable breeding population. Similarly, peregrine falcons, which had been down to 39 breeding pairs in the 1970s, had rebounded to 1,650 pairs by 1999 and were taken off the list. The American alligator was listed as endangered in 1967 because hunting and habitat destruction had reduced populations to precarious levels. Protection has been so effective that the species is now plentiful throughout its entire southern range. Florida alone may have a population of 1 million or more.

Many people are dissatisfied with the slow pace of listing new species, however. Hundreds of species are classified as "warranted but precluded," or deserving of protection but lacking funding or local support. At least 18 species have gone extinct since being nominated for protection.

Part of the reason listing is slow is that political and legal debates can drag on for years. Political opposition is especially fierce when large profits are at stake. An important test of the ESA occurred in 1978 in Tennessee, when construction of the Tellico Dam threatened a tiny fish called the snail darter. The powerful Tennessee Valley Authority, which was building the dam, argued to Supreme Court that the dam was more important than the fish. After this case a new federal committee was given power to override the ESA for economic reasons. (This committee subsequently became known as the God Squad because of its power over the life and death of a species.)

Another important debate over the economics of endangered species protection has been that of the northern spotted owl (chapter 6). Preserving this owl requires the conservation of expansive, undisturbed areas of old-growth temperate rainforest in the Pacific Northwest, where old-growth timber is extremely valuable and increasingly scarce (fig. 5.34). Timber industry economists calculated the cost of conserving a population of 1,600 to 2,400 owls at $33 billion. Ecologists countered that this number was highly inflated; moreover, forest conservation would preserve countless other species and ecosystem services, whose values are almost impossible to calculate.

Sometimes the value of conserving a species is easier to calculate. Salmon and steelhead in the Columbia River are endangered by hydropower dams and water storage reservoirs that block their migration to the sea. Opening the floodgates could allow young fish to run downriver and adults to return to spawning grounds, but at high costs to electricity consumers, barge traffic, and farmers who depend on cheap water and electricity. On the other hand, commercial and sport fishing for salmon is worth over $1 billion per year and employs about 60,000 people directly or indirectly.

Many countries have species protection laws

In the past 25 years or so, many countries have recognized the importance of legal protection for endangered species. Rules for listing and protecting endangered species are established by Canada's Committee on the Status of Endangered Wildlife in Canada (COSEWIC) of 1977, the European Union's Birds Directive (1979) and Habitat Directive (1991), and Australia's Endangered Species Protection Act (1992). International agreements have also been developed, including the Convention on Biological Diversity (1992).

The Convention on International Trade in Endangered Species (CITES) of 1975 provides a critical conservation strategy by blocking the international sale of wildlife and their parts. The Convention makes it illegal to export or import elephant ivory, rhino horns, tiger skins, or live endangered birds, lizards, fish, and orchids. CITES enforcement has been far from perfect: smugglers hide live animals in their clothing and luggage; the volume of international shipping makes it impossible to inspect transport containers and ships; documents may be falsified. The high price of these products in North America and Europe, and increasingly in wealthy cities of China, makes the risk of smuggling worthwhile: a single rare parrot can be worth tens of thousands of dollars,

▲ **FIGURE 5.34** Endangered species often serve as a barometer for the health of an entire ecosystem and as surrogate protector for a myriad of less well-known creatures. A 1990 Herblock Cartoon, "DAMN SPOTTED OWL" copyright by *The Herb Block Foundation*.

even though its sale is illegal. Even so, CITES provides a legal structure for restricting this trade, and it also raises public awareness of the real costs of the trade in endangered species.

A striking example of success in endangered species under CITES is the recovery of the southern white rhino in Africa. A century ago these huge herbivores were near extinction, but now there are at least 17,500 white rhinos in parks and game preserves.

Habitat protection may be better than species protection

Growing numbers of scientists, land managers, policy makers, and developers are arguing that we need a rational, continent-wide preservation of ecosystems that supports maximum biological diversity. They argue that this would be more effective than species-by-species battles for desperate cases. By concentrating on individual species, we spend millions of dollars to breed plants or animals in captivity that have no natural habitat where they can be released. While flagship species, such as mountain gorillas and Indian tigers, are reproducing well in zoos and wild animal parks, the ecosystems that they formerly inhabited have largely disappeared.

A leader of this new form of conservation is J. Michael Scott, who was project leader of the California condor recovery program in the mid-1980s and had previously spent ten years working on endangered species in Hawaii. In making maps of endangered species, Scott discovered that even Hawaii, where more than 50 percent of the land is federally owned, has many vegetation types completely outside of natural preserves (fig. 5.35). The gaps between protected areas may contain more endangered species than are preserved within them.

This observation has led to an approach called **gap analysis**, in which conservationists and wildlife managers look for unprotected landscapes, or gaps in the network of protected lands, that are rich in species. Gap analysis involves mapping protected conservation areas and high-biodiversity areas. Overlaying the two makes it easy to identify priority spots for conservation efforts. Maps also help biologists and land-use planners communicate about threats to biodiversity. This broad-scale, holistic approach seems likely to save more species than a piecemeal approach.

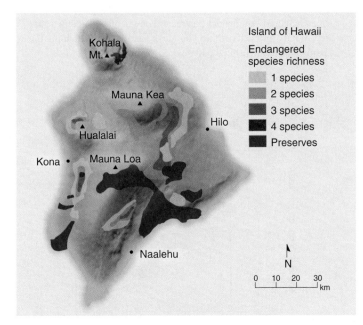

▲ **FIGURE 5.35** Protected lands (*green*) are often different from biologically diverse areas (*red shades*), as shown here on the island of Hawaii.

Conservation biologist R.E. Grumbine suggests four remanagement principles for protecting biodiversity in a large-scale, long-range approach:

1. Protect enough habitat for viable populations of all native species in a given region.

2. Manage at regional scales large enough to accommodate natural disturbances (fire, wind, climate change, etc.).

3. Plan over a period of centuries, so that species and ecosystems can continue to evolve.

4. Allow for human use and occupancy at levels that do not result in significant ecological degradation.

CONCLUSION

Biodiversity can be understood in terms of the environmental conditions where different organisms live, in terms of biomes, and in terms of habitat types. Knowing the influence of climate conditions is an important place to start describing biodiversity and to conserving biological communities. Although we know a great deal about many species (especially mammals, birds, and reptiles), we know very little about most species (especially invertebrates and plants). Of the world's species, many are threatened or endangered. We worry most about threats to mammals, birds, and a few other groups, but a far greater percentage of plants, lichens, mollusks, and other groups are threatened.

Biodiversity is important to us because it can aid ecosystem stability and because we rely on many different organisms for foods, medicines, and other products. We do not know what kinds of undiscovered species may provide medicines and foods in the future. Biodiversity also has important cultural and aesthetic benefits. Pinning a dollar value on these amenities is difficult, but efforts to do so show the wide range of values in conservation areas.

Many factors threaten biodiversity, including habitat loss, invasive species, pollution, population growth, and overharvesting. The acronym HIPPO has been used to describe these threats together as a whole. You can reduce these threats by avoiding exotic pets and unsustainable fisheries or wood products that are harvested unsustainably. Laws to protect biodiversity, including the Endangered Species Act, have been controversial. They have also protected many species, including the bald eagle and gray wolf. Often such laws protect umbrella species, whose habitat also protects many other species. Despite their imperfections, these laws are our only mechanism for preserving biodiversity for future generations.

PRACTICE QUIZ

1. Why did ecologists want to reintroduce wolves to Yellowstone Park? What goals did they have, and have their goals been achieved?

2. Describe nine major types of terrestrial biomes.

3. Explain how climate graphs (as in fig. 5.6) should be read.

4. Describe conditions under which coral reefs, mangroves, estuaries, and tide pools occur.

5. Throughout the central portion of North America is a large biome once dominated by grasses. Describe how physical conditions and other factors control this biome.

6. Explain the difference between swamps, marshes, and bogs.

7. How do elevation (on mountains) and depth (in water) affect environmental conditions and life-forms?

8. Figure 5.15 shows chlorophyll (plant growth) in oceans and on land. Explain why green, photosynthesizing organisms occur in bands at the equator and along the edges of continents. Explain the very dark green areas and yellow/orange areas on the continents.

9. Define *biodiversity* and give three types of biodiversity essential in preserving ecological systems and functions.

10. What is a biodiversity "hot spot"? List several of them (see fig. 5.22).

11. How do humans benefit from biodiversity?

12. What does the acronym HIPPO refer to?

13. Have extinctions occurred in the past? Is there anything unusual about current extinctions?

14. Why are exotic or invasive species a threat to biodiversity? Give several examples of exotic invasive species (see fig. 5.26).

15. What does the Endangered Species Act do?

CRITICAL THINKING AND DISCUSSION

Apply the principles you have learned in this chapter to discuss these questions with other students.

1. Many poor tropical countries point out that a hectare of shrimp ponds can provide 1,000 times as much annual income as the same area in an intact mangrove forest. Debate this point with a friend or classmate. What are the arguments for and against saving mangroves?

2. Genetic diversity, or diversity of genetic types, is believed to enhance stability in a population. Most agricultural crops are genetically very uniform. Why might the usual importance of genetic diversity *not* apply to food crops? Why *might* it apply?

3. Scientists need to be cautious about their theories and assumptions. What arguments could you make for *and* against the statement that humans are causing extinctions unlike any in the history of the earth?

4. A conservation organization has hired you to lead efforts to reduce the loss of biodiversity in a tropical country. Which of the following problems would you focus on first and why: habitat destruction and fragmentation, hunting and fishing activity, harvesting of wild species for commercial sale, or introduction of exotic organisms?

5. Many ecologists and resource scientists work for government agencies to study resources and resource management. Do these scientists serve the public best if they try to do pure science, or if they try to support the political positions of democratically elected representatives, who, after all, represent the positions of their constituents?

6. You are a forest ecologist living and working in a logging community. An endangered salamander has recently been discovered in your area. What arguments would you make for and against adding the salamander to the official endangered species list?

DATA ANALYSIS   Confidence Limits in the Breeding Bird Survey

A central principle of science is the recognition that all knowledge involves uncertainty. No study can observe every possible event in the universe, so there is always missing information. Scientists try to define the limits of their uncertainty, in order to allow a realistic assessment of their results. A corollary of this principle is that the more data we have, the less uncertainty we have. More data increase our confidence that our observations represent the range of possible observations.

One of the most detailed records of wildlife population trends in North America is the Breeding Bird Survey (BBS). Every June, volunteers survey more than 4,000 established 25-mile routes. The accumulated data from thousands of routes, over more than 40 years, indicate population trends, telling which populations are increasing, decreasing, or expanding into new territory. To examine a sample of BBS data, go to Connect, where you can explore the data and explain the importance of uncertainty in data.

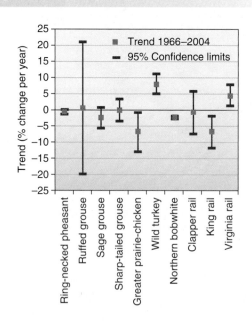

6 Environmental Conservation: Forests, Grasslands, Parks, and Nature Preserves

Approximately 1.7 billion metric tons of carbon are released annually due to land-use change, mainly from tropical deforestation. This is more than all global transportation emissions combined.

LEARNING OUTCOMES

After studying this chapter, you should be able to answer the following questions:

▶ What portion of the world's original forests remain?

▶ What activities threaten global forests? What steps can be taken to preserve them?

▶ Why is road construction a challenge to forest conservation?

▶ Where are the world's most extensive grasslands?

▶ How are the world's grasslands distributed, and what activities degrade grasslands?

▶ What are the original purposes of parks and nature preserves in North America?

▶ What are some steps to help restore natural areas?

Protecting Forests to Prevent Climate Change

In 2010 Norway signed an agreement to support Indonesia's efforts to reduce greenhouse gas emissions from deforestation and forest degradation. Based on Indonesia's performance over the next eight years, Norway will provide up to (U.S.) $1 billion to support this partnership. Indonesia has the third largest area of tropical rainforest in the world (after Brazil and the Democratic Republic of Congo), and because it's an archipelago of more than 17,000 islands, many of which have unique assemblages of plants and animals, Indonesia has some of the highest biological diversity in the world.

Indonesia is an excellent example of the benefits of forest protection. Deforestation, land-use change, and the drying, decomposition, and burning of peatlands cause about 80 percent of the country's current greenhouse gas emissions. This means that Indonesia can make deeper cuts in CO_2 emissions and do it more quickly than most other countries. Reducing deforestation will help preserve biodiversity and protect indigenous forest people. And according to government estimates, up to 80 percent of Indonesia's logging is illegal, so bringing it under control also will increase national revenue and help build civic institutions.

Indonesia recognizes that climate change is one of the greatest challenges facing the world today. In 2009, President Susilo Bambang Yudhoyono committed to reducing Indonesia's CO_2 emissions 26 percent by 2020 compared to a business-as-usual trajectory. This is the largest absolute reduction pledge made by any developing country and could exceed reductions by most industrialized countries as well.

The partnership between Norway and Indonesia is the largest example so far of a new, UN-sponsored program called REDD (Reducing Emissions from Deforestation and Degradation of forests) that aims to slow climate change by paying countries to stop cutting down their forests. One of the few positive steps agreed on at recent UN climate negotiations, REDD could result in a major transfer of financial assistance from wealthy countries to poor ones. It's estimated that it will take about (U.S.) $30 billion per year to fund this program. But it offers a chance to save one of the world's most precious ecosystems. Forests would no longer be viewed merely as timber waiting to be harvested or land awaiting clearance for agriculture.

Many problems need to be solved for the Norway/Indonesia partnership to work. For one thing, it will be necessary to calculate how much carbon is stored in a particular forest as well as

▲ **FIGURE 6.1** Logging valuable hardwoods is generally the first step in tropical forest destruction. Although loggers may take only one or two large trees per hectare, the damage caused by extracting logs exposes the forest to invasive species, poachers, and fires.

how much carbon could be saved by halting or slowing deforestation. Historical forest data, on which these predictions often are based, is often unreliable or nonexistent in tropical countries. Satellite imaging and computer modeling can give answers to these questions, but the technology can be expensive. In the first phase of funding, Norway will support political and institutional reform along with building infrastructure and increasing capacity.

Like other donor nations, Norway is also concerned about how permanent the protections will be. What happens if they pay to protect a forest but a future administration decides to log it? Furthermore, loggers are notoriously mobile and adept at circumventing rules by bribing local authorities, if necessary. What's to prevent them from simply moving to new areas to cut trees? If you avoid deforestation in one place but then cut an equal number of trees somewhere else (sometimes known as "leakage"), carbon emissions won't have gone down at all. Similarly, there's concern that a reduction in logging in one country could lead to pressure on other countries to cut down their forests to meet demand. And there would be a financial incentive to do so if reductions in logging pushed up the price of timber.

Will this partnership protect indigenous people's rights? In theory, yes. Indonesia has more than 500 ethnic groups, and many forest communities lack secure land tenure. Large mining, logging, and palm oil operations often push local people off their traditional lands with little or no compensation. Indonesia has promised a two-year suspension on new projects to convert natural forests. They also have promised to recognize the rights of native people and local communities.

Could having such a sudden influx of money cause corruption? Yes, that's possible. But Indonesia has a good track record of managing foreign donor funds under President Yudhoyono. The Aceh and Nias Rehabilitation and Reconstruction Agency (BRR), established after the 2004 tsunami, managed around (U.S.) $7 billion of donations in line with the best international standards. Indonesia has promised that the same governance principles will be used to manage REDD funds.

In this chapter, we'll look at other examples of how we protect biodiversity and preserve landscapes. For Google Earth™ placemarks that will help you explore these landscapes via satellite images, visit www.mcgrawhillconnect.com. ■

6.1 WORLD FORESTS

Forests, woodlands, pastures, and rangelands together occupy almost 60 percent of global land cover (fig. 6.2). These ecosystems provide many of our essential resources, such as lumber, paper pulp, and grazing for livestock. They also provide essential ecological services, including regulating climate, controlling water runoff, providing wildlife habitat, purifying air and water, and supporting rainfall. Forests and grasslands also have scenic, cultural, and historic values that deserve protection. But these are also among the most heavily disturbed ecosystems (chapter 5).

As the opening case study for this chapter shows, balancing competing land uses and needs can be complicated. Many conservation debates have concerned protection or use of forests, prairies, and rangelands. This chapter examines the ways we use and abuse these biological communities, as well as some of the ways we can protect them and conserve their resources. We discuss forests first, followed by grasslands and then strategies for conservation, restoration, and preservation.

Boreal and tropical forests are most abundant

Forests are widely distributed, but most remaining forests are in the cold boreal ("northern") or taiga regions and the humid tropics (fig. 6.3). Assessing forest distribution is tricky, because forests vary in density and height, and many are inaccessible. The UN Food and Agriculture Organization (FAO) defines "forest" as any area where trees cover more than 10 percent of the land. This definition includes woodlands ranging from open **savannas**, whose trees occupy less than 20 percent of the area, to **closed-canopy forests**, in which tree crowns cover most of the ground. The largest tropical forest is in the Amazon River basin. The highest rates of forest loss are in Africa (fig. 6.4). Some of the world's most biologically diverse regions are undergoing rapid deforestation, including Southeast Asia and Central America (see related stories "Saving an African Eden" and "Protecting Forests to Preserve Rain" at www.mhhe.com/cunningham7e).

▲ **FIGURE 6.3** A tropical rainforest in Queensland, Australia. Primary, or old-growth forests, such as this, aren't necessarily composed entirely of huge, old trees. Instead, they have trees of many sizes and species that contribute to complex ecological cycles and relationships.

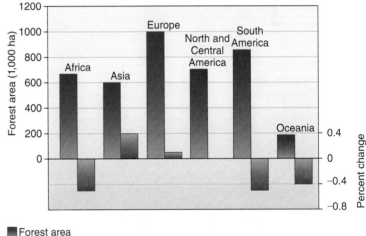

Forest area
% change

▲ **FIGURE 6.4** Forest area and annual net change 2005–2010. The largest annual net deforestation rate in the world is in Africa. Largely because China has planted 50 billion trees in the past decade, Asia has a net increase in forest area. Europe, also, is gaining forest, while North and Central America have had no net change. SOURCE: Data from FAO, 2008.

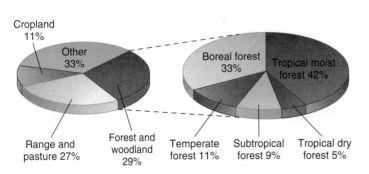

▲ **FIGURE 6.2** World land use and forest types. The "other" category includes tundra, desert, wetlands, and urban areas. SOURCE: UN Food and Agriculture Organization (FAO).

Among the forests of greatest ecological importance are the remnants of primeval forests that are home to much of the world's biodiversity, endangered species, and indigenous human cultures. Sometimes called frontier forests, **old-growth forests** are those that cover a relatively large area and have been undisturbed by human activities long enough that trees can live out a natural life cycle and ecological processes can occur in fairly normal fashion. That doesn't mean that all trees need be enormous or thousands of years old. In some forest systems, such as lodgepole pine forests of the Rocky Mountains, most trees live less than a century before being killed by disease or some natural disturbance, such as a fire. Nor does it mean that humans have never been present. Where human occupation entails relatively little impact, an old-growth forest may have been inhabited by people for a very long

Active LEARNING

Calculating Forest Area

Examine figure 6.4, which shows forest losses between 2005 and 2010. This graph shows only percentage losses of current forest area. How would you evaluate the total losses from these data? To do so, you'd need some additional information about the forest area before humans started deforestation. The following table will help you in these calculations.

Changes in Forest Area:

Region	Original forest (millions of ha)	2010 (millions of ha)
Africa	2,200	674
Asia	1,900	592
Europe	2,200	1,005
South America	1,400	864
Oceania	400	191
North & Cent. America	700	705
World	8,800	4,031

1. Which region has lost the greatest percentage of its original forest, and how much was that?
2. Which region has gained the most forest, and how much was that?
3. Europe gained forest between 2005 and 2010, but how does that compare to total forest lost since the beginning of human history?
4. How much global forest has been lost since the beginnings of human habitation?
5. Working with round numbers makes comparison easy. How important are the details that are lost when you read approximate numbers from the graph? What kinds of generalization might have gone into producing the FAO's original data? What kinds of generalization might have been unavoidable?

ANSWERS: 1. Africa lost 1,526 million ha or nearly 70 percent; 2. North America gained 5 million ha; 3. Europe has lost 55 percent of its original forest, or 1,195 million ha, as a result of deforestation; 4. The world lost about 4,769 million ha of forest, or about 54 percent of its original forest; 5. Usually approximate numbers provide a quick, useful comparison, and additional detail isn't needed until further analysis is done. Defining forests and forest types, or determining extent of either "original" or "current" forest, all involve considerable, usually unavoidable generalization.

---

time. Even forests that have been logged or converted to cropland often can revert to old-growth characteristics if allowed to undergo normal successional processes.

Even though forests still cover about half the area they once did worldwide, only one-quarter of those forests retain old-growth features. The largest remaining areas of old-growth forest are in Russia, Canada, Brazil, Indonesia, and Papua New Guinea. Together, these five countries account for more than three-quarters of all relatively undisturbed forests in the world. In general, remoteness rather than laws protect those forests. Although official data describe only about one-fifth of Russian old-growth forest as threatened, rapid deforestation—both legal and illegal—especially in the Russian Far East, probably puts a much greater area at risk.

Forests provide essential products

Wood plays a part in more activities of the modern economy than does any other commodity. There is hardly any industry that does not use wood or wood products somewhere in its manufacturing and marketing processes. Think about the amount of junk mail, newspapers, photocopies, packaging, and other paper products that each of us in developed countries handles, stores, and disposes of in a single day. Total annual world wood consumption is about 4 billion m³. This is more than steel and plastic consumption combined. International trade in wood and wood products amounts to more than $100 billion each year. Developed coun-

tries produce less than half of all industrial wood but consume about 80 percent of it. Less-developed countries, mainly in the tropics, produce more than half of all industrial wood but use only 20 percent.

Paper pulp, the fastest-growing forest product, accounts for nearly a fifth of all wood consumption. Most of the world's paper is used in the wealthier countries of North America, Europe, and Asia. Global demand for paper is increasing rapidly, however, as other countries develop. The United States, Russia, and Canada are the largest producers of both paper pulp and industrial wood (lumber and panels). Much industrial logging in Europe and North America occurs on managed plantations, rather than in untouched old-growth forest. However, paper production is increasingly blamed for deforestation in Southeast Asia, West Africa, and other regions.

Fuelwood accounts for nearly half of global wood use. At least two billion people depend on firewood or charcoal as a principal source of heating and cooking fuel (fig. 6.5). The average amount of fuelwood used in less-developed countries is about 1 m³ per person per year, roughly equal to the amount that each American consumes each year as paper products alone. Demand for fuelwood, which is increasing at slightly less than the global population growth rate, is causing severe fuelwood shortages and depleting forests in some developing areas, especially around growing cities. About 1.5 billion people have less fuelwood than they need, and many experts expect shortages to worsen as poor urban areas grow. In some countries, firewood harvesting is a major cause of deforestation, but foresters argue that biomass energy could be produced sustainably in most developing countries, with careful management.

Approximately one-quarter of the world's forests are managed for wood production. Ideally, forest management involves scientific planning for sustainable harvests, with particular attention paid to forest regeneration. In temperate regions, according to the UN Food and Agriculture Organization, more land is being replanted or allowed to regenerate naturally than is being permanently deforested. Much of this reforestation, however, is in large plantations of single-species, single-use, intensive cropping called **monoculture forestry**. Although this produces rapid growth and easier harvesting than a more diverse forest, a dense, single-species stand often supports little biodiversity and does poorly in providing the ecological services, such as soil erosion control and clean water production, that may be the greatest value of native forests (fig. 6.6).

Some of the countries with the most successful reforestation programs are in Asia. China, for instance, cut down most of its forests 1,000 years ago and has suffered centuries of erosion and terrible floods as a consequence. Recently, however, timber cutting in the headwaters of major rivers has been outlawed, and a massive reforestation project has begun. In the past 20 years, China planted some 50 billion trees, mainly in Xinjiang Province, to stop the spread of deserts.

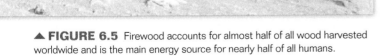

▲ **FIGURE 6.5** Firewood accounts for almost half of all wood harvested worldwide and is the main energy source for nearly half of all humans.

▲ **FIGURE 6.6** Monoculture forestry, such as this Wisconsin tree farm, produces valuable timber and pulpwood, but has little biodiversity.

Korea and Japan also have had very successful forest restoration programs. After being almost totally denuded during World War II, both countries are now about 70 percent forested.

Tropical forests are being cleared rapidly

Tropical forests are among the richest and most diverse terrestrial systems. Although they now occupy less than 10 percent of the earth's land surface, these forests are thought to contain more than two-thirds of all higher plant biomass and at least half of all the plant, animal, and microbial species in the world.

A century ago, an estimated 12.5 million km^2 (an area larger than the entire United States) of the tropics were covered with closed-canopy forest. The FAO estimates that only about 40 percent of that forest remains in its original condition, and that about 10 million ha, or about 0.6 percent, of existing tropical forests are cleared each year (fig. 6.7).

There is considerable debate about current rates of deforestation in the tropics. Some environmental groups, for example, estimate that the Amazon alone loses at least 1 million ha per year, while the government of Brazil claims the amount is only about half that much. There are different definitions of **deforestation**. Some scientists insist that it means a complete change from forest to agriculture, urban areas, or desert. Others include any area that has been logged, even if the cut was selective and regrowth will be rapid. Furthermore, savannas, open woodlands, and succession following natural disturbance are hard to distinguish from logged areas. Consequently, estimates for total tropical forest losses range from about 5 million to more than 20 million ha per year. The FAO estimates of 10 million ha deforested per year are generally the most widely accepted. To put that figure in perspective, it means that about 1 acre—or the area of a football field—is cleared every second, on average, around the clock.

In 2004, Brazil was reported to have lost 2.7 million ha (6.8 million acres) of forest to clearing and fires. This was the highest rate in the world, but Brazil also has by far the largest tropical forests. Much of this destruction was to make room for cattle ranching

1975   1989   2001

▲ **FIGURE 6.7** Forest destruction in Rondomia, Brazil, between 1975 and 2001. Construction of logging roads creates a feather-like pattern that opens forests to settlement by farmers.

(fig. 6.8). In 2011, Brazil claimed its deforestation rate had fallen to 0.6 million ha. It remains to be seen if this progress is sustainable. Indonesia appears to be the current deforestation leader. The FAO estimates that Indonesia is losing about 2 million ha, or about 2 percent of its remaining forest each year. Logging, clearing for oil palm plantations, and fires set to cover up illegal activities account for much of this destruction.

In Africa, the coastal forests of Senegal, Sierra Leone, Ghana, Madagascar, Cameroon, and Liberia already have been mostly demolished. Haiti was once 80 percent forested; today, essentially all that forest has been destroyed, and the land lies barren and eroded. India, Burma, Kampuchea (Cambodia), Thailand, and Vietnam all have little old-growth lowland forest left. In Central America, nearly two-thirds of the original moist tropical forest has been destroyed, mostly within the past 30 years and primarily due to logging and conversion of forest to cattle range. (See related story on "Disappearing Butterfly Forests" at www.mhhe.com/cunningham7e.)

Causes of Deforestation A variety of factors contribute to deforestation, and different forces predominate in various parts of the world. Logging for valuable tropical hardwoods, such as teak and mahogany, is generally the first step. Although loggers might take only one or two of the largest trees per hectare, the canopy of tropical forests is usually so strongly linked by vines and interlocking branches that felling one tree can bring down a dozen others. Building roads to remove logs kills more trees, but even more important, it allows entry to the forest by farmers, miners, hunters, and others who cause further damage.

In Africa, conversion of forest into small-scale agriculture accounts for nearly two-thirds of all tropical forest destruction. In Latin America, poor, landless farmers often start the deforestation but are bought out—or driven out—after a few years by large-scale farmers or ranchers.

Shifting cultivation (sometimes called "slash and burn" or milpa farming) is often blamed for forest destruction. But in many countries, indigenous people have discovered sustainable ways to use complex cycles of mixed polyculture and soil amendment practices to improve soil fertility. Growing non-indigenous populations and logging have destabilized these practices in many areas.

As forests are cleared, rainfall patterns can change. Computer models indicate that as forests are cleared, landscapes dry and temperatures rise. The reason is that forests take up and release moisture, absorbing heat and contributing to regional rainfall in the process. Drought and wild fire become more frequent, and it becomes harder for forests to regrow. Even the Amazon basin, one of world's the greatest rainforests, appears to be drying in the combination of forest clearing and climate change.

Forest Protection What can be done to stop this destruction and encourage tropical forest protection? Although much of the news is discouraging, there are some hopeful signs for forest conservation in the tropics. Many countries now recognize that forests are valuable resources.

About 14 percent of all world forests are in some form of conservation status, but the effectiveness of that protection varies greatly. Costa Rica has one of the best plans in the world for forest guardianship. Attempts are being made there not only to rehabilitate the land (make an area useful to humans) but also to restore the ecosystems to naturally occurring associations. One of the best-known of these projects is Dan Janzen's work in Guanacaste National Park. Like many dry, tropical forests, the northwestern part of Costa Rica had been almost completely converted to ranchland. By controlling fires, however, Janzen and his coworkers are bringing back the forest. One of the keys to this success is involving local people in the project. Janzen also advocates grazing in the park. The original forest evolved, he reasons, together with ancient grazing animals that are now extinct. Horses and cows can play a valuable role as seed dispersers.

▲ **FIGURE 6.8** Cattle ranching can increase pressure for forest destruction, but, in the proper setting, cattle also can assist forest regeneration by dispersing seeds.

Saving forests stabilizes our climate

Wealthier countries have a new incentive to care about global forest losses, as we have come to understand the importance of land conversion in climate change. Forests are a huge carbon sink, storing some 422 billion metric tons of carbon in standing biomass. Clearing and burning of forests is responsible for about 17 percent of all the carbon released by human actions every year—more than all vehicles combined—and is a major factor in global climate change (chapter 10). Moisture released from forests affects rainfall not only locally, but sometimes far away. For example, recent climate studies suggest that deforestation of the Amazon could reduce precipitation in the American Midwest.

As discussed in the opening case study, reducing emissions from deforestation and forest degradation (REDD) programs are increasingly viewed as a way to help developing countries value and conserve standing forests. Agreements to finance this idea has been among the few real successes in recent global climate conferences. The idea, first proposed by Papua New Guinea and Costa Rica at climate talks in 2005, is to recognize the ways wealthy countries depend on climate stablilization and other benefits of forests in developing areas. Part of this recognition is that wealthy countries should pay for some of those services by helping to finance forest protection in poorer countries (see Key Concepts, p. 136). It is calculated that replanting 300 million ha of degraded forest should capture about 1 billion tons of CO_2 over the next 50 years.

Part of the reason REDD has gained traction is that it protects an array of ecological functions that serve a wide variety of interest groups. More than 1.2 billion people depend on forests for their livelihoods. Governments are interested in political stability in these communities. Forests protect water resources that supply cities, as well as biodiversity that supports ecotourism. Often indigenous and tribal groups have specialized knowledge about forest systems, and REDD promotes efforts to involve these people in conservation strategies (see Exploring Science, p. 134). Supplementing their income may also allow them to avoid intensive use of forest resources and yet to remain on the land, where their traditional knowledge and stewardship are valuable resources.

Protection of indigenous rights is an important aim of REDD efforts, but these protections have always been a contentious, and often a dangerous, proposition. In 2011, Joao Claudio Ribeiro da Silva, a rubber tapper and leading forest conservationist, and his wife, Maria do Espirito Santo, were ambushed and killed in the Amazon state of Para. It isn't known who killed them, but loggers and cattle ranchers had made death threats against da Silva. The couple joined a list of others who died defending nature and human rights in the Amazon. In 2009 an American nun named Sister Dorothy Stang was shot by ranchers, and in 1988 Chico Mendez, who founded the rubber tappers union to which da Silva belonged, also was murdered by ranchers. It is hoped that greater incentives for conservation and valuing of non-timber forest services will help reduce the incidence of violence to people as well as to natural resources.

Administering a program as large as this will not be easy. The United Nations estimates that fully funding REDD will cost between $20 billion and $30 billion per year globally. This is a large number but it is just 0.0003 percent of world GDP, or somewhat less than the annual net profit of large companies such as ExxonMobil ($41 billion in 2011) or Apple ($28 billion in 2011). Careful monitoring and good governance (often lacking in developing countries) will be needed to make sure this money is spent wisely and that projects are sustainable. Nevertheless, this represents the largest experiment in tropical conservation in world history.

Temperate forests also are at risk

Tropical countries aren't unique in harvesting forests at an unsustainable rate. Northern countries, such as the United States and Canada, also have allowed controversial forest management practices in many areas. For many years the official policy of the U.S. Forest Service was "multiple use," which implied that the forests could be used for everything that we might want to do there simultaneously. Some uses are incompatible, however. Birdwatching, for example, isn't very enjoyable in an open-pit mine. And protecting species that need unbroken old-growth forest isn't easy when you cut down the forest.

Old-Growth Forests Some of the most contentious forestry issues in the United States and Canada in recent years have centered on logging in old-growth temperate rainforests in the Pacific Northwest. As you've learned in the opening case study for this chapter, such forests have incredibly high levels of biodiversity, and they can accumulate five times as much standing biomass per hectare as a tropical rainforest (fig. 6.9). Many endemic species, such as the northern spotted owl (see What Do

Climate scientists estimate that deforestation and land-use change now account for about 17 percent of our global carbon emissions, or more than all the automobiles, airplanes, and other vehicles combined. There's a growing consensus that we need to protect forests—particularly in the tropics—but for these projects to succeed, tropical nations have to be able to measure and report on the changing state of their forests. Developing countries have a difficult time affording the equipment and expertise to make these challenging assessments.

But new technology is changing that situation. Every day, earth-orbiting satellites gather an incredible amount of environmental data, but storing and publishing all this data is an enormous task, so that often the data is never seen by the people who need it most. Until recently the only people who could access and analyze satellite images of the earth were government officials, the military, well-equipped scientists, and resource extraction companies.

Today, anyone with a computer and an Internet connection can access to Google Earth. Since its introduction in 2005, this powerful application has become a valuable tool for scientists, activists, and ordinary citizens who want to better understand, monitor, and communicate about the environment. But how can we carry out the complex computational analysis to sift through the mass of available data?

Google has recently introduced a new program called Earth Engine to make environmental monitoring accessible to ordinary citizens. Earth Engine provides a dynamic digital model of the whole planet that's updated daily. By using "cloud" computing, calculations and mapping that used to take hours or days to perform with a high-end desktop or expensive, complex software can now be done in minutes or seconds using only a smartphone or other handheld device.

Perhaps even more remarkable is that these tools are being made available to indigenous and traditional communities in remote places, such as Brazil's Amazon Basin. In 2008 a team from Google traveled to Rondonia Province in northwestern Brazil to meet with members of the Surui tribe, whose rainforest homeland is being degraded by illegal logging and mining. The Surui's first encounter with Europeans was only about 40 years ago, and germs introduced in that first contact killed more than 95 percent of the tribe.

The tribe is still very poor, and their educational opportunities are limited. Nevertheless, the Google team taught them how to use android smartphones to take forest measurements and record GPS locations as well as how to enter those data into Earth Engine using laptop computers. The Surui also mapped information about hunting, religious, and cultural sites about their territory on Google Earth so that others can learn about their history.

Working with the Amazon Conservation Team, the Surui have started a project, funded by Norway as part of the REDD program, similar to the opening case study for this chapter, to track and report on the state of the forest. This is one of the first examples where the forest dwellers themselves are measuring carbon storage. Using the smartphones donated by the Google team, they document the size and location of trees in their forest so they can calculate carbon storage in biomass. Then they enter the data they've collected into a database in Earth Engine. This provides forest monitoring, reporting, and verification, which are essential for REDD to succeed.

Having tools for forest assessment—including the daily updates available on Google Earth—also enables the tribe to investigate and report illegal activities in their forest. In the past, if an illegal logger cleared a patch of forest in a remote area, no one might notice it for a long time, and if they did, the culprits would be long gone before authorities could arrive. Today the local people can identify changes online, and then go to the location with their smartphones to take photos and videos. They can find out what's happening and who's doing it. Furthermore, they can post their report immediately for the whole world to see. Officials in far off Brasilia, Oslo, or New York can be made aware of the situation in real-time, and pressure can be applied to stop forest destruction and degradation. Although exposure to Western culture and technology haven't always been beneficial to the Surui tribe, it looks as if this time it may work in their favor. For more information, see www.amazonteam.org.

▶ Members of the Surui tribe use computers and smartphones to map, monitor, and protect their ancestral lands in the Amazon rainforest.

You Think? p. 138), Vaux's swift, and the marbled murrelet, are so highly adapted to the unique conditions of these ancient forests that they live nowhere else.

The U.S. Northwest forest management plan established in 1994 is a model for integrating scientific study, local needs, and best practices in land use. This plan attempts to integrate human

▲ **FIGURE 6.9** The huge trees of the old-growth temperate rainforest accumulate more total biomass in standing vegetation per unit area than any other ecosystem on earth. They provide habitat to many rare and endangered species, but they also are converted by loggers who can sell a single tree for thousands of dollars.

and economic dimensions of issues while also protecting the long-term health of forests, wildlife, and waterways. It focuses on scientifically sound, ecologically credible, and legally responsible strategies and implementation. It aims to produce predictable and sustainable level of timber sales and nontimber resources. And it tries to ensure that federal agencies work together. It may not be enough protection, however, to ensure survival of endangered salmon and trout populations in some rivers. Several salmon and steelhead trout populations have been listed as endangered and more are under consideration. What do you think? How would you balance logging, farming, and cheap hydropower against fishing, native rights, and wildlife protection?

Harvesting Methods Most lumber and pulpwood in the United States and Canada currently are harvested by **clear-cutting**, in which every tree in a given area is cut, regardless of size (fig. 6.10). This method is effective for producing even-age stands of sun-loving species, such as aspen or some pines, but often increases soil erosion and eliminates habitat for many forest species when carried out on large blocks. It was once thought that good forest

▲ **FIGURE 6.10** Large clear-cuts, such as this, threaten species dependent on old-growth forest and expose steep slopes to soil erosion. Restoring something like the original forest will take hundreds of years.

management required immediate removal of all dead trees and logging residue. Research has shown, however, that standing snags and coarse woody debris play important ecological roles, including soil protection, habitat for a variety of organisms, and nutrient recycling.

Some alternatives to clear-cutting include **shelterwood harvesting**, in which mature trees are removed in a series of two or more cuts, and **strip-cutting**, in which all the trees in a narrow corridor are harvested. For many forest types, the least disruptive harvest method is **selective cutting**, in which only a small percentage of the mature trees are taken in each 10- or 20-year rotation. Ponderosa pine, for example, are usually selectively cut to thin stands and improve growth of the remaining trees. A forest managed by selective cutting can retain many of the characteristics of age distribution and groundcover of a mature old-growth forest. (See related story "Forestry for the Seventh Generation" at www.mhhe.com/cunningham7e.)

Roads and Logging An increasing number of people in the United States are calling for an end to all logging on federal lands. They argue that ecological services, such as maintaining water supplies and recreation, generate more revenue at lower costs. Many remote communities depend on logging jobs, but these jobs rely on subsidies. The federal government builds roads, manages forests, fights fires, and sells timber for less than the administrative costs of the sales. How should we weigh these different costs and benefits?

Some argue that logging should be restricted to privately owned lands. Just 4 percent of the nation's timber comes from national forests, and this harvest adds only about $4 billion to the American economy per year. In contrast, recreation, fish and wildlife, clean water, and other ecological services provided by the forest, by their calculations, are worth at least $224 billion

Save a tree, save the climate?

Forest destruction and land conversion produce about 17 percent of all human-caused CO_2 emissions—more than all global transportation emissions. REDD (Reducing Emissions from Deforestation and Forest Degradation) aims to reduce those emissions and help avert a climate catastrophe. Reducing deforestation could accomplish about half of global emission reduction goals. Billions of dollars' worth of ecosystem services, and precious biological diversity, can be saved at the same time. Every day over 30,000 hectares of tropical forest are destroyed by logging and burning; another 30,000 ha are degraded. Each year this adds up to an area twice the size of Alabama.

KC 6.1

How do deforestation and degradation release carbon?

- Trees are burned, releasing carbon (C) stored in wood and leaves.
- Fallen vegetation decays, releasing stored C (see chapter 2).
- Accumulation of C in soil litter declines; exposed soils dry, and C in soil oxidizes to CO_2.
- The forest ecosystem is no longer available to store C.

What drives deforestation?

- Industrial-scale agriculture (soy and palm oil production, cattle ranching)
- Industrial logging driven by international demand for timber
- Poverty and population pressure as people seek farmland and fuelwood
- Road development, oil development, mining, and dams

KC 6.2

KC 6.3

Products from deforested lands

- Oil and gasoline
- Food, cosmetics containing palm oil
- Paper products
- Aluminum (from bauxite ore)
  - Metals, gems, electronic components
  - Many, many others

PALM OIL
100% Pure

OIL

100' ALUMINUM FOIL

Lost value from deforestation

Losses in human welfare are estimated at at least **$2 trillion to $4 trillion each year**[1]. Losses in ecosystem services and the value of carbon storage may be still greater.

[1]Sukdev, P. 2010. Putting a price on nature. *Solutions* 1(6):34-43. www.thesolutionsjournal.org.

KC 6.10

5 mi (11 km)

NASA landsat image reveals parallel clearings on either side of a road near the Amazon River.

What ecological services would be protected under REDD?

We rely on forests for countless goods and services; here are some primary examples:

- Water supplies are maintained by forested areas, which store moisture and release it slowly during a dry season.
- Biodiversity, which provides for wild foods, medicines, building materials, migratory species, and tourism.
- Climate and weather regulation: forested areas have less volatile temperature and humidity changes than do cleared areas.

KC 6.5a

KC 6.5b

KC 6.6

KC 6.4

▼ The world's remaining forest area is about 4 billion hectares. Nearly half of these forests are boreal (northern) forest (purple); about half are tropical forest (green).

KC 6.7

Would REDD cost money?

Yes, but it also represents payment for products and services. Many developing countries rely on exporting tropical timber, or conversion to oil palm and soy farms, for most of their income. To cooperate with REDD, they would want this income replaced to some extent.

Wealthier countries rely on resources and ecosystem services from developing areas. Paying for the timber, oil, paper, and food products is easy, but REDD suggests that now we should also pay to protect some ecosystem services we rely on, including global climate stabilization, biodiversity, and water resources.

The United Nations REDD program estimates it will take $20-30 billion annually from developed countries to pay for forest protection, carbon offsets, and alternative development strategies.

What about human rights?

Some 1.2 billion people rely on forests for their livelihoods. More than 2 billion—a third of the world's population—use firewood to cook and to heat their homes. REDD efforts must recognize the rights of native people and local communities. Channeling money to urban central governments could worsen threats to these communities.

KC 6.8

How can we be sure that REDD projects are sustainable and enduring?

Monitoring, good government, and working at the local level are essential for REDD to succeed. A fascinating and successful example of local involvement is that of the Amazon Conservation Team (ACT), which has been partnering with indigenous peoples to map, monitor, and protect their ancestral lands using Google Earth and GPS (see Exploring Science, p. 134).

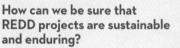

KC 6.9

CAN YOU **EXPLAIN?**

1. **How does deforestation contribute to carbon emissions?**
2. **What are some tropical forest (or formerly forest) resources you use?**

Northern Spotted Owls

What's the most controversial bird in the world? If you count the number of scientists, lawyers, journalists, and activists who have debated its protection, as well as the amount of money, time, and effort spent on research and recovery, the answer must be the northern spotted owl (*Strix occidentalis caurina*). This brown, medium-size owl lives in the complex old-growth forests of North America's Pacific Northwest. It's thought that before European settlement, northern spotted owls occurred throughout the Coastal Ranges and Cascade Mountains from southern British Columbia to the San Francisco Bay.

Spotted owls nest in cavities in the huge old-growth trees of the ancient forest. They depend on flying squirrels and wood rats as their primary prey, but they'll also eat voles, mice, gophers, hares, birds, and occasionally insects. With 90 percent of their preferred habitat destroyed or degraded, northern spotted owl populations are declining throughout their former range. When the U.S. Congress established the Endangered Species Act (ESA) in 1973, the northern spotted owl was identified as potentially endangered. After decades of study—but little action to protect them—northern spotted owls were listed as threatened in 1990 by the U.S. Fish and Wildlife Service. At that time, the population was estimated to contain 5,431 breeding pairs.

Several environmental organizations sued the federal government for its failure to do more to protect the owls. In 1991 a federal district judge agreed that the government wasn't following the requirements of the ESA, and temporarily shut down all logging in old-growth habitat in the Pacific Northwest. Timber sales dropped precipitously, and thousands of loggers and mill workers lost their jobs. Although mechanization and export of whole logs to foreign countries accounted for much of these job losses, many people blamed the owls for the economic woes across the region. Fierce debates broke out between loggers, who hung owls in effigy, and conservationists, who regarded the owls as protectors of the forest and of the whole biological community that lives in it.

In an effort to protect the remaining old growth while still providing timber jobs, President Clinton started a broad planning process for the whole area. After a great deal of study and consultation, a comprehensive Northwest Forest Plan was adopted in 1994 as a management guide for about 9.9 million hectares (24.5 million acres) of federal lands in Oregon, Washington, and northern California. The plan was based on the latest science of ecosystem management and represented compromises on all sides. Nevertheless, loggers complained that this plan

◀ Only about 2,000 pairs of northern spotted owls remain in the old-growth forests of the Pacific Northwest. Cutting old-growth forests threatens the endangered species, but reduced logging threatens the jobs of many timber workers.

locked up forests on which their jobs depended, while environmentalists lamented the fact that millions of hectares of old-growth forest would still be vulnerable to logging.

In spite of the habitat protection provided by the forest plan, northern spotted owl populations continued to decline. By 2004, researchers could find only 2,044 breeding pairs. They reported that 80 percent of the nesting areas occupied two decades earlier no longer had spotted owls, and that 9 of the 13 geographic populations were declining. The courts ordered the Fish and Wildlife Service to establish a recovery plan as required by the ESA. After four more years of study and deliberation, a recovery plan was published in 2008. The plan identified 133 owl conservation areas encompassing 2.6 million hectares (6.4 million acres) of federal lands that will be managed to protect old-growth habitat and, hopefully, stabilize owl populations. Again, both sides complained about the compromise. Loggers accused the government of caring more for owls than people. Conservationists deplored the fact that although less than 10 percent of the original old-growth is left, nearly a third of that remnant is still open to harvesting.

Recently, barred owls (*Strix varia*) have been moving into the Pacific Northwest. These larger and more aggressive cousins of the spotted owl have a wider habitat and prey tolerance, giving them a competitive advantage. When barred owls move in, spotted owls generally move out. In addition, barred owls sometimes interbreed with spotted owls, further diluting the endangered spotted owl gene pool. Some wildlife managers suggest that the only way to rebuild spotted owl populations is to kill barred owls, which are common across most of the middle of North America, but many people oppose this plan.

As you can see, there are a number of thorny ethical issues here. Is it right to kill one species to protect another? And where there are trade-offs between jobs, local economies, and homes for people versus habitat for wildlife and the existence of pristine landscapes, how should we weigh these competing values? Can we coexist with these shy, highly specialized forest creatures? There aren't easy answers for these dilemmas. The solutions depend on your values and worldviews. How would you answer these questions?

each year. Timber industry officials, on the other hand, dispute these claims, arguing that logging not only provides jobs and supports rural communities but also keeps forests healthy. What do you think? Could we make up for decreased timber production from public lands by more intensive management of private

holdings and by substitution or recycling of wood products? Are there alternative ways you could suggest to support communities now dependent on timber harvesting?

Roads on public lands are another controversy. Over the past 40 years, the Forest Service has expanded its system of logging

What Can YOU DO?

Lowering Your Forest Impacts

For most urban residents, forests—especially tropical forests—seem far away and disconnected from everyday life. There are things that each of us can do, however, to protect forests.

- Reuse and recycle paper. Make double-sided copies. Save office paper, and use the back for scratch paper.

- Use email. Store information in digital form, rather than making hard copies of everything.

- If you build, conserve wood. Use wafer board, particle board, laminated beams, or other composites, rather than plywood and timbers made from old-growth trees.

- Buy products made from "good wood" or other certified sustainably harvested wood.

- Don't buy products made from tropical hardwoods, such as ebony, mahogany, rosewood, or teak, unless the manufacturer can guarantee that the hardwoods were harvested from agriforestry plantations or sustainable-harvest programs.

- Don't patronize fast-food restaurants that purchase beef from cattle grazing on deforested rainforest land. Don't buy coffee, bananas, pineapples, or other cash crops if their production contributes to forest destruction.

- Do buy Brazil nuts, cashews, mushrooms, rattan furniture, and other nontimber forest products harvested sustainably by local people from intact forests. Remember that tropical rainforest is not the only biome under attack. Contact the Taiga Rescue Network for information about boreal forests.

- If you hike or camp in forested areas, practice minimum-impact camping. Stay on existing trails, and don't build more or bigger fires than you absolutely need. Use only downed wood for fires. Don't carve on trees or drive nails into them.

- Write to your congressional representatives, and ask them to support forest protection and environmentally responsible government policies. Contact the U.S. Forest Service, and voice your support for recreation and nontimber forest values.

roads more than tenfold, to a current total of nearly 550,000 km (343,000 mi), or more than ten times the length of the interstate highway system. Government economists regard road building as a benefit because it opens up the country to motorized recreation and industrial uses. Wilderness enthusiasts and wildlife supporters, however, see this as an expensive and disruptive program. In 2001 President Clinton established a plan to protect 23.7 million ha (58.5 million acres) of de facto wilderness from roads. Land developers, logging, mining, and energy companies protested this "roadless rule." President G. W. Bush, supported by industry-friendly western judges, overturned the rule and ordered resource managers to expedite logging, mining, and motorized recreation. In 2009, President Obama ordered the rule reinstated. He noted that this measure protects habitat for 1,600 endangered species (including bears and owls) and watersheds for 60 million people. What do you think? How much of the remaining old-growth should be protected as ecological reserves?

Fire Management Following a series of disastrous fire years in the 1930s, in which hundreds of millions of hectares of forest were destroyed, whole towns burned to the ground, and hundreds of people died, the U.S. Forest Service adopted a policy of aggressive fire control in which every blaze on public land was to be out before 10 A.M. Smokey Bear was adopted as the forest mascot and warned us that "only you can prevent forest fires." However, recent studies of fire's ecological role suggest that our attempts to suppress all fires may have been misguided. Many biological communities are fire-adapted and require periodic burning for regeneration. And eliminating fire from these forests has allowed woody debris to accumulate, greatly increasing the chances of a very big fire (fig. 6.11).

Forests that once were characterized by 50 to 100 mature, fire-resistant trees per hectare and an open understory now have a thick tangle of up to 2,000 small, spindly, mostly dead saplings in the same area. The U.S. Forest Service estimates that 33 million ha (73 million acres), or about 40 percent of all federal forestlands, are at risk of severe fires. To make matters worse, Americans increasingly live in remote areas where wildfires are highly likely. Because there haven't been fires in many of these places in living

▲ **FIGURE 6.11** By suppressing fires and allowing fuel to accumulate, we make major fires such as this more likely. The safest and most ecologically sound management policy for some forests may be to allow natural or prescribed fires, which don't threaten property or human life, to burn periodically.

memory, many people assume there is no danger, but by some estimates 40 million U.S. residents now live in areas with high wildfire risk.

A recent prolonged drought in the western United States has heightened fire danger. In 2007 nearly 80,000 wildfires burned 3.6 million ha (8.9 million acres) of forests and grasslands in the United States. Federal agencies spent almost $2 billion to fight these fires, nearly four times the previous ten-year average. And 2011 nearly matched this record.

The dilemma is how to undo years of fire suppression and fuel buildup. Fire ecologists favor small, prescribed burns to clean out debris. Loggers decry this approach as a waste of valuable timber, and local residents of fire-prone areas fear that prescribed fires will escape and threaten them. Recently the Forest Service proposed a massive new program of forest thinning and emergency salvage operations (removing trees and flammable material from mature or recently burned forests) on 16 million ha (40 million acres) of national forest. Carried out over a 20-year period, this program could cost as much as $12 billion and would open up much road-less, de facto wilderness to invasive species and industrial-scale logging. Field evidence shows, moreover, that salvage logging impedes regeneration on burned forest land. Proponents neverthe-less argue that the only way to save the forest is to log it.

Ecosystem Management   In the 1990s many federal agen-cies began to shift their policies from a strictly economic focus to **ecosystem management**, which is very similar to the Northwest Forest Plan in its unified, systems approach. Some of its prin-ciples include:

- Manage across whole landscapes, watersheds, or regions over ecological time scales.

- Depend on scientifically sound, ecologically credible data for decision making.

- Consider human needs and promote sustainable economic development and communities.

- Maintain biological diversity and essential ecosystem processes.

- Utilize cooperative institutional arrangements.

- Generate meaningful stakeholder and public involvement and facilitate collective decision making.

- Adapt management over time, based on conscious experimen-tation and routine monitoring.

Elements of ecosystem management appear in the *National Report on Sustainable Forests,* which suggests goals for sustain-able forest management (table 6.1). Similarly, in 2011, President Obama signed an executive order charging agencies to create a strategic plan for ecosystem-based management of oceans, coasts, and the Great Lakes.

6.2 GRASSLANDS

After forests, grasslands are among the biomes most heavily used by humans. Prairies, savannas, steppes, open woodlands, and other grasslands occupy about one-quarter of the world's land

TABLE 6.1 | Draft Criteria for Sustainable Forestry

1. Conservation of biological diversity
2. Maintenance of productive capacity of forest ecosystems
3. Maintenance of forest ecosystem health and vitality
4. Maintenance of soil and water resources
5. Maintenance of forest contribution to global carbon cycles
6. Maintenance and enhancement of long-term socioeconomic benefits to meet the needs of legal, institutional, and economic framework for forest conservation and sustainable management

SOURCE: Data from USFS, 2002.

surface. Much of the U.S. Great Plains and the Prairie Provinces of Canada fall in this category (fig. 6.12). The 3.8 billion ha (12 million mi^2) of pastures and grazing lands in this biome make up about twice the area of all agricultural crops. When you add to this about 4 billion ha of other lands (forest, desert, tundra, marsh, and thorn scrub) used for raising livestock, more than half of all land is used at least occasionally for grazing. At least 3 billion cattle, sheep, goats, camels, buffalo, and other domestic animals on these lands make a valuable contribution to human nutrition. Sustainable pastoralism can increase productivity while maintain-ing biodiversity in a grassland ecosystem.

Because grasslands, chaparral, and open woodlands are attractive for human occupation, they frequently are converted to cropland, urban areas, or other human-dominated landscapes. Worldwide the rate of grassland disturbance each year is three times that of tropical forest. Although they may appear to be uni-form and monotonous to the untrained eye, native prairies can be highly productive and species-rich. According to the U.S. Depart-ment of Agriculture, more threatened plant species occur in range-lands than in any other major American biome.

▲ **FIGURE 6.12** This short-grass prairie in northern Montana is too dry for trees but nevertheless supports a diverse biological community.

Grazing can be sustainable or damaging

By carefully monitoring the numbers of animals and the condition of the range, ranchers and **pastoralists** (people who live by herding animals) can adjust to variations in rainfall, seasonal plant conditions, and the nutritional quality of forage to keep livestock healthy and avoid overusing any particular area. Conscientious management can actually improve the quality of the range.

When land is abused by overgrazing—especially in arid areas—rain runs off quickly before it can soak into the soil to nourish plants or replenish groundwater. Springs and wells dry up. Seeds can't germinate in the dry, overheated soil. The barren ground reflects more of the sun's heat, changing wind patterns, driving away moisture-laden clouds, and leading to further desiccation. This process of conversion of once-fertile land to desert is called **desertification**.

This process is ancient, but in recent years it has been accelerated by expanding populations and the political conditions that force people to overuse fragile lands. According to the International Soil Reference and Information Centre in the Netherlands, nearly three-quarters of all rangelands in the world show signs of either degraded vegetation or soil erosion. Overgrazing is responsible for about one-third of that degradation (fig. 6.13). The highest percentage of moderate, severe, and extreme land degradation is in Mexico and Central America, while the largest total area is in Asia, where the world's most extensive grasslands occur. Can we reverse this process? In some places, people are reclaiming deserts and repairing the effects of neglect and misuse.

Overgrazing threatens many rangelands

As is the case in many countries, most public grazing lands in the United States are not in good health. Political and economic pressures encourage managers to increase grazing allotments beyond the carrying capacity of the range. Lack of enforcement of existing regulations and limited funds for range improvement have resulted in **overgrazing**, damage to vegetation and soil including loss of native forage species and erosion. The Natural Resources Defense Council claims that only 30 percent of public rangelands are in fair condition, and 55 percent are poor or very poor (fig. 6.14).

Overgrazing has allowed populations of unpalatable or inedible species, such as sage, mesquite, cheatgrass, and cactus, to build up on both public and private rangelands. Wildlife conservation groups regard cattle grazing as the most ubiquitous form of ecosystem degradation and the greatest threat to endangered species in the southwestern United States. They call for a ban on cattle and sheep grazing on all public lands, noting that it provides only 2 percent of the total forage consumed by beef cattle and supports only 2 percent of all livestock producers.

Like federal timber management policy, grazing fees charged for use of public lands often are far below market value and represent an enormous hidden subsidy to western ranchers. Holders of grazing permits generally pay the government less than 25 percent of what it would cost to lease comparable private land. The 31,000 permits on federal range bring in only $11 million in grazing fees

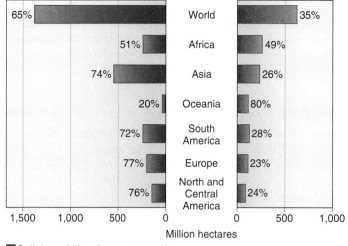

| | Soil degradation due to overgrazing |
| | Soil degradation due to other causes |

▲ **FIGURE 6.13** Percentage of rangeland soil degradation (red bars) due to overgrazing. Notice that, in Europe, Asia, and the Americas, farming, logging, mining, urbanization, and other causes are responsible for about three-quarters of all soil degradation. In Africa and Oceania, where more grazing occurs and desert or semiarid scrub make up much of the range, grazing damage is higher.

but cost $47 million per year for administration and maintenance. The $36 million difference amounts to a massive "cow welfare" system of which few people are aware.

On the other hand, ranchers defend their way of life as an important part of western culture and history. Although few cattle go directly to market from their ranches, they produce almost all the beef calves subsequently shipped to feedlots. And without a

▲ **FIGURE 6.14** More than half of all publicly owned grazing land in the United States is in poor or very poor condition. Overgrazing and invasive weeds are the biggest problems.

viable ranch economy, they claim, even more of the western landscape would be subdivided into small ranchettes to the detriment of both wildlife and environmental quality. What do you think? How much should we subsidize extractive industries to preserve rural communities and traditional occupations?

Ranchers are experimenting with new methods

Where a small number of livestock are free to roam a large area, they generally eat the tender, best-tasting grasses and forbs first, leaving the tough, unpalatable species to flourish and gradually dominate the vegetation. In some places, farmers and ranchers find that short-term, intensive grazing helps maintain forage quality. As South African range specialist Allan Savory observed, wild ungulates (hoofed animals), such as gnus or zebras in Africa or bison (buffalo) in America, often tend to form dense herds that graze briefly but intensively in a particular location before moving on to the next area. Rest alone doesn't necessarily improve pastures and rangelands. Short-duration, **rotational grazing**—confining animals to a small area for a short time (often only a day or two) before shifting them to a new location—simulates the effects of wild herds (fig. 6.15). Forcing livestock to eat everything equally, to trample the ground thoroughly, and to fertilize heavily with manure before moving on helps keep weeds in check and encourages the growth of more desirable forage species. This approach doesn't work everywhere, however. Many plant communities in the U.S. desert Southwest, for example, apparently evolved in the absence of large, hoofed animals and can't withstand intensive grazing.

Restoring fire can be as beneficial to grasslands as it is to forests. In some cases ranchers are cooperating with environmental groups in range management and preservation of a ranching economy.

▲ **FIGURE 6.15** Intensive, rotational grazing encloses livestock in a small area for a short time (often only one day) within a movable electric fence to force them to eat vegetation evenly and fertilize the area heavily.

▲ **FIGURE 6.16** Red deer (*Cervus elaphus*) are raised in New Zealand for antlers and venison.

Another approach to ranching in some areas is to raise wild species, such as impala, wildebeest, oryx, or elk (fig. 6.16). These animals forage more efficiently, tolerate harsh climates, often are more pest- and disease-resistant, and fend off predators better than usual domestic livestock. Native species also may have different feeding preferences and needs for water and shelter than cows, goats, or sheep. The African Sahel, for instance, can provide only enough grass to raise about 20 to 30 kg (44 to 66 lb) of beef per hectare. Ranchers can produce three times as much meat with wild native species in the same area because these animals browse on a wider variety of plant materials.

In the United States, ranchers find that elk, American bison, and a variety of African species take less care and supplemental feeding than cattle or sheep and result in a better financial return because their lean meat can bring a better market price than beef or mutton. Media mogul Ted Turner has become both the biggest private landholder in the United States and the owner of more American bison than anyone other than the government.

6.3 PARKS AND PRESERVES

While most forests and grasslands serve useful, or utilitarian, purposes, most societies also set aside some natural areas for aesthetic or recreational purposes. Natural preserves have existed for thousands of years. Ancient Greeks protected sacred groves for religious purposes. Royal hunting grounds have preserved forests in Europe for centuries. Although these areas were usually reserved for elite classes in society, they have maintained biodiversity and natural landscapes in regions where most lands are heavily used.

The first public parks open to ordinary citizens may have been the tree-sheltered agoras in planned Greek cities. But the idea of providing natural space for recreation, and to preserve natural environments, has really developed in the past 50 years (fig. 6.17). While the first parks were intended mainly for the recreation of growing urban populations, parks have taken on many additional purposes. Today we see our national parks as playgrounds for rest and recreation, as havens for wildlife, as places to experiment with ecological management, and as opportunities to restore ecosystems.

▲ **FIGURE 6.17** Growth of protected areas worldwide, 1907–2007.

Currently, nearly 13 percent of the land area of the earth is protected in some sort of park, preserve, or wildlife management area. This represents about 19 million km² (7.3 million mi²) in 122,000 different preserves. This is an encouraging environmental success story.

Many countries have created nature preserves

Different levels of protection are found in nature preserves. The World Conservation Union divides protected areas into five categories depending on the intended level of allowed human use (table 6.2). In the most stringent category (ecological reserves and wilderness areas), few or no human impacts are allowed. In some strict nature preserves, where particularly sensitive wildlife or natural features are located, human entry may be limited only to scientific research groups that visit on rare occasions. In some wildlife sanctuaries, for example, only a few people per year are allowed to visit, to avoid introducing invasive species or disrupting native

TABLE 6.2 | IUCN Categories of Protected Areas

CATEGORY	ALLOWED HUMAN IMPACT OR INTERVENTION
1. Ecological reserves and wilderness areas	Little or none
2. National parks	Low
3. Natural monuments and archaeological sites	Low to medium
4. Habitat and wildlife management areas	Medium
5. Cultural or scenic landscapes, recreation areas	Medium to high

SOURCE: Data from World Conservation Union, 1990.

species. In the least restrictive categories (national forests and other natural resource management areas), on the other hand, there may be a high level of human use.

Venezuela claims to have the highest proportion of its land area protected (66 percent) of any country in the world. About half this land is designated as preserves for indigenous people or for sustainable resource harvesting. With little formal management, however, there is only minimal protection from poaching by hunters, loggers, and illegal gold hunters. Unfortunately, it's not uncommon in the developing world to have "paper parks" that exist only as a line drawn on a map with no budget for staff, management, or infrastructure. The United States, by contrast, has only about 22 percent of its land area in protected status, and less than one-third of that amount is in IUCN categories I or II (nature reserves, wilderness areas, national parks). The rest is in national forests or wildlife management zones that are designated for sustainable use. With hundreds of thousands of state and federal employees, billions of dollars in public funding, and a high level of public interest and visibility, U.S. public lands are generally well managed.

Currently, Brazil has the largest total area in protected status of any country. More than 2.5 million km² or 29 percent of the nation's land—mostly in the Amazon basin—is in some protected status. In 2006 the northern Brazilian state of Para, in collaboration with Conservation International (CI) and other nongovernmental organizations, announced the establishment of nine new protected areas along the border with Suriname and Guyana. These new areas, about half of which will be strictly protected nature preserves, will link together several existing indigenous areas and nature preserves to create the largest tropical forest reserve in the world. More than 90 percent of the new 15 million ha (58,000 mi², or about the size of Illinois) Guyana Shield Corridor is in pristine natural state. CI president Russ Mittermeir says, "If any tropical rainforest on earth remains intact a century from now, it will be this portion of northern Amazonia." In contrast to this dramatic success, the Pantanal, the world's largest wetland/savanna complex, which lies in southern Brazil and is richer in some biodiversity categories than the Amazon, is almost entirely privately owned. There are efforts to set aside some of this important wetland, but so far little is in protected status.

Some other countries with very large reserved areas include Greenland (with a 980,000 km² national park that covers most of the northern part of the island) and Saudi Arabia (with a 825,000 km² wildlife management area in its Empty Quarter). These areas are relatively easy to set aside, however, being mostly ice-covered (Greenland) or desert (Saudi Arabia). Canada's Quttinirpaaq National Park on Ellesmere Island is an example of a preserve with high wilderness values but little biodiversity. Only 800 km (500 miles) from the North Pole, this remote park gets fewer than 100 human visitors per year during its brief, three-week summer

▲ **FIGURE 6.18** Canada's Quttinirpaaq National Park at the north end of Ellesmere Island has plenty of solitude and pristine landscapes, but little biodiversity.

season (fig. 6.18). With little evidence of human occupation, it has abundant solitude and stark beauty, but very little wildlife and almost no vegetation.

Figure 6.19 compares the percentage of each major biome in protected status. Not surprisingly, there's an inverse relationship between the percentage converted to human use (and where people

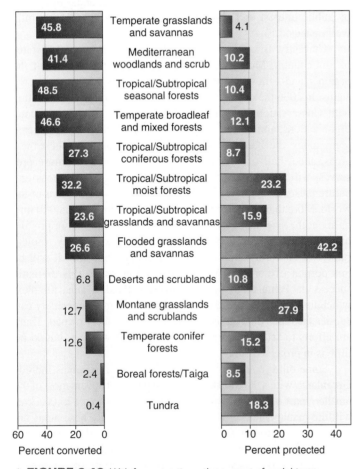

▲ **FIGURE 6.19** With few exceptions, the percent of each biome converted to human use is roughly inverse to the percent protected in parks and preserves. Rock and ice, lakes, and Antarctic ecoregions are excluded. World Database on Protected Areas, 2009.

live) and the percentage protected. Temperate grasslands and savannas (such as the American Midwest) and Mediterranean woodlands and scrub (such as the French Riviera or the coast of southern California) are highly domesticated, and, therefore, expensive to set aside in large areas. Temperate conifer forests (think of Siberia, or Canada's vast expanse of boreal forest) are relatively uninhabited, and therefore easy to put into some protected category.

Not all preserves are preserved

Even parks and preserves designated with a high level of protection aren't always safe from exploitation or changes in political priorities. Serious problems threaten natural resources and environmental quality in many countries. In Greece, the Pindus National Park is threatened by plans to build a hydroelectric dam in the center of the park. Furthermore, excessive stock grazing and forestry exploitation in the peripheral zone are causing erosion and loss of wildlife habitat. In Colombia, dam building also threatens the Paramillo National Park. Ecuador's largest nature preserve, Yasuni National Park, which contains one of the world's most megadiverse regions of lowland Amazonian forest, has been opened to oil drilling, while miners and loggers in Peru have invaded portions of Huascaran National Park. In Palau, coral reefs identified as a potential biosphere reserve are damaged by dynamite fishing, while on some beaches in Indonesia, almost every egg laid by endangered sea turtles is taken by egg hunters. These are just a few of the many problems faced by parks and preserves around the world. Often countries with the most important biomes lack funds, trained personnel, and experience to manage the areas under their control.

Even in rich countries, such as the United States, some of the "crown jewels" of the National Park System suffer from overuse and degradation. Yellowstone and Grand Canyon National Parks, for example, have large budgets and are highly regulated, but are being "loved to death" because they are so popular. When the U.S. National Park Service was established in 1916, Stephen Mather, the first director, reasoned that he needed to make the parks comfortable and entertaining for tourists as a way of building public support. He created an extensive network of roads in the largest parks so that visitors could view famous sights from the windows of their automobiles, and he encouraged construction of grand lodges in which guests could stay in luxury.

His plan was successful; the National Park System is cherished and supported by many American citizens. But sometimes entertainment has superseded nature protection. Visitors were allowed—in some cases even encouraged—to feed wildlife. Bears lost their fear of humans and became dependent on an unhealthy diet of garbage and handouts (fig. 6.20). In Yellowstone and Grand Teton National Parks, the elk herd was allowed to grow to 25,000 animals, or about twice the carrying capacity of the habitat. The excess population overgrazed the vegetation to the detriment of many smaller species and the biological community in general. As we discussed earlier in this chapter, 70 years of fire suppression resulted in changes of forest composition and fuel buildup that made huge fires all but inevitable. In Yosemite, you can stay in a world-class hotel, buy a pizza, play video games, do laundry, play golf or tennis, and shop for curios, but you may find it difficult

▶ **FIGURE 6.20** Wild animals have always been one of the main attractions in national parks. Many people lose all common sense when interacting with big, dangerous animals. This is not a petting zoo.

to experience the solitude or enjoy the natural beauty extolled by John Muir as a prime reason for creating the park.

In many of the most famous parks, traffic congestion and crowds of people stress park resources and detract from the experience of unspoiled nature (fig. 6.21). Some national parks, such as Yosemite and Zion, have banned private automobiles from the most congested areas. Visitors must park in remote lots and take clean, quiet electric or natural-gas-burning buses to popular sites. Other parks are considering limits on the number of visitors admitted each day. How would you feel about a lottery system that might allow you to visit some famous parks only once in your lifetime, but to have an uncrowded, peaceful experience on your one allowed visit? Or would you prefer to be able to visit whenever you wish even if it means fighting crowds and congestion?

Originally the great wilderness parks of Canada and the United States were distant from development and isolated from most human impacts. This has changed in many cases. Forests are clearcut right up to some park boundaries. Mine drainage contaminates streams and groundwater. At least 13 U.S. National Monuments are open to oil and gas drilling, including Texas's Padre Island, the only U.S. breeding ground for endangered Kemps Ridley sea turtles. Even in the dry desert air of the Grand Canyon, where visibility was once up to 150 km, it's often too smoggy now to see across the canyon due to air pollution from power plants just outside the park. Snowmobiles and off-road vehicles (ORV) create pollution and noise and cause erosion while disrupting wildlife in many parks (fig. 6.22).

Chronically underfunded, the U.S. National Park System now has a maintenance backlog estimated to be at least $5 billion. During election campaigns, politicians from both major political parties vow to repair park facilities, but then find other uses for public funds once in office. Ironically, a recent study found that, on average, parks generate $4 in user fees for every $1 they receive in federal subsidies. In other words, they more than pay their own way, and should have a healthy surplus if they were allowed to retain all the money they generate.

In recent years, the U.S. National Park System has begun to emphasize nature protection and environmental education over entertainment. This new agenda is being adopted by other countries as well. The IUCN has developed a **world conservation strategy** for protecting natural resources that includes the following three objectives: (1) to maintain essential ecological processes and life-support systems (such as soil regeneration and protection, nutrient recycling, and water purification) on which human survival and development depend; (2) to preserve genetic diversity essential for breeding programs to improve cultivated plants and domestic animals; and (3) to ensure that any utilization of wild species and ecosystems is sustainable.

Marine ecosystems need greater protection

As ocean fish stocks become increasingly depleted globally, biologists are calling for protected areas where marine organisms are sheltered from destructive harvest methods. Although about

▲ **FIGURE 6.21** Thousands of people wait for an eruption of Old Faithful geyser in Yellowstone National Park. Can you find the ranger who's giving a geology lecture?

▲ **FIGURE 6.22** Off-road vehicles cause severe, long-lasting environmental damage when driven through wetlands.

14 percent of land area is in some conservation status, only about 5 percent of nearshore marine biomes are protected. As the opening case study for chapter 1 describes, limiting the amount and kind of fishing in marine reserves can quickly replenish fish stocks in surrounding areas. In a study of 100 marine refuges around the world, researchers found that, on average, the number of organisms inside no-take preserves was twice as high as surrounding areas where fishing was allowed. In addition, the biomass of organisms was three times as great and individual animals were, on average, 30 percent larger inside the refuge compared to outside. Recent research has shown that closing reserves to fishing even for a few months can have beneficial results in restoring marine populations. The size necessary for a safe haven to protect flora and fauna depends on the species involved, but some marine biologists call on nations to protect at least 20 percent of their nearshore territory as marine refuges.

Coral reefs are among the most threatened marine ecosystems in the world. Surveys show that, worldwide, living coral reefs have declined by about half in the past century, and 90 percent of all reefs face threats from rising sea temperatures, destructive fishing methods, coral mining, sediment runoff, and other human disturbance. In many ways, coral reefs are the old-growth rainforests of the ocean (fig. 6.23). Biologically rich, these sensitive communities can take a century or more to recover from damage. If current trends continue, some researchers predict that in 50 years there will be no viable coral reefs anywhere in the world.

What can be done to reverse this trend? Some countries are establishing large marine reserves specifically to protect coral reefs. Australia has one of the largest marine reserves in the world in its 1.3 million km² Great Barrier Reef/Coral Sea reserve. In his final days in office, President G. W. Bush declared more than 505,000 km² (195,000 mi²) of ocean, including the Mariana trench and atolls around Samoa and some other uninhabited Pacific islands, as national monuments. Altogether, however, aquatic reserves make up less than one-tenth of all the world's protected areas despite the fact that nearly three-quarters of the earth's surface is water. A survey of marine biological resources identified the ten richest and most threatened "hot spots," including the Philippines, the Gulf of Guinea and Cape Verde Islands (off the west coast of Africa), Indonesia's Sunda Islands, the Mascarene Islands in the Indian Ocean, South Africa's coast, southern Japan and the east China Sea, the western Caribbean, and the Red Sea and Gulf of Aden. We urgently need more no-take preserves to protect marine resources.

Conservation and economic development can work together

Many of the most biologically rich communities in the world are in developing countries, especially in the tropics. These countries are the guardians of biological resources important to all of us. Unfortunately, where political and economic systems fail to provide residents with land, jobs, food, and other necessities of life, people do whatever is necessary to meet their own needs. Immediate survival takes precedence over long-term environmental goals. Clearly the struggle to save species and ecosystems can't be divorced from the broader struggle to meet human needs (see Exploring Science, next page).

▲ **FIGURE 6.23** Coral reefs are among both the most biologically rich and endangered ecosystems in the world. Marine reserves are being established in many places to preserve and protect these irreplaceable resources.

People in some developing countries are beginning to realize that their biological resources may be their most valuable assets, and that their preservation is vital for sustainable development. **Ecotourism** (tourism that is ecologically and socially sustainable) can be more beneficial in many places over the long term than extractive industries, such as logging and mining. See What Can You Do? on page 148 for some ways to ensure that your vacations are ecologically responsible.

Native people can play important roles in nature protection

The American ideal of wilderness parks untouched by humans is unrealistic in many parts of the world. As we mentioned earlier, some biological communities are so fragile that human intrusions have to be strictly limited to protect delicate natural features or particularly sensitive wildlife. In many important biomes, however, aboriginal people have been present for thousands of years and have a legitimate right to pursue traditional ways of life. Furthermore, many of the approximately 5,000 indigenous or native cultures that remain today possess ecological knowledge about their ancestral homelands that can be valuable in ecosystem management. According to author Alan Durning, "encoded in indigenous languages, customs, and practices may be as much understanding of nature as is stored in the libraries of modern science."

Some countries have adopted draconian policies to remove native people from parks (fig. 6.24). In South Africa's Kruger National Park, for example, heavily armed soldiers keep intruders out with orders to shoot to kill. This is very effective in protecting wildlife. In all fairness, before this policy was instituted there was a great deal of poaching by mercenaries armed with automatic weapons. But it also means that people who were forcibly displaced from the park could be killed on sight merely for returning to their former homes to collect firewood or to hunt for small game. Similarly, in 2006, thousands of peasant farmers on the edge of the vast Mau Forest in Kenya's Rift Valley were forced

How would you like to use cool technology to help save the forest habitat of the most famous population of chimps in the world? That's what Dr. Lilian Pintea does as vice president of conservation science at the Jane Goodall Institute (JGI). At JGI, Dr. Pintea integrates remote-sensing data with on-the-ground observations to study chimpanzee populations, distribution, and environment. But in addition to high-level science and international wildlife policy issues, Dr. Pintea also works with local Tanzanian villagers to develop land-use and development plans that benefit both wildlife and humans.

Lilian has an interesting history. Born in Moldova, he dreamed from an early age of working with African wildlife. A high-school research paper on snakes won him a place at Moscow State University. After postgraduate zoology studies in Bucharest, Romania, he won a Fulbright Fellowship to learn remote sensing at the University of Delaware. From there, he moved to the University of Minnesota, where he earned a PhD in conservation biology. His dissertation research used Geographic Information Systems (GIS) to map the Gombe National Park in western Tanzania.

▲ Dr. Lilian Pintea takes ground-based measurements in Gombe National Park.

Using the latest high-resolution satellite imagery, Dr Pintea has created a detailed, digital map of the park to identify specific locations where Dr. Jane Goodall and her team made observations over decades of groundbreaking wildlife research. Objects as small as 0.5 m can be seen in the map, visualizing not only of specific trees but streams, footpaths, and other landscape features. This digitized database is a valuable tool for analyzing geographic patterns in animal distribution as well as how habitat type is related to particular chimpanzee behaviors.

In addition to helping carry on the Institute's pioneering research, Dr. Pintea is working with local communities to use their natural resources in a sustainable manner. In the 52 years since chimpanzee research started in Gombe, population growth in the villages around the park, along with an influx of refugees from war zones in nearby countries, has led to rapid deforestation and land degradation right up to park boundaries. Increasingly, the park has become an island of forest surrounded by a barren, eroding landscape. Poaching, illegal logging, and trespassing in the park are increasing problems. It isn't clear that either humans or chimps can survive over the long term if these trends continue.

About 20 years ago, JGI began working with the local communities through the Lake Tanganyika Catchment Reforestation and Education (TACARE, pronounced "take care") project to find ways to protect local natural resources on which both chimps and local people depend. The JGI staff is using its GIS expertise to help villagers develop conservation action plans. They also work with communities on health, economic development, and clean water programs to improve lives and reduce pressures on the park. For example, high-yield hybrid oil palms now produce four times as much oil from a given area as the conventional varieties. Efficient stoves and high-value crops, such as coffee and chocolate, also reduce the need for more forest clearing. Like in Brazil and Indonesia, described earlier in this chapter, this work is supported in part with funds from Norway in the REDD program.

Village land-use planning is essential for sustainability of both wildlife and human communities. GIS is used to analyze deforestation, topography, and human use patterns, and to identify forest conservation areas that can increase chimpanzee viability both inside and outside the park. These conservation areas will also help stabilize watersheds and support human livelihoods.

Recently, 13 villages within the Greater Gombe Ecosystem completed their village land-use plans, which were ratified by the Tanzanian government. Local communities have voluntarily assigned 9,690 hectares, or 26 percent, of their village lands as Village Forest Reserves. These interconnected reserves contain about two-thirds of the priority conservation area identified in the planning process. In the end, it's hoped that both wildlife and local communities will benefit from these conservation efforts. It's a dream come true for a young man from Moldova.

▼ Village land-use plans for the area surrounding Gombe National Park (dark green).

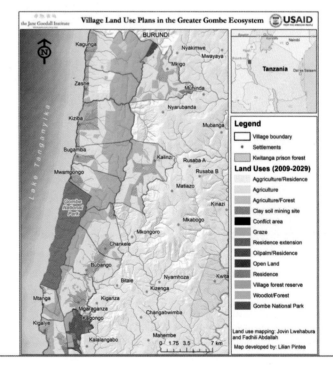

▲ **FIGURE 6.24** Some parks take draconian measures to expel residents and prohibit trespassing. How can we reconcile the rights of local or indigenous people with the need to protect nature?

from their homes at gunpoint by police who claimed that the land needed to be cleared to protect the country's natural resources. Critics claimed that the forced removal amounted to "ethnic cleansing" and was based on tribal politics rather than nature protection.

Other countries recognize that finding ways to integrate local human needs with those of nature is essential for successful conservation. In 1986, UNESCO (United

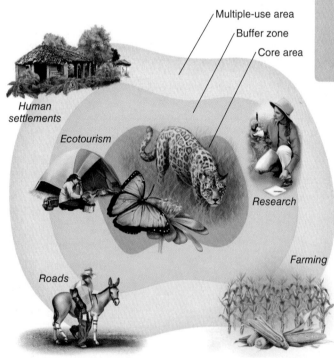

Multiple-use area
Buffer zone
Core area
Human settlements
Ecotourism
Research
Farming
Roads

▲ **FIGURE 6.25** A model biosphere reserve. Traditional parks and wildlife refuges have well-defined boundaries to keep wildlife in and people out. Biosphere reserves, by contrast, recognize the need for people to have access to resources. Critical ecosystem is preserved in the core. Research and tourism are allowed in the buffer zone, while sustainable resource harvesting and permanent habitations are situated in the multiple-use area around the perimeter.

Nations Educational, Scientific, and Cultural Organization) initiated its **Man and Biosphere (MAB) program**, which encourages the designation of **biosphere reserves**, protected areas divided into zones with different purposes. Critical ecosystem functions and endangered wildlife are protected in a central core region, where limited scientific study is the only human access allowed. Ecotourism and research facilities are located in a relatively pristine buffer zone around the core, while sustainable resource harvesting and permanent habitation are allowed in multiple-use peripheral regions (fig. 6.25).

A well-established example of a biosphere reserve is Mexico's 545,000 ha (2,100 mi²) Sian Ka'an Reserve on the Tulum Coast of the Yucatán. The core area includes 528,000 ha (1.3 million acres) of coral reefs, bays, wetlands, and lowland tropical forest. More than 335 bird species have been observed within the reserve, along with endangered manatees, five types of jungle cats, spider and howler monkeys, and four species of increasingly rare sea turtles. Approximately 25,000 people reside in communities and the countryside around Sian Ka'an. In addition to tourism, the economic base of the area includes lobster fishing, small-scale farming, and coconut cultivation.

The Amigos de Sian Ka'an, a local community organization, played a central role in establishing the reserve and is working to protect natural resources while also improving living standards for local people. New intensive farming techniques and sustainable harvesting of forest products enable residents to make a living without harming their ecological base. Better lobster harvesting techniques developed at the reserve have improved the catch without depleting native stocks. Local people now see the reserve as a benefit rather than an imposition from the outside. Similar success stories from many parts of the world show how we can support local people and recognize indigenous rights while still protecting important environmental features.

As is the case in many countries, technology is aiding in protecting parks and involving local communities in land-use planning in Africa (see Exploring Science, p. 147).

Species survival can depend on preserve size and shape

Many natural parks and preserves are increasingly isolated, remnant fragments of ecosystems that once extended over large areas. As park ecosystems are shrinking, however, they are also becoming more and more important for maintaining biological diversity. Principles of landscape design and landscape structure become important in managing and restoring these shrinking islands of habitat.

For years, conservation biologists have disputed whether it is better to have a *single large* or *several small* reserves (the SLOSS debate). Ideally, a reserve should be large enough to support viable populations of endangered species, keep ecosystems intact, and isolate critical core areas from damaging external forces. For some species with small territories, several small, isolated refuges can support viable populations, and having several small reserves provides insurance against a disease, habitat destruction, or other calamity that might wipe out a single population. But small preserves can't support species such as elephants or tigers, which need large amounts of space. Given human needs and pressures, however, big preserves aren't always possible. One proposed solution has been to create **corridors** of natural habitat that can connect to smaller habitat areas (fig. 6.26). Corridors could effectively create a large preserve from several small ones. Corridors could also allow populations to maintain genetic diversity or expand into new breeding territory. The effectiveness of corridors probably depends on how long and wide they are, and on how readily a species will use them.

One of the reasons large preserves are considered better than small preserves is that they have more **core habitat**, areas deep in the interior of a habitat area, and that core habitat has better conditions for specialized species than do edges. **Edge effects** is a term generally used to describe habitat edges: for example, a forest edge is usually more open, bright, and windy than a forest interior, and temperatures and humidity are more varied. For a grassland, on the other hand, edges may be wooded, with more shade, and perhaps more predators, than in the core of the grassland area. As human disturbance fragments an ecosystem, habitat is broken into increasingly isolated islands, with less core and more edge. Small, isolated fragments of habitat often support fewer species, especially

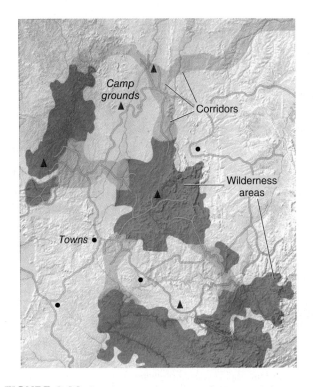

▲ **FIGURE 6.26** Corridors serve as routes of migration, linking isolated populations of plants and animals in scattered nature preserves. Although individual preserves may be too small to sustain viable populations, connecting them through river valleys and coastal corridors can facilitate interbreeding and provide an escape route if local conditions become unfavorable.

fewer rare species, than do extensive, uninterrupted ecosystems. The size and isolation of a wildlife preserve, then, may be critical to the survival of rare species.

A dramatic experiment in reserve size, shape, and isolation is being carried out in the Brazilian rainforest. In a project funded by the World Wildlife Fund and the Smithsonian Institution, loggers left 23 test sites when they clear-cut a forest. Test sites range from 1 ha (2.47 acres) to 10,000 ha. Clear-cuts surround some, and newly created pasture surrounds others (fig. 6.27); others remain connected to the surrounding forest. Selected species are regularly inventoried to monitor their survival after disturbance. As expected, some species disappear very quickly, especially from small areas. Sun-loving species flourish in the newly created forest edges, but deep-forest, shade-loving species disappear, particularly when the size or shape of a reserve reduces the availability of core habitat. This experiment demonstrates the importance of maintaining core habitat in preserves.

CONCLUSION

Forests and grasslands cover nearly 60 percent of global land area. The vast majority of humans live in these biomes, and we obtain many valuable materials from them. And yet these biomes also are the source of much of the world's biodiversity on which we depend for life-supporting ecological services. How we can live

▲ **FIGURE 6.27** How small can a nature preserve be? In an ambitious research project, scientists in the Brazilian rainforest are carefully tracking wildlife in plots of various sizes, either connected to existing forests or surrounded by clear-cuts. As you might expect, the largest and most highly specialized species are the first to disappear.

sustainably on our natural resources while also preserving enough nature so those resources can be replenished represents one of the most important questions in environmental science.

There is some good news in our search for a balance between exploitation and preservation. Although deforestation and land degradation are continuing at unacceptable rates—particularly in some developing countries—many places are more densely forested now than they were two centuries ago. Protection of indigenous preserves in Brazil and Australia's Great Barrier Reef shows that we can choose to protect some biodiverse areas in spite of forces that want to exploit them. Overall, nearly 14 percent of the earth's land area is now in some sort of protected status. Although the level of protection in these preserves varies, the rapid recent increase in number and area in protected status exceeds the goals of the United Nations Millennium Project.

We haven't settled the debate between focusing on individual endangered species versus setting aside representative samples of habitat, but pursuing both strategies seems to be working. Protecting charismatic umbrella organisms, such as the chimps of Gombe National Park in Tanzania, can result in preservation of innumerable unseen species. At the same time, protecting whole landscapes for aesthetic or recreational purposes can also achieve the same end.

PRACTICE QUIZ

1. What do we mean by *closed-canopy* forest and *old-growth* forest?
2. What continent is experiencing the greatest forest losses?
3. What is REDD, and how might it work?
4. Why is fire suppression a controversial strategy?
5. What portion of the United States' public rangelands are in poor or very poor condition due to overgrazing? Why do some groups say grazing fees amount to a "hidden subsidy"?
6. What is *rotational grazing*, and how does it mimic natural processes?
7. How do the size and design of nature preserves influence their effectiveness? What do landscape ecologists mean by *interior habitat* and *edge effects*?
8. What percentage of the earth's land area has some sort of protected status? How has the amount of protected areas changed globally?
9. What is *ecotourism*, and why is it important?
10. What is a *biosphere reserve*, and how does it differ from a wilderness area or wildlife preserve?

CRITICAL THINKING AND DISCUSSION

Apply the principles you have learned in this chapter to discuss these questions with other students.

1. Conservationists argue that watershed protection and other ecological functions of forests are more economically valuable than timber. Timber companies argue that continued production supports stable jobs and local economies. If you were a judge attempting to decide which group was right, what evidence would you need on both sides? How would you gather this evidence?

2. Divide your class into a ranching group, a conservation group, and a suburban home-builders group, and debate the merits of subsidized grazing in the American West. What is the best use of the land? What landscapes are most desirable? Why? How do you propose to maintain these landscapes?

3. Calculating forest area and forest losses is complicated by the difficulty of defining exactly what constitutes a forest. Outline a definition for what counts as forest in your area, in terms of size, density, height, or other characteristics. Compare your definition to those of your colleagues. Is it easy to agree? Would your definition change if you lived in a different region?

4. There is considerable uncertainty about the extent of degradation on grazing lands. Suppose you were a range management scientist, and it was your job to evaluate degradation for the state of Montana. What data would you need? With an infinite budget, how would you gather the data you need? How would you proceed if you had a very small budget?

5. Why do you suppose dry tropical forest and tundra are well represented in protected areas, while grasslands and wetlands are protected relatively rarely? Consider social, cultural, geographic, and economic reasons in your answer.

6. Oil and gas companies want to drill in several parks, monuments, and wildlife refuges. Do you think this should be allowed? Why or why not? Under what conditions would drilling be allowable?

DATA ANALYSIS   Detecting Edge Effects

Edge effects are a fundamental consideration in nature preserves. We usually expect to find dramatic edge effects in pristine habitat with many specialized species. But you may be able to find interior-edge differences on your own college campus, or in a park or other unbuilt area near you. Here are three testable questions you can examine using your own local patch of habitat: (1) Can an edge effect be detected or not? (2) Which species will indicate the difference between edge and interior conditions? (3) At what distance can you detect a difference between edge and interior conditions? Go to Connect to find a field exercise that you can do to form and test a hypothesis regarding edge effects in your own area.

TO ACCESS ADDITIONAL RESOURCES FOR THIS CHAPTER, PLEASE VISIT CONNECT AT www.mcgrawhillconnect.com.

You will find LearnSmart, an adaptive learning system, Google Earth™ exercises, additional Case Studies, Data Analysis exercises, and an interactive ebook.

Food and Agriculture

Soybean harvest in the Brazilian state of
Mato Grosso, whose name means "dense forest."

LEARNING OUTCOMES

After studying this chapter, you should be able to answer the following questions:

▶ How many people are chronically hungry, and why does
hunger persist in a world of surpluses?

▶ What are some health risks of undernourishment, poor diet,
and overeating?

▶ What are our primary food crops?

▶ Describe five components of soil.

▶ What was the green revolution?

▶ What are GMOs, and what traits are most commonly
introduced with GMOs?

▶ Describe some environmental costs of farming, and ways we
can minimize these costs.

Farming the Cerrado

A soybean boom is sweeping across South America, changing the ways people eat and farm from Rio to Beijing. A convergence of factors underlie these changes: inexpensive land, new crop varieties, government policies that allow (or encourage) agricultural expansion into Amazon rainforests, and a booming global export market have made South America the fastest-growing agricultural area in the world.

The center of this rapid expansion is the Cerrado, an area of grassland, savanna, and tropical forest roughly equal in size to the American Midwest. The Cerrado stretches from Bolivia and Paraguay across the center of Brazil almost to the Atlantic Ocean (fig. 7.1). Biologically, this rolling expanse of grasslands and tropical woodland is the richest savanna in the world, with at least 130,000 plant and animal species, many of them declining rapidly as the land is cleared for crops.

Until recently, the Cerrado was considered unsuitable for cultivation. Its red, iron-rich soils are highly acidic and poor in essential plant nutrients. The warm, humid climate harbors countless destructive pests and pathogens. For hundreds of years the Cerrado was primarily cattle country with many poor-quality pastures producing low livestock yields.

In recent decades, however, Brazilian farmers have found that applications of lime and phosphorus can quadruple yields of soybeans, corn (maize), cotton, and other valuable export crops. With rapidly expanding global trade, it also now makes economic sense to use soil amendments. More than 40 varieties of soybeans have been developed, specially adapted for the soils and climate of the Cerrado, using both conventional breeding and genetic modification.

Until the 1970s, soybeans were a relatively minor crop in Brazil, but since then the area planted with soy has exploded, from 1 million ha (2.45 million acres) in 1972 to more than 25 million ha (62 million acres, or 80 percent of U.S. soy acreage) in 2012. Brazil now competes with the United States as the world's top soy exporter, shipping some 36 million metric tons per year. With two crops per year, cheap land, low labor costs, favorable tax rates, and yields per hectare equal to those in the American Midwest, Brazilian farmers can produce soybeans for less than half the cost in America.

Rising income in China is a major driver in Brazil's current soy expansion. With more money to spend, the Chinese are consuming much more pork and chicken meat—for which soy is a dominant feed source. (Chinese still eat less than half as much meat as Americans do.) Between 2002 and 2012 alone, China's soy imports more than tripled, to over 46 million metric tons, or about one-third of total global soy shipments.

Beef production has gone hand in hand with the soy boom. Brazil has long been an exporter of grass-fed beef from the country's vast pasture areas (many of them converted from tropical rainforest), and agriculture ministers now aim to double beef exports. Clearing pasture also makes land available for sale to more lucrative soy operations, so that ranchers dramatically aid the transition from forest to crop field.

Increasing demand for both soybeans and beef create land conflicts in Brazil and threaten the region's biodiversity. Deforestation for cropland and pasture expansion is especially severe in the "arc of destruction" between the Cerrado and the Amazon. Small family farms are being gobbled up. Farmworkers, displaced by mechanization, migrate to cities or cut out new clearings in the retreating forest margin. Increasing conflicts between poor farmers and big landowners have led to violent confrontations.

The Landless Workers Movement claims that over 1,200 rural workers have died in Brazil since 1985 as a result of assassinations and clashes over land rights. The 2011 murders of Joao Claudio Ribeiro da Silva, a rubber tapper and leading forest conservationist, and his wife, Maria do Espirito Santo, in the Amazon state of Para, are among the recently reported deaths. Their assassins are unknown, but Joao had received death threats from loggers and ranchers angered by his efforts to protect forests and worker's rights.

Over the past 20 years, Brazil claims to have resettled 600,000 families from the Cerrado. Still, tens of thousands of landless farmworkers and displaced families live in unauthorized squatter camps and shantytowns across the country, awaiting relocation.

▲ **FIGURE 7.1** Brazil's Cerrado, 2 million ha of savanna, grassland, and open woodland, is the site of the world's fastest-growing soybean production. Cattle ranchers and agricultural workers, displaced by mechanized crop production, are moving northward into the "arc of destruction" at the edge of the Amazon rainforest, where the continent's highest rate of forest clearing is occurring.

(continued)

Rapid growth of beef and soy production in Brazil have both positive and negative aspects. On the one hand, more protein-rich food is available to feed the world. The 2 million km² of the Cerrado represents one of the world's last opportunities to open a large area of new, highly productive cropland. On the other hand, the rapid expansion and mechanization of agriculture in Brazil is destroying globally important biodiversity and creating social conflicts. This case study illustrates many of the major themes in this chapter. Will there be enough food for everyone in the world? What will be the environmental and social consequences of producing the nutrition we need? Are there more sustainable ways to achieve these goals? ■

We can't solve problems by using the same kind of thinking we used when we created them.

—ALBERT EINSTEIN

7.1 GLOBAL TRENDS IN FOOD AND HUNGER

Brazilian agriculture represents one of the most dramatic cases of environmental change today. The Cerrado is a long way from Iowa and Illinois, in the heart of the U.S. corn belt, but farmers in those states recognize the changes in the soy fields of South America. Food production has been transformed from small-scale, diversified operations to vast operations of thousands of hectares, growing one or two genetically modified crops, with abundant inputs of fuel and fertilizer, for a competitive global market.

These changes have dramatically increased production, lowered food prices, and provided affordable meat protein in developing countries from Brazil to China. Food production has increased so dramatically that we now use edible corn and sugar to run our cars (chapter 13). According to the International Monetary Fund, 2005 global food costs (in inflation-adjusted dollars) were the lowest ever recorded, less than one-quarter of the cost in the mid-1970s. In the United States and Europe, overproduction has driven prices low enough that we pay farmers billions of dollars each year to take land out of production.

Despite these changes, food costs have risen recently, which spells disaster for the poorest populations. In general, global food supply problems have more to do with distribution than with supplies. We continue to produce surpluses, but hunger remains an urgent problem.

Revolutions in production have also profoundly altered our environment and our diets. In this chapter we'll examine these changes, as well as the ways farmers have managed to feed more and more of the world's growing population. We'll also consider some of the strategies needed for long-term sustainability in food production.

Food security is unevenly distributed

Fifty years ago, hunger was one of the world's most prominent, persistent problems. In 1960, nearly 60 percent of people in developing countries were chronically undernourished, and the world's population was increasing by more than 2 percent every year. Today, some conditions have changed dramatically, while others have changed very little. The world's population has more than doubled, from 3 billion to over 7 billion, but food production has risen even faster. While the average population growth in the past 50 years has been 1.7 percent per year, food production has increased by an average of 2.2 percent per year. Food availability has increased in most countries to well over 2,200 kilocalories, the amount generally considered necessary for a healthy and productive life (fig. 7.2a). Protein intake has also increased in most countries, including China and India, the two most populous countries (fig. 7.2b). Less than 20 percent of people in developing countries now face chronic food shortages, as compared to 60 percent just 50 years ago.

But hunger is still with us. An estimated 900 million people—almost one in every eight people on earth—suffer chronic hunger (fig. 7.3). This number is up slightly from a few years ago, but because of population growth, the *percentage* of malnourished people is still falling (fig. 7.4).

About 95 percent of hungry people are in developing countries. Hunger is especially serious in sub-Saharan Africa, a region plagued by political instability (figs. 7.2, 7.3). Increasingly, we are coming to understand that **food security**, or the ability to obtain sufficient, healthy food on a day-to-day basis, is a combined problem of economic, environmental, and social conditions. Even in wealthy countries such as the United States, millions lack a sufficient, healthy diet. Poverty, job losses, lack of social services, and other factors lead to persistent hunger—and even more to persistent poor nutrition—despite the fact that we have more, cheaper food (in terms of the work needed to acquire it) than almost any society in history.

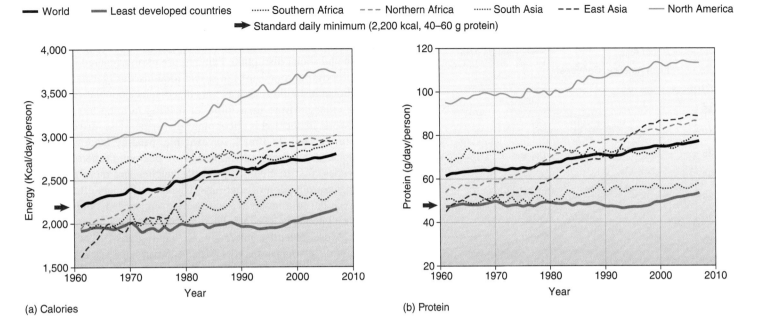

Legend (shared by both graphs):
— World — Least developed countries ⋯⋯ Southern Africa – – – Northern Africa ⋯⋯ South Asia – – – East Asia — North America
→ Standard daily minimum (2,200 kcal, 40–60 g protein)

(a) Calories

(b) Protein

▲ **FIGURE 7.2** Changes in dietary energy (kcal) and protein consumption in selected regions. North America and other developed regions consume more calories and protein than are needed. SOURCE: Data from Food and Agriculture Organization (FAO), 2012.

Food security is important at multiple scales. In the poorest countries, entire national economies can suffer from a severe drought, flood, or insect outbreak. Individual villages also suffer from a lack of food security: a bad crop year can devastate a family or a village, and local economies can collapse if farmers cannot produce enough to eat and sell. Even within families there can be unequal food security. Males often get both the largest share and the most nutritious food, while women and children—who need food most—all too often get the poorest diet. At least 6 million children under 5 years old die every year of diseases exacerbated by hunger and malnutrition. Providing a healthy diet might eliminate as much as 60 percent of all child deaths worldwide. Because

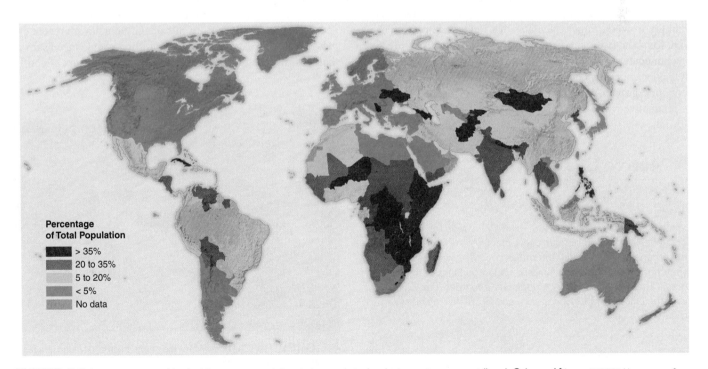

Percentage of Total Population
- > 35%
- 20 to 35%
- 5 to 20%
- < 5%
- No data

▲ **FIGURE 7.3** Hunger rates worldwide. Most severe and chronic hunger is in developing regions, especially sub-Saharan Africa. SOURCE: Hunger map from Food and Agriculture Organization of the United Nations website. Used by permission.

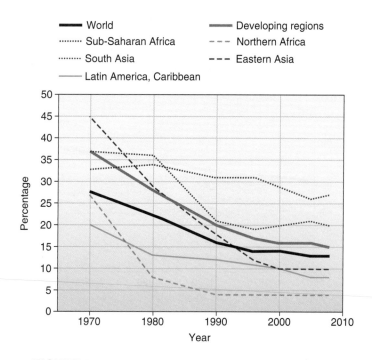

Legend:
- World
- Sub-Saharan Africa
- South Asia
- Latin America, Caribbean
- Developing regions
- Northern Africa
- Eastern Asia

▲ **FIGURE 7.4** Changes in numbers and rates of malnourishment, by region. SOURCE: Data from UN Food and Agriculture Organization, 2012.

high infant mortality is strongly associated with high population growth rates, reducing childhood hunger is an important strategy for slowing global population growth.

Hungry people can't work their way out of poverty. Nobel Prize–winning economist Robert Fogel has estimated that in 1790, 20 percent of the population of England and France were effectively excluded from the labor force because they were too weak and hungry to work. Fogel calculates that improved nutrition can account for about half of all European economic growth during the nineteenth century. This analysis suggests that reducing hunger in today's poor countries could yield more than $120 billion (U.S.) in economic growth, by producing a healthier, longer-lived, and more productive workforce.

Active LEARNING

Mapping Poverty and Plenty

Examine the map in figure 7.3. Using the map of political boundaries at the end of your book, identify ten of the hungriest countries. Then identify ten of the countries with less than 5 percent of people facing chronic undernourishment (green areas). The world's five most populous countries are China, India, the United States, Indonesia, and Brazil. Which classes do these five belong to?

ANSWERS: China, Indonesia, and Brazil have 5–20 percent malnourished; India has 20–35 percent; the United States has <5 percent.

Famines have political and social roots

Famines are large-scale food shortages, with widespread starvation and social disruption. Famines are often triggered by drought or floods, but root causes of severe food insecurity usually include political instability, such as wars that displace populations, removing villagers from their farms or making farming too dangerous to work in the fields. Economic disparities also drive peasants from the land. In Brazil, for example, wealthy landowners have displaced hundreds of thousands of peasant farmers in recent decades, first to create large cattle ranches and more recently to expand production of soybeans. Displaced farmers often have no choice but to migrate to the already overcrowded slums of major cities when they lose their land (chapter 14).

More recently, international "land grabs" have displaced peasants in countries across Africa and parts of Asia. International land speculators, agricultural corporations, and developers have contracted to lease lands occupied by traditional communities but legally owned by governments.

Economist Amartya K. Sen, of Harvard, has shown that while natural disasters often precipitate famines, farmers have almost always managed to survive these events if they aren't thwarted by inept or corrupt governments or greedy elites. Professor Sen points out that armed conflict and political oppression are almost always at the root of famine. No democratic country with a relatively free press, he says, has ever had a major famine.

Famines can lead to long-term disruption, as people are forced to eat their seed grain and slaughter breeding livestock in a desperate attempt to keep themselves and their families alive. Even when better conditions return, they have often sacrificed their productive capacity and will take a long time to recover. Famines often trigger mass migrations to relief camps, where people survive but cannot maintain a healthy and productive life (fig. 7.5). Mass migrations and acute hunger in war-torn Sudan and Somalia are recent examples: in 2011 an estimated 12 to 15 million people faced starvation that was triggered by drought but was rooted in years of conflict.

▲ **FIGURE 7.5** Children wait for their daily ration of porridge at a feeding station in Somalia. When people are driven from their homes by hunger or war, social systems collapse, diseases spread rapidly, and the situation quickly becomes desperate.

China's recovery from the famines of the 1960s is a dramatic example of the relationship between politics and famine (see trends in fig. 7.2). Misguided policies from the central government destabilized farming economies across China. Then two years of bad crops in 1959–1960 precipitated famines that may have killed 30 million people. In recent years, new political and economic policies have transformed access to food. Even though the population has doubled, from 650 million in 1960 to 1.3 billion in 2012, hunger is much less common. China now consumes almost twice as much meat (pork, chicken, and beef) as the United States—and much of this meat comes from livestock fed on soybeans bought from Brazil.

7.2 HOW MUCH FOOD DO WE NEED?

A good diet is essential to keep you healthy. You need a balance of foods to provide the right nutrients, as well as enough calories for a productive and energetic lifestyle. The United Nations Food and Agriculture Organization (FAO) estimates that nearly 3 billion people (almost half the world's population) suffer from vitamin, mineral, or protein deficiencies. These shortages result in devastating illnesses and death, as well as reduced mental capacity, developmental abnormalities, and stunted growth.

A healthy diet includes the right nutrients

Malnourishment is a general term for nutritional imbalances caused by a lack of specific nutrients. In conditions of extreme food shortages, a lack of protein in young children can cause kwashiorkor, which is characterized by a bloated belly and discolored hair and skin. *Kwashiorkor* is a West African word meaning "displaced child." (A young child is displaced—and deprived of nutritious breast milk—when a new baby is born.) Marasmus (from Greek, "to waste away") is another severe condition in children who lack both protein and calories. A child suffering from severe marasmus is generally thin and shriveled (fig. 7.6a). These conditions lower resistance to disease and infections, and children may suffer permanent debilities in mental, as well as physical, development.

Deficiencies in vitamin A, folic acid, and iodine are more widespread problems. Vitamin A and folic acid are found in vegetables, especially dark green leafy vegetables. Deficiencies in folic acid have been linked to neurological problems in babies. Shortages of vitamin A cause an estimated 350,000 people to go blind every year. Dr. Alfred Sommer, an ophthalmologist from Johns Hopkins University, has shown that giving children just two cents' worth of vitamin A twice a year could prevent almost all cases of childhood blindness and premature death associated with shortages of vitamin A. Vitamin supplements also reduced maternal mortality by nearly 40 percent in one study in Nepal.

Iodine deficiencies can cause goiter (fig. 7.6b), a swelling of the thyroid gland. Iodine is essential for synthesis of thyroxin, an endocrine hormone that regulates metabolism and brain development, among other things. The FAO estimates that 740 million people, mostly in Southeast Asia, suffer from iodine deficiency, including 177 million children whose development and growth have been stunted. Developed countries have largely eliminated this problem by adding iodine to our table salt.

(b) Goiter

(a) Marasmus

▲ **FIGURE 7.6** Dietary deficiencies can cause serious illness. (a) Marasmus results from protein and calorie deficiency and gives children a wizened look and dry, flaky skin. (b) Goiter, a swelling of the thyroid gland, results from an iodine deficiency.

Starchy foods, such as maize, polished rice, and manioc (tapioca), form the bulk of the diet for many poor people, but these foods are low in several essential vitamins and minerals. One celebrated effort to deliver crucial nutrients has been through genetic engineering of common foods, such as "golden rice," developed by Monsanto to include a gene for producing vitamin A. This strategy has shown promise, but it also has critics, who argue that genetically modified rice is too expensive for poor populations. In addition, the herbicides needed to grow the golden rice kill the greens that villagers rely on to provide their essential nutrients.

The best human diet is mainly vegetables and grains, with moderate amounts of eggs and dairy products, and sparing amounts of meat and oils. A solid base of regular exercise underpins an ideal diet outlined by Harvard dieticians (fig. 7.7) Modest amounts of fats are essential for healthy skin, cell function, and metabolism. But your body is not designed to process excessive amounts of fats or sugars, and these should be consumed sparingly.

Overeating is a growing world problem

For the first time in history, there are probably more overweight people (more than 1 billion) than underweight people (about 850 million). This trend isn't limited to richer countries. Obesity is spreading around the world (fig. 7.8). Diseases once thought to afflict only wealthy nations, such as heart attack, stroke, and diabetes, are now becoming the most prevalent causes of death and disability everywhere (chapter 8).

▲ FIGURE 7.7 The Harvard food pyramid emphasizes fruits, vegetables, and whole grains as the basis of a healthy diet. Unlike most other representations of a healthy diet, this one rests on a foundation of exercise, and it distinguishes the value of whole grains from white bread and starches.

In the United States, and increasingly in Europe, China, and developing countries, highly processed sugary, fatty foods have become a large part of our diet. Some 64 percent of adult Americans are overweight, up from 40 percent only a decade ago. About

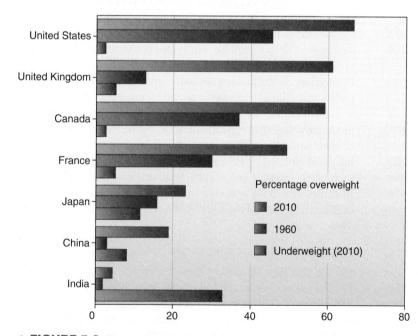

▲ FIGURE 7.8 Chronic obesity is a growing problem worldwide. In wealthier countries and many developing areas, the number overweight vastly exceeds the number underweight. DATA SOURCE: World Health Organization, 2010.

one-third of us are seriously overweight, or **obese**—generally considered to mean more than 20 percent over the ideal weight for a person's height and sex.

Being overweight increases your risk of hypertension, diabetes, heart attacks, stroke, gallbladder disease, osteoarthritis, respiratory problems, and some cancers. Every year about 400,000 people in the United States die from illnesses related to obesity. This number is approaching the number related to smoking (435,000 annually). Paradoxically, food insecurity and poverty can contribute to obesity. In one study, more than half the women who reported not having enough to eat were overweight, compared with one-third of the food-secure women. A lack of time for cooking, limited access to healthy food choices, and ready availability of fast-food snacks and calorie-laden soft drinks, often lead to dangerous dietary imbalances for many people.

More production doesn't necessarily reduce hunger

Most strategies for reducing world hunger have to do with increasing efficiency of farm production, expanding use of fertilizers and improved seeds, and converting more unused land or forest to agriculture. But the prevalence of obesity, and unstable farm economies, suggest that lack of supply is not necessarily the principal cause of world hunger. An overabundance of food supplies in much of the world suggests that answers to global hunger may lie in better use and distribution of food resources.

For most farmers in the developed world, overproduction constantly threatens prices for farm products. To reduce food supplies and stabilize prices in the United States, Canada, and Europe, we send millions of tons of food aid to developing areas every year. Often, however, these shipments of free food destabilize farm economies in receiving areas. Prices for local farm products collapse, and political corruption can expand if war lords control distribution. Even in developing areas, lack of food production is not always the cause of hunger.

There are also inefficiencies in food use. Global food waste amounts to some 30 percent of all food production—1.3 billion tons annually—as food is spoiled in storage and transit, used inefficiently, or thrown away after preparation. We also prefer inefficient foods—particularly meats that require 4 to 10 times as much animal feed as is produced in meat.

In the past decade, global hunger has had two newer causes: international financial speculation on food commodities and biofuel production. Food has long been a globally traded commodity, but a new law passed by the U.S. Congress in 2000, the Commodities Futures Modernization Act, eliminated long-standing rules restricting risky speculation on a wide variety of commodities. Global food products were among these commodities. Freed from restraints, traders gambled on the value of future crops, which pushed up the price of current crops. This deregulation had far-reaching affects for ordinary people. (In the U.S., a widely felt effect of the rule change was the acceleration of trade in home mortgages. This led to a financial "bubble" in home prices, followed by a collapse in which millions of Americans lost their homes.) In 2008, speculative trading in food commodities led to a crisis in food, as a frenzy of speculative trading drove prices of agricultural commodities well beyond their actual value. In response to this bubble in global prices, local food prices rose sharply in towns and villages worldwide. The cost of some basic staples, such as cooking oil and rice, quadrupled. Farm land prices rose as well, often driving struggling peasant farmers from their land and livelihoods. Thus the effects of frenzied Wall Street trading in wheat, food oils, and other commodities were felt in households worldwide. These changes were inconvenient in wealthy countries, where prices rose on grocery store shelves. In poorer countries, food costs can make up as much as 80 percent of household expenses, and there the effect was far worse. Food riots followed in the Philippines, India, Indonesia, and many other countries.

Biofuels have boosted commodity prices

Food prices declined again after 2008, but they were shored up by an additional change: new policies in the United States and Europe promoting biofuels. Using crops such as soy, corn, palm oil, or sugar cane to drive our cars is an important strategy for supporting farm economies in wealthy countries, as the increased demand keeps prices high. These new policies also led to global increases production of these crops reflects these new policies (table 7.1). In the United States, federal ethanol subsidies led to a doubling of corn prices in 2007. But in developing countries, production of soy, palm oil, and other products for export often displaces food production. When biofuel policies compete with food supplies in Asia and Africa, the efficiency of this market becomes less clear.

There has been much debate about the environmental and economic costs and benefits of biofuel production. Some studies have found that biofuels represent a net energy loss, taking more energy to produce than they provide in fuel. In the United States and Europe, the production of biofuels, especially ethanol, has not been economically viable without heavy subsidies for growing and processing crops. On the other hand, plant oils from sunflowers, soybeans, corn, and other oil seed crops can be burned directly in most diesel engines, and may be closer than ethanol to a net energy gain (chapter 12). Brazil produces ethanol more efficiently from its tropical sugarcane crops—although the net energy balance and environmental impacts remain serious questions.

Do we have enough farmland?

The problem of farmland availability has always been a central question in debates about how to feed the world. Because we currently produce more than enough to feed the world we probably do have enough farmland to feed more people than currently live on earth—if we ate according to the recommendations in fig. 7.7, and if we used all farm products for food, rather than fuel and livestock feed. Could we expand production further by clearing new farmlands? That is harder. Tropical soils often are deeply weathered and infertile, so they make poor farmland unless expensive inputs of lime and fertilizer are added. Much of the world's uncultivated land is too steep, sandy, waterlogged, salty, acidic, cold, dry, or nutrient-poor for most crops.

About 11 percent of the earth's land area, some 1,400 million ha, is used for agricultural production. This land area amounts to about 0.2 ha per person (1 ha = 100 m × 100 m). Arable land per person has fallen from about 0.5 ha per person in 1960, mainly because of population growth. Population projections suggest we will have about 0.15 ha by 2050. In Asia, cropland will be even more scarce—0.09 ha per person by 2050.

Much of the world's uncultivated land could potentially be converted to cropland, but this land currently provides essential ecological services on which farmers depend. Forested watersheds help maintain stream flow, regulate climate, and provide refuge for biological and cultural diversity. Wetlands also regulate water supplies, and forests and grasslands support insect pollinators that ensure productive crops.

Despite the importance of these services, the conversion of tropical forests and savannas to farmland continues, spurred by global trade in export crops, such as soy, palm oil, sugar cane (for ethanol), and corn (maize, largely for animal feed), as well as beef and other livestock. The FAO reports that 13 to 16 million ha of forest land is cleared each year, about half of that in tropical Africa and South America. Clearing new cropland has helped increase the world's food production capacity. Brazil, for example, has become a global leader in exports of soy, sugar, beef, poultry, and orange juice, and it is emerging as a leading producer of rice and corn. Most of these export crops serve those who already have enough to eat, however: according to FAO data, 85 percent of global crop exports are bound for Europe, North America, China, Japan, and other wealthy countries. Meanwhile, human rights groups protest that land conversions have displaced traditional communities and subsistence farmers across South America, Africa, and much of Asia. Many displaced farmers end up in the rapidly growing cities of the developing world (chapter 15).

TABLE 7.1 | Key Global Food Sources

CROP	1990**	2010**
Coarse Grains*	841	1,109
Vegetables, melons	466	965
Maize (corn)	483	844
Rice	518	672
Wheat	592	651
Cassava, sweet potatoes	275	335
Potatoes	266	324
Soybeans	108	261
Oil palm fruit	60	210
Oil crops (other)	75	168
Pulses (beans, peas)	59	67
Meat and milk	722	1,013
Fish and seafood	76	144

*Barley, oats, sorghum, rye, millet.; **Production in million metric tons
SOURCE: Data from Food and Agriculture Organization (FAO), 2012.

7.3 WHAT DO WE EAT?

Of the thousands of edible plants and animals in the world, only a few provide almost all our food. About a dozen types of grasses, three root crops, twenty or so fruits and vegetables, six mammals, two domestic fowl, and a few fish species make up almost all the food we eat (table 7.1). Two grasses, wheat and rice, are especially important because they are the staple foods for most of the 5 billion people in developing countries.

In the United States, corn (another grass, also known as maize) and soybeans have become our primary staples. We rarely eat either corn or soybeans directly, but corn provides the corn sweeteners, corn oil, and the livestock feed for producing our beef, chicken, and pork, as well as industrial starches and many synthetic vitamins. Soybeans are also fed to livestock, and soy provides protein and oils for processed foods. Because we have developed so many uses for corn, it now accounts for nearly two-thirds of our bulk commodity crops (which include corn, soy, wheat, rice, and other commodities). Together, corn and soy make up about 85 percent of commodity crops, with an annual production of 268 million tons of corn and 88 million tons of soy (fig. 7.9).

Rising meat production is a sign of wealth

Largely because of dramatic increases in corn and soy production, meat consumption has grown worldwide. In developing countries, meat consumption has risen from just 10 kg per person per year in the 1960s to over 26 kg today (fig. 7.10). In the United States, our meat consumption has risen from 90 kg to 136 kg per person per year in the same 50 years. Meat is a concentrated, high-value source of protein, iron, fats, and other nutrients that give us the energy to lead productive lives. Dairy products are also a key protein source: globally we consume more than twice as much dairy as meat. But dairy production per capita has declined slightly while global meat production has more than doubled in the past 50 years.

▲ **FIGURE 7.10** Meat and dairy consumption has quadrupled in the past 40 years, and China represents about 40 percent of that increased demand.

Meat is a good indicator of wealth because it is expensive to produce in terms of the resources needed to grow an animal (fig. 7.11). As discussed in chapter 2, herbivores use most of the energy they consume in growing muscle and bone, moving around, staying warm, and metabolizing (digesting) food. Only a little food energy is stored for consumption by carnivores, at the next level of the food pyramid. For every 1 kg of beef, a steer consumes over 8 kg of grain. Pigs are more efficient: Producing 1 kg of pork takes just 3 kg of pig feed. Chickens and herbivorous fish (such as catfish) are still more efficient. Globally, some 660 million metric tons of cereals are used as livestock feed each year. This is just over a third of the world cereal use. As figure 7.11 suggests, we could feed at least 8 times as many people by eating those cereals directly.

A number of technological and breeding innovations have made this increased production possible. One of the most important is the **confined animal feeding operation (CAFO),**

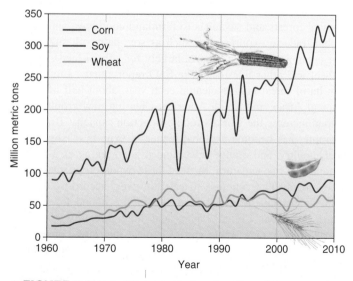

▲ **FIGURE 7.9** United States production of our three dominant crops, corn, soybeans, and wheat. SOURCE: Data from USDA and UN FAO, 2012.

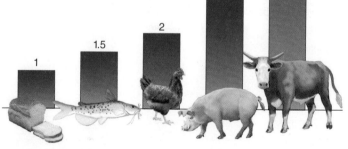

▲ **FIGURE 7.11** Number of kilograms of grain needed to produce 1 kg of bread or 1 kg live weight gain.

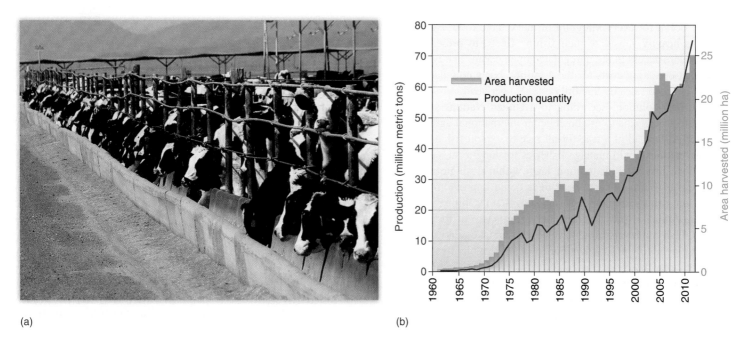

(a) (b)

▲ **FIGURE 7.12** (a) Confined animal feeding operations have expanded the global market for corn and soybeans. (b) Brazil's soy production has grown from near zero to being the country's dominant agricultural product. SOURCE: Data from UN Food and Agriculture Organization 2012.

where animals are housed and fed—mainly soy and corn—for rapid growth (fig. 7.12). These operations dominate livestock raising in the United States, Europe, and increasingly in China and other countries. Animals are housed in giant enclosures, with up to 10,000 hogs or a million chickens in an enormous barn complex (fig. 7.13), or 100,000 cattle in a feed lot. Operators feed the animals specially prepared mixes of corn, soy, and animal protein (parts that can't be sold for human food) that maximizes their growth rate. New breeds of livestock have been developed that produce muscle (meat) rapidly, rather than simply getting fat. The turnaround time is getting shorter, too. A U.S. chicken

▲ **FIGURE 7.13** Most livestock grown in the United States are raised in large-scale, concentrated animal-feeding operations. Up to a million animals can be held in a single facility. High population densities require heavy use of antibiotics and can cause severe local air and water pollution.

producer can turn baby chicks into chicken nuggets after just 8 weeks of growth. Steers reach full size by just 18 months of age. Increased use of antibiotics, which are mixed in daily feed, make it possible to raise large numbers of animals in close quarters. About 90 percent of U.S. hogs receive antibiotics in their feed.

Seafood, both wild and farmed, depends on wild-source inputs

The 140 million metric tons of seafood we eat every year is an important part of our diet. Seafood provides about 15 percent of all animal protein eaten by humans, and it is the main animal protein source for about 1 billion people in developing countries. Over-harvesting and habitat destruction threaten most of the world's wild fisheries, however. Annual catches of ocean fish rose by about 4 percent annually between 1950 and 1988. Since 1989, 13 of 17 major marine fisheries have declined dramatically or become commercially unsustainable. According to the United Nations, three-quarters of the world's edible ocean fish, crustaceans, and mollusks are declining and in urgent need of managed conservation.

The problem is too many boats using efficient but destructive technology to exploit a dwindling resource base. Boats as big as ocean liners travel thousands of kilometers and drag nets large enough to scoop up a dozen jumbo jets, sweeping a large patch of ocean clean of fish in a few hours. Longline fishing boats set cables up to 10 km long with hooks every 2 meters that catch birds, turtles, and other unwanted "by-catch" along with targeted species. Trawlers drag heavy nets across the bottom, scooping up everything indiscriminately and reducing broad swaths of habitat to rubble. One marine biologist compared the technique to harvesting forest mushrooms with a bulldozer. In some operations, up to 15 kg of dead bycatch (nontarget species) are dumped back into the ocean for every kilogram of marketable food. Nearly all countries heavily subsidize their fishing fleets, to preserve jobs and to ensure access in the unregulated open ocean. The FAO estimates that operating costs for the 4 million boats now harvesting wild fish exceed sales by $50 billion (U.S.) per year.

The best solution, according to a United Nations study on ecosystem services, is to establish better international agreements on fisheries. Instead of a free-for-all race to exploit fish as fast as possible, nations could manage fisheries for long-term, sustained production. Just as with hunting laws within a single country, agreement to international fishery rules could improve total food production, fishery employment, and ecosystem stability.

Aquaculture is providing an increasing share of the world's seafood. Fish can be grown in farm ponds that take relatively little space but are highly productive. For cultivation of plant-eating fish such as tilapia, these systems can be very sustainable. Cultivation of high-value carnivorous species such as salmon, however, threatens wild fish populations, which are caught to feed captive fish. A 2012 study of fish catches for fish farms has shown that global declines of seabird populations, such as puffins and albatrosses, closely tracks the increase in farm-raised carnivorous fish.

In tropical areas, coastal fish and shrimp-rearing ponds have replaced hundreds of thousands of hectares of mangrove forests and wetlands, which serve as irreplaceable nurseries for marine species. Net pens anchored in near-shore areas encourage spread of diseases, as they release feces, uneaten food, antibiotics, and other pollutants into surrounding ecosystems (fig. 7.14).

Biohazards arise in industrial production

Increasingly efficient production has a variety of externalized (unaccounted for) costs. Land conversion from forest or grassland to crop fields increases soil erosion, which degrades water quality.

▲ **FIGURE 7.14** Pens for fish-rearing in Thailand.

Bacteria in the manure in the feedlots, or liquid wastes in manure storage lagoons (holding tanks) around hog farms, escapes into the environment—from airborne fecal dust around feedlots or from breaches in the walls of a manure tank. When Hurricane Floyd hit North Carolina's coastal hog production region in 1999, some 10 million m^3 of hog and poultry waste overflowed into local rivers, creating a dead zone in Pamlico Sound. While this was an especially severe disaster, smaller spills and leaks to groundwater are widespread.

Constant use of antibiotics has been associated with antibiotic-resistant infections in hog-producing areas. Eighty percent of all antibiotics used in the United States are administered to livestock. This massive and constant exposure kills off most bacteria, but occasional strains that resist the effects the antibiotics survive, creating new strains that are untreatable in human hosts. Thus overuse is slowly rendering our standard antibiotics ineffective for human health care, and thousands of people have died from antibiotic-resistant infections. The next time you are prescribed an antibiotic by your doctor, you might ask whether she or he worries about antibiotic-resistant bacteria. What would you do if your prescription was ineffective against an illness?

Although the public is increasingly aware of these environmental and health risks of concentrated meat production, we seem to be willing to accept these risks because this production system has made our favorite foods cheaper, bigger, and more available. A fast-food hamburger today is more than twice the size it was in 1960, especially if you buy the kind with multiple patties and special sauce, and Americans love to eat them. At the same time, this larger burger costs less per pound, in inflation-adjusted dollars, than the 1960 version. As a consequence, for much of the world, consumption of protein and calories has climbed beyond what we really need to be healthy (figs. 7.2, 7.8).

As environmental scientists, we are faced with a conundrum: Improved efficiency has great environmental costs; it has also given us the abundant, inexpensive foods that we love. We have more protein, but also more obesity, heart disease, and diabetes than we ever had before. What do you think? Do the environmental risks balance a globally improved quality of life? Should we consider reducing our consumption to lower environmental and health costs? How might we do that?

7.4 LIVING SOIL IS A PRECIOUS RESOURCE

Understanding the limits and opportunities for feeding the world requires an understanding of the soil that supports us. In this section we examine the nature of soils and the ways we use them. Many of us think of soil as just dirt. But healthy soil is a marvelous substance with astonishing complexity. Soil contains mineral grains weathered from rocks, partially decomposed organic molecules, and a host of living organisms. The complex community of bacteria and fungi are primarily responsible for providing nutrients that plants need to grow. Soil bacteria also help filter and purify the water we drink. Soil can be considered a living ecosystem by itself. How can we manage our soils to maintain and build these precious systems?

Building a few millimeters of soil can take anything from a few years (in a healthy grassland) to a few thousand years (in a desert or tundra). Under the best circumstances, topsoil accumulates at about 1 mm per year. With careful husbandry that prevents erosion and adds organic material, soil can be replenished and renewed indefinitely (fig. 7.15). But many farming techniques deplete soil. Crops consume the nutrients; plowing exposes the soil to erosion by wind or water. Severe erosion can carry away 25 mm or more of soil per year, far more than can accumulate under the best of conditions.

What is soil?

Soil is a complex mixture of six components:

1. *sand and gravel* (mineral particles from bedrock, either in place or moved from elsewhere, as in wind-blown sand)
2. *silts and clays* (extremely small mineral particles; many clays are sticky and hold water because of their flat surfaces and ionic charges; others give red color to soil)
3. *dead organic material* (decaying plant matter stores nutrients and gives soils a black or brown color)
4. *soil fauna and flora* (living organisms, including soil bacteria, worms, fungi, roots of plants, and insects, recycle organic compounds and nutrients)
5. *water* (moisture from rainfall or groundwater, essential for soil fauna and plants)
6. *air* (tiny pockets of air help soil bacteria and other organisms survive)

Variations in these six components produce almost infinite variety in the world's soils. Abundant clays make soil sticky and wet. Abundant organic material and sand make the soil soft and easy to dig. Sandy soils drain quickly, often depriving plants of moisture. Silt particles are larger than clays and smaller than sand, so they aren't sticky and soggy, and they don't drain too quickly. Thus silty soils are ideal for growing crops, but they are also light and blow away easily when exposed to wind. Soils with abundant soil fauna quickly decay dead leaves and roots, making nutrients available for new plant growth. Compacted soils have few air spaces, making soil fauna and plants grow poorly. You can see some of these differences just looking at soil. Reddish soils are colored by iron-rich, rust-colored clays, the kind that store few nutrients for plants. Deep black soils are rich in decayed organic material, and thus are generally rich in nutrients.

Healthy soil fauna can determine soil fertility

Soil bacteria, algae, and fungi decompose and recycle leaf litter into plant-available nutrients, as well as helping to give soils structure and loose texture (fig. 7.16). Microscopic worms and nematodes process organic matter and create air spaces as they burrow through soil. These organisms mostly stay near the surface, often within the top few centimeters. The sweet aroma of freshly turned soil is caused by actinomycetes, bacteria that grow in fungus-like strands and give us the antibiotics streptomycin and tetracycline.

The health of the soil ecosystem depends on environmental conditions, including climate, topography, and parent material (the mineral grains or bedrock on which soil is built), and frequency of disturbance. Too much rain washes away nutrients and organic matter, but soil fauna cannot survive with too little rain. In extreme cold, soil fauna recycle nutrients slowly; in extreme heat they may work so fast that leaf litter on the forest floor is taken up by plants in just weeks or months—so that the soil retains little organic matter. Frequent disturbance prevents the development of a healthy soil ecosystem, as does steep topography that allows rain to wash away soils. In the United States, the best farming soils tend to occur where the climate is not too wet or dry, on glacial silt deposits such as those in the upper Midwest, and on silt and clay-rich flood deposits, like those along the Mississippi River.

Most soil fauna occur in the uppermost layers of a soil, where they consume leaf litter. This layer is known as the "O" (organic) horizon. Just below the O horizon is a layer of mixed organic and mineral soil material, called the A horizon (fig. 7.17), also known as **topsoil** or surface soil.

The B horizon, or **subsoil**, lies below most organic activity, and it tends to have more clays than the A layer. The B layer accumulates clays that seep downward from the A horizon with rainwater that percolates through the soil. If you dig a hole, you may be able to tell where the B horizon begins, because the soil tends to become slightly more cohesive. If you squeeze a handful of B soil, it should hold its shape better than a handful of A soil.

▼ **FIGURE 7.15** Terracing, as in these Balinese rice paddies, can control erosion and make steep hillsides productive. These terraced rice paddies have produced two or three crops a year for centuries because the soils are carefully managed and organic nutrients are maintained.

▲ **FIGURE 7.16** Soil ecosystems include countless organisms that consume and decompose organic material, aerate soil, and distribute nutrients through the soil.

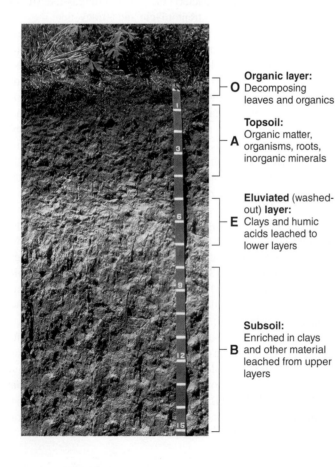

O Organic layer:
Decomposing
leaves and organics

A Topsoil:
Organic matter,
organisms, roots,
inorganic minerals

E Eluviated (washed-out) layer:
Clays and humic
acids leached to
lower layers

B Subsoil:
Enriched in clays
and other material
leached from upper
layers

Sometimes an E (eluviated, or washed-out) layer lies between the A and B horizons. The E layer is loose and light-colored because most of its clays and organic material have been washed down to the B horizon. The C horizon, below the subsoil, is mainly decomposed rock fragments. Parent materials underlie the C layer. Parent material is the sand, wind-blown silt, bedrock, or other mineral material on which the soil is built. About 70 percent of the parent material in the United States was transported to its present site by glaciers, wind, and water, and is not related to the bedrock formations below it.

Your food comes mostly from the A horizon

Ideal farming soils have a thick, organic-rich A horizon. The soils that support the corn belt farm states of the U.S. Midwest have rich, black A horizon that can be more than two meters thick, although a century of farming has washed away much of this soil. Most soils have less than half a meter of A horizon. Desert soils, with slow rates of organic activity, might have almost no O or A horizons (fig. 7.18).

◀ **FIGURE 7.17** A sample soil profile showing common horizons. Soils vary greatly in composition and thickness of layers. Below the B horizon there is generally a C horizon of weathered rock, sand, or other parent material. Depth is marked in 10-cm increments.

(a)

(b)

▲ **FIGURE 7.18** Soils vary dramatically among different climate areas. Temperate grassland soils tend to have a thick, soft, organic-rich A horizon (a). Arid land soils may have little or no workable A horizon, as in this Libyan valley (b).

Because topsoil is so important to our survival, we differentiate soils largely according to the thickness and composition of their upper layers. The U.S. Department of Agriculture classifies the thousands of different soil types into 11 soil orders (http://soils.usda.gov/technical/classification/orders/). Mollisols (*mollic* = soft, *sol* = soil), for example, have a thick, organic-rich A horizon that develops from the deep, dense roots of prairie grasses. Alfisols (*alfa* = first) have a slightly thinner A horizon, with slightly less organic matter. Alfisols develop in deciduous forests, where leaf litter is abundant. In contrast, the aridisols (*arid* = dry) of the desert Southwest have little organic matter, and they often have accumulations of mineral salts. Organic-rich mollisols and alfisols dominate most of the farming regions of the United States (fig. 7.19).

How do we use and abuse soil?

Agriculture both causes and suffers from environmental degradation. The International Soil Reference and Information Centre in the Netherlands estimates that, every year, 3 million ha of cropland are made useless by erosion, 4 million ha are turned into deserts, and 8 million ha are converted to nonagricultural uses, such as homes, highways, shopping centers, factories, and reservoirs. Over the past 50 years, some 1,900 million ha of agricultural land (an area greater than that now in production) have been degraded to some extent. About 300 million ha of this land are strongly degraded, with deep gullies, severe nutrient depletion, or poor crop growth, making restoration difficult and expensive.

Water and wind erosion cause a vast majority of global soil degradation. Chemical degradation, a secondary cause of soil degradation, includes nutrient depletion, salt accumulation, acidification,

and pollution. Physical degradation includes factors such as compaction or loss of soil structure after cultivation by heavy machinery, waterlogging from excess irrigation, and laterization—rock-hard solidification of tropical soils exposed to sun and rain.

Water is the leading cause of soil loss

Erosion is an important natural process that redistributes the products of geologic weathering. Erosion contributes to soil formation, creating regions of wind-blown silt that form much of the U.S.

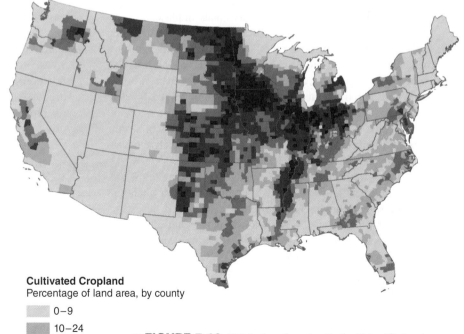

Cultivated Cropland
Percentage of land area, by county

- 0–9
- 10–24
- 25–49
- 50–74
- 75–100

▲ **FIGURE 7.19** Distribution of cropland in the United States, by county. Dark blue areas especially highlight the fertile farm belt of the Midwest and Great Plains, as well as Mississippi river bottom.
SOURCE: U.S. Census of Agriculture.

(a) Sheet erosion   (b) Gully erosion   (c) Wind erosion

▲ **FIGURE 7.20** Land degradation affects more than 1 billion ha yearly, or about two-thirds of all global cropland. Globally, erosion by water accounts for about 56 percent of soil loss from fields (a, b). Wind erosion accounts for another 28 percent (c).

Farm Belt, and to soil loss, as well as rich river bottom soils on which early civilizations were built. Erosion also depletes farmland soils when they are left bare or worked at the wrong time.

Some erosion occurs so rapidly that you can watch it happen. Running water can scour deep gullies, leaving fence posts and trees standing on pedestals as the land erodes away around them. But most erosion is more subtle. **Sheet erosion** removes a thin layer of soil as a sheet of water flows across a nearly level or gently sloping field. When small rivulets of running water form and cut small channels in the soil, the process is called **rill erosion** (fig. 7.20a). Where rills expand to form bigger channels or ravines, too large to be removed by cultivation, we call the process **gully erosion** (fig. 7.20b). **Streambank erosion** also is a problem in farmland streams, where soil washes away from stream banks. Cattle and other livestock often accelerate streambank erosion, as they trample and graze down the vegetation that holds a bank in place.

Sheet and rill erosion cause most soil loss on farmland, even though this process is often too gradual to notice. Studies in Iowa farm fields have found areas with as much as 50 metric tons lost per hectare during winter and spring runoff, more than 12 times the rate considered tolerable or officially acknowledged by state soil scientists. That loss represents just a few millimeters of soil over the whole surface of the field, but repeated years of storms and spring runoff are gradually sending Iowa's productive capacity downstream to the Gulf of Mexico.

All this waterborne sediment damages aquatic systems, too. Increased sediment loads erode rivers and reduce water quality in lakes. Excess nutrients create eutrophic "dead zones" in estuaries and coastal waterways, most famously in the Gulf of Mexico. Even infrastructure such as reservoirs and harbors can be badly impacted by sediment deposits.

Wind is a close second in erosion

Wind can equal or exceed water in erosive force, especially in dry regions, and where vegetation is sparse (fig. 7.20c). When plant cover and surface litter are removed from the land by agriculture or grazing, wind lifts loose soil particles and sweeps them away. In extreme conditions, windblown dunes encroach

on useful land and cover roads and buildings. Over the past 30 years, China has lost 93,000 km² (about the size of Indiana) to desertification, or conversion of productive land to desert. Advancing dunes from the Gobi desert are now only 160 km (100 mi) from Beijing. Every year more than 1 million tons of sand and dust blow from Chinese drylands, often traveling across the Pacific Ocean to the West Coast of North America. Since 1985, China has planted more than 40 billion trees to try to stabilize the soil and hold back deserts.

Intensive farming practices are largely responsible for this situation. Row crops, such as corn and soybeans, leave soil exposed for much of the growing season. Deep plowing and heavy herbicide applications create weed-free fields that look tidy but are subject to erosion. Farmers, under pressure to maximize yields and meet loan obligations, often plow through grass-lined watercourses (low areas where water runs off after a rain) and pull out windbreaks and fencerows to accommodate the large machines and to get every last square meter into production.

An estimated 25 billion metric tons of soil are lost from croplands every year due to wind and water erosion. The net effect, worldwide, of this widespread topsoil erosion is a reduction in crop production equivalent to removing about 1 percent of the world's cropland each year.

7.5 AGRICULTURAL INPUTS

Soil is the foundation of food production, but there are many other critical factors. Reliable water resources, nutrients, favorable temperatures and rainfall, productive crop varieties, and the mechanical energy to tend and harvest the crops are also essential. Strategies for applying these different inputs vary greatly among regions and among contrasting farming strategies (see Key Concepts, p. 168).

High yields usually require irrigation

Agriculture uses at least two-thirds of all fresh water withdrawn from rivers, lakes, and groundwater. Although estimates vary widely, about 15 percent of all cropland, worldwide, is irrigated.

Some countries are water rich and can readily afford to irrigate farmland, while other countries are water poor and must use water very carefully. The efficiency of irrigation water use varies greatly. In some places, high evaporative and seepage losses from unlined and uncovered canals can mean that as much as 80 percent of water withdrawn for irrigation never reaches its intended destination. Poor farmers may over-irrigate because they lack the technology to meter water and distribute just the amount needed. In wealthier countries, farmers can afford sometimes-extravagant uses of water, including center-pivot irrigation systems (fig. 7.21). They can also afford water-saving methods such as drip irrigation, which waters only the base of a crop, reducing evaporative losses.

Excessive use not only wastes water but also often results in **waterlogging**. Waterlogged soil is saturated with water, and plant roots die from lack of oxygen. **Salinization**, in which mineral salts accumulate in the soil, is often a problem when irrigation water dissolves and mobilizes salts in the soil. As the water evaporates, it leaves behind a salty crust on the soil surface that is lethal to most plants. Flushing with excess water can wash away this salt accumulation, but the result is even more saline water for downstream users.

The FAO estimates that 20 percent of all irrigated land is damaged to some extent by waterlogging or salinity. Water conservation techniques can greatly reduce problems arising from excess water use.

Fertilizers boost production

Plants require small amounts of inorganic nutrients from soil. In large-scale farming, fertilizers are used to ensure a sufficient supply of these nutrients. The major elements required by most plants are nitrogen, phosphorus, potassium, and calcium, with lesser amounts of magnesium and sulfur. Nitrogen, a component of all living cells, is the most common limiting factor for plant growth. Phosphorus and potassium can also be limited in supply, so nitrogen, phosphorus, and potassium are our primary fertilizers. In rainy regions such as Brazil, calcium and magnesium often are washed away and must be supplied in the form of lime. Much of the doubling in worldwide crop production since 1950 has come from increased inorganic fertilizer use. In 1950 the average amount of fertilizer used was 20 kg per hectare. In 2000 this had increased to an average of 90 kg per hectare worldwide.

Overfertilizing is also a common problem. Growing plants can use only limited amounts of nutrients, and the nutrients not captured by crops run off of fields or seep into groundwater. Excess fertilizers contaminate drinking water and destabilize aquatic ecosystems. Phosphates and nitrates from farm fields and cattle feedlots are a major cause of aquatic ecosystem pollution. Nitrate levels in groundwater have risen to dangerous levels in many areas where intensive farming is practiced. Young children are especially sensitive to the presence of nitrates. Using nitrate-contaminated water to mix infant formula can be fatal for newborns.

There is considerable potential for expanding the world's food supply by increasing fertilizer use in low-income countries, although this would threaten water supplies. African farmers, for instance, use an average of only 19 kg of fertilizer per hectare, less than one-fourth of the world average. The developing world could triple its crop production by raising fertilizer use to the world average, according to some estimates. Finite global supplies of phosphorus would deplete rapidly with this increase in use, however.

Enriching organic material in soils is an alternative way to fertilize crops. Manure is an important natural source of soil nutrients. Cover crops also can be grown and then plowed into the soil. Nitrogen-fixing bacteria living symbiotically in root nodules of legumes are valuable for making nitrogen available as a plant nutrient (chapter 2). Interplanting and rotating beans or some other leguminous crop with such crops as corn and wheat are traditional ways of increasing nitrogen availability.

Modern agriculture runs on oil

The food system in the United States consumes about 16 percent of the total energy we use. Most of our foods require more energy to produce, process, and get to market than they yield when we eat them. Reliance on fossil fuels began in the 1920s, with the adoption of tractors. Energy use increased sharply after World War II, with the invention of nitrogen fertilizer produced from natural gas. Reliance on diesel and gasoline to run tractors, combines, irrigation pumps, and other equipment has continued to grow since then. David Pimentel, of Cornell University, has calculated that each hectare of corn produced in the United States consumes the equivalent of 800 liters of oil (5 barrels of oil). One third of this energy is used in producing nitrogen fertilizers from natural gas. Another third is inputs for machinery and fuel. The remaining third is used for irrigation, synthetic herbicides, and other fertilizers.

After crops leave the farm, additional energy is used in food processing, distribution, and storage. It has been estimated that the average food item in the American diet travels 2,000 km (1,250 mi) between the farm that grew it and the person who consumes it. For some foods, processing, storage, and distribution use even more energy than growing the crop.

▲ **FIGURE 7.21** Pivot irrigation systems deliver water to most U.S. farm fields. These systems create green circles in the landscape that you can see in aerial photos of the Great Plains.

How can we feed the world?

The world's population has climbed from 3 billion to 7 billion in about two generations (since 1960). Despite this growth, the proportion of chronically hungry people has declined from 60 percent to about 20 percent in developing countries, where most population growth has occurred.

How have we managed to increase food production so rapidly? What are the pros and cons of these strategies? What additional choices do we have? Presented here are three main strategies in food production.

The Green Revolution involved development of high responders—crops that grow and yield well with increased use of fertilizer, irrigation, and pesticides.

Benefits

- Yields have grown dramatically with increased inputs.
- Efficiency is high for large-scale production, which aids in feeding billions.
- Development of pesticides has improved yields by eliminating competition with other plants or predation by insects.
- Labor costs are low: one farmer can work a huge area.

Problems

- Dependence on pesticides and fertilizers has grown; pesticides have sometimes lost effectiveness through overuse.
- Agricultural chemicals have unintended ecological consequences, including lost biodiversity, probable loss of pollinating insects, and contamination of drinking water.
- New varieties can be hard for poor farmers to afford, so that wealthier farmers and wealthier regions gain a relative advantage.
- Increased use of nitrogen fertilizer is a major source of greenhouse gases and consumer of fossil fuels.

KC 7.1

Circular green fields, irrigated by center-pivot sprinklers, produce the corn, soybeans, and other crops that are the backbone of our food system.
BACKGROUND IMAGE SOURCE: NAIP, 2009.

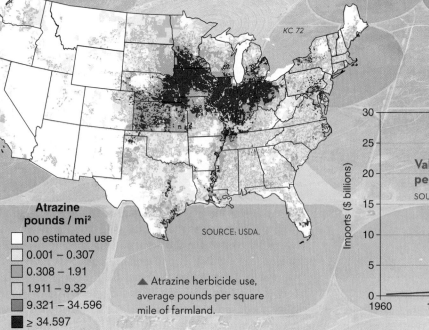

KC 72

Atrazine pounds / mi²

- ☐ no estimated use
- ☐ 0.001 – 0.307
- ☐ 0.308 – 1.91
- ☐ 1.911 – 9.32
- ☐ 9.321 – 34.596
- ■ ≥ 34.597

SOURCE: USDA.

▲ Atrazine herbicide use, average pounds per square mile of farmland.

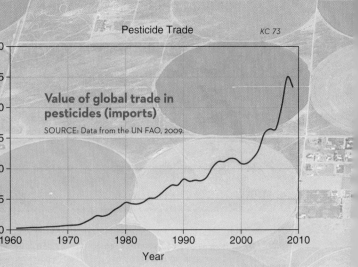

Pesticide Trade   *KC 73*

Value of global trade in pesticides (imports)
SOURCE: Data from the UN FAO, 2009.

Imports ($ billions)

Year

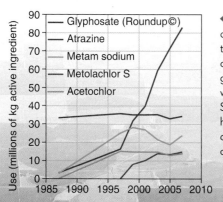

Use (millions of kg active ingredient)

- Glyphosate (Roundup©)
- Atrazine
- Metam sodium
- Metolachlor S
- Acetochlor

90
80
70
60
50
40
30
20
10
0

1985 1990 1995 2000 2005 2010

SOURCE: USDA, 2009.

KC 7.4

◀ The majority of GM crops are designed to tolerate high doses of herbicides such as glyphosate (Roundup), which is now the United States' dominant herbicide. Other GM crops produce their own pesticides.

KC 7.5

◀ **Conflicting evidence**

GM "golden rice" is engineered to provide vitamin A. Indian activist Vandana Shiva charges that golden rice provides less nutrition than traditional greens, which are more affordable for the poor.

Genetically modified (GM) crops have borrowed genes inserted into their DNA, allowing them to produce or tolerate new kinds of organic substances. GM seeds have increased the efficiency of many types of farming.

Benefits

- Yields have increased with GM crops, largely because herbicides help them compete with weeds.
- Bt (*Bacillus thuringiensis*) genes originating from soil bacteria provide a natural insecticide to further protect crops. Many GM crops are designed to produce this insecticide.
- GM crops have allowed expansion of agriculture into formerly unfarmed lands, including Brazil's rainforests and Cerrado region.
- Increased global soy production has raised protein consumption rates in many regions, including China.

Problems

- New varieties are expensive, forcing poor farmers into debt. Wealthier farmers and wealthy regions gain economic advantage.
- In some cases growth hormones (as in GM milk products) are suspected of causing premature puberty and other growth anomalies in humans. These effects remain unclear, however.
- Poorer populations in developing areas often rely on leafy greens to provide much of the nutrition in their diet. Increased herbicide use destroys nontarget crops, increasing malnutrition in some areas.
- Herbicides in drinking water have unknown health effects.

KC 7.6

KC 7.7

Organic production involves mixed strategies: crop rotation retains soil fertility; mixed cropping reduces pest risk; organic fertilizers and pesticides reduce costs of commercial inputs. Organic methods are more sustainable but don't lend themselves to the industrial-scale production of conventional farming, which involves vast expanses of a single crop, such as soy or corn.

Benefits

- Input costs (fertilizers, pesticides, fuel) are minimal.
- Can be sustainable as it preserves or even improves soil, water quality, and biodiversity.
- Crop varieties are usually more mixed, which contributes to healthier diets and more stable farm ecosystems.
- Organic methods are traditional in most areas, and low costs make this approach appropriate for poor farmers, especially in developing regions.
- Integrated pest management (small amounts of pesticides together with other strategies) can protect yields.

Problems

- Labor costs can be high for weed and pest control.
- Careful planning and management may be necessary to ensure good crop management. Creative and innovative problem solving is needed, not simple solutions such as spraying fields.
- Organically farmed food can be more expensive in the United States because it lacks the large-scale distribution networks of conventional products and because conventional methods receive abundant subsidies in tax breaks and price-support payments.
- These methods are unfamiliar to many farmers and can be less efficient in terms of volume of production per unit labor.

KC 7.8

CAN YOU EXPLAIN?

1. What was the Green Revolution?
2. Which are our top three pesticides? Which of these have you heard of previously?
3. Think of what you've eaten today. Which of the three agricultural strategies on this page contributed to your food?

Pesticide use continues to rise

Biological pests reduce crop yields and spoil as much as half of the crops harvested every year in some areas. Modern agriculture depends on toxic chemicals to kill these pests. There are concerns about the types and amounts of pesticides we use, but our reliance on them has grown steadily over the years (fig. 7.22).

Traditional strategies for evading pest damage often involved mixed crops and crop rotation. With many small patches of multiple crops, pest populations tend to remain relatively small. Reliance on a suite of different food sources can also reduce people's vulnerability to a particular crop pest. Rotating crops from one year to another reduces pests' ability to build up populations over time, and traditional crops often include varieties selected for pest resistance.

Modern agriculture, however, involves vast expanses of a single crop, often with little genetic variation, which increases the need for new methods of pest control. The invention of synthetic organic chemicals, such as DDT (dichlorodiphenyltrichloroethane), transformed our approach to pest control. These chemicals have been an important part of our increased crop production and have helped control many disease-causing organisms.

Organophosphates are the most abundantly used synthetic pesticides. These can be designed to prevent growth of broadleaf weeds or to attack the nervous systems of insect pests. Glyphosate, the single most heavily used herbicide in the United States, is also known by a variety of trade names, including Roundup. Glyphosate is applied to 90 percent of U.S. soybeans. "Roundup-ready" soybeans (see discussion below) are one of the most commonly used type of genetically modified crop (see discussion of GM crops, below). **Chlorinated hydrocarbons**, also called organochlorines, are persistent and highly toxic to sensitive organisms. Atrazine, which is applied to 96 percent of all U.S. corn, was the most abundantly used pesticide until about 1998, when Roundup-Ready soy became available. Since then, glyphosate has skyrocketed to first place (see Key Concepts, p. 169).

Other important pesticides include fumigants, highly toxic gases such as methylene bromide, which is used to kill fungus on strawberries and other low-growing crops; inorganic pesticides, such

▲ **FIGURE 7.23** Spraying pesticides by air is quick and cheap, but toxins often drift to nearby fields.

as arsenic and copper; and natural "botanical" pesticides derived from plants, such as nicotinoid alkaloids derived from tobacco.

Pesticides are thus important to us, but indiscriminate use has serious effects. Chemicals drifting or washing away from fields endanger nontarget species (fig. 7.23). Often highly persistent and mobile in the environment, many pesticides move through air, water, and soil and bioaccumulate in food chains. Most famously, bioaccumulation of DDT was responsible for nearly eliminating a variety of predatory birds, including peregrine falcons, brown pelicans, and bald eagles, in the 1960s. Many of our dominant pesticides, including atrazine and glyphosate, are known to cause developmental deformities in aquatic organisms such as frogs, salamanders, and fish. Ecologists and endocrinologists have repeatedly found links between these agricultural chemicals and the disappearance of frogs and salamanders in farming regions.

One of the most urgent consequences of heavy reliance on pesticides has been the evolution of pesticide-resistant varieties of weeds and parasites. As discussed in chapter 2, a species' traits can shift if environmental conditions change. In the presence of pesticides, a few tolerant individuals often survive and reproduce, and tolerance of the pesticide can become dominant in a population. New pesticide-tolerant varieties of weeds or parasites then emerge. Around 400 different cases of pesticide-resistant weed populations have been recorded by the Weed Science Society of America (fig. 7.24). In just a decade of glyphosate use, scores of glyphosate-resistant weeds have appeared, from Iowa to Australia to Brazil. Farmers must use increasingly complex and expensive cocktails of pesticides to combat these new pests. A **pesticide treadmill** results, as new resistant strains arise, causing further increases in pesticide applications, which lead to more resistant strains, which require expanded pesticide use.

Your exposure to pesticides is probably higher than you suspect. A study by the U.S. Department of Agriculture found that 73 percent of conventionally grown foods (out of 94,000 samples assayed) had residues from at least one pesticide. (It is important to note, however, that your greatest risk probably involves household pesticides, the most rapidly growing pesticide market.)

Alternatives for reducing our dependence on chemical pesticides include management changes, such as using cover crops and mechanical cultivation, and planting mixed polycultures rather than vast monoculture fields. Biological controls, such as insect

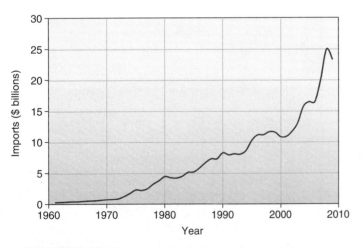

▲ **FIGURE 7.22** Value of global trade in pesticides (imports).
SOURCE: Data from the UN Food and Agriculture Organization, 2012.

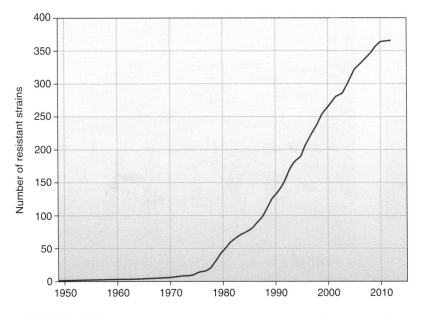

▲ FIGURE 7.24 Pesticide-resistant varieties have increased, resulting in a "pesticide treadmill," in which farmers must use increasingly expensive combinations of chemicals to suppress weeds, diseases, and pests. DATA SOURCES: U.S. EPA; Weed Society of America (http://www.weedscience.com).

predators or pathogens, can help reduce chemical use. Genetic breeding and biotechnology can produce pest-resistant crop and livestock strains as well. Integrated pest management (IPM) combines all of these alternative methods, together with judicious use of synthetic pesticides under precisely controlled conditions. Consumers can also learn to accept less than perfect fruits and vegetables. For ways you can minimize pesticides in your diet, see What Can You Do? (at right).

7.6 HOW HAVE WE MANAGED TO FEED BILLIONS?

In the developed countries, 95 percent of agricultural growth in the twentieth century came from improved crop varieties or increased fertilization, irrigation, and pesticide use, rather than from bringing new land into production. In fact, less land is being cultivated now than 100 years ago in North America, or 600 years ago in Europe. As more effective use of labor, fertilizer, and water and improved seed varieties have increased in the more-developed countries, productivity per unit of land has increased, and much marginal land has been retired, mostly to forests and grazing lands. In developing countries, at least two-thirds of recent production gains have come from new crop varieties and more intense cropping, rather than expansion into new lands.

The green revolution has increased yields

So far, the major improvements in farm production have come from technological advances and modification of a few well-known species. Yield increases often have been spectacular. A century ago, when all maize (corn) in the United States was open pollinated, average yields were about 25 bushels per acre (bu/acre). In 2009 the average yield from rain-fed fields in Iowa was 182 bu/acre, and irrigated maize in Arizona averaged 208 bu/acre. The highest yield ever recorded in field production was 370 bu/acre on an Illinois farm, but theoretical calculations suggest that 500 bu/acre (32 metric tons per hectare) could be possible. Most of this gain was accomplished by use of synthetic fertilizers along with conventional plant breeding: geneticists laboriously hand-pollinating plants and looking for desired characteristics in the progeny.

Starting about 50 years ago, agricultural research stations began to breed tropical wheat and rice varieties that would provide food for growing populations in developing countries. The first of the "miracle" varieties was a dwarf, high-yielding wheat developed by Norman Borlaug (who received a Nobel Peace Prize for his work) at a research center in Mexico (fig. 7.25). At about the same time, the International Rice Institute in the Philippines developed dwarf rice strains with three or four times the production of varieties in use at the time. The spread of these new high-yield varieties around the world has been called the **green revolution**. It is one of the main reasons that world food supplies have more than kept pace with the growing human population over the past few decades.

Most green revolution breeds really are "high responders," meaning that they yield well in response to optimal inputs of fertilizer, water, and chemical protection from pests and diseases (fig. 7.26). With fewer inputs, on the other hand, high responders may not produce as well as traditional varieties. Poor farmers who can't afford

What Can YOU DO?

Reducing Pesticides in Your Food

If you want to reduce the amount of pesticide residues and other toxic chemicals in your diet, follow these simple rules.

- Wash and scrub all fresh fruits and vegetables thoroughly, and peel fruits and vegetables when possible. Pesticides often accumulate on skin or outer leaves. Also trim the fat from meats, since contaminants bioaccumulate in fats.

- Eat lower on the food chain (more vegetables, moderate amounts of meats). This reduces your exposure to bioaccumulated chemicals and will provide you a healthier diet overall.

- Cooking foods can break down some chemical residues.

- Ask for organically grown food at your local grocery store, or shop at a farmers' market or co-op where you can get such food.

- Minimize your use of household disinfectants, toxic cleaning agents, and household pesticides and weed killers.

▲ **FIGURE 7.25** Semidwarf wheat (*right*), bred by Norman Borlaug, has shorter, stiffer stems and is less likely to lodge (fall over) when wet than its conventional cousin (*left*). This "miracle" wheat responds better to water and fertilizer and has played a vital role in feeding a growing human population.

▲ **FIGURE 7.27** A researcher measures growth of genetically engineered rice plants. Superior growth and yield, pest resistance, and other useful traits can be moved from one species to another by molecular biotechnology.

the expensive seed, fertilizer, and water required to become part of this movement usually are left out of the green revolution. In fact, they may be driven out of farming altogether as rising land values and falling commodity prices squeeze them from both sides.

Genetic engineering has benefits and costs

Genetic engineering, splicing a gene from one organism into the chromosome of another, has the potential to greatly increase both the quantity and the quality of our food supply. It is now possible to build entirely new genes, and even new organisms, often called "transgenic" organisms or **genetically modified organisms (GMOs,** or **GM crops)**, by taking a bit of DNA from here, a bit from there, and even synthesizing artificial sequences to create desired characteristics in engineered organisms (fig. 7.27).

Proponents predict dramatic benefits from this new technology. Research is now under way to improve yields and create crops that resist drought, frost, or diseases. Other strains are being

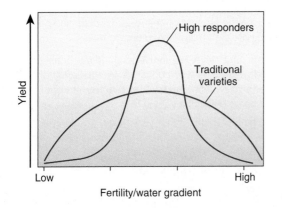

▲ **FIGURE 7.26** Green revolution miracle crops are really high responders, meaning that they have excellent yields under optimum conditions. For poor farmers who can't afford the fertilizer and water needed by high responders, traditional varieties may produce better yields.

developed to tolerate salty, waterlogged, or low-nutrient soils, allowing degraded or marginal farmland to become productive. All of these could be important for reducing hunger in developing countries. Plants that produce their own pesticides might reduce the need for toxic chemicals, while engineering for improved protein or vitamin content could make our food more nutritious. Crops such as bananas and tomatoes have been altered to contain oral vaccines that can be grown in developing countries where refrigeration and sterile needles are unavailable. Plants have been engineered to make industrial oils and plastics. Animals, too, are being genetically modified to grow faster, gain weight on less food, and produce pharmaceuticals, such as insulin, in their milk. Pigs are now being engineered to produce omega-3 fatty acids for a "heart healthy" diet. It may even be possible to create animals with human cell-recognition factors that could serve as organ donors.

Both the number of GM crops and the acreage devoted to growing them is increasing rapidly. Worldwide, over 25 percent of all cropland (72 million ha) was planted with GM crops in 2009. The United States accounted for 63 percent of that acreage, followed by Argentina with 21 percent. Canada, Brazil, and China together make up 22 percent of all transgenic cropland.

Most GMOs are engineered for pesticide production or pesticide tolerance

The dominant transgenic crops are engineered to tolerate high doses of herbicides—allowing fields to be sprayed for weeds without affecting the crops themselves—or to create their own insecticides (fig. 7.28). Leading among herbicide-tolerant (HT) varieties are Monsanto's Roundup-Ready crops—which withstand treatment with Monsanto's best-selling herbicide, Roundup (glyphosate). Another dominant herbicide-tolerant group is AgrEvo's "Liberty Link" crops, which resist that company's Liberty (glufosinate) herbicide. Because crops with these genes can grow in spite of high herbicide doses, farmers can spray fields heavily to exterminate weeds. This allows for conservation tillage and leaving more crop residue on fields to protect topsoil from erosion—both good ideas—but it also can mean using more herbicide in higher doses than might otherwise be used.

Another widespread use of GM technology is to insert genes from a natural soil bacterium, *Bacillus thuringiensis* (Bt), into plants. This bacterium makes toxins lethal to Lepidoptera (butterflies and moths) and Coleoptera (the beetle family). The genes for some of these toxins have been transferred into crops such as corn (to protect against European cutworms), potatoes (to fight potato beetles), and cotton (to protect against boll weevils). Because Bt crops produce their own insecticides, farmers can spray less. Arizona cotton farmers, for example, report reducing their use of chemical insecticides by 75 percent.

As with other widely used pesticides, Bt varieties have begun to produce resistant populations. Bt plants churn out toxin throughout the growing season, regardless of the level of infestation, creating perfect conditions for selection of Bt resistance in pests. Already 500 species of insects, mites, and ticks are resistant to one or more pesticides. Having Bt constantly available can only aggravate this dilemma.

Nearly all soybeans, more than one-quarter of all corn, and nearly 80 percent of all cotton grown in the United States (and possibly worldwide) are now GM varieties, either Bt- or herbicide-tolerant. With this level of use, chances are you have eaten some genetically modified foods. Estimates are that at least 60 percent of all processed food in America contains GM ingredients, chiefly from corn or soy.

Is genetic engineering safe?

Consumers have long worried about unknown health effects of GM foods, but the U.S. Food and Drug Administration has declined to require labeling of foods containing GMOs. The agency argues that these new varieties are "substantially equivalent" to related varieties bred via traditional practices. After all, proponents say, we have been moving genes around for centuries through plant and animal breeding. Biotechnology may just be a more precise way of creating novel organisms than normal breeding procedures.

Despite continuing debates, direct effects of GM foods on human health remain unclear. Official studies of have tended not to find clear effects on humans. Opponents of GM crops argue that studies have been biased or incomplete. The question is likely to be disputed for some time.

The first genetically modified animal designed to be eaten by humans is an Atlantic salmon (*Salmo salar*) containing extra growth hormone genes from an oceanic pout (*Macrozoarces americanus*). The greatest worry from this fish is not that it will introduce extra hormones into our diet—that's already being done by chickens and beef that get extra growth hormone via injections

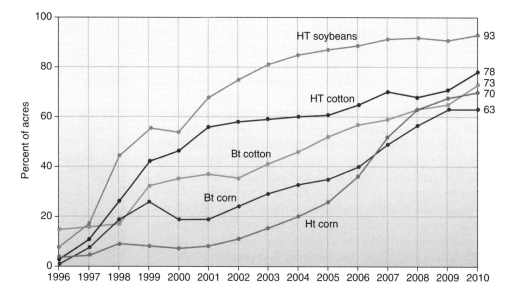

▲ **FIGURE 7.28** Growth of genetically engineered corn, cotton, and soybeans in the United States. HT (herbicide tolerant) varieties tolerate herbicides such as glyphosate (Monsanto's Roundup). Bt crops contain insect-killing proteins from the bacterium *Bacillus thuringiensis*.   SOURCE: USDA Economic Research Service, 2011.

or their diet—but, rather, the ecological effects if the fish escape from captivity. The transgenic fish grow seven times faster and are more attractive to the opposite sex than a normal salmon. If they escape from captivity, they may outcompete already endangered wild relatives for food, mates, and habitat.

Perhaps the greatest unknowns are the social and economic implications of GMOs. Will they help feed the world, or will they lead to a greater consolidation of corporate power and economic disparity? This technology may be available only to the richest countries or the wealthiest corporations, making family farms uncompetitive and driving developing countries even further into poverty. Or higher yields and fewer losses to pests and diseases might allow poor farmers in developing countries to stop using marginal land. Critics suggest that there are simpler and cheaper ways other than high-tech crop varieties to provide vitamin A to children in developing countries or to increase the income of poor rural families. Adding a cow or a fishpond or training people in water harvesting or sustainable farming techniques (see next section) may have a longer-lasting impact than selling them new designer-brand seeds.

On the other hand, if we hope to reduce malnutrition and feed 9 billion people in 50 years, maybe we need all the tools we can get. Where do you stand in this debate? What additional information would you need to reach a reasoned judgment about the risks and benefits of this new technology?

7.7 SUSTAINABLE FARMING STRATEGIES

Innovations in farming have changed the ways we eat and live. Can we ensure that farming and food production also are sustainable and equitable? These are the goals of **sustainable agriculture**, or **regenerative farming**, which aim to reduce or repair the damage caused by destructive practices. Some alternative methods are developed through scientific research; others are discovered in traditional cultures and practices nearly forgotten in our mechanization and industrialization of agriculture.

Recent analysis of decades of crop-monitoring data have found that sustainable farming techniques can provide comparable levels of productivity and economic gain for farmers, compared to conventional monocropping, especially in developing countries. UN studies in Africa, for example, have found that sustainable techniques including integrated pest management, intercropping, and crop rotation can provide income and productivity as good as that of conventional techniques but far more cheaply. Crops grown without synthetic fertilizers or pesticides produced, on average, about a 20 percent lower yield than similar crops grown by conventional methods, but they more than made up the difference with lower operating costs, premium crop prices, less ecological damage, and healthier farm families. For impoverished and hungry populations in developing areas, argues UN researcher Olivier De Schutter, it is foolish not to pursue all available options, including sustainable techniques.

Even in wealthy countries, sustainable practices can provide advantages to farmers. A blend of pest management strategies may be needed, rather than standard applications of one or two pesticides, and production of vast areas of a single crop is not well suited to most sustainable techniques. But food production can be high, and because costs are much lower—with fewer purchases of chemicals and fertilizers—the net gain for farmers can be good even in wealthier countries.

Soil conservation is essential

With careful husbandry, soil is a renewable resource that can be replenished and renewed indefinitely. Because agriculture is the area in which soil is most essential and most often lost through erosion, agriculture offers the greatest potential for soil conservation and rebuilding. Some rice paddies in Southeast Asia have been farmed continuously for a thousand years without any apparent loss of fertility. The rice-growing cultures that depend on these fields have developed management practices that return organic material to the paddies and carefully sustain and build the soil.

While American agriculture hasn't reached that level of sustainability, federal soil conservation programs have reduced erosion considerably since the disastrous years of the 1920s and 1930s. In a study of one Wisconsin watershed, erosion rates were 90 percent less in 1975–1993 than they were in the 1930s.

Among the most important elements in soil conservation are terracing, ground cover, and reduced tillage. Most erosion happens because water runs downhill. The faster it runs, the more soil it carries off the fields. Comparisons of erosion rates in Africa have shown that a 5 percent slope in a plowed field has three times the water runoff volume and eight times the soil erosion rate of a comparable field with a 1 percent slope. Water runoff can be reduced by grass strips in waterways and by **contour plowing**—that is, plowing across the hill rather than up and down. Contour plowing is often combined with **strip-farming**, the planting of different kinds of crops in alternating strips along the land contours (fig. 7.29).

 ▲ **FIGURE 7.29** Contour plowing and strip cropping help prevent soil erosion on hilly terrain, as well as create a beautiful landscape.

When one crop is harvested, the other is still present to protect the soil and keep water from running straight downhill. The ridges created by cultivation trap water, allowing it to seep into the soil rather than running off.

Terracing involves shaping the land to create level shelves of earth to hold water and soil. The edges of the terrace are planted with soil-anchoring plant species. This is an expensive procedure, requiring either much hand labor or expensive machinery, but it makes farming of very steep hillsides possible. The rice terraces in the Chico River Valley in the Philippines rise as much as 300 m (1,000 ft) above the valley floor. They are considered one of the wonders of the world. Some of these terraces have been farmed for 1,000 years without any apparent fertility loss. Will we be able to say the same thing about our farm fields a millennium from now?

Groundcover, reduced tilling protect soil

Annual row crops, such as corn or beans, generally cause the highest erosion rates because they leave soil bare for much of the year (table 7.2). Often the easiest way to provide cover that protects soil from erosion is to leave crop residues on the land after harvest (fig. 7.30). They not only cover the surface to break the erosive effects of wind and water but also reduce evaporation and soil temperature in hot climates and protect ground organisms that help aerate and rebuild soil. In some experiments, crop residue increased water infiltration 99 percent and reduced soil erosion 98 percent.

Leaving crop residues on the field also can increase disease and pest problems, however, and conservation tillage often depends on increased use of pesticides (insecticides, fungicides,

TABLE 7.2 | Soil Cover and Soil Erosion

CROPPING SYSTEM	AVERAGE ANNUAL SOIL LOSS (TONS/HECTARE)	PERCENT RAINFALL RUNOFF
Bare soil (no crop)	41.0	30
Continuous corn	19.7	29
Continuous wheat	10.1	23
Rotation: corn, wheat, clover	2.7	14
Continuous bluegrass	0.3	12

SOURCE: Based on 14 years of data from Missouri Experiment Station, Columbia, Missouri.

and herbicides) to control insects, weeds, and disease. Increased use of agricultural chemicals can diminish the overall environmental benefits of conservation tilling. Increased pesticide use is not always necessary, however. Pesticide use can be minimized through crop rotation and integrated pest management strategies such as trap crops, natural repellents, and biological controls.

Often soil is protected with cover crops as rye, alfalfa, or clover, which can be planted immediately after harvest to hold and protect the soil. These cover crops can be plowed under at planting time to add organic matter and nutrients to the soil.

In some cases, interplanting of two different crops in the same field not only protects the soil but also is a more efficient use of the land, providing double harvests. Native Americans and pioneer farmers, for instance, planted beans or pumpkins between the corn rows. The beans provided nitrogen needed by the corn, the pumpkins crowded out weeds, and both crops provided foods that nutritionally balance corn. Traditional swidden (slash-and-burn) cultivators in Africa and South America often plant as many as 20 different crops together in small plots. The crops mature at different times, so that there is always something to eat, and the soil is never exposed to erosion for very long. Mixed cultivation can include highly valuable crops, as well as helping to conserve biological diversity (see What Do You Think?, p. 176).

Low-input sustainable agriculture can benefit people and the environment

In contrast to the trend toward industrialization and dependence on chemical fertilizers, pesticides, antibiotics, and artificial growth factors common in conventional agriculture, some farmers are going back to a more natural, agroecological farming style. Finding that they can't—or don't want to—compete with factory farms, these folks are making money and staying in farming by returning to small-scale, low-input agriculture. The Minar family, for instance, operates a highly successful 150-cow dairy operation on 97 ha (230 acres) near New Prague, Minnesota. No synthetic chemicals are used on their farm. Cows are rotated every day between 45 pastures or paddocks to reduce erosion and maintain healthy grass. Even in the winter, livestock remain outdoors to avoid the spread of diseases common in confinement (fig. 7.31). Antibiotics are used only to fight diseases. Milk and meat from this operation are marketed through co-ops and a community-supported agriculture (CSA) program. Sand Creek, which flows across the Minar land, has been shown to be cleaner when leaving the farm than when it enters.

Similarly, the Franzen family, who raise livestock on their organic farm near Alta Vista, Iowa, allow their pigs to roam in lush pastures, where they can supplement their diet of corn and soybeans with grasses and legumes. Housing for these happy hogs is in spacious, open-ended hoop structures. As fresh layers of straw are added to the bedding, layers of manure beneath are composted, breaking down into odorless organic fertilizer.

Low-input farms such as these typically don't turn out the quantity of meat or milk that their intensive agriculture neighbors do, but their production costs are lower, and they get higher prices for their crops, so that the all-important net gain is often higher. The Franzens, for example, calculate that they pay 30 percent less

▲ **FIGURE 7.31** On the Minar family's 230-acre dairy farm near New Prague, Minnesota, cows and calves spend the winter outdoors in the snow, bedding down on hay. Dave Minar is part of a growing counterculture that is seeking to keep farmers on the land and bring prosperity to rural areas. ©2008 Star Tribune/Minneapolis-St. Paul.

▲ **FIGURE 7.30** No-till planting involves drilling seeds through debris from last year's crops. Here, soybeans grow through corn mulch. Debris keeps weeds down, reduces wind and water erosion, and keeps moisture in the soil.

Shade-Grown Coffee and Cocoa

Do your purchases of coffee and chocolate help to protect or destroy tropical forests? Coffee and cocoa are two of the many products grown exclusively in developing countries but consumed almost entirely in the wealthier, developed nations. Coffee grows in cool, mountain areas of the tropics, while cocoa is native to the warm, moist lowlands. What sets these two apart is that both come from small trees adapted to grow in low light, in the shady understory of a mature forest. **Shade-grown** coffee and cocoa (grown beneath an understory of taller trees) allow farmers to produce a crop at the same time as forest habitat remains for birds, butterflies, and other wild species.

Until a few decades ago, most of the world's coffee and cocoa were shade-grown. But new varieties of both crops have been developed that can be grown in full sun. Growing in full sun, trees can be crowded together more closely. With more sunshine, photosynthesis and yields increase.

There are costs, however. Sun-grown trees die earlier from stress and diseases common in crowded growing conditions. Crowding also requires increased use of expensive pesticides and fungicides. Shade-grown coffee and cocoa generally require fewer pesticides (or sometimes none) because the birds and insects residing in the forest canopy eat many of the pests. Ornithologists have found as little as 10 percent as many birds in a full-sun plantation, compared to a shade-grown plantation. The number of bird species in a shaded plantation can be twice that of a full-sun plantation. Shade-grown plantations also need less chemical fertilizer because many of the plants in these complex forests add nutrients to the soil. In addition, shade-grown crops rarely need to be irrigated because heavy leaf fall protects the soil, while forest cover reduces evaporation.

Over half the world's coffee and cocoa plantations have been converted to full-sun varieties. Thirteen of the world's 25 biodiversity hot spots occur in coffee or cocoa regions. If all the 20 million ha of

◀ Cocoa pods grow directly on the trunk and large branches of cocoa trees.

coffee and cocoa plantations in these areas are converted to monocultures, an incalculable number of species will be lost.

The Brazilian state of Bahia demonstrates both the ecological importance of these crops and how they might help preserve forest species. At one time, Brazil produced much of the world's cocoa, but in the early 1900s, the crop was introduced into West Africa. Now Côte d'Ivoire alone grows more than 40 percent of the world total. Rapid increases in global supplies have made prices plummet, and the value of Brazil's harvest has dropped by 90 percent. Côte d'Ivoire is aided in this competition by a labor system that reportedly includes widespread child slavery. Even adult workers in Côte d'Ivoire get only about $165 (U.S.) per year (if they get paid at all), compared with a minimum wage of $850 (U.S.) per year in Brazil. As African cocoa production ratchets up, Brazilian landowners are converting their plantations to pastures or other crops.

The area of Bahia where cocoa was once king is part of Brazil's Atlantic Forest, one of the most threatened forest biomes in the world. Only 8 percent of this forest remains undisturbed. Although cocoa plantations don't have the full diversity of intact forests, they do provide an economic rationale for preserving the forest. And Bahia's cocoa plantations protect a surprisingly large sample of the biodiversity that once was there. Brazilian cocoa will probably never be as cheap as that from other areas. There is room in the market, however, for specialty products. If consumers choose to pay a small premium for organic, fair-trade, shade-grown chocolate and coffee, it might provide the incentive needed to preserve biodiversity. Wouldn't you like to know that your chocolate or coffee wasn't grown with child slavery and is helping protect plants and animal species that might otherwise go extinct? What does it take to make that idea spread?

---

for animal feed, 70 percent less for veterinary bills, and half as much for buildings and equipment as their neighboring confinement operations. And on the Minar's farm, erosion after an especially heavy rain was measured to be 400 times lower than on a conventional farm nearby.

Preserving small-scale family farms also helps preserve rural culture. As Marty Strange of the Center for Rural Affairs in Nebraska asks, "Which is better for the enrollment in rural schools, the membership of rural churches, and the fellowship of rural communities—two farms milking 1,000 cows each or twenty farms milking 100 cows each?" Family farms help keep rural towns alive by purchasing machinery at the local implement

dealer, gasoline at the neighborhood filling station, and groceries at the mom-and-pop grocery store.

7.8 CONSUMER ACTION AND FARMING

Since the 1960s, U.S. farm policy and agricultural research has focused on developing large-scale production methods that use fertilizer, pesticides, breeding, and genetic engineering to provide abundant, inexpensive grain, meat, and milk. As a consequence we can afford to eat more calories and more meat than ever before. Hunger still plagues many regions, but increasing nutrition has reached most of the world's population, not just wealthy countries. We now worry about weight-associated illnesses as a cause of mortality, possibly a first in human history.

Cheap-food policies have raised production dramatically to feed growing populations both in the United States and abroad. Farm commodity prices have fallen so low, that we now spend billions of dollars every year to support prices or repay farmers whose production expenses are greater than the value of their crops. The United States buys millions of tons of surplus food every year, often using it as food aid to famine-stricken regions. This social good is also problematic, as cheap or free food donations undermine small farmers in other countries, who cannot compete with free or nearly free food on the local market.

These policies are deeply entrenched in our way of life, politics, and food systems. But there may be some steps that consumers can take to support the beneficial changes in farming methods and farm policies, while reducing their negative effects.

You can be a locavore

A "locavore" is a person who consumes locally produced food. Supporting local farmers can have a variety of benefits, from keeping money in the local economy to ensuring a fresh and healthy diet. Maintaining a viable farm economy can also help slow the conversion of farmland into expanding suburban subdivisions. As many farmers point out, you don't need to eat 100 percent local. Converting just part of your shopping activity to locally produced goods can make a big difference to farmers.

Farmers' markets are usually the easiest way to eat locally (fig. 7.32). The produce is fresh, and profits go directly to the farmer who grows the crop. "Pick your own" farms also let you buy fresh fruit and other products—and they make a fun, social outing. Many conventional grocery stores also now offer locally produced, organic, and pesticide-free foods. Buying these products may (or may not) cost a little more than nonorganic and nonlocal produce, but they can be better for you and they can help keep farming and fresh, local food in the community.

▲ **FIGURE 7.32** Your local farmer's market is a good source of locally grown organic produce.

Many colleges and universities have adopted policies to buy as much locally grown food as possible. Because schools purchase a lot of vegetables, meat, eggs, and milk, this can mean a large amount of income for local and regional farm economies. Although this policy can take more effort and creativity than ordering from centralized, national distributors, many college food service administrators are happy to try to buy locally, if they see that students are interested. If your school doesn't have such a policy, perhaps you could talk to administrators about starting one.

Many areas also have "community supported agriculture" (CSA) projects, farms supported by local residents who pay ahead of time for shares of the farm's products, which can vary from vegetables to flowers to meat and eggs. CSAs require a lump payment early in the season, but the net cost of food by the end of the season is often less than you would have paid at the grocery store. You also get to meet interesting people and learn more about your local area by participating in a CSA.

You can eat low on the food chain

Because there is less energy involved in producing food from plants than producing it from animals, one way you can reduce your impact on the world's soil and water is to eat a little more grains, vegetables, and dairy and a little less meat. This doesn't mean turning vegetarian—unless you choose to do so. Just returning to the level of protein and fat consumption your grandparents had could make a big difference for the environment and for your health.

Low-input, organic foods also reduce the environmental impact of your food choices. When you buy organic food, you support farmers who use no pesticides or artificial fertilizers. Often these farmers use crop rotations to preserve soil nutrients and manage erosion carefully, to prevent loss of their topsoil. Sometimes these farmers preserve diverse varieties of crops, helping to maintain genetic diversity and pest resistance in crops.

Grass-fed beef and free-range poultry or pork can also be excellent low-input foods. By converting grass to protein, they can be an efficient food source where soils or steep hillsides are unsuitable for cropping. With good management, pastures have minimal soil erosion, because vegetation keeps the soil covered year-round.

CONCLUSION

Food production has grown faster than the human population in recent decades, and the percentage of people facing chronic hunger has declined, although the number has increased. Most of us consume more calories and protein than we need, but some 900 million people still are malnourished. Much, or perhaps most, hunger results from political instability, which displaces farmers, inhibits food distribution, and undermines local farming economies.

Increases in food production result from many innovations in agricultural production. The green revolution produced new varieties that yield more crops per hectare, although these crops usually require extra inputs such as fertilizers, irrigation, or pesticides. Genetically modified organisms have also increased yields. Most GMOs are designed to tolerate herbicides, and many produce their own pesticides. Confined animal feeding operations, made possible by large-scale corn and soy production, have greatly increased the efficiency of producing meat. The rise of soy in Brazil, and of meat consumption in the United States, China, and elsewhere, result from these innovations.

These changes bring about important environmental effects, such as soil erosion and degradation, and water contamination from pesticide and fertilizer applications. Health effects are also a concern, including weight-related diseases, such as diabetes, exposure to agricultural chemicals, and antibiotic resistance. There are also many more sustainable approaches, which offer the opportunity to regenerate soils and maintain healthy ecosystems: these systems can offer significant economic advantages to farmers, especially in developing regions, but they are not widely adopted because they are poorly suited to mass production of the type seen in Brazil's soy fields. Consumers can help support sustainable practices by eating locally, eating low on the food chain, buying organic foods or grass-fed meat, and shopping at farmers' markets.

PRACTICE QUIZ

1. What is Brazil's Cerrado, and how is agriculture affecting it?
2. Explain how soybeans grown in Brazil are improving diets in China.
3. What does it mean to be chronically undernourished? How many people in the world currently suffer from this condition?
4. Why do nutritionists worry about food security? Who is most likely to suffer from food insecurity?
5. Describe the conditions that constitute a famine. Why does Amartya Sen say that famines are caused more by politics and economics than by natural disasters?
6. Define *malnutrition* and *obesity*. How many Americans are now considered obese?
7. What three crops provide most human caloric intake?
8. What are confined animal feeding operations, and why are they controversial?
9. What is *soil*? Why are soil organisms so important?
10. What are four dominant types of soil degradation? What is the primary cause of soil erosion?
11. What do we mean by the *green revolution*?
12. What is *genetic engineering*, and how can it help or hurt agriculture?
13. What is *sustainable agriculture*?
14. How could your choices of coffee or cocoa help preserve forests, biodiversity, and local economies in tropical countries?
15. What are the economic advantages of low-input farming?

CRITICAL THINKING AND DISCUSSION

Apply the principles you have learned in this chapter to discuss these questions with other students.

1. Explain the nature of hunger in the world today. How much should we worry about hunger increasing? Why?
2. Review some of the major reasons for global hunger. Which ones are easiest to resolve? How might we approach them?
3. Debate the claim that famines are caused more by human actions (or inactions) than by environmental forces. What evidence would you need to have to settle this question? What hypotheses could you test to help resolve the debate?
4. Should farmers be forced to use ecologically sound techniques that serve their long-term interests, regardless of short-term costs? How might this be done?
5. What aspects of genetically modified food products do you find most beneficial? Which are most worrisome? Are your answers the same as those of your fellow students or family members?
6. Suppose that you were engaged in genetic engineering or pesticide development. What environmental or social safeguards would you impose on your own research, if any? What restrictions would you tolerate from someone else concerned about the effects of your work?

Understanding where your food comes from helps you understand the environmental questions involved with producing the food you eat. Because this information is so important, culturally and economically, the United States Department of Agriculture (USDA) publishes maps and statistical analysis of our major crops. This rich repository of information lets you explore where your food comes from.

Go to Connect to find a link to some of these maps and to demonstrate your understanding of where we produce some of our major foods.

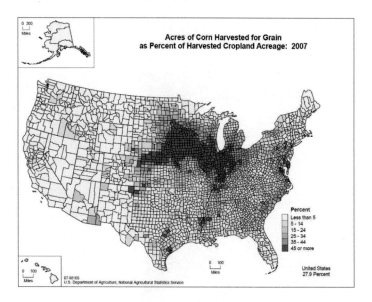

Acres of Corn Harvested for Grain as Percent of Harvested Cropland Acreage: 2007

Percent
- Less than 5
- 5 - 14
- 15 - 24
- 25 - 34
- 35 - 44
- 45 or more

United States
27.9 Percent

07-M165
U.S. Department of Agriculture, National Agricultural Statistics Service

---

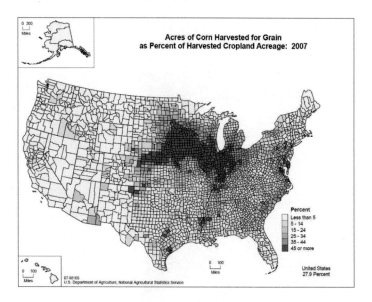

TO ACCESS ADDITIONAL RESOURCES FOR THIS CHAPTER, PLEASE VISIT CONNECT AT www.mcgrawhillconnect.com.

You will find LearnSmart, an adaptive learning system, Google Earth™ exercises, additional Case Studies, Data Analysis exercises, and an interactive ebook.

Environmental Health and Toxicology

Should we be worried about what we're eating? In recent decades, thousands of new, synthetic chemicals have been introduced into our diets and our lives. How dangerous are they? This is an important question in environmental health.

LEARNING OUTCOMES

After studying this chapter, you should be able to answer the following questions:

▶ What is environmental health?

▶ What health risks should worry us most?

▶ Emergent diseases seem to be more frequent now. What human factors may be involved in this trend?

▶ Are there connections between ecology and our health?

▶ When Paracelsus said, "The dose makes the poison," what did he mean?

▶ What makes some chemicals dangerous and others harmless?

▶ How much risk is acceptable, and to whom?

How Dangerous Is BPA?

Bisphenol A (BPA), a key ingredient of both polycarbonate plastics and epoxy resins, is one of the world's most widely used chemical compounds. In 2011, total global production was about 3 million metric tons, and the chemical industry expects use to double by 2015 as China and other developing countries manufacture increasing amounts of plastic or plastic-coated wares. BPA is used in items ranging from baby bottles, automobile headlights, eyeglass lenses, CDs, DVDs, water pipes, the linings of cans and bottles, and tooth-protecting sealants.

Traces of BPA are found in humans nearly everywhere. In one study of several thousand ordinary American adults, 95 percent had measurable amounts of this chemical in their bodies. The most likely source of contamination is from food and beverage containers. During plastic polymerization, not all BPA gets locked up into chemical bonds. Unbound molecules can leach out, especially when plastic is heated, washed with harsh detergents, scratched, or exposed to acidic compounds, such as tomato juice, vinegar, or soft drinks. In one study of canned food from major supermarket chains, half the samples had BPA higher than government-recommended dietary levels.

How dangerous is BPA? In recent years dozens of scientists around the world have linked BPA to myriad health effects in rodents, including mammary and prostate cancer, genital defects in males, early onset of puberty in females, obesity, and even behavior problems such as attention-deficit hyperactivity disorder. Furthermore, epidemiological studies in humans show a correlation between urine concentrations of BPA and cardiovascular disease, type 2 diabetes, and liver-enzyme abnormalities. Scientists found that BPA, like phthalates, dioxins, and PCBs acts as an endocrine hormone disrupter. That is, it upsets the normal function of your body's own hormones. Interestingly, the first use for BPA after it was synthesized in 1891 was as a synthetic estrogen.

But rodents, especially those raised in laboratory conditions, may not be accurate models for how humans react in the real world. We have very different genetics, diet, and physiology. And cross-sectional or retrospective studies of human populations show only correlations, not causality. It could be just a coincidence that people exposed to BPA develop common chronic diseases, such as cardio-

▲ **FIGURE 8.1** What's in our food? How safe are we really?

vascular disease or diabetes. Furthermore, as you'll learn in this chapter, detectable levels aren't always dangerous. New technology allows us to measure tiny amounts of chemicals that may or may not be deleterious. Risk assessment is a complex and difficult task.

Industry-funded scientists point to contradictions and unexplained uncertainties in published studies of BPA toxicity. Some investigators find deleterious effects at low BPA levels; others say they can't reproduce those results. Some of this variability may be linked to funding. In one examination of 115 peer-reviewed, published studies of BPA, 94 of those supported by government agencies found adverse health effects from BPA exposure, whereas none of the 15 financed by industry sources found any problem.

Current federal guidelines put the daily upper limit of safe exposure at 50 micrograms of BPA per kilogram of body weight. But that level is based on a small number of high-dose experiments done years ago, rather than on the hundreds of more recent animal and laboratory studies indicating that serious health risks could result from much lower doses of BPA. Several animal studies show adverse effects, such as abnormal reproductive development, at exposures of 2.4 micrograms of BPA per kilogram of body weight per day, a dose that could be reached by a child eating one or two servings daily of certain canned foods.

BPA can leach into food from many sources. Canada, Japan, and most European countries have restricted use of this chemical in most consumer applications. In 2012, the United States Food and Drug Administration banned BPA in baby bottles and children's drinking cups, but didn't extend the ruling to other food and beverage containers. Not surprisingly, industry representatives emphasize scientific uncertainty and the need for further research, while most scientists and consumer groups demand a more comprehensive ban now.

This case study introduces a number of important themes for this chapter. How dangerous are low-level but widespread exposures to a variety of environmental toxins? What are the effects of disruption of endocrine systems by synthetic (or natural) compounds? And how should we test and evaluate toxic substances? ■

8.1 ENVIRONMENTAL HEALTH

What is health? The World Health Organization (WHO) defines **health** as a state of complete physical, mental, and social well-being, not merely the absence of disease or infirmity. By that definition, we all are ill to some extent. Likewise, we all can improve our health to live happier, longer, more productive, and more satisfying lives if we think about what we do.

What is disease? A **disease** is an abnormal change in the body's condition that impairs important physical or psychological functions. Diet and nutrition, infectious agents, toxic substances, genetics, trauma, and stress all play roles in **morbidity** (illness) and **mortality** (death). **Environmental health** focuses on factors that cause disease, including elements of the natural, social, cultural, and technological worlds in which we live. WHO estimates that 24 percent of all global disease burden and 23 percent of premature mortality are due to environmental factors. Among children (0 to 14 years), deaths attributable to environmental factors may be as high as 36 percent. Figure 8.2 shows some of these environmental risk factors as well as the media through which we encounter them.

In ecological terms, your body is an ecosystem. Of the approximately 100 trillion cells that make up each of us, only about 10 percent are actually human. The others are bacteria, fungi, protozoans, arthropods, and other species. Ideally, the various organisms in this complex system maintain a harmonious balance. Beneficial species help regulate the dangerous ones. The health challenge shouldn't be to try to totally eradicate all these other species; we couldn't live without them. Rather, we need to find ways to live in equilibrium with our environment and our fellow travelers.

Ever since the publication of Rachel Carson's *Silent Spring* in 1962, the discharge, movement, fate, and effects of synthetic chemical toxins have been a special focus of environmental health, but infectious diseases still remain a grave threat. In this chapter, we'll study many environmental risk factors. First, however, let's look at some major causes of illness worldwide.

Global disease burden is changing

In the past, health organizations have focused on the leading causes of death as the best summary of world health. Mortality data, however, fail to capture the impacts of nonfatal outcomes of disease and injury, such as dementia or blindness, on human well-being. When people are ill, work isn't done, crops aren't planted or harvested, meals aren't cooked, and children can't study and learn. Health agencies now calculate **disability-adjusted life years (DALYs)** as a measure of disease burden. DALYs combine premature deaths and loss of a healthy life resulting from illness or disability. This is an attempt to evaluate the total cost of disease, not simply how many people die. Clearly, many more years of expected life are lost when a child dies of neonatal tetanus than when an 80-year-old dies of pneumonia. Similarly, a teenager permanently paralyzed by a traffic accident will have many more years of suffering and lost potential than will a senior citizen who has a stroke. According to WHO, chronic diseases now account for nearly 60 percent of the 56.5 million total deaths worldwide each year and about half of the global disease burden.

The world is now undergoing a dramatic epidemiological transition. Chronic conditions, such as cardiovascular disease and cancer, no longer afflict only wealthy people. Marvelous progress in eliminating communicable diseases, such as smallpox, polio, and malaria, is allowing people nearly everywhere to live longer. As chapter 6 points out, over the past century the average life expectancy worldwide has risen by about two-thirds. In some poorer countries, such as India, life expectancies nearly tripled in the twentieth century. Although the traditional killers in developing countries—infections, maternal and perinatal (birth) complications, and nutritional deficiencies—still take a terrible toll, diseases such as depression and heart attacks that once were thought to occur only in rich countries are rapidly becoming the leading causes of disability and premature death everywhere.

WHO predicts that in 2020, heart disease, which was fifth in the list of causes of global disease burden a decade ago, will be the leading source of disability and deaths worldwide (table 8.1). Most of that increase will be in the poorer parts of the world where people are rapidly adopting the lifestyles and diet of the richer countries. Similarly, global cancer rates will likely increase by 50 percent. It's expected that by 2020, there will be 15 million people who have cancer and 9 million will die from it.

Taking disability as well as death into account in our assessment of disease burden reveals the increasing role of mental health as a worldwide problem. WHO projections suggest that psychiatric

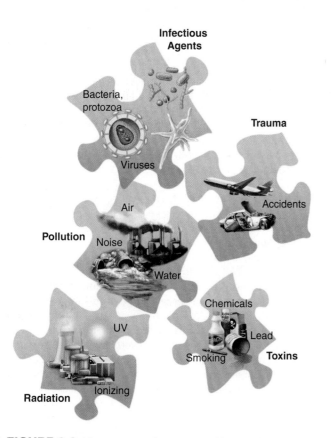

▲ **FIGURE 8.2** Major sources of environmental health risks.

TABLE 8.1 | Leading Causes of Global Disease Burden

RANK	1990	RANK	2020
1	Pneumonia	1	Heart disease
2	Diarrhea	2	Depression
3	Perinatal conditions	3	Traffic accidents
4	Depression	4	Stroke
5	Heart disease	5	Chronic lung disease
6	Stroke	6	Pneumonia
7	Tuberculosis	7	Tuberculosis
8	Measles	8	War
9	Traffic accidents	9	Diarrhea
10	Birth defects	10	HIV/AIDS
11	Chronic lung disease	11	Perinatal conditions
12	Malaria	12	Violence
13	Falls	13	Birth defects
14	Iron anemia	14	Self-inflicted injuries
15	Malnutrition	15	Respiratory cancer

SOURCE: Data from World Health Organization, 2002.

and neurological conditions could increase their share of the global burden from 10 percent currently to 15 percent of the total load by 2020. Again, this isn't just a problem of the developed world. Depression is expected to be the second largest cause of all years lived with disability worldwide, as well as the cause of 1.4 percent of all deaths. For women in both developing and developed regions, depression is the leading cause of disease burden, while suicide, which often is the result of untreated depression, is the fourth largest cause of female deaths.

Notice in table 8.1 that diarrhea, which was the second leading cause of disease burden in 1990, is expected to be ninth on the list in 2020, while measles and malaria are expected to drop out of the top 15 causes of disability. Tuberculosis, which is becoming resistant to antibiotics and is spreading rapidly in many areas (especially in Russia and South Africa), is the only infectious disease whose ranking is not expected to change over the next 20 years. Traffic accidents are now soaring as more people drive. War, violence, and self-inflicted injuries similarly are becoming much more important health risks than ever before.

Chronic obstructive lung diseases (e.g., emphysema, asthma, and lung cancer) are expected to increase from eleventh to fifth in disease burden by 2020. A large part of the increase is due to rising use of tobacco in developing countries, sometimes called "the tobacco epidemic." Every day about 100,000 young people—most of them in poorer countries—become addicted to tobacco. At least 1.1 billion people now smoke, and this number is expected to increase at least 50 percent by 2020. If current patterns persist, about 500 million people alive today will eventually be killed by tobacco. This is expected to be the biggest single cause of death worldwide (because illnesses such as heart attack and depression are triggered by multiple factors). In 2003 the World Health Assembly adopted a historic tobacco-control convention that requires countries to impose restrictions on tobacco advertising, establish clean indoor air controls, and clamp down on tobacco smuggling. If ratified by enough nations, these measures could save billions of lives.

As chapter 7 points out, the world is now experiencing an epidemic of obesity. Poor diet and lack of exercise are now the second leading underlying cause of death in America, causing at least 400,000 deaths per year. Obesity is expected to soon overtake tobacco as the largest single health risk in many countries. However, as we'll see later in this chapter, disruption of endocrine hormones may play a role in obesity as well as many chronic conditions.

Emergent and infectious diseases still kill millions of people

Although the ills of modern life have become the leading killers almost everywhere in the world, communicable diseases still are responsible for about one-third of all disease-related mortality. Diarrhea, acute respiratory illnesses, malaria, measles, tetanus, and a few other infectious diseases kill about 11 million children under age 5 every year in the developing world. Better nutrition, clean water, improved sanitation, and inexpensive inoculations could eliminate most of those deaths (fig. 8.3).

A wide variety of **pathogens** (disease-causing organisms) afflict humans, including viruses, bacteria, protozoans (single-celled animals), parasitic worms, and flukes (fig. 8.4). The greatest loss of life from an individual disease in a single year was the great influenza pandemic of 1918. Epidemiologists now estimate that at least one-third of all humans living at the time were infected, and that between 50 to 100 million died. Businesses, schools, churches, and sport and entertainment events were shut down for months. There were worries that the H1N1 pandemic that spread around the world in 2009 might sicken 2 billion people, kill up to 150 million, and bring the world economy to a standstill. Fortunately, it hasn't been nearly as bad—so far—as the 1918 strain. Influenza is caused by a family of viruses (fig. 8.4a) that mutate rapidly and move from wild and domestic animals to humans, making control of this disease very difficult.

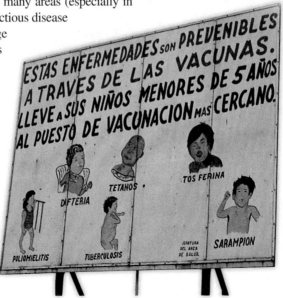

◀ **FIGURE 8.3** At least 3 million children die every year from easily preventable diseases. This billboard in Guatemala encourages parents to have their children vaccinated against polio, diphtheria, TB, tetanus, pertussis (whooping cough), and scarlet fever.

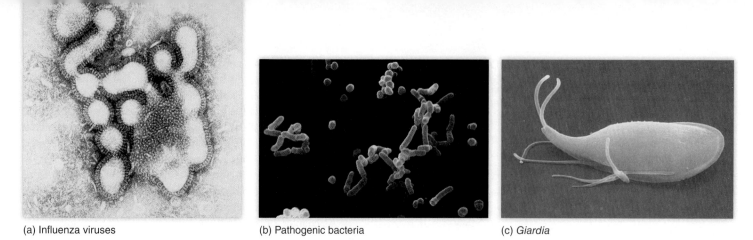

(a) Influenza viruses   (b) Pathogenic bacteria   (c) *Giardia*

▲ **FIGURE 8.4** (a) A group of influenza viruses magnified about 300,000 times. (b) Pathogenic bacteria magnified about 50,000 times. (c) *Giardia*, a parasitic intestinal protozoan, magnified about 10,000 times.

Every year there are 76 million cases of foodborne illnesses in the United States, resulting in 300,000 hospitalizations and 5,000 deaths. Both bacteria and intestinal protozoa cause these illnesses (fig. 8.4b and c). They are spread from feces through food and water. In 2010 nearly 6 million pounds (about 2,700 metric tons) of ground beef were recalled in the United States because of contamination by *E. coli* strain O157:H7.

At any given time, around 2 billion people—nearly one-third of the world population—suffer from worms, flukes, and other internal parasites. Though parasites rarely kill people, they can be extremely debilitating, and can cause poverty that leads to other, more deadly, diseases.

Malaria is one of the most prevalent remaining infectious diseases. Every year about 500 million new cases of this disease occur, and about one million people die from it. The territory infected by this disease is expanding as global climate change allows mosquito vectors to move into new territory. Simply providing insecticide-treated bed nets and a few dollars' worth of anti-malarial pills could prevent tens of millions of cases of this debilitating disease every year. Tragically, some of the countries where malaria is most widespread tax both bed nets and medicine as luxuries, placing them out of reach for ordinary people.

Emergent diseases are those not previously known or that have been absent for at least 20 years. The H1N1 flu that spread around the world in 2009 is a good example. There have been at least 40 outbreaks of emergent diseases over the past two decades, including the extremely deadly Ebola and Marburg fevers, which have afflicted Central Africa in at least six different locations in the past decade. Similarly, cholera, which had been absent from South America for more than a century, reemerged in Peru in 1992 (fig. 8.5). Some other examples include a new drug-resistant form of tuberculosis, now spreading in South Africa; dengue fever,

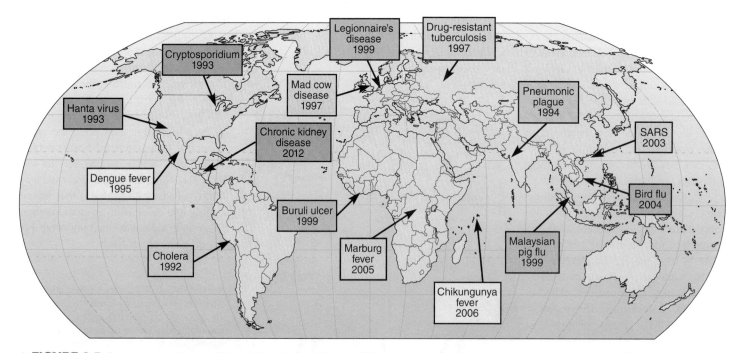

▲ **FIGURE 8.5** Some recent outbreaks of highly lethal infectious diseases. Why are supercontagious organisms emerging in so many different places?

which is spreading through Southeast Asia and the Caribbean; and a new human lymphotropic virus (HTLV), which is thought to have jumped from monkeys into people in Cameroon who handled or ate bushmeat. These HTLV strains are now thought to infect 25 million people.

Growing human populations push people into remote areas where they encounter diseases that may have existed for a long time but only now are exposed to humans. Rapid international travel makes it possible for these new diseases to spread around the world at jet speed. Epidemiologists warn that the next deadly epidemic is only a plane ride away.

West Nile virus shows how fast new diseases can travel. West Nile belongs to a family of mosquito-transmitted viruses that cause encephalitis (brain inflammation). Although recognized in Africa in 1937, the West Nile virus was absent from North America until 1999, when it apparently was introduced by an imported bird or mosquito. The disease spread rapidly from New York, where it was first reported, throughout the eastern United States in only two years (fig. 8.6). Within five years, it was found almost everywhere in the lower 48 states. The virus infects at least 250 bird species and 18 mammalian species. In 2007, about 4,000 people contracted West Nile and about 100 died.

The largest recent human death toll from an emergent disease is due to HIV/AIDS. Although it was first recognized in the early 1980s, acquired immune-deficiency syndrome has now become the fifth greatest cause of contagious deaths. WHO estimates that about 33 million people are now infected with the human immune-deficiency virus and that 3 million die every year from AIDS complications. Although two-thirds of all current HIV infections are now in sub-Saharan Africa, the disease is spreading rapidly in South and East Asia. Over the next 20 years, there could be an additional 65 million AIDS deaths. In Swaziland, health officials estimate

nearly 40 percent of all adults are HIV-positive and that two-thirds of all current 15-year-olds will die of AIDS before age 50. Without AIDS, the life expectancy in Swaziland would be expected to be 55.3 years. With AIDS, Swaziland's average life expectancy is now 35.7 years. Worldwide, more than 15 million children—the equivalent of every child under age 5 in America—have lost one or both parents to AIDS. The economic costs of treating patients and lost productivity from premature deaths resulting from this disease are estimated to be at least $35 billion (U.S.) per year, or about one-tenth of the total GDP of sub-Saharan Africa.

Conservation medicine combines ecology and health care

Humans aren't the only ones to suffer from new and devastating diseases. Domestic animals and wildlife also experience sudden and widespread epidemics, which are sometimes called ecological diseases. We are coming to recognize that the delicate ecological balances that we value so highly—and disrupt so frequently—are important to our own health. **Conservation medicine** is an emerging discipline that attempts to understand how our environmental changes threaten our own health as well as that of the natural communities on which we depend for ecological services. Although it is still small, this new field is gaining recognition from mainstream funding sources such as the World Bank, the World Health Organization, and the U.S. National Institutes of Health.

Ebola hemorrhagic fever, for example, is one of the most virulent viruses ever known, killing up to 90 percent of its victims. In 2002 an outbreak of Ebola fever began killing humans along the Gabon-Congo border. A few months later, researchers found that 221 of the 235 western lowland gorillas they had been studying in this area disappeared in just a few months. Many chimpanzees also died. Although the study team could find only a few of the dead gorillas, 75 percent of those tested positive for Ebola. Altogether, researchers estimate that 5,000 gorillas died in this small area of the Congo. Extrapolating to all of central Africa, it's possible that Ebola has killed one-quarter of all the gorillas in the world. It's thought that the spread of this disease in humans resulted from the practice of hunting and eating primates.

In 2006, people living near a cave west of Albany, New York, reported something peculiar: little brown bats (*Myotis lucifugus*) were flying outside during daylight in the middle of the winter. Inspection of the cave by the Department of Conservation found numerous dead bats near the cave mouth. Most had white fuzz on their faces and wings, a condition now known as white-nose syndrome (WNS). Little brown bats are tiny creatures, about the size of your thumb. They depend on about 2 grams of stored fat to get them through the winter. Hibernation is essential to making their energy resources last. Being awakened just once can cost a bat a month's worth of fat.

The white fuzz has now been identified as filamentous fungus (*Geomyces destructans*), which thrives in the cool, moist conditions where bats hibernate. We don't know where the fungus came from, but, true to its name, the pathogen has spread like wildfire through half a dozen bat species in 14 states along the Appalachian Mountains and in two Canadian provinces. Biologists estimate

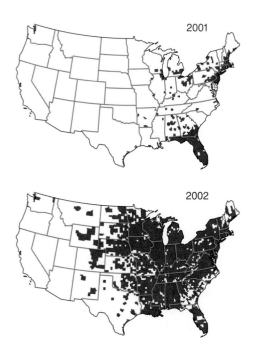

▲ **FIGURE 8.6** The spread of West Nile virus in birds, 2001–2002.

that at least 7 million bats already have died from this disease. It isn't known how the pathogen spreads. Perhaps it moves from animal to animal through physical contact. It's also possible that humans introduce fungal spores on their shoes and clothing when they go from one cave to another.

One mammalogist calls WNS "the chestnut blight of bats." So far, six species of bat are known to be susceptible to this plague. In infected colonies, mortality can be 100 percent. Some researchers fear that bats could be extinct in 20 years in the eastern United States. Losing these important species would have devastating ecological consequences.

An even more widespread epidemic is currently sweeping through amphibians worldwide. A disease called Chytridiomycosis is causing dramatic losses or even extinctions of frogs and toads throughout the world (fig. 8.7). A fungus called *Batrachochytrium dendrobatidis* causes the disease. It was first recognized in 1993 in dead and dying frogs in Queensland, Australia, and now is spreading rapidly, perhaps because the fungus has become more virulent or amphibians are more susceptible due to environmental change. Most of the world's approximately 6,000 amphibian species appear to be susceptible to the disease, and around 2,000 species have declined or even become extinct in their native habitats as part of this global epidemic.

Temperatures above 28°C (82°F) kill the fungus, and treating frogs with warm water can cure the disease in some species. Topical application of the drug chloramphenicol also has successfully cured some frogs. And certain skin bacteria seem to confer immunity to fungal infections. In some places, refuges have been established in which frogs can be maintained under antiseptic conditions until a cure is found. It's hoped that survivors can eventually be reintroduced back to their native habitat and species will be preserved.

Climate change also facilitates expansion of parasites and diseases into new territories. Tropical diseases, such as malaria, cholera, yellow fever, and dengue fever, have been moving into areas from which they were formerly absent as mosquitoes, rodents, and other vectors expand into new habitat. This affects other species besides humans. A disease called Dermo is spreading northward through oyster populations along the Atlantic coast of North America. This disease is caused by a protozoan parasite (*Perkinsus marinus*) that was first recognized in the Gulf of Mexico about 70 years ago. In the 1950s the disease was found in Chesapeake Bay. Since then the parasite has been moving northward, probably assisted by higher sea temperatures caused by global warming. It is now found as far north as Maine. This disease doesn't appear to be harmful to humans, but it is devastating oyster populations.

A puzzling condition is killing honey bees across America. Called colony collapse disorder, this epidemic is reducing bee populations by about 30 percent per year. At least 70 crops worth at least $15 billion annually are at risk. Many factors have been suggested to explain the mysterious bee die-off, including pesticides, parasites, viruses, fungi, malnutrition, genetically modified crops, and long-distance shipping of bees to pollinate crops. Although many authors claim to have discovered the cause of this disorder, it remains a mystery.

◀ **FIGURE 8.7** Frogs and toads throughout the world are succumbing to a deadly disease called Chytridiomycosis. Is this a newly virulent fungal disease, or are amphibians more susceptible because of other environmental stresses?

Resistance to antibiotics and pesticides is increasing

In recent years, health workers have become increasingly alarmed about the rapid spread of methicillin-resistant *Staphylococcus aureus* (MRSA). Staphylococcus (or staph) is very common. Most people have at least some of these bacteria. They are a common cause of sore throats and skin infections, but are usually easily controlled. This new strain, however, is resistant to penicillin and related antibiotics, and can cause deadly infections, especially in people with weak immune systems. MRSA is most frequent in hospitals, nursing homes, correctional facilities, and other places where people are in close contact. It's generally spread through direct skin contact. School locker rooms, gymnasiums, and contact sports also are sources of infections. Several states have closed schools as a result of MRSA contamination. In 2006 the U.S. Centers for Disease Control and Prevention estimated that at least 100,000 MRSA infections in the United States resulted in about 19,000 deaths. Since then a campaign for better hygiene in hospitals and schools appears to have brought infection rates down. A much worse situation is reported in China, where about half of the 5 million annual staph infections are thought to be methicillin-resistant.

Why have vectors, such as mosquitoes, and pathogens, such as bacteria or the malaria parasite, become resistant to pesticides and antibiotics? Part of the answer is natural selection and the ability of many organisms to evolve rapidly. Another factor is the human tendency to use control measures carelessly. Many doctors prescribe penicillin and other antibiotics just in case they might do some good. Similarly, when we discovered that DDT and other insecticides could control mosquito populations, we spread them everywhere. This not only harmed wildlife and beneficial insects but also created perfect conditions for natural selection.

Many pests and pathogens are exposed only to low levels of antibiotics in these widespread releases, allowing those with natural resistance to survive and spread their genes through the population (fig. 8.8). After repeated cycles of exposure and selection, many microorganisms and their vectors have become insensitive to almost all our weapons against them.

Raising huge numbers of cattle, hogs, and poultry in densely packed barns and feedlots is another reason for widespread antibiotic resistance in pathogens. Confined animals are dosed constantly with antibiotics and steroid hormones to keep them disease-free and to

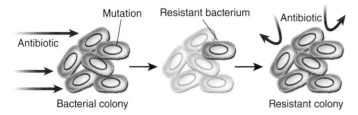

(a) Mutation and selection create drug-resistant strains

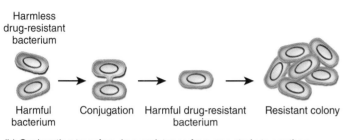

(b) Conjugation transfers drug resistance from one strain to another

▲ **FIGURE 8.8** How microbes acquire antibiotic resistance. (a) Random mutations make a few cells resistant. When challenged by antibiotics, only those cells survive to give rise to a resistant colony. (b) Sexual reproduction (conjugation) or plasmid transfer moves genes from one strain or species to another.

pursue drugs to treat baldness and obesity, depression in dogs, and erectile dysfunction, billions of people are sick or dying from treatable infections and parasitic diseases to which little attention is paid. Worldwide, only 2 percent of the people with AIDS have access to modern medicines. Every year some 600,000 infants acquire HIV—almost all of them through mother-to-child transmission during birth or breast-feeding. Antiretroviral therapy costing only a few dollars can prevent most of this transmission. The Bill and Melinda Gates Foundation has pledged $200 million for medical aid to developing countries to help fight AIDS, TB, and malaria.

Dr. Jeffrey Sachs of the Columbia University Earth Institute says that disease is as much a cause as a consequence of poverty and political unrest, yet the world's richest countries now spend just $1 per person per year on global health. He predicts that raising our commitment to about $25 billion annually (about 0.1 percent of the annual GDP of the 20 richest countries) not only would save about 8 million lives each year but also would boost the world economy by billions of dollars. There also would be huge benefits for rich countries in living in a world less endangered by mass social instability, the spread of pathogens, terrorism, and drug trafficking. Sachs also argues that reducing disease burden would

make them gain weight faster. Up to 70 percent of all antibiotics used in the United States each year are fed to livestock. A significant amount of these antibiotics and hormones are excreted in urine and feces, which are spread, untreated, on the land or discharged into surface water, where they contribute further to the evolution of supervirulent pathogens.

At least half of the 100 million antibiotic doses prescribed for humans every year in the United States are unnecessary or are the wrong ones. Furthermore, many people who start a course of antibiotic treatment fail to carry it out for the time prescribed. For your own health and that of the people around you, if you are taking an antibiotic, follow your doctor's orders. Finish your prescribed doses and don't stop taking the medicine as soon as you start feeling better.

Who should pay for health care?

The heaviest burden of illness is borne by the poorest people, who can afford neither a healthy environment nor adequate health care. WHO estimates that 90 percent of all disease burden occurs in developing countries, where less than one-tenth of all health care dollars is spent. The group Médecins Sans Frontières (MSF, or Doctors without Borders) calls this the 10/90 gap. While wealthy nations

What Can YOU DO?

Tips for Staying Healthy

- Eat a balanced diet with plenty of fresh fruits, vegetables, legumes, and whole grains. Wash fruits and vegetables carefully; they may have come from a country where pesticide and sanitation laws are lax.
- Use unsaturated oils, such as olive or canola, rather than hydrogenated or semisolid fats, such as margarine.
- Cook meats and other foods at temperatures high enough to kill pathogens; clean utensils and cutting surfaces; store food properly.
- Wash your hands frequently. You transfer more germs from hand to mouth than any other means of transmission.
- When you have a cold or flu, don't demand antibiotics from your doctor—they aren't effective against viruses.
- If you're taking antibiotics, continue for the entire time prescribed—quitting as soon as you feel well is an ideal way to select for antibiotic-resistant germs.
- Practice safe sex.
- Don't smoke; avoid smoky places.
- If you drink alcohol, do so in moderation. Never drive when your reflexes or judgment are impaired.
- Exercise regularly: walk, swim, jog, dance, garden. Do something you enjoy that burns calories and maintains flexibility.
- Get enough sleep. Practice meditation, prayer, or some other form of stress reduction. Get a pet.
- Make a list of friends and family who make you feel more alive and happy. Spend time with one of them at least once a week.

help reduce population growth. When parents believe their offspring will survive, they have fewer children and invest more in food, health, and education for smaller families.

What do you think? Would you spend $1 per year for a healthier world?

8.2 TOXICOLOGY

Toxic means poisonous. Toxicology is the study of the adverse effects of external factors on an organism or a system. This includes environmental chemicals, drugs, and diet as well as physical factors, such as ionizing radiation, UV light, and electromagnetic forces. In addition to studying agents that cause toxicity, scientists in this field are concerned with movement and fate of poisons in the environment, routes of entry into the body, and effects of exposure to these agents. Toxic substances damage or kill living organisms because they react with cellular components to disrupt metabolic functions. Toxins often are harmful even in extremely dilute concentrations. In some cases, billionths, or even trillionths, of a gram can cause irreversible damage.

Many toxicologists limit the term "toxin" to proteins or other molecules synthesized by living organisms. Nonbiological noxious substances are called toxicants (from Latin *toxicum,* or poison). The modes of action of organic or inorganic as well as synthetic or natural materials are so similar, however, that we'll use the generic terms "toxins" and "toxics" for poisons in this chapter regardless of their origin.

Hazardous materials aren't necessarily toxic. Some substances are dangerous because they're flammable, explosive, acidic, caustic, irritants, or sensitizers. Many of these materials must be handled carefully in large doses or high concentrations, but they can be rendered relatively innocuous by dilution, neutralization, or other physical treatment.

Environmental toxicology, or ecotoxicology, specifically deals with the interactions, transformation, fate, and effects of toxic materials in the biosphere, including individual organisms, populations, and whole ecosystems. In aquatic systems the fate of the pollutants is primarily studied in relation to mechanisms and processes at interfaces of the ecosystem components. Special attention is devoted to the sediment/water, water/organisms, and water/air interfaces. In terrestrial environments, the emphasis tends to be on effects of metals on soil fauna community and population characteristics.

Table 8.2 is a list of the top 20 toxic and hazardous substances considered the highest risk by the U.S. Environmental Protection Agency. Compiled from the 275 substances regulated by the Comprehensive Environmental Response, Compensation, and Liability Act (CERCLA), commonly known as the Superfund Act, these materials are listed in order of assessed importance in terms of human and environmental health.

How do toxics affect us?

Allergens are substances that activate the immune system. Some allergens act directly as **antigens**; that is, they are recognized as foreign by white blood cells and stimulate the production of specific

TABLE 8.2 | Top 20 Toxic and Hazardous Substances

MATERIAL	MAJOR SOURCES
1. Arsenic	Treated lumber
2. Lead	Paint, gasoline
3. Mercury	Coal combustion
4. Vinyl chloride	Plastics, industrial uses
5. Polychlorinated biphenyls (PCBs)	Electric insulation
6. Benzene	Gasoline, industrial use
7. Cadmium	Batteries
8. Benzo(a)pyrene	Waste incineration
9. Polycyclic aromatic hydrocarbons	Combustion
10. Benzo(b)fluoranthene	Fuels
11. Chloroform	Water purification, industry
12. DDT	Pesticide use
13. Aroclor 1254	Plastics
14. Aroclor 1260	Plastics
15. Trichloroethylene	Solvents
16. Dibenz (a, h)anthracene	Incineration
17. Dieldrin	Pesticides
18. Chromium, hexavalent	Paints, coatings, welding, anticorrosion agents
19. Chlordane	Pesticides
20. Hexachlorobutadiene	Pesticides

SOURCE: Data from U.S. Environmental Protection Agency.

antibodies (proteins that recognize and bind to foreign cells or chemicals). Other allergens act indirectly by binding to and changing the chemistry of foreign materials so they become antigenic and cause an immune response.

Formaldehyde is a good example of a widely used chemical that is a powerful sensitizer of the immune system. It is directly allergenic and can trigger reactions to other substances. Widely used in plastics, wood products, insulation, glue, and fabrics, formaldehyde concentrations in indoor air can be thousands of times higher than in normal outdoor air.

Some people suffer from what is called **sick building syndrome**: headaches, allergies, and chronic fatigue caused by poorly vented indoor air contaminated by molds, carbon monoxide, nitrogen oxides, formaldehyde, and other toxic chemicals released by carpets, insulation, plastics, building materials, and other sources. The Environmental Protection Agency estimates that poor indoor air quality may cost the nation $60 billion a year in absenteeism and reduced productivity (see Key Concepts, p. 190).

Neurotoxins are a special class of metabolic poisons that specifically attack nerve cells (neurons). The nervous system is so important in regulating body activities that disruption of its activities is especially fast-acting and devastating. Different types of neurotoxics act in different ways. Heavy metals, such as lead and mercury, kill nerve cells and cause permanent neurological damage. Anesthetics (ether, chloroform, halothane, etc.) and chlorinated hydrocarbons (DDT, Dieldrin, Aldrin) disrupt nerve cell membranes necessary for nerve action. Organophosphates (Malathion, Parathion) and carbamates (carbaryl, zeneb, maneb) inhibit acetylcholinesterase, an enzyme that regulates signal transmission between nerve cells and the tissues or organs they innervate (for example, muscle). Most neurotoxics are both acute and extremely toxic. More than 850 compounds are now recognized as neurotoxics.

In December 2011, the Environmental Protection Agency issued standards (which had been due for more than 20 years) on mercury and fine particles from power plants. The agency calculated this new rule would provide about $90 billion per year in public health benefits compared to costs of about $10 billion to power companies. The benefits were based mostly on increased lifetime wages for people who would otherwise suffer from mercury-caused brain damage in childhood. Many other benefits (reduced heart attacks, asthma, and lung diseases) are real, but hard to quantify. Opponents in Congress tried to kill the measure because it was "unfriendly to business." What do you think, would you raise business expenses by $10 billion if it saves public costs of $90 billion?

Mutagens are agents, such as chemicals and radiation, that damage or alter genetic material (DNA) in cells. This can lead to birth defects if the damage occurs during embryonic or fetal growth. Later in life, genetic damage may trigger neoplastic (tumor) growth. When damage occurs in reproductive cells, the results can be passed on to future generations. Cells have repair mechanisms to detect and restore damaged genetic material, but some changes may be hidden, and the repair process itself can be flawed. It is generally accepted that there is no "safe" threshold for exposure to mutagens. Any exposure has some possibility of causing damage.

Teratogens are chemicals or other factors that specifically cause abnormalities during embryonic growth and development. Some compounds that are not otherwise harmful can cause tragic problems in these sensitive stages of life. Perhaps the most prevalent teratogen in the world is alcohol. Drinking during pregnancy can lead to **fetal alcohol syndrome**—a cluster of symptoms including craniofacial abnormalities, developmental delays, behavioral problems, and mental defects, that last throughout a child's life. Even one alcoholic drink a day during pregnancy has been associated with decreased birth weight.

Carcinogens are substances that cause **cancer**—invasive, out-of-control cell growth that results in malignant tumors. Cancer rates rose in most industrialized countries during the twentieth century, and cancer is now the second leading cause of death in the United States, killing about a half a million people per year. Sixteen of the 20 compounds listed by the U.S. EPA as the greatest risk to human health are probable or possible human carcinogens. More than 200 million Americans live in areas where the combined upper limit lifetime cancer risk from these carcinogens exceeds 10 in 1 million, or 10 times the risk normally considered acceptable.

In 2010, the U.S. President's Cancer Panel warned that exposure to carcinogens and hormone-disrupting environmental chemicals may be a much greater threat than previously recognized. In particular, the panel was concerned about exposure during pregnancy, when the risk seems greatest. Noting that 300 contaminants have been detected in umbilical cord blood, they said that babies are being born "pre-polluted." The panel called for more rigorous regulation of chemicals, noting that only a few hundred of the more than 80,000 chemicals in use in the United States have been tested thoroughly for human toxicity.

Endocrine hormone disrupters are of special concern

One of the most recently recognized environmental health threats are **endocrine hormone disrupters**, chemicals that interrupt the normal endocrine hormone functions. Hormones are chemicals released into the bloodstream by glands in one part of the body to regulate the development and function of tissues and organs elsewhere in the body (fig. 8.9). You undoubtedly have heard about sex hormones and their powerful effects on how we look and behave, but these are only one example of the many regulatory hormones that rule our lives.

We now know that some of the most insidious effects of persistent chemicals, such as BPA, DDT and PCBs, are that they interfere with the normal growth, development, and physiology of a variety of animals—presumably including humans—at very low doses. In some cases, picogram concentrations (trillionths of a gram per liter) may be enough to cause developmental abnormalities in sensitive organisms. These chemicals are sometimes called environmental estrogens or androgens, because they often cause sexual dysfunction (reproductive health problems in females or feminization of males, for example). They are just as likely, however, to disrupt thyroxin functions or those of other important regulatory molecules as they are to obstruct sex hormones.

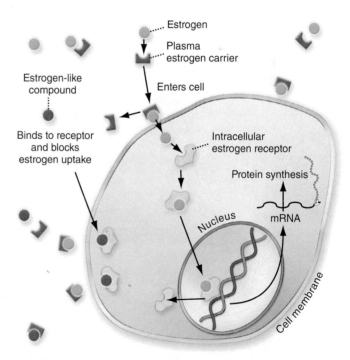

▲ **FIGURE 8.9** Steroid hormone action. Plasma hormone carriers deliver regulatory molecules to the cell surface, where they cross the cell membrane. Intracellular receptors deliver hormones to the nucleus, where they bind to and regulate expression of DNA. Estrogen-like compounds bind to receptors and either block uptake of endogenous hormones or act as a substitute hormone to disrupt gene expression.

What toxins and hazards are present in your home?

The EPA warns that indoor air can be much more polluted than outdoor air. Many illnesses can be linked to poor air quality and exposure to toxins in the home. Since 1950, at least 70,000 new chemical compounds have been invented and dispersed into our environment. Only a fraction of these have been tested for human toxicity, but it's suspected that many may contribute to allergies, birth defects, cancer, and other disorders. Which of the following materials can be found in your home? What toxins and hazards are present in your home?

Garage
- Antifreeze
- Automotive polishes and waxes
- Batteries
- Insecticides, herbicides, fungicides, pesticides
- Gasoline and solvents
- Paints, stains
- Pool supplies
- Rust remover
- Wood preservatives

Kitchen/Laundry Area
- Bleach
- Carbon monoxide and fine particulates
- Cleansers, disinfectants
- Laundry detergents
- Drain cleaners
- Floor polishes
- Oven cleaners
- Nonstick by-products
- Window cleaners

Basement
- Carbon monoxide from furnace and water heaters
- Epoxy glues
- Gasoline, kerosene, and other flammable solvents
- Lye and other caustics
- Mold, bacteria, and other pathogens or allergens
- Paint and paint remover
- PVC and other plastics
- Radon gas from subsoil

Bathroom

- Chloroform from showers and bath
- Leftover drugs and medications
- Fingernail polish and remover
- Makeup
- Mold
- Mouthwash
- Toilet bowl cleaner

Attic

- Asbestos
- Fiberglass insulation
- PBDE-treated cellulose

Bedroom

- Aerosols
- Bisphenol A, lead, cadmium in toys and jewelry
- Flame retardants, fungicides, and insecticides in carpets and bedding
- Mothballs

Living Room

- Asbestos from floor or ceiling tiles
- Benzepyrenes from smoking
- Flame retardants
- Freons from air conditioners
- Furniture and metal polishes
- Lead or cadmium from toys
- Paints, fabrics
- Plastics

CAN YOU EXPLAIN?

1. Which space has the largest number of toxic materials?

2. In which room do you spend the most time?

3. Which space has the greatest number of toxins to which you're likely to be exposed on a regular basis?

8.3 MOVEMENT, DISTRIBUTION, AND FATE OF TOXINS

There are many sources of toxic and hazardous chemicals in the environment. The danger of a chemical is determined by many factors related to the chemical itself, its route or method of exposure, and its persistence in the environment, as well as characteristics of the target organism (table 8.3). We can think of both individuals and an ecosystem as sets of interacting compartments between which chemicals move, based on molecular size, solubility, stability, and reactivity (fig. 8.10). The dose (amount), route of entry, timing of exposure, and sensitivity of the organism all play important roles in determining toxicity. In this section, we will consider some of these characteristics and how they affect environmental health.

Solubility and mobility determine when and where chemicals move

Solubility is one of the most important characteristics in determining how, where, and when a toxic material will move through the environment or through the body to its site of action. Chemicals can be divided into two major groups: those that dissolve more readily in water and those that dissolve more readily in oil. Water-soluble compounds move rapidly and widely through the environment because water is ubiquitous. They also tend to have ready access to most cells in the body because aqueous solutions bathe all our cells. Molecules that are oil- or fat-soluble (usually organic molecules) generally need a carrier to move through the environment and into or within the body. Once inside the body,

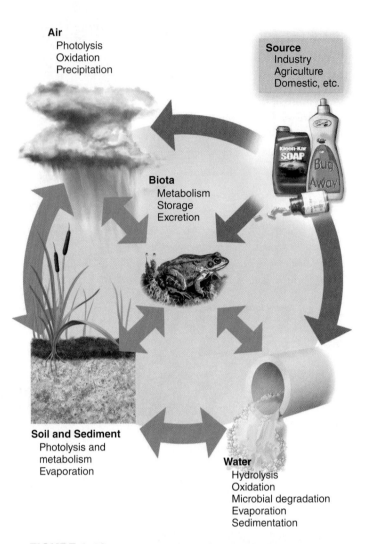

Air
Photolysis
Oxidation
Precipitation

Source
Industry
Agriculture
Domestic, etc.

Biota
Metabolism
Storage
Excretion

Soil and Sediment
Photolysis and metabolism
Evaporation

Water
Hydrolysis
Oxidation
Microbial degradation
Evaporation
Sedimentation

▲ **FIGURE 8.10** Movement and fate of chemicals in the environment. Toxins also move directly from a source to soil and sediment.

TABLE 8.3 | Factors in Environmental Toxicity

FACTORS RELATED TO THE TOXIC AGENT
1. Chemical composition and reactivity
2. Physical characteristics (such as solubility, state)
3. Presence of impurities or contaminants
4. Stability and storage characteristics of toxic agent
5. Availability of vehicle (such as solvent) to carry agent
6. Movement of agent through environment and into cells

FACTORS RELATED TO EXPOSURE
1. Dose (concentration and volume of exposure)
2. Route, rate, and site of exposure
3. Duration and frequency of exposure
4. Time of exposure (time of day, season, year)

FACTORS RELATED TO THE ORGANISM
1. Resistance to uptake, storage, or cell permeability of agent
2. Ability to metabolize, inactivate, sequester, or eliminate agent
3. Tendency to activate or alter nontoxic substances so they become toxic
4. Concurrent infections or physical or chemical stress
5. Species and genetic characteristics of organism
6. Nutritional status of subject
7. Age, sex, body weight, immunological status, and maturity

however, oil-soluble toxics penetrate readily into tissues and cells because the membranes that enclose cells are themselves made of similar oil-soluble chemicals. Once inside cells, oil-soluble materials are likely to accumulate and to be stored in lipid deposits, where they may be protected from metabolic breakdown and persist for many years.

Exposure and susceptibility determine how we respond

Just as there are many sources of toxic materials in our environment, there are many routes for entry into our bodies (fig. 8.11). Airborne toxics generally cause more ill health than toxics from any other exposure source. We breathe far more air every day than the volume of food we eat or water we drink. Furthermore, the lining of our lungs, which is designed to exchange gases very efficiently, also absorbs toxics very well. Epidemiologists estimate that 3 million people—two-thirds of them children—die each year from diseases caused or exacerbated by air pollution.

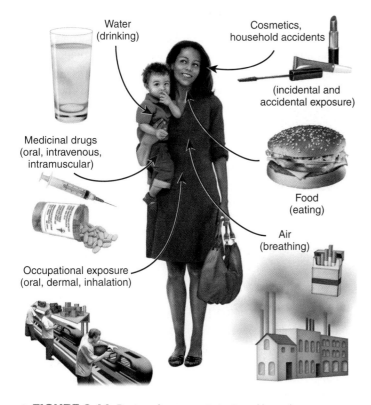

Water
(drinking)

Cosmetics,
household accidents

(incidental and
accidental exposure)

Medicinal drugs
(oral, intravenous,
intramuscular)

Food
(eating)

Air
(breathing)

Occupational exposure
(oral, dermal, inhalation)

▲ **FIGURE 8.11** Routes of exposure to toxic and hazardous environmental factors.

But food, water, and skin contact also can expose us to a wide variety of hazards. The largest exposures for many toxics are found in industrial settings, where workers may encounter doses thousands of times higher than would be found anywhere else. The European Agency for Safety and Health at Work warns that 32 million people (20 percent of all employees) in the European Union are exposed to unacceptable levels of carcinogens and other chemicals in their workplaces.

Condition of the organism and timing of exposure also have strong influences on toxicity. Healthy adults, for example, may be relatively insensitive to doses that are very dangerous to young children or to someone already weakened by other diseases. Pound for pound, children drink more water, eat more food, and breathe more air than do adults. Putting fingers, toys, and other objects into their mouths increases children's exposure to toxics in dust or soil.

Furthermore, children generally have less-developed immune systems or processes to degrade or excrete toxics. The developing brain is especially sensitive to damage. Obviously, disrupting the complex and sensitive process of brain growth and development can have tragic long-term consequences. Researchers estimate that one in six children in America has a developmental disability, usually involving the nervous system.

The best-known example of an environmental risk for children is lead poisoning. Before lead paint and leaded gasoline were banned in the 1970s, at least 4 million American children had dangerous levels of lead in their blood. Banning these products has been one of the greatest successes in environmental health.

Blood lead levels in children have fallen more than 90 percent in the past three decades. Unfortunately, this tremendous success hasn't yet extended to developing countries. In 2009 the *China Daily* reported that 1.1 million children are born in China every year with birth defects attributed to environmental factors.

The notorious teratogen thalidomide is a prime example of differences in sensitivity between species and within stages of fetal development. A single dose of this teratogen taken in the third week of human pregnancy (a time when many women aren't aware they're pregnant) can cause severe abnormalities in fetal limb development. Thalidomide was tested on a number of laboratory animals without showing any deleterious effects. Unfortunately, however, it's a powerful teratogen in humans.

Bioaccumulation and biomagnification increase chemical concentrations

Cells have mechanisms for **bioaccumulation**, the selective absorption and storage of a great variety of molecules. This allows them to accumulate nutrients and essential minerals, but at the same time they also may absorb and store harmful substances through the same mechanisms. Materials that are rather dilute in the environment can reach dangerous levels inside cells and tissues through this process of bioaccumulation.

Toxic substances also can be magnified through food webs. **Biomagnification** occurs when the toxic burden of a large number of organisms at a lower trophic level is accumulated and concentrated by a predator in a higher trophic level. Phytoplankton and bacteria in aquatic ecosystems, for instance, take up heavy metals or toxic organic molecules from water or sediments (fig. 8.12). Their predators—zooplankton and small fish—collect and retain the toxics from many prey organisms, building up higher concentrations of toxics. The top carnivores in the food chain—game fish, fish-eating birds, and humans—can accumulate such high toxic levels that they suffer adverse health effects.

One of the first well-known examples of bioaccumulation and biomagnification was DDT, which accumulated through food chains, so that by the 1960s it was shown to be interfering with reproduction of peregrine falcons, brown pelicans, and other predatory birds at the top of their food chains.

Persistence makes some materials a greater threat

Many toxic substances degrade when exposed to sun, air, and water. This can destroy them or convert them to inactive forms, but some materials are persistent and can last for years or even centuries as they cycle through ecosystems. Even if released in minute concentrations, they can bioaccumulate in food webs to reach dangerous levels. Heavy metals, such as lead and mercury, are classic examples. Mercury, like lead, can destroy nerve cells and is particularly dangerous to children. The largest source of mercury in the United States is from burning coal. Every year, power plants in the United States release 48 tons of this toxic metal into the air. It works its way through food chains and is concentrated to dangerous levels in fish. Mercury contamination is the most

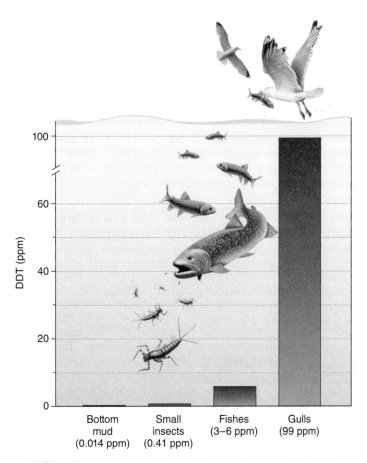

DDT (ppm)

| Bottom mud (0.014 ppm) | Small insects (0.41 ppm) | Fishes (3–6 ppm) | Gulls (99 ppm) |

▲ **FIGURE 8.12** Bioaccumulation and biomagnification in a Lake Michigan food chain. The DDT tissue concentration in gulls, a tertiary consumer, was about 240 times that in the small insects sharing the same environment.

common cause of lakes and rivers failing to meet pollution regulation standards. Forty states have issued warnings that children and pregnant women should not eat local fish. In a nationwide survey of lakes and rivers in 2007, the Environmental Protection Agency found that 55 percent of the 2,500 fish sampled had mercury levels that exceeded dietary recommendations.

Many organic compounds, such as bisphenol A (BPA) and chlorinated hydrocarbon pesticides, are highly resistant to degradation. This makes them very useful, but it also allows them to accumulate in the environment and have unexpected effects far from the site of their original use. Some of these **persistent organic pollutants (POPs)** have become extremely widespread, being found now from the tropics to the Arctic. They often accumulate in food webs and reach toxic concentrations in long-living top predators, such as humans, sharks, raptors, swordfish, and bears. These are some of the greatest current concerns:

• Polybrominated diphenyl ethers (PBDEs) are widely used as flame-retardants in textiles, foam in upholstery, and plastic in appliances and computers. These compounds were first reported accumulating in women's breast milk in Sweden in the 1990s. They were subsequently found in humans and other species everywhere from Canada to Israel. Nearly 150 million metric tons (330 million lb) of PBDEs are used every year worldwide. The toxicity and environmental persistence of PBDEs are much like those of PCBs, to which they are

closely related chemically. The dust at ground zero in New York City after September 11 was heavily laden with PBDEs. The European Union has already banned these compounds.

• Perfluorooctane sulfonate (PFOS) and perfluorooctanoic acid (PFOA, also known as C8) are members of a chemical family used to make nonstick, waterproof, and stain-resistant products such as Teflon, Gortex, Scotchguard, and Stainmaster. Industry makes use of their slippery, heat-stable properties to manufacture everything from airplanes and computers to cosmetics and household cleaners. Now these chemicals—which are reported to be infinitely persistent in the environment—are found throughout the world, even the most remote and seemingly pristine sites.

   Almost all Americans have one or more perfluorinated compounds in their blood. In one long-term study, workers exposed to high levels of PFOA were twice as likely to die of prostate cancer or stroke than colleagues with little or no exposure to the chemical. Heating some nonstick cooking pans above 260°C (500°F) can release enough PFOA to kill pet birds. This chemical family has been shown to cause liver damage as well as various cancers and reproductive and developmental problems in rats. Exposure may be especially dangerous to women and girls, who may be 100 times more sensitive than men to these chemicals.

• Perchlorate is a waterborne contaminant left over from propellants and rocket fuels. About 12,000 sites in the United States were used by the military for live munition testing and are contaminated with perchlorate. Polluted water used to irrigate crops such as alfalfa and lettuce has introduced the chemical into the human food chain. Tests of cow's milk and human breast milk detected perchlorate in nearly every sample from throughout the United States. Perchlorate can interfere with iodine uptake in the thyroid gland, disrupting adult metabolism and childhood development.

• Phthalates (pronounced *thal ates*) are found in cosmetics, deodorants, and many plastics (such as soft polyvinyl chloride, or PVC) used for food packaging, children's toys, and medical devices. Some members of this chemical family are known to be toxic to laboratory animals, causing kidney and liver damage and possibly some cancers. In addition, many phthalates act as endocrine hormone disrupters and have been linked to reproductive abnormalities and decreased fertility. A correlation has been found between phthalate levels in urine and low sperm numbers and decreased sperm motility in men. Nearly everyone in the United States has phthalates in his or her body at levels reported to cause these problems. While not yet conclusive, these results could help explain a 50-year decline in semen quality in most industrialized countries.

   It's also possible that phthalates and other endocrine-disrupting chemicals are contributing to the obesity crisis. Rather than being simply a matter of "greed and sloth," metabolic disturbances caused by chemicals in shampoos, soaps, cosmetics. and other products could be playing a role in abnormal weight gain. In 2007, California banned phthalates in products designed for children.

- Atrazine is the most widely used herbicide in America. More than 60 million pounds of this compound are applied per year, mainly on corn and cereal grains, but also on golf courses, sugarcane, and Christmas trees. It has long been known to disrupt endocrine hormone functions in mammals, resulting in spontaneous abortions, low birth weights, and neurological disorders. Studies of families in corn-producing areas in the American Midwest have found higher rates of developmental defects among infants, and certain cancers in families with elevated atrazine levels in their drinking water. University of California professor Tyrone Hayes has shown that atrazine levels as low as 0.1 ppb (30 times less than the EPA maximum contaminant level) caused severe reproductive effects in amphibians, including abnormal gonadal development and hermaphroditism.

  Atrazine now is found in rain and surface waters nearly everywhere in the United States at levels that could cause abnormal development in frogs. In 2003 the European Union withdrew regulatory approval for this herbicide, and several countries banned its use altogether. Some toxicologists have suggested a similar rule in the United States.

Chemical interactions can increase toxicity

Some materials produce *antagonistic* reactions. That is, they interfere with the effects or stimulate the breakdown of other chemicals. For instance, vitamins E and A can reduce the response to some carcinogens. Other materials are *additive* when they occur together in exposures. Rats exposed to both lead and arsenic show twice the toxicity of only one of these elements. Perhaps the greatest concern is synergistic effects. **Synergism** is an interaction in which one substance exacerbates the effects of another. For example, occupational asbestos exposure increases lung cancer rates 20-fold. Smoking increases lung cancer rates by the same amount. Asbestos workers who also smoke, however, have a 400-fold increase in cancer rates. How many other toxic chemicals are we exposed to that are below threshold limits individually but combine to give toxic results?

8.4 MECHANISMS FOR MINIMIZING TOXIC EFFECTS

A fundamental concept in toxicology is that every material can be poisonous under some conditions, but most chemicals have a safe level or threshold below which their effects are undetectable or insignificant. Each of us consumes lethal doses of many chemicals over the course of a lifetime. One hundred cups of strong coffee, for instance, contain a lethal dose of caffeine. Similarly, 100 aspirin tablets, 10 kg (22 lb) of spinach or rhubarb, or a liter of alcohol would be deadly if consumed all at once. Taken in small doses, however, these materials can be broken down or excreted before they do much harm. Furthermore, the damage they cause can be repaired. Sometimes, however, mechanisms that protect us from a particular toxin at one stage in the life cycle become deleterious with another substance or in another stage of development. Let's look at how these processes help protect us from harmful substances, as well as how they can go awry.

Metabolic degradation and excretion eliminate toxics

Most organisms have enzymes that process waste products and environmental poisons to reduce their toxicity. In mammals, most of these enzymes are located in the liver, the primary site of detoxification of both natural wastes and introduced poisons. Sometimes, however, these reactions work to our disadvantage. Compounds such as benzepyrene, for example, that aren't toxic in their original form are processed by the same liver enzymes into cancer-causing carcinogens. Why would we have a system that makes a chemical more dangerous? Evolution and natural selection are expressed through reproductive success or failure. Defense mechanisms that protect us from toxins and hazards early in life are "selected for" by evolution. Factors or conditions that affect postreproductive ages (such as cancer or premature senility) usually don't affect reproductive success or exert "selective pressure."

We also reduce the effects of waste products and environmental toxics by eliminating them from the body through excretion. Volatile molecules, such as carbon dioxide, hydrogen cyanide, and ketones, are excreted via breathing. Some excess salts and other substances are excreted in sweat. Primarily, however, excretion is a function of the kidneys, which can eliminate significant amounts of soluble materials through urine formation. Accumulated toxins in the urine can damage this vital system, however. In the same way, the stomach, intestine, and colon often suffer damage from materials concentrated in the digestive system and may be afflicted by diseases and tumors.

Repair mechanisms mend damage

In the same way that individual cells have enzymes to repair damage to DNA and protein at the molecular level, tissues and organs that are exposed regularly to physical wear-and-tear or to toxic or hazardous materials often have mechanisms for damage repair. Our skin and the epithelial linings of the gastrointestinal tract, blood vessels, lungs, and urogenital system have high cellular reproduction rates to replace injured cells. With each reproduction cycle, however, there is a chance that some cells will lose normal growth controls and run amok, creating a tumor. Thus any agent, such as smoking or drinking, that irritates tissues is likely to be carcinogenic. And tissues with high cell-replacement rates are among the most likely to develop cancers.

8.5 MEASURING TOXICITY

In 1540 the Swiss scientist Paracelsus said, "The dose makes the poison," by which he meant that almost everything is toxic at very high levels, but can be safe if diluted enough. This remains the most basic principle of toxicology. Sodium chloride (table salt), for instance, is essential for human life in small doses. If you were forced to eat a kilogram of salt all at once, however, it would make you very sick. A similar amount injected into your bloodstream would be lethal. How a material is delivered—at what rate, through which route of entry, and in what medium—plays a vital role in determining toxicity.

This doesn't mean that all toxics are identical, however. Some are so poisonous that a single drop on your skin can kill you. Others require massive amounts injected directly into the blood to be lethal. It is difficult to measure and compare the toxicity of various materials, because species differ in sensitivity and individuals within a species respond differently to a given exposure. In this section, we will look at methods of toxicity testing and at how results are analyzed and reported.

We usually test toxic effects on lab animals

The most commonly used and widely accepted toxicity test is to expose a population of laboratory animals to measured doses of a specific substance under controlled conditions. This procedure is expensive, time-consuming, and often painful and debilitating to the animals being tested. It commonly takes hundreds—or even thousands—of animals, several years of hard work, and hundreds of thousands of dollars to thoroughly test the effects of a toxic at very low doses. More humane toxicity tests using computer simulations of model reactions, cell cultures, and other substitutes for whole living animals are being developed. However, conventional large-scale animal testing is the method in which scientists have the most confidence and on which most public policies about pollution and environmental or occupational health hazards are based.

In addition to humanitarian concerns, several other problems in laboratory animal testing trouble both toxicologists and policy makers. One problem is differences in toxic sensitivity among the members of a specific population. Figure 8.13 shows a typical dose/response curve for exposure to a hypothetical chemical. Some individuals are very sensitive to the material, while others are insensitive. Most, however, fall in a middle category, forming a bell-shaped curve. The question for regulators and politicians is whether we should set pollution levels that will protect everyone, including the most sensitive people, or only aim to protect the average person. It might cost billions of extra dollars to protect a very small number of individuals at the extreme end of the curve. What do you think, is that a good use of resources?

Dose/response curves aren't always symmetrical, making it difficult to compare toxicity of unlike chemicals or different species of organisms. A convenient way to describe toxicity of a chemical is to determine the dose to which 50 percent of the test population is sensitive. In the case of a lethal dose (LD), this is called the **LD50** (fig. 8.14).

Unrelated species can react very differently to the same poison, not only because body sizes vary but also because physiology and metabolism differ. Even closely related species can have very dissimilar reactions to a particular chemical. Hamsters, for instance, are nearly 5,000 times less sensitive to some dioxins than are guinea pigs. Of 226 chemicals found to be carcinogenic in either rats or mice, 95 cause cancer in one species but not the other. These variations make it difficult to estimate the risks for humans, because we don't consider it ethical to perform controlled experiments in which we deliberately expose people to toxins.

Even within a single species, there can be variations in responses between different genetic lines. A current controversy in determining the toxicity of bisphenol A (BPA) concerns the type of rats used for toxicology studies. In most labs, a sturdy strain called the Sprague-Dawley rat is standard. It turns out, however, that these animals, which were bred to grow fast and breed prolifically in lab conditions, are thousands of times less sensitive to endocrine disrupters than ordinary rats. Industry reports that declare BPA to be harmless based on Sprague-Dawley rats are highly suspect.

There is a wide range of toxicity

It's useful to group materials according to their relative toxicity. A moderately harmful substance takes about 1 g per kg of body weight (about 2 oz for an average human) to make a lethal dose.

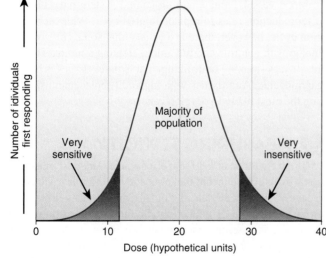

▲ **FIGURE 8.13** Probable variations in sensitivity to a toxin within a population. Some members of a population may be very sensitive to a given toxin, while others are much less sensitive. The majority of the population falls somewhere between the two extremes.

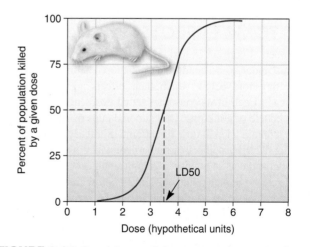

▲ **FIGURE 8.14** Cumulative population response to increasing doses of a toxin. The LD50 is the dose that is lethal to half the population.

Very toxic materials take about one-tenth that amount, while extremely poisonous materials take one-hundredth as much (only a few drops) to kill most people. Supertoxic chemicals are extremely potent; for some, a few micrograms (millionths of a gram—an amount invisible to the naked eye) make a lethal dose. These materials aren't all synthetic. One of the most toxic chemicals known, for instance, is ricin, a protein found in castor bean seeds. It is so poisonous that 0.3 billionths of a gram given intravenously will kill a mouse. If aspirin were this toxic for humans, a single tablet, divided evenly, could kill 1 million people.

Many carcinogens, mutagens, and teratogens are dangerous at levels far below their direct toxic effect because abnormal cell growth exerts a kind of biological amplification. A single cell, perhaps altered by a molecular event, such as methylation, can multiply into millions of tumor cells or an entire organism. Just as there are different levels of direct toxicity, however, there are different degrees of carcinogenicity, mutagenicity, and teratogenicity. Methanesulfonic acid, for instance, is highly carcinogenic, while the sweetener saccharin is a possible carcinogen whose effects may be vanishingly small.

Acute versus chronic doses and effects

Most of the toxic effects that we have discussed so far have been **acute effects**. That is, they are caused by a single exposure to the threat and result in an immediate health crisis. Often, if the individual experiencing an acute reaction survives this immediate crisis, the effects are reversible. **Chronic effects**, on the other hand, are long-lasting, perhaps even permanent. A chronic effect can result from a single dose of a very toxic substance, or it can be the result of a continuous or repeated sublethal exposure.

We also describe long-lasting exposures as chronic, although their effects may or may not persist after the toxic agent is removed. It usually is difficult to assess the specific health risks of chronic exposures because other factors, such as aging or normal diseases, act simultaneously with the factor under study. It often requires very large populations of experimental animals to obtain statistically significant results for low-level chronic exposures. Toxicologists talk about "megarat" experiments in which it might take a million rats to determine the health risks of some supertoxic chemicals at very low doses. Such an experiment would be terribly expensive for even a single chemical, let alone for the thousands of chemicals and factors suspected of being dangerous.

An alternative to enormous studies involving millions of animals is to give massive amounts—usually the maximum tolerable dose—of a compound being studied to a smaller number of individuals and then to extrapolate what the effects of lower doses might have been. This is a controversial approach because it is not clear that responses to toxics are linear or uniform across a wide range of doses.

Figure 8.15 shows three possible results from low doses of a toxic material. Curve (a) shows a baseline level of response in the population, even at zero dose. This suggests that some other factor in the environment also causes this response. Curve (b) shows a straight-line relationship from the highest doses to zero exposure. Many carcinogens and mutagens show this kind of response. Any

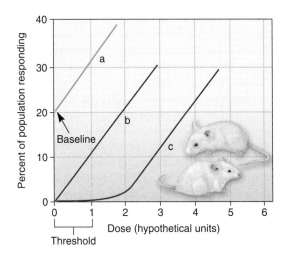

▲ **FIGURE 8.15** Three possible dose–response curves at low doses. (a) Some individuals respond, even at zero dose, indicating that some other factor must be involved. (b) Response is linear down to the lowest possible dose. (c) Threshold must be passed before any response is seen.

exposure to such agents, no matter how small, carries some risks. Curve (c) shows a threshold for the response where some minimal dose is necessary before any effect can be observed. This generally suggests the presence of a defense mechanism that prevents the toxin from reaching its target in an active form or repairs the damage that the toxic causes. Low levels of exposure to the material in question may have no deleterious effects, and it might not be necessary to try to keep exposures to zero.

Which, if any, environmental health hazards have thresholds is an important but difficult question. The 1958 Delaney Clause to the U.S. Food and Drug Act forbids the addition of any amount of known carcinogens to food and drugs, based on the assumption that any exposure to these substances represents unacceptable risks. This standard was replaced in 1996 by a "no reasonable harm" requirement, defined as less than one cancer for every million people exposed over a lifetime. This change was supported by a report from the National Academy of Sciences concluding that synthetic chemicals in our diet are unlikely to represent an appreciable cancer risk. We'll discuss risk analysis in section 8.6.

Detectable levels aren't always dangerous

You may have seen or heard dire warnings about toxic materials detected in samples of air, water, or food. A typical headline announced recently that 23 pesticides were found in 16 food samples. What does that mean? The implication seems to be that any amount of dangerous materials is unacceptable and that counting the numbers of compounds detected is a reliable way to establish danger. We have seen, however, that the dose makes the poison. It matters not only what is there but also how much, where it's located, how accessible it is, and who's exposed. At some level, the mere presence of a substance is insignificant.

Noxious materials may seem to be more widespread now than in the past, and this is surely a valid perception for many substances (fig. 8.16). The daily reports we hear of new chemicals found in new places, however, are also due, in part, to our more sensitive measuring techniques. Twenty years ago, parts per million were generally the limits of detection for most materials. Anything below that amount was often reported as "zero" or "absent," rather than more accurately as "undetected." A decade ago, new machines and techniques were developed to measure parts per billion. Suddenly, substances were found where none had been suspected. Now we can detect parts per trillion or even parts per quadrillion in some cases. Increasingly sophisticated measuring capabilities may lead us to believe that toxic materials have become more prevalent. In fact, our environment may be no more dangerous; we're just better at finding trace amounts.

What accounts for the seemingly enormous increase in conditions, such as asthma, autism, food allergies, and behavioral disorders in American children? Is this evidence for higher levels of environmental pollutants or merely, overdiagnosis or increased public paranoia? Perhaps these conditions always existed but weren't so often given labels. In one surprising study, the only factor that appeared to correlate with increased childhood autism was the educational level of the parents. And how can it be that if asthma is caused by air pollution or other environmental con-

▲ **FIGURE 8.16** "Do you want to stop reading those ingredients while we're trying to eat?" SOURCE: Reprinted with permission of the Star-Tribune, Minneapolis-St. Paul.

taminants, the disease is almost unknown in China, where conditions are far worse than in America? One interesting suggestion is that when we aren't exposed to common pathogens early in life, our immune systems began to attack normal tissues and organs. Clearly, humans are complex organisms and we still have much to learn about our interactions with toxins and our environment.

Low doses can have variable effects

A complication in assessing risk is that the effects of low doses of some toxics and health hazards can be nonlinear. They may be either more or less dangerous than would be predicted from exposure to higher doses. For example, low doses of DHEP suppress activity of an enzyme essential for rat brain development. This is surprising because higher doses stimulated this enzyme. Thus low doses can be more damaging to brain development than expected.

On the other hand, very low amounts of radiation seem to be protective against certain cancers. This is perplexing, because ionizing radiation has long been recognized as a human carcinogen. It's thought now, however, that very low radiation exposure may stimulate DNA repair along with enzymes that destroy free radicals (atoms with unpaired, reactive electrons in their outer shells). Activating these repair mechanisms may defend us from other, unrelated hazards. These nonlinear effects are called **hormesis**.

Another complication is that some substances can have long-lasting effects on genetic expression. For example, researchers found that exposure of pregnant rats to certain chemicals can have effects not only on the exposed rats, but on their daughters and granddaughters. A single dose given on a specific day in pregnancy can be expressed several generations later, even if those offspring have never been exposed to the chemical (see Exploring Science, p. 199).

This effect doesn't require a permanent mutation in genes, but it can result in changes, both positive and negative, in expression of whole groups of critical genes over multiple generations. It also can have different outcomes in variants of the same gene. Thus exposure to a particular toxin could be very harmful to you

Could your diet, behavior, or environment affect the lives of your children or grandchildren? For a century or more, scientists assumed that the genes you receive from your parents irreversibly fix your destiny, and that factors, such as stress, habits, toxic exposure or parenting have no effect on future generations.

Now, however, a series of startling discoveries are making us reexamine those ideas. Scientists are finding that a complex set of chemical markers and genetic switches—called the **epigenome**—consisting of DNA and its associated proteins and other small molecules, regulates gene function in ways that can affect numerous functions simultaneously and persist for multiple generations. "Epi" means above, and the epigenome is above ordinary genes in that it regulates their functions. Understanding how this system works helps us see how many environmental factors affect health, and may become useful in treating a variety of diseases.

One of the most striking epigenetic experiments was carried out a decade ago by researchers at Duke University. They were studying the affects of diet on an a strain of mice carrying a gene called "agouti" that makes them obese, yellow, and prone to cancer and diabetes. Starting just before conception, mother agouti mice were fed a diet rich in B vitamins (folic acid and B_{12}). Amazingly, this simple dietary change resulted in baby mice that were sleek, brown, and healthy. The vitamins somehow had turned off the agouti gene in the offspring.

We know now that B vitamins as well as vegetables, such as onions, garlic, and beets, are methyl donors—that is, they can add a carbon atom and three hydrogens to proteins and nucleic acids. Attaching an extra methyl group can switch genes on or off by changing the way proteins and nucleic acids translate the DNA. Similarly, acetylating DNA (addition of an acetyl group: CH_3CO) can also either stimulate or inhibit gene expression. Both of these reactions are key methods of regulating gene expression.

These reactions involve not only the genes themselves, but also a huge set of what we once thought was useless, or junk DNA in chromosomes as well as a large amount of protein that once seemed to be merely packing material. We now know that both this extra DNA and the proteins around which genes are wrapped play vital roles in gene expression. And methylating or acetylating these proteins or nucleic acid sequences can have lasting effects on whole families of genes.

More remarkable is that changes in the epigenome can carry through multiple generations. In 2004, Michael Skinner, a geneticist at Washington State University was studying the effects on rats of exposure to a commonly used fungicide. He found that male rats exposed in utero had lower sperm counts later in life. It only

took a single exposure to cause this effect. Amazingly, the effect lasted for at least four generations even though those subsequent offspring were never exposed to the fungicide. Somehow, the changes in the switching system can be passed from one generation to another along with the DNA it controls.

The way a mother rodent nurtures her young also can cause changes in methylation patterns in her babies' brains that are somewhat like the prenatal vitamins and nutrients that affected the agouti gene. It's thought that licking and grooming activate serotonin receptors that turn on genes to reduce stress responses, resulting in profound brain changes. In another study, rats given extra attention, diet, and mental stimulation (toys) did better at memory tests than did environmentally deprived controls. Altered methylation patterns in the hippocampus—the part of the brain that controls memory—were detected in both these cases. Subsequent generations maintained this methylation pattern.

Epigenetic effects have also been found in humans. One of the most compelling studies involved comparison of two centuries of health records, climate, and food supply in a remote village in northern Sweden. The village of Overkalix was so isolated that when bad weather caused crop failures, famine struck everyone. In good years, on the other hand, there was plenty of food and people stuffed themselves. A remarkable pattern emerged. When other social factors were factored in, grandfathers who were preteens during lean years had grandsons who lived an amazing 32 years longer than those whose grandfathers had gorged themselves as preteens. Similarly, women whose mothers had an access to a rich diet while they were pregnant were much more likely to have daughters and granddaughters with health problems and shortened lives.

In an another surprising human health study, researchers found, in a long-term analysis of couples in Bristol, England, that fathers who started smoking before they were eleven years old (just as they were starting puberty and sperm formation was beginning) were much more likely to have sons and grandsons who were overweight and who lived significantly shortened lives than those of nonsmokers. Both these results are attributed to epigenetic effects.

A wide variety of factors can cause epigenetic changes. Smoking, for example, leaves a host of persistent methylation markers in your DNA. So does exposure to a number of pesticides, toxics, drugs, and stressors. At the same time, polyphenols in green tea and deeply colored fruit, B vitamins, as well as healthy foods, such as garlic, onions, and turmeric, can help prevent deleterious methylations. Not surprisingly, epigenetic changes are implicated

▲ Agouti mice have a gene that makes them obese, yellow, and prone to cancer and diabetes. If a mother agouti mouse (*left*) is given B vitamins during pregnancy, the gene is turned off and its baby (*right*) is sleek, brown, and healthy. Amazingly, this genetic repair lasts for several generations before the gene resumes its deleterious effects.

Continued ▶

◀ *Continued*

in many cancers, including colon, prostate, breast, and blood. This may explain many confusing cases in which our environment seems to have long-lasting effects on health and development that can't be explained by ordinary metabolic effects.

Unlike mutations, epigenetic changes aren't permanent. Eventually the epigenome returns to normal if the exposure isn't repeated. This makes them candidates for drug therapy. Currently the U.S. Food and Drug Administration has approved two drugs, Vidaza and Dacogen, that inhibit methylation and are used to treat

a precursor to leukemia. Another drug, Zolinza, which enhances acetylation, is approved to treat another form of leukemia. Dozens of other drugs that may treat a variety of diseases including rheumatoid arthritis, neurodegenerative diseases, and diabetes are under development.

So, your diet, behavior, and environment can have a much stronger impact on both your health and that of your descendents than we previously understood. What you ate, drank, smoked, or did last night may have profound effects on future generations.

---

but have no detectable effects in someone who has slightly different forms of the same genes. This may explain why, in a group of people exposed to the same carcinogen, some will get cancer while others don't. Or it could explain why a particular diet protects the health of some people but not others. As scientists are increasing our knowledge of this intricate network of switches and controls, they're helping explain much about our environment and health.

8.6 RISK ASSESSMENT AND ACCEPTANCE

Even if we know with some certainty how toxic a specific chemical is in laboratory tests, it's still difficult to determine **risk** (the probability of harm multiplied by the probability of exposure) if that chemical is released into the environment. As we have seen, many factors complicate the movement and fate of chemicals both around us and within our bodies. Furthermore, public perception of relative dangers from environmental hazards can be skewed so that some risks seem much more important than others.

Our perception of risks isn't always rational

A number of factors influence how we perceive relative risks associated with different situations.

- People with social, political, or economic interests—including environmentalists—tend to downplay certain risks and emphasize others that suit their own agendas. We do this individually as well, building up the dangers of things that don't benefit us, while diminishing or ignoring the negative aspects of activities we enjoy or profit from.

- Most people have difficulty understanding and believing probabilities. We feel that there must be patterns and connections in events, even though statistical theory says otherwise. If the coin turned up heads last time, we feel certain that it will turn up tails next time. In the same way, it is difficult to understand the meaning of a 1-in-10,000 risk of being poisoned by a chemical.

- Our personal experiences often are misleading. When we have not personally experienced a bad outcome, we feel it

is more rare and unlikely to occur than it actually may be. Furthermore, the anxieties generated by life's gambles make us want to deny uncertainty and to misjudge many risks (fig. 8.17).

- We have an exaggerated view of our own abilities to control our fate. We generally consider ourselves above-average drivers, safer than most when using appliances or power tools, and less likely than others to suffer medical problems, such as heart attacks. People often feel they can avoid hazards because they are wiser or luckier than others.

- News media give us a biased perspective on the frequency of certain kinds of health hazards, overreporting some accidents or diseases, while downplaying or underreporting others. Sensational, gory, or especially frightful causes of death, such as murders, plane crashes, fires, or terrible accidents, receive a disproportionate amount of attention in the public media. Heart disease, cancer, and stroke kill nearly 15 times as many people in the United States as do accidents and 75 times as many people as do homicides, but the emphasis placed by the media on accidents and homicides is nearly inversely proportional to their relative frequency, compared with either cardiovascular disease or cancer. This gives us an inaccurate picture of the real risks to which we are exposed.

◀ **FIGURE 8.17** How dangerous is trick skating? Many parents regard them as extremely risky, while many students—expecially males—believe the risks are accceptable. Perhaps the more important question is whether the benefits outweigh the risks.

- We tend to have an irrational fear or distrust of certain technologies or activities that leads us to overestimate their dangers. Nuclear power, for instance, is viewed as very risky, while coal-burning power plants seem to be familiar and relatively benign; in fact, coal mining, shipping, and combustion cause an estimated 10,000 deaths each year in the United States, compared with none known so far for nuclear power generation. An old, familiar technology seems safer and more acceptable than does a new, unknown one.

- Alarmist myths and fallacies spread through society, often fueled by xenophobia, politics, or religion. For example, the World Health Organization campaign to eradicate polio worldwide has been thwarted by religious leaders in northern Nigeria—the last country where the disease remains widespread—who claim that oral vaccination is a U.S. plot to spread AIDS or infertility among Muslims.

How much risk is acceptable?

How much is it worth to minimize and avoid exposure to certain risks? Most people will tolerate a higher probability of occurrence of an event if the harm caused by that event is low. Conversely, harm of greater severity is acceptable only at low levels of frequency. A 1-in-10,000 chance of being killed might be of more concern to you than a 1-in-100 chance of being injured. For most people, a 1-in-100,000 chance of dying from some event or some factor is a threshold for changing what they do. That is, if the chance of death is less than 1 in 100,000, we are not likely to be worried enough to change our ways. If the risk is greater, we will probably do something about it. The Environmental Protection Agency generally assumes that a risk of 1 in 1 million is acceptable for most environmental hazards. Critics of this policy ask, acceptable to whom?

For activities that we enjoy or find profitable, we are often willing to accept far greater risks than this general threshold. Conversely, for risks that benefit someone else, we demand far higher protection. For instance, your chance of dying in a motor vehicle accident in any given year are about 1 in 5,000, but that doesn't deter many people from riding in automobiles. Your chances of dying from lung cancer if you smoke one pack of cigarettes per day are about 1 in 1,000. By comparison, the risk from drinking water with the EPA limit of trichloroethylene is about 2 in 1 billion. Strangely, many people demand water with zero levels of trichloroethylene while continuing to smoke cigarettes.

More than 1 million Americans are diagnosed with skin cancer each year. Some of these cancers are lethal, and most are disfiguring, yet only one-third of teenagers routinely use sunscreen. Tanning beds more than double your chances of cancer, especially if you're young, but about 10 percent of all teenagers admit regularly using these devices.

Table 8.4 lists lifetime odds of dying from some leading causes. These are statistical averages, of course, and there clearly are differences in where one lives and how one behaves that affect the danger level of these activities. Although the average lifetime chance of dying in an automobile accident is 1 in 100, there are

TABLE 8.4	Lifetime Chances of Dying in the United States	
SOURCE		**ODDS (1 IN x)**
Heart disease		2
Cancer		3
Smoking		4
Lung disease		15
Pneumonia		30
Automobile accident		100
Suicide		100
Falls		200
Firearms		200
Fires		1,000
Airplane accident		5,000
Jumping from high places		6,000
Drowning		10,000
Lightning		56,000
Hornets, wasps, bees		76,000
Dog bite		230,000
Poisonous snakes, spiders		700,000
Botulism		1 million
Falling space debris		5 million
Drinking water with EPA limit of trichloroethylene		10 million

SOURCE: Data from U.S. National Safety Council, 2003.

clearly things you can do—such as wearing a seat belt, driving defensively, and avoiding risky situations—to improve your odds. Still, it is interesting how we readily accept some risks while shunning others.

Active LEARNING

Calculating Probabilities

You can calculate the statistical danger of a risky activity by multiplying the probability of danger by the frequency of the activity. For example, in the United States, 1 person in 3 will be injured in a car accident in their lifetime (so the probability of injury is 1 per 3 persons, or ⅓). In a population of 30 car-riding people, the cumulative risk of injury is 30 people × (1 injury/ 3 people) = 10 injuries over 30 lifetimes.

1. If the average person takes 50,000 trips in a lifetime, and the accident risk is ⅓ per lifetime, what is the probability of an accident per trip?

2. If you have been riding safely for 20 years, what is the probability of an accident during your next trip?

ANSWERS: 1. Probability of injury per trip = (1 injury/3 lifetimes) × (1 lifetime/50,000 trips) = 1 injury/150,000 trips. 2. 1 in 150,000. Statistically, you have the same chance each time.

Our perception of relative risks is strongly affected by whether risks are known or unknown, whether we feel in control of the outcome, and how dreadful the results are. Risks that are unknown or unpredictable and results that are particularly gruesome or disgusting seem far worse than those that are familiar and socially acceptable.

Studies of public risk perception show that most people react more to emotion than to statistics. We go to great lengths to avoid some dangers while gladly accepting others. Factors that are involuntary, unfamiliar, undetectable to those exposed or catastrophic; those that have delayed effects; and those that are a threat to future generations are especially feared. Factors that are voluntary, familiar, detectable, or immediate cause less anxiety. Even though the actual number of deaths from automobile accidents, smoking, or alcohol, for instance, is thousands of times greater than those from pesticides, nuclear energy, or genetic engineering, the latter group preoccupies us far more than the former.

8.7 ESTABLISHING PUBLIC POLICY

Risk management combines principles of environmental health and toxicology with regulatory decisions based on socioeconomic, technical, and political considerations (fig. 8.18). The biggest problem in making regulatory decisions is that we are usually exposed to many sources of harm, often unknowingly. It is difficult to separate the effects of all these different hazards and to evaluate their risks accurately, especially when the exposures are near the threshold of measurement and response. In spite of often vague and contradictory data, public policy makers must make decisions.

The struggle over whether to vaccinate children against common illnesses is a good example of the difficulties in risk assessment. In 1998 a British physician published a paper suggesting that the measles, mumps, and rubella (MMR) vaccine is linked to autism. In 2010 the U.K. General Medical Council found the author of that study guilty of "dishonesty and misleading conduct" for failing to disclose his personal interest in this research and scientific and ethical errors led to retraction of the original paper by the *Lancet*. At least 20 subsequent studies have failed to find any link between vaccines and autism, but none of that scientific evidence is reassuring to thousands of angry and frightened parents who demand answers for why their children are autistic. Many of them remain convinced that vaccines cause this distressing condition and they refuse to allow their children to be vaccinated. Physicians argue that this is a danger not only to the children but also to the population at large, which is at risk from epidemics when there's a large pool of nonimmune children. Nevertheless, the absence of a clear, convincing explanation for what does cause autism simply fuels many people's suspicions. This is a good example of the power of anecdotal evidence and personal bias versus scientific evidence.

In setting standards for environmental toxics, we need to consider (1) combined effects of exposure to many different sources of damage, (2) different sensitivities of members of the population, and (3) effects of chronic as well as acute exposures. Some people argue that pollution levels should be set at the highest amount that does *not* cause measurable effects. Others demand that pollution be reduced to zero if possible, or as low as is technologically feasible. It may not be reasonable to demand that we be protected from every potentially harmful contaminant in our environment, no matter how small the risk. As we have seen, our bodies have mechanisms that enable us to avoid or repair many kinds of damage, so that most of us can withstand a minimal level of exposure without harm.

On the other hand, each challenge to our cells by toxic substances represents stress on our bodies. Although each individual stress may not be life-threatening, the cumulative effects of all the environmental stresses, both natural and human-caused, to which we are exposed may seriously shorten or restrict our lives. Furthermore, some individuals in any population are more susceptible to those stresses than others. Should we set pollution standards so that no one is adversely affected, even the most sensitive individuals, or should the acceptable level of risk be based on the average member of the population?

Finally, policy decisions about hazardous and toxic materials also need to be based on information about how such materials affect the plants, animals, and other organisms that define and maintain our environment. In some cases, pollution can harm or destroy whole ecosystems with devastating effects on the life-supporting cycles on which we depend. In other cases, only the most sensitive species are threatened. Table 8.5 shows the Environmental Protection Agency's assessment of relative risks to human welfare. This ranking reflects a concern that our exclusive focus on reducing pollution to protect human health has neglected risks to natural ecological systems. While there have been many benefits from a case-by-case approach in which we evaluate the health risks of individual chemicals, we have often missed broader ecological problems that may be of greater ultimate importance.

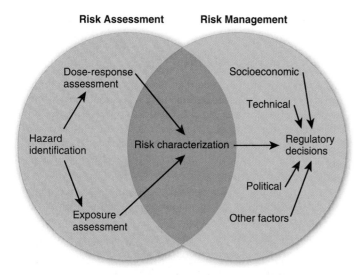

▲ **FIGURE 8.18** Risk assessment organizes and analyzes data to determine relative risk. Risk management sets priorities and evaluates relevant factors to make regulatory decisions.

TABLE 8.5 | Relative Risks to Human Welfare

RELATIVELY HIGH-RISK PROBLEMS
Habitat alteration and destruction
Species extinction and loss of biological diversity
Stratospheric ozone depletion
Global climate change
RELATIVELY MEDIUM-RISK PROBLEMS
Herbicides/pesticides
Toxins and pollutants in surface waters
Acid deposition
Airborne toxins
RELATIVELY LOW-RISK PROBLEMS
Oil spills
Groundwater pollution
Radionuclides
Thermal pollution

SOURCE: Data from U.S. Environmental Protection Agency.

CONCLUSION

We have made marvelous progress in reducing some of the worst diseases that have long plagued humans. Smallpox is the first major disease to be completely eliminated. Guinea worms and polio are nearly eradicated worldwide; typhoid fever, cholera, yellow fever, tuberculosis, mumps, and other highly communicable diseases are rarely encountered in advanced countries. Childhood mortality has decreased 90 percent globally, and people almost everywhere are living twice as long, on average, as they did a century ago.

But the technological innovations and affluence that have diminished many terrible diseases, have also introduced new risks. Chronic conditions, such as cardiovascular disease, cancer, depression, dementia, diabetes, and traffic accidents, that once were confined to richer countries, now have become leading health problems nearly everywhere. Part of this change is that we no longer die at an early age of infectious disease, so we live long enough to develop the infirmities of old age. Another factor is that affluent lifestyles, lack of exercise, and unhealthy diets aggravate these chronic conditions.

New, emergent diseases are appearing at an increasing rate. With increased international travel, diseases can spread around the globe in a few days. Epidemiologists warn that the next deadly epidemic may be only a plane ride away. In addition, modern industry is introducing thousands of new chemical substances every year, most of which aren't studied thoroughly for long-term health effects. Endocrine disrupters, neurotoxics, carcinogens, mutagens, teratogens, and other toxics can have tragic outcomes. The effects of lead on children's mental development is an example of both how we have introduced materials with unintended consequences, and a success story of controlling a serious health risk. Many other industrial chemicals could be having similar harmful effects.

PRACTICE QUIZ

1. Define the terms *health* and *disease*.
2. Name the five leading causes of global disease burden expected by 2020.
3. Define *emergent diseases* and give some recent examples.
4. What is *conservation medicine*?
5. What is the difference between toxic and hazardous? Give some examples of materials in each category.
6. What are *endocrine disrupters*, and why are they of concern?
7. What are *bioaccumulation* and *biomagnification*?
8. Why is atrazine a concern?
9. What is an *LD50*?
10. Distinguish between acute and chronic toxicity.

CRITICAL THINKING AND DISCUSSION

Apply the principles you have learned in this chapter to discuss these questions with other students.

1. Is it ever possible to be completely healthy?
2. How much would be appropriate for wealthy countries to contribute to global health? Why should we do more than we do now? What's in it for us?
3. Why do we spend more money on heart diseases or cancer than childhood diseases?
4. Why do we tend to assume that natural chemicals are safe while industrial chemicals are evil? Is this correct?
5. In the list of reasons why people have a flawed perception of certain risks, do you see things that you or someone you know sometimes do?
6. Do you agree that 1 in 1 million risk of death is an acceptable risk? Notice that almost everything in table 8.4 carries a greater risk than this. Does this make you want to change your habits?

A central question in environmental health is how we perceive different risks around us. When we evaluate environmental hazards, how do we assess known factors, uncertain risks, and the unfamiliarity of new factors we encounter? Which considerations weigh most heavily in our decisions and our actions as we try to avoid environmental risks?

The graph shown here shows one set of answers to this question, using aggregate responses of many people to risk. Go to Connect to find further discussion and a set of questions regarding risk, uncertainty, and fear.

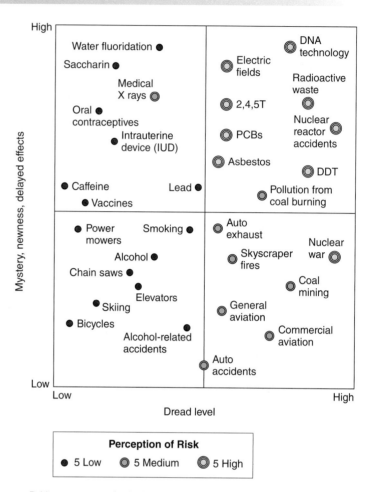

▲ Public perception of risk, depending on the familiarity, apparent potential for harm, and personal control over the risk.

CHAPTER
9
Climate

Cattle have suffered badly in Texas's recent droughts. Climatologists predict that drought will soon be the new normal for the southern and central United States.

LEARNING OUTCOMES

After studying this chapter, you should be able to answer the following questions:

▶ How do the troposphere and stratosphere differ?

▶ What are some factors in natural climate variability?

▶ Explain the greenhouse effect and how it is changing our climate.

▶ How do we know the nature and cause of recent climate change?

▶ List some effects of climate change.

▶ What are some strategies for minimizing global climate change?

CASE STUDY

Weird Weather: The New Normal?

For cattle ranchers in Texas, 2011 was a terrible year—the hottest and driest year observed since record keeping began in 1895. In 2011, Texas saw more than 90 days with temperatures over 100°F (38°C)—roughly double the normal number. Lakes, streams, and stock-watering ponds dried up and disappeared along with the grass on baking pastures, leaving cattle thirsty and starving. Drought also meant no harvest of hay—critical for winter feed. By the end of the year, Texas ranchers were forced to sell or slaughter 1.4 million cattle, or roughly 11 percent of the state total, possibly the biggest sell-off in American history.

Because Texas is the nation's largest producer of beef cattle, impacts were felt in beef prices across the country. This was good news for ranchers in other regions, where the value of herds rose sharply, but it wasn't easy on meat packers and consumers.

With no soil moisture and with spring rains at half their normal level, field crops suffered, too. Just over half of the state's 20 million acres of cropland were harvested. More than half the cotton crop failed entirely. Many farmers didn't even bother planting wheat on their bone-dry fields. The federal government stepped in to ease the pain, however: U.S. taxpayers covered more than $5 billion in crop insurance, disaster payments, and other supports to farmers who lost their cotton, sorghum, wheat, and other crops. Total economic losses associated with that one year of drought are estimated at more than $10 billion.

Federal assistance also helped in fighting the worst fires in Texas history. Nearly 4 million acres of desiccated forests, grasslands, and suburban neighborhoods burned in 2011, in 30,000 separate fires that destroyed 3,000 homes. All these fires were expensive to fight, as over 16,000 firefighters and support crews, with 1,500 airplanes, bulldozers, and other heavy equipment were mobilized from all 50 states.

City-dwellers suffered, too, as they watched their trees and lawns die under mandatory watering bans. City parks lost hundreds of thousands of trees. Industries such as petroleum refineries, a cornerstone of the Texas economy, were badly threatened by shortages of water essential for production (fig. 9.1).

This drought was extreme even for a region accustomed to heat and drought, but the state climatologist has warned that these dry conditions could last for a decade or longer. Much of the South and the West have seen similar extremes. In Arizona half a million acres of forest burned, another state record. Springtime is arriving earlier and becoming warmer: in an average spring week in 2012, for example, U.S. weather stations recorded 1,692 new extreme high temperature records, compared to 19 new low records. Storms have been becoming more intense, too. Freakish concentrations of tornadoes have ripped across the Great Plains, causing billions of dollars in damages and untold misery.

Is all this just weird weather? Or is it part of a larger pattern? This is the question policy makers, scientists, and the public are all asking. In a complex and rapidly shifting system like weather, most of us tend to see data selectively, focusing on pieces of evidence that confirm our expectations. For some observers, several bad years of drought are confirming proof of human-caused climate change predicted by climatologists, ecologists, marine scientists, and others. For others, a string of bad years is simply more variation in the complex system we call weather. How do we know what's really going on?

Climate scientists—like all scientists—seek their answers in data—that is, in collected observations. The longer the data record, the more years and the more sources of evidence or places considered, the better. Weather changes by the hour, but long-term shifts toward warmer weather, or patterns of increasing storms or droughts, indicate that something is happening outside the normal daily variations (fig. 9.2). In the data, climatologists have shown not just patterns of rising temperature but also a strong correlation with emissions of industrial and transportation-associated CO_2. No other known factor can explain the observed patterns. Atmospheric chemistry and physics do explain these changes well (see section 9.3).

Climatologists conclude that the droughts seen across much of the southern and central United States, and many other regions of the world, represent a long-term shift toward increased heat and energy storage in the atmosphere. And they warn that if we don't act soon—in the next few years at the most—those changes will shift us to an entirely new climate regime. The evidence indicates that extreme summer heat, with drying soil and crops, is the new normal for much of the southern and central United States. Texas is expected to look like the Arizona desert does today. Illinois and Iowa, the center of our farm economy, will look like Texas today.

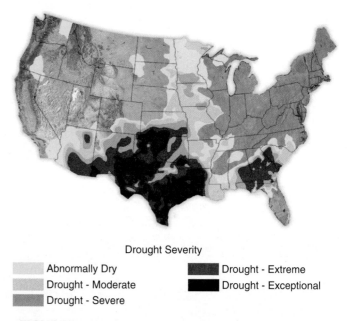

Drought Severity

Abnormally Dry	Drought - Extreme
Drought - Moderate	Drought - Exceptional
Drought - Severe	

▲ **FIGURE 9.1** Extreme drought persisted in Texas and much of the South through most of 2011. SOURCE: U.S. Drought Monitor (droughtmonitor.unl.edu).

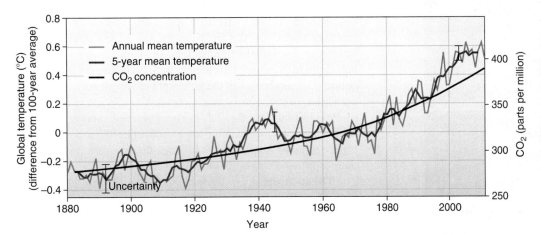

◀ **FIGURE 9.2** Global average temperatures have closely tracked CO_2 concentrations in our atmosphere. CO_2 has increased from 285 parts per million (ppm) in 1880 to nearly 400 parts per million, with the fastest change in the past 50 years. Today's average temperature is about 0.7 degrees warmer than a century ago. SOURCE: NASA Goodard Institute for Space Studies.

Is it too late to change this future? Almost, but not quite. Across the globe, countries, states, cities, and regions are making plans to cut greenhouse gas emissions. They are discovering that they can both serve self-interest—a more secure economy, and a healthier environment with lower health care costs—and earn the respect of their neighbors by reducing greenhouse gas emissions. Energy security becomes possible when we aren't dependent on distant sources of conventional fuel. Smarter planning, transportation, and land use can improve our economy and health as well as our climate.

In this chapter we'll examine the nature of climate, weather, and climate change. We'll also look at promising solutions for reducing our climate impacts. ■

> *The next decade is critical. If emissions do not peak by around 2020, . . . the needed 50% reduction by 2050 will become much more costly. In fact, the opportunity may be lost completely.*
> —INTERNATIONAL ENERGY AGENCY, 2010

9.1 WHAT IS THE ATMOSPHERE?

Earth's atmosphere consists of gas molecules, relatively densely packed near the surface and thinning gradually to about 500 km (300 mi) above the earth's surface. In the lowest layer of the atmosphere, air moves ceaselessly—flowing, swirling, and continually redistributing heat and moisture from one part of the globe to another. The daily temperatures, wind, and precipitation that we call **weather** occur in the troposphere. Long-term temperatures and precipitation trends we refer to as **climate**.

The earliest atmosphere on earth probably consisted mainly of hydrogen and helium. Over billions of years, most of that hydrogen and helium diffused into space. Volcanic emissions added carbon, nitrogen, oxygen, sulfur, and other elements to the atmosphere. Virtually all of the molecular oxygen (O_2) we breathe was probably produced by photosynthesis in blue-green bacteria, algae, and green plants.

Clean, dry air is 78 percent nitrogen and almost 21 percent oxygen, with the remaining 1 percent composed of argon, carbon dioxide (CO_2), and a variety of other gases. Water vapor (H_2O in gas form) varies from near 0 to 4 percent, depending on air temperature and available moisture. Minute particles and liquid droplets—collectively called **aerosols**—also are suspended in the air. Atmospheric aerosols and water vapor play important roles in the earth's energy budget and in rain production.

The atmosphere has four distinct zones of contrasting temperature, due to differences in absorption of solar energy (fig. 9.3). The layer immediately adjacent to the earth's surface is called the **troposphere** (*tropein* means "to turn or change," in Greek). Within the troposphere, air circulates in great vertical and horizontal **convection currents**, constantly redistributing heat and moisture around the globe (fig. 9.4). The troposphere ranges in depth from about 18 km (11 mi) over the equator to about

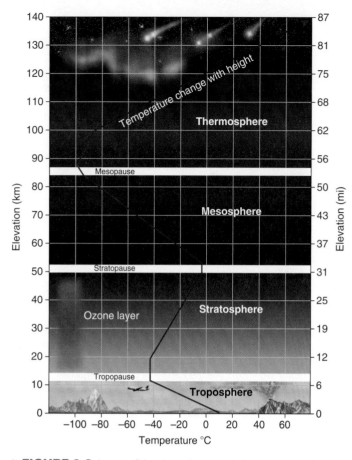

▲ **FIGURE 9.3** Layers of the atmosphere vary in temperature and composition. Our weather happens in the troposphere. Stratospheric ozone is important for blocking ultraviolet solar energy.

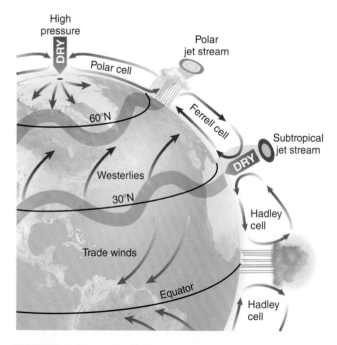

▲ **FIGURE 9.4** Convection cells circulate air, moisture, and heat around the globe. Jet streams develop where cells meet, and surface winds result from convection. Convection cells expand and shift seasonally.

8 km (5 mi) over the poles, where air is cold and dense. Because gravity holds most air molecules close to the earth's surface, the troposphere is much denser than the other layers: it contains about 75 percent of the total mass of the atmosphere. Air temperature drops rapidly with increasing altitude in this layer, reaching about –60°C (–76°F) at the top of the troposphere. A sudden reversal of this temperature gradient creates a boundary called the tropopause. This temperature boundary occurs because **ozone** (O_3) molecules in the stratosphere absorb solar energy. In particular, ozone absorbs ultraviolet (UV) radiation (wavelengths of 290–330 nm; see fig. 2.13). This absorbed energy makes the stratosphere warmer than the upper troposphere. Tropospheric air cannot continue to rise when it is cooler than the surrounding air, so there is little mixing across this boundary.

The **stratosphere** extends about 50 km (31 mi) out from the tropopause. It is far more dilute than the troposphere, but it has a similar composition—except that it has almost no water vapor and nearly 1,000 times more ozone.

Because UV radiation damages living tissues, UV absorption in the stratosphere is also essential for life on earth. Depletion of stratospheric ozone by chemical pollutants has been a major public health concern. Increased UV radiation reaching the earth's surface can increase skin cancer rates and damage biological communities. Fortunately, global cooperation to restrict key pollutants is beginning to reduce the loss of stratospheric ozone.

Unlike the troposphere, the stratosphere is relatively calm. There is so little mixing in the stratosphere that volcanic ash and human-caused contaminants can remain in suspension there for many years.

Above the stratosphere, the temperature diminishes again, creating the mesosphere, or middle layer. The thermosphere (heated layer) begins at about 50 km. This is a region of highly ionized (electrically charged) gases, heated by a steady flow of high-energy solar and cosmic radiation. In the lower part of the thermosphere, intense pulses of high-energy radiation cause electrically charged particles (ions) to glow. This phenomenon is what we know as the *aurora borealis* and *aurora australis*, or northern and southern lights.

No sharp boundary marks the end of the atmosphere. Pressure and density decrease with distance from the earth until they become indistinguishable from the near-vacuum of interstellar space.

The atmosphere captures energy selectively

The sun supplies the earth with abundant energy, especially near the equator. Of the solar energy that reaches the outer atmosphere, about one-quarter is reflected by clouds and atmospheric gases, and another quarter is absorbed by carbon dioxide, water vapor, ozone, methane, and a few other gases (fig. 9.5). This absorbed energy warms the atmosphere slightly. About half of incoming solar radiation (insolation) reaches the earth's surface. Most of this energy is in the form of light or infrared (heat) energy.

Some incoming solar energy is reflected by bright surfaces, such as snow, ice, and sand. The rest is absorbed by the earth's surface and by water. Surfaces that *reflect* energy have a high **albedo** (reflectivity). Fresh snow and dense clouds, for instance, can reflect as much as 85 to 90 percent of the light falling on them

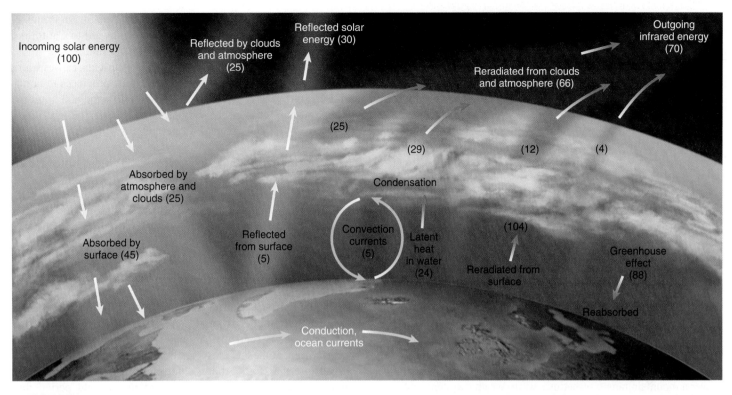

▲ FIGURE 9.5 Energy balance between incoming and outgoing radiation. The atmosphere absorbs or reflects about half of the solar energy reaching the earth. Most of the energy reemitted from the earth's surface is long-wave, infrared energy. Gases and aerosols in the atmosphere absorb and re-radiate most of this energy, keeping the surface much warmer than it would otherwise be. This absorption is known as the greenhouse effect.

(table 9.1). Surfaces that absorb energy have a low albedo and generally appear dark. Black soil, asphalt pavement, and water, for example, have low albedo, with reflectivity as low as 3 to 5 percent.

Absorbed energy heats materials (such as an asphalt parking lot in summer), evaporates water, and provides the energy for photosynthesis in plants. Following the second law of thermodynamics, absorbed energy is gradually reemitted as lower-quality heat energy. The walls of a brick building, for example, absorb light (high-intensity energy) and reemit that energy as heat (low-intensity energy).

The change in energy intensity is important because the gases that make up our atmosphere let light energy pass through—this is why it's bright during the day—but these gases absorb or reflect the lower-intensity heat energy that is reemitted from the earth

(fig. 9.5). Several trace gases in the atmosphere are especially effective at trapping reradiated heat energy. The most effective and abundant of these gases are water vapor (H_2O), carbon dioxide (CO_2), methane (CH_4), and nitrous oxide (N_2O).

If our atmosphere didn't capture this reemitted energy, the earth's average surface temperature would be about –6°C (21°F), rather than the current 14°C (57°F) average. Thus energy capture is necessary for liquid water on earth, and for life as we know it. The "**greenhouse effect**" is a common term to describe the capture of energy by gases in the atmosphere. Something like the glass of a greenhouse, the atmosphere transmits sunlight but traps some heat inside. Also like a greenhouse, the atmosphere lets that energy dissipate gradually to space (see Key Concepts, p. 218).

The balance of the rate of incoming energy and outgoing energy determines the temperature inside the greenhouse. The policy issue that faces us is that we are slowing the rate of heat loss, thus increasing heat storage in our "greenhouse." We are doing this by adding CO_2, CH_4, and N_2O to the atmosphere at levels the earth has not seen since before the appearance of humans as a species. The question is whether we will be able to agree on changing this trend.

Evaporated water stores and redistributes heat

Much of the incoming solar energy is used up in evaporating water. Every gram of evaporating water absorbs 580 calories of energy as it transforms from liquid to gas. Water vapor in the air stores that

TABLE 9.1 | Albedo (reflectivity) of Earth Surfaces

SURFACE	ALBEDO (%)
Fresh snow	80–85
Dense clouds	70–90
Water (low sun)	50–80
Sand	20–30
Forest	5–10
Water (sun overhead)	5
Dark soil	3

580 calories per gram, and we call that stored heat **latent heat**. Later, when the water vapor condenses, it releases 580 calories per gram. Globally, latent heat contains huge amounts of energy, enough to power thunderstorms, hurricanes, and tornadoes. Imagine the sun shining on the Gulf of Mexico in the winter. Warm sunshine and plenty of water allow continuous evaporation that converts an immense amount of solar (light) energy into latent heat stored in evaporated water. Now imagine a wind blowing the humid air north from the Gulf toward Canada. The air cools as it rises and moves north. Eventually, cooling causes the water vapor to condense. Rain (or snow) falls as a consequence. Note that it is not only water that has moved from the Gulf to the Midwest: 580 calories of heat have also moved with every gram of moisture. The heat and water have now moved from the sunny Gulf to the colder Midwest. This redistribution of heat and water around the globe is essential to life on earth.

Why does it rain? Understanding this will help you understand the distribution of latent heat, and water resources, around the globe. Rain falls when there are two conditions: (1) a moisture source, such as an ocean, from which water can evaporate into the atmosphere; (2) a lifting mechanism. Lifting is important because air cools at high elevations. You may have observed this cooling if you have driven over a mountain pass. Sometimes air is lifted as winds push it over a mountain range; sometimes warm weather systems collide with cooler weather systems, and the warm air is forced up over the cooler air. Sometimes hot air, warmed near the earth's surface on a sunny day, rises in convection currents. Any of these three mechanisms can cause air to rise and cool. Moisture in the cooling air then condenses. We see the moisture falling as rain or snow.

Next time you watch the weather report, see if you can find references to these processes in predicted rain or snowfall.

Ocean currents also redistribute heat

Warm and cold ocean currents strongly influence climate conditions on land. Surface ocean currents result from wind pushing on the ocean surface. As surface water moves, deep water wells up to replace it, creating deeper ocean currents. Differences in water density—depending on the temperature and saltiness of the water—also drive ocean circulation. Huge cycling currents called gyres carry water north and south, redistributing heat from low latitudes to high

latitudes (see appendix 3, p. A-4, global climate map). For example, the Alaska current, flowing from Alaska southward to California, keeps San Francisco cool and foggy during the summer.

The Gulf Stream, one of the best-known currents, carries warm Caribbean water north past Canada's maritime provinces to northern Europe (fig. 9.6). This current is immense, some 800 times the volume of the Amazon, the world's largest river. The heat transported from the Gulf keeps Europe much warmer than it should be for its latitude. Stockholm, Sweden, for example, where temperatures rarely fall much below freezing, is at the same latitude as Churchill, Manitoba, where polar bears live in summer. As the warm Gulf Stream passes Scandinavia and swirls around Iceland, the water cools and evaporates, becomes dense and salty, and plunges downward, creating a strong, deep, southward current.

Together, this surface and deep-water circulation system is called the **thermohaline** (temperature and salinity-related) **circulation**, because both temperature and salt concentrations control the density of water, and contrasts in density drive its movement. Dr. Wallace Broecker of the Lamont Doherty Earth Observatory, who first described this great conveyor system, also found that it can shut down suddenly. About 11,000 years ago, as the earth was warming at the end of the last ice age, cold glacial melt-water surged into the North Atlantic and interrupted the thermohaline circulation cycle. Europe was plunged into a cold period that lasted for 1,300 years. Temperatures may have changed dramatically in just a few years.

Could this happen again? Some climatologists suggest that the melting of the Greenland ice sheet, which contains 10 percent of the globe's glacial ice, could cause sudden changes in ocean circulation.

9.2 CLIMATE CHANGES OVER TIME

Climatologist Wallace Broecker has said that "climate is an angry beast, and we are poking it with sticks." He meant that we assume our climate is stable, but our thoughtless actions may be stirring it to sudden and dramatic changes. How stable is climate? That depends upon the time frame you consider. Over centuries and millennia, we know that climate shifts somewhat, but

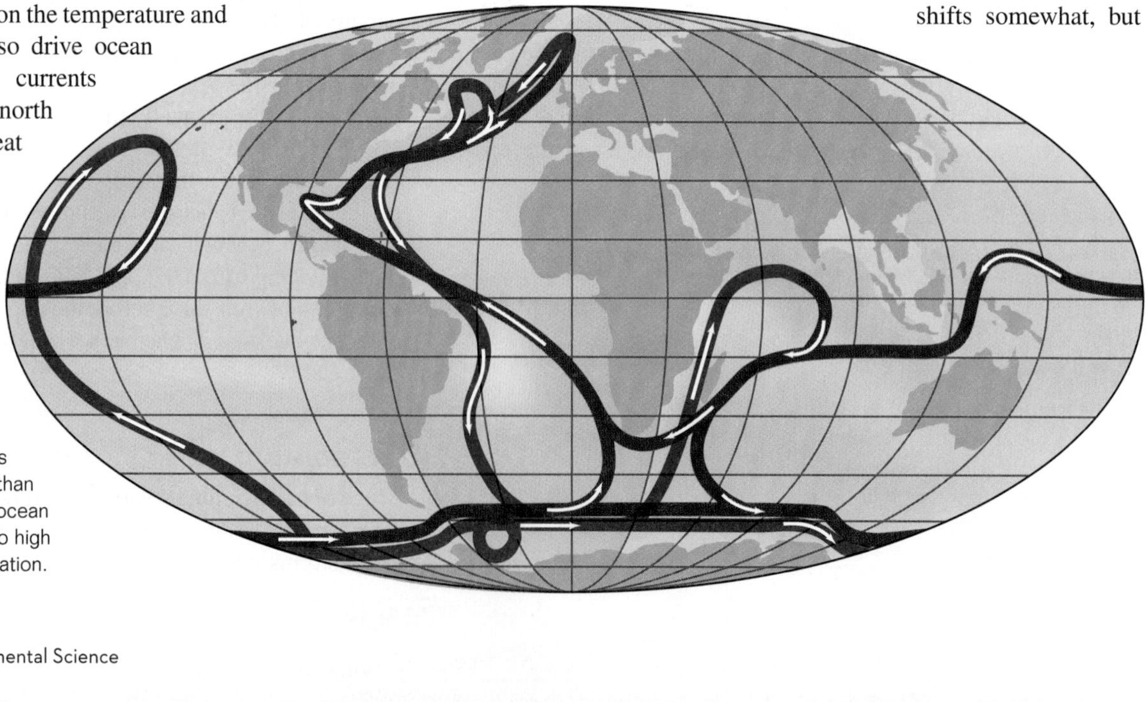

▶ **FIGURE 9.6** Ocean currents act as a global conveyor system, redistributing warm and cold water around the globe. These currents moderate our climate: for example, the Gulf Stream keeps northern Europe much warmer than northern Canada. Variations in ocean salinity and density, low (*blue*) to high (*yellow*), help drive ocean circulation.

usually we expect little change on the scale of a human lifetime. The question now is whether that is a reasonable expectation. If climate does shift, how fast might it change, and what will those changes mean for the environmental systems on which we depend?

Ice cores tell us about climate history

Every time it snows, small amounts of air are trapped in the snow layers. In Greenland and Antarctica and other places where cold is persistent, yearly snows slowly accumulate over the centuries. New layers compress lower layers into ice, but still tiny air bubbles remain, even thousands of meters deep into glacial ice. Each bubble is a tiny sample of the atmosphere at the time that snow fell.

Climatologists have discovered that by drilling deep into an ice sheet, they can extract ice cores, from which they can collect air-bubble samples. Samples taken every few centimeters show how the atmosphere has changed over time. Ice core records have revolutionized our understanding of climate history (fig. 9.7). We can now see how concentrations of atmospheric CO_2 have varied. We can detect ash layers and spikes in sulfate concentrations that record volcanic eruptions.

Most important, we learn about ancient temperatures by comparing oxygen isotopes (atoms of different mass) in these air samples. In cold years, water molecules containing slightly lighter oxygen atoms evaporate more easily than water with slightly heavier atoms. By comparing the proportions of heavier and lighter oxygen atoms, climatologists can reconstruct temperatures over time, and plot temperature changes against CO_2 concentrations and other atmospheric components.

The first very long record was from the Vostok ice core, which reached 3,100 m into the Antarctic ice and which gives us a record of temperatures and atmospheric CO_2 over the past 420,000 years. A team of Russian scientists worked for 37 years at the Vostok site, about 1,000 km from the South Pole,

▲ **FIGURE 9.7** Dr. Mark Twickler, of the University of New Hampshire, holds a section of the 3,000 m Greenland ice sheet core, which records 250,000 years of climate history.

to extract this ice core. A similar core has been drilled from the Greenland ice sheet. More recently the European Project for Ice Coring in Antarctica (EPICA) has produced a record reaching back over 800,000 years (fig. 9.8). All these cores show that climate has varied dramatically over time but that there is a close correlation between atmospheric temperatures and CO_2 concentrations.

From these ice cores, we know that CO_2 concentrations have varied between 180 to 300 ppm (parts per million) in the past 800,000 years. Therefore, we know that today's concentrations of approximately nearly 400 ppm (and rising about 2 ppm per year) are about one-third higher than the earth has seen in nearly a million years. Concentrations of methane and nitrous oxide, two other important greenhouse gases, are also higher than in any records in the EPICA core. We also know from oxygen isotopes that present temperatures are nearly as warm as any in the ice core records. Further warming in the coming decades is likely to exceed anything ever seen by our species, *Homo sapiens*, which appeared just 200,000 years ago.

What causes natural climatic swings?

Ice core records also show that there have been repeated, cyclical climate changes over time. What causes these periodic (repeated) changes? Modest changes correspond to a cycle in the sun's intensity, which peaks about every 11 years. More dramatic changes are associated with periodic shifts in the earth's orbit and tilt (fig. 9.9). These are known as the **Milankovitch cycles,** after the Serbian scientist Milutin Milankovitch, who first described them in the 1920s. There are three of these cycles: (1) the earth's elliptical orbit stretches and shortens in a 100,000-year cycle; (2) the earth's axis changes its angle of tilt in a 40,000-year cycle; (3) over a 26,000-year period, the axis wobbles like an out-of-balance spinning top. These cycles cause variation in the intensity of incoming solar energy at different latitudes, and in the intensity of summer heating or winter cooling.

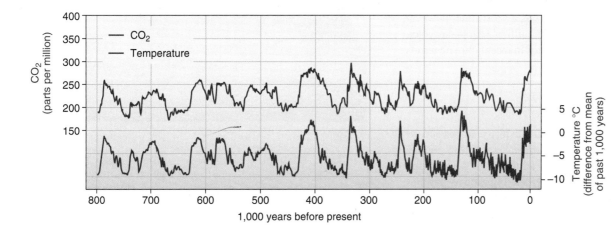

◀ **FIGURE 9.8** Air bubbles in ice cores provide samples of ancient atmospheric composition, going back 800,000 years in this record from the EPICA ice core. Concentrations of CO_2 (*red line*) map closely to temperatures (*blue*, derived from oxygen isotopes). Recent temperatures lag behind rising CO_2, possibly because oceans have been absorbing heat.

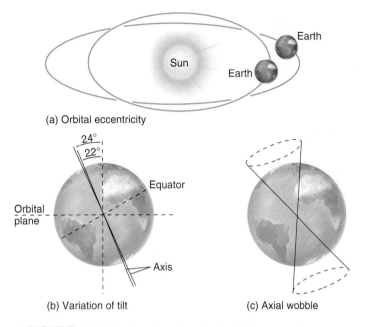

(a) Orbital eccentricity

(b) Variation of tilt

(c) Axial wobble

▲ **FIGURE 9.9** Milankovitch cycles including (a) changes in the eccentricity of the earth's orbit, (b) shifting tilt of the axis, and (c) wobble of the earth, explain some long-term climate variability.

The interplay of these cycles also seems to explain the glacial periods of the last 800,000 years, which you can see in the cold/warm cycles in figure 9.8. For example, when the wobble orients the north pole toward the sun in summer, then there is substantial summer warming and overall warming of the earth. When the wobble points toward the north pole away from the sun, then northern summers are cold, and there is global cooling. Similarly, when the axis tilts toward the sun, the poles warm, and when the axis is more parallel to the sun, the poles warm less.

Volcanic eruptions can cause sudden climate shifts, but usually only for a few years. One exception was the explosion of Mount Toba in western Sumatra about 73,000 years ago. This was the largest volcanic cataclysm in the past 28 million years. The eruption ejected at least 2,800 km³ of material, compared to only 1 km³ emitted by Mount St. Helens in Washington State in 1980. Sulfuric acid and particulate material ejected into the atmosphere from Mount Toba are estimated to have dimmed incoming sunlight by 75 percent and to have cooled the whole planet by as much as 16°C for more than 160 years. In climate history, 160 years is a short time, however, and this was the largest eruption in 28 million years—so volcanoes are a notable but not dominant factor in recent climate trends.

El Niño/Southern Oscillation is one of many regional cycles

On the scale of years or decades, the climate also changes according to oscillations in the ocean and atmosphere. These coupled ocean–atmosphere oscillations occur in all the world's oceans, but the **El Niño/Southern Oscillation** (ENSO) is probably the best known. ENSO affects weather across the Pacific and adjacent continents, causing heavy monsoons or serious droughts.

The core of this system is a huge pool of warm surface water in the Pacific Ocean that sloshes slowly back and forth between Indonesia and South America like water in a giant bathtub. Most years, steady equatorial trade winds hold this pool in place in the western Pacific (fig. 9.10). From Southeast Asia to Australia, this concentration of warm equatorial water provides latent heat (water vapor) that drives strong upward convection (low pressure) in the atmosphere. Resulting heavy rains in Indonesia support dense tropical forests.

On the American side of the Pacific, cold upwelling water along the South American coast replaces westward-flowing surface waters. This upwelling deep water is rich in nutrients. It supports dense schools of anchovies and other fish. In the atmosphere, dry, sinking air in Mexico and California replaces the air moving steadily westward in the trade winds. Normally dry conditions in the southwestern United States are a result.

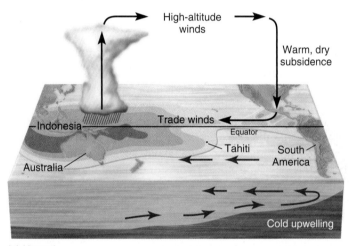

(a) Normal

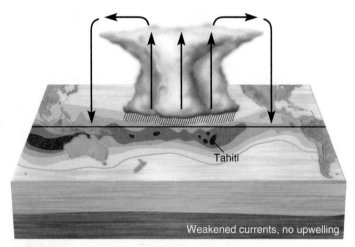

(b) El Niño

▲ **FIGURE 9.10** The El Niño/La Niña Southern Oscillation. (a) Normally surface trade winds drive cold currents from South America toward Indonesia, and cold, deep water wells up near Peru. (b) During El Niño years, winds and currents weaken, and warm, low-pressure conditions shift eastward, bringing storms to the Americas.

Every three to five years, for reasons that we don't fully understand, Indonesian convection (rising air currents) weakens, and westward wind and ocean currents fail. Warm surface water surges back east across the Pacific. Rains increase in the western United States and Mexico, and drought occurs in Indonesia. Upwelling currents that support South American fisheries also fail.

Fishermen in Peru were the first to notice irregular cycles of rising ocean temperatures because the fish disappeared when the water warmed. They named this event El Niño (Spanish for "the Christ child") because it often occurs around Christmastime. The counterpart to El Niño, when the eastern tropical Pacific cools, has come to be called La Niña (little girl).

ENSO cycles have far-reaching effects. During an El Niño year, the northern jet stream—which is normally over Canada—splits and is drawn south over the United States. This pulls moist air from the Pacific and Gulf of Mexico inland, bringing intense storms and heavy rains from California across the Midwestern states. The intervening La Niña years bring hot, dry weather to the same areas. Oregon, Washington, and British Columbia, on the other hand, tend to have warm, sunny weather in El Niño years rather than their usual rain. Droughts in Australia and Indonesia during El Niño episodes cause disastrous crop failures and forest fires, including one in Borneo in 1983 that burned 3.3 million ha (8 million acres).

Some climatologists believe that El Niño conditions are becoming stronger or more frequent because of global climate change. There is evidence that warm ocean surface temperatures are spreading, which could contribute to the intensity of El Niño convection patterns.

9.3 HOW DO WE KNOW THE CLIMATE IS CHANGING FASTER THAN USUAL?

Many scientists consider anthropogenic (human-caused) global climate change to be the most important environmental issue of our time. The possibility that humans might alter world climate is not a new idea. In 1859 John Tyndall measured the infrared absorption of various gases and described the greenhouse effect. In 1895, Svante Arrhenius, who subsequently received a Nobel Prize for his work in chemistry, predicted that CO_2 released by coal burning could cause global warming.

Pivotal evidence that human activities are increasing atmospheric CO_2 came from an observatory on top of the Mauna Loa volcano in Hawaii. The observatory was established in 1957 as part of an International Geophysical Year, to provide atmospheric data in a remote, pristine environment. Measurements taken by atmospheric chemist Charles David Keeling showed CO_2 levels increasing about 0.5 percent per year. The "Keeling curve" (fig. 9.11), which has been continuously updated, shows that CO_2 concentrations have

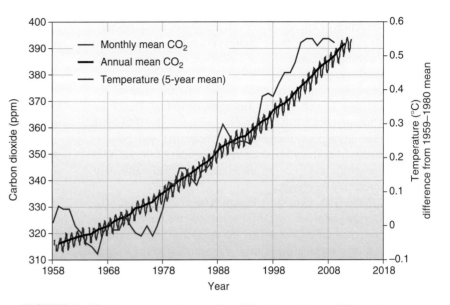

▲ **FIGURE 9.11** Measurements of atmospheric CO_2 taken at the top of Mauna Loa, Hawaii, show an increase of about 2.2 percent per year in recent years. Monthly mean CO_2 (*red*) and annual average CO_2 (*black*) track closely with the general trend in global temperatures (*blue*).

risen from 315 ppm in 1958 to 393 ppm in 2012. The line is jagged because of annual fluctuations: concentrations decline each May as plants in the Northern Hemisphere (which has far greater continental area than the Southern Hemisphere) begin photosynthesizing and capturing CO_2. During the northern winter, plant decay releases CO_2, causing levels to rise.

Scientific consensus is clear

Because the climate is so complex, climate scientists worldwide have collaborated in collecting and sharing data, and in programming models to describe how the climate system works. Evidence shows regional variation in warming and cooling trends, and there are minor differences among models. But among climate scientists who work with the data and models, there is no disagreement

Active LEARNING

The IPCC's Fourth Assessment Report (AR4)

Open a web browser and find the IPCC's AR4 report at http://www.ipcc.ch. Because there are thousands of pages of PDF documents on this site, start with the Synthesis Report, which combines the findings of several working groups.

Choose a figure in the synthesis report, and explain to your class what it tells you, using the ideas you have learned in this chapter and the explanations that accompany the figure. Be patient, and work with a colleague if that makes it easier. Present your explanation to the class, and see if they can understand your presentation.

about the direction of change. The evidence shows unequivocally, as in the Mauna Loa graph, that climate is changing, and the global average is warming because of increased retention of energy in the lower atmosphere.

The most comprehensive effort to describe the state of climate knowledge is that of the **Intergovernmental Panel on Climate Change (IPCC)**. As the name indicates, the IPCC is a collaboration among governments, with scientists and government representatives from 130 countries. The aim of the IPCC is to review scientific evidence on the causes and likely effects of human-caused climate change.

In 2007 the IPCC issued its Fourth Assessment Report. The result of 6 years of work by 2,500 scientists, the four volumes of the report represent a consensus by more than 90 percent of all the scientists working on climate change. The conclusion is a 90 percent certainty that observed climate change is caused by human activity. Subsequent reports have raised that to a 99 percent certainty. You can view the report, with figures and related documents, at the IPCC's website: http://www.ipcc.ch. The IPCC's Fifth Assessment Report is due out in 2013.

Though scientific consensus has long been clear, in a few places, notably the United States, a public debate has developed about climate evidence, a debate unrelated to science. Popular media commentators and a few politicians have recently converted "belief" in climate evidence into a matter of identity and political philosophy, rather than a question of what the evidence shows.

This political positioning is new. In 1997 and 2002, just five years apart, both the Clinton administration and President George W. Bush awarded Charles Keeling medals of honor for his studies of atmospheric CO_2 and climate change. At that time, members of both major parties considered Keeling's work to be of national interest and global importance. A decade later, views on climate had become a polarizing political issue between parties. Some commentators trumpeted marginal evidence for change, while others dismissed the data and the science altogether. Acknowledging climate change had become a matter of group identity and worldview, rather than a question of science and observation. This contrast in views reminds us how important it is that voters have some understanding of data and the scientific process. The apparent divide between political commentators also greatly oversimplifies the spectrum of actual views. As the meteorologist, businessman, and political conservative Paul Douglas has written, you don't have to be a liberal to believe in climate change. He argues that fiscal conservatives should be advocating for new growth opportunities in climate-saving technologies, rather than defending reliance on old-fashioned and inefficient uses of fossil fuels.

Rising heat waves, sea level, and storms are expected

The IPCC's Fourth Assessment Report presents a variety of climate scenarios for predicted emissions of greenhouse gases. For each scenario, the IPCC modeled future emissions, starting in 2000. Scenarios differed in expected population growth, economic growth, energy conservation and efficiency, and adoption of greenhouse gas controls (or lack thereof). The different scenarios

project a temperature increase by 2100 of 1–6°C (2–11°F) compared to temperatures at the end of the twentieth century (fig. 9.12, colored trend lines).

According to the IPCC, the "best estimate" for temperature rise is now about 2–4°C (about 3–8°F). A change of 4°C is just slightly less than the difference in global temperature between now and the middle of the last glacial period, which was about 5°C cooler.

Observations since 2007 show that all the IPCC scenarios were too conservative. Greenhouse gas emissions, temperatures, sea level, and energy use have accelerated faster than even the worst IPCC projections (fig. 9.12, bold line). There is serious concern that increased heat stress and drought could cause increased deaths as well as crop failure and new waves of refugees from drought-stricken regions.

The IPCC projected in 2007 that sea levels should rise 17–57 cm (7–23 in.) by the end of this century. More-recent estimates have raised this estimate to 1–2 m of sea-level rise by 2100. If recent rapid melting of polar ice sheets and Greenland glaciers continues, this change will be higher still. Complete melting of Greenland's ice sheet would raise sea level by more than 6 m (nearly 20 ft). This would flood most of Florida, a broad swath of the Gulf Coast, most of Manhattan Island, Shanghai, Hong Kong, Tokyo, Kolkata, Mumbai, and about two-thirds of the other largest cities in the world (fig. 9.13).

Tipping points are a concern for many climatologists. Projections indicate that if we don't control emissions in the next few decades we will pass points of no return in melting of permafrost, in the loss of Greenland's ice cap, and other factors.

The United States military is concerned about global warming. In 2007 the U.S. Military Advisory Board said, "Climate change, national security, and energy dependence are a related set of global challenges that will lead to tensions even in stable regions of the world." Some humanitarian aid agencies point to unusual drought as a factor in recent civil wars and famines in southern Sudan and in Somalia. These crises are rooted in drought and food shortages caused by changing weather patterns that have

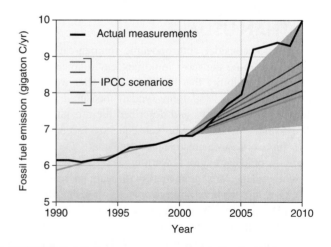

▲ **FIGURE 9.12** Emissions scenarios modeled by the IPCC (colored lines) are considerably lower than observed emissions data (*black*), indicating the IPCC projections have underestimated current and future climate change. Values represent fossil fuels, cement production, and land-use change.

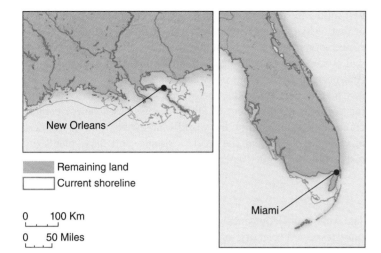

▲ **FIGURE 9.13** Approximate change in land surface with 1 m (3 ft) of sea level rise, a change that is likely by 2100. If no action is taken, sea level change may be 2 m (6 ft) or more.

led to years of below-normal rainfall and desertification. Global climate change may bring more such conflicts along with the millions of refugees and tragic suffering that we see now in the African Sahel.

Policy makers have made little progress in finding solutions. Climate control is a classic free-rider problem, in which nobody wants to take action for fear that someone else might benefit from their sacrifices.

The question is whether the sacrifices need necessarily be as big as some policy makers suggest. Increasingly, economists point out that shifting our energy strategy from coal (our largest emitter of greenhouse gases and other pollutants) to wind, solar, and greater efficiency could produce millions of new jobs and save billions in health care costs associated with coal burning.

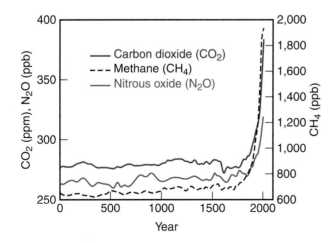

▲ **FIGURE 9.14** Methane (CH_4, *black*) and nitrous oxide (N_2O, *blue*) have risen with CO_2 (*red*) since industrialization began in about 1750. Concentrations are shown in parts per million (ppm) and parts per billion (ppb). SOURCE: USGS 2009.

The main greenhouse gases are CO_2, CH_4, and N_2O

Since preindustrial times atmospheric concentrations of CO_2, methane (CH_4), and nitrous oxide (N_2O) have climbed by over 31 percent, 151 percent, and 17 percent, respectively (fig. 9.14). Carbon dioxide is by far the most important of these because of its abundance and because it lasts for decades or centuries in the atmosphere (fig. 9.15a). Fossil fuel use is responsible for 80 percent of CO_2 emissions. Cement production furnaces and burning of forests and grasslands are also major sources. Together these release more than 33 billion tons of CO_2 every year, on average (fig. 9.15a). About 3 billion tons of this excess carbon is taken up by terrestrial ecosystems, and around 2 billion tons are absorbed by the oceans, leaving an annual atmospheric increase of some

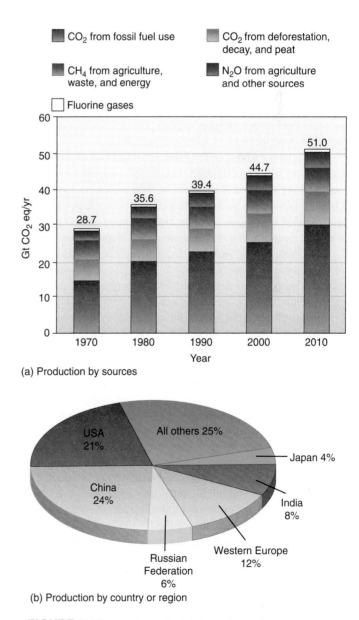

▲ **FIGURE 9.15** Contribution to global warming by different gases and activities (a) and by countries (b).   SOURCE: IPCC.

4 billion tons per year. If current trends continue, CO_2 concentrations could reach about 500 ppm (approaching twice the preindustrial level of 280 ppm) by the end of the twenty-first century.

Methane (CH_4) is much less abundant than CO_2, but it absorbs 23 times as much infrared energy per molecule and is accumulating in the atmosphere about twice as fast as CO_2. Methane can be produced anywhere organic matter decays without oxygen, especially under water. Natural gas, which is mainly CH_4, derives from ancient swamp material. Methane also is released by ruminant animals, wet-rice paddies, coal mines, landfills, wetlands, and pipeline leaks.

Reservoirs for hydroelectricity, usually promoted as a clean power source, are an important source of methane because they capture submerged, decaying vegetation. Philip Fearnside, an ecologist at Brazil's National Institute for Amazon Research, calculates that rotting vegetation in the reservoir behind the Cura-Una dam in Para Province emits so much carbon dioxide and methane every year that it causes three and a half times as much global warming as would generating the same amount of energy by burning fossil fuels. Tropical dams produce roughly 3 percent of global CH_4 emissions.

Nitrous oxide (N_2O) is produced mainly by chemical reactions between atmospheric N and O, which combine in the presence of heat from internal combustion engines. Other sources are burning of organic material and soil microbial activity.

Chlorofluorocarbons (CFCs) and other gases containing fluorine also store heat from infrared energy. CFC releases in developed countries have declined since many of their uses were banned, but increasing production in developing countries, such as China and India, remains a problem. Together, fluorine gases and N_2O account for about 17 percent of human-caused global warming (fig. 9.15a).

The United States, with less than 5 percent of the world's population, releases one-quarter or more of the global CO_2 emissions. In 2007 China passed the United States in total CO_2 emissions (fig. 9.15b), but China's per capita emissions remain less than one-fifth those of the United States. India, with only one ton of CO_2 per person, has only one-twentieth as much as the United States. Oil-rich countries, such as the Middle Eastern oil emirates, have the highest per capita CO_2 output. Qatar, for example, produces more than three times as much CO_2 per person as Australia. But because these countries are small, their overall impact is relatively modest. To examine some of the larger per-capita emitters, see chapter 4, Key Concepts (p. 82).

Africa, in contrast, produces just over one ton of CO_2 per person per year. The lowest emissions in the world are in Chad, where per capita production is only one-thousandth that of the United States.

Some countries with high standards of living release relatively little CO_2. Sweden, for example, produces only 6.5 tons per person per year, or about one-third that of the United States. Remarkably, by adopting renewable energy and conservation measures, Sweden has reduced its carbon emissions by 40 percent over the past 30 years. At the same time, Sweden has seen dramatic increases in both personal income and quality of life measures.

Perhaps the biggest question in environmental science today is whether China and India, which now have the world's largest populations and also are among the fastest-growing economies, will follow the development path of the United States and Canada or that of Sweden and Switzerland. Rising affluence in China has fueled a rapidly growing demand for energy, the vast majority of which comes from coal. China is now building at least one large coal-burning power plant per week. Another large source of CO_2 in China is cement production. Worldwide, cement manufacturing accounts for 4 percent of all CO_2 emissions. Chinese cement companies, stimulated by the world's largest building boom, now produce nearly half of the world's supply, and these plants are responsible for about 10 percent of all Chinese CO_2 emissions.

What consequences do we see?

The American Geophysical Union, one of the nation's largest and most respected scientific organizations, has stated that, as best as can be determined, the world is now warmer than it has been in the last two thousand years—since the beginning of civilization as we know it. If current trends continue, by the end of the century it will be hotter than at any point in the last two million years. The average global temperature has climbed about 0.6°C (1°F) in the past century, and 19 of the 20 warmest years in the past 150 have occurred since 1980. This warming has resulted in some important changes.

- Global sea level has risen approximately 20 cm (8 in.) in the past century. Half of this rise is due to thermal expansion of seawater (because warming water expands). Melting glaciers are responsible for about one-quarter. Melting ice sheets from Greenland and Antarctica account for the remainder (fig. 9.16a).

- Polar regions have warmed much faster than the rest of the world. Permafrost is melting; houses, roads, pipelines, sewage systems, and transmission lines are collapsing as the ground sinks beneath them. Spruce beetle infestations in Alaska (made possible by warmer winters) have killed 250 million spruce trees on the Kenai Peninsula alone. Declining sea ice is leading to coastal erosion and relocation of coastal towns. In Alaska, western Canada, and eastern Russia, average temperatures have increased as much as 4°C (7°F) over the past 50 years.

- Arctic sea ice is less than half as thick and half as extensive in summer as it was 30 years ago. As more heat-absorbing water is exposed, and less heat-reflecting ice persists, the Arctic Ocean is warming at an accelerating rate. Some populations of arctic seals, whose pups must be born on ice, are collapsing. Polar bears, which hunt seals on sea ice, were added to the Endangered Species List in 2008. An aerial survey in 2005 found bears swimming across 260 km (160 mi) of open water to reach the pack ice. Loss of sea ice is also devastating for Inuit people whose traditional lifestyle depends on ice for travel and hunting.

- Ice shelves on the Antarctic Peninsula are breaking up and disappearing rapidly: 90 percent of the glaciers on the peninsula are now retreating. Emperor and Adélie penguin populations have declined by half over the past 50 years as the ice shelves melt (fig. 9.16b). Greenland's ice is melting

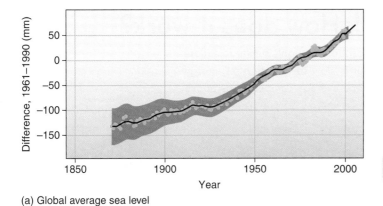

(a) Global average sea level

(b)

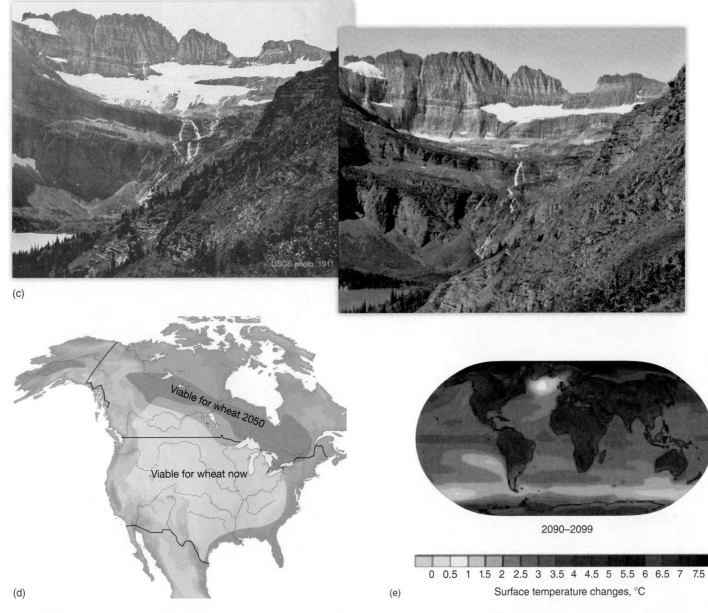

(c)

(d)

(e)

2090–2099

Surface temperature changes, °C

0 0.5 1 1.5 2 2.5 3 3.5 4 4.5 5 5.5 6 6.5 7 7.5

▲ **FIGURE 9.16** Consequences of climate change include a sea level rise of about 20 cm (a) and sharp declines in ice-dependent species, some of which have lost half or more of their population (b). Mountain glaciers are retreating, except where snowfall has increased, including Montana's Glacier National Park (c). Climatic conditions for farming are shifting, and by 2050 wheat production may have shifted to central Canada (d). Surface temperature is projected to rise sharply, even under IPCC scenario B1 (e), which assumes rapid adoption of new energy technology, reduction of fossil fuel use, and a declining population after 2050.

Climate change in a nutshell: How does it work?

The greenhouse effect describes the heating of the earth's atmosphere. Roughly similar to a glass greenhouse, our atmosphere is transparent to light energy but is slow to release heat energy (or infrared radiation). In general, this "greenhouse effect" keeps average temperatures above freezing and supports life, but too much heating can be harmful in a greenhouse or in our atmosphere. Over the past 200 years, we have been emitting heat-absorbing gases (CO_2, CH_4, N_2O, CFCs) at a dramatically increased rate. As a consequence, more heat is retained in the atmosphere. Effects include shorter winters, more heat waves, melting glaciers, declining polar ice, increased droughts in some areas, and increased storms in other areas.

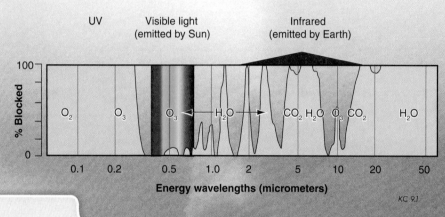

KC 9.1

▲ Different molecules block different wavelengths. CO_2 and H_2O especially prevent infrared energy from escaping the atmosphere.

What are GHGs?

Greenhouses gases (GHGs) are molecules in the atmosphere that block long-wave energy from escaping to space. Water vapor is our most abundant GHG, but human activities have not caused as much change in atmospheric water vapor as other GHGs. We have dramatically increased CO_2, CH_4, N_2O, and other gases since industrialization began in about 1800.

These gases naturally keep our planet warm, but recent increases in GHGs are warming the planet enough to destabilize our economies and resource uses.

GAS	% OF CLIMATE FORCING*
Carbon dioxide (CO_2)	60%
Methane (CH_4)	20%
Nitrous oxide (N_2O)	10%
Aerosols, other gases	10%

*Percentage of anthropogenic change; depends on (1) amount emitted, (2) energy-capture effectiveness, and (3) persistence in the atmosphere.

KC 9.3

KC 9.4

KC 9.5

KC 9.2

CO₂

KC 9.10

N₂O

Where do GHGs come from?

Fossil fuel burning produces about 60% of GHG emissions, followed by deforestation (17%) and industrial and agricultural N_2O (14%). Rice paddies, belching livestock, and tropical dams produce CH_4 (9% of emissions).

KC 9.6

CH₄

KC 9.9

KC 9.8

KC 9.7

How do we know that recent climate changes are caused by human activity?

IPCC models show that observed temperature trends (black line) do not fit mathematical models built without human-caused factors (blue shaded area shows range of model predictions). Observed trends do fit models built with factors such as fossil fuel use and forest clearing (pink shaded area). ▶

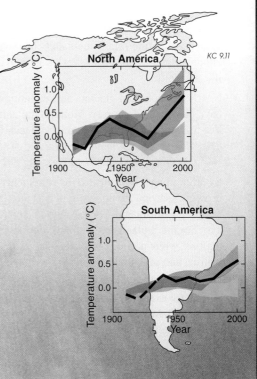

Hasn't the climate changed in geologic time?

Climate conditions have always changed, but never so dramatically since the beginning of civilization, and usually more slowly than now. Our current course is set for higher temperatures by 2100 than have occurred in at least 800,000 years. Current CO_2 concentrations are about 30 percent higher than at any time in the past 800,000 years, according to ice core data. The rate of current climate change is also new: changes now occurring in 100 years took 800 to 5,000 years at the end of the ice ages. ▼

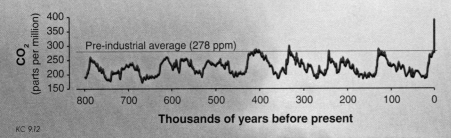

KC 9.12

What effects are observed and expected?

Disappearing ice: Arctic ice, which helps stabilize climate, has declined by nearly half in summer. Mountain glaciers and snow, which provide water to about 75 percent of the western United States and over 1 billion people in Asia, are disappearing worldwide.

Wildfire and pests: Increased fire frequency and severity, aided by expanding parasites, is causing ecosystem change and even human mortality.

Early spring: Early onset of warm weather has led to early flowering, migrations, and hotter summers.

Rising sea level: We are committed to about 0.5 m rise. Without rapid CO_2 reductions, we may soon be committed to 2 m or more.

More storms: More energetic atmospheric circulation is likely to bring more, heavier storms. Heavier rain and snow in the eastern U.S. may already be evident.

Cumulative costs of climate change: $5-90 trillion* by 2100, in damaged infrastructure, lost property values, health costs.

*Pew Environment Group, 2010.

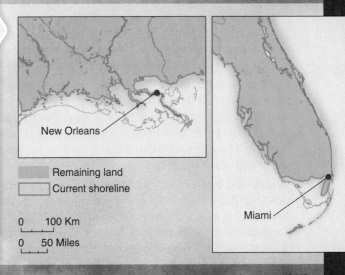

Remaining land
Current shoreline

0   100 Km
0   50 Miles

KC 9.13

Observed changes are greater than models anticipated.

Observed GHG emissions (black lines) have accelerated faster than all the IPCC's projected scenarios (colored lines). Changes in temperature and sea level therefore may be greater than anticipated. ▶

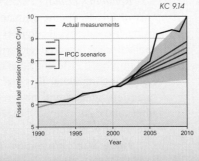

KC 9.14

CAN YOU EXPLAIN?

1. What is a greenhouse gas? What are three main anthropogenic gases?

2. Explain the pink and blue bands in the maps at top right. What do the black lines show?

3. Examine the New Orleans and Miami maps. Identify some strategies to protect these cities against rising sea levels and storm frequencies.

4. In this chapter, what are some strategies we have to reduce climate change?

Are we committed to this path?

Not necessarily. If we adopt alternative plans, especially for energy production, transportation, and more efficient use of energy, we could constrain temperature change to only 2-4 degrees globally. We are committed to at least 20-40 cm of sea-level change, however, as the ocean continues to absorb excess heat from the atmosphere.

at an accelerating rate. Greenland's massive ice cap holds enough water to raise sea level by about 6 m (about 20 ft) if it all melts.

- Nearly all mountain glaciers are retreating rapidly, and many have disappeared entirely. Mount Kilimanjaro has nearly lost its famous ice cap. Montana's Glacier National Park had 150 glaciers when it was created in 1910. Soon it will be Glacierless National Park (fig. 9.16c).

- Satellite images and surface measurements show that growing seasons are now as much as three weeks longer in northern Eurasia and North America, compared to 30 years ago. Southern plants and animals are moving into arctic regions, while arctic species such as musk ox, caribou, walrus, and seal are declining.

- Droughts are becoming more frequent and widespread. In Africa, for example, droughts have increased about 30 percent since 1970. Marginal farmlands are becoming less productive as warmer temperatures evaporate available moisture.

- Plants and animals are breeding earlier or extending their range into new territory. In Europe and North America, 57 butterfly species have disappeared from the southern end of their range or extended the northern limits. Many plants may be unable to migrate as rapidly as conditions are changing: we now are forcing them to move much faster than they did at the end of the last ice age (fig. 9.16d).

- Coral reefs worldwide are "bleaching" (losing the colorful algae they rely on for survival) as water temperatures rise above 30°C (85°F). Marine ecologists worry that warming oceans could be the final blow for reefs, which are already threatened by pollution and overfishing, and for the fisheries and ecosystems they support.

- Oceans have been absorbing CO_2 and heat. This absorption slows warming, but it will take centuries to dissipate the stored heat even if we reduce our greenhouse gas emissions immediately. Absorption of CO_2 is also acidifying the oceans. Acidity dissolves calcium carbonate, making it harder for mollusks to grow shells and for corals to build reefs.

- Storms are becoming stronger and more damaging. Rising energy retention in the atmosphere (fig. 9.16e) leads to more vigorous circulation, including storms, extreme rainfall, snowfall, and hurricane frequency. Insurance companies are dropping storm coverage and raising premium rates, as they plan for more extreme disaster costs (see Exploring Science, p. 221).

Controlling emissions is cheap compared to climate change

A 2010 study by the Pew Trust evaluated estimates of the cost of lost ecological services by 2100. Costs included factors such as lost agricultural productivity from drought, damage to infrastructure from flooding and storms, lost biological productivity, health costs from heat stress, and lost water supplies to the billion or so people who depend on snowmelt for drinking and irrigation.

The Pew report found that climate change is likely to cost between $5 trillion and $90 trillion by 2100, depending on how economic discount rates and other factors are calculated. In a study issued on behalf of the British government, Sir Nicholas Stern, former chief economist of the World Bank, estimated the cumulative costs of climate change equal between 5 and 20 percent of the annual global gross domestic product (GDP).

In contrast, the Stern report estimated it would cost only about 1 percent of global GDP to reduce greenhouse gas emissions now to avoid the worst impacts of climate change. The IPCC says it would cost even less, only 0.12 percent of annual global GDP to reduce carbon emissions the 2 percent per year necessary to stabilize world climate.

For many people, global climate change is a moral and ethical issue as well as a practical one. Religious leaders are joining with scientists and business leaders to campaign for measures to reduce greenhouse gas emissions. Ultimately, the people likely to suffer most from global warming are the poorest in Africa, Asia, and Latin America, who have contributed least to the problem. There is also a question of intergenerational equity. The actions we take—or fail to take—in the next 10 to 20 years will have a profound effect on those living in the second half of this century and in the next. What kind of world are we leaving to our children and grandchildren? What price will they pay if we fail to act?

Why are there disputes over climate evidence?

Scientific studies have long been unanimous about the direction of climate trends, but commentators on television, newspapers, and radio continue to fiercely dispute the evidence. Why is this? Part of the reason may be that change is threatening, and many of us would rather ignore it or dispute it than acknowledge it. Part of the reason may be a lack of information. Another reason is that while scientists tend to look at trends in data, the public might be more impressed by a few recent and memorable events, such as an especially snowy winter in their local area. And on talk radio and TV, colorful opinions sell better than evidence. Climate scientists offer the following responses to some of the claims in the popular media:

Reducing climate change requires abandoning our current way of life. Reducing climate change doesn't necessarily require using *less* energy, it requires that we use *different* energy. If we replace coal-powered electricity with wind, solar, natural gas, and improved efficiency, we can drastically cut our emissions but keep our computers, TVs, cars, and other conveniences. Reducing coal dependence will also reduce air pollution, health expenditures, and destruction of vegetation and buildings (see discussion later in this chapter).

There is no alternative to current energy systems. Without investments in alternative energy sources, this would be true, but Chinese and European energy companies are demonstrating that this claim is false. European and Chinese businesses are showing that alternative energy and improved efficiency can already provide what we need and that there's a great deal of money to be made in new technology. In the coming years there is likely to be more profit in new technologies than in the traditional energy

Our climate system is one vast manipulative experiment: we are injecting greenhouse gases into the atmosphere and observing the changes that result. In most manipulative experiments, though, we have controls, which we can compare to treatments to be certain of the effects. Since we have only one earth, we have no controls in this experiment. So how do we test a hypothesis in an uncontrolled experiment?

One approach is to use models. You build a computer model, a complex set of equations, that include all the known natural causes of climate fluctuation, such as Milankovitch cycles and solar variation. You also include the known human-caused inputs (fossil fuel emissions, methane, aerosols, soot, and so on). Then you run the model and see if it can reproduce observed past changes in temperatures.

Continued ▶

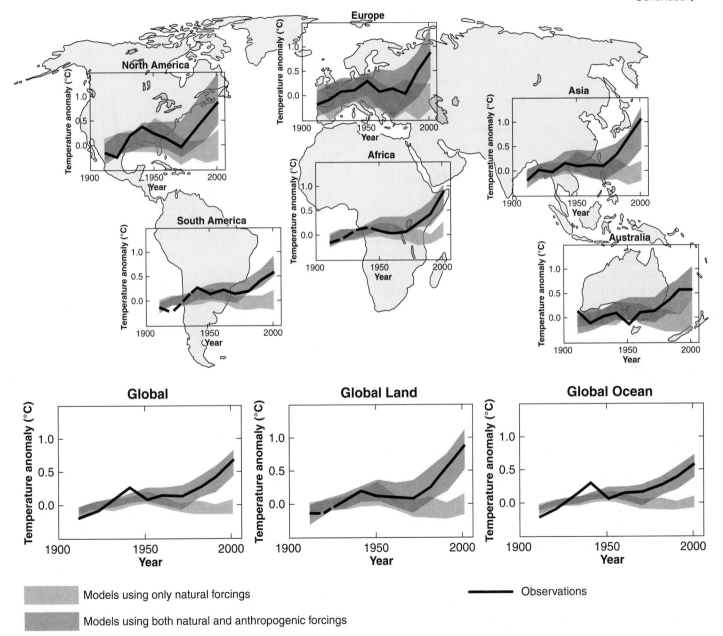

▲ Comparison of observed continental and global changes in surface temperature with results simulated by climate models using either natural factors (*blue*) or combined human and natural factors (*pink*). Observed changes are shown in black. (Dotted lines indicate low density of data record.) Blue shaped bands show the 6–95 percent range for 19 simulations using only natural forcings, including solar energy and volcanoes. Pink shaded bands show the 5–95 percent range for 58 simulations using both natural and anthropogenic forcings. SOURCE: IPCC 2008.

◀ *Continued*

If you can accurately "predict" past changes, then your model is a good description of how the system works. You've done a good job of representing how the atmosphere responds to CO_2 inputs, how oceans absorb heat, how changes in snow cover accelerate energy absorption, and so on.

If you can create a model that represents the system quite well, then you can rerun the model, but this time you leave out all the anthropogenic inputs. If the model *without* human inputs is *inconsistent* with observed changes in temperature, and if the model *with* human inputs is *consistent* with observations, then you can be extremely confident, beyond the shadow of a reasonable doubt, that human inputs had made the difference and caused temperature changes.

This modeling approach is precisely what climate scientists have done. As you can see in the IPCC's summary of model results, observed trends do not fit the models run with only natural changes (in blue). Observed trends do fit models that include human inputs, however.

---

and transportation technologies of the 1940s. Much of the technological innovation in energy is occurring in the United States, although adoption here has been slower than elsewhere.

A comfortable lifestyle requires high CO_2 output. The data show this claim is incorrect. Most northern European countries have higher standards of living (in terms of education, health care, life span, vacation time, financial security) than residents of the United States and Canada, yet their CO_2 emissions are as low as half those of North Americans. Residents of San Francisco consume about 1/6 as much energy as residents of Kansas City, yet quality of life is not necessarily six times greater in Kansas City than in San Francisco.

Natural changes such as solar variation can explain observed warming. Solar input fluctuates, but changes are slight and do not coincide with the direction of changes in temperatures (fig. 9.17). Milankovitch cycles also cannot explain the rapid changes in the past few decades. Increased GHG emissions, however, do correspond closely with observed temperature and sea-level changes (see fig. 9.13).

The climate has changed before, so this is nothing new. Today's CO_2 level of almost 400 ppm exceeds by at least 30 percent anything the earth has seen for nearly a million years, and perhaps as long as 15 million years. Recent change is also far more rapid than natural fluctuations. Antarctic ice cores indicate that CO_2 concentrations for the past 800,000 years have varied from 180 to 300 ppm (see fig. 9.8). This natural variation in CO_2 appears to be a feedback in glacial cycles, resulting from changes in biotic activity in warm periods. Because temperature has closely tracked CO_2 over time, it is likely that temperatures by 2100 will exceed anything in the past million years. The rate of change is probably also unprecedented. Changes that took 1,000 to 5,000 years at the end of ice ages are now occurring on the scale of a human lifetime.

Temperature changes are leveling off. Short-term variation in trends can always be found if we view the data selectively (fig. 9.17), but over decades the trends in surface air temperatures and in sea level continue to rise. Climatologists don't fully understand the slight slowing of temperature changes, but some evidence suggests that heat absorption in deeper ocean layers may account for the slowing rate of increase in several recent years.

We had cool temperatures and snowstorms last year, not heat and drought. Climate models predict regional and seasonal variation in temperature and precipitation trends, as increasingly

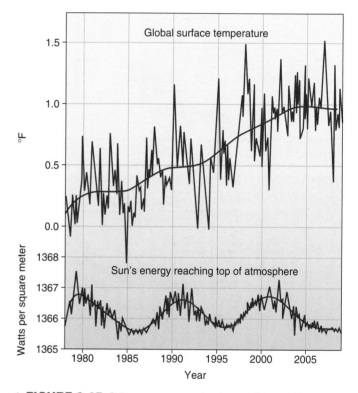

▲ **FIGURE 9.17** Solar energy received at the top of the earth's atmosphere has been measured by satellites since 1978. It has followed a natural 11-year cycle of small shifts but no overall increase (*bottom*). Over the same period, surface temperatures have increased markedly (*top*).   SOURCE: Climate Change Compendium, 2009.

intense atmospheric circulation leads to greater storm activity and precipitation in some regions (fig. 9.18). The global average of these trends, however, points toward warmer conditions overall, associated with widespread drought.

Climate scientists don't know everything, and they have made errors and misstatements. The gaps and uncertainties in climate data are minute compared to the evident trends. There are many unknowns, such as details of precipitation change or interaction of long-term cycles such as El Niño, but the trends are unequivocal. Climatologist James Hansen has noted that while most people make occasional honest mistakes, fraud in data collection is almost unheard of. The scientific process ensures transparency and eventual

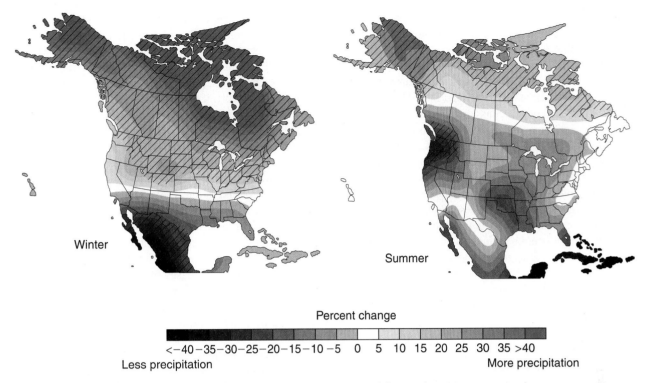

Percent change

<−40 −35 −30 −25 −20 −15 −10 −5 0 5 10 15 20 25 30 35 >40
Less precipitation   More precipitation

▲ **FIGURE 9.18** Precipitation in 2100 is expected to produce wetter northern winters and drier southern winters, compared to recent averages; Drier summers are expected for most of North America. Values are averages of 15 climate models. SOURCE: Climate Change Compendium, 2009.

exposure of errors. Despite this effort, Hansen notes, prominent climate scientists are regularly subjected to personal attacks from climate-change deniers who, lacking evidence for their arguments, resort to harassment to suppress discussion.

9.4 ENVISIONING SOLUTIONS

Former president Bill Clinton has argued that combating climate change doesn't have to mean economic hardship. It could be the biggest development stimulus since World War II, creating millions of jobs and saving trillions of dollars in foreign fuel imports. What are some of the strategies we have available?

We have many options. We can reduce dependence on coal, which produces more CO_2 per unit energy than any other fuel. We can reconsider the $120 billion per year in tax credits for oil and coal development, which transfer operating costs to taxpayers and hide the real price of these fuels. We can plan cities and transportation systems to provide better alternatives to private automobiles, which are the main way individuals contribute to our climate impacts. We can account for the hidden costs (such as health costs, biodiversity and recreation costs, military expenditures to protect oil fields); when these costs are no longer hidden, it will be clear that fossil fuels are not cheaper than alternatives. We can invest in new technologies—the price of solar power has plummeted in recent years, making the cost of new solar, wind, and coal power equivalent. We can help developing countries invest in alternative energy and low-carbon futures. We can reduce the emissions associated with deforestation, chiefly decomposition, burning, and lost carbon sinks, our largest nonfuel source of CO_2 emissions (see chapter 6).

A favored plan thus far has been trading of carbon emission credits. By instituting a legal cap on emissions, then allowing companies to buy and sell shares of that total cap, many strategists think we can keep emissions down and make money all at once. A vigorous market for emissions trading has emerged, growing from a global value of about $15 billion in 2005 to $130 billion in 2012. Other strategists are not so sure this is a real solution. Climate scientists James Hansen argues that European carbon markets have made traders rich but have failed to reduce carbon emissions.

International protocols have sought to establish common rules

One of the centerpieces of the 1992 United Nations Earth Summit meeting in Rio de Janeiro was the Framework Convention on Climate Change, which set an objective of stabilizing greenhouse gas emissions to reduce the threats of global warming. At a follow-up conference in Kyoto, Japan, in 1997, 160 nations agreed to roll back CO_2, methane, and nitrous oxide emissions to about 5 percent below 1990 levels by 2012. Three other greenhouse gases, hydrofluorocarbons, perfluorocarbons, and sulfur hexafluoride, would also be reduced, although by what level was not decided. Known as the **Kyoto Protocol**, this treaty set different limits for individual nations, depending on their output before 1990. Poorer nations, such as China and India, were exempted from emission limits to allow development to increase their standard of living. Wealthy countries created the problem, the poorer nations argued, so wealthy countries should solve it.

Although the United States took a lead role in negotiating the Kyoto protocol, U.S. administrations have declined to participate in the accord. Claiming that reducing carbon emissions would be too costly for the U.S. economy, President George H. W. Bush stated the general policy: "We're going to put the interests of our own country first and foremost." The United States has opted for voluntary limits to greenhouse gas emissions, and has seen emissions rise by approximately 25 percent above 1990 emissions by 2012.

Many of the largest business conglomerates in America have joined environmental groups to call for strong national legislation to achieve significant reductions of greenhouse gas emissions. Those companies would prefer a single national standard rather than a jumble of conflicting local and state rules, which cost them time, money, and sometimes contracts with potential trading partners.

International climate accords continue to be stymied by the largest economies. However, the largest European economies overshot their Kyoto targets by 2012, at the same time as their economies have grown. International recognition of the costs of climate change seems to be growing. What progress we will make remains unclear.

A wedge approach could fix the problem

There have been many proposals for dramatic new inventions that can fix the problem all at once—nuclear fusion, space-based solar energy, or giant mirrors that would reflect solar energy away from the earth's surface. These are intriguing ideas, but all are still in the distant future, and climate scientists warn that we can't wait if we hope to avoid disaster. An alternative approach is wedge analysis, breaking down a large problem into smaller, bite-size pieces. Each one might start small now, but their impacts will grow over time, producing a larger wedgelike impact in 50 years (fig. 9.19) By calculating the contribution of each wedge, we can add them up, see the magnitude of their collective effect, and decide which part to work on first. This idea was proposed by Stephen Pacala and Robert Socolow, of Princeton University's Climate Mitigation Initiative, who calculated that currently available technologies—efficient vehicles, buildings, power plants, alternative fuels—could solve our problems quickly, if we just take them seriously.

Pacala and Socolow's paper described 14 "wedges" (table 9.2). Each wedge represents 1 GT (1 billion tons) of carbon emissions avoided in 2058, compared to a "business as usual" scenario. Accomplishing just half of these wedges could level off our emissions. Accomplishing all of them could return to levels well below those envisioned in the Kyoto Protocol. A "stabilization triangle" shows the difference between our current path of rising CO_2 and alternative strategies to lower emissions. The "business as usual" scenario follows the current pattern of constantly increasing CO_2 output. This trajectory heads toward a tripling of CO_2 by 2100, with temperature increases of around 5°C (9°F) (see fig. 9.2). A "stabilization scenario" would prevent further increases in CO_2 emissions, but atmospheric CO_2 in the atmosphere would still double by 2100 because of processes already in place. Temperatures increase by about 2–3°C. A third trajectory would produce declining CO_2 emissions.

To achieve stabilization, we need to reduce our annual carbon emissions by about 7 billion tons (or 7 gigatons, GT) per year within 50 years. None of the 14 wedge options is going to please everybody, and some will be more popular than others, but the potential benefits of the entire portfolio are large enough that not every option must be used.

TABLE 9.2 | Actions to Reduce Global CO_2 Emissions by 1 Billion Tons over 50 Years

1. Double the fuel economy for 2 billion cars from 30 to 60 mpg.

2. Cut average annual travel per car from 10,000 to 5,000 miles.

3. Improve efficiency in heating, cooling, lighting, and appliances by 25 percent.

4. Update all building insulation, windows, and weather stripping to modern standards.

5. Boost efficiency of all coal-fired power plants from 32 percent today to 60 percent (through co-generation of steam and electricity).

6. Replace 800 large coal-fired power plants with an equal amount of gas-fired power (four times current capacity).

7. Capture CO_2 from 800 large coal-fired, or 1,600 gas-fired, power plants and store it securely.

8. Replace 800 large coal-fired power plants with an equal amount of nuclear power (twice the current level).

9. Add 2 million 1-MW windmills (50 times current capacity).

10. Generate enough hydrogen from wind to fuel a billion cars (4 million 1-MW windmills).

11. Install 2,000 GW of photovoltaic energy (700 times current capacity).

12. Expand ethanol production to 2 trillion liters per year (50 times current levels).

13. Stop all tropical deforestation and replant 300 million ha of forest.

14. Apply conservation tillage to all cropland (10 times current levels).

SOURCE: Data from Pacala and Socolow, 2004.

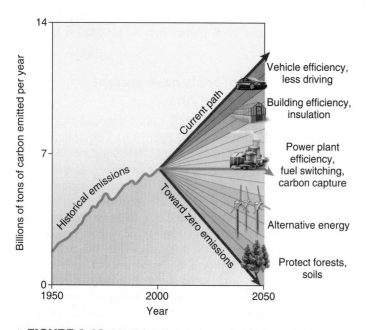

▲ **FIGURE 9.19** A wedge approach could use multiple strategies to reduce or stabilize carbon emissions rapidly and relatively cheaply.

Because most of our CO_2 emissions come from fossil fuel combustion, energy conservation and a switch to renewable fuels probably are the first places we should look. One of the 7 GT reductions could be accomplished by increasing our vehicle fuel efficiency from the expected 30 miles per gallon in 2058 to 60 mpg. For another GT we could reduce reliance on cars by half (with more public transit or less suburban sprawl, for example), helping drivers drop from an average 10,000 miles to 5,000 miles per year. We could save another 2 GT simply by installing the most efficient lighting and appliances available, along with improved insulation in buildings. These steps add up to 4/7 of the stabilization triangle, using currently available technologies. The remaining 3/7 can be accomplished by capturing and storing carbon at power plants, by changing the way power plants operate, and by reducing reliance on coal power. Much of this CO_2 could be injected into oil wells to improve crude oil production.

Another set of seven wedges, including alternative energy, preventing deforestation, and reducing soil loss, could put us on a trajectory to reduce our CO_2 emissions and prevent disastrous rates of climate change. If we used all 14 available options, we could reverse the present trajectory and move toward zero greenhouse emissions.

The net effect of these strategies is likely to be economic gain, which contradicts many traditional fears of economists and politicians that we cannot afford climate mitigation. Many of the needed changes involve efficiency, which means long-term cost savings. Employment is likely to increase as new cars and appliances replace old ones, and as we insulate more buildings.

There are other potential benefits, too. Efficient cars save household income. Cleaner power plants reduce asthma and other respiratory illnesses, saving health care costs as well as improving quality of life. Less reliance on coal reduces toxic mercury in our food chain, because coal burning is the largest single source of airborne mercury emissions. Individuals can make many contributions to this effort (see What Can You Do?, at right).

Local initiatives are everywhere

Many countries are working to reduce greenhouse emissions. The United Kingdom, for example, had already rolled CO_2 emissions back to 1990 levels by 2000 and vowed to reduce them 60 percent by 2050. Britain already has started to substitute natural gas for coal, promote energy efficiency in homes and industry, and raise its already high gasoline tax. Plans are to "decarbonize" British society and to decouple GNP growth from CO_2 emissions. New carbon taxes are expected to lower CO_2 releases and trigger a transition to renewable energy over the next five decades. New Zealand Prime Minister Helen Clark pledged that her country will be the first to be "**carbon neutral**," that is, to reduce net greenhouse gas emissions to zero, although she didn't say when this will occur. Copenhagen, Denmark, is aiming to be the first carbon-neutral capital, a goal they hope to achieve by improved transportation planning, housing efficiency, and alternative energy sources.

Germany, also, has reduced its CO_2 emissions by more than 10 percent by switching from coal to gas and by encouraging energy efficiency throughout society. Atmospheric scientist Steve Schneider calls this a "no regrets" policy; even if we didn't need

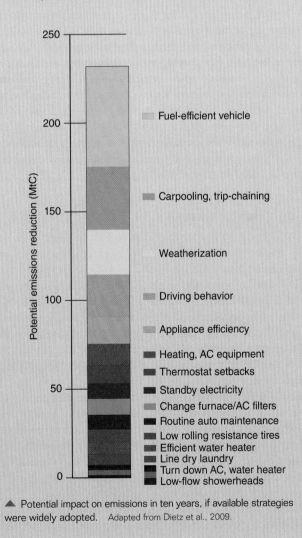

What Can YOU DO?

Reducing Individual CO_2 Emissions

Individual actions can have tremendous impacts on climate change, because our actions are multiplied by millions of others who make similar decisions. Many of our options save money as well as reducing pollution and resource consumption. A recent study of behavior and household options found that we can reduce U.S. emissions by 233 metric tons of carbon with simple changes in driving patterns, vehicle efficiency, household weatherization, and minor changes in behavior. This graph shows a generalized summary of which steps would make the most impact. If they were widely adopted, these steps could reduce total emissions by 7.4 percent in ten years without any new regulations, technology, or reduction in well-being.

To read more, see T. Dietz et al., 2009. Household actions can provide a behavioral wedge to rapidly reduce U.S. carbon emissions. *Proceedings of the National Academy of Sciences* 106(44): 18452–56.

Potential emissions reduction (MtC)

- Fuel-efficient vehicle
- Carpooling, trip-chaining
- Weatherization
- Driving behavior
- Appliance efficiency
- Heating, AC equipment
- Thermostat setbacks
- Standby electricity
- Change furnace/AC filters
- Routine auto maintenance
- Low rolling resistance tires
- Efficient water heater
- Line dry laundry
- Turn down AC, water heater
- Low-flow showerheads

▲ Potential impact on emissions in ten years, if available strategies were widely adopted. Adapted from Dietz et al., 2009.

to stabilize our climate, many of these steps save money, conserve resources, and have other environmental benefits. Nuclear power also is being promoted as an alternative to fossil fuels. It's true that nuclear reactors don't produce greenhouse gases, but it is increasingly unclear whether the economics of nuclear power make sense as alternatives become cheaper.

Renewable energy has been a leading strategy for reducing climate change and for long-term energy stability, because global oil supplies are unstable, economically volatile, and declining in abundance (see chapter 12). Sweden now produces 50 percent of its energy from renewable sources, including wind, geothermal, biomass, biogas (methane from garbage), fuel cells, and especially hydropower. Denmark, the world's leader in wind power, has no hydropower and little sunshine, but now gets 20 percent of its electricity from windmills. Plans are to generate half of the nation's electricity from offshore wind farms by 2030. China, the world's fastest-growing economy, has promised to cut 10 percent of its CO_2 emissions per unit of economic output by shifting to renewable energy and conservation.

In the United States, nearly 1,000 cities and 39 states have announced their own plans to combat global warming. And over 450 college campuses have pledged to reduce greenhouse emissions. Part of the motivation for these steps is that alternative solutions often are advantageous in their own right. Making buildings more energy efficient and buying high-mileage vehicles saves money in the long run. Buying better light bulbs saves money in the near term. Planning cities for better transit, biking, and walking saves tax dollars we now spend on far-flung road and service networks. Walking, biking, and climbing stairs are good for your health, and they help reduce traffic congestion and energy consumption. As the Irish statesman and philosopher Edmund Burke said, "Nobody made a greater mistake than he who did nothing because he could do only a little."

Carbon capture saves CO_2 but is expensive

It is possible, though expensive, to store CO_2 by injecting it deep into geologic formations. Since 1996, Norway's Statoil has been pumping more than 1 million metric tons of CO_2 per year into an aquifer 1,000 m below the seafloor in the North Sea. The pressurized CO_2 enhances oil recovery. It also saves money because otherwise the company would have to pay a $50 per ton carbon tax on its emissions. Around the world, deep, salty aquifers could store a century's output of CO_2 at current fossil fuel consumption rates. A number of companies have started, or are now planning, similar schemes (fig. 9.20).

Coal-fired power plants, our greatest emitters, can prevent CO_2 emissions using integrated gasification combined cycle (IGCC) technology (see chapter 12). Utilities eager to continue business as usual are touting "clean-coal" techniques and carbon sequestration as the answer to global warming. Because of the high cost of these systems, and of capturing, shipping, and pumping carbon into the ground, few of these facilities have been completed.

We focus most attention on CO_2 because it is abundant and long-lasting, but we can address other greenhouse gases, too. Methane can be captured if we maintain pipelines to prevent leakage.

Active LEARNING

Calculate Your Carbon Reductions

How much global warming can you avoid through personal efforts? It's difficult to compare actions as different as walking to work and planting a tree, but most of our CO_2 emissions result from burning fossil fuels. The exact amount of CO_2 you save depends on the source of your energy and how you use it. Coal, for instance, produces about twice as much CO_2 per unit of energy delivered as natural gas. Wind, solar, and hydropower, on the other hand, don't emit any CO_2 at all. It's often easier to find energy consumption data for appliances.

The Tufts University Climate Initiative provides the following information:

The average desktop computer uses about 120 watts (the monitor uses 75 watts, and the CPU uses 45 watts). Laptops use considerably less, around 30 watts total.

1. Suppose you keep your computer on 24 hours a day. How much energy does it use every year?

2. If electricity costs 11 cents per kWh, how much does it cost per year to keep your computer running all the time?

3. If your electricity source produces 1.45 lb of CO_2 per kWh delivered to your home (the average for coal-fired power), how much CO_2 is released every year to keep your computer always on?

4. Suppose you turn your computer off for 12 hours per day when you're not using it. How much energy would you save over a year?

5. How much CO_2 would that save per year?

6. How much money would you save per year?

ANSWERS: 1. 120 watts/hour = 0.12 kWh × 24 hr/day × 365 days/yr = 1,051 kWh/yr. 2. 1,051 kWh × $0.11 = $115.63/yr. 3. 1,051 kWh × 1.45 lb/kWh = 1,524 lb/yr. 4. 0.12 kWh × 12 hr/day × 365 days/yr = 525.6 kWh/yr 5. 525.6 kWh/yr × 1.45 lb/kWh = 762 lb/yr. 6. 525.6 kWh × $0.11 = $57.81/yr.

Unmeasured but significant leakage of methane also happens during gas drilling and extraction, especially when hydraulic fracturing is used to release gas from tight geologic formations. Capturing this gas for sale is good business, too.

Capturing methane from landfills, oil wells, and coal mines would make an important contribution to both fuel and climate problems. Rice paddies, with submerged, decaying plant matter, are also important methane sources: changing flooding schedules and fertilization techniques can reduce methane production in paddies.

Ruminant animals (such as cows, camels, and buffalo) create large amounts of methane in their digestive systems. Feeding them more grass and less corn can reduce these emissions. Some analysts have suggested that eating less beef—skipping just one day per week—could reduce our individual contributions to climate change more than driving a hybrid Toyota Prius would.

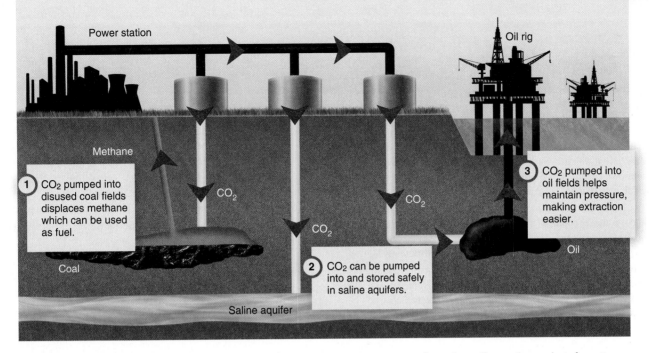

▲ **FIGURE 9.20** Carbon capture and storage involves pumping CO_2 into permanent storage, usually a salty aquifer or other geologic formation. Captured CO_2 can also be used to increase pressure on oil and gas wells, increasing recovery.

CONCLUSION

Climate change may be the most far-reaching issue in environmental science today. Although the challenge is almost inconceivably large, solutions are possible if we choose to act, as individuals and as a society. Temperatures are now higher than they have been in thousands of years, and climate scientists say that if we don't reduce greenhouse gas emissions soon, drought, flooding of cities, and conflict may be inevitable.

Understanding the climate system is essential to understanding the ways in which changing composition of the atmosphere (more carbon dioxide, methane, and nitrous oxide, in particular) matters to us. Basic concepts to remember about the climate system include how the earth's surfaces absorb solar heat, how atmospheric convection transfers heat, and that different gases in the atmosphere absorb and store heat that is reemitted from the earth. Increasing heat storage in the lower atmosphere can cause increasingly vigorous convection, more extreme storms and droughts, melting ice caps, and rising sea levels. Changing patterns of monsoons, cyclonic storms, frontal weather, and other precipitation patterns could have extreme consequences for humans and ecosystems.

Despite the importance of natural climate variation, observed trends in temperature and sea level are more rapid and extreme than other changes in the climate record. Exhaustive modeling and data analysis by climate scientists show that these changes can only be explained by human activity. Increasing use of fossil fuels is our most important effect, but forest clearing, decomposition of agricultural soils, and increased methane production are also extremely important.

The "stabilization wedge" proposal is a list of immediate and relatively modest steps that could be taken to accomplish needed reductions in greenhouse gases. International organizations, national governments, and local communities have all begun to do their part. Individual "wedges" are likely to strengthen jobs and economic growth. Although individual producers of fossil fuels might lose income, economists estimate that the rest of society would benefit in economy, health, and environmental quality.

PRACTICE QUIZ

1. What are the dominant gases that make up clean, dry air?
2. Name and describe four layers of the atmosphere.
3. What is the greenhouse effect? What is a greenhouse gas?
4. What are some factors that influence natural climate variation?
5. Explain the following: Hadley cells, jet streams, Coriolis effect.
6. What is a monsoon, and why is it seasonal?
7. What is a cyclonic storm?
8. What is the IPCC, and what is its function?
9. What method has the IPCC used to demonstrate a human cause for recent climate changes? Why can't we do a proper manipulative study to prove a human cause?
10. What is the Kyoto Protocol?
11. List 5 to 10 effects of changing climate.
12. What is a climate stabilization wedge? Why is it an important concept?
13. What are the "stabilization wedges" that could help flatten or reduce our CO_2 emissions?

Apply the principles you have learned in this chapter to discuss these questions with other students.

1. Weather patterns change constantly over time. From your own memory, what weather events can you recall? Can you find evidence in your own experience of climate change? What does your ability to recall climate changes tell you about the importance of data collection?

2. One of the problems with forming climate policies such as the Kyoto Protocol is that economists and scientists define problems differently and have contrasting priorities. How would an economist and an ecologist explain disputes over the Kyoto Protocol differently?

3. Economists and scientists often have difficulty reaching common terms for defining and solving issues such as the Clean Air Act renewal. How might their conflicting definitions be reshaped to make the discussion more successful?

4. Why do you think controlling greenhouse gases is such a difficult problem? List some of the technological, economic, political, emotional, and other factors involved. Whose responsibility is it to reduce our impacts on climate?

5. How does the decades-long, global-scale nature of climate change make it hard for new policies to be enacted? What factors might be influential in people's perception of the severity of the problem?

6. Would you favor building more nuclear power plants to reduce CO_2 emissions? Why or why not?

7. Of the climate wedges shown in table 9.2, which would you find most palatable? Least tolerable? Why? Can you think of any additional wedges that should be included?

DATA ANALYSIS   Examining the IPCC Fourth Assessment Report (AR4)

The Intergovernmental Panel on Climate Change (IPCC) has a rich repository of figures and data, and because these data are likely to influence some policy actions in your future, it's worthwhile to examine the IPCC reports. The data and conclusions, as well as the points of uncertainty, are presented there to help the public understand the issues with the best available data.

An excellent overview is in the Summary for Policy Makers (SPM) that accompanies the Fourth Assessment Report. Find a copy of this report on Connect. Examine the figures in the report and answer questions in Connect to demonstrate your understanding of the ideas and the issues.

▶ See the evidence: view the IPCC report at http://www.ipcc.ch/ipccreports/ar4-syr.htm.

Change in temperature, sea level, and Northern Hemisphere snow cover

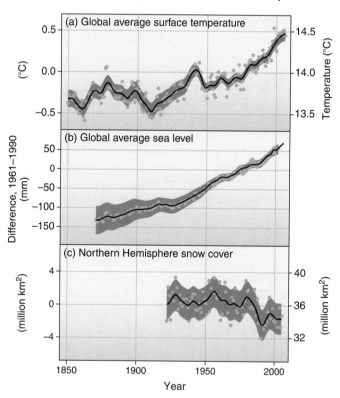

The Great London Smog of 1952 killed thousands and helped change the way we see air population.

LEARNING OUTCOMES

After studying this chapter, you should be able to answer the following questions:

▶ What are the main types and sources of conventional or "criteria" pollutants?

▶ Describe several hazardous air pollutants and their effects.

▶ How do air pollutants affect the climate and stratospheric ozone?

▶ In what ways can air pollution affect human health?

▶ What policies and strategies do we have for reducing air pollution?

▶ Has world air quality been getting better or worse? Why?

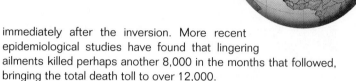

The Great London Smog

London was once legendary for its pea-soup fogs. In the days of Charles Dickens and Sherlock Holmes, darkened skies and blackened buildings, saturated with soot from hundreds of thousands of coal-burning fireplaces, were a fact of life. Londoners had been accustomed to filthy air since the beginning of the industrial revolution, but over a period of four days in 1952, days just 60 years ago, they experienced the worst air pollution disaster on record. Smoke, soot, and acidic droplets of fog made the air opaque. Thousands died, thousands more became ill, and our view of air pollution changed forever.

In early December 1952, a dense blanket of coal smoke and fog settled on the city. Under normal conditions, winds keep polluted air moving, away from the city and out over the countryside. These winds occur, broadly speaking, because air near the earth's surface is usually warmed by the sun-heated ground, while air aloft is cool. Turbulence develops as the warmed air rises and cool air sinks, and the turbulence tends to circulate pollutants away from their sources. But from time to time an inversion develops. As the name implies, an inversion occurs when layers of air are out of order: Still, cold air settles near the ground, trapped by warmer layers above. On a cold, dark December day in London, the cold air can settle in to stay.

Inversions can be unpleasantly chilly and damp, but in a city where coal is the primary fuel, burned in countless low-efficiency stoves and furnaces, the stable inversion conditions also trap smoke, particulates of coal dust, and tiny droplets of sulfuric acid (from sulfur in coal) in the city.

The "killer smog" of 1952 came on suddenly, on Friday, December 5. Home heaters and industrial furnaces were working in full force, pumping out smoke on the cold winter day. During the afternoon, visibility plummeted, and traffic came to a halt as drivers were blinded by the smoke and fog. Hundreds of cattle at a cattle market were the first to go. With lungs blackened by soot, they suffocated while standing in their pens. People could cover their faces and go indoors, but the soot soon reached inside buildings, as well. Concerts were cancelled because of blackened air in the halls, and books in the British Museum were tainted with soot. Visibility fell to a foot in some places by the third day of the inversion.

The ill and elderly, especially those with lung or heart ailments and heavy smokers, were the next to go. Hospitals filled with victims of bronchitis, pneumonia, lung inflammations, and heart failure. Like the cattle in the market, victims' lungs were clogged by smoke and microscopic soot particles, their lips turned blue, and they asphyxiated due to lack of oxygen. Healthy people tried to stay indoors and keep quiet, and children were kept home from school—so they would not get lost in the dark, as much as because of the air quality.

Four days later, a change in the weather brought fresh winds into London, and the inversion dissipated. Studies showed that at least 4,700 deaths were attributable to air pollution during and immediately after the inversion. More recent epidemiological studies have found that lingering ailments killed perhaps another 8,000 in the months that followed, bringing the total death toll to over 12,000.

Air pollution wasn't generally treated as a problem at the time. The weekend of December 5 just had worse than usual smoky fog. Everyone understood that coal smoke was unhealthy, of course, but grimy air and illness were a cost of living in the city. Controls on smoke had been proposed for centuries, since at least 1300, but pollution was too normal, and too pervasive, to change. We rarely consider normal conditions a problem, or imagine alternatives, until a crisis makes us start to question the costs of customary ways of doing business.

The smog of 1952 turned out to be one such crisis. Alarming death counts caught the attention of politicians and the public alike, and gradually led to changes in expectations and practices. New government policies gradually began to phase out coal fireplaces, replacing them with oil burners and other forms of heat. New efforts were made to monitor air quality and to put limits on industrial pollutants. These changes were solidified in the United Kingdom's Clean Air Act of 1956, which established health standards and helped homeowners convert to other heat sources. A decade later, the 1968 Clean Air Act expanded these rules to address industrial emissions. In the United States, a similar Clean Air Act was adopted in 1963, with major amendments in 1970 and again in 1990.

While air quality in cities (and often in the countryside) is frequently worse than we would like, extreme conditions like the killer smog are mainly historical curiosities today. We now have higher expectations for air pollution control, and we no longer find it acceptable—at least in principle—for private citizens or industries to emit pollutants that cause illness or death. Air quality standards exist for smoke, particulate matter, sulfuric acid, heavy metals, and other contaminants, all of which can now be captured before they leave the smokestack. New practices and rules keep our environment cleaner, reduce health care costs, and protect buildings, forests and farms from the effects of air pollution.

Ironically, coal burning remains one of our greatest challenges in air pollution control, though the source is now electricity-producing power plants. But there are many other sources. Every year millions of additional cars create more nitrogen oxides and carbon dioxide. Industry produces new classes of hazardous organic air pollutants, airborne metals are emitted from coal burning and mining, dust remains a serious problem, and asthma and cardiovascular conditions remain elevated in cities with poor air quality. But things are nowhere near as bad as they were in London in 1952. In this chapter we'll examine major types and sources of air pollutants. We'll also consider policies and technology that help ensure that events like the smog of London doesn't happen in your town. ■

> *If you think education is expensive, try ignorance.*
>
> —DEREK BOK

10.1 AIR POLLUTION AND HEALTH

Clean air is something we take for granted when we have it, but once a region's air pollution becomes severe, removing those pollutants presents an enormous challenge. Pollution comes in many forms. Smoke, haze, dust, odors, corrosive gases, noise, and toxic compounds are among our most widespread pollutants. Some pollutants, such as sulfur dioxide, irritate our eyes and lungs; fine particulates penetrate deep into our lungs; airborne metals enter our blood stream when we breathe them, then damage nerves and brain function.

Worldwide, these air pollution emissions add up to about 2 billion metric tons per year (Table 10.1). Sometimes the health and environmental costs of air pollution are shocking, as in London in 1952. But most of the time the costs are more subtle. Chronic illness, resulting from low-level, ongoing exposure is likely to cause more deaths and higher health-care costs than serious but infrequent deadly smog events. The Organization for Economic Cooperation and Development has projected that by 2050, chronic exposure to ground-level ozone, fine particulate matter, sulfur dioxide, and other pollutants will cause 3.6 million premature deaths every year.

Most of those deaths are unlikely to be in developed countries. Although we often don't realize it, air quality today is vastly better than it was a generation ago in most of the U.S. and other wealthy countries. We have cleaned up many of the worst pollution sources, especially those that are large, centralized, and easy to monitor. The Environmental Protection Agency (EPA) estimates that emissions have declined more than 1 million tons per year, since 1990, when regulation of the most hazardous materials began. Since the 1970s, the levels of major pollutants have decreased in the United States, despite population growth of more than 30 percent and economic growth of 1,500 percent.

Pollution reductions have resulted from better health standards for pollution, better enforcement of those standards, better monitoring, and efficiency and pollution-control technologies in factories,

▲ **FIGURE 10.1** While air quality is improving in many industrialized countries, newly developing countries have growing pollution problems. Xi'an, China, often has particulate levels above 300 μg/m³.

power plants, and vehicles. These improvements demonstrate the dramatic improvements that we can make when we set our minds to it. But continued public attention is always needed to protect the safeguards we depend on.

In the burgeoning megacities of rapidly industrializing countries, pollution controls are absent or poorly enforced, and air quality has been getting much worse. In many Chinese cities, for example, airborne dust, smoke, and soot often are ten times higher than levels considered safe for human health (fig. 10.1). Of the 20 smoggiest cities in the world, 16 are in China. China's city dwellers are four to six times more likely than rural people to die of lung cancer. Poorly regulated industrial cities of India, Russia, Pakistan, and many other countries cause similar hazards. Respiratory ailments, cardiovascular diseases, lung cancer, infant mortality, and miscarriages are as much as 50 percent higher than in countries with high pollution levels than in those with cleaner air.

TABLE 10.1 | Estimated Fluxes of Pollutants and Trace Gases to the Atmosphere

SPECIES	MAJOR SOURCES	APPROXIMATE ANNUAL FLUX (MILLIONS OF METRIC TONS/YR)	
		NATURAL	ANTHROPOGENIC
CO_2 (carbon dioxide)	Respiration, fossil fuel burning, land clearing, industry	370,000	29,600*
CH_4 (methane)	Rice paddies, wetlands, gas drilling, landfills, cattle, termites	155	350
CO (carbon monoxide)	Incomplete combustion, CH_4 oxidation, plant metabolism	1,580	930
Non-methane hydrocarbons	Fossil fuels, industrial uses, plant isoprenes, other biogenics	860	92
NO_x (nitrogen oxides)	Fossil fuel burning, lightning, biomass burning, soil microbes	90	140
SO_x (sulfur oxides)	Fossil fuel burning, industry, biomass burning, volcanoes, oceans	35	79
SPM (suspended particulate matter)	Fossil fuels, industry, mining, biomass burning, dust, sea salt	583	362

*Only 27.3 percent of this amount—or 8 billion tons—is carbon.
SOURCE: UNEP

Studies of air pollutants over southern Asia reveal that a 3 km (2 mi) thick cloud of ash, acids, aerosols, dust, and smog covers the entire Indian subcontinent for much of the year. Nobel laureate Paul Crutzen estimates that up to 2 million people in India alone die each year from atmospheric pollution. Produced by forest fires, the burning of agricultural wastes, and dramatic increases in the use of fossil fuels, the Asian smog layer cuts the amount of solar energy reaching the earth's surface beneath it by up to 15 percent. Meteorologists suggest that the cloud—80 percent of which is human-made—could disrupt monsoon weather patterns and cut rainfall over northern Pakistan, Afghanistan, western China, and central Asia by up to 40 percent.

When this "Asian Brown Cloud" drifts out over the Indian Ocean at the end of the monsoon season, it cools sea temperatures and may be changing regional climate patterns in the Pacific Ocean as well. This plume of soot and gases can travel half-way around the globe in a week, with unknown, but probably huge, impacts on the world's climate and environmental quality.

The Clean Air Act regulates major pollutants

Air pollution control has evolved gradually. The Clean Air Act of 1963 was the first national legislation in the United States aimed at air quality. The act provided federal grants to aid states in pollution control but was careful to preserve states' rights to set or enforce air quality regulations. It soon became obvious that piecemeal, local standards did not resolve the problem, because neither pollutants nor the markets for energy and industrial products are contained within state boundaries.

Amendments to the law in 1970 designated new standards, to be applied equally across the country, for six major pollutants:

sulfur dioxide, nitrogen oxides, carbon monoxide, ozone (and its precursor volatile organic compounds), lead, and particulate matter. These six are referred to as **conventional** or **criteria pollutants**, and they were addressed first because they contributed the largest volume of air quality degradation and also are considered the most serious threat to human health and welfare. Transportation and power plants are the dominant sources of most criteria pollutants (fig. 10.2). National ambient air quality standards (NAAQS) designated allowable levels for these pollutants in the **ambient air** (the air around us).

In addition to the six conventional pollutants, the Clean Air Act regulates an array of unconventional pollutants, compounds that are produced in less volume than conventional pollutants but that are especially toxic or hazardous, such as asbestos, benzene, mercury, polychlorinated biphenyls (PCBs), and vinyl chloride. Most of these are uncommon in nature or have no natural sources (fig. 10.3).

Many pollutants come from a **point source**, such as a smokestack. **Fugitive**, or **nonpoint-source**, **emissions** are those that do not go through a smokestack. Leaking valves and pipe joints contribute as much as 90 percent of the hydrocarbons and volatile organic chemicals emitted from oil refineries and chemical plants, and increasingly from natural gas wells. Dust, as from mining, agriculture, and building construction and demolition, is also considered fugitive emissions.

Primary pollutants are substances that are harmful when released. Secondary pollutants, by contrast, become harmful after they react with other gases or substances in the air. In particular, **photochemical oxidants** (compounds created by reactions driven by solar energy) and atmospheric acids are probably the most important secondary pollutants.

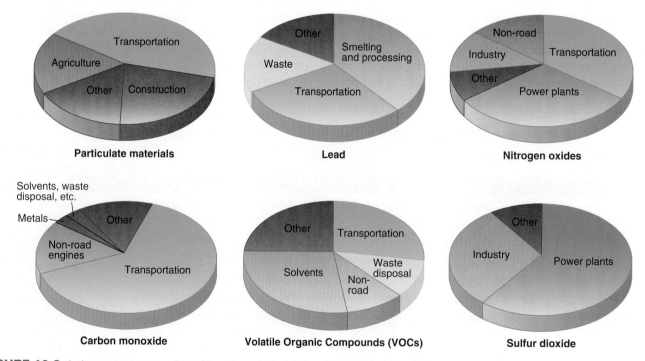

▲ **FIGURE 10.2** Anthropogenic sources of six of the primary "criteria" air pollutants in the United States. SOURCE: UNEP

◀ **FIGURE 10.3** Many of our most serious pollutants are fugitive emissions from petrochemical facilities such as this one in Baton Rouge, Louisiana.

Active LEARNING

Compare Sources of Pollutants

Getting a handle on the nature of air pollution can be difficult because there are many pollutants, all originating from a variety of sources. A good place to start is to remember the identity and sources of the 6 major "criteria" pollutants first targeted by the Clean Air Act. These pollutants aren't as dangerous in minute doses as mercury or other hazardous pollutants, but they are both serious and abundant.

Examine closely the pollutant sources shown in Fig. 10.2, to answer these questions:

1. Close your eyes and list the 6 pollutants shown.

2. Volatile organic compounds (VOCs) are often evaluated together with ozone (O_3), because together these contribute to a variety of photochemical oxidants and other pollutants. In this context, what does "volatile" mean? What does "organic compound" mean?

3. Examine the chart for sulfur dioxide. What is its single largest source? What is another pollutant produced in abundance from that source?

4. Which pollutants have transportation as a major contributor? How many of the six are these? As you read this chapter, what are some of our strategies for minimizing these pollutants?

5. What is the main source of nitrogen in nitrogen oxides from your car engine? (Examine your text and figures for the answer.)

ANSWERS: 1. particulate material, carbon monoxide, lead, volatile organic compounds, nitrogen oxides, sulfur dioxide; 2. "volatile" means easily evaporated, and organic compound means a base of carbon atoms or rings; 3. industry; 4. five of the six pollutants: all but sulfur dioxide—pollution control strategies include better combustion, cleaner fuels, and other efforts; 5. atmospheric nitrogen, which binds to atmospheric oxygen in the heat of fuel combustion.

Conventional pollutants are abundant and serious

Most conventional pollutants are produced primarily by burning fossil fuels, especially in coal-powered electric plants and in cars and trucks, as well as in processing natural gas and oil. Others, especially sulfur and metals, are by-products of mining and manufacturing processes. Of the 188 air toxics listed in the Clean Air Act, about two-thirds are volatile organic compounds, and most of the rest are metal compounds. In this section we will discuss the characteristics and origin of the major pollutants.

Sulfur dioxide (SO_2) is a colorless, corrosive gas that damages both plants and animals. Once in the atmosphere, it can be further oxidized to sulfur trioxide (SO_3), which reacts with water vapor or dissolves in water droplets to form sulfuric acid (H_2SO_4), a major component of acid rain. Sulfur dioxide and sulfate ions are probably second only to smoking as causes of air pollution-related health damage. Sulfate particles and droplets also reduce visibility in the United States by as much as 80 percent. Minute droplets of sulfuric acid can penetrate deep into lungs, causing permanent damage, as well as irritating eyes and corroding buildings. In plants, sulfur dioxide destroys chlorophyll, eventually killing tissues. Sulfur dioxide from coal smoke, along with particulate matter, was a principal component of London's deadly smog event of 1952 (fig. 10.4).

Nitrogen oxides (NO_x) are highly reactive gases formed when the heat of combustion initiates reactions between atmospheric nitrogen (N_2) and oxygen (O_2). The initial product, nitric oxide (NO), oxidizes further in the atmosphere to nitrogen dioxide (NO_2), a reddish-brown gas that gives photochemical smog its distinctive color. Because these gases convert readily from one form to the other, the general term NO_x (with x indicating an unspecified number) is used to describe these gases. Nitrogen oxides combine with water to form nitric acid (HNO_3), which is also a major component of acid precipitation (fig. 10.5). Excess nitrogen in water is also causing eutrophication of inland waters and coastal seas.

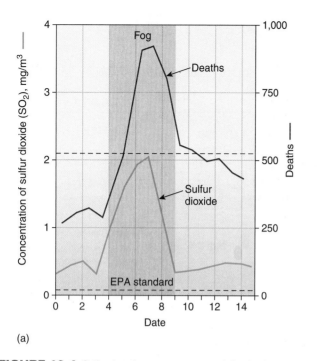

(a)

(b)

▲ **FIGURE 10.4** Sulfur dioxide concentrations and deaths during the London smog of December 1952. The EPA standard limit is 0.08 mg/m³ (dashed line, a). The soybean leaf at right (b) was exposed to 2.1 mg/m³ sulfur dioxide for 24 hours. White patches show where chlorophyll has been destroyed.

Carbon monoxide (CO) is less common but more dangerous than the principal form of atmospheric carbon, carbon dioxide (CO_2). CO is a colorless, odorless, but highly toxic gas produced mainly by incomplete combustion of fuel (coal, oil, charcoal, wood, or gas). CO inhibits respiration in animals by binding irreversibly to hemoglobin in blood. In the United States, two-thirds of the CO emissions are created by internal combustion engines in transportation. Land-clearing fires and cooking fires also are major sources. About 90 percent of the CO in the air is consumed in photochemical reactions that produce ozone.

Ozone (O_3) is important in the upper atmosphere, where it shields us against ultraviolet radiation from the sun (see chapter 9), but at the ground level ozone is a highly reactive oxidizing agent

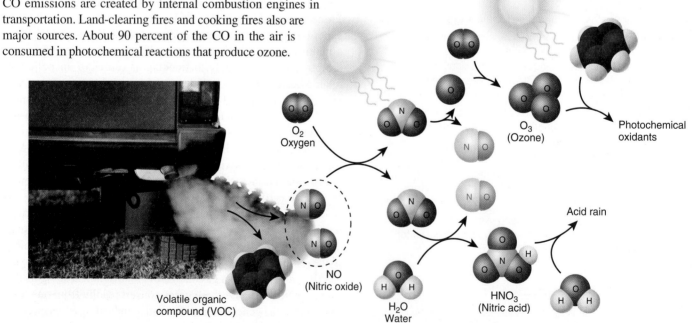

▲ **FIGURE 10.5** The heat of fuel combustion causes nitrogen oxides to form from atmospheric N_2 and O_2. NO_2 interacts with water (H_2O) to form HNO_3, a component of acid rain. In addition, solar radiation can force NO_2 to release a free oxygen atom, which joins to atmospheric O_2, creating ozone (O_3). Fuel combustion also produces incompletely burned hydrocarbons (including volatile organic compounds). Both O_3 and VOCs contribute to photochemical oxidants, in reactions activated by sunlight. The VOC shown here is benzene, a ring of 6 carbon atoms with a hydrogen atom attached to each carbon.

that damages eyes, lungs, and plant tissues, as well as paint, rubber, and plastics. Ground-level ozone is a secondary pollutant, created by chemical reactions that are initiated by solar energy (table 10.2). In general, pollutants created by these light-initiated reactions are known as photochemical oxidants (that is, oxidizing agents that irritate tissues and damage materials). One of the most important of these reactions involves formation of single atoms of oxygen by splitting nitrogen dioxide (NO_2). This atomic oxygen then binds to a molecule of O_2 to make ozone (O_3). The acrid, biting odor of ozone is a distinctive characteristic of photochemical smog.

A variety of **volatile organic compounds (VOCs)** interact with ozone to produce photochemical oxidants in smog. A wide array of these organic (carbon-based), volatile (easily evaporated) chemicals derive from industrial processes such as refining of oil and gas, or plastics and chemical manufacturing. Some of the most dangerous and common of these industrial VOCs are benzene, toluene, formaldehyde, vinyl chloride, phenols, chloroform, and trichloroethylene. Principal sources are incompletely burned fuels from vehicles, power plants, chemical plants, and petroleum refineries.

Lead, our most abundantly produced metal air pollutant, impairs nerve and brain functions. It does this by binding to essential enzymes and cellular components and disabling them. A wide range of industrial and mining processes produce lead, especially smelting of metal ores, mining, and burning of coal and municipal waste, in which lead is a trace element, and burning of gasoline to which lead has been added. Historically, leaded gasoline was the main source of lead in the United States, but leaded gas was phased out in the 1980s. Since 1986, when the ban was enforced, children's average blood lead levels have dropped 90 percent and average IQs have risen three points. Banning leaded gasoline in the United States was one of the most successful pollution-control measures in American history. Now, 50 nations have renounced leaded gasoline. The global economic benefit of this step is estimated to be more than $200 billion per year.

Particulate material includes dust, ash, soot, lint, smoke, pollen, spores, algal cells, and many other suspended materials. **Aerosols**, or extremely minute particles or liquid droplets suspended in the air, are included in this class. Particulates often are the most apparent form of air pollution, since they reduce visibility and leave dirty deposits on windows, painted surfaces, and textiles. Breathable particles smaller than 2.5 micrometers are among the most dangerous of this group because they can damage lung tissues. Asbestos fibers and cigarette smoke are among the most dangerous respirable particles in urban and indoor air because they are carcinogenic.

TABLE 10.2 | Photochemical Oxidant Production

STEPS	PHOTOCHEMICAL PRODUCTS
1. NO + VOC →	NO_2 (nitrogen dioxide)
2. NO_2 + UV sunlight →	NO + O (nitric oxide + atomic oxygen)
3. O + O_2 →	O_3 (ozone)
4. NO_2 + VOC →	PAN (peroxyacetyl nitrate)

Hazardous air pollutants can cause cancer and nerve damage

A special category of toxins is monitored by the U.S. EPA because they are particularly dangerous even in low concentrations. Called **hazardous air pollutants (HAPs)**, these chemicals cause cancer, nerve damage, disrupt hormone function, and fetal development. These persistent substances remain in ecosystems for long periods of time, and accumulate in animal and human tissues. Most of these chemicals are either metal compounds, chlorinated hydrocarbons, or volatile organic compounds. Gasoline vapors, solvents, and components of plastics are all HAPs that you may encounter on a daily basis.

Many HAPs are emitted by chemical-processing factories that produce gasoline, plastics, solvents, pharmaceuticals, and other organic compounds. Benzene, toluene, xylene, and other volatile organic compounds are among these. Dioxins, carbon-based compounds containing chlorine, are released mainly by burning plastics and medical waste containing chlorine. The EPA reports that 100 million Americans (one third of us) live in areas where the cancer rate from HAPs is ten times the normally accepted standard for action (1 in 1 million). Benzene, formaldehyde, acetaldehyde, and 1,3 butadiene are responsible for most of this HAP cancer risk. To help the public track local air quality levels, the EPA recently estimated the concentration of HAPs in localities across the continental United States. You can check pollutant levels and types in your own community by looking online for the Environmental Defense Fund HAP scorecard web page (www.scorecard.org/env-releases/hap).

To help inform communities about toxic substances produced and handled in their area, Congress established the **Toxic Release Inventory (TRI)** in 1986. This inventory collects self-reported statistics from 23,000 factories, refineries, hard rock mines, power plants, and chemical manufacturers to report on toxin releases (above certain minimum amounts) and waste management methods for 667 toxic chemicals. Although this total is less than 1 percent of all chemicals registered for use, and represents a limited range of sources, the TRI is widely considered the most comprehensive source of information about toxic pollution in the United States (fig. 10.6).

Mercury is a key neurotoxin

Airborne metals originate mainly from combustion of fuel, especially coal, which contains traces of mercury, arsenic, cadmium, and other metals, as well as sulfur and other trace elements. Airborne mercury has received special attention because it is a widespread and persistent neurotoxin (a substance that damages the brain and nervous system). Minute doses can cause nerve damage and other impairments, especially in young children and developing fetuses. Some 70 percent of airborne mercury is released by coal-burning power plants. Metal ore smelting and waste combustion also produce airborne mercury and other metals.

About 75 percent of human exposure to mercury comes from eating fish. This is because aquatic bacteria are mainly responsible for converting airborne mercury into methyl mercury, a form that accumulates in living animal tissues. Once methyl mercury enters

▲ **FIGURE 10.6** A variety of hazardous emissions from some 23,000 facilities in the United States are monitored by the Toxic Release Inventory (TRI).

The U.S. National Institutes of Health (NIH) estimates that between 300,000 and 600,000 of the 4 million children born each year in the United States are exposed in the womb to mercury levels that could cause diminished intelligence or developmental impairments. According to the NIH, elevated mercury levels cost the U.S. economy $8.7 billion each year in higher medical and educational costs and in lost workforce productivity.

Mercury became fully regulated by the Clean Air Act in 2000, after decades of debate (see What Do You Think, p. 237). Since then emissions have declined in many areas, as the metal is captured before it leaves the smokestack, but globally mercury is still a growing problem.

Indoor air can be worse than outdoor air

We have spent a considerable amount of effort and money to control the major outdoor air pollutants, but we have only recently become aware of the dangers of indoor air pollutants. The U.S. EPA has found that indoor concentrations of toxic air pollutants are often higher than outdoors. Because people generally spend more time inside than out, they are exposed to higher doses of these pollutants. In some cases, indoor air in homes has chemical concentrations that would be illegal outside or in the workplace. Under some circumstances, compounds such as chloroform, benzene, carbon tetrachloride, formaldehyde, and styrene can be 70 times higher in indoor air than in outdoor air. Molds, pathogens, and other biohazards also represent serious indoor pollutants.

the food web, it bioaccumulates in the flesh and blood stream of predators. As a consequence, large, long-lived, predatory fish contain especially high levels of mercury in their tissues. Contaminated tuna fish alone is responsible for about 40 percent of all U.S. exposure to mercury (fig. 10.7). Swordfish, shrimp, and other seafood are also important mercury sources in our diet.

▲ **FIGURE 10.7** Canned tuna is our dominant source of mercury, because it is a top predator fish that we eat in abundance.

A 2009 report by the United States Geological Survey found that mercury levels in Pacific Ocean tuna have risen 30 percent in the past 20 years, with another 50 percent rise projected by 2050. Increased coal burning in China, which is building two new coal-burning power plants every week, is understood to be the main cause of growing mercury emissions in the Pacific. Similarly, U.S. coal plants produce mercury that is deposited across North America, in the Atlantic, and across Europe. Long-range transport of mercury through the air is even causing bioaccumulation in aquatic ecosystems in remote, high-Arctic areas. There, mercury poisoning can be a serious risk for people and wildlife in whose food chain is based on fish.

Freshwater fish also carry risks. Mercury contamination is the most common cause of impairment of U.S. rivers and lakes, and 45 states have issued warnings against frequent consumption of fresh-caught fish. A 2007 study tested more than 2,700 fish from 636 rivers and streams in 12 western states, and mercury was found in every one of them.

Cigarette smoke is without doubt the most important air contaminant in developed countries in terms of human health. The U.S. surgeon general has estimated that 400,000 people die each year in the United States from emphysema, heart attacks, strokes, lung cancer, or other diseases caused by smoking. These diseases are responsible for 20 percent of all mortality in the United States, or four times as much as infectious agents. Total costs for early deaths and smoking-related illnesses are estimated to be $100 billion per year. Eliminating smoking probably would save more lives than any other pollution-control measure.

In the less-developed countries of Africa, Asia, and Latin America, where such organic fuels as firewood, charcoal, dried dung, and agricultural wastes make up the majority of household energy, smoky, poorly ventilated heating and cooking fires are the worst sources of indoor air pollution (fig. 10.8). The World Health Organization (WHO) estimates that 2.5 billion people—more than one-third of the world's population—are adversely affected by pollution from this source. In particular, this affects women and small children, who spend long hours each day around open fires or unventilated stoves in enclosed spaces.

Cap and Trade for Mercury Pollution?

Often referred to as quicksilver, mercury is used in a host of products including paints, batteries, fluorescent light bulbs, electrical switches, pesticides, skin creams, antifungal agents, and old thermometers. Mercury also is a powerful neurotoxin that destroys the brain and central nervous system at high doses. Minute amounts can cause nerve damage and developmental defects in children. Mercury exposure results mainly from burning garbage, coal, or other mercury-laden materials. It falls to the ground and washes into lakes and wetlands, where it enters the food web. In a survey of freshwater fish from 260 lakes across the United States, the EPA found that every fish sampled contained some level of mercury.

In 1994, the EPA declared mercury a hazardous pollutant regulated under the Clean Air Act. Municipal and medical incinerators were required to reduce their mercury emissions by 90 percent. Industrial and mining operations also agreed to cut emissions. However, the law did not address the 1,032 coal-burning power plants, which emit nearly half of total annual U.S. emissions, some 48 tons per year.

Finally, in 2000, the EPA declared mercury from power plants, like that from other sources, a public health risk. The agency could have applied existing air-toxin regulations and reduced power plant emissions 90 percent in 5 years with existing control technology. But the EPA in 2000 opted instead for a "cap and trade" market mechanism, which should reduce mercury releases 70 percent in about 30 years.

Cap-and-trade approaches set limits (caps) and allow utilities to buy and sell unused pollution credits. This strategy is widely supported because it uses a profit motive rather than rules, and it allows industries to make their own decisions about emission controls. It also allows continued emissions if credits are cheaper than emission controls, and traders have the opportunity to make money on the exchanges.

On the other hand, public health advocates argue that while cap-and-trade systems work well for some pollutants, they are inappropriate for a substance that is toxic at very low levels. Health advocates also object that utilities are allowed to continue emitting mercury for years longer than necessary. Many eastern states are especially concerned because they suffer from high mercury pollution generated in the Midwest and blown east by prevailing winds (fig. 1).

Meanwhile, in the Allegheny Mountains of West Virginia, a huge coal-fired power plant is adding fuel to the mercury debate. The enormous 1,600-megawatt Mount Storm plant ranked second in the nation in mercury emissions just a few years ago. When Mount Storm installed new controls to capture sulfur and nitrogen oxides from its stack, this equipment also caught 95 percent of its mercury emissions, at no extra cost. This is excellent news, but it also raises a policy question: If existing technology can cut mercury economically, why wait 30 years to impose similarly cost-effective limits on other power plants?

This case illustrates the complexity of regulating air pollution. Highly mobile, widely dispersed, produced by a variety of sources, and having diverse impacts, air pollutants can be challenging to regulate. Often air quality controversies—such as mercury control—pit a diffuse public interest (improving general health levels or child development) against a very specific private interest (utilities which must pay millions of dollars per year to control pollutants). How would you set the rules if you were in charge? Would you impose rules or allow for trading of mercury emission permits? Why? How would you negotiate the responsibility for controlling pollutants?

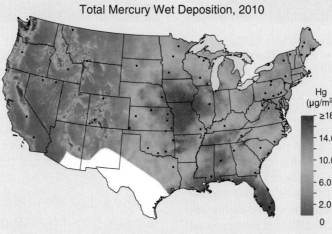

Total Mercury Wet Deposition, 2010

▲ **FIGURE 1** Atmospheric deposition of mercury with precipitation (wet deposition), 2010. Values derive from point sampling and national precipitation maps.   SOURCE: National Atmospheric Deposition Program/Mercury Deposition Network (http://nadp.isws.illinois.edu), 2012.

10.2 AIR POLLUTION AND THE CLIMATE

Physical processes in the atmosphere transport, concentrate, and disperse air pollutants. Global warming, in which pollutants are altering the earth's energy budget, is the best-known case of interaction between anthropogenic pollutants and the atmosphere. But there are many other important interactions. In this section we survey other important climate-pollution interactions.

Air pollutants travel the globe

Dust and fine aerosols can be carried great distances by the wind. Pollution from the industrial belt between the Great Lakes and the Ohio River Valley regularly contaminates the Canadian Maritime Provinces and sometimes can be traced as far as Ireland. Similarly, dust storms from China's Gobi and Takla Makan deserts routinely close schools, factories, and airports in Japan and Korea, and often reach western North America. In one particularly

▲ **FIGURE 10.8** Indoor air pollution affects some 2.5 billion people, mainly women and children, who spend days in poorly ventilated kitchens where carbon monoxide, particulates, and cancer-causing hydrocarbons often reach dangerous levels.

▲ **FIGURE 10.9** A massive dust storm extends more than 1,600 km from the coast of Western Sahara and Morocco. Long-distance transport of dust and pollutants is an important source of contaminants worldwide.

severe dust storm in 1998, chemical analysis showed that 75 percent of the particulate pollution in Seattle, Washington, air came from China. Similarly, dust from North Africa regularly crosses the Atlantic and contaminates the air in Florida and the Caribbean Islands (fig. 10.9). This dust can carry pathogens and is thought to be the source of diseases attacking Caribbean corals. Soil scientists estimate that 3 billion tons of sand and dust are blown around the world every year.

Increasingly sensitive monitoring equipment has begun to reveal industrial contaminants in places usually considered among the cleanest in the world. Samoa, Greenland, and even Antarctica and the North Pole all have heavy metals, pesticides, and radioactive elements in their air. Since the 1950s, pilots flying in the high Arctic have reported dense layers of reddish-brown haze clouding the arctic atmosphere. Aerosols of sulfates, soot, dust, and toxic heavy metals, such as vanadium, manganese, and lead, travel to the pole from the industrialized parts of Europe and Russia.

Circulation of the atmosphere tends to transport contaminants toward the poles. Like mercury (discussed above), volatile compounds (VOCs) evaporate from warm areas, travel through the atmosphere, then condense and precipitate in cooler regions. Over several years, contaminants migrate to the coldest places, generally at high latitudes, where they bioaccumulate in food chains. Whales, polar bears, sharks, and other top carnivores in polar regions have been shown to have dangerously high levels of pesticides, metals, and other hazardous air pollutants in their bodies. The Inuit people of Broughton Island, well above the Arctic Circle, have higher levels of polychlorinated biphenyls (PCBs) in their blood than any other known population except

victims of industrial accidents. Far from any source of this industrial by-product, these people accumulate PCBs from the flesh of fish, caribou, and other animals they eat.

Carbon dioxide and halogens are key greenhouse gases

Some 370 billion tons of CO_2 are emitted each year is from respiration (oxidation of organic compounds by plant and animal cells). These releases are usually balanced by an equal uptake by photosynthesis in green plants. At normal concentrations, CO_2 is nontoxic and innocuous, but steadily increasing atmospheric levels (about 0.5 percent per year) due to human activities are now causing global climate change, with serious implications for both human and natural communities (chapter 9).

Regulating CO_2 has been a subject of intense debate since the 1990s. On the one hand, policy makers have widely acknowledged that climate change is likely to have disastrous effects. On the other hand, CO_2 is difficult to consider limiting because we produce abundant quantities, because reductions involve changes to both technology and behavior, and because its production historically has been closely tied to our economic productivity. Although future economic growth is likely to depend on efficiencies and new technologies, these concerns remain an important part of the debate.

In recent years, many members of Congress have been intent on eliminating this and other pollution regulation, arguing that it is too costly for industry and the economy (see further discussion of this below). Energy companies and their representatives, in particular, have lobbied to prevent legal limits on greenhouse gases. In its 2011 budget, Congress proposed to slash EPA funding by one-third, in part to reduce pollution monitoring and regulation.

The question of whether the EPA should regulate greenhouse gases was so contentious that it went to the Supreme Court in 2007. The Court ruled that it was the EPA's responsibility to limit these gases, on the grounds that greenhouse gases endanger public health and welfare within the meaning of the Clean Air Act. The Court, and subsequent EPA documents, noted that these risks include increased drought, more frequent and intense heat waves and wildfires, sea level rise, and harm to water resources, agriculture, wildlife and ecosystems. In addition to these risks, the U.S. military has cited climate change as a growing security threat. A coalition of generals and admirals signed a report from the Center for Naval Analyses stating that climate change "presents significant national security challenges" including violence resulting from scarcity of water, and migration from sea level rise and crop failure.

Since the Supreme Court ruling, the EPA is charged with regulating six greenhouse gases: carbon dioxide, methane, nitrous oxide, hydrofluorocarbons, perfluorocarbons and sulfur hexafluoride. These are gases whose emissions have grown dramatically in recent decades.

Three of the six gases noted above contain halogens, a group of lightweight, highly reactive elements (fluorine, chlorine, bromine, and iodine). These gases are far more potent greenhouse gases per molecule than CO_2 (fig. 10.10) Because they are generally toxic in their elemental form, they are commonly used as fumigants and disinfectants, but they also have hundreds of uses in industrial and commercial products. Chlorofluorocarbons (CFCs) have been banned for most uses in industrialized countries, but about 600 million tons of these compounds are used annually worldwide in spray propellants, refrigeration compressors, and for foam blowing. They diffuse into the stratosphere where they release chlorine and fluorine atoms that destroy the ozone shield that protects the earth from ultraviolet radiation.

How do we reduce emissions of greenhouse gases? Reducing fuel use through conservation and alternative energy are a first step. Changing subsidy systems that support coal burning is another. A cap-and-trade system, involving a market for trading in emission rights, or "credits," has been the most popular strategy (see section 10.4). It remains unclear how much this approach has actually reduced greenhouse gas emissions, however.

CFCs also destroy ozone in the stratosphere

Long-range pollution transport and the chemical reactions of atmospheric gases and pollution produce the phenomenon known as the ozone "hole," a thinning of ozone concentrations in the stratosphere (fig. 10.11). This phenomenon was discovered in 1985 but has probably been developing since at least the 1960s. Chlorine-based aerosols, such as chlorofluorocarbons (CFCs) and hydrochlorofluorocarbons (HCFCs), are the principal agents of ozone depletion. Nontoxic, nonflammable, chemically inert, longlasting, and cheaply produced, these compounds were extremely useful as industrial gases and in refrigerators, air conditioners, Styrofoam insulation, and aerosol spray cans for many years. From the 1930s until the 1980s, CFCs were used all over the world and widely dispersed through the atmosphere.

Ozone (O_3) is a pollutant near the ground because it irritates skin and plant tissues, but in the stratosphere ozone is valuable. The O_3 molecule is especially effective at absorbing ultraviolet (UV) radiation as it enters the atmosphere from space. UV radiation damages plant and animal cells, potentially causing mutations that

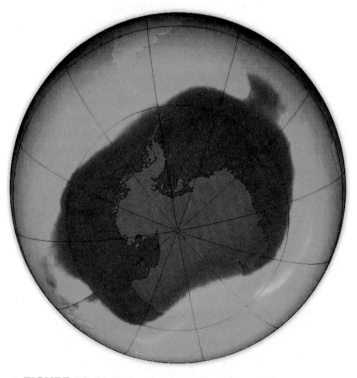

▲ FIGURE 10.11 Reduced concentrations of stratospheric ozone, which occur in the southern springtime, are shown by purple shading. This 2006 ozone "hole" was the largest recorded, although CFC production has been declining since 1990.

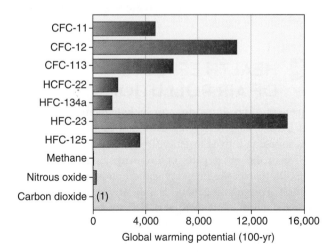

▲ FIGURE 10.10 The Montreal Protocol has contributed to reducing greenhouse gases, as well as preserving stratospheric ozone, because CFCs have high global warming potential and longevity compared to CO_2.

produce cancer. A 1 percent loss of ozone could result in about a million extra human skin cancers per year worldwide. Excessive UV exposure could reduce agricultural production and disrupt ecosystems. Scientists worry, for example, that high UV levels in Antarctica could reduce populations of plankton, the tiny floating organisms that form the base of a food chain that includes fish, seals, penguins, and whales in Antarctic seas.

Antarctica's exceptionally cold winter temperatures (−85° to −90°C) help break down ozone. During the long, dark, winter months, strong winds known as the circumpolar vortex circle the pole. These winds isolate Antarctic air and allow stratospheric temperatures to drop low enough to create ice crystals at high altitudes—something that rarely happens elsewhere in the world. Ozone and chlorine-containing molecules are absorbed on the surfaces of these ice particles. When the sun returns in the spring, it provides energy to liberate chlorine ions, which readily bond with ozone, breaking it down to molecular oxygen (table 10.3). It is only during the Antarctic spring (September through December) that conditions are ideal for rapid ozone destruction. During that season, temperatures are still cold enough for high-altitude ice crystals, but the sun gradually becomes strong enough to drive photochemical reactions.

As the Antarctic summer arrives, temperatures warm slightly. The circumpolar vortex weakens, and air from warmer latitudes mixes with Antarctic air, replenishing ozone concentrations in the ozone hole. Slight decreases worldwide result from this mixing, however. Ozone re-forms naturally, but not nearly as fast as it is destroyed. Since the chlorine atoms are not themselves consumed in reactions with ozone, they continue to destroy ozone for years, until they finally precipitate or are washed out of the air. In 2000 the region of ozone depletion covered 29.8 million km^2 (about the size of North America).

CFC control has had remarkable success

The discovery of stratospheric ozone losses brought about a remarkably quick international response. In 1987 an international meeting in Montreal, Canada, produced the Montreal Protocol, the first of several major international agreements on phasing out most use of CFCs by 2000. As evidence accumulated, showing that losses were larger and more widespread than previously thought, the deadline for the elimination of all CFCs (halons, carbon tetrachloride, and methyl chloroform) was moved up to 1996, and a $500 million fund was established to assist poorer countries

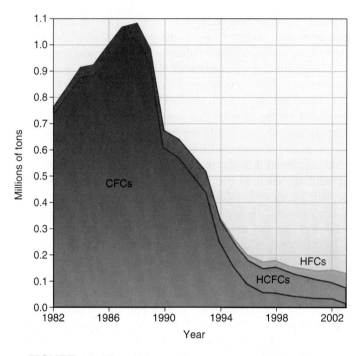

▲ **FIGURE 10.12** The Montreal Protocol has been remarkably successful in eliminating CFC production. The remaining HFC and HCFC use is primarily in recently industrialized countries, such as China and India.

in switching to non-CFC technologies. Fortunately, alternatives to CFCs for most uses already exist. The first substitutes are hydrochlorofluorocarbons (HCFCs), which release much less chlorine per molecule. Scientists hope to eventually develop halogen-free molecules that work as well and are no more expensive than CFCs.

There is evidence that the CFC ban is already having an effect. CFC production in most industrialized countries has fallen sharply since 1988 (fig. 10.12), and CFCs are now being removed from the atmosphere more rapidly than they are being added. In 50 years or so, stratospheric ozone levels are expected to be back to normal.

The Montreal Protocol has also helped reduce climate change, because CFCs and related compounds are such powerful and long-lasting greenhouse gases.

10.3 HEALTH EFFECTS OF AIR POLLUTION

Consequences of breathing dirty air include increased probability of heart attacks, respiratory diseases, and lung cancer. This can mean as much as a five- to ten-year decrease in life expectancy if you live in the worst parts of Los Angeles or Baltimore, for example. Of course, the intensity and duration of your exposure, as well as your age and general health, are extremely important: you are much more likely to be at risk if you are very young, very old, or already suffering from some respiratory or cardiovascular disease. Bronchitis and emphysema are common chronic conditions resulting from air pollution. The U.S. Office of Technology Assessment estimates that 250,000 people suffer from pollution-related bronchitis and emphysema in the United States, and some

TABLE 10.3	Stratospheric Ozone Destruction by Chlorine Atoms and UV Radiation	
STEPS		**PRODUCT**
1. CFCl$_3$ (chlorofluorocarbon) + UV energy		CFCl$_2$ + Cl
2. Cl + O$_3$		ClO + O$_2$
3. O$_2$ + UV energy		2O
4. ClO + 2O		O$_2$ + Cl
5. Return to step 2		

50,000 excess deaths each year are attributable to complications of these diseases, which are probably second only to heart attack as a cause of death.

How does air pollution cause these health effects? Because they are strong oxidizing agents, sulfates, SO_2, NO_x, and O_3 irritate and damage delicate tissues in the eyes and lungs. Fine, suspended particulate materials penetrate deep into the lungs, causing irritation, scarring, and even tumor growth. Heart stress results from impaired lung functions. Carbon monoxide binds to hemoglobin, reducing oxygen flow to the brain. Headaches, dizziness, and heart stress result. Lead also binds to hemoglobin, damaging critical neurons in the brain and resulting in mental and physical impairment and developmental retardation.

Environmental effects of air pollution can involve similarly complex, and often gradually-accumulating, effects.

Acid deposition results from SO_4 and NO_x

Deposition of acidic droplets or particles, from rain, fog, snow, or aerosols in the atmosphere, became recognized as a widespread pollution problem only since the 1980s. But the effects of air pollution on plants, especially sulfuric acid deposition, have been known in industrial areas since at least the 1850s. In the early days of industrialization, fumes from furnaces, smelters, refineries, and chemical plants destroyed vegetation and created desolate, barren landscapes around mining and manufacturing centers. The copper-nickel smelter at Sudbury, Ontario, is a spectacular example. Starting in 1886, open-bed roasting was used to purify sulfide ores of nickel and copper. The resulting sulfur dioxide and sulfuric acid destroyed nearly all plant life within about 30 km of the smelter. Rains washed away the exposed soil, leaving a barren moonscape of blackened bedrock. This pattern has been widespread in mining and smelting regions around the world.

Pollutant levels too low to produce visible symptoms of damage may still have important effects. Field studies show that yields in some crops, such as soybeans, may be reduced as much as 50 percent by currently existing levels of oxidants in ambient air. Some plant pathologists suggest that ozone and photochemical oxidants are responsible for as much as 90 percent of agricultural, ornamental and forest losses from air pollution. The total costs of this damage may be as much as $10 billion per year in North America alone.

Acidic deposition is now understood to affect forests and croplands far from industrial centers. Rain is normally slightly acidic (pH 5.6: see chapter 2), owing to reactions of CO_2 and rainwater, which produce a mild carbonic acid. Industrial emissions of sulfur dioxide (SO_2), sulfate (SO_4), and nitrogen oxides (NO_x) can acidify rain, fog, snow, and mist to pH 4 or lower. Ongoing exposure to acid fog, snow, mist, and dew cause permanent damage to plants, lake ecosystems, and buildings. Acidity causes forest decline partly by damaging leaf tissues and weakening seedlings. Acidity also reduces nutrient availability in forest soils, and it mobilizes toxic concentrations of metals in soils,

especially aluminum. Weakened trees become susceptible other stressors such as diseases and insect pests.

Lakes in Scandinavia were among the first aquatic ecosystems discovered to be damaged by acid precipitation. Prevailing winds from Germany, Poland, and other parts of Europe deliver acids generated by industrial and automobile emissions—principally H_2SO_4 and HNO_3. The thin, acidic soils and nutrient-poor lakes and streams in the mountains of southern Norway and Sweden have been severely affected by this acid deposition. Most noticeable is the reduction of trout, salmon, and other game fish, whose eggs and young die below pH 5. Aquatic plants, insects, and invertebrates also suffer. Many lakes in Sweden are now so acidic that they will no longer support game fish or other sensitive aquatic organisms. Large parts of Europe and eastern North America have also been damaged by acid precipitation.

High-elevation forests are most severely affected. Mountain tops often have thin, often acidic soils under normal conditions, with little ability to neutralize acidic rain, snow, and mist. On Mount Mitchell in North Carolina, nearly all the trees above 2,000 m (6,000 ft) have lost needles, and about half are dead (fig. 10.13). Damage has been reported throughout Europe, from the Netherlands to Switzerland, as well as in China and the states of the former Soviet Union. In 1985 West German foresters estimated that about half the total forest area in West Germany (more than 4 million ha) was declining. The loss to the forest industry is estimated to be about 1 billion Euros per year.

A vigorous program of pollution control has been undertaken by Canada, the United States, and several European countries since the widespread recognition of acid rain. SO_2 and NO_x emissions from power plants have decreased dramatically over the past three decades over much of Europe and eastern North America as a result of pollution-control measures. However, rain falling in these areas remains acidic (fig. 10.14).

▲ **FIGURE 10.13** A Fraser fir forest on Mount Mitchell, North Carolina, killed by acid rain, insect pests, and other stressors.

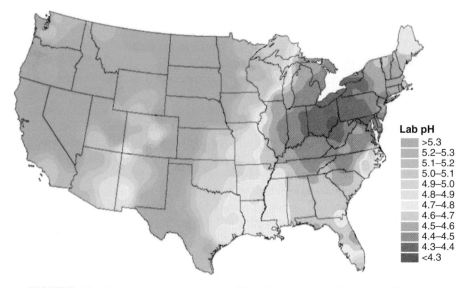

▲ **FIGURE 10.14** Acid precipitation over the United States in 2000. Many areas have improved markedly since then, because of the new rules on sulfur dioxide emissions and other pollutants.   SOURCE: National Acid Depositions Program, 2001

Lab pH

	>5.3
	5.2–5.3
	5.1–5.2
	5.0–5.1
	4.9–5.0
	4.8–4.9
	4.7–4.8
	4.6–4.7
	4.5–4.6
	4.4–4.5
	4.3–4.4
	<4.3

▶ **FIGURE 10.15**
Atmospheric acids, especially sulfuric and nitric acids, have almost eaten away the face of this medieval statue. Annual losses from air pollution damage to buildings and materials amount to billions of dollars.

In cities throughout the world, air pollution is destroying some of the oldest and most glorious buildings and works of art. Smoke and soot coat buildings, paintings, and textiles. Acids dissolve limestone and marble, destroying features and structures of historic buildings (fig. 10.15). The Parthenon in Athens, the Taj Mahal in Agra, the Coliseum in Rome, medieval cathedrals in Europe, and the Washington Monument in Washington, D.C., are slowly dissolving and flaking away because of acidic fumes in the air. Acid deposition also speeds corrosion of steel in reinforced concrete, weakening buildings, roads, and bridges. Limestone, marble, and some kinds of sandstone flake and crumble. The Council on Environmental Quality estimates that U.S. economic losses from architectural and property damage from air pollution amount to about $10 billion every year.

Urban areas endure inversions and heat islands

In urban areas, pollution is most extreme when temperature inversions develop, concentrating dangerous levels of pollutants within cities (as in the opening case study). A temperature inversion is a

situation in which stable, cold air rests near the ground, with warm layers above (fig. 10.16). This situation reverses the normal conditions: usually air is warmed by heat re-emitted from the ground surface, and it cools with elevation above the earth's surface. Warming air rises, mixing the atmospheric layers and helping pollution to disperse from its sources. When cool, dense air lies below a warmer, lighter layer, air remains stable and still, and pollutants accumulate near the ground, where they irritate our lungs and eyes. Stable inversion conditions are usually created by rapid nighttime cooling, especially in a valley, where air movement is restricted.

Los Angeles has ideal conditions for inversions. Mountains surround the city on three sides, reducing wind movement; heavy traffic and industry create a supply of pollutants; skies are generally clear at night, allowing rapid radiant heat loss, and the ground cools quickly. Surface air layers are cooled by contact with the cool ground surface, while upper layers remain relatively warm, and an inversion results. As long as the atmosphere is still and stable, pollutants accumulate near the ground where they are produced and where we breathe them.

Abundant sunlight in Los Angeles initiates photochemical oxidation in the concentrated aerosols and gaseous chemicals in the inversion layer. A brown haze of ozone and nitrogen dioxide,

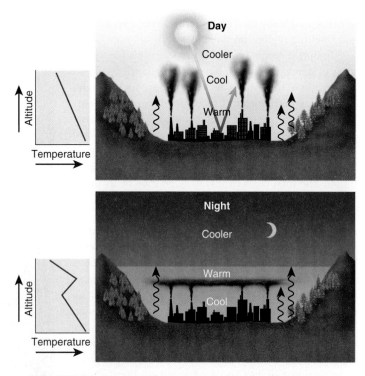

▲ **FIGURE 10.16** Atmospheric temperature inversions occur where ground-level air cools more quickly than upper air. With cold air resting below the warmer air, there is little mixing, and pollutants are trapped near the ground.

quickly develops. Although recent air quality regulations have helped tremendously, on summer days, ozone concentrations in the Los Angeles basin still can reach unhealthy levels.

Heat islands and dust domes occur in cities even without inversion conditions. With their low albedo, concrete and brick surfaces in cities absorb large amounts of solar energy. A lack of vegetation or water results in very slight evaporation (latent heat production); instead, available solar energy is turned into heat. As a result, temperatures in cities are frequently 3° to 5°C (5° to 9°F) warmer than in the surrounding countryside, a condition known as an urban heat island. Tall buildings create convective updrafts that sweep pollutants into the air. Stable air masses created by this heat island over the city concentrate pollutants in a dust dome.

Smog and haze reduce visibility

We have only recently realized that pollution affects rural areas as well as cities. Even supposedly pristine places such as our national parks are suffering from air pollution. Grand Canyon National Park, where maximum visibility used to be 300 km (185 mi), is now so smoggy on some winter days that visitors can't see the opposite rim only 20 km (12.5 mi) across the canyon. Mining operations, smelters, and power plants (some of which were moved to the desert to improve air quality in cities such as Los Angeles) are the main culprits (fig. 10.17) .

▲ **FIGURE 10.17** Reduced visibility is a widespread impact of air pollutants, as on this Los Angeles afternoon.

Huge regions are affected by pollution. A gigantic "haze blob" as much as 3,000 km (about 2,000 mi) across covers much of the eastern United States in the summer, cutting visibility as much as 80 percent. People become accustomed to these conditions and don't realize that the air once was clear. Studies indicate, however, that if all human-made sources of air pollution were shut down, the air would clear up in a few days, and there would be about 150-km (90-mi) visibility nearly everywhere, rather than the 15 km to which we have become accustomed.

Recent studies have found that rural sources also impact urban areas. In some parts of Los Angeles, for example, ammonia (NH_3) drifting from nearby dairy feedlots cause as much smog as cars do. Dust, pulverized manure, and methane are additional airborne pollutants that drift from dairy feedlots into the city. Windblown dust from cultivated fields is also a dominant cause of impaired visibility in some regions, especially during spring and summer.

10.4   AIR POLLUTION CONTROL

"Dilution is the solution to pollution:" this catch phrase has long characterized our main approach to air pollution control. Tall smokestacks were built to send emissions far from the source, where they became difficult to detect or trace to their source. With increasing global industrialization, though, dilution is no longer an effective strategy. We have needed to find different strategies for pollution control.

The best strategy is reducing production

Since most air pollution in the developed world is associated with transportation and energy production, the most effective strategy would be conservation: reducing electricity consumption, insulating homes and offices, and developing better public transportation, and more alternative energy all greatly reduce air pollution at the source. Pollutants can also be captured from effluent after burning.

Particulate removal involves filtering air emissions. Filters trap particulates in a mesh of cotton cloth, spun glass fibers, or asbestos-cellulose. Industrial air filters are generally giant bags through which effluent gas is blown, much like the bag on a vacuum cleaner. Bags can be huge, 10 to 15 m long and 2 to 3 m wide. In power plants, electrostatic precipitators are the most common particulate controls (fig. 10.18). Ash particles pick up an electrostatic surface charge as they pass between large electrodes. The electrically charged particles then precipitate (collect) on an oppositely charged collecting plate. These precipitators consume a large amount of electricity, but maintenance is relatively simple, and collection efficiency can be as high as 99 percent. The ash collected by both of these techniques is sometimes reusable as construction material, but often it is hazardous waste because it contains mercury, lead, or arsenic captured from coal smoke. This waste must be buried in landfills.

Sulfur removal is important because sulfur oxides are among the most damaging of all air pollutants for human health, infrastructure, and ecosystems. Switching from soft coal with a high sulfur content to low-sulfur coal is the surest way to reduce sulfur emissions. But high-sulfur coal is often used for political reasons

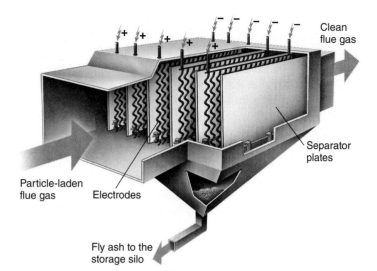

▲ **FIGURE 10.18** An electrostatic precipitator can remove 99 percent unburned particulates in the effluent (smoke) from power plants. Electrodes transfer a static electric charge to dust and smoke particles, which then adhere to collector plates. Particles are then shaken off of the plates and collected for re-use or disposal.

Labels on figure: Clean flue gas; Separator plates; Particle-laden flue gas; Electrodes; Fly ash to the storage silo

or because it is cheap. In the United States, high-sulfur coal comes mainly from Appalachia, a region with political leverage because of its chronic poverty and its powerful coal interests (which now use the controversial process of mountaintop removal to mine coal: see discussion in chapter 12). In China, much domestic coal is rich in sulfur, and companies use it because it is both cheap and convenient. Coal can also be cleaned: it can be crushed, washed, and gasified to remove sulfur and metals before combustion. These measures improve heat efficiency, but they also replace air pollution with solid waste and water pollution, and they are expensive.

Sulfur can be extracted after combustion with catalytic converters, which oxidize or reduce sulfur in effluent gas. Residues, including elemental sulfur, sulfuric acid, and ammonium sulfate can be marketable products. This approach can help companies make money instead of waste, but markets must be reasonably close, and the product must be pure enough for easy reuse.

Nitrogen oxides (NO_x) can be reduced in both internal combustion engines and industrial boilers by as much as 50 percent by carefully controlling the flow of air and fuel. Staged burners, for example, control burning temperatures and oxygen flow to prevent formation of NO_x. The **catalytic converter** on your car uses platinum-palladium and rhodium catalysts to remove up to 90 percent of NO_x, hydrocarbons, and carbon monoxide at the same time.

Hydrocarbon controls mainly involve complete combustion or the control of evaporation. Hydrocarbons and volatile organic compounds are produced by incomplete combustion of fuels or by solvent evaporation from chemical factories, paints, dry cleaning, plastic manufacturing, printing, and other industrial processes. Closed systems that prevent escape of fugitive gases can reduce many of these emissions. Controlling leaks from industrial valves, pipes, and storage tanks can have a significant impact on air quality. Afterburners are often the best method for destroying volatile organic chemicals in industrial exhaust stacks.

Clean air legislation is controversial but extremely successful

Through most of human history, the costs of pollution have been absorbed by the public, which breathes or grows crops in polluted air, rather than by polluters themselves. Rules to control pollution have often sought to make polluters pay for pollution control, in order to reduce public losses in sickness, death, degraded resources, and other costs associated with pollution. Naturally, emitters of pollutants have objected to these rules. It is easier and more profitable to externalize the costs of pollution, and let the public absorb the expenses. Health advocates, on the other hand, argue that industries should pay for pollution prevention. Because of these contrasting interests, clean air laws have always been controversial.

Over time countless ordinances have prohibited objectionable smoke and odors. As far back as 1306, England's King Edward tried to ban the burning of smoky coal in London. The ban failed but was attempted repeatedly over the centuries. As noted earlier in this chapter, it was not until 1956 that better enforcement and alternative fuels finally made clean air rules effective in London.

In the United States, the Clean Air Act of 1963 was the first national law for air pollution control. The act provided federal grants to states to combat pollution but was careful to preserve states' rights to set and enforce their own standards for air quality and enforcement. Because this approach was uneven and difficult to enforce, the act was largely rewritten in the 1970 Clean Air Act amendments. These amendments identified the "criteria" pollutants discussed earlier. The 1970 rules also established primary standards, intended to protect human health, and secondary standards, to protect materials, crops, climate, visibility, and personal comfort.

One of the most contested aspects of the act is the "new source review," which was established in 1977. This provision was originally adopted because industry argued that it would be intolerably expensive to install new pollution-control equipment on old power plants and factories that were about to close down anyway. Congress agreed to "grandfather" or exempt existing equipment from new pollution limits with the stipulation that when they were upgraded or replaced, more stringent rules would apply. The result has been that owners have kept old facilities operating precisely because they were exempted from pollution control. In fact, corporations have poured millions into aging power plants and factories, expanding their capacity rather than build new ones. Decades later, most of those grandfathered plants are still going strong, and continue to be among the biggest contributors to smog and acid rain.

The 1990 amendments included major changes in incentives as well as rules for additional pollutants. Among the major provisions were establishment of new controls for ozone-depleting CFCs, new rules for controlling emissions of benzene, chloroform, and other hazardous air pollutants, and a requirement that comprehensive federal and state standards be set for both industrial and transportation-based sources of common pollutants. The 1990 amendments also provided incentives and rules to support development of alternative fuels and technology, These amendments also established the EPA's right to fine violators of air pollution standards.

Trading pollution credits is one approach

The 1990 revisions also created new incentives for pollution control. One of these is a market-based "cap-and-trade" system. In this approach, the EPA sets maximum emission levels for pollutants. Facilities can then buy and sell emission "credits," or permitted allotments of pollutants. Companies can decide if it's cheaper to install pollution control equipment or to simply buy someone else's credits.

Cap-and-trade has worked well for sulfur dioxide. When trading began in 1990, economists estimated that eliminating 10 million tons of sulfur dioxide would cost $15 billion per year. Left to find the most economical ways to reduce emissions, however, utilities have been able to reach clean air goals for one-tenth that price. A serious shortcoming of this approach is that while trading has resulted in overall pollution reduction, some local "hot spots" remain where owners have found it cheaper to pay someone else to reduce pollution than to do it themselves. This has been the approach adopted for CO_2 and for mercury, among other pollutants.

Presidents and EPA administrators have varied greatly in their enthusiasm for enforcing rules. Business-friendly administrations tend to rely on voluntary emissions controls and a trading program for air pollution allowances. Administrations more focused on public health have sought enforcement of rules, because voluntary action alone rarely reduces pollution.

Amendments have involved acrimonious debate, with bills sometimes languishing in Congress for years because of disputes over burdens of responsibility and cost and definitions of risk. A 2002 report concluded that simply by enforcing existing clean air legislation, the United States could save at least 6,000 lives per year and prevent 140,000 asthma attacks.

Despite controversies, the Clean Air Act has been tremendously successful. Measured only in economic terms, a comparison of the costs of regulation (about $50 billion) and the economic benefits of reduced illness, property damage, and increased productivity (about $1,300 billion), the economic benefits had outweighed costs by more than 25 to 1 by 2010 (see Key Concepts, p. 245). Like automobile industry officials who argued that seat belts in cars were technologically infeasible, unnecessary, and costly, opponents of pollution regulations have often argued that regulations would be technologically infeasible and would hinder economic growth. The evidence shows that these arguments are in error, when cumulative costs and benefits are taken into account.

10.5 THE ONGOING CHALLENGE

Although the United States has not yet achieved the Clean Air Act goals in many parts of the country, air quality has improved dramatically in the last decade in terms of the major large-volume pollutants. For 23 of the largest U.S. cities, the number of days each year in which air quality reached the hazardous level is down 93 percent from a decade ago. Of 97 metropolitan areas that failed to meet clean air standards in the 1980s, 41 are now in compliance. Eighty percent of the United States now meets the National Ambient Air Quality Standards.

Most pollutants have declined sharply since the introduction of Clean Air Act rules. Lead has been reduced by 98 percent, SO_2 by 35 percent, and CO by 32 percent. Filters, scrubbers, and precipitators on power plants and other large stationary sources are responsible for most of the particulate and SO_2 reductions. Catalytic converters on automobiles are responsible for most of the CO and O_3 reductions. The only conventional "criteria" pollutants that have not dropped significantly are particulates and NO_x and particulate matter. Because automobiles are the main source of NO_x, this pollutant has remained high or grown in many areas. Particulate matter, mostly dust and soot produced by agriculture, fuel combustion, metal smelting, concrete manufacturing, and other activities, has also grown in some areas.

Among the areas of sharply increasing air pollutants are rural natural gas producing regions, such as central Wyoming and Colorado. Tens of thousands of gas-producing wells, many of them leaking small amounts of gas, and unregulated waste storage tanks produce ozone, volatile organic compounds, and other contaminants that can exceed some of the worst urban levels in the United States.

Pollution persists in developing areas

The major metropolitan areas of many developing countries are growing at explosive rates to incredible sizes (chapter 15), and environmental quality is abysmal in many of them. Mexico City is notorious for bad air. Its pollution levels exceed World Health Organization (WHO) health standards 350 days per year, and more than half of all its children have lead levels in their blood high enough to lower intelligence and retard development. Mexico City's 131,000 industries and 2.5 million vehicles spew out more than 5,500 tons of air pollutants daily. Santiago, Chile, averages 299 days per year on which suspended particulates exceed WHO standards of 90 mg/m³.

While China is making efforts to control air and water pollution (chapter 1), many of China's 400,000 factories have no air pollution controls. Experts estimate that home coal burners and factories emit 10 million tons of soot and 15 million tons of sulfur dioxide annually and that emissions have increased rapidly over the past 20 years. Seven of the ten cities in the world with the worst air quality are in China. Shenyang, an industrial city in northern China, is thought to have the world's worst continuing particulate problem, with peak winter concentrations over 700 mg/m³ (nine times U.S. maximum standards). Airborne particulates in Shenyang exceed WHO standards on 347 days per year. It's estimated that air pollution is responsible for 400,000 premature deaths every year in China. The high incidence of cancer in Shanghai is thought to be linked to air pollution.

Many places have improved

Low-income countries can control air pollution, too. Delhi, India, was once considered one of the world's ten most polluted cities. Visibility often was less than 1 km on smoggy days. Health experts warned that breathing Delhi's air was equivalent to smoking two packs of cigarettes per day. Pollution levels exceeded WHO standards by nearly five times. The biggest problem was vehicle emissions, which contributed about 70 percent of air pollutants (followed by industrial emissions, at 20 percent, and burning of garbage and firewood).

Can we afford clean air?

Designed to protect human health, crops, and buildings the Clean Air Act requires the control of conventional (criteria) pollutants, metals, organic compounds, and other substances.

As part of the CAA, Congress directed the EPA to evaluate the economic costs and benefits of enforcing the act's provisions. Emitters of air pollutants have charged that emission controls are expensive, and that those costs reduce economic productivity and threaten job creation.

In the most recent of these reports, published in 2011, the EPA calculate the economic costs and benefits of the 1990 revisions to the CAA. The report is available at http://www.epa.gov/air/sect812/prospective2.html. The EPA compared the economic costs that would have been incurred (costs of health care, lost productivity, infrastructure degradation, and other factors) *if we had not implemented the 1990 air pollution controls*. These costs were compared to *observed costs of implementation* of those rules.

> Controlling air pollution after it's produced is expensive. Electrostatic precipitators, better seals, careful monitoring and inspections all cost money.
>
> The EPA study found that by 2020, cumulative public and private costs of the 1990 rules amounted to **$65 billion**.
>
> Cumulative savings amounted to **$2,000 billion**.
>
> At this rate, can we afford *not* to keep our air clean?

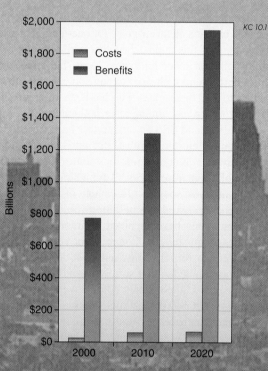

Direct costs (red) and benefits (blue) of Clean Air Act provisions by 2000, 2010, and 2020, in billions of 2006 dollars.   SOURCE: EPA 2011 Clean Air Impacts Summary Report.

What are the 1990 rules?

Among the major provisions were establishment of:

- controls for ozone-depleting CFCs
- a market-based "cap and trade" system, with emissions permits that can be sold, to reduce acid rain-producing SO_2 and NO_x
- requirement of federal and state rules for both industrial and transportation sources
- new rules for controlling emissions of benzene, chloroform, and other hazardous air pollutants
- "new source review" for newly constructed power plants and other major emitters
- the EPA's right to fine violators of air pollution standards
- a framework for developing alternative fuels and technology

What are the costs?

Direct costs of compliance by 2020 for five major source categories (red bars) and for additional minor categories (purple). Vehicles and electricity generation, our dominant sources, bear most costs. ▶

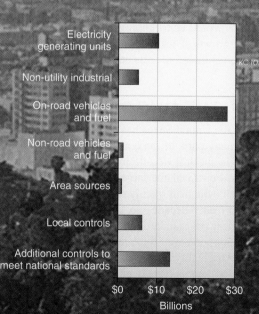

SOURCE: 2011 Clean Air Impacts Summary Report

What are the benefits of the 1990 Clean Air Act amendments?

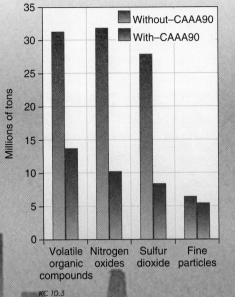

KC 10.3

Without the Clean Air Act amendments (CAAA) of 1990:

◀ Production of key pollutants would be greater in 2020.

Mortality, bronchitis, heart disease, and lost work days would all be greater. ▶

Visibility would be worse. ▼

HEALTH EFFECT REDUCTIONS (PM2.5 & OZONE ONLY)	POLLUTANT(S)	YEAR 2010	YEAR 2020
PM2.5 Adult Mortality	PM	160,000	230,000
PM2.5 Infant Mortality	PM	230	280
Ozone Mortality	Ozone	4,300	7,100
Chronic Bronchitis	PM	54,000	75,000
Acute Bronchitis	PM	130,000	180,000
Heart Disease	PM	130,000	200,000
Asthma Excerbation	PM	1,700,000	2,400,000
Hospital Admissions	PM, Ozone	86,000	135,000
Emergency Room Visits	PM, Ozone	86,000	120,000
Restricted Activity Days	PM, Ozone	84,000,000	110,000,000
School Loss Days	Ozone	3,200,000	5,400,000
Lost Work Days	PM	13,000,000	17,000,000

Without CAAA

KC 10.4

With CAAA

Impacts on the economy?

Since 1970, the six commonly found air pollutants have decreased by more than 50 percent. Hazardous air pollutants from large industrial sources, such as chemical plants, petroleum refineries, and paper mills have declined by nearly 70 percent. Production of most ozone-depleting chemicals has ceased. New cars are more than 90 percent cleaner and will be even cleaner in the future.

At the same time, the U.S. gross domestic product (GDP) has tripled, vehicle use has doubled, and energy consumption has risen 50 percent.

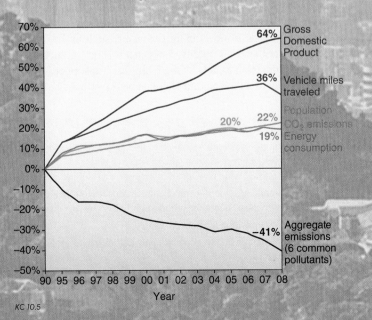

KC 10.5

Visibility in Deciviews, 2020

0 3 6 9 12 15 18 21 24 27 30
Best   Worst

KC 10.6

CAN YOU EXPLAIN?

1. What are several key provisions of the 1990 Clean Air Act amendments?

2. What is the idea of "cap and trade" market mechanisms?

3. In what ways have the amendments saved money?

In the 1990s, catalytic converters were required for automobiles, and unleaded gasoline and low-sulfur diesel fuel were introduced. In 2000, private automobiles were required to meet European standards, and in 2002, more than 80,000 buses, auto-rickshaws, and taxis were required to switch from liquid fuels to compressed natural gas (fig. 10.19). Sulfur dioxide and carbon monoxide levels have dropped 80 percent and 70 percent, respectively, since 1997. Particulate emissions dropped by about 50 percent. Residents report that the air is dramatically clearer and healthier. Unfortunately, rising prosperity, driven by globalization of information management, has doubled the number of vehicles on the roads, threatening this progress. Still, the gains made in New Delhi are encouraging for people everywhere.

Twenty years ago, Cubatao, Brazil, was described as the "Valley of Death," one of the most dangerously polluted places in the world. A steel plant, a huge oil refinery, and fertilizer and chemical factories churned out thousands of tons of air pollutants every year that were trapped between onshore winds and the uplifted plateau on which São Paulo sits. Trees died on the surrounding hills. Birth defects and respiratory diseases were alarmingly high. Since then, however, the citizens of Cubatao have made remarkable progress in cleaning up their environment. The end of military rule and restoration of democracy allowed residents to publicize their complaints. The environment became an important political issue. The state of São Paulo invested about $100 million and the private sector spent twice as much to clean up most pollution sources in the valley. Particulate pollution was reduced 75 percent, ammonia emissions were reduced 97 percent, hydrocarbons that cause ozone and smog were cut 86 percent, and sulfur dioxide production fell 84 percent. Fish are returning to the rivers, and forests are regrowing on the mountains.

CONCLUSION

Air pollution is often the most obvious and widespread type of pollution. Everywhere on earth, from the most remote island in the Pacific, to the highest peak in the Himalayas, to the frigid ice cap over the North Pole, there are traces of human-made contaminants, remnants of the 2 billion metric tons of pollutants released into the air worldwide every year by human activities.

Health effects of these pollutants include respiratory diseases, birth defects, heart attacks, cancer, and developmental disabilities in children. Environmental impacts include destruction of stratospheric ozone, poisoning of forests and waters by acid rain, and corrosion of building materials. Damages to health and environ-

▲ **FIGURE 10.19** Air quality in Delhi, India has improved dramatically since buses, auto-rickshaws, and taxis were required to switch to compressed natural gas. This is an excellent example of success in controlling pollution in the developing world.

ment in turn have economic costs that vastly outweigh the costs of pollution prevention.

We have made encouraging progress in controlling air pollution, progress that has economic benefits as well as health benefits. Many people aren't aware of how much worse air quality was in the industrial centers of North America and Europe a century or two ago than they are now. Cities such as London, Pittsburgh, Chicago, Baltimore, and New York had air quality as bad as or worse than most megacities of the developing world now.

The success of the Montreal Protocol and the Clean Air act show that real progress is achievable if we choose to act. Though the stratospheric ozone hole persists because of the residual chlorine in the air released decades ago, we expect the ozone depletion to end in about 50 years. The Clean Air Act, similarly, has dramatically improved air quality and reduced health costs. Local pollution reductions in developing countries, such as Brazil and India, show that better air quality is not just for the wealthy. What is needed is a decision to act. As the Chinese philosopher Lao-tzu wrote, a journey of a thousand miles must begin with a single step.

PRACTICE QUIZ

1. Define *primary* and *secondary air pollutants.*
2. What are the six "criteria" pollutants in the original Clean Air Act? Why were they chosen? What are some additional hazardous air toxins have been added to the list regulated by the Clean Air Act?
3. What pollutants in indoor air may be hazardous to your health? What is the greatest indoor air problem globally?
4. What is acid deposition? What causes it?
5. What is an atmospheric inversion and how does it trap air pollutants?

6. What is the difference between ambient and stratospheric ozone? What is destroying stratospheric ozone?

7. What is long-range air pollution transport? Give two examples.

8. What is the ratio of direct costs and benefits of the Clean Air Act? What costs are mainly saved?

9. Which of the conventional pollutants has decreased most in the recent past and which has decreased least?

10. Give one example of current air quality problems and one success in controlling pollution in a developing country.

CRITICAL THINKING AND DISCUSSION

Apply the principles you have learned in this chapter to discuss these questions with other students.

1. How would you choose between government regulations and market-based trading programs for air pollution control? Are there situations where one approach would work better than the other?

2. Debate the following proposition: Our air pollution blows into someone else's territory; therefore, it is uneconomical to install pollution controls, because they will bring no direct economic benefit to those of us who have to pay for them.

3. Utility managers once claimed that it would cost $1,000 per fish to control acid precipitation in the Adirondack lakes and that it would be cheaper to buy fish for anglers than to put scrubbers on power plants. Does that justify continuing pollution? Why or why not?

4. Economists and scientists often have difficulty reaching common terms for defining and solving issues such as the Clean Air Act renewal. How might their conflicting definitions be reshaped to make the discussion more successful?

5. Why do you think controlling pollutants like mercury is such a difficult problem? List some of the technological, economic, political, emotional, and other factors involved. Whose responsibility is it to reduce these emissions?

DATA ANALYSIS   How Polluted Is Your Hometown?

How does air quality in your area compare to that in other places? You can examine trends in major air pollutants, both national and local trends in your area, on the EPA's website. The EPA is the principal agency in charge of protecting air quality and informing the public about the air we breathe and how healthy it is.

Go to Connect to find a link to data and maps showing trends in SO_2 emissions since 1980. At the same site you can see trends in NO_x, CO, lead, and other criteria pollutants. Examine national trends, then look at your local area on the map on the same page to answer questions about trends in your area, and to compare your area to others.

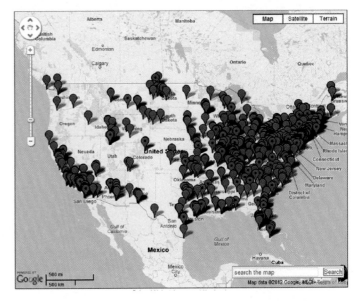

▲ Examine pollutant trends in your area on the EPA website.

TO ACCESS ADDITIONAL RESOURCES FOR THIS CHAPTER, PLEASE VISIT CONNECT AT www.mcgrawhillconnect.com.

You will find LearnSmart, an adaptive learning system, Google Earth™ exercises, additional Case Studies, Data Analysis exercises, and an interactive ebook.

Water: Resources and Pollution

Between 2000 and 2010, the surface level of Lake Mead, the largest reservoir on the Colorado River, fell more than 100 ft (30.5 m) during the worst drought in recorded history. If water levels fall another 100 ft, the reservoir will provide neither the water nor the electrical power on which millions of people depend.

LEARNING OUTCOMES

After studying this chapter, you should be able to answer the following questions:

▶ Where does our water come from? How do we use it?

▶ Where and why do water shortages occur?

▶ How can we increase water supplies? What are some costs of these methods?

▶ How can *you* conserve water?

▶ What is *water pollution*? What are its sources and effects?

▶ Why are sewage treatment and clean water important in developing countries?

▶ How can we control water pollution?

CASE STUDY

When Will Lake Mead Go Dry?

The Colorado River is the lifeblood of the American Southwest. More than 30 million people and a $1.2 trillion regional economy in cities, such as Los Angeles, Phoenix, Las Vegas, and Denver, depend on its water (fig. 11.1). But the sustainability of this essential resource is in doubt. Drought, climate change, and rapid urban growth are creating worries about the future of the entire watershed.

In 2008, Tim Barnett and David Pierce from the Scripps Institute in California published a provocative article suggesting that both Lake Mead and Lake Powell could reach levels within a decade or so at which neither would be able to either produce power or provide water for urban or agricultural use if no changes are made in current water allocations. Representing more than 85 percent of the water storage for the entire Colorado system, reaching "dead pool" levels in these huge lakes would be a catastrophe for the whole region. This warning is based on both historical records and climate models that suggest a 10 to 30 percent run off reduction in the area over the next 50 years.

The roots of this problem can be traced to the Colorado Compact of 1922, which allocated water rights for the seven states that adjoin the river. The previous decade had been the wettest in more than a thousand years. The estimated annual river flow of 18 million acre-feet (22 billion m³) negotiators thought they could allocate was about 20 percent higher than the twentieth-century average. The error didn't matter much at the time, because none of the states were able to withdraw their full share of water.

As cities have grown, however, and agriculture has expanded over the past century, competing claims for water have repeatedly caused tensions and disputes. Cumulatively, massive water diversion projects, such as the Colorado River Aqueduct, which provides water for Los Angeles, the All-American Canal, which irrigates California's Imperial Valley, or the Central Arizona Project, which transports water over the mountains and across the desert to Phoenix and Tucson, are capable of diverting the entire river flow. In 1944, the United States agreed to provide 1.5 million acre-feet to Mexico so there would be at least a little water (although of dubious quality) in the river when it crossed the border.

To make matters worse, climate change is expected to decrease river flows by 10 to 30 percent over the next 50 years. There may be only half as much water in the river in a few decades than negotiators once thought they had to distribute between the states. The Southwest is currently in its eighth year of drought, which may be the first hint of that change. The maximum water level in Lake Mead (an elevation of 1,220 feet, or 372 m) was last reached in 2000. Since then, the lake level has been dropping about 12 feet (3.6 m) per year, reaching 1,097 feet in 2010. The minimum power level (the height at which electricity can be produced) is 1,050 feet (320 m). In 2011, heavy snow in the Rockies raised the surface of the lake 40 feet, but there are fears this may have been a singular event. The minimum level at which water can be drawn off by gravity is 900 feet (274 m). Barnett and Pierce estimate that without changes in current management plans, there's a 50 percent

chance minimum power pool levels in both Lakes Mead and Powell will be reached by 2017 and that there's an equal chance that live storage in both lakes will be gone by about 2021.

Already, we're at or beyond the sustainable limits of the river. Currently, Lake Powell is only 58 percent full, and Lake Mead holds only 43 percent of its maximum volume. The shores of both lakes now display a wide "bath-tub ring" of deposited minerals left by the receding water. One suggestion has been to drain Lake Powell in order to ensure a water supply for Lake Mead. This solution is strenuously opposed by many of the 3 million people per year who recreate in its red rock canyons and sparkling blue water. On the other hand, think of the cost and disruption if Los Angeles, Phoenix, Las Vegas, and other major metropolitan areas of the region were to run out of water and power.

The American Southwest isn't alone in facing this problem. The United Nations warns that water supplies are likely to become one of the most pressing environmental issues of the twenty-first

(continued)

▲ **FIGURE 11.1** The Colorado River flows 2,330 km (1,450 mi) through seven western states. Its water supports 30 million people and a $1.2 trillion regional economy, but drought, climate change, and rapid urban growth threaten the sustainability of this resource.

century. By 2025, two-thirds of all humans could be living in places where water resources are inadequate. In this chapter, we'll look at the sources of our fresh water, what we do with it, and how we might protect its quality and extend its usefulness.

For further reading, see:

Barnett, T. P., and D. W. Pierce. 2008. When will Lake Mead go dry? *Journal of Water Resources Research*, vol. 44, W03201.

Powell, James L. 2009. *Dead Pool: Lake Powell, Global Warming, and the Future of Water in the West.* University of California Press. ∎

I tell you gentlemen; you are piling up a heritage of conflict and litigation of water rights, for there is not sufficient water to supply the land.

—JOHN WESLEY POWELL

11.1 WATER RESOURCES

Water is a marvelous substance—flowing, swirling, seeping, constantly moving from sea to land and back again. It shapes the earth's surface and moderates our climate. Water is essential for life. It is the medium in which all living processes occur (chapter 2). Water dissolves nutrients and distributes them to cells, regulates body temperature, supports structures, and removes waste products. About 60 percent of your body is water. You could survive for weeks without food, but only a few days without water. Water also is needed for agriculture, industry, transportation, and a host of other human uses. In short, clean fresh water is one of our most vital natural resources.

The hydrologic cycle constantly redistributes water

The water we use cycles endlessly through the environment. The total amount of water on our planet is immense—more than 1,404 million km^3 (370 billion billion gal) (table 11.1). This water evaporates from moist surfaces, falls as rain or snow, passes through living organisms, and returns to the ocean in a process known as the **hydrologic cycle** (see fig. 2.17). Every year, about 500,000 km^3, or a layer 1.4 m thick, evaporates from the oceans. More than 90 percent of that moisture falls back on the ocean. The 47,000 km^3 carried onshore joins some 72,000 km^3 evaporated from lakes, rivers, soil, and plants to become our annual, renewable supply of fresh water. Plants play a major role in the hydrologic cycle, absorbing groundwater and pumping it into the atmosphere by transpiration (transport plus evaporation). In tropical forests, as much as 75 percent of annual precipitation is returned to the atmosphere by plants.

TABLE 11.1 | Units of Water Measurement

One cubic kilometer (km^3) equals 1 billion cubic meters (m^3), 1 trillion liters, or 264 billion gal.

One acre-foot is the amount of water required to cover an acre of ground 1 ft deep. This is equivalent to 325,851 gal, or 1.2 million liters, or 1,234 m^3, approximately the amount consumed annually by a family of four in the United States.

One cubic foot per second of river flow equals 28.3 liters per second, or 449 gal per minute.

Solar energy drives the hydrologic cycle by evaporating surface water, which becomes rain and snow. Because water and sunlight are unevenly distributed around the globe, water resources are very uneven. At Iquique in the Chilean desert, for instance, no rain has fallen in recorded history. At the other end of the scale, 26.5 m (86.8 ft) of rain was recorded in 1860 in Cherrapunji in India. Figure 11.2 shows current patterns of precipitation around the world, but climate change is altering that map. Cherrapunji, for example, only receives about one-third as much rain today as it did a century ago.

Most of the world's rainiest regions are tropical, where heavy rainy seasons occur, or in coastal mountain regions. Deserts occur on every continent just outside the tropics (the Sahara, the Namib, the Gobi, the Sonoran, and many others). Rainfall is also slight at very high latitudes, another high-pressure region.

Mountains also influence moisture distribution. The windward sides of mountain ranges, including the Pacific Northwest and the flanks of the Himalayas, typically are wet and have large rivers; on the leeward sides of mountains, in areas known as the rain shadow, dry conditions dominate, and water can be very scarce. The windward side of Mount Waialeale on the island of Kauai, for example, is extremely wet, with an annual rainfall around 1,200 cm (460 in.). The leeward side, only a few kilometers away, has an average yearly rainfall of only 46 cm (18 in.).

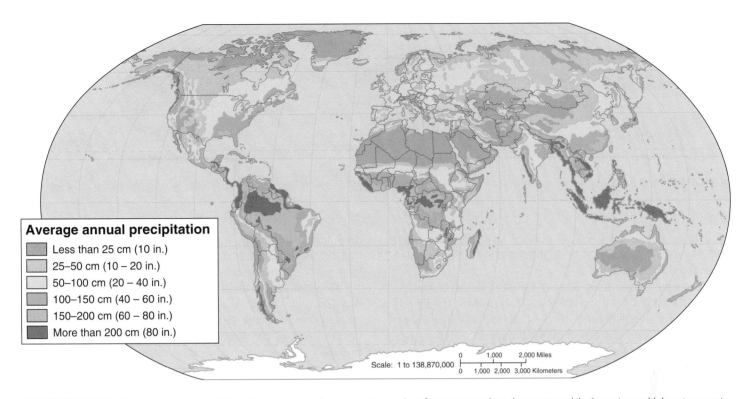

▲ **FIGURE 11.2** Average annual precipitation. Note wet areas that support tropical rainforests occur along the equator, while the major world deserts occur in zones of dry, descending air between 20° and 40° north and south.

An important consideration in water availability is whether moisture is evenly distributed, especially in the growing season. Another question is whether hot, dry weather evaporates available moisture. These factors help determine the amount and variety of biological activity in a place (chapter 5).

11.2 MAJOR WATER COMPARTMENTS

The distribution of water often is described in terms of interacting compartments in which water resides, sometimes briefly and other times for eons (table 11.2). The length of time water typically stays in a compartment is its **residence time**. On average, a water molecule stays in the ocean for about 3,000 years, for example, before it evaporates and starts through the hydrologic cycle again. Nearly all the world's water is in the oceans (fig. 11.3). Oceans play a crucial role in moderating the earth's temperature, and over 90 percent of the world's living biomass is contained in the oceans. What we mainly need, though, is fresh water. Amazingly, only about .02 percent of the world's water is in a form accessible to us and to other organisms that rely on fresh water.

Glaciers, ice, and snow contain most surface, fresh water

Of the 2.4 percent of all water that is fresh, nearly 90 percent is tied up in glaciers, ice caps, and snowfields. Although most of this ice is located in Antarctica, Greenland, and the floating ice cap in

TABLE 11.2 | Earth's Water Compartments

COMPARTMENT	VOLUME (1,000 km³)	PERCENT OF TOTAL WATER	AVERAGE RESIDENCE TIME
Total	1,386,000	100	2,800 years
Oceans	1,338,000	96.5	3,000 to 30,000 years*
Ice and snow	24,364	1.76	1 to 100,000 years*
Saline groundwater	12,870	0.93	Days to thousands of years*
Fresh groundwater	10,530	0.76	Days to thousands of years*
Fresh lakes	91	0.007	1 to 500 years*
Saline lakes	85	0.006	1 to 1,000 years*
Soil moisture	16.5	0.001	2 weeks to 1 year*
Atmosphere	12.9	0.001	1 week
Marshes, wetlands	11.5	0.001	Months to years
Rivers, streams	2.12	0.0002	1 week to 1 month
Living organisms	1.12	0.0001	1 week

*Depends on depth and other factors.
SOURCE: Data from UNEP, 2002.

the Arctic, alpine glaciers and snowfields supply water to billions of people. The winter snowpack on the western slope of the Rocky Mountains, for example, provides 75 percent of the flow in the Colorado River described in the opening case study of this chapter.

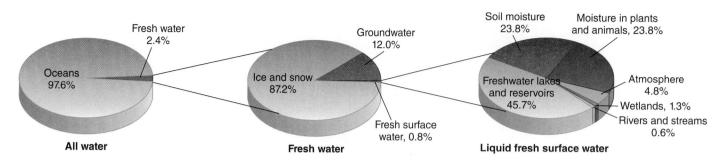

FIGURE 11.3 Less than 1 percent of fresh water, and less than 0.02 percent of all water, is fresh, liquid surface water on which terrestrial life depends.

Drought conditions already have reduced snowfall (and runoff) in the western United States, and global warming is projected to cause even further declines.

As chapter 9 discusses, climate change is shrinking glaciers and snowfields nearly everywhere (fig. 11.4). In Asia, the Tibetan glaciers, which are the source of six of the world's largest rivers and supply drinking water for 3 billion people, are shrinking rapidly. There are warnings that these glaciers could vanish in a few decades, which would bring enormous suffering and economic losses to that continent.

Groundwater stores large resources

Originating as precipitation that percolates into layers of soil and rock, groundwater makes up the largest compartment of liquid, fresh water. The groundwater within 1 km of the surface is more than 100 times the volume of all the freshwater lakes, rivers, and reservoirs combined.

Plants get moisture from a relatively shallow layer of soil containing both air and water, known as the **zone of aeration** (fig. 11.5). Depending on rainfall amount, soil type, and surface topography, the zone of aeration may be a few centimeters or many meters deep. Lower soil layers, where all soil pores are filled with water, make up the **zone of saturation**, the source of water in most wells; the top of this zone is the **water table**.

Geologic layers that contain water are known as **aquifers**. Aquifers may consist of porous layers of sand or gravel or of cracked or porous rock. Below an aquifer, relatively impermeable layers of rock or clay keep water from seeping out at the bottom. Instead, water seeps more or less horizontally through the porous layer. Depending on geology, it can take from a few hours to several years for water to move a few hundred meters through an aquifer. If impermeable layers lie above an aquifer, pressure can develop within the water-bearing layer. Pressure in the aquifer can make a well flow freely at the surface. These free-flowing wells and springs are known as *artesian* wells or springs.

Areas where surface water filters into an aquifer are **recharge zones** (fig. 11.6). Most aquifers recharge extremely slowly, and road and house construction or water use at the surface can further

FIGURE 11.4 Glaciers and snowfields provide much of the water on which billions of people rely. The snowpack in the western Rocky Mountains, for example, supplies about 75 percent of the annual flow of the Colorado River. Global climate change is shrinking glaciers and causing snowmelt to come earlier in the year, disrupting this vital water source.

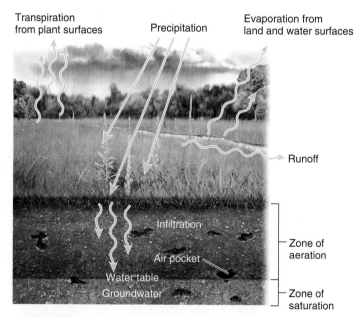

FIGURE 11.5 Precipitation that does not evaporate or run off over the surface percolates through the soil in a process called infiltration. The upper layers of soil hold droplets of moisture between air-filled spaces (usually much smaller than shown here). Lower layers, where all spaces are filled with water, make up the zone of saturation, or groundwater.

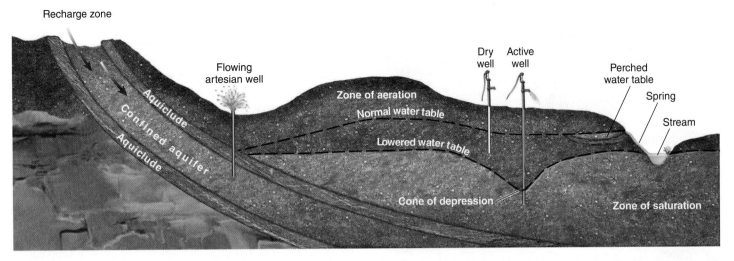

▲ **FIGURE 11.6** An aquifer is a porous or cracked layer of rock. Impervious rock layers (aquicludes) keep water within a confined aquifer. Pressure from uphill makes an artesian well flow freely. Pumping can create a cone of depression, which leaves shallower wells dry.

slow recharge rates. Contaminants can also enter aquifers through recharge zones. Urban or agricultural runoff in recharge zones is often a serious problem. The people who don't have access to clean surface water generally depend on groundwater for drinking and other uses. Every year 700 km³ are withdrawn by humans, mostly from shallow, easily polluted aquifers.

Rivers, lakes, and wetlands cycle quickly

Fresh, flowing surface water is one of our most precious resources. Rivers contain a relatively small amount of water at any one time. Most rivers would begin to dry up in weeks or days if they were not constantly replenished by precipitation, snowmelt, or groundwater seepage.

We compare the size of rivers in terms of **discharge**, or the amount of water that passes a fixed point in a given amount of time. This is usually expressed as liters or cubic feet of water per second. The 16 largest rivers in the world carry nearly half of all surface runoff on the earth, and a large fraction of that occurs in a single river, the Amazon, which carries nearly as much water as the next seven biggest rivers together.

Lakes contain nearly 100 times as much water as all rivers and streams combined, but much of this water is in a few of the world's largest lakes. Lake Baikal in Siberia, the Great Lakes of North America, the Great Rift Lakes of Africa, and a few other lakes contain vast amounts of water. Worldwide, lakes are almost as important as rivers in terms of water supplies, food, transportation, and settlement.

Wetlands—bogs, swamps, wet meadows, and marshes—play a vital and often unappreciated role in the hydrologic cycle. Their lush plant growth stabilizes soil and holds back surface runoff, allowing time for infiltration into aquifers and producing even, year-long stream flow. When wetlands are disturbed, their natural water-absorbing capacity is reduced, and surface waters run off quickly, resulting in floods and erosion during the rainy season and low stream flow the rest of the year.

The atmosphere is one of the smallest compartments

The atmosphere contains only 0.001 percent of the total water supply, but it is the most important mechanism for redistributing water around the world. An individual water molecule resides in the atmosphere for about ten days, on average. Some water evaporates and falls within hours. Water can also travel halfway around the world before it falls, replenishing streams and aquifers on land.

Active **LEARNING**

Mapping the Water-Rich and Water-Poor Countries

The top ten water-rich countries, in terms of water availability per capita, and the ten most water-poor countries are listed below. Locate these countries on the fold-out map at the end of your book. Describe the patterns. Where are the water-rich countries concentrated? (Hint: does latitude matter?) Where are the water-poor countries most concentrated?

Water-rich countries: Iceland, Surinam, Guyana, Papua New Guinea, Gabon, Solomon Islands, Canada, Norway, Panama, Brazil

Water-poor countries: Kuwait, Egypt, United Arab Emirates, Malta, Jordan, Saudi Arabia, Singapore, Moldavia, Israel, Oman

ANSWER: Water-rich countries (per capita) are either in the far north, where populations and evaporation are low, or in the tropics. Water-poor countries are in the desert belt at about 15° to 25° latitude or are densely populated island nations (e.g., Malta, Singapore).

▲ **FIGURE 11.7** Water has always been the key to survival. Who has access to this precious resource and who doesn't has long been a source of tension and conflict.

11.3 WATER AVAILABILITY AND USE

Perhaps more than any other environmental factor, water availability determines the location and activities of humans on the earth (fig. 11.7). Renewable water supplies—mainly surface water and shallow groundwater—are cleansed and replenished by the hydrologic cycle (see fig. 2.17). Water is a renewable resource, but renewal takes time. The rate at which we now use water makes it essential that we protect and conserve local water supplies.

As you've seen in figure 11.2, South America, West Central Africa, and South and Southeast Asia all have areas of very high rainfall. Brazil and the Democratic Republic of Congo, because they have both high precipitation and large land areas, are among the most water-rich countries on earth. Canada and Russia, which both have large areas with high rain and snowfall, also have large annual water supplies.

The highest per capita water supplies occur in countries with wet climates and low population densities. Iceland, for example, has about 160 million gallons (605,000 m³) per person per year. In contrast, Bahrain, where temperatures are extremely high and rain almost never falls, has essentially no natural fresh water. Almost all of Bahrain's water comes from imports and desalinized seawater. Egypt, in spite of the fact that the Nile River flows through it, has only about 11,000 gallons of water annually per capita, or about 15,000 times less than Iceland.

We use water for many purposes

Clean, fresh water is essential for nearly every human endeavor. Water can be recycled if its quality isn't degraded by use. Water withdrawal is the total amount of water taken from a water body. Much of this water can be returned to circulation in a reusable form. Water consumption, on the other hand, is loss of water due to evaporation, absorption, or contamination.

Worldwide, agriculture claims about 70 percent of total water withdrawal, and that use is increasing rapidly worldwide. Agricultural use ranges from 93 percent of all water withdrawn in India to only 4 percent in Kuwait, which cannot afford to spend its limited water on crops. In many developing countries and in parts of the United States, the most common type of irrigation is to simply flood the whole field or run water in rows between crops. As much as half this water can be lost through evaporation or seepage from unlined irrigation canals bringing water to fields. Sprinklers are more efficient in distributing water, but they are more costly and energy-intensive. Water-efficient drip irrigation is even more efficient but currently is used on only about 1 percent of the world's croplands.

Most of us never think about the water required to grow and prepare the food we eat (fig. 11.8). Raising a kilogram of beef in a concentrated feeding operation, for example, takes more than 15,000 liters of water because cattle in these facilities are fed grain, which they process very inefficiently, that often was grown with water-intensive irrigation systems. The result is that the average hamburger takes 2,400 liters (630 gal) to produce. Raising goats requires at least 90 percent less water from all sources than cattle because they eat a much wider diet and metabolize what they do eat more effectively.

Rice cultivation (which generally occurs in wet paddies) takes about three times as much water as raising potatoes or wheat. For

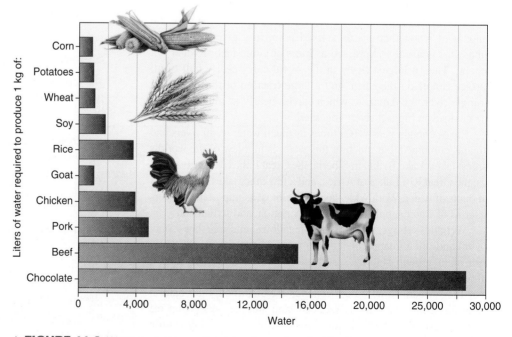

▲ **FIGURE 11.8** Water required to produce 1 kg of some important foods.

some foods, the greatest water use isn't in growing the crop, but in preparing it for our consumption. Chocolate, for example, is highly processed before we eat it.

Industrial and domestic uses tend to be far lower than agricultural use

Industry uses represent about one-fourth of all water withdrawals, worldwide. Some European countries use 70 percent of water for industry; less-industrialized countries use as little as 5 percent. Cooling water for power plants is by far the largest single industrial use of water, typically accounting for 50 to 75 percent of industrial withdrawal. If treated properly, much of the cooling water can be recycled for other uses. A rapidly growing demand for water is for biofuel production. It currently takes 4 to 5 liters of water to produce 1 liter of ethanol. Water scarcity may limit our biofuel use.

Domestic, or household, water use accounts for only about 6 percent of world water use. This includes water for drinking, cooking, and washing. The amount of water used per household varies enormously, however, depending on a country's wealth. The United Nations reports that people in developed countries consume on average, about ten times more water daily than those in developing nations. Poorer countries often can't afford the infrastructure to obtain and deliver water to citizens. There's vicious cycle between water stress and poverty. Each contributes to the other.

Many attempts have been made to enhance local supplies and redistribute water. Creating rain by cloud seeding—distributing condensation nuclei in humid air to help form raindrops—is sometimes effective but it's essentially taking moisture from one area for the benefit of another. Desalination is locally important: in the arid Middle East, where energy and money are available but water is scarce, desalination is sometimes the principal source of water. Some American cities, such as Tampa, Florida, and San Diego, California, also depend partly on energy-intensive desalination.

11.4 FRESHWATER SHORTAGES

Clean drinking water and basic sanitation are necessary to prevent communicable diseases and to maintain a healthy life. The United Nations estimates that at least a billion people lack access to safe drinking water and 2.5 billion don't have adequate sanitation. As populations grow, more people move into cities, and agriculture and industry compete for increasingly scarce water supplies, water shortages are expected to become even worse.

Water scarcity is a growing problem

The World Health Organization considers an average of 1,000 m^3 (264,000 gal) per person per year to be a necessary amount of water for modern domestic, industrial, and agricultural uses. It is fairly common in some areas to use 70 percent of annual river flows and, in a few cases, to use as much as 120 percent of annual renewable resources by drawing on fossil groundwater. This obviously isn't sustainable in the long run.

About a third of the world's current population lives in countries where water supplies don't meet everyone's minimum essential water needs (fig. 11.9). Compare figure 11.9 with figure 2.17. Notice that the scarcity map shows average conditions for each country. Local conditions, such as those in the American Southwest described in the opening case study, can be much worse than the country-wide mean. Similarly, China is vulnerable to water shortages due to drought in its northern and western regions even though its southeast has been suffering catastrophic floods. Not surprisingly, the biggest problems in water stress and scarcity at a human level occur in Africa and Asia where rainfall is low and poor countries can't afford to adapt. Some affluent countries are very dry, but have the money and political capacity to adjust to drought (see What Do You Think? on p. 259).

Water use has been increasing about twice as fast as population growth over the past century. And water withdrawals are expected to continue to grow as more land is irrigated to feed an expanding population. The United Nations Food and Agriculture Organization predicts that, by 2025, 1.8 billion people will be living with severe water scarcity and as many as two-thirds of all humans will experience water-stressed conditions. Conflicts are likely to increase as different countries, economic sectors, and other stakeholders compete for the same limited water supply. Water wars may well be the major source of hostilities in the twenty-first century.

In some countries, the problem isn't the total amount of water but access to clean water. In Mali, for example, 88 percent of the population lacks safe water; in Ethiopia, it is 94 percent. Rural people often have less access to good water than do city dwellers. This often results in diarrhea and a variety of other diseases linked to contaminated water and lack of sanitation. Every year about 1.6 million people, 90 percent of them children under 5 years old, die from these diseases. More children die from diarrhea than malaria, measles, and HIV/AIDS combined.

More than two-thirds of the world's households have to fetch water from outside the home (fig. 11.10). This is heavy work, done mainly by women and children and sometimes taking several hours a day. Improved public systems bring many benefits to these poor families. The hours once spent carrying water can devoted to study or money-making enterprises. The quality of local surface water is often poor, and increased sanitation can dramatically reduce infectious diseases.

Living in an age of thirst

Climate change is expected to exacerbate water scarcity. A general rule is that with global warming, dry areas will generally get drier, while wet places will be wetter. One of the United Nations Millennium Development Goals is to reduce by one-half the proportion of people without reliable access to clean water and improved sanitation, but that will be difficult if water supplies become more stressed.

There is no simple definition of drought. What would be an amazingly wet year in a desert might be considered a catastrophic drought elsewhere. In general, a drought is an extended period of consistently below average precipitation that has a substantial impact on ecosystems, agriculture, and economies.

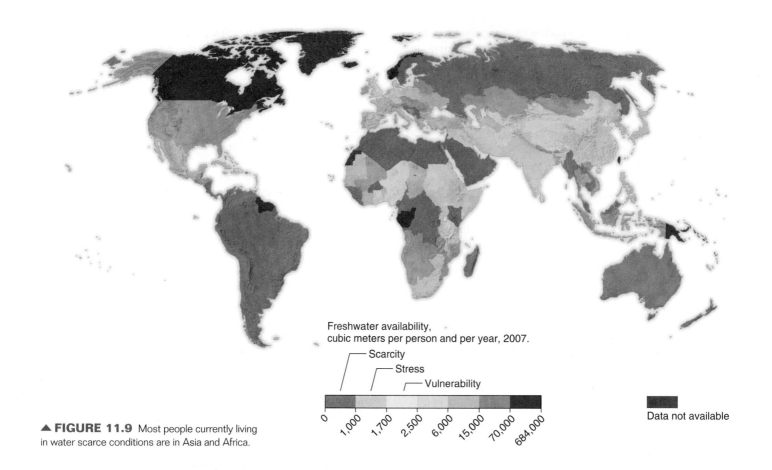

Freshwater availability,
cubic meters per person and per year, 2007.

Scarcity
Stress
Vulnerability

0 1,000 1,700 2,500 6,000 15,000 70,000 684,000

Data not available

▲ **FIGURE 11.9** Most people currently living in water scarce conditions are in Asia and Africa.

▲ **FIGURE 11.10** Women and children often spend hours every day collecting water—which often is unsafe for drinking—from local water sources.

The worst drought in recent United States history, in economic and social terms, was in the 1930s. Poor soil conservation practices and a series of dry years in the Great Plains combined to create the "dust bowl." Wind stripped topsoil from millions of hectares of land, and billowing dust clouds turned day into night. Thousands of families were forced to leave farms and migrate to other areas.

As the opening case study shows, much of the western United States has been exceptionally dry over the past decade. In many states, conditions are now worse than during the 1930s dust bowl.

Droughts in the American Southwest aren't a new phenomenon. In fact, the dry spells in recent years have been relatively short duration compared to historic events. The dust bowl of the 1930s, for example, only lasted about six years. By contrast, the megadroughts that destroyed the Anasazi (or Ancient Pueblo) cultures in the twelfth and thirteenth centuries (fig. 11.11) lasted between 25 and 50 years.

If the government had listened to Major John Wesley Powell, the settlement patterns in the western United States would be very different from what they are today, and we would be less worried that the Colorado River might dry up. Powell, who led the first expedition down the Colorado, went on to be the first head of the U.S. Geological Survey. In that capacity, he did a survey of the agricultural and settlement potential of the western desert. His conclusion, quoted at the beginning of this chapter, was that there isn't enough water to support a large human population.

Powell recommended that the political organization of the West be based on watersheds so that everyone in a given jurisdiction would be bound together by the available water. He thought that farms should be limited to local surface-water supplies, and that cities should be small, oasis settlements. Instead, we've built huge metropolitan areas, such as Los Angeles, Phoenix, Las Vegas, and Denver, in places where there is little or no natural water supply. Will they survive impending shortages?

Australia Adapts to Drought

Could this be a future many of us will face? Over the past decade, Australia has experienced the worst droughts and heat waves in its recorded history. Enormous fires swept huge swaths of the country, and dust storms blanked major cities for days. Elderly people died from heat exhaustion and asthmatic attacks, while Australia's sheep population fell by half. In some years rice and cotton production collapsed almost completely.

Australia is the driest inhabited continent (Antarctica is drier but has no permanent settlements) and has always had large extremes in rainfall from place to place and from year to year. The north

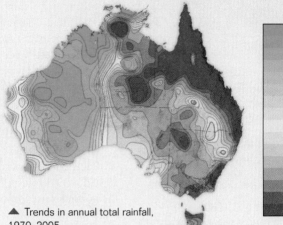

▲ Trends in annual total rainfall, 1970–2005.

coast has torrential rains and floods for part of the year, followed by searing heat and drought for months more, while the center and south are always hot and dry. But current trends suggest that global climate change is going to require new adaptations from all Australians. Since 1950, average rainfall has decreased 15 percent and average temperatures, which were already warm, have increased by about 0.6°C (1°F). The combination of higher evaporation and lower precipitation depletes soil moisture and reduces runoff, making droughts more intense and more frequent.

One of the regions hit hardest by these changing conditions is the Murray-Darling Basin in New South Wales. This fertile region, which covers about 1 million km² (about the size of Germany and France combined), produces one-third of the country's food and agricultural income and is considered Australia's food bowl. Stream flows have been falling in this watershed for decades due to drought and excess withdrawals, and water quality has suffered correspondingly. In 2002 the Murray River was completely dry in its lower reaches for the first time ever. Not only was this a crisis for farms and cities that depended on the river for their water supplies, but it was also a calamity for nature. The collapse of the Coorong wetlands at the mouth of the Murray is considered one of the country's worst environmental disasters. Shorebird and freshwater turtle populations fell by at least 50 percent. Old-growth forests of red river gums are dying, and along with them a unique biological community is disappearing.

This so-called Millennium Drought has gotten Australia's attention. A wide range of educational, technical, and regulatory strategies have been proposed to combat drought and climate change. Cities have adopted aggressive policies to reduce water use and to expand new and unconventional supplies. State and local governments now subsidize efficient appliances, such as dual-flush toilets, and have campaigns to educate the public about the need for water conservation. Outdoor car washing and lawn watering are banned during dry months. Between 2002 and 2008, per capita urban water use, which was already low compared to the United States, dropped by 37 percent.

Among the unconventional water sources being explored are cisterns to harvest rooftop rain runoff, systems that reuse gray water, and recycling of sewage water. The country's five largest cities are spending $13 billion to build desalination plants that will meet 30 percent of current urban water needs. In a few cases, treated sewage effluent is being used as drinking water. Former Australian environment minister Malcolm Turnbull said, "It may sound yucky, but we're not getting rain; we've got no choice."

The government is also reducing subsidies and rules that encourage inefficient or unproductive water use by cities and farms. The federal government has committed $3.1 billion over 10 years to purchase water rights in the Murray-Darling Basin. This water will be used to restore natural flows in rivers and wetlands. Regulators have introduced water markets in hopes of encouraging farmers to install efficient irrigation systems that will produce more food with less water.

Programs have been set up to help farmers plan for drought and adapt to changing conditions. Help is available for assessing conditions and mapping out responses. Farmers are urged to make difficult but needed judgments on how many animals or crops their land can sustain. Although the impulse is to hold on to flocks ("mobs" in Australia) as long as possible and to replant seed in hopes that next season will be better, advisors suggest that the first priority is health of the land. What's the best thing to do to ensure that the soil will still be there and the land will be able to support crops if the drought ends?

Notice that in figure 11.9, Australia, in spite of having so little natural rainfall over most of its territory, isn't considered likely to fall below the minimum requirements for clean water for human health. With enough money and public support, it's possible to adapt to extremely dry conditions. What do you think? Will other countries, such as the United States, rise to this challenge as well?

Groundwater supplies are being depleted

Groundwater provides nearly 40 percent of the fresh water for agricultural and domestic use in the United States. Nearly half of all Americans and about 95 percent of the rural population depend on groundwater for drinking and other domestic purposes. Overuse of these supplies dries up wells, natural springs, and even groundwater-fed wetlands, rivers, and lakes. Pollution of aquifers through dumping of contaminants on recharge zones, leaks through abandoned wells, or deliberate injection of toxic wastes can make this valuable resource unfit for use.

▲ **FIGURE 11.11** The stunning cliff dwelling of Mesa Verde National Park were abandoned by the Anasazi, or ancient Puebloan people, during a severe megadrought between 1275 and 1299. Will some of our modern cities face the same fate?

In many areas of the United States, groundwater is being withdrawn from aquifers faster than natural recharge can replace it. On a local level, this causes a cone of depression in the water table. On a broader scale, heavy pumping can deplete a whole aquifer. The Ogallala Aquifer underlies eight Great Plains states from Texas to North Dakota. This porous bed of sand, gravel, and sandstone once held more water than all the freshwater lakes, streams, and rivers on the earth. Excessive pumping for irrigation has removed so much water that wells have dried up in many places, and farms, ranches, even whole towns are being abandoned. Recharging many such aquifers will take thousands of years. Using "fossil" water like this is essentially water mining. For all practical purposes, these aquifers are nonrenewable resources.

Water withdrawal also allows aquifers to collapse. Subsidence, or sinking of the ground surface, follows. The San Joaquin Valley in California has sunk more than 10 m (33 ft) in the past 50 years because of excessive groundwater pumping. Where aquifers become compressed, recharge becomes impossible.

Another consequence of aquifer depletion is saltwater intrusion. Along coastlines and in areas where saltwater deposits are left from ancient oceans, overuse of freshwater reservoirs often allows saltwater to intrude into aquifers used for domestic and agricultural purposes.

Diversion projects redistribute water

Dams and canals are a foundation of civilization because they store and redistribute water for farms and cities. Many great civilizations have been organized around large-scale canal systems, including ancient empires of Sumeria, Egypt, and India. As modern dams and water diversion projects have grown in scale and number, though, their environmental costs have raised serious questions about efficiency, costs, and the loss of river ecosystems.

More than half of the world's 227 largest rivers have been dammed or diverted. Of the 50,000 large dams in the world, 90 percent were built in the twentieth century. Half of those are in China, and China continues to build and plan dams on its remaining rivers. Dams are justified in terms of flood control, water storage, and electricity production. However, the costs of relocating villages, as well as lost fishing, farming, and water losses to evaporation, are enormous. Economically speaking, at least one-third of the world's large dams should never have been built.

The largest water diversion project in the world is now being built in China. It's projected to move more than twice the flow of the American Colorado River. The initial cost estimate of this scheme is estimated to be (U.S.) $62 billion, but it could easily be twice that much. But without more water, Beijing, the national capital and home to about 20 million people, might have to be moved.

Las Vegas, Nevada, facing a similar situation with the drying of Lake Mead—which supplies 40 percent of its water—has started a $3.5 billion, 525 km (326 m) pipeline to tap aquifers in the northeastern part of the state. Local ranchers fear that groundwater pumping will decimate the range, destroy native vegetation, and cause massive dust storms. They point to Owens Valley in California, where a similar water grab by Los Angeles in 1913 dried up the river and destroyed both ranching and economic development. Las Vegas also has suggested that if local water supplies fail, they may ask states east of the Mississippi to share some of their water. If you live in a moist area, how would you feel about sharing your resources?

Las Vegas is also digging a $3.5 billion tunnel that will burrow into Lake Mead, 100 m (300 ft) below the normal outlet (fig. 11.12). Even if the lake reaches the "dead pool" level as warned in the beginning of this chapter, the city will still be able to draw off water. Of course this might prevent refilling the reservoir to provide water and power to downstream users. If you lived downstream, how would you feel about this outcome?

▲ **FIGURE 11.12** Hoover Dam powers Las Vegas, Nevada. Lake Mead, behind the dam, loses about 1.3 billion m³ per year to evaporation. Reduced inflows now threaten the viability of this system.

One of the most disastrous diversions in world history is that of the Aral Sea. Situated in arid Central Asia, on the border of Kazakhstan and Uzbekistan, the Aral Sea is a shallow inland sea fed by rivers from distant mountains. Starting in the 1950s, the Soviet Union began diverting these rivers to water cotton fields and rice fields. Gradually the Aral Sea has evaporated, leaving vast, toxic salt flats (fig. 11.13). The economic value of the cotton and rice has probably never equalled the cost of lost fisheries, villages, and health.

Recently some river flow has been restored to the "Small Aral" or northern lobe of the once-great sea. Water levels have risen 8 m and native fish are being reintroduced. It's hoped that one day commercial fishing may be resumed. The fate of the larger, southern remnant is more uncertain. There may never be enough water to refill it, and if there were, the toxins left in the lake bed could make it unusable anyway.

Questions of justice often surround dam projects

While dams provide hydroelectric power and water to distant cities, local residents often suffer economic and cultural losses. In some cases, dam builders have been charged with using public money to increase the value of privately held farmlands, as well as encouraging inappropriate farming and urban growth in arid lands.

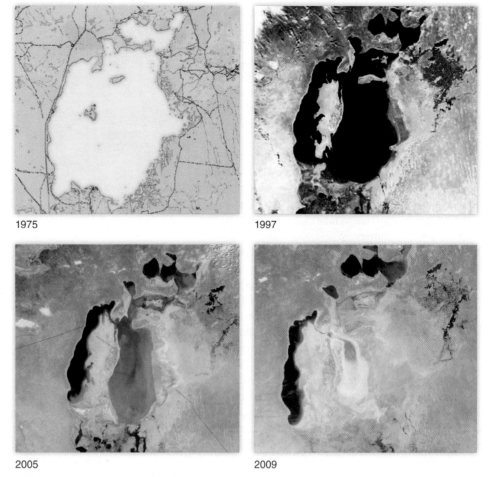

1975   1997

2005   2009

▲ **FIGURE 11.13** For nearly 40 years, rivers feeding the Aral Sea have been diverted to irrigate cotton and rice fields. The main body of the sea has lost more than 90 percent of its volume. Toxic dust storms from remaining salt flats now contaminate the region.

International Rivers, an environmental and human rights organization, reports that dam projects have forced more than 23 million people from their homes and land, and many are still suffering the impacts of dislocation years after it occurred. Often the people being displaced are ethnic minorities. Currently at least 144 dams on eight rivers in Southeast Asia have been proposed or are under construction. This includes the Lancang (Upper Mekong), the Nu (Upper Salween), and the Jinsha (Upper Yangtze). Several of these projects are in or adjacent to the Three Rivers World Heritage Site, threatening the ecological and cultural integrity of one of the most spectacular and biologically rich areas in the world.

There's increasing concern that big dams in seismically active areas can trigger earthquakes. In more than 70 cases worldwide, large dams have been linked with increased seismic activity. Geologists suggest that filling the reservoir behind the nearby Zipingpu Dam on the Min River caused the devastating 7.9-magnitude Sichuan earthquake that killed an estimated 90,000 people in 2008. If true, it would be the world's deadliest dam-induced earthquake ever. But it pales in comparison to the potential catastrophe if the Three Gorges Dam on the Yangtze were to collapse. As one engineer says, "It would be a flood of biblical proportions for the 100 million people who live downstream."

Would you fight for water?

Many environmental scientists have warned that water shortages could lead to wars between nations. *Fortune* magazine wrote, "Water will be to the 21st century what oil was to the 20th." With one-third of all humans living in areas with water stress now, the situation could become much worse as the population grows and climate change dries up some areas and brings more severe storms to others. Already we've seen skirmishes—if not outright warfare—over insufficient water. In Kenya, for instance, nomadic tribes have fought over dwindling water and grazing. An underlying cause of the genocide now occurring in the Darfur region of Sudan is water scarcity. When rain was plentiful, Arab pastoralists and African farmers coexisted peacefully. Drought—perhaps caused by global warming—has upset that truce. The hundreds of thousands who have fled to Chad could be considered climate refugees as well as war victims.

Although they haven't usually risen to the level of war, there have been at least 37 military confrontations in the past 50 years in which water has been at least one of the motivating factors. Thirty of those conflicts have been between Israel and its neighbors. India, Pakistan, and Bangladesh also have confronted each other over water rights, and Turkey and Iraq threatened to send their armies to protect access to the water in the Tigris and Euphrates Rivers. Water can even be used as a weapon. Saddam Hussein cut off water flow to the massive Iraq marshes as a way of punishing his enemies among the marsh Arabs. Drying of the marshes drove 140,000 people from their homes and destroyed a unique way of life. It also caused severe ecological damage to what is regarded by many as the original Garden of Eden.

Public anger over privatization of the public water supply in Bolivia sparked a revolution that overthrew the government in 2000. Water sales are already a $400-billion-a-year business. Multinational corporations are moving to take control of water systems in many countries. Who owns water and how much they are able to charge for it could become the question of the century. Investors are now betting on scarce water resources by buying future water rights. One Canadian water company, Global Water Corporation, puts it best: "Water has moved from being an endless commodity that may be taken for granted to a rationed necessity that may be taken by force."

Freshwater shortages may become much worse in the future because of global climate change. Figure 11.14 shows the best estimate from the Intergovernmental Panel on Climate Change (IPCC) on likely changes in global precipitation for the period 2090–2099 compared to 1980–1989. White areas are where models are inconclusive. How does this map compare to figure 11.2? Which areas are most likely to suffer from water shortages by the end of this century? How do these projections impact the case study on Lake Mead?

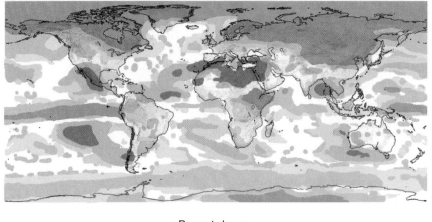

Percent change

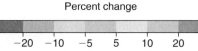

▲ FIGURE 11.14 Relative changes in precipitation (in percentage) for the period 2090–2099 compared to 1980–1989, predicted by the Intergovernmental Panel on Climate Change. SOURCE: IPCC, 2007.

11.5 WATER MANAGEMENT AND CONSERVATION

Watershed management and conservation are often more economical and environmentally sound ways to prevent flood damage and store water for future use than building huge dams and reservoirs. A **watershed**, or catchment, is all the land drained by a stream or river. It has long been recognized that retaining vegetation and groundcover in a watershed helps hold back rainwater and lessens downstream floods. In 1998 Chinese officials acknowledged that unregulated timber cutting upstream on the Yangtze contributed to massive floods that killed 30,000 people. Similarly, after disastrous floods in the upper Mississippi Valley in 1993, it was suggested that, rather than allowing residential, commercial, or industrial development on floodplains, these areas should be reserved for water storage, aquifer recharge, wildlife habitat, and agriculture. Unfortunately, this advice has been ignored in most places. Further discussion of flooding hazards can be found in chapter 12.

Sound farming and forestry practices can reduce runoff. Retaining crop residue on fields reduces flooding, and minimizing plowing and forest cutting on steep slopes protects watersheds. Effects of deforestation on weather and water supplies are discussed in chapter 6. Wetlands conservation preserves natural water storage

capacity and aquifer recharge zones. A river fed by marshes and wet meadows tends to run consistently clear and steady, rather than in violent floods.

A series of small dams on tributary streams can hold back water before it becomes a great flood. Ponds formed by these dams provide useful wildlife habitat and stock-watering facilities. They also catch soil where it could be returned to the fields. Small dams can be built with simple equipment and local labor, eliminating the need for massive construction projects and huge dams.

In 1998 U.S. Forest Service chief Mike Dombeck announced a major shift in his agency's priorities. "Water," he said, "is the most valuable and least appreciated resource the national forests provide. More than 60 million people in 33 states obtain their drinking water from national forest lands. Protecting watersheds is far more economically important than logging or mining, and will be given the highest priority in forest planning."

Everyone can help conserve water

We could probably save as much as half of the water we now use for domestic purposes without great sacrifice or serious changes in our lifestyles. Simple steps, such as taking shorter showers, fixing leaks, and washing cars, dishes, and clothes as efficiently as possible, can go a long way toward forestalling the water shortages that many authorities predict. Isn't it better to adapt to more conservative uses now when we have a choice than to be forced to do it by scarcity in the future?

Water-conserving appliances, such as low-volume showerheads and efficient dishwashers, can reduce water consumption greatly (see What Can You Do? p. 263). If you live in an arid part of the country, you might consider whether you really need a lush, green lawn that requires constant watering, feeding, and care. Planting native vegetation can be both ecologically sound and aesthetically pleasing (fig. 11.15). As part of its water conservation

▲ **FIGURE 11.15** By using native plants in a natural setting, residents of Phoenix save water and fit into the surrounding landscape.

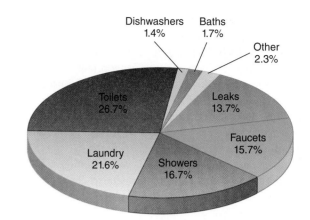

▲ **FIGURE 11.16** Typical household water use in the United States. SOURCE: Data from the American Water Works Association, 2010.

efforts, Las Vegas is paying residents to replace turf with natural vegetation, has asked golf courses to rip up fairways, is encouraging hotels to use recycled water in fountains, and water cops patrol the streets to identify illegal watering or car washing.

Toilets are our greatest domestic water user (fig. 11.16). Usually each flush uses several gallons of water to dispose of a few ounces of waste. On average, each person in the United States uses about 50,000 L (13,000 gal) of drinking-quality water annually to flush toilets. Low-flush toilets can drastically reduce this water use. Gray water (recycled from other uses) could be used for flushing, but installing separated plumbing systems is expensive.

California already uses more than 555 million m³ (450,000 acre-feet) of recycled water annually—mostly for irrigation. That's equivalent to about two-thirds of the water consumed by Los Angeles every year.

Efficiency is reducing water use in many areas

Growing recognition that water is a precious and finite resource has changed policies and encouraged conservation across the United States. Despite a growing population, the United States is now saving some 144 million liters (38 million gal) per day—a tenth the volume of Lake Erie—compared with per capita consumption rates of 30 years ago. With nearly 90 million more people in the United States now than in 1980, we get by with 10 percent less water. New requirements for water-efficient fixtures in many cities help conserve water on the home front.

Pricing has an effect on our water usage. Ironically, water from Lake Mead, which is facing a supply crisis, currently costs Las Vegas residents 33 cents per m³. By comparison, the same amount costs $3 in Atlanta and $7 in Copenhagen, where water is abundant. What do you think those prices do to motivate conservation?

What Can YOU DO?

Saving Water and Preventing Pollution

Each of us can conserve much of the water we use and avoid water pollution in many simple ways.

- Don't flush every time you use the toilet. Take shorter showers, and shower instead of taking baths.
- Don't let the faucet run while brushing your teeth or washing dishes. Draw a basin of water for washing and another for rinsing dishes. Don't run the dishwasher when it's half full.
- Use water-conserving appliances: low-flow showers, low-flush toilets, and aerated faucets.
- Fix leaking faucets, tubs, and toilets. A leaky toilet can waste 50 gal per day. To check your toilet, add a few drops of dark food coloring to the tank and wait 15 minutes. If the tank is leaking, the water in the bowl will change color.
- Put a brick or full water bottle in your toilet tank to reduce the volume of water in each flush.
- Dispose of used motor oil, household hazardous waste, batteries, and so on, responsibly. Don't dump anything down a storm sewer that you wouldn't want to drink.
- Avoid using toxic or hazardous chemicals for simple cleaning or plumbing jobs. A plunger or plumber's snake will often unclog a drain just as well as caustic acids or lye. Hot water and soap can accomplish most cleaning tasks.
- If you have a lawn, use water, fertilizer, and pesticides sparingly. Plant native, low-maintenance plants that have low water needs.
- If possible, use recycled (gray) water for lawns, house plants, and car washing.

Charging a higher proportion of real costs to users of public water projects can rationalize use patterns as will water marketing policies that allow prospective users to bid on water rights. Some countries already have effective water pricing and allocation policies that encourage the most socially beneficial uses and discourage wasteful water uses. It will be important, as water markets develop, to be sure that environmental, recreational, and wildlife values are not sacrificed to the lure of high-bidding industrial and domestic users.

Several countries with desperate water shortages, including Singapore, Australia, and parts of the United States, are using recycled water for drinking (see What Do You Think? p. 259).

11.6 WATER POLLUTION

Any physical, biological, or chemical change in water quality that adversely affects living organisms or makes water unsuitable for desired uses can be considered pollution. There are natural sources of water contamination, such as poison springs, oil seeps, and sedimentation from erosion, but here we'll focus primarily on human-caused changes that affect water quality or usability.

▲ **FIGURE 11.17** Sewer outfalls, industrial effluent pipes, acid draining out of abandoned mines, and other point sources of pollution are generally easy to recognize. Pollution-control laws have made this sight less common today than it once was.

Pollution includes point sources and nonpoint sources

Pollution-control standards and regulations usually distinguish between point and nonpoint pollution sources. Factories, power plants, sewage treatment plants, underground coal mines, and oil wells are classified as **point sources** because they discharge pollution from specific locations, such as drainpipes, ditches, or sewer outfalls (fig. 11.17). These sources are discrete and identifiable, so they are relatively easy to monitor and regulate. It is generally possible to divert effluent from the waste streams of these sources and treat it before it enters the environment.

In contrast, **nonpoint sources** of water pollution are diffuse, having no specific location where they discharge into a particular body of water. They are much harder to monitor and regulate than point sources because their origins are hard to identify. Nonpoint sources include runoff from farm fields and feedlots, golf courses, lawns and gardens, construction sites, logging areas, roads, streets, and parking lots. While point sources may be fairly uniform and predictable throughout the year, nonpoint sources are often highly episodic. The first heavy rainfall after a dry period may flush high concentrations of gasoline, lead, oil, and rubber residues off city streets, for instance, while subsequent runoff may be much cleaner.

Perhaps the ultimate in diffuse, nonpoint pollution is atmospheric deposition of contaminants carried by air currents and precipitated into watersheds or directly onto surface waters as rain, snow, or dry particles. The Great Lakes, for example, have been found to be accumulating industrial chemicals, such as PCBs (polychlorinated biphenyls) and dioxins, as well as agricultural toxins, such as the insecticide toxaphene, that cannot be accounted for by local sources alone. The nearest sources for many of these chemicals are sometimes thousands of kilometers away.

Biological pollution includes pathogens and waste

Although the types, sources, and effects of water pollutants are often interrelated, it is convenient to divide them into major categories for discussion (table 11.3). Here, we look at some of the important sources and effects of different pollutants.

Pathogens The most serious water pollutants in terms of human health worldwide are pathogenic (disease-causing) organisms (chapter 8). Among the most important waterborne diseases are typhoid, cholera, bacterial and amoebic dysentery, enteritis, polio, infectious hepatitis, and schistosomiasis. Malaria, yellow fever, and filariasis are transmitted by insects that have aquatic larvae. Altogether, at least 25 million deaths each year are blamed on water-related diseases. Nearly two-thirds of the mortalities of children under 5 years old in poorer countries are linked to these diseases.

The main source of these pathogens is untreated or improperly treated human wastes. Animal wastes from feedlots or fields near waterways and food-processing factories with inadequate waste treatment facilities also are sources of disease-causing organisms.

In developed countries, sewage treatment plants and other pollution-control techniques have reduced or eliminated most of the worst sources of pathogens in inland surface waters. Furthermore, drinking water is generally disinfected by chlorination, so epidemics of waterborne diseases are rare in these countries. The United Nations estimates that 90 percent of the people in developed countries have adequate (safe) sewage disposal, and 95 percent have clean drinking water.

The situation is quite different in less-developed countries, where billions of people lack adequate sanitation and access to clean drinking water. Conditions are especially bad in remote, rural areas, where sewage treatment is usually primitive or nonexistent

TABLE 11.3 | Major Categories of Water Pollutants

CATEGORY	EXAMPLES	SOURCES
CAUSE OF ECOSYSTEM DISRUPTION		
1. Oxygen-demanding wastes	Animal manure, plant residues	Sewage, agricultural runoff, paper mills, food processing
2. Plant nutrients	Nitrates, phosphates, ammonium	Agricultural and urban fertilizers, sewage, manure
3. Sediment	Soil, silt	Land erosion
4. Thermal changes	Heat	Power plants, industrial cooling
CAUSE OF HEALTH PROBLEMS		
1. Pathogens	Bacteria, viruses, parasites	Human and animal excreta
2. Inorganic chemicals	Salts, acids, caustics, metals	Industrial effluents, household cleansers, surface runoff
3. Organic chemicals	Pesticides, plastics, detergents, oil, gasoline	Industrial, household, and farm use
4. Radioactive materials	Uranium, thorium, cesium, iodine, radon	Mining and processing of ores, power plants, weapons production, natural sources

and purified water is either unavailable or too expensive to obtain. The World Health Organization estimates that 80 percent of all sickness and disease in less-developed countries can be attributed to waterborne infectious agents and inadequate sanitation.

Detecting specific pathogens in water is difficult, time-consuming, and costly, so water quality is usually described in terms of concentrations of **coliform bacteria**—any of the many types that commonly live in the colon, or intestines, of humans and other animals. The most common of these is *Escherichia coli* (or *E. coli*). Other bacteria, such as *Shigella*, *Salmonella*, or *Listeria*, can also cause serious, even fatal, illness. If any coliform bacteria are present in a water sample, infectious pathogens are assumed to be present as well, and the Environmental Protection Agency (EPA) considers the water unsafe for drinking.

Biological Oxygen Demand   The amount of oxygen dissolved in water is a good indicator of water quality and of the kinds of life it will support. An oxygen content above 6 parts per million (ppm) will support game fish and other desirable forms of aquatic life. At oxygen levels below 2 ppm, water will support mainly worms, bacteria, fungi, and other detritus feeders and decomposers. Oxygen is added to water by diffusion from the air, especially when turbulence and mixing rates are high, and by photosynthesis of green plants, algae, and cyanobacteria. Turbulent, rapidly flowing water is constantly aerated, so it often recovers quickly from oxygen-depleting processes. Oxygen is removed from water by respiration and chemical processes that consume oxygen. Because oxygen is so important in water, **dissolved oxygen (DO)** levels are often measured to compare water quality in different places.

Adding organic materials, such as sewage or paper pulp, to water stimulates activity and oxygen consumption by decomposers. Consequently, **biochemical oxygen demand (BOD)**, or the amount of dissolved oxygen consumed by aquatic microorganisms, is another standard measure of water contamination. Alternatively, chemical oxygen demand (COD) is a measure of all organic matter in water.

Downstream from a point source, such as a municipal sewage plant discharge, a characteristic decline and restoration of water quality can be detected either by measuring DO content or by observing the types of flora and fauna that live in successive sections of the river. The oxygen decline downstream is called the **oxygen sag** (fig. 11.18). Upstream from the pollution source, oxygen levels support normal populations of clean-water organisms. Immediately below the source of pollution, oxygen levels begin to fall as decomposers metabolize waste materials. Rough fish, such as carp, bullheads, and gar, are able to survive in this oxygen-poor environment, where they eat both decomposer organisms and the waste itself.

Farther downstream, the water may become so oxygen depleted that only the most resistant microorganisms and invertebrates can survive. Eventually most of the nutrients are used up, decomposer populations are smaller, and the water becomes oxygenated once again. Depending on the volumes and flow rates of the effluent plume and the river receiving it, normal communities may not appear for several miles downstream.

Plant Nutrients and Cultural Eutrophication   Water clarity (transparency) is affected by sediments, chemicals, and the abundance of plankton organisms; clarity is a useful measure of water quality and water pollution. Rivers and lakes that have clear water and low biological productivity are said to be **oligotrophic** (*oligo* = little + *trophic* = nutrition). By contrast, **eutrophic** (*eu* + *trophic* = well-nourished) waters are rich in organisms and organic materials. Eutrophication, an increase in nutrient levels and biological productivity, often accompanies successional changes (chapter 3) in lakes. Tributary streams bring in sediments and nutrients that stimulate plant growth. Over time, ponds and lakes often fill in, becoming marshes or even terrestrial biomes. The rate of eutrophication depends on water chemistry and depth, volume of inflow, mineral content of the surrounding watershed, and biota of the lake itself.

Human activities can greatly accelerate eutrophication, an effect called **cultural eutrophication**. Cultural eutrophication is caused mainly by increased nutrient input into a water body. Increased productivity in an aquatic system sometimes can be beneficial. Fish and other desirable species may grow faster, providing a welcome food source. Often, however, eutrophication

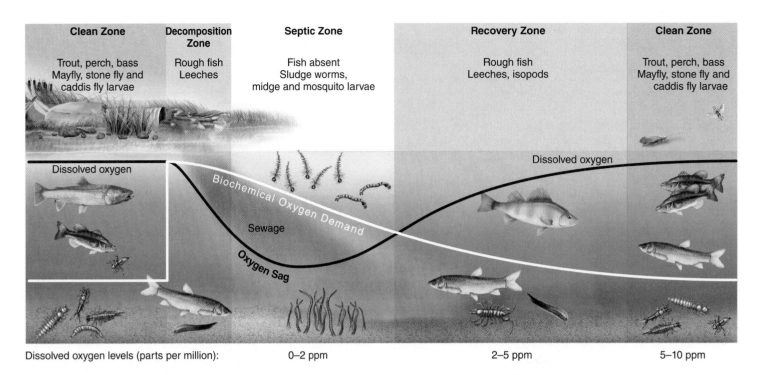

Clean Zone	Decomposition Zone	Septic Zone	Recovery Zone	Clean Zone
Trout, perch, bass Mayfly, stone fly and caddis fly larvae	Rough fish Leeches	Fish absent Sludge worms, midge and mosquito larvae	Rough fish Leeches, isopods	Trout, perch, bass Mayfly, stone fly and caddis fly larvae

Dissolved oxygen

Biochemical Oxygen Demand

Sewage

Oxygen Sag

Dissolved oxygen

Dissolved oxygen levels (parts per million): 0–2 ppm 2–5 ppm 5–10 ppm

▲ **FIGURE 11.18** Oxygen sag downstream of an organic source. A great deal of time and distance may be required for the stream and its inhabitants to recover.

produces "blooms" of algae or thick growths of aquatic plants stimulated by elevated phosphorus or nitrogen levels (fig. 11.19). Bacterial populations then increase, fed by larger amounts of organic matter. The water often becomes cloudy, or turbid, and has unpleasant tastes and odors. Cultural eutrophication can accelerate the "aging" of a water body enormously over natural rates. Lakes and reservoirs that normally might exist for hundreds or thousands of years can be filled in a matter of decades.

▼ **FIGURE 11.19** Eutrophic lake. Nutrients from agriculture and domestic sources have stimulated growth of algae and aquatic plants. This reduces water quality, alters species composition, and lowers the lake's recreational and aesthetic values.

Eutrophication also occurs in marine ecosystems, especially in nearshore waters and partially enclosed bays or estuaries. Partially enclosed seas, such as the Black, Baltic, and Mediterranean Seas, tend to be in especially critical condition. During the tourist season, the coastal population of the Mediterranean, for example, swells to 200 million people. Eighty-five percent of the effluents from nearby large cities go untreated into the sea. Beach pollution, fish kills, and contaminated shellfish result. Extensive "dead zones" often form where rivers dump nutrients into estuaries and shallow seas. (See related story "Studying the Gulf Dead Zone" at www.mhhe.com/cunningham7e). A federal study of the condition of U.S. coastal waters found that 28 percent of estuaries are impaired for aquatic life, and 80 percent of all coastal water is in fair to poor condition.

Marine animals in hypoxic zones die not only because of depleted oxygen, but also because of high concentrations of harmful organisms, including toxic algae, pathogenic fungi, and parasitic protists. Excessive nutrients support blooms of these deadly aquatic microorganisms in polluted nearshore waters. Red tides—and tides of other colors, depending on the species involved—have become increasingly common where nutrients and wastes wash down rivers.

Inorganic pollutants include metals, salts, and acids

Some toxic inorganic chemicals are naturally released into water from rocks by weathering processes (chapter 12). Humans accelerate the transfer rates in these cycles thousands of times above natural background levels by mining, processing, using, and discarding minerals.

When Ashok Gadgil was a child in Bombay, India, five of his cousins died in infancy from diarrhea spread by contaminated water. Although he didn't understand the implications of those deaths at the time, as an adult he realized how heartbreaking and preventable those deaths were. After earning a degree in physics from the University of Bombay, Gadgil moved to the University of California at Berkeley, where he was awarded a PhD in 1979. He's now senior staff scientist in the Environmental Energy Technology Division, where he works on solar energy and indoor air pollution.

But Dr. Gadgil wanted to do something about the problem of waterborne diseases in India and other developing countries. Although progress has been made in bringing clean water to poor people in many countries, about a billion people still lack access to safe drinking water. After studying ways to sterilize water, he decided that UV light treatment had the greatest potential for poor countries. It requires far less energy than boiling, and it takes less sophisticated chemical monitoring than chlorination.

▲ Women carry water from the village WaterHealth kiosk. More than 6 million people's lives have been improved by this innovative system of water purification.

There are many existing UV water treatment systems, but they generally involve water flowing around an unshielded fluorescent lamp. However, minerals in the water collect on the glass lamp and must be removed regularly to maintain effectiveness. Regular disassembly, cleaning, and reassembly of the apparatus are difficult in primitive conditions. The solution, Gadgil realized, was to mount the UV source above the water where it couldn't

develop mineral deposits. He designed a system in which water flows through a shallow, stainless steel trough. The apparatus can be gravity fed and requires only a car battery as an energy source.

The system can disinfect 15 liters (4 gallons) of water per minute, killing more than 99.9 percent of all bacteria and viruses. This produces enough clean water for a village of 1,000 people. This simple system costs only about 5 cents per ton (950 liter). Of course, removing pathogens doesn't do anything about minerals, such as arsenic, or dangerous organic chemicals, so UV sterilization is often combined with filtering systems to remove those contaminants.

WaterHealth International, the company founded to bring this technology to market, now makes several versions of Gadgil's disinfection apparatus for different applications. A popular version provides a complete water purification system including a small kiosk, jugs for water distribution, and training on how to operate everything.

A village-size system costs about $5,000. Grants and loans are available for construction, but villagers own and run the facility to ensure there's local responsibility. Each family in the cooperative pays about $1 per month for pure water. These systems have been installed in thousands of villages in India, Bangladesh, Africa, and the Philippines. Currently, about 6.6 million people are getting clean, healthy water at an easily affordable price from the simple system Dr. Gadgil invented.

Among the chemicals of greatest concern are heavy metals, such as mercury, lead, tin, and cadmium. Supertoxic elements, such as selenium and arsenic, also have reached hazardous levels in some waters. Other inorganic materials, such as acids, salts, nitrates, and chlorine, that are nontoxic at low concentrations may become concentrated enough to lower water quality and adversely affect biological communities.

Metals Many metals, such as mercury, lead, cadmium, and nickel, are highly toxic in minute concentrations. Because metals are highly persistent, they accumulate in food chains and have a cumulative effect in humans.

Currently the most widespread toxic metal contamination in North America is mercury released from incinerators and coal-

burning power plants. Transported through the air, mercury precipitates in water supplies, where it bioconcentrates in food webs to reach dangerous levels in top predators. As a general rule, Americans are warned not to eat more than one meal of wild-caught fish per week (fig. 11.20). Top marine predators, such as shark, swordfish, bluefin tuna, and king mackerel, tend to have especially high mercury content. Pregnant women and small children should avoid these species entirely. Public health officials estimate that 600,000 American children now have mercury levels in their bodies high enough to cause mental and developmental problems, while one woman in six in the United States has blood-mercury concentrations that would endanger a fetus.

Mine drainage and leaching of mining wastes are serious sources of metal pollution in water. A survey of water quality in

▲ **FIGURE 11.20** Mercury contamination is the most common cause of impairment of U.S. rivers and lakes. Forty states have issued warnings about eating locally caught freshwater fish. Long-lived, top predators are especially likely to bioaccumulate toxic concentrations of mercury.

eastern Tennessee found that 43 percent of all surface streams and lakes and more than half of all groundwater used for drinking supplies were contaminated by acids and metals from mine drainage. In some cases, metal levels were 200 times higher than what is considered safe for drinking water.

Nonmetallic Salts Some soils contain high concentrations of soluble salts, including toxic selenium and arsenic (see related story "Arsenic in Drinking Water" at www.mhhe.com/cunningham7e). Tens of millions of people are at risk in India and Bangladesh where groundwater is polluted with arsenic. Irrigation and drainage of desert soils can mobilize toxic salts and result in serious pollution problems, as in Kesterson Marsh in California, where selenium poisoning killed thousands of migratory birds in the 1980s.

Ordinarily nontoxic salts, such as sodium chloride (table salt), that are harmless at low concentrations also can be mobilized by irrigation and concentrated by evaporation, reaching levels that are dangerous for plants and animals. Salinity levels in the Colorado River and surrounding farm fields have become so high in recent years that millions of hectares of valuable croplands have had to be abandoned. In northern states, millions of tons of sodium chloride and calcium chloride are used to melt road ice in the winter. Leaching of road salts into surface waters has deleterious effects on aquatic ecosystems.

Acids and Bases Acids are released as by-products of industrial processes, such as leather tanning, metal smelting and plating, petroleum distillation, and organic chemical synthesis. Coal mining is an especially important source of acid water pollution. Sulfur compounds in coal react with oxygen and water to make sulfuric acid. Thousands of kilometers of streams in the United States have been acidified by acid mine drainage, some so severely that they are essentially lifeless.

Acid precipitation (chapter 10) also acidifies surface-water systems. In addition to damaging living organisms directly, these acids leach aluminum and other elements from soil and rock, further destabilizing ecosystems.

Organic chemicals include pesticides and industrial substances

Thousands of different natural and synthetic organic chemicals are used in the chemical industry to make pesticides, plastics, pharmaceuticals, pigments, and other products that we use in everyday life. Many of these chemicals are highly toxic (chapter 8). Exposure to very low concentrations (perhaps even parts per quadrillion, in the case of dioxins) can cause birth defects, genetic disorders, and cancer. Some can persist in the environment because they are resistant to degradation and toxic to organisms that ingest them.

The two principal sources of toxic organic chemicals in water are (1) improper disposal of industrial and household wastes and (2) pesticide runoff from farm fields, forests, roadsides, golf courses, and private lawns. The EPA estimates that about 500,000 metric tons of pesticides are used in the United States each year. Much of this washes into the nearest waterway, where it passes through ecosystems and may accumulate in high levels in nontarget organisms. The bioaccumulation of DDT in aquatic ecosystems was one of the first of these pathways to be understood (chapter 8). Dioxins and other chlorinated hydrocarbons (hydrocarbon molecules that contain chlorine atoms) have been shown to accumulate to dangerous levels in the fat of salmon, fish-eating birds, and humans and to cause health problems similar to those resulting from toxic metal compounds.

Hundreds of millions of tons of hazardous organic wastes are thought to be stored in dumps, landfills, lagoons, and underground tanks in the United States (chapter 14). Many, perhaps most, of these sites have leaked toxic chemicals into surface waters, groundwater, or both. The EPA estimates that about 26,000 hazardous waste sites will require cleanup because they pose an imminent threat to public health, mostly through water pollution.

Is bottled water safer?

It has become trendy to drink bottled water. Every year, Americans buy about 28 billion bottles of water at a cost of about $15 billion with the mistaken belief that it's safer than tap water. Worldwide, some 160 billion liters (42 billion gallons) of bottled water are consumed annually. Public health experts say that municipal water is often safer than bottled water because most large cities test their water supplies every hour for up to 25 different chemicals and pathogens, while the requirements for bottled water are much less rigorous. About one-quarter of all bottled water in the United States is simply reprocessed municipal water, and much of the rest is drawn from groundwater aquifers, which may or may not be safe. A recent survey of bottled water in China found that two-thirds of the samples tested had dangerous levels of pathogens and toxins.

Though the plastics used for bottling water are easily recycled, 80 percent of the bottles purchased in the United States end up in a landfill (the recycling rate is even poorer in most other countries). Overall, the average energy cost to make the plastic, fill the bottle, transport it to market, and then deal with the waste would be "like filling up a quarter of every bottle with oil," says water-expert Peter Gleick. Furthermore, it takes 3 to 5 times as much water to make the bottles as they hold. In blind tasting tests, most adults either can't tell the difference between municipal and bottled water, or they actually prefer municipal water. Furthermore, if water is held in a plastic bottle for weeks or months, plasticizers and other toxic chemicals can leach from the bottle into the water.

In most cases, bottled water is expensive, wasteful, and often less safe than most municipal water. Drink tap water and do a favor for your environment, your budget, and, possibly, your health.

Sediment and heat also degrade water

Sediment is a natural and necessary part of river systems. Sediment fertilizes floodplains and creates fertile deltas. But human activities, chiefly farming and urbanization, greatly accelerate erosion and increase sediment loads in rivers. Silt and sediment are considered the largest source of water pollution in the United States, being responsible for 40 percent of the impaired river miles in EPA water quality surveys. Cropland erosion contributes about 25 billion metric tons of soil, sediment, and suspended solids to world surface waters each year. Forest disturbance, road building, urban construction sites, and other sources add at least 50 billion additional tons.

This sediment fills lakes and reservoirs, obstructs shipping channels, clogs hydroelectric turbines, and makes purification of drinking water more costly. Sediments smother gravel beds in which insects take refuge and fish lay their eggs. Sunlight is blocked, so that plants cannot carry out photosynthesis, and oxygen levels decline. Murky, cloudy water also is less attractive for swimming, boating, fishing, and other recreational uses (fig. 11.21). Sediment washed into the ocean clogs estuaries and coral reefs.

Thermal pollution, usually effluent from cooling systems of power plants or other industries, alters water temperature. Raising or lowering water temperatures from normal levels can adversely affect water quality and aquatic life. Water temperatures are usually much more stable than air temperatures, so aquatic organisms tend to be poorly adapted to rapid temperature changes. Lowering the temperature of tropical oceans by even 1°C can be lethal to some corals and other reef species. Raising water temperatures can have similar devastating effects on sensitive organisms. Oxygen solubility in water decreases as temperatures increase, so species requiring high oxygen levels are adversely affected by warming water.

Humans also cause thermal pollution by altering vegetation cover and runoff patterns. Reducing water flow, clearing streamside trees, and adding sediment all make water warmer and alter the ecosystems in a lake or stream.

Warm-water plumes from power plants often attract fish and birds, which find food and refuge there, especially in cold weather. This artificial environment can be a fatal trap, however. Florida's manatees, an endangered mammal, are attracted to the abundant food supply and warm water in power plant thermal plumes. Often they are enticed into spending the winter much farther north than they normally would. On several occasions, a midwinter power plant breakdown has exposed a dozen or more of these rare animals to a sudden, deadly thermal shock.

▲ **FIGURE 11.21** Sediment and industrial waste flow from this drainage canal into Lake Erie.

11.7 WATER QUALITY TODAY

Surface-water pollution is often both highly visible and one of the most common threats to environmental quality. In more-developed countries, reducing water pollution has been a high priority over the past few decades. Billions of dollars have been spent on control programs, and considerable progress has been made. Still, much remains to be done.

The 1972 Clean Water Act protects our water

Like most developed countries, the United States and Canada have made encouraging progress in protecting and restoring water quality in rivers and lakes over the past 40 years. In 1948 only about one-third of Americans were served by municipal sewage systems, and most of those systems discharged sewage without any treatment or with only primary treatment (the bigger lumps of waste are removed). Most people depended on cesspools and septic systems to dispose of domestic wastes.

Areas of Progress The 1972 Clean Water Act established a National Pollution Discharge Elimination System (NPDES), which requires an easily revoked permit for any industry, municipality, or other entity dumping wastes in surface waters. The permit requires disclosure of what is being dumped and gives regulators valuable data and evidence for litigation. As a consequence, only about 10 percent of our water pollution now comes from industrial and municipal point sources. One of the biggest improvements has been in sewage treatment.

Since the Clean Water Act was passed in 1972, the United States has spent more than $180 billion in public funds and perhaps ten times as much in private investments on water pollution control. Most of that effort has been aimed at point sources, especially to build or upgrade thousands of municipal sewage treatment plants. As a result, nearly everyone in urban areas is now served by municipal sewage systems, and no major city discharges raw sewage into a river or lake except as overflow during heavy rainstorms.

This campaign has led to significant improvements in surface-water quality in many places. Fish and aquatic insects have returned to waters that formerly were depleted of life-giving oxygen. Swimming and other water-contact sports are again permitted in rivers, in lakes, and at ocean beaches that once were closed by health officials.

The Clean Water Act goal of making all U.S. surface waters "fishable and swimmable" has not been fully met, but currently the EPA reports that 91 percent of all monitored river miles and 88 percent of all assessed lake acres are suitable for their designated uses (fig. 11.22). This sounds good, but you have to remember that not all water bodies are monitored. Furthermore, the designated goal for some rivers and lakes is merely to be "boatable." Water quality doesn't have to be very high to allow boating. Even in "fishable" rivers and lakes, there isn't a guarantee that you can catch anything other than rough fish, such as carp or bullheads, nor can you be sure that what you catch is safe to eat. Even with billions of dollars of investment in sewage treatment plants, elimination of much of the industrial dumping and other gross sources of pollutants, and a general improvement in water quality, the EPA reports that 21,000 water bodies still do not meet their designated uses. According to the EPA, an overwhelming majority of the American people—almost 218 million—live within 16 km (10 mi) of an impaired water body.

In 1998 a new regulatory approach to water quality assurance was instituted by the EPA. Rather than issue standards on a river-by-river approach or factory-by-factory permit discharge, the focus was changed to watershed-level monitoring and protection. Some 4,000 watersheds are now monitored for water quality. You can find information about your watershed at www.epa.gov/owow/tmdl/. The intention of this program is to give the public more and better information about the health of their watersheds. In addition, states can have greater flexibility as they identify impaired water bodies and set priorities, and new tools can be used to achieve goals. States are required to identify waters not meeting water quality goals and to develop **total maximum daily loads**

▲ **FIGURE 11.22** Not all rivers and lakes are "fishable or swimmable," but we've made substantial progress since the Clean Water Act was passed in 1972.

(TMDL) for each pollutant and each listed water body. A TMDL is the amount of a particular pollutant that a water body can receive from both point and nonpoint sources. It considers seasonal variation and includes a margin of safety.

Currently, all 56 U.S. states and territories have submitted TMDL lists, and the EPA has approved most of them. Of the 5.6 million km of rivers monitored, only 480,000 km fail to meet their clean water goals. Similarly, of 40 million lake hectares, only 12.5 percent (in about 20,000 lakes) fail to meet their goal. To give states more flexibility in planning, the EPA has proposed new rules that include allowances for reasonably foreseeable increases in pollutant loadings to encourage "Smart Growth." In the future, TMDLs also will include load allocations from all nonpoint sources, including air deposition and natural background levels.

An encouraging example of improved water quality is seen in Lake Erie. Although widely regarded as "dead" in the 1960s, the lake today is promoted as the "walleye capital of the world." Bacteria counts and algae blooms have decreased more than 90 percent since 1962. Water that once was murky brown is now clear. Interestingly, part of the improved water quality is due to immense numbers of exotic zebra mussels, which filter the lake water very efficiently. Swimming is now officially safe along 96 percent of the lake's shoreline. Nearly 40,000 nesting pairs of double-crested cormorants nest in the Great Lakes region, up from only about 100 in the 1970s.

Canada's 1970 Water Act has produced comparable results. Seventy percent of all Canadians in towns over 1,000 population are now served by some form of municipal sewage treatment. In Ontario, the vast majority of those systems include tertiary treatment. After ten years of controls, phosphorus levels in the Bay of Quinte in the northeast corner of Lake Ontario have dropped nearly by half, and algal blooms that once turned waters green are less frequent and less intense than they once were. Elimination of mercury discharges from a pulp and paper mill on the Wabigoon-English River system in western Ontario has resulted in a dramatic decrease in mercury contamination. Twenty years ago this mercury contamination was causing developmental retardation in local residents. Extensive flooding associated with hydropower projects has raised mercury levels in fish to dangerous levels elsewhere, however.

Remaining Problems The greatest impediments to achieving national goals in water quality in both the United States and Canada are sediment, nutrients, and pathogens, especially from nonpoint discharges of pollutants (fig. 11.23). These sources are harder to identify and to reduce or treat than are specific point sources. About three-fourths of the water pollution in the United States comes from soil erosion, fallout of air pollutants, and surface runoff from urban areas, farm fields, and feedlots. In the United States, as much as 25 percent of the 46,800,000 metric tons (52 million tons) of fertilizer spread on farmland each year is carried away by runoff.

Cattle in feedlots produce some 129,600,000 metric tons (144 million tons) of manure each year, and the runoff from these sites is rich in viruses, bacteria, nitrates, phosphates, and other contaminants. A single cow produces about 30 kg (66 lb) of manure per day. Some feedlots have 100,000 animals with no provision for capturing or treating runoff water. Imagine drawing

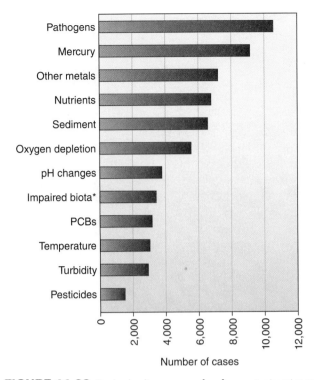

▲ **FIGURE 11.23** Twelve leading causes of surface-water impairment in the United States. *Undetermined causes. SOURCE: Data EPA, 2009.

▲ **FIGURE 11.24** Half of the water in China's major rivers is too polluted to be suitable for any human use. Although the government has spent billions of yuan in recent years, dumping of industrial and domestic waste continues at dangerous levels.

your drinking water downstream from such a facility. Pets also can be a problem. It is estimated that the wastes from about a half million dogs in New York City are disposed of primarily through storm sewers and therefore do not go through sewage treatment.

Loading of both nitrates and phosphates in surface water have decreased from point sources but have increased about fourfold since 1972 from nonpoint sources. Fossil fuel combustion has become a major source of nitrates, sulfates, arsenic, cadmium, mercury, and other toxic pollutants that find their way into water. Carried to remote areas by atmospheric transport, these combustion products now are found nearly everywhere in the world. Toxic organic compounds, such as DDT, PCBs, and dioxins, also are transported long distances by wind currents.

Developing countries often have serious water pollution

Japan, Australia, and most of western Europe also have improved surface-water quality in recent years. Sewage treatment in the wealthier countries of Europe generally equals or surpasses that in the United States. Sweden, for instance, serves 98 percent of its population with at least secondary sewage treatment (compared with 70 percent in the United States), and the other 2 percent have primary treatment. Poorer countries have much less to spend on sanitation. Spain serves only 18 percent of its population with even primary sewage treatment. In Ireland, it is only 11 percent, and in Greece, less than 1 percent of the people have even primary treatment. Most of the sewage, both domestic and industrial, is dumped directly into the ocean.

The fall of the "iron curtain" in 1989 revealed appalling environmental conditions in much of the former Soviet Union and its satellite states in eastern and central Europe. The countries geographically and socially closest to western Europe—the Czech Republic, Hungary, East Germany, and Poland—have made massive investments and encouraging progress toward cleaning up environmental problems. Parts of Russia itself, however, along with former socialist states in the Balkans and Central Asia, remain some of the most polluted places on earth. In Russia, for example, only about half the tap water is fit to drink. In cities like St. Petersburg, even boiling and filtering aren't enough to make municipal water safe.

As we saw earlier in this chapter, at least 200 million Chinese live in areas without sufficient fresh water. Sadly, pollution makes much of the limited water unusable (fig. 11.24). It's estimated that 70 percent of China's surface water is unsafe for human consumption, and that the water in half the country's major rivers is so contaminated that it's unsuited for any use, even agriculture. The situation in Shanxi Province exemplifies the problems of water pollution in China. An industrial powerhouse, in the north-central part of the country, Shanxi has about one-third of China's known coal resources and currently produces about two-thirds of the country's energy. In addition to power plants, major industries include steel mills, tar factories, and chemical plants.

Economic growth has been pursued in recent decades at the expense of environmental quality. According to the Chinese Environmental Protection Agency, the country's ten worst polluted cities are all in Shanxi. Factories have been allowed to exceed pollution discharges with impunity. For example, 3 million tons of waste-water is produced every day in the province with two-thirds of it discharged directly into local rivers without any treatment. Locals complain that the rivers, which once were clean and fresh, now run black with industrial waste. Among the 26 rivers in the province, 80 percent were rated Grade V (unfit for any human use) or higher in 2006. More than half the wells in Shanxi are reported to have dangerously high arsenic levels. Many of the 85,000 reported public protests in China in 2006 involved complaints about air and water pollution.

However, there is also evidence of progress in pollution control. In 1997 Minamata Bay in Japan, long synonymous with mercury poisoning, was declared officially clean again. Another important success is found in Europe, where one of its most important rivers has been cleaned up significantly through international cooperation. The Rhine, which starts in the rugged Swiss Alps and winds 1,320 km through five countries before emptying through a Dutch delta into the North Sea, has long been a major commercial artery into the heart of Europe. More than 50 million people live in its catchment basin, and nearly 20 million get their drinking water from the river or its tributaries. By the 1970s, the Rhine had become so polluted that dozens of fish species disappeared and swimming was discouraged along most of its length.

Efforts to clean up this historic and economically important waterway began in the 1950s, but a disastrous fire at a chemical warehouse near Basel, Switzerland, in 1986 provided the impetus for major changes. Through a long and sometimes painful series of international conventions and compromises, land-use practices, waste disposal, urban runoff, and industrial dumping have been changed and water quality has significantly improved. Oxygen concentrations have gone up fivefold since 1970 (from less than 2 mg/l to nearly 10 mg/l, or about 90 percent of saturation) in long stretches of the river. Chemical oxygen demand has fallen fivefold during the same period, and organochlorine levels have decreased as much as tenfold. Many species of fish and aquatic invertebrates have returned to the river. In 1992, for the first time in decades, mature salmon were caught in the Rhine.

The less-developed countries of South America, Africa, and Asia have even worse water quality than do the poorer countries of Europe. Sewage treatment is usually either totally lacking or woefully inadequate. In urban areas, 95 percent of all sewage is discharged untreated into rivers, lakes, or the ocean. Low technological capabilities and little money for pollution control are made even worse by burgeoning populations, rapid urbanization, and the shift of much heavy industry (especially the dirtier ones) from developed countries where pollution laws are strict to less-developed countries where regulations are more lenient.

Appalling environmental conditions often result from these combined factors. In many countries, open sewers run through urban areas (fig. 11.25). Two-thirds of India's surface waters are contaminated sufficiently to be considered dangerous to human health. The Yamuna River in New Delhi has 7,500 coliform bacteria per 100 ml (37 times the level considered safe for swimming in the United States) *before* entering the city. The coliform count increases to an incredible 24 *million* cells per 100 ml as the river leaves the city! At the same time, the river picks up some 20 million liters of

▲ **FIGURE 11.25** Ditches in this Haitian slum serve as open sewers into which all manner of refuse and waste are dumped. The health risks of living under these conditions are severe.

industrial effluents every day from New Delhi. It's no wonder that disease rates are high and life expectancy is low in this area. Only 1 percent of India's towns and cities have any sewage treatment, and only eight cities have anything beyond primary treatment.

Groundwater is especially hard to clean up

About half the people in the United States, including 95 percent of those in rural areas, depend on underground aquifers for their drinking water. This vital resource is threatened in many areas by overuse and pollution and by a wide variety of industrial, agricultural, and domestic contaminants. For decades it was widely assumed that groundwater was impervious to pollution because soil would bind chemicals and cleanse water as it percolated through. Springwater or artesian well water was considered to be the definitive standard of water purity, but that is no longer true in many areas.

A recent source of aquifer pollution in many areas comes from hydraulic fracturing (or fracking) to release gas and oil from tight shale formations (see chapter 12 for more discussion). The U.S. EPA has found high levels of methane and hydraulic fracturing fluids in water wells near fracking operations. In 2010, drilling companies used 32 million gallons (121 million liters) of diesel fuel along with other toxic chemicals, such as benzene, toluene, ethylbenzene, and xylene, in fracking mixtures.

The EPA estimates that every day some 4.5 trillion liters (1.2 trillion gal) of contaminated water seep into the ground in the United States from septic tanks, leaking underground storage tanks at gas stations, municipal and industrial landfills and waste disposal sites, surface impoundments, agricultural fields, forests, and abandoned wells (fig. 11.26). The most toxic of these are probably waste disposal sites. Agricultural chemicals and wastes

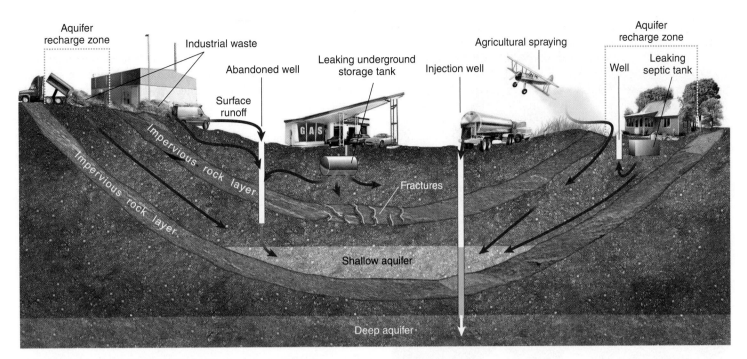

▲ **FIGURE 11.26** Sources of groundwater pollution. Septic systems, landfills, and industrial activities on aquifer recharge zones leach contaminants into aquifers. Wells provide a direct route for injection of pollutants into aquifers.

are responsible for the largest total volume of pollutants and area affected. Because deep underground aquifers often have residence times of thousands of years, many contaminants are extremely stable once underground. It is possible, but expensive, to pump water out of aquifers, clean it, and then pump it back.

In farm country, especially in the Midwest's corn belt, fertilizers and pesticides commonly contaminate aquifers and wells. Herbicides such as atrazine and alachlor are widely used on corn and soybeans and show up in about half of all wells in Iowa, for example. Nitrates from fertilizers often exceed safety standards in rural drinking water. These high nitrate levels are dangerous to infants (nitrates combine with hemoglobin in the blood and result in "blue-baby" syndrome).

Every year, epidemiologists estimate that around 1.5 million Americans fall ill from infections caused by fecal contamination. In 1993, for instance, a pathogen called cryptosporidium got into the Milwaukee public water system, making 400,000 people sick and killing at least 100 people. The total costs of these diseases amount to billions of dollars per year. Preventative measures, such as protecting water sources and aquifer recharge zones and updating treatment and distribution systems, would cost far less.

Ocean pollution has few controls

Although we don't use ocean waters directly, ocean pollution is serious and one of the fastest-growing water pollution problems. Coastal bays, estuaries, shoals, and reefs are often overwhelmed by pollution. Dead zones and poisonous algal blooms are increasingly widespread. Toxic chemicals, heavy metals, oil, sediment, and plastic refuse affect some of the most attractive and productive ocean regions. The potential losses caused by this pollution

amount to billions of dollars each year. In terms of quality of life, the costs are incalculable.

Discarded plastic flotsam and jetsam are becoming a ubiquitous mark of human impact on the oceans. Even the most remote beaches of distant islands are likely to have mounds of trash. It's been estimated that 6 million metric tons of plastic bottles, packaging material, and other litter are tossed from ships every year into the ocean, where they ensnare and choke seabirds, mammals, and even fish (fig. 11.27).

Researchers have recently discovered a vast swath of the Pacific Ocean filled with a soup of plastic refuse. Dubbed the "Great Pacific Garbage Patch," this slowly swirling vortex fills two "convergence zones" that collect trash from all over the world. One of these gyres occurs between Hawaii and California; the other one is closer to Japan. Each is larger than Texas. Currents sweep all sorts of refuse into these huge vortices that have been called the world's biggest garbage dumps. Much of this plastic consists of tiny particles suspended at or just below the water surface. The smallest particles may be more dangerous biologically than bigger pieces. Plankton and small fish ingest the plastic bits along with the contaminants to their surface, and introduce them into the marine food chain. See chapter 14 for further discussion of this topic.

Oil pollution affects beaches and open seas around the world. Oceanographers estimate that between 3 million and 6 million metric tons of oil are discharged into the world's oceans each year from oil tankers, fuel leaks, intentional discharges of fuel oil, and coastal industries. About half of this amount is due to maritime transport. Of this portion, most is not from dramatic, headline-making accidents such as the 2010 *Deepwater Horizon* spill in the Gulf of Mexico. Rather, routine open-sea bilge pumping and tank cleaning are the primary source. These activities are illegal but very common.

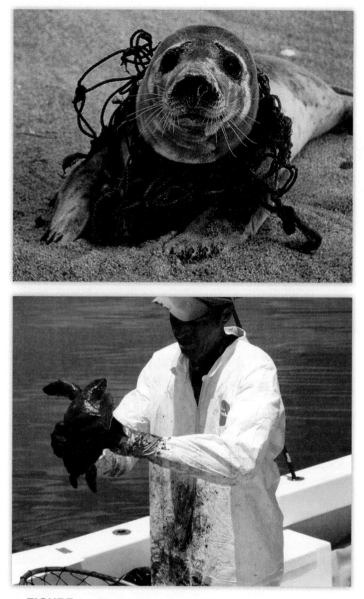

▲ **FIGURE 11.27** A deadly necklace. Marine biologists estimate that castoff nets, plastic beverage yokes, and other packing residue kill hundreds of thousands of birds, mammals, and fish each year.

Fortunately, awareness of ocean pollution is growing. Oil spill cleanup technologies and response teams are improving, although most oil is eventually decomposed by natural bacteria. Efforts are growing to control waste plastic. Criminals who ship toxic waste to poor countries are being prosecuted. Volunteer efforts also are helping to reduce beach pollution locally: in one day, volunteers in Texas gathered more than 300 tons of plastic refuse from Gulf Coast beaches.

11.8 POLLUTION CONTROL

The cheapest and most effective way to reduce pollution is to avoid producing it or releasing it in the first place. Eliminating lead from gasoline has resulted in a dramatic decrease in the amount of lead in U.S. surface waters. Studies have shown that as much as 90 percent less road deicing salt can be used in many areas without significantly affecting the safety of winter roads. Careful handling of oil and petroleum products can greatly reduce the amount of water pollution caused by these materials. Although we still have problems with persistent chlorinated hydrocarbons spread widely in the environment, the banning of DDT and PCBs in the 1970s has resulted in significant reductions in levels in wildlife.

Industry can reduce pollution by recycling or reclaiming materials that otherwise might be discarded in the waste stream. These approaches usually have economic as well as environmental benefits. Companies can extract valuable metals and chemicals and sell them, instead of releasing them as toxic contaminants into the water system. Both markets and reclamation technologies are improving as awareness of these opportunities grows. In addition, modifying land use is an important component of reducing pollution.

Nonpoint sources are often harder to control than point sources

Farmers have long contributed a huge share of water pollution, including sediment, fertilizers, and pesticides that flow from fields. Soil conservation practices on farmlands (chapter 7) aim to keep soil and contaminants on fields, where they are needed. Precise application of fertilizer, irrigation water, and pesticides saves money and reduces water contamination. Preserving wetlands which help capture sediment and contaminants, also helps protect surface and groundwaters.

In urban areas, reducing waste that enters storm sewers is essential. It is getting easier for city residents to recycle waste oil and to properly dispose of paint and other household chemicals that they once dumped into storm sewers or the garbage. Urbanites can also minimize use of fertilizers and pesticides. Regular street sweeping greatly reduces nutrient loads (from decomposing leaves and debris) in rivers and lakes.

The tremendous challenge of managing these sources is seen in Chesapeake Bay, America's largest estuary. Once fabled for its abundant oysters, crabs, shad, striped bass, and other valuable fisheries, the bay had deteriorated seriously by the early 1970s. Citizens' groups, local communities, state legislatures, and the federal government together established an innovative pollution-control program that made the bay the first estuary in America targeted for protection and restoration.

Among the principal objectives of this plan is reducing nutrient loading through land-use regulations in the bay's six watershed states to control agricultural and urban runoff. Pollution-prevention measures, such as banning phosphate detergents, also are important, as are upgrading wastewater treatment plants and improving compliance with discharge and filling permits. Efforts are underway to replant thousands of hectares of sea grasses and to restore wetlands that filter out pollutants.

Since the 1980s, annual phosphorous discharges into Chesapeake Bay have dropped 40 percent. Nitrogen levels, however, have remained constant or have even risen in some tributaries. Although progress has been made, the goals of reducing both nitrogen and phosphate levels by 40 percent and restoring viable fish and shellfish populations are still decades away. Still, as former EPA Administrator

Carol Browner said, it demonstrates the "power of cooperation" in environmental protection. (See related story "Watershed Protection in the Catskills" at www.mhhe.com/cunningham7e.)

How do we treat municipal waste?

Under natural conditions, water purification occurs constantly in soils and water. Bacteria take up and transform nutrients or break down oils. Sand and soil filter water; plant roots and fungi use nutrients in the water and simultaneously capture metals and other components. When water is cool and moving, oxygen from the air mixes in, eliminating stagnant conditions where harmful organisms can grow.

The high population densities of cities, however, produce much more waste than natural systems can process. As we have already seen, human and animal wastes usually create the most serious health-related water pollution problems. More than 500 types of disease-causing (pathogenic) bacteria, viruses, and parasites can travel from human or animal excrement through water.

Most developed countries require that cities and towns build municipal water treatment systems to purify the human and household waste. Most rural households use septic systems, which allow solids to settle in a tank, where bacteria decompose them. Liquids percolate through soil, where soil bacteria presumably purify them. Where population densities are not too high, this can be an effective method of waste disposal. With urban sprawl, however, groundwater pollution often becomes a problem.

Municipal treatment has three levels of quality

Over the past 100 years, sanitary engineers have developed ingenious and effective municipal wastewater treatment systems to protect human health, ecosystem stability, and water quality (fig. 11.28). This topic is an important part of pollution control, and is a principal responsibility of every municipal government.

Primary treatment physically separates large solids from the waste stream with screens and settling tanks. Settling tanks allow grit and some dissolved (suspended) organic solids to fall out as sludge. Water drained from the top of settling tanks still carries up to 75 percent of the organic matter, including many pathogens. These pathogens and organics are removed by **secondary treatment**, in which aerobic bacteria break down dissolved organic compounds. In secondary treatment, effluent is aerated, often with sprayers or in an aeration tank, in which air is pumped through the microorganism-rich slurry. Fluids can also be stored in a sewage lagoon, where sunlight, algae, and air process waste more cheaply but more slowly. Effluent from secondary treatment processes is usually disinfected with chlorine, UV light, or ozone to kill harmful bacteria before it is released to a nearby waterway.

Tertiary treatment removes dissolved metals and nutrients, especially nitrates and phosphates, from the secondary effluent. Although wastewater is usually free of pathogens and organic material after secondary treatment, it still contains high levels of these inorganic nutrients. If discharged into surface waters, these nutrients stimulate algal blooms and eutrophication. Allowing effluent to flow through a wetland or lagoon can remove nitrates and phosphates. Alternatively, chemicals often are used to bind and precipitate nutrients.

Sewage sludge can be a valuable fertilizer, but it can be unsafe if it contains metals and toxic chemicals. Some cities spread sludge on farms and forest lands, while others convert it to methane (natural gas). Many cities, however, incinerate or landfill sludge, both expensive options. Often, sanitary sewers are connected to storm sewers, which carry contaminated runoff from streets, parking lots, and yards. This allows treatment to remove oil, gasoline, fertilizers, and pesticides. Heavy storms, however, often overload municipal systems, resulting in large volumes of raw sewage and toxic surface runoff being dumped directly into rivers or lakes.

Conventional Treatment Misses New Pollutants In 2002 the USGS released the first-ever study of pharmaceuticals and hormones in streams. Scientists sampled 130 streams, looking for 95 contaminants, including antibiotics, natural and synthetic hormones, detergents, plasticizers, insecticides, and fire retardants (fig. 11.29). All these substances were found, usually in low concentrations. One stream had 38 of the compounds tested. Drinking-water standards exist for only 14 of the 95 substances. A similar study found the same substances in groundwater, which is much harder to clean than surface waters. What are the effects of these widely used chemicals on our environment or on people consuming the water? Nobody knows. This study is a first step toward filling huge gaps in our knowledge about their distribution, though.

Natural wastewater treatment can be an answer

Natural wastewater treatment systems offer a promising alternative to remote locations, developing countries, and small factories that can't afford conventional treatment. These systems are still unfamiliar and unconventional, so they are relatively uncommon, but they offer many advantages. Natural wastewater treatment systems are normally cheaper to build and operate than conventional systems. They use less energy and less chlorine or other purifiers, because gravity moves water, and plants and bacteria do most (or all) disinfection. With fewer pumps and filters to manage, less staff time is needed. Plants remove nutrients, metals, and other contaminants that are not captured by most conventional systems.

Constructed wetlands are a complex of artificial marshes designed to filter and decompose

◀ **FIGURE 11.28** In conventional sewage treatment, aerobic bacteria digest organic materials in high-pressure aeration tanks. This is described as secondary treatment.

Could natural systems treat our wastewater?

Conventional sewage treatment systems are designed to treat large volumes of effluent quickly and efficiently. Water treatment is necessary for public health and environmental quality, but it is expensive. Industrial-scale installations, high energy inputs, and caustic chemicals are needed. Huge quantities of sludge must be incinerated or trucked off-site for disposal.

KC 11.2

KC 11.1

▲ An aeration tank helps aerobic (oxygen-using) bacteria digest organic compounds.

Conventional treatment misses new pollutants. Pharmaceuticals and hormones, detergents, plasticizers, insecticides, and fire retardants are released freely into surface waters, because these systems are not designed for those contaminants.

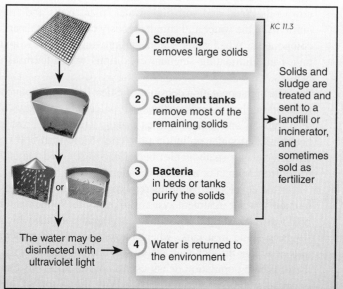

KC 11.3

1 **Screening** removes large solids

2 **Settlement tanks** remove most of the remaining solids

3 **Bacteria** in beds or tanks purify the solids

or

The water may be disinfected with ultraviolet light →

4 Water is returned to the environment

Solids and sludge are treated and sent to a landfill or incinerator, and sometimes sold as fertilizer

▲ The process of conventional sewage treatment

Natural wastewater treatment is unfamiliar but usually cheaper

We depend on ecological systems—natural bacteria and plants in water and soil—to finish off conventional treatment. Can we use these systems for the entire treatment process? Although they remain unfamiliar to most cities and towns, **wetland-based treatment systems** have operated successfully for decades—at least as long as the lifetime of a conventional plant. Because they incorporate healthy bacteria and plant communities, there is potential for uptake of novel contaminants and metals as well as organic contaminants. These systems also remove nutrients better than most conventional systems do. These systems can be half as expensive as conventional systems because they have

- few sprayers, electrical systems, and pumps → cheaper installation
- gravity water movement → low energy consumption
- few moving parts or chemicals → low maintenance
- biotic treatment → little or no chlorine use
- nutrient uptake → more complete removal of nutrients, metals, and possibly organic compounds

KC 11.4

KC 11.5

1 **ANAEROBIC TANKS** In the absence of oxygen, anaerobic bacteria decompose waste.

◀ Drinkable quality water is produced by a well-designed natural system. This photo shows before and after treatment. Most people are squeamish about the prospect of drinking treated wastewater, so recycled water is generally used for other purposes such as toilets, washing, or irrigation. Since these uses make up about 95 percent of many municipal water supplies, they can represent a significant savings.

Constructed wetland systems can be designed with endless varieties, but all filter water through a combination of beneficial microorganisms and plants. Here are common components:

- **Anaerobic (oxygen-free) tanks:** here anaerobic bacteria convert nitrate (NO_3) to nitrogen gas (N_2), and organic molecules to methane (CH_4). In some systems, methane can be captured for fuel.

- **Aerobic (oxygen-available) tanks:** aerobic bacteria convert ammonium (NH_4) to nitrate (NO_3); green plants and algae take up nutrients.

- **Gravel-bedded wetland:** beneficial microorganisms and plants growing in a gravel bed capture nutrients and organic material. In some systems, the wetland provides wildlife habitat and recreational space.

- **Presumable disinfection:** water is clean leaving the system, but rules usually require that chlorine be added to ensure disinfection. Ozone or ultraviolet light can also be used.

A constructed wetland outside can be an attractive landscaping feature that further purifies water. ▶

KC 11.6

KC 11.7

◀ Where space is available, a larger constructed wetland can serve as recreational space, a wildlife refuge, a living ecosystem, and a recharge area for groundwater or streamflow.

4 DISINFECTION
Ozone, chlorine, UV light, or other methods ensure that no harmful bacteria remain. Water can then be reused or released.

In this system, after passing through the growing tanks, the effluent water runs over a waterfall and into a small fish pond for additional oxygenation and nutrient removal. This verdant greenhouse is open to the public and adds an appealing indoor space in a cold, dry climate. ▶

KC 11.8

3 CONSTRUCTED WETLANDS
Plants take up remaining nutrients. Remaining nitrate is converted to nitrogen gas.

KC 11.9

◀ The growing tanks need to be in a greenhouse or other sunny space to provide light for plants.

2 AEROBIC TANKS
Oxygen is mixed into water, supporting plants and bacteria that further break down and decontaminate waste. Remaining solids settle out.

CAN YOU EXPLAIN?

1. Based on your reading of this chapter, what are the primary contaminants for which water is treated?
2. What is the role of bacteria in a system like this?
3. What factors make conventional treatment expensive?
4. Why is conventional treatment more widely used?

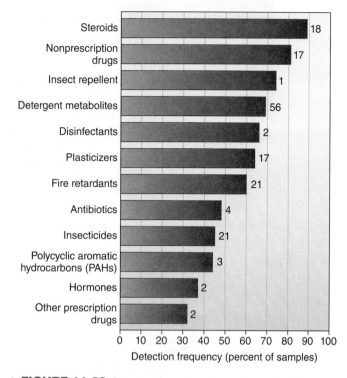

▲ **FIGURE 11.30** This Wisconsin cheese factory treats its effluent by passing it through a series of large tanks in which plants bacteria, algae, and other aquatic organisms filter the water and remove organic material, suspended solids, and nutrients.

▲ **FIGURE 11.29** Detection frequency of organic, wastewater contaminants in a recent USGS survey. Maximum concentrations in water samples are shown above the bars in micrograms per liter. Dominant substances included DEET insect repellent, caffeine, and triclosan, which comes from antibacterial soaps.

waste. One of the best-known of these is in Arcata, California, which was required to build a new and very expensive sewer system upgrade 30 years ago. As an alternative, the city transformed a 65-ha garbage dump into a series of ponds and marshes that serve as a simple, low-cost, waste-treatment facility. Arcata saved millions of dollars and improved its environment simultaneously. The marsh is a haven for wildlife and has become a prized recreation area for the city.

Similar wetland waste treatment systems are now operating in many developing countries. Effluent from these operations can be used to irrigate crops or raise fish for human consumption if care is taken first to destroy pathogens. Usually 20 to 30 days of exposure to sun, air, and aquatic plants is enough to make the water safe. These systems make an important contribution to human food supplies. A 2,500-ha waste-fed aquaculture facility in Kolkata (Calcutta), for example, supplies about 7,000 metric tons of fish annually to local markets.

Many institutions don't have the space for a constructed wetland. One of these is the Cedar Grove Cheese Factory in southern Wisconsin, which has built a "**Living Machine**®," a sequence of tanks, bacteria, algae, and small artificial wetlands (fig. 11.30). This system converts factory effluent to nearly pure water and vegetation. It removes 99 percent of the biological oxygen demand, 98 percent of the suspended solids, 93 percent of the nitrogen, and 57 percent of the phosphorus.

Systems like this can be built adjacent to, or even inside of, buildings. Combinations of plants and animals, including algae, rooted aquatic plants, clams, snails, and fish are present, each chosen to provide a particular service in a contained environment. The water leaving such a system is of drinkable quality, and it's cleaner than the water received by the facility. Often the final effluent is used to flush toilets or for irrigation, because most people are squeamish about the idea of drinking treated water. This novel approach can save resources and money, and it can serve as a valuable educational tool.

Remediation can involve containment, extraction, or biological treatment

Just as there are many sources of water contamination, there are many ways to clean it up. New developments in environmental engineering are providing promising solutions to many water pollution problems. Containment methods keep dirty water from spreading. Many pollutants can be destroyed or detoxified by chemical reactions that oxidize, reduce, neutralize, hydrolyze, precipitate, or otherwise change their chemical composition. Where chemical techniques are ineffective, physical methods may work. Solvents and other volatile organic compounds, for instance, can be stripped from solution by aeration and then burned in an incinerator.

Often, living organisms can clean contaminated water effectively and inexpensively. We call this **bioremediation**. Restored wetlands, for instance, along stream banks or lake margins can effectively filter out sediment and remove pollutants. Some plants are very efficient at taking up heavy metals and organic contaminants. Bioremediation offers exciting and inexpensive alternatives to conventional cleanup.

TABLE 11.4 | Some Important Water Quality Legislation

1. *Federal Water Pollution Control Act (1972).* Establishes uniform nationwide controls for each category of major polluting industries.

2. *Marine Protection Research and Sanctuaries Act (1972).* Regulates ocean dumping and established sanctuaries for protection of endangered marine species.

3. *Ports and Waterways Safety Act (1972).* Regulates oil transport and the operation of oil-handling facilities.

4. *Safe Drinking Water Act (1974).* Requires minimum safety standards for every community water supply. Among the contaminants regulated are bacteria, nitrates, arsenic, barium, cadmium, chromium, fluoride, lead, mercury, silver, and pesticides; radioactivity and turbidity also are regulated. This act also contains provisions to protect groundwater aquifers.

5. *Resource Conservation and Recovery Act (RCRA) (1976).* Regulates the storage, shipping, processing, and disposal of hazardous wastes and sets limits on the sewering of toxic chemicals.

6. *Toxic Substances Control Act (TOSCA) (1976).* Categorizes toxic and hazardous substances, establishes a research program, and regulates the use and disposal of poisonous chemicals.

7. *Comprehensive Environmental Response, Compensation, and Liability Act (CERCLA) (1980)* and *Superfund Amendments and Reauthorization Act (SARA) (1984).* Provide for sealing, excavation, or remediation of toxic and hazardous waste dumps.

8. *Clean Water Act (1985) (amending the 1972 Water Pollution Control Act).* Sets as a national goal the attainment of "fishable and swimmable" quality for all surface waters in the United States.

9. *London Dumping Convention (1972).* Calls for an end to all ocean dumping of industrial wastes, tank-washing effluents, and plastic trash. The United States is a signatory to this international convention.

11.9 WATER LEGISLATION

Water pollution control has been among the most broadly popular and effective of all environmental legislation in the United States. It has not been without controversy, however. Table 11.4 describes some of the most important water legislation in the United States.

The Clean Water Act was ambitious, popular, and largely successful

Passage of the U.S. Clean Water Act of 1972 was a bold, bipartisan step that made clean water a national priority. Along with the Endangered Species Act and the Clean Air Act, this is one of the most significant and effective pieces of environmental legislation ever passed by the U.S. Congress. It also is an immense and complex law, with more than 500 sections regulating everything from urban runoff, industrial discharges, and municipal sewage treatment to land-use practices and wetland drainage.

The ambitious goal of the Clean Water Act was to return all U.S. surface waters to "fishable and swimmable" conditions. For point sources, the act requires discharge permits and use of the best practicable control technology (BPT). For toxic substances, the act sets national goals of best available, economically achievable technology (BAT) and zero discharge goals for 126 priority toxic pollutants. As discussed earlier, these regulations have had a positive effect on water quality. Although surface water is not yet swimmable or fishable everywhere, its quality in the United States has significantly improved on average over the past quarter century. Perhaps the most important result of the act has been investment of $54 billion in federal funds and more than $128 billion in state and local funds for municipal sewage treatment facilities.

Opponents of federal regulation have tried repeatedly to weaken or eliminate the Clean Water Act. They regard restriction of their "right" to dump toxic chemicals and wastes into wetlands and waterways to be an infringement of personal freedom. They resent being forced to clean up municipal water supplies, and call for cost–benefit analysis that places greater weight on economic interests in all environmental planning.

Supporters of the Clean Water Act would like to see a shift away from an "end-of-the-pipe" focus on effluent removal and more attention to changing industrial processes, so that toxic substances aren't produced in the first place. Many people also would like to see stricter enforcement of existing regulations, mandatory minimum penalties for violations, more effective community right-to-know provisions, and increased powers for citizen lawsuits against polluters.

CONCLUSION

Water is a precious resource. As human populations grow and climate change affects rainfall patterns, water is likely to become even more scarce in the future. Already, about 2 billion people live in water-stressed countries (where there are inadequate supplies to meet all demands), and at least half those people don't have access to clean drinking water. Depending on population growth rates and climate change, it's possible that by 2050 there could be 7 billion people (about 60 percent of the world population) living in areas with water stress or scarcity. Conflicts over water rights are becoming more common between groups within countries and between neighboring countries that share water resources. This is made more likely by the fact that most major rivers cross two or more countries before reaching the sea, and droughts, such as the one in the southeastern United States, may become more frequent and severe with global warming. Many experts agree with *Fortune* magazine that "water will be to the 21st century what oil was to the 20th."

Forty years ago, rivers in the United States were so polluted that some caught fire while others ran red, black, orange, or other unnatural colors with toxic industrial wastes. Many cities still dumped raw sewage into local rivers and lakes, so that warnings had to be posted to avoid any bodily contact. We've made huge progress since that time. Not all rivers and lakes are "fishable or swimmable," but federal, state, and local pollution controls have greatly improved our water quality in most places.

In rapidly developing countries, such as China and India, water pollution remains a serious threat to human health and ecosystem well-being. It will take a massive investment to correct this growing problem. But there are relatively low-cost solutions to many pollution issues. Constructed wetlands for ecological sewage treatment provide low-tech, inexpensive ways to reduce pollution. "Living Machines®" for water treatment in individual buildings or communities also offer hope for better ways to treat our wastes. Perhaps you can use the information you've learned by studying environmental science to plan one for your own community.

1. Describe the path a molecule of water might follow through the hydrologic cycle from the ocean to land and back again.
2. About what percent of the world's water is liquid, fresh, surface water that supports most terrestrial life?
3. What is an *aquifer*? How does water get into an aquifer? Explain the idea of an *artesian well* and a *cone of depression*.
4. What is the difference between water *withdrawal* and *consumption*? Which sector of water use consumes most globally? Overall, has water use increased in the past century? Has efficiency increased or decreased in the three main use sectors?
5. Describe at least one example of the environmental costs of water diversion from rivers to farms or cities.
6. Explain the difference between point and nonpoint pollution. Which is harder to control? Why?
7. Why are nutrients considered pollution? Explain the ideas of *eutrophication* and an *oxygen sag*.
8. Describe primary, secondary, and tertiary water treatment.
9. What are some sources of groundwater contamination? Why is groundwater pollution such a difficult problem?
10. What is a "Living Machine®," and how does it work?

CRITICAL THINKING AND DISCUSSION

Apply the principles you have learned in this chapter to discuss these questions with other students.

1. What changes might occur in the hydrologic cycle if our climate were to warm or cool significantly?
2. Why does it take so long for deep ocean waters to circulate through the hydrologic cycle? What happens to substances that contaminate deep ocean water or deep aquifers in the ground?
3. If you were a judge responsible for allocating the dwindling water supply in the Colorado River among the various stakeholders, how would you assign water rights? What would be your criteria for needs and rights?
4. Do you think that water pollution is worse now than it was in the past? What considerations go into a judgment such as this? How do your personal experiences influence your opinion?
5. What additional information would you need to make a judgment about whether conditions are getting better or worse? How would you weigh different sources, types, and effects of water pollution?
6. Under what conditions might sediment in water or cultural eutrophication be beneficial? How should we balance positive and negative effects?

DATA ANALYSIS   Graphing Global Water Stress and Scarcity

According to the United Nations, **water stress** is when annual water supplies drop below 1,700 m³ per person. **Water scarcity** is defined as annual water supplies below 1,000 m³ per person. More than 2.8 billion people in 48 countries will face either water stress or scarcity conditions by 2025. Of these 48 countries, 40 are expected to be in West Asia or Africa. By 2050, far more people could be facing water shortages, depending both on population projections and scenarios for water supplies based on global warming and consumption patterns.

To explore some of the issues and questions about future water scarcity, go to Connect, and answer questions about this figure and others from this chapter.

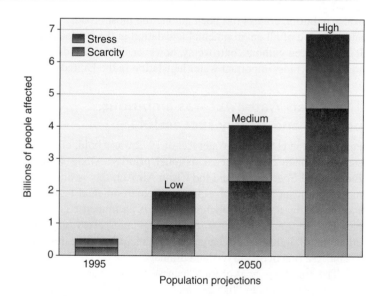

12 Environmental Geology and Earth Resources

The tops of at least 500 mountains in the southern Appalachians have been sheared off to access buried coal seams. More than a million acres (0.5 million ha) of forest have been destroyed and about 2,000 miles (3,200 km) of streams have been buried by waste rock pushed down into "valley fills" by mountaintop removal mining.

LEARNING OUTCOMES

After studying this chapter, you should be able to answer the following questions:

▶ What are tectonic plates and how does their movement shape our world?

▶ Where and why do volcanoes and earthquakes occur?

▶ What are some of the environmental and social costs of mining and oil- and gas-drilling?

▶ How can we reduce our consumption of geologic resources?

▶ Explain why floods and mass wasting are problems.

Mountaintop Removal Mining

Many mountains in Appalachia are structured like layer cakes with multiple horizontal coal seams separated by layers of sedimentary rock. The earliest approach for mining this coal was to bore into a coal seam from its exposed face on the mountainside, either by hand or with mechanical boring machines. As earthmoving equipment has become larger and more powerful, however, mining companies have found it cheaper to simply blast and scrape off the rock (overburden) lying on top of coal deposits, allowing the coal to be scooped up by giant shovels. This produces far more coal per man-hour than traditional underground mining, and brings high profits to mining companies.

Mountaintop removal mining (MTR), as this technique is called, is controversial, however, because of the destruction it wreaks on the mountain, as well as the water pollution, destroyed vegetation, and damage it causes to streams, farms, and villages in surrounding valleys. The first step in MTR is to clear-cut forests (fig. 12.1). Wildlife habitat is destroyed and barren slopes are prone to erosion, floods, and landslides. Next, thousands of tons of high-powered explosives rip open the mountain. Every week, the explosive equivalent of a Hiroshima-sized atomic bomb (15,000 tons of TNT) is detonated in MTR mines in the four states (Kentucky, Tennessee, Virginia, and West Virginia) where this technique is most common. The noise, dust, and vibrations disrupt life in nearby communities, and flying boulders have destroyed property, even killing some people.

As much as 800 feet (244 m) of rock may be removed from the mountaintop to get to underlying coal. Sometimes the overburden can be piled back on top of the mountain or ridge, but often it's simply shoved into the valley below. Nearly 1.2 million acres (0.5 million ha) of land in Appalachia on more than 500 mountain peaks and ridgelines have been turned into mining wastelands, and about 2,000 mi (3,200 km) of headwater streams have been buried or severely polluted by the "valley fills" created by MRT. The broken rock left from mining has high concentrations of sulfur, arsenic, mercury, and other toxic metals. As rainwater percolates through the rock rubble, it leaches toxins and becomes highly acidic as sulfur compounds are oxidized to sulfuric acid.

Coal must be crushed and washed before being sent to market. The wastewater is a slurry of coal dust and poisonous chemicals that can't be dumped into local rivers. Millions of gallons of toxic waste are stored in huge ponds behind earthen dams in coal-mining country. A number of these dams have failed, sending floods of toxic sludge onto communities below.

Many years ago, underground mineral rights were separated from surface ownership across much of Appalachia. Property owners had no idea that what lay below their land had any value, and they often sold mineral resources for a pittance. Subsequently, laws were rewritten to give mineral rights priority over surface rights. In theory, mining companies are required to restore land to its original condition when mining is finished, but when a mountain has been flattened and the rock has been dumped into the valley below, it's impossible to recreate former conditions. One

survey of former surface mines in Appalachia found that less than 3 percent are currently used for any economic activity.

There have been many public protests against MTR mining, but jobs are scarce and the economy is depressed across Appalachia. Mine workers don't want to give up their jobs. The U.S. EPA and the Federal Courts have tried repeatedly to limit or block

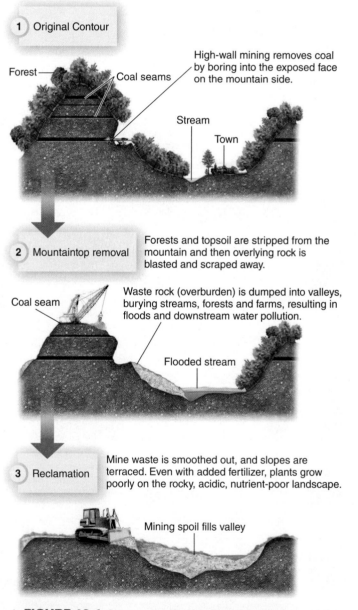

1 Original Contour

High-wall mining removes coal by boring into the exposed face on the mountain side.

Forest

Coal seams

Stream

Town

2 Mountaintop removal

Forests and topsoil are stripped from the mountain and then overlying rock is blasted and scraped away.

Coal seam

Waste rock (overburden) is dumped into valleys, burying streams, forests and farms, resulting in floods and downstream water pollution.

Flooded stream

3 Reclamation

Mine waste is smoothed out, and slopes are terraced. Even with added fertilizer, plants grow poorly on the rocky, acidic, nutrient-poor landscape.

Mining spoil fills valley

▲ **FIGURE 12.1** Steps in mountaintop removal mining.

MTR mining, but the mining companies and their supporters, so far, have always found loopholes that allow them to continue. Although America's use of coal to generate electricity is dropping dramatically, coal production remains fairly constant. Some MTR mines export 90 percent of their output. We're destroying our landscape to sell coal to China. Is that right?

In this chapter, we'll look at the principles of geology and how they affect our lives. But we'll also look at ways in which we humans have become geologic agents. MTR mining is a prime example of how we're causing long-term, massive-scale damage to the earth. For related resources, including Google Earth placemarks that show locations discussed in this chapter, visit www.mcgrawhillconnect.com. ■

> When we heal the earth, we heal ourselves.
> —DAVID ORR

12.1 EARTH PROCESSES SHAPE OUR RESOURCES

Many people are exposed to geologic hazards of one type or another, but all of us benefit from the earth's geological resources. Right now you are undoubtedly using products made from these resources: plastics, of many types are made from petroleum; iron, copper, and aluminum mines produce the materials for the electrical wiring that brings you power; rare earth metals are essential in your cell phone, your MP3 player, and your computer. All of us also share responsibility for the environmental and social devastation that often results from mining, drilling, and processing materials.

Fortunately, there are many promising solutions to reduce these costs, including recycling and alternative materials. But why are these risks and resources distributed as they are? To understand how and where geological risks and resources are created, we need to learn about the earth's structure and the processes that shape it.

Earth is a dynamic planet

Although we think of the ground under our feet as solid and stable, the earth is a dynamic and constantly changing structure. Titanic forces inside the earth cause continents to split, move apart, and then crash into each other in slow but inexorable collisions.

The earth is a layered sphere. The **core**, or interior, is composed of a dense, intensely hot mass of metal—mostly iron—thousands of kilometers in diameter (fig. 12.2). Solid in the center but more fluid in the outer core, this immense mass generates the magnetic field that envelops the earth.

Surrounding the molten outer core is a hot, pliable layer of rock called the **mantle**. The mantle is much less dense than the core because it contains a high concentration of lighter elements, such as oxygen, silicon, and magnesium.

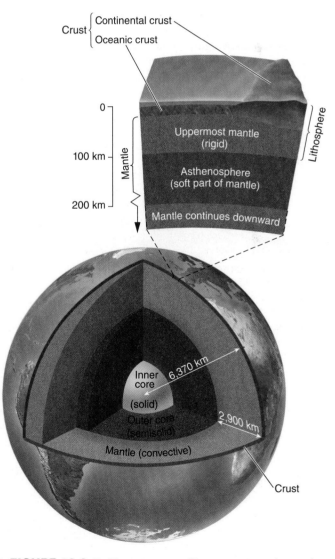

▲ **FIGURE 12.2** Earth's cross section. Slow convection in the mantle causes the thin, brittle crust to move.

The outermost layer of the earth is the cool, lightweight, brittle rock **crust**. The crust below oceans is relatively thin (8–15 km), dense, and young (less than 200 million years old) because of constant recycling. Crust under continents is relatively thick (25–75 km) and light, and as old as 3.8 billion years, with new material being added continually. It also is predominantly granitic, whereas oceanic crust is mainly dense basaltic rock. Table 12.1 compares the composition of the whole earth (dominated by the dense core) and the crust.

Tectonic processes reshape continents and cause earthquakes

The huge convection currents in the mantle are thought to break the overlying crust into a mosaic of huge blocks called **tectonic plates** (fig. 12.3). These plates slide slowly across the earth's surface like wind-driven ice floes on water, in some places breaking up into smaller pieces, in other places crashing ponderously into each other to create new, larger landmasses. Ocean basins form where continents crack and pull apart. The Atlantic Ocean, for example, is growing slowly as Europe and Africa move away from the Americas. **Magma** (molten rock) forced up through the cracks forms new oceanic crust that piles up under water in **mid-ocean ridges**. Creating the largest mountain range in the world, these ridges wind around the earth for 74,000 km (46,000 mi) (see fig. 12.3). Although concealed from our view, this jagged range

boasts higher peaks, deeper canyons, and sheerer cliffs than any continental mountains. Slowly spreading from these fracture zones, ocean plates push against continental plates.

Earthquakes, such as the ones that struck Japan in 2011, are caused by jerking as plates grind past each other. Mountain ranges like those on the west coasts of North and South America are pushed up at the margins of colliding continental plates. The Himalayas are still rising as the Indian subcontinent collides inexorably with Asia. Southern California is sailing very slowly north toward Alaska. In about 30 million years, Los Angeles will pass San Francisco, if both still exist by then.

TABLE 12.1 | Eight Most Common Chemical Elements (Percent) in Whole Earth and Crust

WHOLE EARTH		CRUST	
Iron	33.3	Oxygen	45.2
Oxygen	29.8	Silicon	27.2
Silicon	15.6	Aluminum	8.2
Magnesium	13.9	Iron	5.8
Nickel	2.0	Calcium	5.1
Calcium	1.8	Magnesium	2.8
Aluminum	1.5	Sodium	2.3
Sodium	0.2	Potassium	1.7

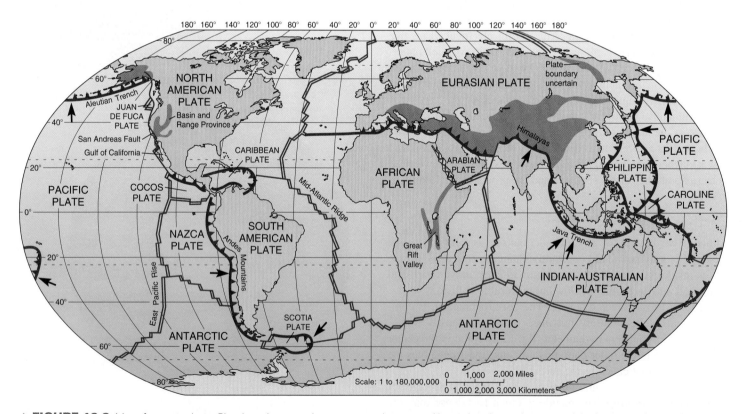

▲ **FIGURE 12.3** Map of tectonic plates. Plate boundaries are dynamic zones, characterized by earthquakes, volcanism, and the formation of great rifts and mountain ranges (shaded areas). Arrows indicate the direction of subduction, where one plate is diving beneath another. These zones are sites of deep trenches in the ocean floor and high levels of seismic and volcanic activity.

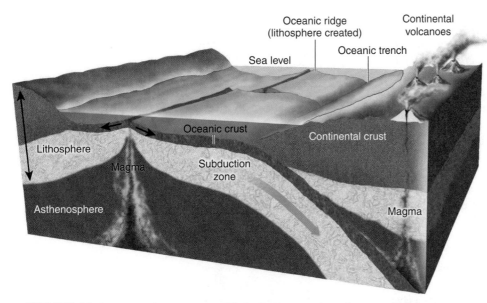

▲ **FIGURE 12.4** Tectonic plate movement. Where thin, oceanic plates diverge, upwelling magma forms mid-ocean ridges. A chain of volcanoes, like the Hawaiian Islands, may form as plates pass over a "hot spot." Where plates converge, melting can cause volcanoes, such as the Cascades.

When an oceanic plate collides with a continental landmass, the continental plate usually rides up over the seafloor, while the oceanic plate is **subducted**, or pushed down into the mantle, where it melts and rises back to the surface as magma (fig. 12.4). Deep ocean trenches mark these subduction zones, and volcanoes form where the magma erupts through vents and fissures in the overlying crust. Trenches and volcanic mountains ring the Pacific Ocean rim from Indonesia to Japan to Alaska and down the west coast of the Americas, forming a so-called ring of fire where oceanic plates are being subducted under the continental plates. This ring is the source of more earthquakes and volcanic activity than any other region on the earth.

Over millions of years, continents can drift long distances. Antarctica and Australia once were connected to Africa, for instance, somewhere near the equator and supported luxuriant forests. Geologists suggest that several times in the earth's history, most or all of the continents have gathered to form supercontinents, which have broken up and re-formed over hundreds of millions of years (fig. 12.5). The redistribution of continents has profound effects on the earth's climate and may help explain the periodic mass extinctions of organisms marking the divisions between many major geologic periods (fig. 12.6).

As the opening case study for this chapter shows, we humans have now become a major geological force. We may move more dirt and rock and cause more species extinction every year than any natural forces. Some geologists believe this new era should be called the Anthropocene after its most important driving force—us.

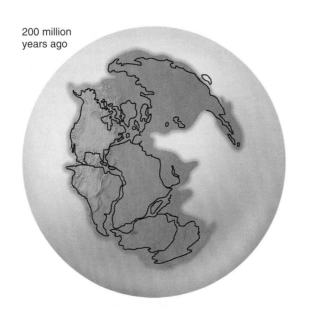

200 million years ago

▲ **FIGURE 12.5** Pangaea, an ancient supercontinent of 200 million years ago, combined all the world's continents in a single landmass. Continents have combined and separated repeatedly.

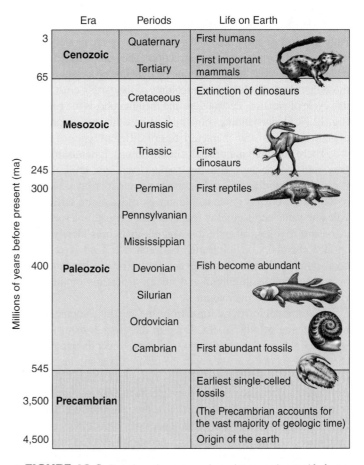

Era	Periods	Life on Earth
Cenozoic	Quaternary	First humans
	Tertiary	First important mammals
Mesozoic	Cretaceous	Extinction of dinosaurs
	Jurassic	
	Triassic	First dinosaurs
Paleozoic	Permian	First reptiles
	Pennsylvanian	
	Mississippian	
	Devonian	Fish become abundant
	Silurian	
	Ordovician	
	Cambrian	First abundant fossils
Precambrian		Earliest single-celled fossils (The Precambrian accounts for the vast majority of geologic time)
		Origin of the earth

Millions of years before present (ma): 3, 65, 245, 300, 400, 545, 3,500, 4,500

▲ **FIGURE 12.6** Periods and eras in geological time, and major life-forms that mark some periods. Some authors claim we've now entered a a new era, the Anthropocene, dominated by humans.

12.2 MINERALS AND ROCKS

A **mineral** is a naturally occurring, inorganic solid with a specific chemical composition and a specific internal crystal structure. A mineral is solid; therefore, ice is a mineral (with a distinct composition and crystal structure), but liquid water is not. Similarly molten lava is not crystalline, although it generally hardens to create distinct minerals. Metals (such as iron, copper, aluminum, or gold) come from mineral ores, but once purified, metals are no longer crystalline and thus are not minerals. Depending on the conditions in which they were formed, mineral crystals can be microscopically small, such as asbestos fibers, or huge, such as the tree-size selenite crystals recently discovered in a Chihuahua, Mexico mine.

A **rock** is a solid, cohesive aggregate of one or more minerals. Within the rock, individual mineral crystals (or grains) are mixed together and held firmly in a solid mass. The grains may be large or small, depending on how the rock was formed, but each grain retains its own unique mineral qualities. Each rock type has a characteristic mixture of minerals, grain sizes, and ways in which the grains are mixed and held together. Granite, for example, is a mixture of quartz, feldspar, and mica crystals. Rocks with a granite-like mineral content but much finer crystals are called rhyolite; chemically similar rocks with large crystals are called pegmatite.

The rock cycle creates and recycles rocks

Although rocks appear hard and permanent, they are part of a relentless cycle of formation and destruction. They are crushed, folded, melted, and recrystallized over time by dynamic processes related to those that shape the large-scale features of the earth's crust. We call this cycle of creation, destruction, and metamorphosis the **rock cycle** (fig. 12.7). Understanding something of how this cycle works helps explain the origin and characteristics of different types of rocks.

There are three major rock classifications: igneous, metamorphic, and sedimentary. **Igneous rocks** (from *igni*, the Latin word for fire) are solidified from hot, molten magma or lava. Most rock in the earth's crust is igneous. Magma extruded to the surface from volcanic vents cools quickly to make finely crystalline rocks, such as basalt, rhyolite, or andesite. Magma that cools slowly in subsurface chambers or is intruded between overlying layers makes coarsely crystalline rocks, such as gabbro (rich in iron and silica) or granite (rich in aluminum and silica), depending on the chemical composition of the magma.

Metamorphic rocks form from the melting, contorting, and recrystallizing of other rocks. Deep in the ground, tectonic forces squeeze, fold, heat, and transform solid rock. Under these conditions, chemical reactions can alter both the composition and the structure of the component minerals. Metamorphic rocks are classified by their chemical composition and by the degree of recrystallization: some minerals form only under extreme pressure and heat (diamonds or jade, for example); others form under more moderate conditions (graphite or talc). Some common metamorphic rocks are marble (from limestone), quartzite (from sandstone), and slate (from mudstone and shale). Metamorphic rocks often have beautiful colors and patterns left by the twisting and folding that created them.

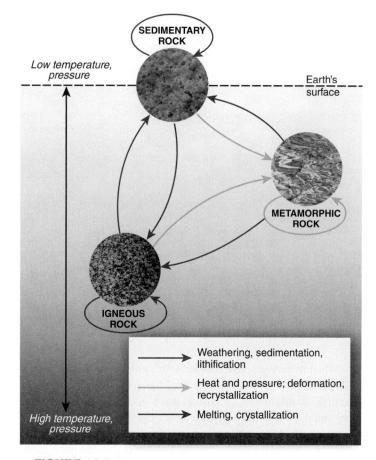

▲ **FIGURE 12.7** The rock cycle includes a variety of geologic processes that can transform any rock.

Sedimentary rocks are formed when loose grains of other rocks are consolidated by time and pressure. Sandstone, for example, is solidified from layers of sand, and mudstone consists of extremely hardened mud and clay. Tuff is formed from volcanic ash, and conglomerates are aggregates of sand and gravel. Some sedimentary rocks develop from crystals that precipitate out of extremely salty water. Rock salt, made of the mineral halite, is ground up to produce ordinary table salt (sodium chloride). Salt deposits often form when a body of saltwater dries up, leaving salt crystals behind. Limestone is a rock composed of cemented remains of marine organisms. You can often see the shapes of shells and corals in a piece of limestone. Sedimentary formations often have distinctive layers that show different conditions when they were laid down. Erosion can reveal these layers and inform us of their history (fig. 12.8).

Weathering and sedimentation

Most crystalline rocks are extremely hard and durable, but exposure to air, water, changing temperatures, and reactive chemical agents slowly breaks them down in a process called **weathering** (fig. 12.9). Mechanical weathering is the physical breakup of rocks into smaller particles without a change in chemical composition

▲ **FIGURE 12.8** Different colors of soft sedimentary rocks deposited in ancient seas during the Tertiary period 63 to 40 million years ago have been carved by erosion into the fluted spires and hoodoos of the Pink Cliffs of Bryce Canyon National Park.

▲ **FIGURE 12.9** Weathering slowly reduces an igneous rock to loose sediment. Here, exposure to moisture expands minerals in the rock, and frost may also force the rock apart.

of the constituent minerals. You have probably seen the results of mechanical weathering in rounded rocks in rivers or on shorelines, smoothed by constant tumbling in waves or currents. On a larger scale, mountain valleys are carved by rivers and glaciers.

Chemical weathering is the selective removal or alteration of specific minerals in rocks. This alteration leads to weakening and disintegration of rock. Among the more important chemical weathering processes are oxidation (combination of oxygen with an element to form an oxide or a hydroxide mineral) and hydrolysis (hydrogen atoms from water molecules combine with other chemicals to form acids). The products of these reactions are more susceptible to both mechanical weathering and dissolving in water. For instance, when carbonic acid (formed when rainwater absorbs CO_2) percolates through porous limestone layers in the ground, it dissolves the rock and creates caves.

Particles of rock loosened by wind, water, ice, and other weathering forces are carried downhill, downwind, or downstream until they come to rest again in a new location. The deposition of these materials is called **sedimentation**. Water, wind, and glaciers deposit particles of sand, clay, and silt far from their source. Much of the American Midwest, for instance, is covered with hundreds of meters of sedimentary material left by glaciers (till, or rock debris deposited by glacial ice), wind (loess, or fine dust deposits), river deposits of sand and gravel, and ocean deposits of sand, silt, clay, and limestone.

12.3 ECONOMIC GEOLOGY AND MINERALOGY

The earth is unusually rich in mineral variety. Mineralogists have identified some 4,400 different mineral species, far more, we believe, than any of our neighboring planets. What makes the difference? The processes of plate tectonics and the rock cycle on this planet have gradually concentrated uncommon elements and allowed them to crystallize into new minerals. But this accounts for only about one-third of our geologic legacy. The biggest difference is life. Most of our minerals are oxides, but there was little free oxygen in the atmosphere until it was released by photosynthetic organisms, thus triggering evolution of our great variety of minerals.

Economic mineralogy is the study of resources that are valuable for manufacturing and trade. Most economic minerals are metal ores, minerals with unusually high concentrations of metals. Lead, for example, generally comes from the mineral galena (PbS), and copper comes from sulfide ores, such as bornite (Cu_5FeS_4). Nonmetallic geologic resources include graphite, feldspar, quartz crystals, diamonds, and other crystals that are valued for their usefulness or beauty. Metals have been so important in human affairs that major epochs of human history are commonly known by their dominant materials and the technology involved in using those materials (Stone Age, Bronze Age, Iron Age, etc.). The mining, processing, and distribution of these materials have broad implications for both our culture and our environment. Most economically valuable crustal resources exist everywhere in small amounts; the important thing is to find them concentrated in economically recoverable levels.

The U.S. mining law passed in 1872 encourages mining on public lands as a way of boosting the economy and utilizing natural resources. There have been repeated efforts to update this law and to recover public revenue from publicly owned resources, but powerful friends in Congress together with a tradition of supporting extractive industries in many states have continually blocked these reforms.

Metals are essential to our economy

Metals are malleable substances that are useful and valuable because they are strong, relatively light, and can be reshaped for many purposes. The availability of metals and the methods to extract and use them have determined technological developments, as well as economic and political power for individuals and nations (fig. 12.10).

▲ **FIGURE 12.10** The availability of metals and minerals and the ways we extract and use them have profound effects on our society and environment.

The metals consumed in greatest quantity by world industry include iron (740 million metric tons annually), aluminum (40 million metric tons), manganese (22.4 million metric tons), copper and chromium (8 million metric tons each), and nickel (0.7 million metric tons). The complex, worldwide network that extracts, processes and distributes these metals has become crucially important to the economic and social stability of all nations. Table 12.2 shows some important uses of these metals.

The largest countries (in surface area) have the greatest likelihood of mountain building (orogeny), volcanism, or other events

TABLE 12.2 | Primary Uses of Some Major Metals

METAL	USE
Aluminum	Packaging foods and beverages (38%), transportation, electronics
Chromium	High-strength steel alloys
Copper	Building construction, electric and electronic industries
Iron	Heavy machinery, steel production
Lead	Leaded gasoline, car batteries, paints, ammunition
Manganese	High-strength, heat-resistant steel alloys
Nickel	Chemical industry, steel alloys
Platinum group	Automobile catalytic converters, electronics, medical uses
Gold	Medical, aerospace, electronic uses; accumulation as monetary standard
Silver	Photography, electronics, jewelry

that create economic mineral deposits. Russia, China, Canada, the United States, and Australia, for example, are especially rich in metal resources. Africa has relatively few geologic resources, except for South Africa, which is unusually rich in diamonds, gold, and other valuable minerals. In 2011, geologists announced the discovery of an estimated $1 trillion (U.S.) in metals and other valuable minerals in Afghanistan. Critics claimed this announcement was merely an excuse for continued military occupation of this troubled country. In any case, difficult terrain and political instability will make these deposits difficult to access.

The rapid growth of green technologies, such as renewable energy and electric vehicles has made a group of rare earth metals especially important. Worries about impending shortages of these minerals complicate future developments in this sector (see Exploring Science, p. 289 and Key Concepts, pp. 290–291).

Nonmetal mineral resources include gravel, clay, glass, and salts

Nonmetal minerals constitute a broad class that covers resources from gemstones to sand, gravel, salts, limestone, and soils. Sand and gravel production for road and building construction comprise by far the greatest volume and dollar value of all nonmetal mineral resources and a far greater volume than all metal ores. Sand and gravel are used mainly in brick and concrete construction, in paving, as loose road filler, and for sandblasting. High-purity silica sand is our source of glass. These materials usually are retrieved from surface pit mines and quarries, where they were deposited by glaciers, winds, or ancient oceans. Some industry sources predict a building crisis in the United States as supplies of easily accessible sand and gravel are exhausted.

Limestone, like sand and gravel, is mined and quarried for concrete and crushed for road rock. It also is cut for building stone, pulverized for use as an agricultural soil additive that neutralizes acidic soil, and roasted in lime kilns and cement plants to make plaster (hydrated lime) and cement.

Evaporites (materials deposited by evaporation of chemical solutions) are mined for halite, gypsum, and potash. These are often found at or above 97 percent purity. Halite, or rock salt, is used for water softening and ice melting on winter roads in some northern areas. When refined, it is a source of table salt. Gypsum (calcium sulfate) now makes our plaster wallboard, but it has been used to cover walls ever since the Egyptians plastered their frescoed tombs along the Nile River some 5,000 years ago. Potash is an evaporite composed of a variety of potassium chlorides and potassium sulfates. These highly soluble potassium salts have long been used as a soil fertilizer. Sulfur deposits are mined mainly for sulfuric acid production. In the United States, sulfuric acid use amounts to more than 200 lb per person per year, mostly because of its use in industry, car batteries, and some medicinal products. Interestingly, the world's largest supply of lithium, which is essential for lightweight batteries for electric cars, cell phones, and other consumer goods, is a vast salt flat, called the Salar de Uyuni, in southwestern Bolivia. The lithium, which was deposited when a huge salt lake evaporated eons ago, could make this impoverished country the Saudi Arabia of the electronic age.

Could shortages of a group of obscure minerals limit the growth of alternative energy supplies and green technology? A recent decision by China to limit exports of rare earth elements is seen by some experts as a serious threat to the global clean tech industry.

"Rare earth" elements are a collection of metallic elements including yttrium, scandium, and 15 lanthanides, such as neodymium, dysprosium, and gadolinium, that are essential in modern electronics. These metals are used in cell phones, high-efficiency lights, hybrid cars, superconductors, high-strength magnets, lightweight batteries, lasers, energy-conserving lamps, and a variety of medical devices. Because of their unusual properties, small amounts of these metals can make motors 90 percent lighter and lights 80 percent more efficient. Without these materials, MP3 players, hybrid vehicles, high-capacity wind turbines, and much other high-tech equipment would be impossible. A Toyota Prius, for example, uses about a kilogram of neodymium and dysprosium for its electric motor and as much as 15 kg of lanthanum for its battery pack.

Despite their name, these elements occur widely in the earth's crust, but commercially viable concentrations are found in only a few locations. China produces about 95 percent of all rare earth metals, an increase from about 30 percent two decades ago. China's dominance in mining these metals results partly because China uses these materials in electronics production, partly because of low labor costs in mining, and partly because the government has been willing to overlook the high environmental costs of extracting these metals from the ground. About half of all Chinese production of rare earth metals occurs in a single mine in Baotou in Inner Mongolia; most of the rest come from small, often unlicensed mines in southern China where pollution levels are high.

Like gold, silver, and other precious metals, rare earth elements are often separated from ore by crushing ore-bearing rocks and washing the ore in strong acids. Acids release metals from the ore, but when the metals are later separated from the acid slurry, tremendous amounts of toxic wastewater are produced. Often acids are pumped directly into a borehole drilled in the ground, and metals are dissolved from ores in place. The resulting slurry is then pumped to the surface for processing. Acidic wastewater is frequently stored behind earthen dams, which can leak into surface and ground waters. Processing also releases sulfur and radioactive uranium and thorium that frequently occur with rare earth elements. Establishing better control on illegal mines is one reason for China's interest in controlling export and production.

For China, maintaining control of supplies, as well as a near monopoly on production, ensures that domestic electronic needs will be met. Outside of China, there is concern about supplies for both strategic needs (such as military guidance systems), consumer electronics, and alternative energy supplies. Having a near-monopoly of rare earth metals production has helped China become a center of technology innovation, and other countries now wonder how to keep up in the high-tech race. Many firms are simply moving to China. The division of General Motors that deals with miniaturized magnet research, for example, shut down its U.S. office and moved its entire staff to China in 2006. The Danish wind turbine company Vestas moved much of its production to China in 2009.

In response to expected shortages and rising prices, several companies are working to reopen mines in North America and Australia. Molycorp Minerals is reopening its mine in Mountain Pass, California, and hopes to meet about 10 percent of global demand. Similarly, Avalon Rare Metals of Toronto is working on a mine in Canada's Northwest Territories. Greenland is also jumping into this new gold rush, with hopes to produce up to 25 percent of rare earth metals from recently discovered ore bodies.

It remains to be seen whether new environmental controls will be in place for this coming expansion. Australia is now building a $200 million (U.S.) rare earth facility in Malaysia. The ore will be mined near Mt. Weld in the Australian desert, but for processing it will be shipped to Malaysia where water is abundant, wages are low, and environmental rules are more lenient than in Australia. Japan supports this project as a way of breaking the Chinese monopoly on strategic metals, but locals have held up the project because of worries about pollution. An earlier rare earth facility in Malaysia was closed when high levels of cancer and birth defects were found in people working and living near the plant.

Meanwhile, the Toyota Corporation announced in 2012 that it has found ways to produce batteries, electric motors, and other high-tech components without the use of rare earth alloys.

▼ Rare earths are valuable in a wide variety of high-tech and energy-saving applications. Clockwise from top center: praseodymium, cerium, lanthanum, neodymium, samarium, and gadolinium.

Currently, the earth provides almost all our fuel

At present, modern society functions largely on energy produced from geologic deposits of oil, coal, and natural gas. Nuclear energy, which runs on uranium, provides about 20 percent of our electricity. But, as chapter 13 shows, renewable sources, such as sun, wind, hydropower, biomass, and geothermal energy, could replace all the fossil fuels and nuclear power we now use, while reducing pollution and reducing global climate change.

Where does your cell phone come from?

Mobile phones, computers, and other electronic gadgets have transformed our lives, but few of us think about the geologic resources that went into making those devices. We enjoy these items for their useful lives, but we're always eager to trade up to the next newer and better models. While each individual gadget may be small and contain only tiny amounts of precious metals, rare earths, fossil fuels, and other materials, collectively they make a big impact.

Currently there are at least 1 billion personal computers and 5 billion mobile phones in use around the world, and the numbers are climbing rapidly. In the United States, most people change their cell phone every 18 to 24 months. Computers last only 2 to 3 years, contributing to the mountains of eWaste discarded in the United States every year.

Here are a few examples of the many source areas that contribute to our cell phones.

We depend increasingly on electronics, but where do they originate?

KC 12.1

KC 12.2

Although they may be tiny, cell phones contain a surprising amount of metal that could be recycled, and yet hundreds of millions are thrown away every year or lie unused in drawers and cupboards. **A typical cell phone can contain gold, silver, copper, palladium, platinum, lead, zinc, mercury, chromium, cadmium, rhodium, beryllium, arsenic, and lithium** among other metals and chemical compounds.

Specialized lightweight metals are often mined in remote, hard-to-monitor locations. One of these is "coltan," an ore containing the metals columbium, tantalum, and niobium, which are essential in many electronics, including cell phones. Congo contains 80 percent of the world's supply of coltan. Inhumane conditions and child labor have been involved, and sales of these metals have helped finance wars in central Africa.

Depending on the ore quality, a metric ton of ore may yield only 0.3 g of gold, and the miners may have to move 2 to 5 tons of overburden (unwanted rock) to get to that ore.

Abandoned mines often leak strong acids laced with arsenic, mercury, and other toxic metals into local groundwater and surface waters. By some calculations, humans now move more earth than the glaciers did.

KC 12.3

Workers mine coltan

Open-pit copper mine

KC 12.4

As we've seen elsewhere in this chapter, supplies of rare earth metals, such as neodymium, dysprosium, lanthanum, and yttrium are also critical for modern electronics. Monopolies in the sources of some of these materials, together with adverse environmental impacts in their extraction and purification, could constrain some technologies.

Oil refineries produce our plastics from oil. KC 12.5

Every year, billions of cell phones and computers are discarded around the world. Added to the piles of refrigerators, air conditioners, TV sets, and other unwanted appliances, disposal of electronic waste is a huge problem. In the United States, where 3 million tons of electronics are tossed out every year, 70 percent of the heavy metals in landfills are from eWaste. Increasingly, this electronic garbage is shipped to developing countries where scavengers disassemble it under dangerous conditions in an effort to recover valuable metals. Modern recycling facilities can recover 99 percent of the contents much more safely and efficiently.

An offshore oil well provides energy and plastics.

KC 12.7

The plastic case for your phone or computer is made from petroleum. Energy is also used to manufacture and ship all the gadgets we use. Extracting, shipping, and refining fossil fuels are among our biggest geologic impacts. Obtaining the 20 billion barrels of petroleum or the 7 billion short tons of coal we use globally every year has uncalculated environmental, social, and economic impacts.

KC 12.6

eWaste collection and monitoring is increasingly important.

KC 12.8

Gold smelter

Smelting (extracting metals from ore by baking it) takes a huge amount of energy and often releases large amounts of air and water pollution, especially in countries where environmental regulation is weak.

CAN YOU EXPLAIN?

1. If each cell phone contains 0.03 g of gold, how much would be in 5 billion phones?
2. List 15 of the elements or earth materials found in cell phones.
3. Do we have an ethical responsibility for what happens to our waste once we're through with it?

Oil, coal, and gas are organic, created over millions of years as extreme heat and pressure transformed the remains of ancient organisms. They are not minerals, because they have no crystalline structure, but they can be considered part of economic mineralogy because they are such important geologic resources. In addition to providing energy, oil is the source material for plastics, and natural gas is used to make agricultural fertilizers. We'll discuss fossil fuels further in chapter 13.

12.4 ENVIRONMENTAL EFFECTS OF RESOURCE EXTRACTION

Each of us depends daily on geologic resources mined or pumped from sites around the world. We use scores of metals and minerals, many of which we've never even heard of, in our lights, computers, watches, fertilizers, and cars. Extracting and purifying these resources can have severe environmental and social consequences. The most obvious effect of mining and well drilling is often the disturbance or removal of the land surface. Farther-reaching effects, though, include air and water pollution. The EPA lists more than 100 toxic air pollutants, from acetone to xylene, released from U.S. mines and wells every year. Nearly 80,000 metric tons of particulate matter (dust) and 11,000 tons of sulfur dioxide are released from nonmetal mining alone. Pollution from chemical and sediment runoff is a major problem in many local watersheds. Acidic mine runoff has damaged or destroyed aquatic ecosystems in many places (fig. 12.11).

▼ **FIGURE 12.11** Thousands of abandoned mines on public lands poison streams and groundwater with acid, metal-laced drainage.

Gold and other metals are often found in sulfide ores that produce sulfuric acid when exposed to air and water. In addition, metal elements often occur in very low concentrations—10 to 20 parts per billion may be economically extractable for gold, platinum, and other metals. Consequently, vast quantities of ore must be crushed and washed to extract these metals. Cyanide, mercury, and other toxic substances are used to chemically separate metals from the minerals that contain them, and these substances can easily contaminate lakes and streams. Furthermore, a great deal of water is used in washing crushed ore with cyanide and other solutions. The USGS estimates that in arid Nevada, mining consumes about 230,000 m^3 (60 million gal) per day. After use in ore processing, much of this water contains sulfuric acid, arsenic, heavy metals, and other contaminants and is unsuitable for any other use.

Mining and drilling can degrade water quality

There are many techniques for extracting geologic materials. The most common methods are open-pit mining, strip-mining, and underground mining. An ancient method of accumulating gold, diamonds, and coal is placer mining, in which pure flakes or nuggets are washed from stream sediments. Since the California gold rush of 1849, placer miners have used water cannons to blast away hillsides. This method, which chokes stream ecosystems with sediment, is still used in Alaska, Canada, and many other regions. Another ancient and much more dangerous method is underground mining. Ancient Roman, European, and Chinese miners tunneled deep into tin, lead, copper, coal, and other mineral seams. Mine tunnels occasionally collapse, and natural gas in coal mines can explode. Water seeping into mine shafts also dissolves toxic minerals. Contaminated water seeps into groundwater; it is also pumped to the surface, where it enters streams and lakes.

A current controversy in the United States involves extraction of methane gas from coal deposits that are too deep or too dispersed for mining. Vast deposits of coal-bearing shale underlie both the Rocky Mountains and the Appalachian Mountains. Because the gas doesn't migrate easily through tight shales, it often takes many closely spaced wells to extract this methane (fig. 12.12). In Wyoming's Powder River basin, for example, 140,000 wells have been proposed for methane extraction. Together with the vast network

▲ **FIGURE 12.12** Hundreds of thousands of natural gas wells have been (or are being) drilled in the United States. Often, these wells require a controversial procedure called hydraulic fracturing, or fracking, to release gas from sedimentary formations.

▲ **FIGURE 12.13** Some giant mining machines stand as tall as a 20-story building and can scoop up thousands of cubic meters of rock per hour.

of roads, pipelines, pumping stations, and service facilities, this industry is having serious impacts on ranching, wildlife, and recreation in formerly remote areas.

Perhaps even worse is the effect on water supplies. Each well produces up to 75,000 liters of salty water per day. Dumping this toxic waste into streams causes widespread pollution. To boost well output, mining companies rely on hydraulic fracturing (or "fracking"). A mixture of water, sand, and toxic chemicals is pumped into the ground and rock formations at extremely high pressure. The pressurized fluid cracks sediments and releases the gas. This often disrupts aquifers, however, and contaminates wells. And in some areas, deep injection of fracking liquids has been correlated with increased earthquake frequency.

For decades, coal-bed methane extraction was a problem only in western states, but this controversial technology is now moving to the East Coast as well. The Marcellus and Devonian Shales underlie parts of ten eastern states, ranging from northern Georgia to Upstate New York. It has long been recognized that methane can be extracted from these formations, but estimates of economically recoverable amounts were relatively small. New developments in horizontal drilling and hydraulic fracturing along with increased exploratory drilling have now made this deposit a potentially "supergiant" gas field. The U.S. Geological Survey now estimates that the Marcellus/Devonian formation may contain 500 trillion ft^3 (13 trillion m^3) of methane. If all of it were accessible, it would make a 100-year supply for the United States at current consumption rates. But the same issues, concerning a multitude of wells, water pollution, and threats to water supplies on which millions of people depend, raise thorny problems.

Surface mining destroys landscapes

Open-pit mines are used to extract massive beds of metal ores and other minerals. The size of modern open pits can be hard to comprehend. The Bingham Canyon mine, near Salt Lake City, Utah, is 800 m (2,640 ft) deep and nearly 4 km (2.5 mi) wide at the top. More than 5 billion tons of copper ore and waste material have been removed from the hole since 1906. A chief environmental challenge of open-pit mining is that groundwater accumulates in the pit. In metal mines, a toxic soup results. No one yet knows how to detoxify these lakes, which endanger wildlife and nearby watersheds.

Half the coal used in the United States comes from strip mines. Because coal is often found in expansive, horizontal beds, the entire land surface can be stripped away to cheaply and quickly expose the coal (fig. 12.13). Up to 100 m (328 ft) of overburden, or surface material, may be removed in relatively flat areas. The waste rock is generally placed back into the mine, but usually in long ridges called spoil banks. Spoil banks are very susceptible to erosion and chemical weathering. Because the spoil banks have no topsoil (the complex organic mixture that supports vegetation—see chapter 7), revegetation occurs very slowly (fig. 12.14).

▲ **FIGURE 12.14** This surface coal mine in Saskatchewan was abandoned a century ago. Plants are slowly vegetating the waste rock piles, but the land has little economic or biological value.

The 1977 federal Surface Mining Control and Reclamation Act (SMCRA) requires better restoration of strip-mined lands, especially where mines replaced prime farmland. Since then, the record of strip-mine reclamation has improved substantially. Complete mine restoration is expensive, often more than $10,000 per hectare. Restoration is also difficult because the developing soil is usually acidic and compacted by the heavy machinery used to reshape the land surface.

As the opening case study for this chapter shows, bitter controversy has grown over mountaintop removal mining. In 2011 the EPA denied permits for one of the largest of these operations ever proposed. The Spruce Top #1 mine in West Virginia would have dumped 110 million yd^3 (84 million m^3) of mine waste directly into streams. Although the mine was approved in 2007, it was held up by litigation, and now may never be opened. Opponents hope this marks the end of an especially destructive mining method.

The Mineral Policy Center in Washington, D.C., estimates that 19,000 km (12,000 mi) of rivers and streams in the United States are contaminated by mine drainage. The EPA estimates that cleaning up impaired streams, along with 550,000 abandoned mines in the United States, may cost $70 billion. Worldwide, mine closing and rehabilitation costs are estimated in the trillions of dollars. Because of the volatile prices of metals and coal, many mining companies have gone bankrupt before restoring mine sites, leaving the public responsible for cleanup.

In 2012, Interior Secretary Ken Salazar declared nearly 1 million acres (400,000 ha) of public land near the Grand Canyon off-limits to uranium mining because of worries about the potential for pollution in the nearby national park.

Processing contaminates air, water, and soil

Metals are extracted from ores by heating or by using chemical solvents. Both processes release large quantities of toxic materials that can be even more environmentally hazardous than mining. **Smelting**—roasting ore to release metals—is a major source of air pollution. One of the most notorious examples of ecological devastation from smelting in the United States is a wasteland near Ducktown, Tennessee. In the mid-1800s, mining companies began excavating the rich copper deposits in the area. To extract copper from the ore, they built huge, open-air wood fires, using timber from the surrounding forest. Dense clouds of sulfur dioxide released from sulfide ores poisoned the vegetation and acidified the soil over a 50 mi^2 (13,000 ha) area. Rains washed the soil off the denuded land, creating a barren moonscape.

Sulfur emissions from Ducktown smelters were reduced in 1907 after Georgia sued Tennessee over air pollution. In the 1930s the Tennessee Valley Authority (TVA) began treating the soil and replanting trees to cut down on erosion. Recently, upward of $250,000 per year has been spent on this effort. The trees and other plants are still spindly and feeble, but more than two-thirds of the area is considered "adequately" covered with vegetation. Similarly, smelting of copper-nickel ore in Sudbury, Ontario, a century ago caused widespread ecological destruction that is slowly being repaired following pollution-control measures.

Chemical extraction is used to dissolve or mobilize pulverized ore, but it uses and pollutes a great deal of water. A widely used method is **heap-leach extraction**, which involves piling crushed ore in huge heaps and spraying it with a dilute alkaline-cyanide solution. The solution percolates through the pile and dissolves gold. The gold-containing solution is then pumped to a processing plant that removes the gold by electrolysis. A thick clay pad and plastic liner beneath the ore heap is supposed to keep the poisonous cyanide solution from contaminating surface or groundwater, but leaks are common.

Once all the gold is recovered, mine operators have simply walked away from the operation, leaving vast amounts of toxic effluent in open ponds behind earthen dams. A case in point is the Summitville Mine near Alamosa, Colorado. After extracting $98 million in gold, the absentee owners declared bankruptcy in 1992, abandoning millions of tons of mine waste and huge, leaking ponds of cyanide. The Environmental Protection Agency may spend more than $100 million trying to clean up the mess and keep the cyanide pool from spilling into the Alamosa River.

12.5 CONSERVING GEOLOGIC RESOURCES

Conservation offers great potential for extending our supplies of economic minerals and reducing the effects of mining and processing. The advantages of conservation are significant: less waste to dispose of, less land lost to mining, and less consumption of money, energy, and water resources.

Recycling saves energy as well as materials

Some waste products already are being exploited, especially for scarce or valuable metals. Aluminum, for instance, must be extracted from bauxite by electrolysis, an expensive, energy-intensive process. Recycling waste aluminum, such as beverage cans, on the other hand, consumes one-twentieth of the energy of extracting new aluminum. Today, nearly two-thirds of all aluminum beverage cans in the United States are recycled, up from only 15 percent 20 years ago. The high value of aluminum scrap ($2,000 a ton versus $200 for iron, and $45 for glass) gives consumers plenty of incentive to deliver their cans for collection. Recycling is so rapid and effective that half of all the aluminum cans now on a grocer's shelf will be made into another can within two months. Table 12.3 shows the energy cost of extracting other materials.

Platinum, the catalyst in automobile catalytic exhaust converters, is valuable enough to be regularly retrieved and recycled from used cars. Other commonly recycled metals are gold, silver, copper, lead, iron, and steel. The last four are readily available in a pure and massive form, including copper pipes, lead batteries, and steel and iron auto parts. Gold and silver are valuable enough to warrant recovery, even through more difficult means. Nearly all scrapped automobiles and car batteries are now recycled in the United States (fig. 12.15).

Although total U.S. steel production has fallen in recent decades—largely because of inexpensive supplies from new and efficient Japanese steel mills—a new type of mill subsisting entirely on a readily available supply of scrap/waste steel and iron

TABLE 12.3	Energy Requirements in Producing Various Materials from Ore and Raw Source Materials	
	ENERGY REQUIREMENT (MJ/KG)[1]	
PRODUCT	NEW	FROM SCRAP
Glass	25	25
Steel	50	26
Plastics	162	n.a.[2]
Aluminum	250	8
Titanium	400	n.a.[2]
Copper	60	7
Paper	24	15

[1] Megajoules per kilogram.
[2] Not available.
SOURCE: Data from E. T. Hayes, *Implications of Materials Processing*, 1997.

is a growing industry. **Minimills**, which remelt and reshape scrap iron and steel, are smaller and cheaper to operate than traditional integrated mills that perform every process from preparing raw ore to finishing iron and steel products. Minimills use less than half as much energy per ton of steel produced, compared to integrated mill furnaces. Minimills now produce about half of all U.S. steel. Some minimills use as much as 90 percent recycled iron and steel. Now that all new steel made in North America must contain at least 28 percent recycled contents, iron and steel recycling in the United States reached 83 percent in 2010.

New materials can replace mined resources

Mineral and metal consumption can be reduced by new materials or new technologies developed to replace traditional uses. This is a long-standing tradition; for example, bronze replaced stone technology, and iron replaced bronze. More recently, the introduction of plastic pipe has decreased our consumption of copper,

▼ **FIGURE 12.15** The richest metal source we have—our mountains of scrapped cars—offers a rich, inexpensive, and ecologically beneficial resource that can be "mined" for a number of metals.

lead, and steel pipes. In the same way, the development of fiber-optic technology and satellite communication reduces the need for copper telephone wires.

Iron and steel have been the backbone of heavy industry, but we are now moving toward other materials. One of our primary uses for iron and steel has been machinery and vehicle parts. In automobile production, steel is being replaced by polymers (long-chain organic molecules similar to plastics), aluminum, ceramics, and new high-technology alloys. All of these reduce vehicle weight and cost, while increasing fuel efficiency. Some of the newer alloys that combine steel with titanium, vanadium, or other metals wear much better than traditional steel. Ceramic engine parts provide heat insulation around pistons, bearings, and cylinders, keeping the rest of the engine cool and operating efficiently. Plastics and glass fiber–reinforced polymers are used in body parts and some engine components.

Electronics and communications (telephone) technology, once major consumers of copper and aluminum, now use ultrahigh-purity glass cables to transmit pulses of light, instead of metal wires carrying electron pulses. Once again, this technology has been developed for its greater efficiency and lower cost, but it also reduces consumption of our most basic metals.

12.6 GEOLOGIC HAZARDS

Earthquakes, along with volcanoes, floods, and landslides, are normal earth processes, events that have made our earth what it is today. However, when they affect human populations, their consequences can be among the worst and most feared disasters that befall us.

Earthquakes are frequent and deadly hazards

Earthquakes are among our planet's most destructive geological disasters. In 2004 an earthquake and tsunami just off the coast of Banda Aceh, Indonesia, for example, killed over 230,000 people and caused damage as far away as Africa. A far larger toll is thought to have been caused by a 1976 earthquake in Tangshan, China. Government officials reported 655,000 deaths, although some geologists doubt it was that high.

Earthquakes are sudden movements in the earth's crust that occur along faults (planes of weakness), where one rock mass slides past another one. When movement along faults occurs gradually and relatively smoothly, it is called creep, or seismic slip, and may be undetectable to the casual observer. When friction prevents rocks from slipping easily, stress builds up until it is finally released with a sudden jerk. The point on a fault at which the first movement occurs during an earthquake is called the epicenter.

Earthquakes have always seemed mysterious, sudden, and violent, coming without warning and leaving ruined cities and dislocated landscapes in their wake. Cities built on poorly consolidated alluvial soil, often suffer the greatest damage from earthquakes. Water-saturated soil can liquefy when shaken. Buildings sometimes sink out of sight or fall down like a row of dominoes under these conditions.

Earthquakes frequently occur along the edges of tectonic plates, especially where one plate is being subducted, or pushed down, beneath another. Earthquakes also occur in the centers of continents, however. In fact, the largest earthquake in recorded history in North America was one of an estimated magnitude 8.8 that struck the area around New Madrid, Missouri, in 1812 (fig. 12.16). Fortunately, few people lived there at the time, and the damage was minimal.

Modern building codes in earthquake zones attempt to prevent damage and casualties by constructing buildings that can withstand tremors. The primary methods used are heavily reinforced structures, strategically placed weak spots in the building that can absorb vibration from the rest of the structure, and pads or floats beneath the building on which it can shift harmlessly with ground motion.

There's evidence that human activities can trigger earthquakes. Raising and lowering water levels in reservoirs behind large dams often correlates with increased seismic activity. Chinese geologists, for example, suspect that the 2008 earthquake that killed 69,000 people in Sichuan Province may have been triggered by recent building of the Zipingpu dam on the Min River only a few kilometers from the earthquake epicenter. Similarly, injection of fluids into deep wells also correlates with increased earth tremors. In 2009, two geothermal deep-well projects—one in Switzerland and the other in California—were abruptly canceled when earthquakes in the immediate vicinity suddenly intensified.

One of the most notorious effects of earthquakes is the **tsunami** (Japanese for "harbor wave.") These giant sea swells (sometimes called tidal waves), such as the one that struck Japan in 2011, can move at 1,000 kph (600 mph), or faster, away from the center of an earthquake. About 25,000 people were killed and four nuclear reactors were destroyed by the 2011 tsunami (fig. 12.17). When these swells approach the shore, they can create breakers as high as 65 m (nearly 200 ft). Tsunamis also can be caused by underwater volcanic explosions or massive seafloor slumping.

Volcanoes eject deadly gases and ash

Volcanoes and undersea magma vents are the sources of most of the earth's crust. Over hundreds of millions of years, gaseous emissions from these sources formed the earth's earliest oceans

▲ **FIGURE 12.17** In 2011 a magnitude 9.0 earthquake just off the coast of Japan created a massive tsunami that destoyed homes, killed at least 25,000 people, and destroyed four nuclear power plants.

and atmosphere. Many of the world's fertile soils are weathered volcanic materials. Volcanoes have also been an ever-present threat to human populations (fig. 12.18). One of the most famous historic volcanic eruptions was that of Mount Vesuvius in southern Italy, which buried the cities of Herculaneum and Pompeii in A.D. 79. The mountain had been showing signs of activity before it erupted, but many citizens chose to stay and take a chance on survival. On August 24, the mountain buried the two towns in ash. Thousands were killed by the dense, hot, toxic gases that accompanied the ash flowing down from the volcano's mouth. It continues to erupt from time to time.

Nuees ardentes (French for "glowing clouds") are deadly, denser-than-air mixtures of hot gases and ash like those that inundated Pompeii and Herculaneum. Temperatures in these clouds may exceed 1,000°C, and they move at more than 100 kph (60 mph). *Nuees ardentes* destroyed the town of St. Pierre on the Caribbean

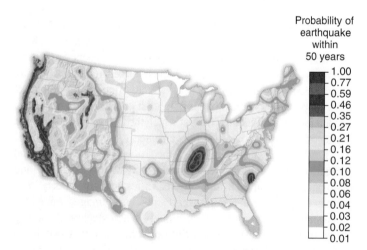

Probability of earthquake within 50 years

1.00	
0.77	
0.59	
0.46	
0.35	
0.27	
0.21	
0.16	
0.12	
0.10	
0.08	
0.06	
0.04	
0.03	
0.02	
0.01	

▲ **FIGURE 12.16** A seismic map of the lower 48 states shows the risk of earthquakes over the next 50 years (1.00 = a 2 percent chance). You might be surprised to learn that some of the highest risk is along the Mississippi River near New Madrid, Missouri. SOURCE: USGS 2010.

▲ **FIGURE 12.18** Volcanic ash covers damaged houses and dead vegetation after the November 2010 eruption of Mt. Merapi (background) in central Java. More than 300,000 residents were displaced and at least 325 were killed by this multiday eruption.

island of Martinique on May 8, 1902. Mount Pelee released a cloud of *nuees ardentes* that rolled down through the town, killing between 25,000 and 40,000 people within a few minutes. All of the town's residents died except for a single prisoner being held in the town dungeon.

Disastrous mudslides are also associated with volcanoes. The 1985 eruption of Nevado del Ruíz, 130 km (85 mi) northwest of Bogotá, Colombia, caused mudslides that buried most of the town of Armero and devastated the town of Chinchina. An estimated 25,000 people were killed. Heavy mudslides also accompanied the eruption of Mount St. Helens in Washington in 1980. Sediments mixed with melted snow destroyed roads, bridges, and property, but because of sufficient advance warning, there were few casualties. Geologists worry that similar mudflows from an eruption at Mount Rainier would threaten much larger populations (fig. 12.19).

Volcanic eruptions often release large volumes of ash and dust into the air. Mount St. Helens expelled 3 km³ of dust and ash, causing ash fall across much of North America. This was only a minor eruption. An eruption in a bigger class of volcanoes was that of Tambora in Indonesia in 1815, which expelled 175 km³ of dust and ash, more than 58 times that of Mount St. Helens. These dust clouds circled the globe and reduced sunlight and air temperatures enough so that 1815 was known as the year without a summer.

It is not just a volcano's dust that blocks sunlight. Sulfur emissions from volcanic eruptions combine with rain and atmospheric moisture to produce sulfuric acid (H_2SO_4). Droplets of H_2SO_4 interfere with solar radiation and can significantly cool the

▲ **FIGURE 12.20** In 2010, unusually heavy monsoon rains flooded about one-quarter of Pakistan, forcing about 20 million people from their homes and killing at least 2,000.

world climate. In 1991 Mount Pinatubo in the Philippines emitted 20 million tons of sulfur dioxide aerosols, which remained in the stratosphere for two years. This thin haze cooled the entire earth by 1°C for about two years. It also caused a 10 to 15 percent reduction in stratospheric ozone, allowing increased ultraviolet light to reach the earth's surface.

Floods are part of a river's land-shaping processes

Like earthquakes and volcanoes, **floods** are normal events that cause damage when people get in the way. As rivers carve and shape the landscape, they build broad **floodplains**, level expanses that are periodically inundated. Large rivers often have huge floodplains. Many cities have been built on these flat, fertile plains, to be convenient to the river. Floodplains may appear safe for many years, but eventually they do flood. The severity of floods can be described by the height of water above the normal stream banks or by how frequently a similar event occurs—on average—for a given area. Note that these are statistical averages. A "ten-year flood" would be expected to occur once in every ten years; a "100-year flood" would be expected to occur once every century. But two 100-year floods can occur in successive years or even in the same year.

Among direct natural disasters, floods take the largest number of human lives and cause the most property damage (fig. 12.20). A flood on the Yangtze River in China in 1931, for example, killed 3.7 million people, making it the most deadly natural disaster in recorded history. In another flood on China's Yellow River in 1959, about 2 million people died, mostly due to famine and disease.

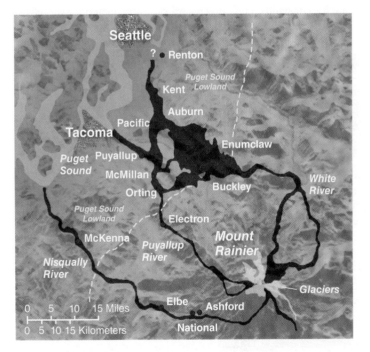

▲ **FIGURE 12.19** The highest risk for earthquakes, tsunamis, and volcanic activity in the contiguous 48 states is in the Pacific Northwest, where the Juan de Fuca plate is diving under the North American plate. Historic mudflows from Mount Rainier (brown areas) cover large areas that are now settled newar Seattle and Tacoma, Washington.

The biggest flood in U.S. history occurred in 1927 when the Mississippi River and its tributaries overflowed their banks from Minnesota to the Gulf of Mexico. Exceptional rain and snowfalls across the central U.S. in 1926 set the stage for this catastrophe. In April 1927, Mississippi levees broke in more than 145 places. Below St. Louis, the river spread out up to 80 miles (128 km) from side to side. More than 17 million acres (almost 7 million ha) were flooded up to 20 m (65 ft) deep, causing at least $400 million in damages. More than 200,000 people were displaced. Some languished in refugee camps with little food, shelter, or medical help for months. White authorities' refusals to help African-Americans during this time played a role in the diaspora to northern cities—particularly Chicago—in the 1930s.

Will floods get worse due to global climate change? Many climate scientists predict that global warming will cause more extreme weather events, including both severe droughts in some places and more intense rainfall in others. In addition, many other human activities increase both the severity and the frequency of floods. Covering the land with hardened surfaces, such as roads, parking lots, and roofs, reduces water infiltration into the soil and speeds the rate of runoff into streams and lakes. Clearing forests for agriculture and destroying natural wetlands also increases both the volume and the rate of water discharge after a storm. In Iowa, for example, at least 99 percent of the natural wetlands that existed before settlement have been filled for farmland and urban development.

Similarly, forest clearing and floodplain development coupled with torrential rains in 2011 flooded nearly half of Thailand. Although only about 400 people died, economic losses may have been $50 billion (U.S.), because at least 14,000 factories were closed and 660,000 people were out of work. Shortages in computer parts, batteries, and other high-tech components had severe effects on world trade.

Flood control

After the 1927 floods the U.S. Army Corps of Engineers was ordered to control the Mississippi River. The world's longest system of levees and flood walls were built to contain water within riverbanks, and river channels were dredged and deepened to allow water to recede faster. But every levee simply transfers the problem downstream. The water has to go somewhere. If it doesn't soak into the ground upstream, it will exacerbate floods somewhere else—leading to more levee development, and then to more flooding farther downstream, and so on.

More than $25 billion of river-control systems have been built on the Mississippi and its tributaries since 1928. These systems have protected

many communities over the past century. In recent years, however, this elaborate system helped turn large floods into major disasters. Deprived of the ability to spill out over floodplains, the river is pushed downstream to create faster currents and deeper floods until eventually a levee gives way somewhere. Hydrologists calculate that Mississippi floods in 1993 were about 3 m (10 ft) higher than they would have been, given the same rainfall in 1900, before so many wetlands were filled in and flood-control structures were built.

Under current rules, the government is obligated to finance most levees and flood-control structures. Many people think that it would be much better to spend this money to restore wetlands, replace groundcover on water courses, build check dams on small streams, move buildings off the floodplain, and undertake other nonstructural ways of reducing flood danger. According to this view, floodplains should be used for wildlife habitat, parks, recreation areas, and other uses not susceptible to flood damage.

The National Flood Insurance Program administered by the Federal Emergency Management Agency (FEMA) was intended to aid people who cannot buy insurance at reasonable rates, but its effects have been to encourage building on the floodplains by making people feel that, whatever happens, the government will take care of them. Many people would like to relocate homes and businesses out of harm's way after the recent floods or to improve them so they will be less susceptible to flooding, but owners of damaged property can collect only if they rebuild in the same place and in the same way as before. This perpetuates problems rather than solves them.

Mass wasting includes slides and slumps

Gravity constantly pulls downward on every material everywhere on earth. Hillsides, beaches, even relatively flat farm fields can lose material to erosion. Often water helps mobilize loose material, and catastrophic slumping, beach erosion, and gully development can occur in a storm. A general term for downhill slides of earth is "mass wasting."

Landslides are sudden collapses of hillsides. In the United States alone, landslides and related mass wasting cause over $1 billion in property damage every year. When unconsolidated sediments on a hillside are saturated by a storm or exposed by logging, road building, or house construction, slopes are especially susceptible to sudden landslides (fig. 12.21).

◀ **FIGURE 12.21** Damaged homes sit on a hillside after a landslide in Laguna Beach, California.

Often people are unaware of the risks they face by locating on or under unstable hillsides. Sometimes they simply ignore clear and obvious danger. In southern California, where land prices are high, people often build expensive houses on steep hills and in narrow canyons. Most of the time, this dry environment appears quite stable, but in fact, steep hillsides slip and slump frequently. Especially when soil is exposed or rainfall is heavy, mudslides and debris flows can destroy whole neighborhoods. In developing countries, mudslides have buried entire villages in minutes. **Soil creep**, on the other hand, moves material inexorably downhill at an imperceptibly slow pace.

Erosion destroys fields and undermines buildings

Gullying is the development of deep trenches on relatively flat ground. Especially on farm fields, which have a great deal of loose soil unprotected by plant roots, rainwater running across the surface can dig deep gullies. Sometimes land becomes useless for farming because gullying is so severe and because erosion has removed the fertile topsoil. Agricultural soil erosion has been described as an invisible crisis. Erosion has reduced the fertility of millions of acres of prime farmland in the United States alone.

Beach erosion occurs on all sandy shorelines because the motion of the waves is constantly redistributing sand and other sediments. One of the world's longest and most spectacular beach complexes runs down the Atlantic coast of North America from New England to Florida and around the Gulf of Mexico. Much of this beach lies on some 350 long, thin **barrier islands** that stand between the mainland and the open sea. Behind these barrier islands lie shallow bays or brackish lagoons fringed by marshes or swamps.

Early inhabitants recognized that exposed, sandy shores were hazardous places to live, and they settled on the bay side of barrier islands or as far upstream on coastal rivers as was practical. Modern residents, however, place a high value on living where they have an ocean view and ready access to the beach. And they assume that modern technology makes them immune to natural forces. The most valuable and prestigious property is closest to the shore. Over the past 50 years, more than 1 million acres (400,000 ha) of estuaries and coastal marshes have been filled to make way for housing or recreational developments.

Construction directly on beaches and barrier islands can cause irreparable damage to the whole ecosystem. Under normal circumstances, fragile vegetative cover holds the shifting sand in place. Damaging this vegetation with construction, building roads, and breaching dunes with roads can destabilize barrier islands. Storms then wash away beaches or even whole islands. Hurricane Katrina in 2005 caused $100 billion in property damage along the Gulf Coast of the United States, mostly from the storm washing over barrier islands and coastlines (fig. 12.22). FEMA estimates that 25 percent of all coastal homes in the United States could have the ground washed out from under them by 2060 as intensified storms and rising sea levels caused by global warming make barrier islands and low-lying areas even riskier places to live.

Cities and individual property owners often spend millions of dollars to protect beaches from erosion and repair damage after

▲ **FIGURE 12.22** The aftermath of Hurricane Katrina on Dauphin Island, Alabama. Since 1970, this barrier island at the mouth of Mobile Bay has been overwashed at least five times by storms. More than 20 million yd³ (15 million m³) of sand have been dredged or trucked in to restore the island. Some beach houses have been rebuilt, mostly at public expense, five times. Does it make sense to keep rebuilding in such an exposed place?

storms. Sand is dredged from the ocean floor or hauled in by the truckload, only to wash away again in the next storm. Building artificial barriers, such as groins or jetties, can trap migrating sand and build beaches in one area, but they often starve downstream beaches and make erosion there even worse.

As is the case for inland floodplains, government policies often encourage people to build where they probably shouldn't. Subsidies for road building and bridges, support for water and sewer projects, tax exemptions for second homes, flood insurance, and disaster relief are all good for the real estate and construction businesses but invite people to build in risky places. Flood insurance typically costs $400 per year for $100,000 of coverage. In 2011 FEMA paid out $17 billion in disaster claims, 80 percent of which were flood- or storm-related. Settlement usually requires that structures be rebuilt exactly where and as they were before. There is no restriction on how many claims can be made, and policies are rarely canceled, no matter what the risk. Some beach houses have been rebuilt—at public expense—five times in two decades. The General Accounting Office reports that 2 percent of federal flood policies are responsible for 30 percent of all claims.

The Coastal Barrier Resources Act of 1982 prohibited federal support, including flood insurance, for development on sensitive islands and beaches. In 1992, however, the U.S. Supreme Court ruled that ordinances forbidding floodplain development amount to an unconstitutional "taking," or confiscation, of private property.

CONCLUSION

Geologic hazards, including earthquakes, volcanic eruptions, tsunamis, floods, and landslides, represent major threats. Devastating events have altered human history many times in the past, sending geopolitical, economic, genetic, and even artistic repercussions around the planet. But the same processes that cause threats to humans also create resources, such as fossil fuels, metals, and building materials. The study of geology has allowed us to predict where these threats and resources will occur. The extraction of earth resources often carries severe environmental costs, however, including water contamination, habitat destruction, and air pollution.

The earth's surface is shaped by shifting pieces of crust, which split and collide slowly but continuously. Earthquakes, volcanoes, and mountains generally occur on plate margins. Rocks are composed of minerals, and rocks can be described according to their origins in molten material (igneous rocks), eroded or deposited sediments (sedimentary rocks), or materials transformed by heat and pressure deep in the earth (metamorphic rocks).

Earth resources, including oil, gas, and coal, are the foundation of our economy. Metals are expensive to extract, but they are extremely valuable because they are ductile (bendable) yet strong, and because they carry electricity (e.g., copper). Mining and refining of metals and other geologic resources can cause severe environmental damage. In some cases, land can be restored to something close to its original state, but when you dig a mile-deep hole in the ground, or lop the whole top off a mountain and dump the waste into a nearby valley, it's unlikely that the damage will ever be undone.

Many materials can be recycled, saving money, energy, and environmental quality. Recycling aluminum consumes one-twentieth of the energy to extract new aluminum, for example, and recycling copper takes about one-eighth as much energy. We can also save energy and resources by replacing traditional materials with newer, more efficient ones. Fiber-optic communication lines have replaced much of our copper wiring, adding speed and efficiency while saving copper use.

Understanding the forces and processes that create both geologic hazards and resources is important for human existence. As the famous geologist Louis Agassiz said, "Learn geology or die."

PRACTICE QUIZ

1. How does tectonic plate movement create ocean basins, mid-ocean ridges, and volcanoes?
2. What is the "ring of fire"?
3. Describe the processes and components of the rock cycle.
4. What is the difference between metals and nonmetal mineral resources?
5. What is a *mineral* and a *rock*? Why are pure metals not minerals?
6. Which countries are the single greatest producers of our major metals?
7. Describe some of the mining, processing, and drilling methods that can degrade water or air quality.
8. Compare the different mining methods of underground, open-pit, strip, and placer mining, as well as mountaintop removal.

9. What resources, aside from minerals themselves, can be saved by recycling?
10. Describe the most deadly risks of volcanoes.
11. What is *mass wasting*? Give three examples and explain why they are a problem.
12. Why is building on barrier islands risky?
13. What is a *floodplain*? Why is building on floodplains controversial?
14. Describe the processes of chemical weathering and mechanical weathering. How do these processes contribute to the recycling of rocks?
15. The Mesozoic period begins and ends with the appearance and disappearance of dinosaurs. What fossils mark the other geologic eras?

CRITICAL THINKING AND DISCUSSION

Apply the principles you have learned in this chapter to discuss these questions with other students.

1. Understanding and solving the environmental problems of mining are basically geologic problems, but geologists need information from a variety of environmental and scientific fields. What are some of the other sciences (or disciplines) that could contribute to solving mine contamination problems?
2. Geologists are responsible for identifying and mapping mineral resources, but mineral resources are buried below the soil and covered with vegetation. How do you suppose geologists in the field find clues about the distribution of rock types?
3. If you had an igneous rock with very fine crystals and another with very large crystals, which would you expect to have formed deep in the ground, and why?

4. Heat and pressure tend to help concentrate metal ores. Explain why such ores might often occur in mountains such as the Andes in South America.
5. The idea of tectonic plates shifting across the earth's surface is central to explanations of geologic processes. Why is this idea still called the "theory" of plate tectonic movement?
6. Geologic data from fossils and sediments provided important evidence for past climate change. What sorts of evidence in the rocks and landscape around you suggest that the place where you live once looked much different than it does today?
7. We all use metals and fossil fuels obtained by mining. What responsibility does each of us have for the methods that produce the goods and services we enjoy?

The USGS Earthquake Center is a site that gives you access to global earthquake monitoring data. Understanding the distribution of earthquakes will help you understand the patterns of earth movement, volcanoes, and mountain building processes. Earthquakes are also one of our most important geologic hazards worldwide. Go to Connect to find data and questions that let you examine real-time earthquake records and to demonstrate your understanding of this geologic phenomenon.

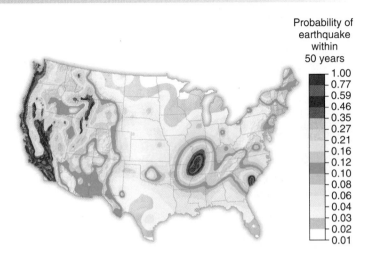

Probability of earthquake within 50 years

- 1.00
- 0.77
- 0.59
- 0.46
- 0.35
- 0.27
- 0.21
- 0.16
- 0.12
- 0.10
- 0.08
- 0.06
- 0.04
- 0.03
- 0.02
- 0.01

▲ **FIGURE 1** Probability of earthquakes in the coming half-century.

TO ACCESS ADDITIONAL RESOURCES FOR THIS CHAPTER, PLEASE VISIT CONNECT AT www.mcgrawhillconnect.com.

You will find LearnSmart, an adaptive learning system, Google Earth™ exercises, additional Case Studies, Data Analysis exercises, and an interactive ebook.

Concentrating solar power (CSP) plants in desert regions could supply more electricity than all current world use.

LEARNING OUTCOMES

After studying this chapter, you should be able to answer the following questions:

▶ What are our dominant sources of energy?

▶ What is peak oil production? Why is it hard to evaluate future oil production?

▶ How important is coal in domestic energy production?

▶ What are the environmental effects of coal burning?

▶ How do nuclear reactors work? What are some of their advantages and disadvantages?

▶ What are our main renewable forms of energy?

▶ Could solar, wind, hydropower, and other renewables eliminate the need for fossil fuels?

▶ What are photovoltaic cells, and how do they work?

▶ What are biofuels? What are arguments for and against their use?

CASE STUDY

Renewable Energy in Europe

Northern Europe has a problem. They'd like to be environmentally responsible and wean themselves away from fossil fuels. Coastal regions generally have good wind power resources, and Great Britain, Germany, the Netherlands, and Scandinavia lead the world in offshore wind farms. But the most abundant renewable energy supply—solar—is often sorely lacking in the notoriously dark, cloudy, northern regions. Look at the location of northern Europe on a globe. Stockholm, Oslo, and Helsinki, for example, are all at about the same latitude as Anchorage, Alaska.

A great solar resource exists, however, just across the Mediterranean Sea in the Sahara desert, where the skies are cloudless and the sun shines fiercely nearly every day. An area about 125 × 125 km—or about 0.3 percent of North Africa—receives enough sunlight to supply all the current electrical consumption in Europe. And high-voltage, direct-current (HVDC) transmission lines have advanced, so it's economically and technically feasible to ship electrical current from Africa to Europe. Transmission losses are only 3 percent per 1,000 km and add just 1–2 cents per kilowatt-hour, an insignificant amount when you consider that the fuel is free.

A consortium led by the German Aerospace Center has been studying this issue for a decade. Operating under the name Desertech, about a dozen German banks and energy companies, together with other interested parties in more than 20 countries, have begun building a giant network of renewable energy facilities and a HVDC supergrid they hope will eventually link Europe, the Middle East, and North Africa (EU-MENA) to make a significant contribution both to regional development and to combating global climate change.

Some three dozen concentrating solar power (CSP) plants, spread across North Africa and the Middle East, together with about 20 offshore wind farms, a dozen hydroelectric dams, and a few biomass or geothermal facilities (fig. 13.1) linked together by HVDC "electric highways" form the heart of this ambitious plan. We'll discuss details of CSP later in this chapter, but basically it captures solar energy to generate steam to produce electricity. This technology is already competitive with fossil fuels. In fact, in 2008, when oil hit $140 per barrel, CSP was less than half the price of an equivalent amount of oil energy.

Why would oil-rich Arab countries want to help Europe kick their fossil fuel habit? Perhaps because the world is approaching—or may have already passed—peak oil production. And remaining supplies are becoming increasingly expensive and difficult to reach. Many formerly oil-rich countries are facing the prospect of life without oil. Why not sell an endless supply of solar power, and save your remaining oil for your own use or to sell for higher prices at a later date?

For Europe, wouldn't this just mean trading dependency on unstable Middle Eastern countries for oil to dependency on their solar electricity? Perhaps. But if Desertech leads to local economic development (it's expected that building and operating all those power plants will add 80,000 well- paying local jobs), and if local economies become dependent on the power and water from renewable energy, mutual benefits from the system may help make it safe from political threats and civil unrest.

The first steps in Desertech implementation are now taking place. In 2011, contracts were issued for 65 km of HVDC to connect Spain and France—the first link in the Supergrid. And at the same time, Morocco, which has been selected for the first CSP plant, announced it had chosen both the site for the facility and four consortia partners to design, finance, build, and operate it.

Many other parts of the world are following this development with interest. China, Australia, South Africa, and western North America also have vast solar potential. The Desertech Consortium points out that within 6 hours, world deserts receive more energy from the sun than humankind consumes in an entire year. Perhaps many others of us could benefit from a similar system.

In this chapter we'll look at our options for finding environmentally and socially sustainable ways to meet our energy needs. ∎

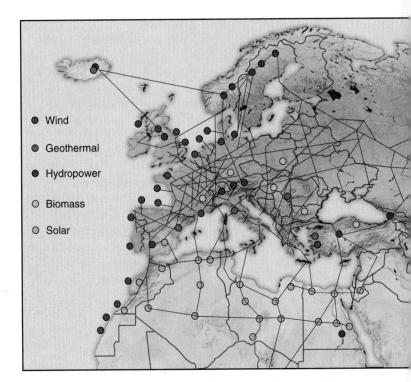

▲ **FIGURE 13.1** A supergrid of HVDC transmission lines may link a network of renewable energy facilities in Europe, North Africa, and the Middle East and provide both a substantial percentage of electricity for the region as well as drinking water for desert nations. SOURCE: German Aerospace Center, 2010.

13.1 ENERGY RESOURCES AND USES

Using external energy sources is one of the main things that sets us apart from other species and makes us human. Fire was probably our first external energy source. We learned long ago to use fire to heat and light our encampments, cook our food, and keep predators at bay. At least 10,000 years ago, we domesticated animals and trained them to carry us and our belongings as well as to pull plows and carts. About the same time, we learned how to use wind and water power to move boats, grind grain, pump water, and do other useful tasks. When James Watt invented his steam engine 250 years ago, he unleashed an age of industrialization that greatly magnified our ability to transform our world.

The **fossil fuels**—coal, oil, and natural gas—that have powered the industrial age have brought us many benefits, but have also caused huge social, political, and environmental problems. As we discussed in chapter 9, perhaps the most threatening of these problems is that the burning of fossil fuels emits carbon dioxide (CO_2), which is changing our global climate. We now get nearly 90 percent of all commercial energy from fossil fuels. How we'll end our dependence on—some would say addiction to—fossil fuels is one of the most important problems that face us today. In this chapter we'll look at the costs and consequences of various energy sources as well as our options for the future. We'll start with the fossil fuels and nuclear power that provide most of our energy today, and then turn to renewable sources that could supply all the energy we will need in the not-too-distant future.

How do we measure energy?

To understand the magnitude of energy use, it is helpful to know the units used to measure it. **Work** is the application of force over distance, and we measure work in **joules** (table 13.1). **Energy** is the capacity to do work. **Power** is the rate of energy flow or the rate of work done: for example, one **watt** (W) is one joule per second. If you use a 100-watt light bulb for 10 hours, you have used 1,000 watt-hours, or one kilowatt-hour (kWh). Most American households use about 11,000 kWh per year (table 13.2).

TABLE 13.1 | Energy Units

1 joule (J) = work needed to accelerate 1 kg 1 m/sec² for 1 m (or 1 amp/sec flowing through 1 ohm resistance)
1 watt (W) = 1 J per second
1 terawatt (TW) = 1 trillion watts
1 kilowatt hour (kWh) = 1,000 W exerted for 1 hour (or 3.6 million J)
1 megawatt (MW) = 1 million (10^6) W
1 gigajoule (GJ) = 1 billion (10^9) J
1 standard barrel (bbl) of oil = 42 gal (160 liters)

TABLE 13.2 | Energy Uses

USES	kWh/YEAR*
Computer	100
Television	125
100 W light bulb	250
15 W fluorescent bulb	40
Dehumidifier	400
Dishwasher	600
Electric stove/oven	650
Clothes dryer	900
Refrigerator	1,100

*Averages shown; actual rates vary greatly.
SOURCE: U.S. Department of Energy.

Fossil fuels supply most of our energy

Like most industrialized nations, the United States gets a vast majority of its energy from fossil fuels. According to the U.S. Energy Information Agency, oil currently provides 37 percent of this supply, followed by natural gas (25 percent) and coal (21 percent) (fig. 13.2). Renewables (hydro, wind, solar, biomass) provide 11 percent and nuclear power supplies 9 percent. In the twentieth century, although the rich countries of the world made up less than 5 percent of the total population, they consumed more than half the commercial energy. That situation is now changing, however. Rising incomes in China are leading to more energy consumption. China now consumes as much primary energy as all of Europe, and 85 percent as much as the United States. And because so much of China's energy comes from coal, it has now passed the United States in total CO_2 production.

How we, and the other countries of the world, can transition to sustainable energy is one of the most challenging issues we face. The scenario in figure 13.2 suggests that by 2050 we may be getting about one-quarter of our energy from renewables and another 15 percent or so from nuclear power. But many people

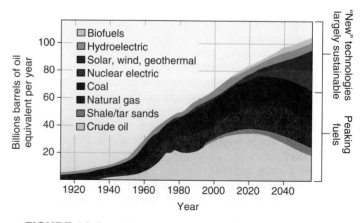

▲ **FIGURE 13.2** Fossil fuels, which now supply about 88 percent of all commerical energy in the world, are likely to decline as their costs increase and renewable energy gets cheaper.

believe that both fossil fuels and nuclear power are unacceptable, and that we should move much more quickly to renewables. Is that possible? As you'll learn in this chapter, there's more than enough energy from solar, wind, geothermal, and biomass power to provide everyone with a healthy, productive life.

How much energy do you use every year? Most of us don't think about it much, but maintaining the lifestyle we enjoy requires an enormous energy input. On average, each person in the United States and Canada uses more than 300 gigajoules (GJ) (equivalent to about 60 barrels of oil) per year. By contrast, in some of the poorest countries of the world, such as Ethiopia, Nepal, and Bhutan, each person generally consumes less than 1 GJ per year. This means that each each U.S. citizen consumes, on average, almost as much energy in a single day as a person in one of these countries consumes in a year.

Clearly, energy consumption is linked to the comfort and convenience of our lives. Those of us in the richer countries enjoy many amenities not available to most people in the world. The link isn't absolute, however. Several European countries, including Sweden, Denmark, and Finland, have higher standards of living than does the United States by almost any measure but use about half as much energy.

How do we use energy?

The largest share of the energy used in the United States is consumed by industry (fig. 13.3). Mining, milling, smelting, and forging of primary metals consume about one-quarter of that industrial energy share. The chemical industry is the second largest industrial user of fossil fuels, but only half of that use is for energy generation. The remainder is raw material for plastics, fertilizers, solvents, lubricants, and hundreds of thousands of organic chemicals in commercial use.

Residential and commercial customers use roughly 41 percent of the primary energy consumed in the United States, mostly for space heating, air conditioning, lighting, and water heating. Transportation requires about 28 percent of all energy used in the United States each year, almost all of that comes from petroleum. About three-quarters of all transport energy is used by motor vehicles. Nearly 3 trillion passenger miles and 600 billion ton miles of freight are carried annually by motor vehicles in the United States. About 75 percent of all freight traffic in the United States is carried by trains, barges, ships, and pipelines, but because they are very efficient, they use only 12 percent of all transportation fuel.

Producing and transporting energy also consumes and wastes energy. About half of all the energy in primary fuels is lost during conversion to more useful forms, while being shipped to the site of end use, or during use. Electricity is generally promoted as a clean, efficient source of energy because, when it is used to run a resistance heater or an electrical appliance, almost 100 percent of its energy is converted to useful work and no pollution is given off.

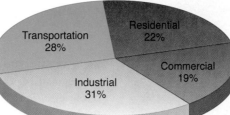

◀ FIGURE 13.3 U.S. energy consumption by sector in 2012. SOURCE: U.S. EIA.

What happens, however, before electricity reaches us? It takes large amounts of energy to mine, clean, and ship coal. Then nearly two-thirds of the energy in the coal we mine is lost in thermal conversion in the power plant. Finally, about 10 percent more is lost during conventional transmission and stepping down to household voltages. We need to take the whole fuel cycle into account when determining efficiency or the footprint of a particular source.

13.2 FOSSIL FUELS

Fossil fuels are organic (carbon-based) compounds derived from decomposed plants, algae, and other organisms buried in rock layers for hundreds of millions of years. Most of the richest deposits date to about 286 million to 360 million years ago (the Mississippian, Pennsylvanian, and Permian periods: see chapter 12), when the earth's climate was much warmer and wetter than it is now.

Coal resources are vast

World coal deposits are enormous, ten times greater than conventional oil and gas resources combined. Almost all the world's coal is in North America, Europe, and Asia (fig. 13.4), and just three countries, the United States, Russia, and China, account for two-thirds of all proven reserves. Coal seams can be 100 m thick and can extend across tens of thousands of square kilometers that were vast, swampy forests in prehistoric times. The total resource is estimated to be 10 trillion metric tons. If all this coal could be extracted, and we could find environmentally benign ways to use it, this would amount to several thousand years' supply. But do we really want to use all that coal? Coal mining is a dirty, dangerous activity. Underground mines are notorious for cave-ins, explosions, and lung diseases, such as black lung suffered by miners. Surface mines (called strip mines, where large machines scrape off overlying sediment to expose coal seams) are cheaper and generally safer for workers than tunneling, but leave huge holes where coal has been removed and vast piles of discarded rock and soil.

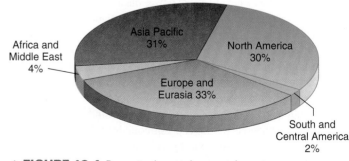

▲ FIGURE 13.4 Proven-in-place coal reserves by region. SOURCE: U.S. CIA Factbook, 2012.

▲ **FIGURE 13.5** One of the most environmentally destructive methods of coal mining is mountaintop removal. Up to 250 m of the mountain is scraped off and pushed into the valley below, burying forests, streams, farms, and sometimes whole towns.

An especially damaging technique employed in Appalachia is called mountaintop removal. Typically, the whole top of a mountain ridge is scraped off to access buried coal (fig. 13.5). In 2010 the EPA announced it would ban "valley fill," in which waste rock is pushed into nearby valleys, but existing operations are "grandfathered in" (see chapter 12 for further discussion). Mine reclamation is now mandated in the United States, but efforts often are only partially successful.

Coal burning releases huge amounts of air pollution. Every year the roughly 1 billion tons of coal burned in the United States (83 percent for electric power generation) releases close to a trillion metric tons of CO_2. This is about half of the industrial CO_2 released by the United States each year.

Coal also contains toxic impurities, such as mercury, arsenic, chromium, lead, and uranium, which are released into the air during combustion. The coal burned every year in the United States releases 18 million metric tons of sulfur dioxide (SO_2), 5 million metric tons of nitrogen oxides (NO_x), 4 million metric tons of airborne particulates, 600,000 metric tons of hydrocarbons and carbon monoxide, and 40 tons of mercury. This is about three-quarters of the SO_2 and one-third of the NO_x released by the United States each year. Sulfur and nitrogen oxides combine with water in the air to form sulfuric and nitric acids, making coal burning the largest single source of acid rain in many areas (chapter 9).

Most people aren't aware of it, but coal-burning plants emit radioactivity from uranium and thorium. You'd get more radioactivity living 70 years next to a coal power plant than next to a nuclear plant—assuming no accidents at the nuclear plant. It's possible to make either gas or liquid fuels out of coal, but these processes are even dirtier and more expensive than burning the coal directly. Both coal-to-liquid and coal-to-gas are environmentally disastrous.

Another problem with coal combustion was revealed in 2009 when an earthen dam broke in eastern Tennessee and released a billion gallons (3.8 billion liters) of coal ash sludge into a tributary of the Tennessee River. The ash contained dangerous levels of arsenic, mercury, and toxic hydrocarbons. After the spill, the U.S. EPA revealed that this impoundment was only one of hundreds of equally risky coal ash dumps across the country.

Coal may be on the way out

In 2010 the U.S. Energy Information Agency (EIA) predicted that coal would drop to 44 percent of America's electrical generation by 2035. Actually, we reached that level in 2011. Currently the government is projecting that coal will provide only 39 percent of our electricity by 2035, but that estimate appears to still be far too high. In reality, coal is fading quickly from our energy picture. Only half a dozen new coal-fired power plants are now under construction in the U.S. or in the planning stage. And when the last of those plants is finished about five years from now, no other new projects are proposed for the foreseeable future.

Federal regulations are part of this decline. The Mercury and Air Toxics Standards announced by the Environmental Protection Agency in 2012 will slash the allowable mercury emissions from coal-fired power plants. This was required by the 1970 Clean Air Act, but it was delayed for decades by owners of old power plants, who argued that their facilities were about to be closed anyway and so they shouldn't have to install expensive pollution control equipment. Forty years later, many of those plants are still in operation and still emitting dangerous pollutants.

The EPA estimates the new rules will cost utilities about $9 billion, but will save $90 billion in health care costs by 2016 by reducing our exposure to mercury, arsenic, chromium, and fine particulates that cause mental retardation, cardiovascular diseases, asthma, and other disorders. In 2012 the EPA also proposed limiting carbon emissions from power plants. If this rule goes into effect, new facilities will be allowed to emit no more than 1,000 lb (454 kg) of CO_2 per megawatt hour of electricity produced. Natural gas plants can easily meet that standard, but it's about half the amount released by the average coal-fired power plant. The only way to meet this limit with coal is to install expensive carbon capture and storage equipment.

Another problem for coal is that its prices are going up rapidly while solar costs are falling (fig. 13.6). By some accounts, solar, which cost almost twice as much as coal-fired electricity in 2012, could be one-third cheaper by 2020. One reason for this expected

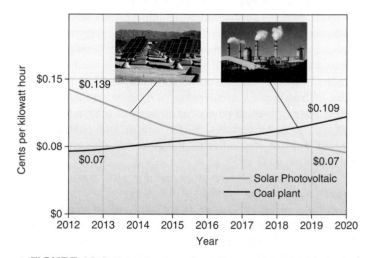

▲ **FIGURE 13.6** Solar electricity is becoming cost competitive with coal, and by 2020 solar should be cheaper than coal, experts predict.
SOURCE: Beyond Coal 2012.

price reversal is that solar panel prices fell by half in 2011 and are projected to come down more, while mercury limits and carbon-mitigation policies are driving up the price of coal.

Rapidly growing supplies and falling prices of natural gas also represent a challenge for coal. Gas is more versatile and cleaner burning than coal, but, as we'll discuss shortly, there are concerns that leakage from natural gas wells may negate its advantages.

There are cleaner ways to generate energy from coal. Gasification involves heating a coal slurry at high pressure in the presence of almost pure oxygen. The coal doesn't burn, instead it reacts with the oxygen and breaks down into a variety of gases, mostly hydrogen and carbon dioxide. The gases are cooled, separated, and purified of contaminants, such as sulfur, mercury, and arsenic. Carbon dioxide also can be captured and used for industrial processes or stored in geological formations. This technology has been proven in small-scale operations, but despite their enthusiasm for "Clean Coal," no utilities have opted for full-scale deployment. China claims to have about a dozen carbon-capture coal gasification plants, but observers report that although these facilities are capturing CO_2, none are actually storing it because they aren't required to by the central government. Instead, they simply vent it to the air.

China and India, both of which have very large coal resources, now burn about half of all coal mined annually in the world. Both of these countries have been increasing coal production greatly in the recent past to fuel their rapidly growing economies. Continuing to do so could cause runaway global climate change, so it's good news that China, at least, seems determined to move quickly to renewable energy.

Have we passed peak oil?

In the 1940s Dr. M. King Hubbert, a Shell Oil geophysicist, predicted that oil production in the United States would peak in the 1970s, based on estimates of U.S. reserves at the time. Hubbert's predicted peak was correct, and subsequent calculations have estimated a similar peak in global oil production in about 2005–2010 (fig. 13.7). Global production has not yet slowed significantly, but many oil experts expect that we will pass this peak in the next few years.

About half of the world's original 4 trillion bbl (600 billion metric tons) of liquid oil are thought to be ultimately recoverable. (The rest is too diffuse, too tightly bound in rock formations, or too deep to be extracted.) Of the 2 trillion recoverable barrels, roughly 1.26 trillion bbl are in proven reserves (commercially extractable using currently available technology). We have already used more than 0.5 trillion bbl—almost half of proven reserves—and the remainder is expected to last 41 years at current consumption rates of 30.7 billion bbl per year. Middle Eastern countries have more than half of the proven world supplies (fig. 13.8).

Consumption rates continue to climb, however, both in developed countries and in the fast-growing economies, such as China, India, and Brazil. China's energy demands have more than tripled in the past 35 years (much of this energy is used to produce goods for the U.S. and European markets), and China anticipates another doubling of energy demands in the next 15 years. Although renewables are supplying a growing share of China's energy, it's clear that competition is growing for global oil and gas supplies.

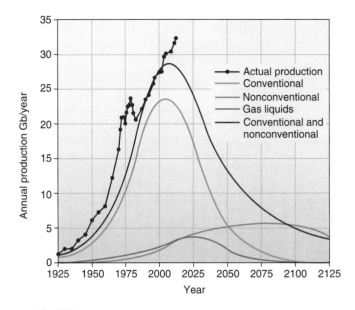

▲ **FIGURE 13.7** Worldwide production of crude oil with predicted Hubbert production. Gb = billion barrels. SOURCE: Jean Laherrère, www.hubbertpeak.org; International Energy Agency 2011.

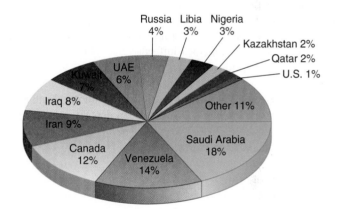

▲ **FIGURE 13.8** Proven oil reserves. Twelve countries (7 of them in the Greater Middle East) account for 89 percent of all known, economically recoverable oil. Numbers add to more than 100 percent due to rounding. SOURCE: CIA Factbook, 2012.

Extreme oil and tar sands have extended our supplies

Estimates of our recoverable oil supplies have expanded dramatically as we've developed techniques for obtaining oil from ever-more extreme places. Some countries, such as Canada, which wasn't even in the list of top ten oil-rich countries, and Venezuela, which was only seventh a few years ago, have suddenly vaulted to second and third in terms of their proven oil reserves (see fig. 13.8). Even the United States, which has been an oil importer for decades, is producing more of its own oil. Still, just 12 countries control 88 percent of this strategic resource.

Most of us hadn't thought much about the dangers of deep ocean oil wells in remote places until the 2010 explosion and sinking of the *Deepwater Horizon* in the Gulf of Mexico (fig. 13.9).

▲ **FIGURE 13.9** In 2010, the oil drilling rig *Deepwater Horizon* exploded and sank, spilling 6 million barrels (800 million liters) of crude oil into the Gulf of Mexico. It was drilling in water 1 mile (1.6 km) deep, but other wells are now more than twice as deep.

▲ **FIGURE 13.10** Alberta tar sands are now the largest single source of oil for the United States, but water pollution, forest destruction, and the energy used to liquify the sticky tar are among the many costs of for extracting this oil.

At least 5 million barrels (800 million liters) of oil were spilled during the four months it took to plug the leak. The well was being drilled in about 1 mi deep (1.6 km) water, but that isn't very deep by current standards. For the Gulf of Mexico, the current record is held by the Perdido Spar rig, which is drilling in more than 3,000 m (9,627 ft) of water and then to a depth of more than 6 km below the seafloor. Brazil is drilling at a similar depth about 300 km (186 mi) offshore. This ultradeep deposit, which Brazil estimates could hold 50 to 100 billion barrels, could make that country fifth or sixth in the world in oil resources.

By some estimates, Venezuela could have more than 300 billion barrels of oil (more even than Saudi Arabia) accessible with current technology, but much of Canada's and Venezuela's new oil resources are from tar sands. Canadian deposits in northern Alberta are estimated to be equivalent to 1.7 trillion bbl of oil, and Venezuela has nearly as much. Together these deposits are three times as large as all conventional liquid oil reserves. **Tar sands** are composed of sand and shale particles coated with bitumen, a viscous mixture of long-chain hydrocarbons. Shallow tar sands are excavated and mixed with hot water and steam to extract the bitumen. For deeper deposits, superheated steam can be injected to melt the bitumen, which is then pumped to the surface like liquid crude. Once the oil has been retrieved, it still must be cleaned and refined to be useful. Depending on how much energy is used to extract and refine oil from tar sands, this resource may emit more CO_2 than coal.

In 2012, Canada produced more than 3 million bbl per day, or twice the maximum projected output of the Arctic National Wildlife Refuge (ANWR). There are severe environmental costs, however, in producing this oil (fig. 13.10). A typical facility producing 125,000 bbl of oil per day creates about 15 million m^3 of toxic sludge, releases 5,000 metric tons of greenhouse gases, and consumes or contaminates billions of liters of water each year. Surface mining in Canada could destroy millions of hectares of boreal forest. Native Cree, Chipewyan, and Métis people worry about the effects on their traditional ways of life if forests are destroyed and wildlife and water are contaminated.

A battle over a proposed pipeline to carry tar-sands oil from mines in Alberta to Houston, Texas, has brought this fuel source to public attention. Supporters of the Keystone XL pipeline claimed it would bring energy security to the United States and provide 20,000 jobs. Opponents countered that the pipeline wouldn't help the United States very much because the oil would be shipped to Texas and sold abroad. This could raise U.S. oil prices rather than lower them. Critics also say that the pipeline supporters' estimates of job creation are not based just on pipeline workers but include everyone who would play a role supporting those workers. A more realistic number, according to opponents, is only 50 permanent jobs on the pipeline. Furthermore, a rupture in the Keystone pipeline could contaminate the Ogallala aquifer, which supplies drinking water and irrigation for much of the Great Plains.

TransCanada, the company behind the pipeline, is also pursuing a northern "Gateway" route that would cross the Canadian Rockies on its way to a terminal in the fjords of British Columbia's Great Bear Rainforest. First Nations people fear that heavy tanker traffic through the narrow, twisting fjords could result in another *Exxon Valdez*–size accident in this pristine wilderness. Pipelines carrying tar sands oil have a much higher rupture rate than those for conventional oil. The residual sand is more abrasive, the oil is more acidic and corrosive, and heavy oil must be heated to higher temperatures to be shipped, all making tar sands pipelines more accident prone.

The United States also has large supplies of unconventional oil. **Oil shales** are fine-grained sedimentary rock rich in solid organic material called kerogen. Like tar sands, the kerogen can be heated, liquefied, and pumped out like liquid crude oil. Oil shale beds up to 600 m (1,800 ft) thick underlie much of Colorado, Utah, and Wyoming. If these deposits could be extracted at a reasonable price and with acceptable environmental impacts, they might yield the equivalent of several trillion barrels of oil.

However, mining and extraction of oil shale—like tar sands—uses vast amounts of water (a scarce resource in the arid western United States), releases much more carbon dioxide than burning an equivalent amount of coal, and creates enormous quantities of waste. The rock matrix expands when heated, resulting in two or three times the volume that was dug out of the ground. Billions of dollars were spent in the 1980s on pilot projects to produce shale oil. When oil prices dropped, these schemes were abandoned. With rapidly rising crude oil prices in recent years, interest in this resource has rekindled. The Bureau of Land Management has just approved the first tar sands mine in the U.S. and has designated 1 million ha (2.5 million acres) for tar sands and oil shale production in western mountain states.

Natural gas is growing in importance

Natural gas (mostly methane) is the world's second-largest commercial fuel, making up about one-quarter of global energy consumption. Gas burns more cleanly than either coal or oil, and it generally produces only half as much CO_2 as an equivalent amount of coal. Many people hope that switching from coal to gas will help reduce global warming.

More than half of all the world's proven natural gas reserves are in the Middle East and the former Soviet Union (fig. 13.11). Both eastern and western Europe are highly dependent on imported gas. The total ultimately recoverable natural gas resources are thought to be 10,000 trillion ft^3, corresponding to about 80 percent as much energy as the estimated recoverable reserves of crude oil. The proven world reserves of natural gas are 6,200 trillion ft^3 (176 million metric tons). Because gas consumption rates are only about half of those for oil, current gas proven reserves represent roughly a 60-year supply at present usage rates.

Large amounts of methane are known to occur in many relatively shallow sedimentary beds. For the past decade or so, there has been intense drilling activity in the western United States. It often takes many closely spaced wells and directional drilling to extract methane from these "coal-bed" methane deposits. In Wyoming's Powder River basin, for example, 140,000 wells have been proposed for methane extraction. Together with the vast network of roads, pipelines, pumping stations, and service facilities, this

 ▲ **FIGURE 13.12** Some of the thousands of gas wells in the Jonah Field in Wyoming's Upper Green River Basin.

industry is having serious impacts on ranching, wildlife, and recreation in formerly remote areas (fig. 13.12).

Now attention is shifting to eastern states. It has long been recognized that methane is present in the Marcellus and Devonian Shales, which underlie much of the Appalachian mountains (fig. 13.13), but the economically recoverable resource was thought to be relatively small. New drilling techniques, however, have now made this deposit a potentially "supergiant" gas field. The U.S. Geological Survey estimates that the Marcellus/Devonian formation may contain 500 trillion ft^3 (13 trillion m^3) of methane. If all of it were recoverable, it would make a 100-year supply for the United States at current consumption rates. Large amounts of gas now available have driven prices down sharply in recent years.

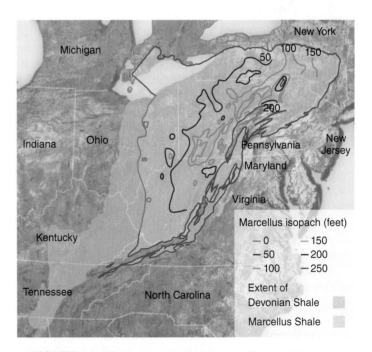

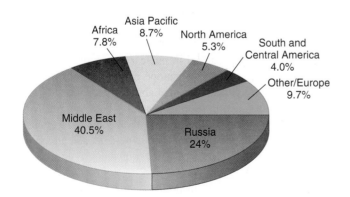

▲ **FIGURE 13.11** Proven natural gas reserves by region, 2011.
SOURCE: Data from British Petroleum, 2012.

▲ **FIGURE 13.13** The Marcellus and Devonian Shales, which underlie much of the Appalachian Mountain chain, contain a "supergiant" gas field.

But these shale deposits are generally "tight" formations through which gas doesn't flow easily. To boost well output, mining companies rely on hydraulic fracturing (or "fracking"). A mixture of water, sand, and various chemicals is pumped into the ground and rock formations at extremely high pressure. The pressurized fluid cracks sediments and releases the gas. Fracturing rock formations often disrupts aquifers, however, and contaminates water wells.

Drilling companies generally refuse to reveal the chemical composition of the fluids used in fracking. They claim it's a proprietary secret, but it's well known that a number of petroleum distillates, such as diesel fuel, benzene, toluene, xylene, polycyclic aromatic hydrocarbons, glycol ethers, as well as hydrochloric acid or sodium hydroxide, are often used. Many of these chemicals are known to be toxic to humans and wildlife. The U.S. EPA recently forced mining companies to reveal the contents, but not specific fractional composition, of their fracking fluids used on public land.

A study released in 2011 by the National Academy of Sciences reported that drinking water samples from shallow wells near methane drilling sites in Pennsylvania and New York had 17 times as much methane as those from sites far from drilling. And a study by researchers at Dartmouth concluded that 3.6 to 7.9 percent of the methane from shale-gas wells escapes to the atmosphere in leaks and venting over the life of the well. These methane emissions are up to twice those from conventional gas wells. Compared to coal, the climate footprint of shale gas is at least 20 percent greater for a comparable amount of energy, and may be twice as much over 100 years. This new gas supply may not help combat global climate change after all. Nevertheless, the U.S. EIA predicts that by 2020, about two-thirds of U.S. natural gas will come from shale beds and tight gas formations.

13.3 NUCLEAR POWER

In 1953 President Dwight Eisenhower presented his "Atoms for Peace" speech to the United Nations. He announced that the United States would build nuclear-powered electrical generators to provide clean, abundant energy. He predicted that nuclear energy would fill the deficit caused by predicted shortages of oil and natural gas. It would provide power "too cheap to meter" for continued industrial expansion of both the developed and the developing world. Today there are about 440 reactors in use worldwide, 104 of them in the United States. Half of the U.S. plants (52) are more than 30 years old and are approaching the end of their expected operational life (fig. 13.14). Cracking pipes, leaking valves, and other parts increasingly require repair or replacement as a plant ages. Nuclear power now amounts to about 9 percent of U.S. energy supply (almost twice the world average). All of it is used to generate electricity.

Rapidly increasing construction costs, safety concerns, and the difficulty of finding permanent storage sites for radioactive waste have made nuclear energy less attractive than promoters expected in the 1950s. Of the 140 reactors on order in 1975, 100 were subsequently canceled. The costs of decommissioning old reactors is a serious concern, because demolishing a worn-out plant may cost ten times as much as building it in the first place.

▲ **FIGURE 13.14** New York's Indian Point nuclear power plant is ranked the riskiest in the country by the U.S. Nuclear Regulatory Commission due to its age and location on the Hudson River just 24 miles (38 km) north of Manhattan Island. What would it cost to evacuate New York City if these reactors melt down?

Ten nuclear reactors have been shut down in the United States and deconstruction of most of them is now under way. Although these plants were generally small, costs have averaged several hundred million dollars each.

The nuclear power industry has been campaigning for greater acceptance, arguing that reactors don't release greenhouse gases that cause global warming. That's true during ordinary operation of the reactor, but the mining, processing, and shipping of nuclear fuel, together with decommissioning of old reactors and perpetual storage of wastes, result in up to 25 times more carbon emissions than an equal amount of wind energy.

Nevertheless, a number of prominent environmentalists have endorsed nuclear power as a solution to global climate change. In 2012 the Nuclear Regulatory Commission approved permits for two new nuclear reactors to be built in Georgia by the Southern Company. These reactors will be supported by $8 billion in loan guarantees from the federal government in addition to insurance caps on catastrophic accidents. If completed, these will be the first new nuclear power plants built in three decades in the United States.

How do nuclear reactors work?

The most commonly used fuel in nuclear power plants is U^{235}, a naturally occurring radioactive isotope of uranium. Uranium ore must be purified to a concentration of about 3 percent U^{235}, enough to sustain a chain reaction in most reactors. The uranium is then formed into cylindrical pellets slightly thicker than a pencil and about 1.5 cm long. Although small, these pellets pack an amazing amount of energy. Each 8.5 g pellet is equivalent to a ton of coal or 4 bbl of crude oil.

The pellets are stacked in hollow metal rods approximately 4 m (13 ft) long. About 100 of these rods are bundled together to make a **fuel assembly**. Thousands of fuel assemblies containing about

100 tons of uranium are bundled in a heavy steel vessel called the reactor core. Radioactive uranium atoms are unstable—that is, when struck by a high-energy subatomic particle called a neutron, they undergo **nuclear fission** (splitting), releasing energy and more neutrons. When uranium is packed tightly in the reactor core, the neutrons released by one atom will trigger the fission of another uranium atom and the release of still more neutrons (fig. 13.15). Thus a self-sustaining **chain reaction** is set in motion, and vast amounts of energy are released.

The chain reaction is moderated (slowed) in a power plant by a neutron-absorbing cooling solution that circulates between the fuel rods. In addition, **control rods** of neutron-absorbing material, such as cadmium or boron, are inserted into spaces between fuel assemblies to shut down the fission reaction or are withdrawn to allow it to proceed. Water or some other coolant is circulated between the fuel rods to remove excess heat. The greatest danger in one of these complex machines is a cooling system failure. If the pumps fail or pipes break during operation, the nuclear fuel quickly overheats, and a "meltdown" can result that releases deadly radioactive mate-rial. Although nuclear power plants cannot explode like a nuclear bomb, the radioactive releases from a worst-case disaster, such as the meltdown of the Fukushima reactors in Japan in 2011, can entail enormous costs (see What Do You Think?, p. 312).

Nuclear reactor design

Seventy percent of the world's nuclear plants are pressurized water reactors (PWR). Water circulates through the core, absorbing heat as it cools the fuel rods (fig. 13.16). This primary cooling water is heated to 317°C (600°F) and reaches a pressure of 2,235 psi. It then is pumped to a steam generator, where it heats a secondary water-cooling loop. Steam from the secondary loop drives a high-speed turbine generator that produces electricity. Both the reactor vessel and the steam generator are contained in a thick-walled, concrete-and-steel containment building that prevents radiation from escaping and is designed to withstand high pressures and temperatures in case of accidents.

Overlapping layers of safety mechanisms are designed to prevent accidents, but these fail-safe controls make reactors both expensive and complex. A typical nuclear power plant has 40,000 valves, compared with only 4,000 in a fossil fuel-fired plant of similar size. In some cases, the controls are so complex that they confuse operators and cause accidents rather than prevent them. Under normal operating conditions, however, a PWR releases very little radioactivity and is probably less dangerous for nearby residents than a coal-fired power plant.

We lack safe storage for radioactive waste

One of the most difficult problems associated with nuclear power is the disposal of wastes produced during mining, fuel production, and reactor operation. How these wastes are managed may ultimately be the overriding obstacle to nuclear power.

Enormous piles of mine wastes and abandoned mill tailings in uranium-producing countries represent another serious waste disposal problem. Production of 1,000 tons of uranium fuel typically generates 100,000 tons of tailings and 3.5 million liters of liquid waste. There now are approximately 200 million tons of radioactive waste in piles around mines and processing plants in the United States. This material is carried by the wind or washes into streams, contaminating areas far from its original source. Canada has even more radioactive mine waste on the surface than does the United States.

In addition to the leftovers from fuel production, the United States has about 77,000 tons of high-level (very radioactive) wastes. The high-level wastes consist mainly of spent fuel rods from commercial nuclear power plants and assorted wastes from nuclear weapons production. While they're still intensely radioactive, spent fuel assemblies are being stored in deep, water-filled pools at the power plants. These pools were originally intended only as temporary storage until the wastes were shipped to reprocessing centers or permanent disposal sites.

In 1987 the U.S. Department of Energy announced plans to build the first high-level waste repository in the desert under Yucca Mountain in Nevada. Waste was to be buried deep in the ground, where it was hoped it would remain unexposed

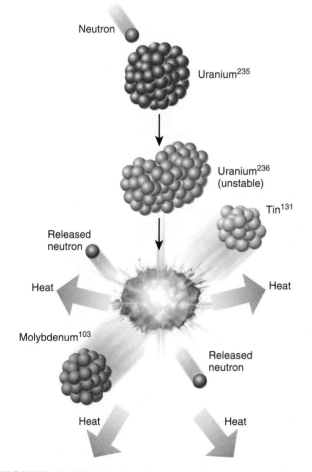

▲ **FIGURE 13.15** The nuclear fission carried out in the core of a nuclear reactor. The unstable isotope uranium-235, absorbs a neutron and splits to form tin-131 and molybdenum-103. Two or three neutrons are released per fission event and continue the chain reaction. A tiny amount of mass is converted to energy (mostly heat).

Twilight for Nuclear Power?

On Friday, March 11, 2011, at 2:46 p.m. Tokyo time, a magnitude 9.0 earthquake hit northern Japan. The largest earthquake in Japan's recorded history damaged buildings and roads in its own right, but even worse, it generated tsunami waves up to 30 m (98 ft.) high that crushed buildings, toppled power lines, and washed away cars, boats, and millions of tons of debris. Authorities reported 15,846 deaths, 6,011 injuries, 3,320 people missing, and 125,000 buildings damaged or destroyed by waves.

Perhaps the worst result of this catastrophe was the destruction of four of the six nuclear reactors at the Fukushima Daiichi power station 170 mi (273 km) north of Tokyo. The reactors shut down, as they were designed to do, when the earthquake hit, but that eliminated the electricity needed to pump cooling water through the intensely hot reactor core. Backup generators and connections to the regional power grid that would have provided emergency power were destroyed by the tsunami. The reactors quickly overheated, and the fuel rods began to melt in three of the six reactors cores. Hydrogen explosions in the reactor buildings at the complex destroyed roofs and walls and scattered radioactive debris around the area. In addition, spent fuel rods in storage pools of two units also overheated and caught fire, releasing even more radiation.

The plant operators sprayed seawater onto reactors to cool the reactors and put out fires, but that washed radioactive pollution into the ocean and contaminated seafood on which Japan depends. In 2012 Japan began tunneling under the nuclear complex to install a giant concrete diaper to try to contain radioactive drainage from the site. High radioactivity caused authorities to order evacuation of 140,000 people living within a 12-mile (20 km) radius around the facility. But the toll could have been worse. If the melting fuel and fires hadn't been contained, the radiation release could have been ten times greater than the 1986 disaster at Chernobyl in Ukraine.

At one point, government officials seriously considered evacuating the Tokyo metropolitan area. That might have meant moving up to 40 million people, which would have been the largest mass relocation in world history. However, westerly winds blew most of the radiation out to sea and abandoning Tokyo wasn't necessary.

Still, cleanup will take decades, and some areas near the reactors may never be habitable again. Altogether, Japanese officials estimate that losses may be $300 (U.S.) billion. This disaster is causing people in many countries to reconsider nuclear power. In Japan, which once got about one-third of its electricity from nuclear plants and had plans to expand that share to more than half the nation's power supply,

▲ Three of the four nuclear reactors at Japan's Fukushima Diiachi that were destroyed by fuel melting and hydrogen explosions after the 2011 tsunami knocked out emergency cooling systems.

more than 80 percent of the population now say they are anti-nuclear. After the disaster, all the nation's 54 reactors were shut down. An intense debate occurred about whether to restart any of them.

After Fukushima, Germany immediately shut down eight reactors and promised to close all the rest of its nuclear plants by 2022. China has suspended approvals for new reactors. Italy, Switzerland, and Spain voted to keep their countries nonnuclear. And in France, which gets three-quarters of its electricity from nuclear power, 62 percent of the population favored a phase-out of this energy source.

Could this be the death knell for nuclear power? Although public opinion swung strongly against this technology after Chernobyl, some people have been arguing that we need nuclear power at least as a temporary stopgap to replace fossil fuels in an effort to stop global climate change. What do you think? Are the risks of other disasters like Fukushima and Chernobyl worth the benefits of eliminating fossil fuels? And if we are going to make this Faustian bargain, what safeguards could we employ to reduce our risks?

to groundwater and earthquakes for the tens of thousands of years required for the radioactive materials to decay to a safe level. But continuing worries about the stability of the site led the Obama administration to cut off funding for the project in 2009 after 20 years of research and $100 billion in exploratory drilling and development.

For the foreseeable future, the high-level wastes that were to go to Yucca Mountain will be held in large casks in temporary surface storage facilities located at 131 sites in 39 states (fig. 13.17). But local residents living near these sites fear casks will leak.

Most nuclear power plants are built near rivers, lakes, or seacoasts. Radioactive materials could spread quickly over large areas if leaks occur. A hydrogen gas explosion and fire in 1997 in a dry storage cask at Wisconsin's Point Beach nuclear plant intensified opponents' suspicions about this form of waste storage.

If the owners of nuclear facilities had to pay the full cost for fuel, waste storage, and insurance against catastrophic accidents, no one would be interested in this energy source. Rather than be too cheap to meter, it would be too expensive to matter.

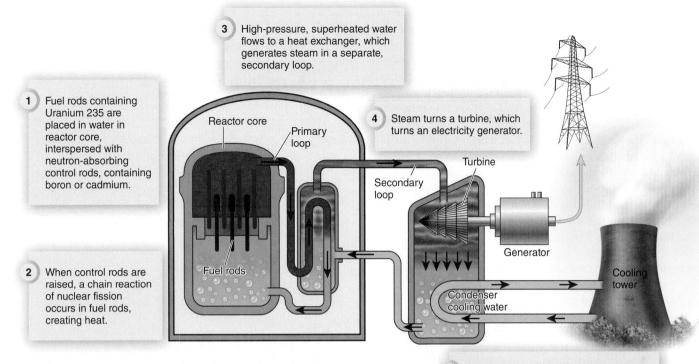

3 High-pressure, superheated water flows to a heat exchanger, which generates steam in a separate, secondary loop.

1 Fuel rods containing Uranium 235 are placed in water in reactor core, interspersed with neutron-absorbing control rods, containing boron or cadmium.

Reactor core

Primary loop

4 Steam turns a turbine, which turns an electricity generator.

Turbine

Secondary loop

Generator

2 When control rods are raised, a chain reaction of nuclear fission occurs in fuel rods, creating heat.

Fuel rods

Condenser cooling water

Cooling tower

▲ FIGURE 13.16 Pressurized water nuclear reactor. Water is superheated and pressurized as it flows through the reactor core. Heat is transferred to nonpressurized water in the steam generator, which drives the turbine to produce electricity.

5 Cooling water condenses steam to water, which returns to the heat exchanger.

▲ FIGURE 13.17 Spent fuel is being stored temporarily in large, aboveground "dry casks" at many nuclear power plants.

13.4 RENEWABLE ENERGY

In his 2011 State of the Union speech, President Barack Obama said, "To truly transform our economy, protect our security, and save our planet from the ravages of climate change, we need to ultimately make clean, renewable energy the profitable kind of energy. . . . So tonight, I challenge you to join me in setting a new goal: By 2035, 80 percent of America's electricity will come from clean energy sources." The good news is that using currently available technology and only those sites where energy facilities are socially,

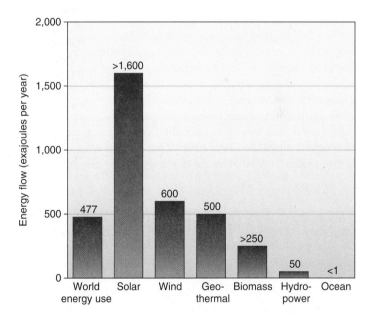

▲ FIGURE 13.18 Potential energy available from renewable resources using currently available technology in presently accessible sites. Together, these sources could supply more than six times current world energy use.   SOURCE: Adapted from UNDP and International Energy Agency.

economically, and politically acceptable, there's more than enough renewable energy from the sun, wind, geothermal, biomass, and other sources to meet all our present energy needs (fig. 13.18).

One of the best ways to avoid energy shortages and to relieve environmental and health effects of our current energy technologies is simply to use less (see What Can You Do?, this page). Much of the energy we consume is wasted. Our ways of using energy are so inefficient that most potential energy in fuel is lost as waste heat, becoming a form of environmental pollution. Conservation involves technology innovation as well as changes in behavior, but we have met these challenges in the past.

Oil price shocks in the 1970s led to rapid improvements in industrial and household energy use. Although population and GDP have continued to grow since then, the **energy intensity**, or amount of energy needed to provide goods and services, has declined while prices have risen sharply. In response to federal regulations and high gasoline prices, average U.S. automobile fuel economy more than doubled from 13 mpg in 1975 to 28.8 mpg in 1988.

Unfortunately, an oil glut and falling fuel prices in the 1990s discouraged further conservation. In fact, over the next decade, average fuel economy decreased to only 20.4 mpg, or less than Henry Ford's Model T got nearly a century earlier. In 2012, however, the Obama administration announced a plan to increase national fuel economy standards to 54.5 miles per gallon by the year 2025. This will reduce U.S. oil consumption by 2.2 million barrels a day (see Active Learning, p. 315). But you don't have to wait until 2025. High-efficiency automobiles are already available. Low-emission, hybrid gas-electric vehicles get up to 72 mpg (30.3 km/l) on the highway. And walking, biking, or taking public transport can lower your personal energy footprint far more.

Many improvements in domestic energy efficiency also have occurred in recent decades. Today's average new home uses one-half the fuel required in a house built in 1974, but much more can be done. Reducing air infiltration is usually the cheapest, quickest, and most effective way of saving energy because it is the largest source of losses in a typical house. An energy audit can tell you where your greatest energy losses are (fig. 13.19). It doesn't take much skill or investment to caulk around doors, windows, foundation joints, electrical outlets, and other sources of air leakage. Mechanical ventilation is needed to prevent moisture buildup in tightly sealed homes. Household energy losses can be reduced

by one-half to three-fourths by using better insulation, installing double- or triple-glazed windows, purchasing thermally efficient curtains or window coverings, and sealing cracks and loose joints.

Green building can cut energy costs by half

Innovations in "green" building have been stirring interest in both commercial and household construction. Much of the innovation has occurred in large commercial structures, which have larger budgets—and more to save through efficiency—than most homeowners have. Elements of green building are evolving rapidly, but they include extra insulation in walls and roofs, coated windows to keep summer heat out and winter heat in, and recycled materials, which save energy in production (fig. 13.20). Orienting windows toward the sun, or providing roof overhangs for shade, are important for comfort as well as for saving money.

Many appliances, such as dishwashers and coffee makers, already have timers you can program to operate at specific times. Suppose your whole house or apartment had similar capacities. Several utilities are experimenting with **smart metering**, in which you get information not only on how much energy any particular appliance is using at a given time, but also the source of that energy and how much it costs. Using one of these systems, you might program your water heater to operate only after midnight when electricity is cheapest or surplus wind power is available. These systems can be controlled remotely. You might turn on your heating system or air conditioning with your telephone as you're on your way home. Or the utility might turn off those same systems for short periods to avoid bringing expensive peak power online.

New houses can also be built with extra-thick, superinsulated walls and roofs. Windows can be oriented to let in sunlight, and

▲ **FIGURE 13.19** Infrared photography shows heat loss in a building.

Active LEARNING

Driving Down Gas Costs

Most Americans drive at least 1,000 miles per month in vehicles that get about 20 miles per gallon. Suppose gasoline costs $4.00 per gallon.

1. At these rates, how much does driving cost in a year? You can calculate the annual cost of driving with the following equation. Before multiplying the numbers, cross out the units that appear on both the bottom and the top of the fractions—if the units cancel out and give you $/year, then you know your equation is set up right. Then use a paper and pencil, and multiply the top numbers and divide by the bottom numbers. What is your annual cost?

$$\frac{12 \text{ months}}{\text{year}} \times \frac{1{,}000 \text{ mi}}{\text{month}} \times \frac{1 \text{ gal}}{20 \text{ mi}} \times \frac{\$4.00}{\text{gal}} =$$

2. Driving fast lowers your mileage by about 25 percent. At 75 mph, you get about three-fourths as many miles per gallon as you get driving 60 mph. This would drop your 20 mpg rate to 15 mpg. Recalculate the equation above, but replace the 20 with a 15. What is your annual cost now?

3. Efficiency also declines by about 33 percent if you drive aggressively, because rapid acceleration and braking cost energy. Aggressive driving would drop your 20 mpg mileage to about 13.4 mpg. How much would your yearly gas cost at 13.4 mpg?

4. In 2012, the Obama administration announced a goal of an average fleet efficiency of 54.5 mpg by 2025. How much would your yearly gas cost at 54.5 mpg? What is the difference between that cost and your cost at 20 mpg?

ANSWERS: 1. $2,400/year; 2. $3,200/year; 3. $3,592/year; 4. $881/year, or $1,519 less than at 20 mpg.

eaves can be used to provide shade. Double-glazed windows that have internal reflective coatings and that are filled with an inert gas (argon or xenon) have an insulation factor of R11, the same as a standard 4-inch-thick insulated wall, or ten times as efficient as a single-pane window. Superinsulated houses now being built in Sweden require 90 percent less energy for heating and cooling than the average American home. President Obama's "Cash for Caulkers" initiative planned to retrofit 100 million American homes and generate a million green jobs while cutting greenhouse gas emissions 5 percent over the next 20 years.

Improved industrial design has also cut our national energy budget. More efficient electric motors and pumps, new sensors and control devices, advanced heat-recovery systems, and material recycling have reduced industrial energy requirements significantly. In the early 1980s, U.S. businesses saved $160 billion per year through conservation. When oil prices collapsed, however, many businesses returned to wasteful ways.

Cities can make surprising contributions to energy conservation. New York City has become a leader in this effort, replacing 11,000 traffic signals with more-efficient LEDs (light-emitting diodes), and 180,000 old refrigerators with energy-saving models. Ann Arbor, Michigan, replaced 1,000 streetlights with LED models. These lights saved the city over $80,000 in the first year, and will pay for themselves in just over two years.

Cogeneration makes electricity from waste heat

One of the fastest growing sources of new energy is **cogeneration**, the simultaneous production of both electricity and steam or hot water in the same plant. By producing two kinds of useful energy in the same facility, the net energy yield from the primary fuel is increased from 30–35 percent to 80–90 percent. In 1900 half the electricity generated in the United States came from plants that also provided industrial steam or district heating. As power plants became larger, dirtier, and less acceptable as neighbors, they were forced to move away from their customers. Waste heat from the turbine generators became an unwanted pollutant to be disposed of in the environment. Furthermore, long transmission lines, which are unsightly and lose up to 20 percent of the electricity they carry, became necessary.

By the 1970s, cogeneration had fallen to less than 5 percent of our power supplies, but interest in this technology is growing. District heating systems are being rejuvenated, and the EPA estimates that cogeneration could produce almost 20 percent of U.S. electrical use, or the equivalent of 400 coal-fired plants.

Shade trees
Cut cooling cost

High-efficiency furnace and air conditioner

Photovoltaic roofing Solar water heating

Roof
12 in. (30 cm) insulation = R36-48

Overhang blocks summer sun

Winter sun warms house

Weather stripping
Doors, windows, vents, fixtures
Prevents air leaks

Windows
Double or triple panes;
Well-placed for light

SIPs
Structural insulated panels for extra insulation

Heat exchanger
Saves on heating/cooling;
prevents moisture buildup

Advanced framing
Prevents conductive heat loss

Geothermal
Heating and cooling

◀ FIGURE 13.20 Energy-efficient buildings can lower energy costs dramatically. Many features can be added to older structures. New buildings that start with energy-saving features (such as SIPs or advanced framing) can save even more money.

13.5 ENERGY FROM BIOMASS

Plants capture immense amounts of solar energy by storing it in the chemical bonds of plant cells. Firewood is probably our first fuel source. As recently as 1850, wood supplied 90 percent of the heat used in the United States. For more than a billion people in developing countries, burning **biomass** (plant materials) remains the principal energy source for heating and cooking. An estimated 1,500 m³ of fuelwood is gathered each year globally. This amounts to half of all wood harvested. Wood gathering and charcoal burning are important causes of deforestation in many rural areas. Providing efficient wood stoves can improve people's lives while also saving forests. In developed countries, where we depend on fossil fuels for most energy, wood burning is a minor heat source.

In developed countries, biomass burning can make a significant contribution to renewable energy supplies. Both agricultural wastes (such as straw and corn stalks) and biomass crops, such as reeds and elephant grass growing on land unsuitable for food crops, can be highly sustainable. These crops are carbon neutral because they absorb as much CO_2 in growing as they emit when burned.

Ethanol and biodiesel can contribute to fuel supplies

Biofuels, such as ethanol and biodiesel, are by far the biggest recent news in biomass energy. Globally, production of these two fuels is booming—from Brazil, which gets about 40 percent of its transportation energy from ethanol generated from sugarcane, to Southeast Asia, where millions of hectares of tropical forest have been cleared for palm oil plantations, to the United States, where about one-fifth of the corn (maize) crop currently is used to make ethanol. In 2009 President Obama proposed spending $150 billion over 10 years to develop renewable fuels and create 5 million "green collar" jobs. He proposed increasing ethanol production in the United States from 9 billion to 36 billion gallons per year (30 billion to 136 billion liters) by 2022. However, it would take the entire U.S. corn crop to produce that much ethanol from corn. We need to find other ways to create biofuels.

Crops with a high oil content, such as soybeans, sunflower seed, rape seed (usually called canola in America), and palm oil fruits are relatively easy to make into biodiesel. In some cases the oil needs only minimal cleaning to be used in a standard diesel engine. However, it would take a very large land area to meet our transportation needs with soy or sunflowers. Furthermore, diversion of these oils for vehicles deprives humans of important sources of edible oils.

Oil palms are considerably more productive per unit area than soy or sunflower (although palm fruit is more expensive to harvest and transport). Currently millions of hectares of species-rich forests in Southeast Asia are being destroyed to create palm oil plantations. Indonesia already has 6 million ha (15 million acres) of palm oil plantations, and Malaysia has nearly as much. Together these two countries produce nearly 90 percent of the world's palm oil.

Cellulosic ethanol may offer hope for the future

Crops such as sugarcane and sugar beets have a high sugar content that can be fermented into ethanol, but sugar is expensive and the yields from these crops are generally low, especially in temperate climates. Starches in grains, such as corn, have higher yields and can be converted into sugars that can be turned by yeast into ethanol or other alcohols. The idea of burning alcohol in vehicles isn't new. Henry Ford designed his 1908 Model T to run on ethanol.

Since 1980 more than 100 new refineries have been built, and U.S. ethanol production has grown from about 500 million liters to 30 billion liters per year. Most scientists calculate that corn ethanol contains only slightly more energy than you put into producing it, but everyone agrees that producing ethanol from **cellulosic** (woody) **crops** would have considerable environmental, social, and economic advantages over using edible grains or sugar crops for transportation fuel (fig. 13.21).

A number of techniques have been proposed for extracting sugars from cellulosic materials. Most involve mechanical chopping or shredding followed by treatment with bacteria or fungi to break down cellulose into soluble sugars (fig. 13.22).

So far, there are no commercial-scale cellulosic ethanol factories operating in North America, but the Department of Energy has provided $385 million in grants for six cellulosic biorefinery plants. These pilot projects will test a wide variety of feedstocks, including rice and wheat straw, milo stubble, switchgrass hay, almond hulls, corn stover (stalks, leaves, and cobs), and wood chips.

One of the most exciting biofuel crops is *Miscanthus x giganteus*, a perennial grass from Asia. Often called elephant grass (although this name is also used for other species), Miscanthus is a sterile, hybrid grass that grows 3 or 4 m in a single season

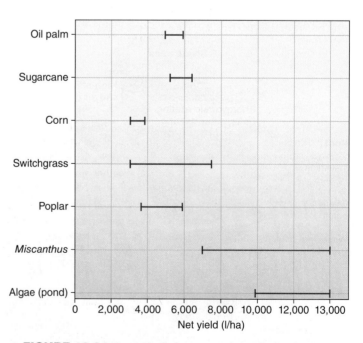

▲ **FIGURE 13.21** Proven biofuel sources include oil palms, sugarcane, and corn grain (maize). Other experimental sources may produce better yields, however. SOURCE: Data from E. Marris, 2006. *Nature* 444:670–678.

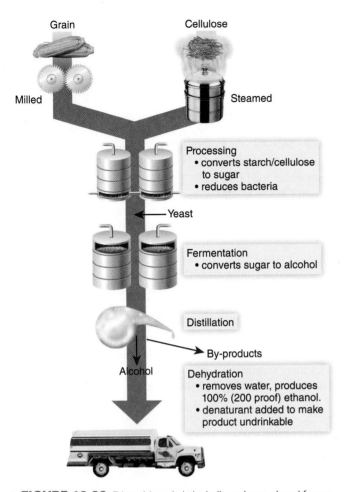

▲ FIGURE 13.22 Ethanol (or ethyl alcohol) can be produced from a wide variety of sources. Maize (corn) and other starchy grains are milled (ground) and then processed to convert starch to sugar, which can be fermented by yeast into alcohol. Distillation removes contaminants and yields pure alcohol. Cellulosic crops, such as wood or grasses, can also be converted into sugars, but the process is more difficult. Steam blasting, alkaline hydrolysis, enzymatic conditioning, and acid pretreatment are a few of the methods for breaking up woody material. Once sugars are released, the processes are similar.

(fig. 13.23). Miscanthus can produce at least five times as much dry biomass per hectare as corn. Its perennial growth and long-lasting canopy also protect the soil from erosion and require much less fuel for cultivation.

Where using corn to produce enough ethanol to replace 20 percent of U.S. gasoline consumption would take about one-quarter of all current U.S. cropland out of food production, Miscanthus could produce the same amount on less than half that much area. And it wouldn't need to be prime farm fields. Miscanthus can grow on marginal soil with far less fertilizer than corn needs. In the fall, Miscanthus moves nutrients into underground rhizomes. This means that the standing stalks are almost entirely cellulose and next year's crop needs very little fertilizer.

Harvesting, storing, and shipping biomass crops remains a problem. The low energy content of straw or wood chips, compared to oils or sugars, makes it prohibitively expensive to ship them more

▲ FIGURE 13.23 *Miscanthus x giganteus* is a perennial grass that can grow 3 or 4 m in a single season. It thrives on marginal land with little fertilizer or water and can produce five times as much biomass as corn.

than about 50 km (31 mi) to a refinery. We might need to have a very large number of small refineries if we depend on cellulosic ethanol. Interestingly, some authors claim that you could drive a hybrid automobile about twice as far on the electricity generated by burning a ton of dry biomass than you could on the ethanol fermented from that same ton. So, burning biomass may still be a better solution than fermentation if we have electric vehicles.

Methane from biomass is efficient and clean

Just about any organic waste, but especially sewage and manure, can be used to produce methane. Methane gas, the main component of natural gas, is produced when anaerobic bacteria (bacteria living in an oxygen-free space) digest organic matter (fig. 13.24).

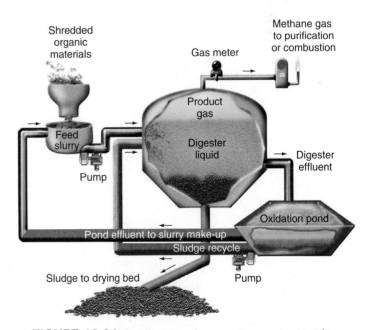

▲ FIGURE 13.24 Continuous unit for converting organic material to methane by anaerobic fermentation. One kilogram of dry organic matter will produce 1–1.5 m^3 of methane, or 2,500–3,600 million calories per metric ton.

The main by-product of this digestion, CH_4, has no oxygen atoms because no oxygen was available in digestion. But this molecule oxidizes, or burns, easily, producing CO_2 and H_2O (water vapor). Consequently, methane is a relatively clean, efficient fuel. Today, as more cities struggle to manage urban sewage and feedlot manure, methane could be a rich source of energy. In China, in addition to solar and wind power, more than 6 million households use methane, also known as biogas, for cooking and lighting. Two large municipal facilities in Nanyang, China, for example, provide fuel for more than 20,000 families.

Methane is a promising resource, but it has not been adopted as widely as it could be. Gas is harder to store than liquid fuels like ethanol, and low prices for natural gas and other fuels have reduced incentives for building methane production systems. However, concerns about greenhouse gases may lead to further development, because methane is a powerful agent of atmospheric warming (chapter 9). Especially around livestock facilities, such as poultry or hog barns, large lagoons of liquid manure release a constant flow of methane to the atmosphere. These lagoons are also a threat to water bodies, because they occasionally overflow. But trapping this methane would provide energy, save money, and reduce atmospheric impacts. City sewage treatment plants and landfills also offer rich, and mostly untapped, potential for methane generation (chapter 14).

Could algae be a hope for the future?

Algae might be an even more productive biofuel crop than any we've discussed so far. While Miscanthus can yield up to 13,000 liters (3,500 gal) of ethanol per hectare, some algal species growing in a photobioreactor (high-tech greenhouse) might theoretically produce 30 times as much high-quality oil. This is partly because single-celled algae can grow 30 times as fast as higher plants. Furthermore, some algae store up to half their total mass as oil. Photobioreactors are much more expensive to build and operate than planting crops, but they could be placed on land unsuitable for agriculture and they could use recycled water. Open ponds are much cheaper than photobioreactors, but they also produce far less biomass per unit area. So far, the actual yield from algal growth facilities is actually about the same as Miscanthus.

One of the most intriguing benefits of algal growth facilities is that they could be placed next to conventional power plants, where CO_2 from burning either fossil fuels or biomass could be captured and used for algal growth. Thus they'd actually be carbon negative: providing a net reduction in atmospheric carbon while also creating useful fuel.

An algal bioreactor started producing biodiesel in South Africa in 2006, and one in Brazil aims to soon start trapping CO_2 from a coal-fired power plant. A number of U.S. companies, including Solix Biofuels, Sapphire Energy, OriginOil, PetroAlgae, and Shell Oil, are exploring algal biofuels. In 2009 Japan Airlines made a test flight using a combination of jet fuel and algal oils. Another tantalizing fact is that some algae produce hydrogen gas as a photosynthetic by-product. If fuel cells ever become economically feasible, algae might provide them with a energy source that doesn't depend on fossil fuels.

13.6 WIND AND SOLAR ENERGY

Renewable sources could supply all the energy we need (see Key Concepts, p. 320). Solar energy is our most abundant and ubiquitous renewable resource, followed by wind power. Engineering developments and mass production have brought prices for wind and solar down so they now compete with fossil fuels almost everywhere. Renewables provide more jobs, take less land, and keep more money at home, where it can benefit local communities rather than support dictatorial regimes in unstable countries as we now do with our payments for fossil fuels. Perhaps most important, wind and solar don't emit CO_2 that disrupts our global climate.

As the opening case study for this chapter shows, wind and solar energy can be tied together over a wide geographical area to create a steady, dependable, affordable electrical supply. Within 6 hours, deserts of the world receive more energy from the sun than humankind consumes within an entire year. If we can capture just a fraction of that energy, we could stop burning fossil fuels almost entirely. Wind has a number of advantages over most other power sources. Wind farms and solar collectors have much shorter planning and construction times than fossil fuel or nuclear power plants. Furthermore these renewable facilities are modular (a few or a lot more turbines or solar panels can be added if loads grow), and neither has ongoing fuel costs or air emissions.

Wind could meet all our energy needs

Wind power is the world's fastest-growing energy source and could replace all the commercial energy we now use. With 250,000 MW of globally installed capacity in 2012, wind power is producing about 500 TWhr of electricity annually. The Wind Energy Association predicts that 1.5 million MW of capacity could be possible by 2020.

China is now the world's largest producer of wind turbines. Clean technology provides more than 1 million Chinese jobs building equipment, much of which is exported. China now has at least 63 GW of wind power, or about one-quarter of the world total (fig. 13.25). The biggest wind turbines now being built have

▼ **FIGURE 13.25** China is now the world's largest producer of wind turbines and has about one-quarter of all installed wind power capacity. Together with solar, this clean technology provides more than 1 million jobs in China.

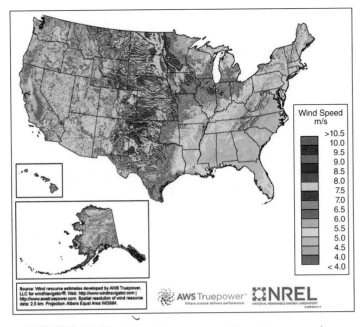

▲ **FIGURE 13.26** U.S. wind resource map. Mountain ranges and areas of the High Plains have the highest wind potential, but much of the country has fair to good wind supply. SOURCE: Data from U.S. Department of Energy.

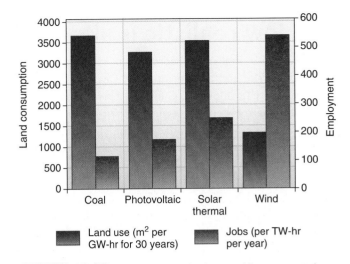

▲ **FIGURE 13.27** If you include the land required for mining, wind power takes about one-third as much area and creates about five times as many jobs to create the same amount of electrical energy as coal.

towers up to 150 m tall with 62 m long blades that reach as high as a 45-story building. Each can generate 5 MW of electricity, or enough for 2,500 typical American homes. Out of commission for maintenance only about three days per year, many turbines can produce power 90 percent of the time. Theoretically up to 60 percent efficient, modern windmills typically produce about 35 percent of peak capacity under field conditions.

Prices for wind power have fallen sharply in the past few years, and this is currently the cheapest source of new electrical generation, costing as little as 3 cents/kWh compared to 4 to 5 cents/kWh for coal and five times that much for nuclear fuel. If the carbon "cap and trade" program proposed by President Obama becomes law, wind energy could be cheaper than fossil fuels in many places.

Like many countries, the United States has a tremendous potential for wind power. Large areas of the Great Plains and mountain states have persistent winds suitable for commercial development (fig. 13.26). Thirty-seven states now have utility-scale wind farms. Texas, with 10,377 MW, leads the nation, followed by Iowa, California, Illinois, and Minnesota, but there is a huge potential waiting to be tapped.

As figure 13.27 shows, wind power takes about one-third as much area and creates about five times as many jobs to create the same amount of electrical energy as coal when the land consumed by mining is taken into account. Furthermore, with each tower taking only about a 0.1 ha (0.25 acre) of cropland, farmers can continue to cultivate 90 percent of their land while getting $2,000 or more in annual rent for each wind machine. An even better return results if the landowner builds and operates the wind generator, selling the electricity to the local utility. Annual profits can be as much as $100,000 per turbine, a wonderful bonus for use of 10 percent of your land.

Cooperatives are springing up to help landowners and communities finance, build, and operate their own wind generators. One thousand megawatts of wind power (equivalent to one large nuclear or fossil fuel plant) can create more than 3,000 permanent jobs, while paying about $4 million in rent to landowners and $3.6 million in tax payments to local governments. About 20 Native American tribes, for example, have formed a coalition to study wind power. Together their reservations (which are in the windiest, least productive parts of the Great Plains) could generate at least 350,000 MW of electrical power, equivalent to about half of the current total U.S. installed electrical capacity.

Wind does have limitations, however. Like solar energy, it's an intermittent source. Not every place has strong enough or steady enough wind to make this an economical resource. As the opening case study for this chapter shows, part of Europe's renewable energy plan is a network of offshore wind farms. Although the United States does have good offshore wind potential, installation and operating costs are much higher for ocean-based facilities, compared to those based on land.

There are problems with wind energy. In some places, high bird and bat mortality has been reported around wind farms. This seems to be particularly true in California, where rows of generators were placed at the summit of mountain passes where wind velocities are high but where migrating birds and bats are likely to fly into rotating blades. New generator designs and more careful tower placement seems to have reduced this problem in most areas. Although national polls in the United States show that 82 percent of the public supports additional wind power, the rate of support is often considerably less among people who live close to the towers.

Some people object to the sight of large machines looming over the landscape, and there are claims that low-frequency sound waves and flickering shadows produced by moving blades cause headaches, insomnia, digestive problems, panic attacks, and other health issues. There is no medical evidence to connect these symptoms to wind turbines, however, and researchers point out

How realistic is alternative energy?

It's very realistic, according to studies from Stanford University and the University of California at Davis.* With existing technology, renewable sources could provide all the energy we need, including the fossil fuels we use now. And we could save money at the same time. **Land-based wind, water power, and solar potential exceed all global energy consumption.** Renewable energy supplies over the oceans are even larger, since oceans cover two-thirds of the earth's surface. Many studies suggest that renewables could meet future demand more economically and more safely than fossil-based energy plans. How would this energy future look?

Balance of global energy supply using only alternative sources (estimated demand for 2030 = 11.4 TW)

- ← Additional demand if we depend on fossil/nuclear energy. Because of fuel production and transportation costs plus lower efficiency overall, we need about 25–30% more energy from coal, oil, gas, and nuclear sources than from renewables.
- **9%** ← Hydro, tidal, geothermal, wave
- **41%** ← Solar
- **50%** ← Wind

KC 13.2

GLOBAL ENERGY DEMAND, 2030
in terawatts (TW)
(1 TW = 1,000,000,000,000 watts)

KC 13.1

1. **Wind** could supply 50 percent of our energy, according to this plan. It would take 3.8 million large wind turbines to supply electricity to the whole world. Isn't that an impossible task? Not necessarily: we build that many cars and trucks every year worldwide.

2. **Solar energy** could provide 41 percent of our total energy supply. It would take 1.7 billion rooftop photovoltaic systems and nearly 100,000 concentrated solar power plants to provide 4.6 TW. Rooftop collectors can be located where energy is used, so they don't lose energy in transmission and don't compete with other land uses.

KC 13.3

◄ Solar thermal collectors already are price competitive with fossil fuels, but they generally can't be located close to consumers, and they may require scarce cooling water in arid lands where sunshine is plentiful.

3. **Hydropower (dams, tidal, geothermal, wave energy)** could supply about 9 percent of our energy. Most major rivers are already dammed, but underwater turbines in rivers and tidal areas could be effective. Deep wells could tap geothermal energy, but there are worries about triggering earthquakes and contaminating aquifers.

KC 13.4

*For more information: see Jacobson, M. Z., and M. A. Delucchi. 2009. A path to sustainable energy. *Scientific American* 301(5) 58–65.

Geothermal plant ▶

Wouldn't we have problems with unreliable supplies and a need for expensive storage?

Fortunately, the wind blows more at night to complement sunshine during the day. By balancing renewable sources, we can have just as reliable supplies as we now have with fossil fuels. Renewable sources also have a much better service record. Coal-burning power plants are out of production 46 days per year for maintenance. Solar panels and wind turbines average only 7 days down for repairs per year.

Solar, wind, and water power also solve two of our most pressing global problems: (1) the problem of climate change, perhaps the most serious and costly problem we face currently, as water shortages, crop failures, and refugee migrations destabilize developing regions; and (2) political conflict over fuel supplies, as in the oil fields of Iraq, Nigeria, and Ecuador, or nuclear fuel processing in Iran.

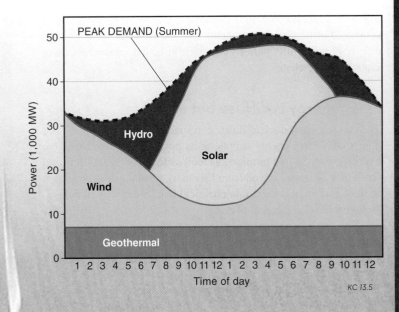

KC 13.5

In addition to the energy we can obtain from renewable sources, conservation measures could save up to half the energy we now use. Mass transit, weather-proofing, urban in-fill, and efficient appliances are among the available strategies that can save money in the near term and in the long term.

What would renewable energy cost?

By 2020, wind and hydroelectricity should cost about half as much as fossil fuels or nuclear power, and because renewable energy sources are inherently more efficient than fossil fuels, it should take about one-third less energy to supply the same services with sun, wind, and water.

▲ Light Rail   KC 13.6

CAN YOU EXPLAIN?

1. What would be the greatest benefits of switching to renewable energy?
2. Which of these sources is forecast to produce the most energy?
3. How many windmills would we need in this plan?
4. Who would benefit most and least from an alternative energy future? From a fossil and nuclear future? Why?

that such general complaints could be caused by a wide variety of sources. Furthermore, proponents say, the aesthetic or economic effects of windmills pale in comparison to the consequences of global climate change.

Solar energy is diffuse but abundant

The sun is a giant nuclear furnace in space, constantly bathing our planet with a free energy supply. Solar power drives winds and the hydrologic cycle. All biomass, as well as fossil fuels and our food (both of which are derived from biomass), results from conversion of light energy (photons) into chemical bond energy by photosynthetic bacteria, algae, and plants.

The average amount of solar energy arriving at the top of the atmosphere is 1,330 watts per square meter. About half of this energy is absorbed or reflected by the atmosphere (more at high latitudes than at the equator), but the amount reaching the earth's surface is still about 10,000 times all the commercial energy used each year. However, this tremendous infusion of energy comes in a form that, until recently, has been too diffuse and low in intensity to be used except for environmental heating and photosynthesis. But as the opening case study for this chapter shows, there are now ways to use this vast power source, so we might never again have to burn fossil fuels. Figure 13.28 shows world solar energy potential.

Solar collectors can be passive or active

Our simplest and oldest use of solar energy is passive heat absorption, using natural materials or absorptive structures with no moving parts to simply gather and hold heat. For thousands of years people have built thick-walled stone and adobe dwellings that slowly collect heat during the day and gradually release that heat at night. After cooling at night, these massive building materials maintain a comfortable daytime temperature within the house, even as they absorb external warmth.

▲ **FIGURE 13.29** Solar water heaters can be scaled up to provide hot water and space heating for whole cities.

A modern adaptation of this principle is a glass-walled "sun space" or greenhouse on the south side of a building. Incorporating massive energy-storing materials, such as brick walls, stone floors, or barrels of heat-absorbing water, into buildings also collects heat to be released slowly at night. Active solar systems generally pump a heat-absorbing, fluid medium (air, water, or an antifreeze solution) through a relatively small collector, rather than passively collecting heat in a stationary medium like masonry. Active collectors can be located adjacent to or on top of buildings rather than being built into the structure. Because they are relatively small and structurally independent, active systems can be retrofitted to existing buildings.

A flat, black surface sealed with a double layer of glass makes a good solar collector. Water pumped through the collector picks up heat for space heating or to provide hot water. A collector with about 5 m² of surface can reach 95°C (200°F) and can provide enough hot water for an average family. China currently produces about 80 percent of the world's solar water heaters, which cost less than $200 each. At least 30 million Chinese homes get hot water and/or space heat from solar energy. In Europe, municipal solar systems provide district heating for whole cities (fig. 13.29).

In a symbolic act to illustrate his commitment to solar energy, Barack Obama restored to the White House roof the solar water heating panels that were removed 30 years ago by Ronald Reagan.

High-temperature solar energy

Solar high-temperature solar thermal plants are suitable for industrial-size facilities. The solar farms being built in North Africa for the Desertech project, for example, are **concentrating solar power** (CSP) systems. They use long, trough-shaped, parabolic mirrors to reflect and concentrate sunlight on a central tube containing a heat-absorbing fluid (see p. 302). Reaching much higher temperatures than possible in a basic flat panel collector, the fluid passes

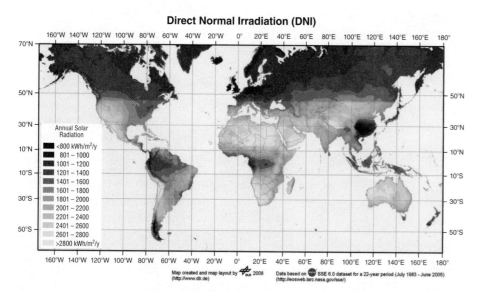

Direct Normal Irradiation (DNI)

Annual Solar Radiation

■ <800 kWh/m²/y
■ 801 – 1000
■ 1001 – 1200
■ 1201 – 1400
■ 1401 – 1600
■ 1601 – 1800
■ 1801 – 2000
■ 2001 – 2200
■ 2201 – 2400
■ 2401 – 2600
■ 2601 – 2800
□ >2800 kWh/m²/y

Map created and map layout by DLR 2008 (http://www.dlr.de)   Data based on SSE 6.0 dataset for a 22-year period (July 1983 - June 2005) (http://eosweb.larc.nasa.gov/sse/)

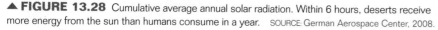

▲ **FIGURE 13.28** Cumulative average annual solar radiation. Within 6 hours, deserts receive more energy from the sun than humans consume in a year.   SOURCE: German Aerospace Center, 2008.

through a heat exchanger, where it generates steam to turn a turbine to produce electricity. Research by the German Aerospace Center suggests that CSP plants in North Africa and the Middle East should be able to provide 470,000 MW by 2050, or about 17 percent of the power used by the European Union. Costs, they estimate, should be equal to or lower than nuclear or fossil fuel power.

There are several advantages for a CSP plant besides fuel cost. Heat from the transfer fluid can be stored in a medium, such as molten salt, for later use. This allows the system to continue to generate electricity on cloudy days or at night. Desertech expects to be able to produce power nearly around the clock. In addition, those plants located near coastlines (see fig. 13.1) can use seawater to cool the power cycle (necessary to keep turbines operating). But the heat absorbed from turbines isn't all wasted. Much of it can be used to flash-evaporate water to create pure drinking water—something sorely lacking in most of North Africa and the Middle East. A 250 MW collector field is expected to provide 200 MW of electricity plus 100,000 m³ (about 26 million gal) of distilled water per day.

But wouldn't highly polished mirrors in a CSP plant be damaged by desert sand storms? The parabolic troughs follow the sun to maximize solar energy absorption. On days when storms are forecast, the mirrors can be rotated into a protective position.

Solar-thermal power plants in California's Mojave Desert have been operating for over 20 years, and have withstood hailstorms, sandstorms, and gale-force winds. Wouldn't it take huge areas of land to capture solar energy? According to the German Aerospace Center, supplying 17 percent of Europe's energy requirements will take 2,500 km², or less than 0.3 percent of the Sahara desert. Interestingly, the nuclear disaster at Fukushima has made about 0.3 percent of Japan permanently uninhabitable. Which energy source do you think is a better choice?

CSP's relatively small footprint doesn't necessarily mollify critics, however. While some people regard deserts as useless, barren wastelands, others view them as beautiful, biologically rich, and captivating. In 2010, state regulators approved 13 large solar thermal facilities and wind farms for California's Mojave Desert. Subsequently, however, most of these projects were canceled or delayed by protests over land use. Some people argued that these plants would harm rare or endangered species, such as the desert tortoise or the fringe-toed lizard. Native American groups protested that some areas were sacred cultural sites, while others simply love the solitude and mystery of the desert and believe that large industrial facilities are an unwelcome intrusion. In response to this challenge, millions of hectares of desert have been added to new or existing protected areas to forestall energy development.

Another high-temperature system uses thousands of smaller mirrors arranged in concentric rings around a tall central tower (fig. 13.30). The mirrors, driven by electric motors, track the sun and focus light on a heat absorber at the top of the "power tower" where a transfer medium is heated to temperatures as high as 500°C (1,000°F), which then drives a steam-turbine electric generator. Under optimum conditions, a 50 ha (130 acre) mirror array should be able to generate 100 MW of clean, renewable power. Southern California Edison's Solar II plant in the Mojave Desert

▲ **FIGURE 13.30** A "Power Tower" is a form of concentrated solar thermal electrical generation. Thousands of movable mirrors focus intense energy on the tower, where fluid is heated to drive a steam turbine.

east of Los Angeles is an example. Its 2,000 mirrors focused on a 100 m (300 ft) tall tower generate 10 MW, or enough electricity for 5,000 homes at an operating cost far below that of nuclear power or oil.

Because all the mirrors are focused on a single point, the heat transfer medium has to be capable of absorbing much higher energy levels than in solar troughs. So far most of these plants use liquid sodium or molten nitrate salt for heat absorption. These materials are much more corrosive and difficult to handle than the lower-temperature fluids suitable for a solar trough. The Worldwatch Institute estimates that U.S. deserts could provide more than 7,000 GW of solar energy—nearly seven times the potential in Europe.

Photovoltaic cells generate electricity directly

Photovoltaic (PV) cells capture solar energy and convert it directly to electrical current by separating electrons from their parent atoms and accelerating them across a one-way electrostatic barrier formed by the junction between two different types of semiconductor material (fig. 13.31). The first photovoltaic cells were made by slicing thin wafers from giant crystals of extremely pure silicon.

Over the past 25 years, the efficiency of energy captured by photovoltaic cells has increased from less than 1 percent of incident light to more than 10 percent under field conditions and over 75 percent in the laboratory. Promising experiments are under way using exotic metal alloys, such as gallium arsenide, and semiconducting polymers of polyvinyl alcohol, which are more efficient in energy conversion than silicon crystals.

In 2010, thin-film photovoltaic cells finally broke the $1-per-watt barrier, a price that begins to make them competitive with fossil fuels and nuclear power in many situations. As further research improves their efficiency and lifespan, industry experts believe they could produce electricity for about 7¢ per kilowatt-hour by 2020 (see fig. 13.6). This makes photovoltaic solar competitive with fossil fuels in many places for utility-scale baseload power arrays (fig. 13.32a).

Systems that generate electricity closer to the end user, on the other hand, have many advantages. Mounting a photovoltaic system in your yard or on your rooftop delivers electricity without

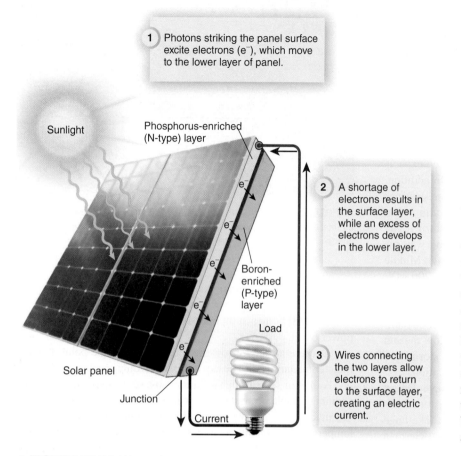

1 | Photons striking the panel surface excite electrons (e⁻), which move to the lower layer of panel.

Sunlight

Phosphorus-enriched (N-type) layer

2 | A shortage of electrons results in the surface layer, while an excess of electrons develops in the lower layer.

Boron-enriched (P-type) layer

Load

Solar panel

Junction

Current

3 | Wires connecting the two layers allow electrons to return to the surface layer, creating an electric current.

▲ **FIGURE 13.31** When solar energy strikes a photovoltaic (PV) cell, an electron is dislodged from atoms in the p-layer in the silicon crystal. These electrons cross an electrostatic junction between different semiconductor materials. This creates a surplus of electrons in the n-layer and a shortage of electrons (or a positive charge) in the p-layer. The difference in charge creates an electric current in a circuit connecting the two layers.

the losses inherent in long-distance distribution. A photovoltaic array of about 30 to 40 m² will generate enough electricity for an efficient house.

One of the most promising developments in photovoltaic cell technology in recent years is the invention of **amorphous silicon collectors**. First described in 1968 by Stanford Ovshinky, these non-crystalline silicon semiconductors can be made into lightweight, paper-thin sheets that require much less material than conventional crystalline silicon cells. They also are cheaper to manufacture and can be made in a variety of shapes and sizes, permitting ingenious applications. Roof tiles with amorphous silicon collectors layered on their surface already are available (fig. 13.32b). Even flexible films can be coated with these materials.

There's a huge potential for rooftop solar energy. One study estimated that more than 1,000 mi² (2,590 km²) of roofs suitable for photovoltaic systems in the United States could generate about three-quarters of present electrical consumption. In 2010, Southern California Edison started construction of photovoltaic arrays on roofs of warehouses and big-box retail stores (fig. 13.32c). Over the next five years, the utility expects to install a total of 250 MW of solar voltaic power. Overall, the 1 million solar roofs project aims to install 3,000 MW of photovoltaic energy on homes and apartments in California by 2016. More than $2.8 billion in incentives are available to homeowners to cover costs.

Innovative financing programs are helping make this dream a reality. First introduced in Berkeley, California, **Property Assessed Clean Energy (PACE)** uses city bonds to pay for renewable energy and conservation expenses. The bonds are paid off through a 20-year assessment on property taxes. Decreased utility bills often offset tax increases, so that switching to renewable energy is relatively painless for the property owner. And after 20 years you own the system and never have to pay another electric bill.

(a) Base-load power facility

(b) Flexible, thin-film solar tiles

(c) Roof-top solar array

▲ **FIGURE 13.32** Solar photovoltaic energy is highly versatile and can be used in a variety of dispersed settings. (a) Utility-scale PV arrays can provide base-load power. (b) Thin-film PV collectors can be printed on flexible backing and used like oridinary roof tiles. (c) Millions of square meters of roof tops on schools and commercial buildings could be fitted with solar panels.

Some other financing arrangements that help overcome the high up-front costs of renewable energy are power purchasing agreements and solar leasing programs. In both cases an investor builds a certain amount of solar or wind energy in return for a contract to buy the energy produced at a specific rate for a fixed length of time. This frees property owners from large capital expenses, while giving investors a secure return on their investment. Feed-in tariffs that require utilities to buy excess power from homeowners at a fair price also help make solar photovoltaics economically feasible.

An intriguing option for storing electricity is in plug-in hybrid vehicles, which could provide an enormous, distributed battery array. You'd recharge your auto battery at night when power plants have excess generating capacity. During the day, your car would be plugged into a smart meter that could sell electricity back to your utility if prices rise. A few million mobile battery arrays could greatly help smooth out power peaks and valleys.

▲ **FIGURE 13.33** Hydropower dams produce clean, renewable energy but can be socially and ecologically damaging. China's Three Gorges Dam, shown here under construction, displaced 1.5 million people.

13.7 HYDROPOWER

Moving water is one of our oldest power sources. In early American settlements, water-powered gristmills and sawmills were essential, and most early industrial cities were built where falling water could run mills. The invention of water turbines in the nineteenth century greatly increased the efficiency of electricity-producing **hydropower dams**. By 1925 falling water generated 40 percent of the world's electric power. Since then, hydroelectric production capacity has grown 15-fold, but fossil fuel use has risen so rapidly that water power is now only one-quarter of total electrical generation. Still, many countries produce most of their electricity from falling water. Norway, for instance, depends on hydropower for 99 percent of its electricity; Brazil, New Zealand, and Switzerland all produce at least three-quarters of their electricity with water power. Canada is the world's leading producer of hydroelectricity, running 400 power stations with a combined capacity exceeding 60,000 MW. First Nations people protest, however, that their rivers are being diverted and lands flooded to generate electricity, most of which is sold to the United States.

The total world potential for hydropower is estimated to be about 3 million MW. If all of this capacity were put to use, the available water supply could provide between 8 and 10 terawatt hours (10^{12} watt-hours) of electrical energy. Currently we use only about 10 percent of the potential hydropower supply. The energy derived from this source in 1994 was equivalent to about 500 million tons of oil, or 8 percent of the total world commercial energy consumption.

Most hydropower comes from large dams

Much of the hydropower development since the 1930s has focused on enormous dams. There is a certain efficiency of scale in giant dams, and they bring pride and prestige to the countries that build them, but they can have unwanted social and environmental effects that spark protests in many countries. China's Three Gorges Dam on the Yangtze River, for instance, spans 2.0 km (1.2 mi) and is 185 m (600 ft) tall (fig. 13.33). The reservoir it creates is 644 km (400 mi) long and has displaced more than 1 million people.

In warm, dry climates, large reservoirs often suffer enormous water losses. Lake Nasser, behind the Aswan High Dam in Egypt, loses 15 billion m^3 each year to evaporation and seepage. Unlined canals lose another 1.5 billion m^3. Together, these losses represent one-half of the Nile River flow, or enough water to irrigate 2 million ha of land. The silt trapped by the Aswan High Dam formerly fertilized farmland during seasonal flooding and provided nutrients that supported a rich fishery in the delta region. Farmers now must buy expensive chemical fertilizers, and the fish catch has dropped almost to zero. Schistosomiasis, spread by snails that flourish in the reservoir, is an increasingly serious problem.

Large dams also destroy biodiversity. In 2010, Brazil announced approval of a controversial Belo Monte Dam on the Xingu River (a major tributary of the Amazon) in Para State. The $17 billion dam would be the third largest in the world. It will fuel development in this remote area, but it will flood 250 km^2 (96.5 mi^2) of tropical rainforest. Indigenous Kayapo people protested the loss of traditional hunting lands.

Dam promoters claim that the area to be flooded is less than the 5,000 km^2 originally planned, and equal to the forest flooded every year during the rainy season. Dam opponents, on the other hand, point out that the seasonally flooded forest is a unique ecosystem in which plants and animals are exquisitely adapted to changing water levels. The reservoir created by the dam will irreversibly change local ecology and eliminate many endemic species. Furthermore, decaying vegetation in the drowned forest will emit methane that could cause more global climate change than burning an equivalent amount of coal.

Unconventional hydropower comes from tides and waves

Ocean tides and waves also contain enormous amounts of energy that can be harnessed to do useful work. A **tidal station** works like a hydropower dam, with its turbines spinning as the tide flows through them. A high-tide/low-tide differential of several meters is required to spin the turbines. Unfortunately, variable tidal periods often cause problems in integrating this energy source into the electric utility grid. Nevertheless, some of these plants have operated for many decades.

Ocean wave energy can easily be seen and felt on any seashore. The energy that waves expend as millions of tons of water are picked up and hurled against the land, over and over, day after day, can far exceed the combined energy budget for both insolation (solar energy) and wind power in localized areas. Captured and turned into useful forms, that energy could make a substantial contribution to meeting local energy needs.

Dutch researchers estimate that 20,000 km of ocean coastline are suitable for harnessing wave power. Among the best places in the world for doing this are the west coasts of Scotland, Canada, the United States (including Hawaii), South Africa, and Australia. Wave energy specialists rate these areas at 40 to 70 kW per meter of shoreline. Altogether, it's calculated, if the technologies being studied today become widely used, wave power could amount to as much as 16 percent of the world's current electrical output.

Some of the **wave power** designs being explored include oscillating water columns that push or pull air through a turbine, as well as a variety of floating buoys, barges, and cylinders that bob up and down as waves pass, using a generator to convert mechanical motion into electricity. However, it's difficult to design a mechanism that can survive the worst storms.

An interesting new development in this field is the Pelamis generator developed by the Scottish company Ocean Power Delivery (fig. 13.34). The first application of this technology is now in operation 5 km off the coast of Portugal, with three units producing 2.25 MW of electricity, or enough to supply 1,500 Portuguese households. Another 28 units are currently being installed. Each of the units consists of four cylindrical steel sections linked by hinged joints. Anchored to the seafloor at its nose, the snakelike machine points into the waves and undulates up and down and side to side as swells move along its 125 m length. This motion pumps fluid to hydraulic motors that drive electrical generators to

▲ **FIGURE 13.34** The Pelamis wave converter (named after a sea snake) is a 125 m long and 3.5 m diameter tube, hinged so it undulates as ocean swells pass along it. This motion drives pistons that turn electrical generators. Energy experts calculate that capturing just 1 to 2 percent of global wave power could supply at least 16 percent of the world's electrical demand.

produce electricity, which is carried to shore by underwater cables. Portugal considers wave energy one of its most promising sources of renewable energy.

Pelamis's inventor, Richard Yemm, says that survivability is the most important feature of a wave-power device. Being offshore, the Pelamis isn't exposed to the pounding breakers that destroy shore-based wave-power devices. If waves get too steep, the Pelamis simply dives under them, much as a surfer dives under a breaker. These wave converters lie flat in the water and are positioned far offshore, so they are unlikely to stir up as much opposition as do the tall towers of wind generators.

Geothermal heat could supply substantial amounts of energy

The earth's internal temperature can provide a useful source of energy in some places. High-pressure, high-temperature steam fields exist below the earth's surface. Around the edges of continental plates or where the earth's crust overlays magma (molten rock) pools close to the surface, this **geothermal energy** is expressed in the form of hot springs, geysers, and fumaroles. Yellowstone National Park is the largest geothermal region in the United States. Iceland, Japan, and New Zealand also have high concentrations of geothermal springs and vents.

Depending on the shape, heat content, and access to groundwater, these sources produce wet steam, dry steam, or hot water. Iceland, which sits on a mid-ocean ridge (chapter 12), has abundant geothermal energy. Iceland has ambitious plans to be the first carbon-neutral country, largely because the earth's heat provides steam for heat and electric energy. Even places that don't naturally have geysers or hot springs may have hot spots close enough to the surface to be tapped by deep wells. In 2010, however, two large deep-well projects in Switzerland and California were shut down abruptly when evidence surfaced that they might trigger earthquakes.

Although few places have geothermal steam, the earth's warmth can help reduce energy costs nearly everywhere. Pumping water through deeply buried pipes can extract enough heat so that a heat pump will operate more efficiently. Similarly, the relatively uniform temperature of the ground can be used to augment air conditioning in the summer (fig. 13.35).

13.8 FUEL CELLS

Rather than store and transport energy, another alternative is to generate it locally, on demand. **Fuel cells** are devices that use ongoing electrochemical reactions to produce an electrical current. They are very similar to batteries except that, rather than recharging them with an electrical current, you add more fuel for the chemical reaction. Depending on the environmental costs of input fuels, fuel cells can be a clean energy source for office buildings, hospitals, or even homes.

All fuel cells consist of a positive electrode (the cathode) and a negative electrode (the anode) separated by an electrolyte, a material that allows the passage of charged atoms, called ions but is impermeable to electrons (fig. 13.36). In the most common

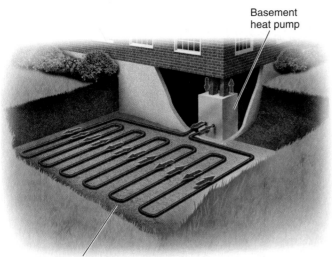

Basement
heat pump

Horizontal earth loop

▲ **FIGURE 13.35** Geothermal energy can cut heating and cooling costs by half in many areas. In summer (shown here), warm water is pumped through buried tubing (earth loops), where it is cooled by constant underground temperatures. In winter, the system reverses and the relativity warm soil helps the house.

systems, hydrogen or a hydrogen-containing fuel is passed over the anode while oxygen is passed over the cathode. At the anode, a reactive catalyst, such as platinum, strips an electron from each hydrogen atom, creating a positively charged hydrogen ion (a proton). The hydrogen ion can migrate through the electrolyte to the cathode, but the electron is excluded. Electrons flow through an external circuit, and the electrical current generated by their passage can be used to do useful work. At the cathode, the electrons and protons are reunited and combined with oxygen to make water.

Fuel cells provide direct-current electricity as long as they are supplied with hydrogen and oxygen. For most uses, oxygen is provided by ambient air. Hydrogen can be supplied as a pure gas, but storing hydrogen gas is difficult and dangerous because it's highly explosive. An alternative is a device called a **reformer** or converter that strips hydrogen from fuels such as natural gas, methanol, ammonia, gasoline, ethanol, or even vegetable oil. Even methane effluents from landfills and wastewater treatment plants can be used as a fuel source. Or hydrogen gas could be provided by solar, wind, or geothermal facilities that generate electricity to hydrolyze (split) water.

A fuel cell that runs on pure oxygen and hydrogen produces no waste products except drinkable water and radiant heat. Other fuels create some pollutants (most commonly carbon dioxide), but the levels are typically far less than conventional fossil fuel combustion in a power plant or an automobile engine. Although the theoretical efficiency of electrical generation of a fuel cell can be as high as 70 percent, the actual yield is closer to 40 or 45 percent. The quiet, clean operation and variable size of fuel cells make them useful in certain places (fig. 13.37), such as buildings where waste heat can be captured for water heating or space heating. A 45-story office building at 4 Times Square, for example, has two 200 kW fuel cells that provide both electricity and heat. The same

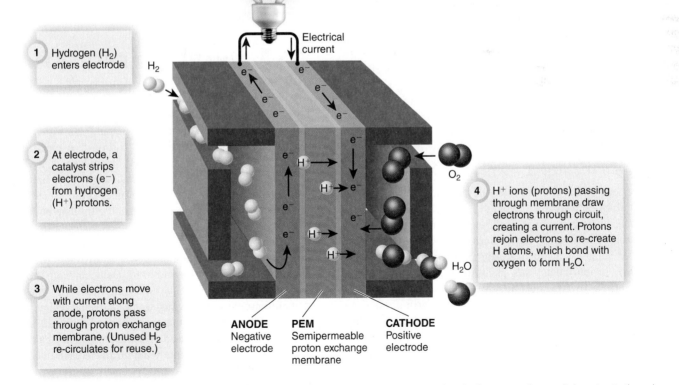

1 Hydrogen (H_2) enters electrode

H_2

Electrical current

2 At electrode, a catalyst strips electrons (e^-) from hydrogen (H^+) protons.

O_2

4 H^+ ions (protons) passing through membrane draw electrons through circuit, creating a current. Protons rejoin electrons to re-create H atoms, which bond with oxygen to form H_2O.

H_2O

3 While electrons move with current along anode, protons pass through proton exchange membrane. (Unused H_2 re-circulates for reuse.)

ANODE
Negative electrode

PEM
Semipermeable proton exchange membrane

CATHODE
Positive electrode

▲ **FIGURE 13.36** Fuel cell operation. Electrons are removed from hydrogen atoms at the anode to produce hydrogen ions (protons) that migrate through a semipermeable electrolyte medium to the cathode, where they reunite with electrons from an external circuit and oxygen atoms to make water. Electrons flowing through the circuit connecting the electrodes create useful electrical current.

◀ **FIGURE 13.37** The Long Island Power Authority has installed 75 stationary fuel cells to provide reliable backup power.

building has photovoltaic panels on its façade, natural lighting, fresh-air intakes to reduce air conditioning, and a number of other energy conservation features.

Utilities are promoting renewable energy

Utility restructuring currently being planned in the United States could include policies to encourage conservation and alternative energy sources. Among the proposed policies are (1) "distributional surcharges" in which a small per kilowatt-hour charge is levied on all utility customers to help finance renewable energy research and development, (2) "renewables portfolio" standards to require power suppliers to obtain a minimum percentage of their energy from sustainable sources, and (3) **green pricing** that allows utilities to profit from conservation programs and charge premium prices for energy from renewable sources.

Some states already are pursuing these policies. For example, Iowa has a Revolving Loan Fund supported by a surcharge on investor-owned gas and electric utilities. This fund provides low-interest loans for renewable energy and conservation. Several states have initiated green pricing programs as a way to encourage a transition to sustainable energy. One of the first was in Colorado, where 1,000 customers agreed to pay $2.50 per month above their regular electric rates to help finance a 10 MW wind farm on the Colorado–Wyoming border. Buying a 100 kW "block" of wind power provides the same environmental benefits as planting a half acre of trees or not driving an automobile 4,000 km (2,500 mi) per year.

We need a supergrid

As you've seen in this chapter, many of the places with the greatest potential for both solar and wind development are far from the urban centers where power is needed. This means we'll need a vastly increased network of power lines if we're going to depend on wind or solar for a much greater proportion of our energy (fig. 13.38). In introducing his plans to double the amount of renewable energy over the next three years, President Obama said, "Today, the electricity we use is carried along a grid of lines and wires that dates back to Thomas Edison—a grid that can't support the demands of clean energy." He designated $4.5 billion to modernize and expand the transmission grid as part of the $86 billion in clean-energy investments in the economic recovery bill.

Fortunately, as we've seen in the case study for this chapter, high-voltage direct-current lines make it possible to transmit electricity over long distances with relatively minor loses. Interestingly,

▲ **FIGURE 13.38** New high-voltage power lines (orange) will be needed to tie together regional networks (green) if the United States is to make effective use of its renewable energy potential. The pink area served by the Eastern electrical grid needs to be connected to the west by interlinks (black dots) for maximum efficiency.

studies in California show that integration of renewable resources can smooth out daily variations. The wind blows more strongly at night, and the sun shines (obviously) during the day. And because hydropower can start up quickly, it easily fills in gaps. Even though the wind doesn't blow every day in most locations, linking together wind farms even a few hundred kilometers apart can give a steadier electrical supply than does a single site. A supergrid, such as the one proposed for Desertech, could make our entire energy supply more robust, reliable, and sustainable.

13.9 WHAT'S OUR ENERGY FUTURE?

In 2008, former vice president Al Gore issued a bold and inspiring challenge to the United States. Currently, he said, "We're borrowing money from China to buy oil from the Persian Gulf to burn in ways that destroy the planet." He urged America to repower itself with 100 percent carbon-free electricity within a decade. Doing so, he proposed, would solve the three biggest crises we face—environmental, economic, and security—simultaneously. This ambitious project could create millions of jobs, spur economic development, and eliminate our addiction to imported fossil fuels.

But could we get all our electricity from renewable, environmentally friendly sources in such a short time? Mark Jacobson from Stanford University and Mark Delucchi from the University of California–Davis believe we can. Moreover, they calculate that currently available wind, water, and solar technologies could supply 100 percent of the world's energy by 2030 and completely eliminate all our use of fossil fuels. They calculate that it would take 3.8 million large wind turbines (each rated at 5 MW), 1.7 billion rooftop photovoltaic systems, 720,000 wave converters, half a million tidal turbines, 89,000 concentrated solar power plants and industrial-sized photovoltaic arrays, 5,350 geothermal plants, and 900 hydroelectric plants, worldwide.

Wouldn't it be an overwhelming job to build and install all that technology? It would be a huge effort, but it's not impossible. Jacobson and Delucchi point out that society has achieved massive transformations before. In 1956 the United Sates began building the Interstate Highway System, which now extends 47,000 mi (75,600 km) and has changed commerce, landscapes, and society. And every year roughly 60 million new cars and trucks are added to the world's highways.

Is there enough clean energy to meet our needs? Yes, there is. In 2012, the U.S. National Renewable Energy Laboratory concluded that using currently available, affordable technology, renewable energy could supply 80 percent of total U.S. electricity generation in 2050, while meeting electricity demand on an hourly basis in every region of the country (fig 13.39). By the end of the twenty-first century, renewable sources could provide all our energy needs if we take the necessary steps to make this happen. After the meltdown of four nuclear reactors in Japan in 2011, Germany, Japan, Switzerland, Sweden, and several other countries announced intentions to move away from both nuclear and fossil fuels and to emphasize renewable energy sources in the future.

Interestingly, it would take about 30 percent less total energy to meet our needs with sun, wind, and water than to continue using fossil fuels. That's because electricity is a more efficient way to use energy than burning dead plants and animals. For example, only about 20 percent of the energy in gasoline is used to move a vehicle (the rest is wasted as heat). An electric vehicle, on the other hand, uses about three-quarters of the energy in electricity for motion. Furthermore, much of the energy from renewable sources could often be produced closer to where it's used, so there are fewer losses in transmission and processing.

Won't it be expensive to install so much new technology? Yes it will be, but the costs of continuing our current dependence on fossil fuels would be much higher. It's estimated that investing $700 billion per year now in clean energy will avoid 20 times that much in a few decades from the damages of climate change.

One of the biggest challenges in moving to clean energy is that the wind doesn't blow all the time and the sun doesn't always shine in a given location. But a smart balance of sources can even out shortages. We'll need a large investment in the electric transmission grid—including some high-voltage interchange lines—to tie together the areas with abundant sun and wind with the cities where most people live. A smart grid that transmits energy more efficiently and safely is a good investment in any case.

China is taking bold steps to develop and employ wind, hydro, and solar energy. Let's hope that other developing countries follow their lead. Even some richer countries may see the benefits of this path. A decade ago it wasn't clear that clean energy would be technically or economically feasible. Now that it is, we all need to work to make it politically feasible as well. The energy choices we make now will have profound effects in the future on our lives and our environment.

CONCLUSION

Fossil fuels—oil, coal, and natural gas—remain our dominant energy sources. Coal is extremely abundant, especially in North America, but extracting and burning coal have been major causes of environmental damage and air pollution. Oil (petroleum) currently provides most of our transportation energy, but we're running out of cheap, easily extracted oil. And burning oil also releases greenhouse gases. Natural gas is more abundant than oil and cleaner than coal, but fracking, the most common method for extracting natural gas, may release so much methane into the atmosphere (in addition to contaminating surface and ground water) that the fuel it produces is actually worse for global climate change than coal.

Nuclear power doesn't create CO_2 while operating, but mining, processing, and shipping fuel, together with perpetual storage of wastes, results in far more greenhouse gases than does wind energy. Furthermore, the danger of accidents and the problem of storing the highly dangerous wastes are expensive and unresolved problems.

Conservation is a key factor in a sustainable energy future. New designs in housing, office buildings, industrial production, and transportation can all save huge amounts of energy. Biofuels, including ethanol and oil (biodiesel), vary greatly in their net energy yield and environmental effects, but cellulosic feedstocks and algae may provide useful energy. Hydropower can be clean and reliable, but a focus on huge dams has led to many environmental and social problems. Rapid innovations in solar, wind, wave power, and other renewable energy sources now make it possible to get all our energy from clean technologies. The choices we make about our energy sources and uses will have profound effects on our environment and society.

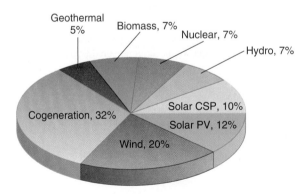

▲ **FIGURE 13.39** A renewable energy scenario for 2050. Cogeneration would mostly burn natural gas to generate both electricity and space heating. SOURCE: 2008 Worldwatch Report.

PRACTICE QUIZ

1. What is Desertech, and how will it help Europe meet its energy needs?
2. Define *energy*, *power*, and *kilowatt-hour* (kWh).
3. What are the major sources of global commercial energy?
4. How does energy consumption in the United States compare to that in other countries?
5. Why don't we want to use all the coal in the ground?
6. Where is most liquid oil located? How long are supplies likely to last?
7. What are *tar sands* and *oil shales*? What are the environmental costs of their extraction?
8. How are nuclear wastes now being stored?
9. Explain active and passive solar energy.
10. How do photovoltaic cells work?

CRITICAL THINKING AND DISCUSSION

Apply the principles you have learned in this chapter to discuss these questions with other students.

1. If you were the energy czar of your state or country, where would you invest your budget? Why?
2. We have discussed a number of different energy sources and energy technologies in this chapter. Each has advantages and disadvantages. If you were an energy policy analyst, how would you compare such different problems as the risk of a nuclear accident versus air pollution effects from burning coal?
3. If your local utility company were going to build a new power plant in your community, what kind would you prefer? Why?
4. The nuclear industry is placing ads in popular magazines and newspapers, claiming that nuclear power is environmentally friendly because it doesn't contribute to the greenhouse effect. How do you respond to that claim?
5. How would you evaluate the debate about net energy loss or gain in biofuels? What questions would you ask the experts on each side of this question? What worldviews or hidden agendas do you think might be implicit in this argument?
6. It clearly will cost a lot of money to switch from fossil fuels to renewables. How would you respond to someone who says that future costs from climate change are no concern of theirs?

DATA ANALYSIS Personal Energy Use

For many college students, a car and a computer are essentials of life. Cars are also one of our most important single uses of energy, so differences in efficiency can greatly affect your energy consumption (and energy costs). This exercise asks you to modify an Excel spreadsheet in order to evaluate the impact of vehicle efficiency on energy use.

Suppose you were to buy a very efficient car, such as the Honda Insight, rather than a sport utility vehicle, such as a Ford Excursion. How much energy would that save, and how long could you run your computer with that energy? Go to Connect to find a spreadsheet that explores these questions, and to answer questions about personal energy use.

"Biogas for a greener Kristianstad": this bus runs on methane from local organic waste products.

LEARNING OUTCOMES

After studying this chapter, you should be able to answer the following questions:

▸ What are the major components of the waste stream?

▸ How does a sanitary landfill operate? Why are we searching for alternatives to landfills?

▸ Why is ocean dumping a problem?

▸ What are the "three Rs" of waste reduction, and which is most important?

▸ How can biomass waste be converted to natural gas?

▸ What are toxic and hazardous waste? How do we dispose of them?

▸ What is bioremediation?

▸ What is the Superfund, and has it shown progress?

A Waste-free City

Garbage is something most of us don't want to think about. It's messy and smelly, a worthless liability. It has also been with us since the dawn of civilization. Unpleasant, disease-generating middens of food waste, broken crockery, and animal bones have always occurred just on the edge of town. Or waste was dumped into rivers or into the sea, where we couldn't see or smell it. (In the United States, the earliest clean water legislation was an effort to protect commercial navigation from the hazards of burgeoning mounds of garbage lurking below the surface of American rivers and harbors.) Modern civilization improved on this system by burying its garbage piles. We concentrate disposal into fewer, larger, better-built "sanitary" landfills, which are vast, expensive to manage and, because nobody wants them around, often hundreds of miles from garbage-producing cities.

The city of Kristianstad has embraced its garbage. This southern Swedish city has turned an unpleasant liability into a money-making resource. Twenty years ago, Kristianstad faced all the usual waste-disposal challenges of midsize cities: growing volumes and rising costs of trucking, concerns about odors and air quality; and declining landfill space. At the same time, Sweden had committed to reducing greenhouse gases and fossil fuel dependence, both to reduce climate change and to create a more diversified and less vulnerable energy future.

Kristianstad residents decided to focus on multiple solutions. The city actively promotes recycling and waste reduction. It provides assistance and facilities to help households and businesses reduce, reuse, and recycle trash.

Most notably, the city built a biogas plant on the edge of town. Biogas digesters take organic waste—household food waste, effluent from food processing and slaughterhouses, manure from local farms—mix it with water, and let the slurry digest for about 3 weeks. Submerged in the liquid mixture, bacteria decompose the organic material without oxygen. Decomposition produces methane, CH_4, instead of CO_2 (the product of decomposition when oxygen is present). Methane is a high-quality fuel that can power vehicles, run electricity-generating plants, and provide heat to the town's households. Undigested solids are redistributed on farm fields as sterile, disease-free, odor-free, and weed-free fertilizer (fig. 14.1).

With a biogas plant, farm manure and food-processing wastes become valuable resources. Local food-processing industries, including pork processors, provide over half of the waste in Kristianstad's biogas plant; another third is manure from local farms; about 6 percent is household solid waste.

Kristianstad also produces methane from the city sewage plant and from the old landfill at the edge of town. All landfills emit methane, as decades-old garbage gradually decomposes without oxygen. Most landfills and sewage plants simply let methane to leak into the atmosphere, where it is a potent, climate-changing greenhouse gas. The Kristianstad landfill and sewage plant pipe methane to a district heating plant, which provides hot water for household heating. These plants also produce 75 MW of electricity (about as much as 75 standard wind turbines or one small coal-fired power plant). Local households get most of their heat from this

▲ **FIGURE 14.1** Biogas plants convert all types of organic waste into usable methane and fertilizer.

system. The district heating plant ensures efficiency and flexibility by accepting multiple fuel sources: in addition to burning landfill gas, it burns wood chips and sawdust from local forests, urban trees, farms, and other sources.

The regional bus fleet is probably the most visible aspect of the waste-to-energy system. Over two-thirds of the 1,034 city and regional buses now run on biogas. The city vehicle fleet also burns methane, as do a growing number of private vehicles. Biogas has become a moneymaker, as well as a moneysaver, because Kristianstad is selling a growing amount of biogas to other cities in the region. Across the country, biogas generates some 1.5 TW of energy and makes up more than half the country's total use of natural gas.

Kristianstad is one of many European cities that have closed their landfills and shifted to recycling, waste reduction, and biogas. Incineration is also widely used for a small proportion of waste (often less than 15 percent in cities with strong recycling programs). Strict rules on emissions make new incinerators far cleaner and safer than most in the United States.

Integrated waste disposal systems serve many needs. They save money and produce marketable goods. They protect environmental quality and human health. They reduce dependence on imported fuel and lower the climate impacts of waste disposal and energy. In just two decades Kristianstad has become nearly carbon neutral, thanks largely to its waste management systems.

We are living in a false economy where the price of goods and services does not include the cost of waste and pollution.

—LYNN LANDES, FOUNDER AND
DIRECTOR OF ZERO WASTE AMERICA

14.1 WHAT WASTE DO WE PRODUCE?

Waste is everyone's business, even though we don't think about it every day. We all produce unwanted by-products in nearly everything we do. According to the Environmental Protection Agency (EPA), the United States produces 11 billion tons of solid waste each year. That's roughly 3.6 tons per person. Nearly half of that amount consists of agricultural waste, such as crop residues and animal manure, which are generally recycled into the soil on the farms where they are produced. Agricultural wastes provide groundcover to reduce erosion, and they nourish new crops, but they are also the single largest source of nonpoint air and water pollution in the country. Another one-third of all solid wastes are mine tailings, overburden from strip mines, smelter slag, and other industrial waste from mining and metal processing. Much of this material is stored in or near its source of production. Improper disposal practices, however, can result in serious and widespread pollution.

Industrial waste—other than mining and mineral production—amounts to some 400 million metric tons per year in the United States. Most of this material is recycled, converted to other forms, sent to landfills, or disposed of in injection wells, which should be deep enough not to interact with usable groundwater. About 60 million metric tons of industrial waste fall into a special category of hazardous and toxic waste, which we will discuss later in this chapter.

Municipal solid waste, the garbage we produce in our houses, offices, and cities, accounts for a small percentage of total waste by weight, but it is one of our most important challenges in waste management. Municipal solid waste is hard to reuse and recycle because it contains many different kinds of materials, yet it amounts to about 250 million metric tons per year in the United States (fig. 14.2). That's just over 2 kg (4.6 lb) per person per day—twice as much per capita as Europe or Japan, and five to ten times as much as most developing countries.

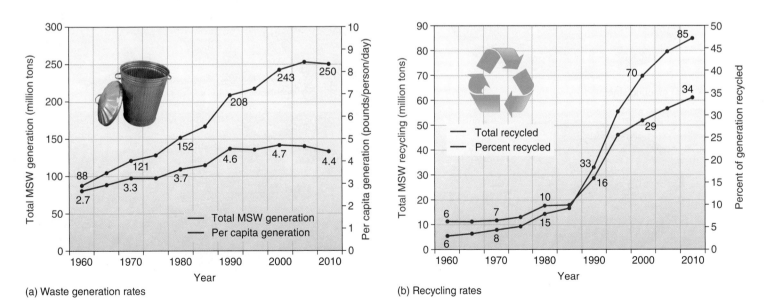

(a) Waste generation rates

(b) Recycling rates

▲ **FIGURE 14.2** Bad news and good news in solid waste production. Per capita waste has risen steadily to more than 2 kg/person/day (a). Recycling rates are also rising, however (b).

Despite considerable progress in the past 20 years, we still recycle only about 30 percent of our glass bottles and jars, less than 50 percent of aluminum drink cans, and less than 7 percent of our plastic food and beverage containers. We could save money, energy, land, and many other resources if we could improve on these rates.

The waste stream is everything we throw away

What kinds of materials are in all that waste? There are organic materials, such as yard and garden wastes, food wastes, and sewage sludge from sewage treatment plants; junked cars; worn-out furniture; and consumer products of all types. Newspapers, magazines, packaging, and office refuse make paper one of our major wastes (fig. 14.3).

The **waste stream** is a term for the steady flow of varied wastes that we all produce, from domestic garbage and yard wastes to industrial, commercial, and construction refuse. Many of the materials in our waste stream would be valuable resources if they were not mixed with other garbage. Unfortunately, our collecting and dumping processes mix and crush everything together, making separation an expensive and sometimes impossible task. In a landfill or incinerator, much of the value of recyclable materials is lost.

When hazardous materials get mixed into the waste stream, they get dispersed through thousands of tons of miscellaneous garbage. This mixing makes the disposal or burning of what might have been rather innocuous stuff a difficult, expensive, and risky business. Spray-paint cans, pesticides, batteries (zinc, lead, or mercury), cleaning solvents, smoke detectors containing radioactive material, and plastics that produce dioxins and PCBs (polychlorinated biphenyls) when burned are mixed with paper, table scraps, and other nontoxic materials. The best thing to do with household toxic and hazardous materials is to separate them for safe disposal or recycling, as we will see later in this chapter.

14.2 WASTE DISPOSAL METHODS

Where do our wastes go? In this section, we will examine some historic methods of waste disposal, as well as some future options. We begin with the least desirable but most commonly used methods. We'll end with the most important strategies, the "three Rs" of Reduction, Reuse, and Recycling.

Open dumps release hazardous substances into the air and water

Often people dispose of waste by simply dropping it someplace. Open, unregulated dumps are still the predominant method of waste disposal in most developing countries, where government infrastructure, including waste collection, has difficulty serving growing populations. Giant megacities in the developing world have enormous garbage problems (fig. 14.4). Mexico City, one of the largest cities in the world, generates some 10,000 tons of trash each day. Until recently, most of this torrent of waste was left in giant piles, exposed to the wind and rain, as well as rats, flies, and other vermin. Manila, in the Philippines, has at least ten huge open dumps. The most notorious is called "Smoky Mountain" because of its constant smoldering fires. Thousands of people live and work on this 30 m high heap of refuse. They spend their days sorting

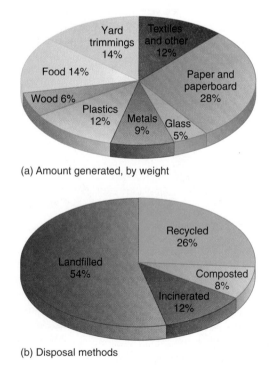

(a) Amount generated, by weight

(b) Disposal methods

▲ **FIGURE 14.3** Composition of municipal solid waste in the United States, by weight and by disposal methods. SOURCE: U.S. EPA Office of Solid Waste Management, 2011.

▲ **FIGURE 14.4** Trash disposal has become a crisis in the developing worlds, where people have adopted cheap plastic goods and packaging but lack good recycling and disposal options.

through the garbage for edible or recyclable materials. Health conditions are abysmal, but these people have nowhere else to go, and the city has no current alternatives for waste disposal.

Most developed countries forbid open dumping, at least in urban areas, but illegal dumping is still a problem. You have undoubtedly seen trash accumulating along roadsides and in vacant, weedy lots. Is this just an aesthetic problem? No. Much of this trash washes into sewers and then into the ocean (see next section). Often illegally dumped garbage includes waste oil and solvents. An estimated 200 million liters of waste motor oil are poured into the sewers or allowed to soak into the ground every year in the United States. This is about five times as much as was spilled by the *Exxon Valdez* in Alaska in 1989! No one knows the volume of solvents and other chemicals disposed of by similar methods.

Increasingly, these toxic chemicals are showing up in groundwater, on which nearly half of Americans depend for drinking (chapter 11). An alarmingly small amount of oil or other solvents can pollute large quantities of drinking or irrigation water. One liter of gasoline, for instance, can make a million liters of water undrinkable.

Ocean dumping is mostly uncontrolled

The oceans are vast, but they're not large enough to absorb our waste without harm. Every year some 25,000 metric tons (55 million lb) of packaging, including millions of bottles, cans, and plastic containers, are dumped at sea. Even in remote regions, beaches are littered with the nondegradable flotsam and jetsam (fig. 14.5a). About 150,000 tons (330 million lb) of fishing gear—including more than 1,000 km (660 mi) of nets—are lost or discarded at sea each year. An estimated 50,000 northern fur seals are entangled in this refuse and drown or starve to death every year in the North Pacific alone.

Until recently, many cities in the United States dumped municipal refuse, industrial waste, sewage, and sewage sludge into the ocean. Federal legislation now prohibits this dumping. New York City, the last to stop offshore sewage sludge disposal, finally ended this practice in 1992.

Plastic debris is a growing problem in all the world's oceans. Millions of tons of plastic drink bottles, bottle caps, plastic shopping bags, and other debris end up at sea. Most is probably carelessly discarded litter and uncontained garbage, but there is also deliberate disposal at sea, especially from cruise ships and container ships. All this debris, floating just below the surface, accumulates in vast regions of slowly swirling ocean currents.

The **Great Pacific Garbage Patch**, discovered in 1997 by sailing captain Charles Moore, is the best-known of these plastic debris fields. In all the world's oceans, vast circulating currents known as gyres are driven by the earth's rotation. These currents collect floating plastic debris, much of it tiny fragments, in regions thousands of km wide. The North Pacific gyre has captured at least 100 million tons of plastic. An estimated 80 percent of this debris originates from improper or accidental disposal of plastics on land. The remaining 20 percent is dumped or lost by ships at sea.

All this plastic flotsam outweighs the living biomass in large parts of the Pacific and Atlantic oceans. Fish have been found with stomachs full of plastic fragments. Seabirds gulp down plastic debris, then regurgitate it for their chicks. With stomachs

(a)

(b)

▲ **FIGURE 14.5** (a) Plastic trash dumped on land and at sea ends up on remote beaches and (b) in the bellies of seabirds. Here an autopsy of a young Laysan albatross shows the plastic its parents accidentally swallowed and then regurgitated to feed the chick. SOURCE: NOAA.

blocked by undigestible bottle caps, disposable lighters, and other fragments, chicks slowly starve to death. In one study of Laysan albatrosses, 90 percent of the carcasses of dead albatross chicks contained plastic (fig. 14.5b).

Oceanographers are trying to find ways of collecting or controlling this debris that is slowly starving ocean ecosystems. Most material is too fine-grained to capture easily in nets, and it is distributed widely around the world's oceans. Growing awareness, however, is a first step toward resolving the problem. Increasingly images and information about the Pacific garbage gyre, and other garbage gyres, can be found online.

Landfills receive most of our waste

Currently, 54 percent of all municipal solid waste in the United States goes to landfills, 33 percent is recycled, and 13 percent is incinerated. While we have a long way to go in controlling waste, this is a dramatic change from 1960, when 94 percent was landfilled and only 6 percent was recycled.

A modern **sanitary landfill** is designed to contain waste. Operators are required to compact the refuse and cover it every day with a layer of dirt, to decrease smells and litter and to discourage insects and rats. This method helps control pollution, but the dirt fill also takes up as much as 20 percent of landfill space. Since 1994, all operating landfills in the United States have been required to control such hazardous substances as oil, chemical compounds, and toxic metals, which seep through piles of waste along with rain water. To prevent leakage to groundwater and streams, landfills require an impermeable clay and/or plastic lining (fig. 14.6). Drainage systems are installed in and around the liner to catch drainage and to help monitor chemicals that leak out. Modern municipal solid waste landfills now have many of the safeguards of hazardous waste repositories described later in this chapter.

Sanitary landfills also must manage methane, a greenhouse gas produced when organic material decomposes in the anaerobic conditions deep inside a landfill. Landfills are the single largest anthropogenic source of methane in the United States. Globally, landfills are estimated to produce more than 700 million metric tons of methane annually. Because methane is 20 times as potent at absorbing heat as CO_2, this represents about 12 percent of all greenhouse gas emissions. Until recently almost all this landfill methane was simply vented into the air. Now about half of all landfill gas in the United States is either flared (burned) on site or is collected and used as fuel to generate electricity. Methane recovery in the United States produces 440 trillion Btu per year, and is equivalent to removing 25 million vehicles from the highway. Some landfill operators are deliberately pumping water into their waste as a way of speeding up production of this valuable fuel.

Historically, landfills were convenient and cheap. This was because we had much less waste to deal with—we produced only

▲ **FIGURE 14.6** A plastic liner being installed in a sanitary landfill. This liner and a bentonite clay layer below it prevent leakage to groundwater. Trash is also compacted and covered with earth fill every day.

a third as much in 1960 as today—and because there were few regulations about disposal sites and methods. Since 1984, when stricter financial and environmental protection requirements for landfills took effect, roughly 90 percent of landfills in the United States have closed. Nearly all of those landfills lacked environmental controls to keep toxic materials from leaking into groundwater and surface waters. Many areas now suffer groundwater and stream contamination from decades-old unregulated dump sites.

With new rules for public health protection, landfills are becoming fewer, larger, and more expensive. Cities often truck their garbage hundreds of miles for disposal, a growing portion of the $10 billion per year that we spend to dispose of trash. A decade from now, it may cost us $100 billion per year to dispose of our trash and garbage. On the other hand, rising landfill costs make it more economical to pursue alternatives strategies, including waste reduction and recycling.

We often export waste to countries ill-equipped to handle it

Most industrialized nations agreed to stop shipping hazardous and toxic waste to less-developed countries in 1989, but the practice still continues. In 2006, for example, 400 tons of toxic waste were illegally dumped at 14 open dumps in Abidjan, the capital of the Ivory Coast. The black sludge—petroleum wastes containing hydrogen sulfide and volatile hydrocarbons—killed ten people and injured many others. At least 100,000 city residents sought medical treatment for vomiting, stomach pains, nausea, breathing difficulties, nosebleeds, and headaches. The sludge—which had been refused entry at European ports—was transported by an Amsterdam-based multinational company on a Panamanian-registered ship and handed over to an Ivoirian firm (thought to be connected to corrupt government officials) to be dumped in the Ivory Coast. The Dutch company agreed to clean up the waste and pay the equivalent of (U.S.) $198 million to settle claims.

Most of the world's obsolete ships are now dismantled and recycled in poor countries. The work is dangerous, and old ships often are full of toxic and hazardous materials, such as oil, diesel fuel, asbestos, and heavy metals. On India's Alang Beach, for example, more than 40,000 workers tear apart outdated vessels using crowbars, cutting torches, and even their bare hands. Metal is dragged away and sold for recycling. Organic waste is often simply burned on the beach, where ashes and oily residue wash back into the water.

Discarded electronics, or **e-waste**, is one of the greatest sources of toxic material currently going to developing countries. There are at least 2 billion television sets and personal computers in use globally. Televisions often are discarded after only about five years, while computers, PlayStations, cellular telephones, and other electronics become obsolete even faster. It's estimated that 50 million tons of e-waste are discarded every year worldwide. Only about 20 percent of the components are currently recycled. The rest generally goes to open dumps or landfills. This waste stream contains at least 2.5 billion kg of lead, as well as mercury, gallium, germanium, nickel, palladium, beryllium, selenium, arsenic, and valuable metals, such as gold, silver, copper, and steel.

Until recently, most of this e-waste went to China, where villagers, including young children, would break it apart to retrieve valuable metals. Often this scrap recovery was done under primitive conditions where workers had little or no protective gear (fig. 14.7a). Health risks in this work are severe, especially for growing children. Soil, groundwater, and surface-water contamination at these sites is extremely high. Food grown in contaminated soils often contains toxic levels of lead and other metals.

Shipping e-waste to China is now officially banned, but illegal smuggling continues. With tighter regulation in China, informal e-waste recycling has shifted to India, Congo, and other areas with weak environmental regulation. Adding to the difficulty of this problem, these developing areas will soon be producing more e-waste than wealthier countries with better regulation (fig. 14.7b). Will those developing areas be able to defend public health with this increase in waste production?

The Basel Action Network is an international network of activists seeking better controls on global trade in toxic materials. This group, named after the city where an international agreement was made banning the practice, tracks international e-waste shipments and working conditions. Exporting waste to poor communities also occurs within countries (see What Do You Think?, p. 338).

Incineration produces energy from trash

Faced with growing piles of garbage and a lack of available landfills at any price, many cities have built waste incinerators to burn municipal waste. Incineration reduces the volume of waste by 80 to 90 percent. Residual ash and unburnable residues representing 10 to 20 percent of the original volume are usually taken to a landfill for disposal, but landfilling costs are greatly reduced.

Most incinerators do some degree of **energy recovery**, using the heat derived from incinerated refuse to heat nearby buildings or to produce steam and generate electricity. Internationally, well over 1,000 waste-to-energy plants in Brazil, Japan, and Western Europe generate energy. In the United States more than 110 waste

(a)

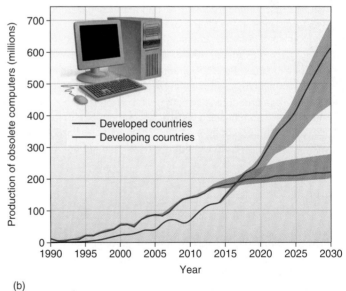

(b)

▲ **FIGURE 14.7** A Chinese woman breaks up e-waste to extract valuable metals (a). This kind of unprotected demanufacturing is hazardous to workers and the environment, but production of e-waste is rising in both developed and developing areas (b). Most waste will be produced in developing areas after about 2015.
SOURCE: Modified from Yu et al., 2010, *Environmental Science and Technology.*

Environmental Justice

When a new landfill, petrochemical factory, incinerator, or other unwanted industrial facility is proposed for a minority neighborhood, charges of environmental racism often are raised by those who oppose this siting. Everyday experiences tell us that minority neighborhoods are much more likely to have high pollution levels and facilities that you wouldn't want to live near than are middle- or upper-class white neighborhoods. But does this prove that land-use decisions are racist, or just that minorities are less politically powerful than middle- or upper-class residents? Could it be that land prices are simply cheaper and public resistance to locating a polluting facility in a place that's already polluted is less than putting it in a cleaner environment? Or does this distinction matter? Perhaps showing that a disproportionate number of minorities live in dirtier places is evidence enough of racism. How would you decide?

▲ Native Americans march in protest of toxic waste dumping on tribal lands.

One of the first systematic studies showing this inequitable distribution of environmental hazards based on race in the United States was conducted by Robert D. Bullard in 1978. Asked for help by a predominantly black community in Houston that was slated for a waste incinerator, Bullard discovered that all five of the city's existing landfills and six of eight incinerators were located in African-American neighborhoods. In a book entitled Dumping on Dixie, Bullard showed that this pattern of risk exposure in minority communities is common throughout the United States. Here are some of his findings:

- Three of the five largest commercial hazardous waste landfills, accounting for about 40 percent of all hazardous waste disposal in the United States, are located in predominantly African-American or Hispanic communities.
- Sixty percent of African-Americans and Latinos and nearly half of all Asians, Pacific Islanders, and Native Americans live in communities with uncontrolled toxic waste sites.
- The average percentage of the population made up by minorities in communities without a hazardous waste facility is 12 percent. By contrast, communities with one hazardous waste facility have, on average, twice as high (24 percent) a

minority population, while those with two or more such facilities average three times as high a minority population (38 percent) as those without one.

But does this prove that race, not class or income, is the strongest determinant of who is exposed to environmental hazards? What additional information might you look for to make this distinction? One of the lines of evidence Dr. Bullard raises is the fact that the discrepancy between the pollution exposure of middle-class blacks and that of middle-class whites is even greater than the difference between poorer whites and blacks. While upper-class whites can "vote with their feet" and move out of polluted and dangerous neighborhoods, Bullard argues, minorities are restricted by color barriers and prejudice to less desirable locations.

Some additional evidence uncovered by this research is variation in the way toxic waste sites are cleaned up and how polluters are punished in different neighborhoods. White communities see faster responses and get better results once toxic wastes are discovered than do minority communities. For instance, the EPA takes 20 percent longer to place a hazardous waste site in a minority community on the Superfund National Priority List than it does for one in a white community. Penalties assessed against polluters of white communities average six times higher than those against polluters of minority communities. Cleanup is more thorough in white communities as well. Most toxic wastes in white communities are treated—that is, removed or destroyed. By contrast, waste sites in minority neighborhoods are generally only "contained" by putting a cap over them, leaving contaminants in place to potentially resurface or leak into groundwater at a later date.

How would you evaluate these findings? Do they convince you that racism is at work, or do you think that other explanations might be equally likely? Which of these arguments do you find most persuasive, or what other evidence would you need to make a reasoned judgment about whether or not environmental racism is a factor in determining who gets exposed to pollution and who enjoys a cleaner, more pleasant environment?

incinerators burn 45,000 metric tons of garbage daily. Some of these are simple incinerators, but newer facilities usually provide heating and/or electricity.

Municipal incinerators are specially designed plants capable of burning thousands of tons of waste per day. In some plants, refuse is sorted as it comes in to remove unburnable or recyclable materials before combustion. This sorted waste is called **refuse-**

derived fuel because the enriched burnable fraction has a higher energy content than the raw trash. Another approach, called a **mass burn**, is to dump everything into a giant furnace, unsorted, and burn as much as possible (fig. 14.8). This technique avoids the expensive and unpleasant job of sorting, but it produces more unburned ash, and often it produces more air pollution as plastics, batteries, and other mixed substances are burned.

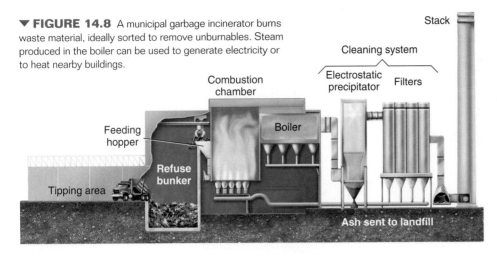

▼ **FIGURE 14.8** A municipal garbage incinerator burns waste material, ideally sorted to remove unburnables. Steam produced in the boiler can be used to generate electricity or to heat nearby buildings.

The cost-effectiveness of garbage incinerators is the subject of heated debates. Initial construction costs are high—usually between $100 million and $300 million for a typical municipal facility. Tipping fees at an incinerator (the fee charged to haulers for each ton of garbage dumped) are often much higher than those at a landfill. Ironically, one worry about incinerators is whether enough garbage will be available to feed them. Incinerators also compete with recyclers for paper and plastics, which make high-energy fuel. Cities usually have contracts guaranteeing certain amounts of waste daily. Some communities in which recycling has been really successful have had to buy garbage from neighbors to meet contractual obligations to waste-to-energy facilities.

If they are not well built and well managed, incinerators can produce large amounts of ash and hazardous airborne emissions. Residual from a mass burn often contains a variety of toxic components, including dioxins and furans (from incompletely burned plastics), lead, and cadmium (from batteries, toys, and other products) in incinerator ash. These toxic materials are more concentrated in the fly ash (lighter, airborne particles, which can penetrate deep into the lungs) than in heavy bottom ash, which collects inside the burner system. In one EPA study, all of the incinerators examined exceeded cadmium standards, and 80 percent exceeded lead standards. One way to reduce these dangerous emissions is to remove batteries containing heavy metals and plastics containing chlorine before wastes are burned.

European countries rely heavily on incineration, in part because they lack landfill space (fig. 14.9). Studies of hazardous air pollutants in Europe have shown that although municipal incinerators were major sources of hazardous emissions in 1990, modern incinerators are among the smallest sources. They now produce 1/80 as much emissions as metal processors do, and 1/20 as much as household fireplaces emit, because of tighter permitting processes, better inspections, and high burn temperatures.

In addition, municipal waste systems in many European countries sort waste to remove both recyclables and toxic materials. Increasingly, European cities are banning plastics from incinerator waste and requiring households to separate plastics from other garbage. This is expected to eliminate nearly all dioxins and other combustion by-products. Separation also helps avoid the expense of installing costly pollution-control equipment.

In places with weaker oversight, including many U.S. cities, it is often less clear that pollution-control equipment is correctly used or maintained. Opponents of incineration argue that we should be investing in more recycling rather than building expensive incinerators to deal with waste.

14.3 SHRINKING THE WASTE STREAM

Compared to landfilling and incineration, recycling saves money, energy, raw materials, and land space, and it reduces pollution. **Recycling**, as the term is used in solid waste management, is the reprocessing of discarded materials into new products. Usually recycling is easiest when materials are separated. Kristianstad, Sweden, is one of many cities that now expect citizens to separate organic waste, as well as more conventional recyclables, in order to maintain a high-quality material source for biogas. Keeping food waste out of paper, plastic, and metals also greatly facilitates processing and reuse of those recyclables (fig. 14.10).

Sometimes recycling reproduces products: old aluminum cans and glass bottles are usually melted and recast into new cans and bottles, and the lead from car batteries can be made into new batteries. Sometimes entirely new products are made. Old tires are shredded and turned into rubberized playground or road surfacing. Newspapers become cellulose insulation, and steel cans become new automobiles and construction materials.

There have been some dramatic successes in recycling in recent years. Nationally, the United States recycles or composts one-third of municipal solid waste. Minneapolis and Seattle claim a 60 percent recycling rate, something thought unattainable a

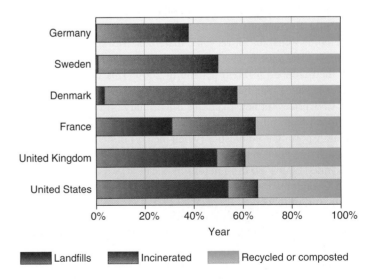

▲ **FIGURE 14.9** Reliance on landfills, recycling/composting, and incineration varies considerably among countries. DATA SOURCE: Eurostat, U.S. EPA, 2012.

▲ **FIGURE 14.10** Kristianstad, Sweden, provides separate bins for food waste (*left*) and other recyclables (*right*). These sidewalk bins are provided for use by pedestrians. This system keeps both waste streams pure and easy to reuse.

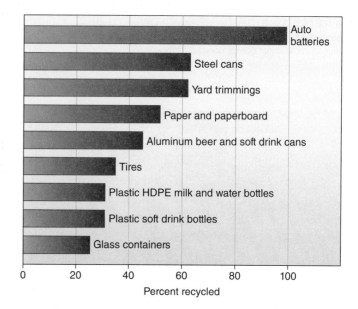

▲ **FIGURE 14.11** Recycling rates for selected materials in the United States. Car battery recycling is required by law, to keep lead out of the waste stream.   SOURCE: U.S. EPA.

decade ago. New Jersey, renowned for its waste sites, is a national leader at more than 60 percent recycling. San Francisco is aiming for 100 percent recycling. All this makes good environmental sense, but it also saves San Francisco the cost of waste disposal.

Making recycling pay for itself is often the critical challenge. Some materials are heavy and low-value, so they can be difficult to ship economically to recycling facilities. Aluminum is normally the easiest and most valuable material to recycle. Lightweight, high-value, and expensive to produce from raw materials, aluminum can be reused for thousands of purposes. Even so, only half of aluminum cans are recycled in the United States. This rate is up from only 15 percent 20 years ago, but Americans still throw away nearly 350,000 metric tons of aluminum beverage containers each year (fig. 14.11). That is enough to make 3,800 Boeing 747 airplanes. This is especially unfortunate because producing new aluminum is extraordinarily energy-intensive, while recycling is relatively easy.

Wild fluctuations in commodity prices are a challenge in developing a market for recycled materials. Newsprint, for example, cost $160 a ton in 1995; by 1999 it dropped to just $42 per ton and then climbed to $650 per ton in 2009. This unpredictability makes business planning difficult.

Low prices for new materials is a primary obstacle. New plastic, made from oil, is usually cheaper than the cost of collecting and transporting used plastics (when the cost of disposal and other expenses are not considered). Consequently, less than 7 percent of the United States' 30 million tons of plastic waste is recycled each year. Contamination is a major obstacle in plastics recycling. Most plastic soft drink bottles are made of PET (polyethylene terephthalate), which can be remanufactured into carpet, fleece clothing, plastic strapping, and nonfood packaging. However, even a trace of vinyl—a single PVC (polyvinyl chloride) bottle in a truckload, for example—can make PET useless. Because single-use beverage containers are so costly to recycle, they have been outlawed in Denmark and Finland, so that taxpayers don't have to bear the burden of cleanup and disposal.

The growing popularity of bottled water has created a serious waste disposal problem. Of the 300 billion bottles of water consumed each year globally, less than 20 percent are recycled. It takes around 75 billion liters (500 million barrels) of oil to manufacture and ship these bottles. In most American cities, tap water is safe and is subjected to more rigorous testing than bottled water. The best way to control this problem is through bottle deposits. States with deposit laws recover about 78 percent of all beverage containers, while those without generally have recycling rates of 20 percent or less.

Bottle deposits also help prevent the costs—and nuisance—of roadside litter. Americans pay an estimated 32¢ for each piece of litter picked up by crews along state highways, which adds up to $500 million every year. Bottle deposits have reduced littering in many states.

Recycling saves money, energy, and space

Some recycling programs cover their own expenses with materials sales; others have difficulty paying for themselves. Yet recycling is usually cheaper than other disposal methods, which aren't expected to pay for themselves. Curbside pickup of recyclables costs around $35 per ton, as opposed to the $80 paid to dispose of them at an average metropolitan landfill.

Philadelphia is investing in neighborhood collection centers that will recycle 600 tons a day, enough to eliminate the need for a previously planned, high-priced incinerator. New York City, still producing 27,000 tons of garbage a day despite the closing of its only large landfill more than a decade ago, has set a target of 50 percent waste reduction to be accomplished by recycling office paper and household and commercial waste. In 2002 Mayor Michael Bloomberg discontinued most recycling, arguing that the program was too expensive. The city quickly found that disposing of waste was more expensive than recycling, and most programs were reinstated.

Japan probably has the most successful recycling program in the world. Half of all household and commercial wastes in Japan are recycled, while the rest are about equally incinerated or landfilled. Some communities have raised recycling rates to 80 percent, and others aim to reduce waste altogether by 2020. This level of recycling takes a high level of participation and commitment. In Yokohama, a city of 3.5 million, there are now 10 categories of recyclables, including used clothing and sorted plastics. Some communities have 30 or 40 categories for sorting recyclables.

Recycling lowers demand for raw resources. The United States cuts down 2 million trees every day to produce newsprint and paper products, a heavy drain on its forests. Recycling the print run of a single Sunday issue of the *New York Times* would spare 75,000 trees (fig. 14.12). Every piece of plastic made in the United States reduces the reserve supply of petroleum and makes the country more dependent on foreign oil. Recycling 1 ton of aluminum saves 4 tons of bauxite (aluminum ore) and 700 kg of petroleum coke and pitch, as well as keeping 35 kg of aluminum fluoride out of the air.

Recycling also reduces energy consumption and air pollution. Plastic bottle recycling could save 50 to 60 percent of the energy needed to make new bottles. Making new steel from old scrap offers up to 75 percent energy savings. Producing aluminum from scrap instead of bauxite ore cuts energy use by 95 percent, yet the United States still throws away more than a million tons of aluminum every year. If aluminum recovery were doubled worldwide, more than a million tons of air pollutants would be eliminated every year (see Key Concepts, p. 342).

Composting recycles organic waste

Pressed for landfill space, many cities have banned yard waste from municipal garbage. Rather than bury this valuable organic material, they are turning it into a useful product through **composting**: biological degradation or breakdown of organic matter under aerobic (oxygen-rich) conditions. The organic compost resulting from this process makes a nutrient-rich soil amendment that aids water retention, slows soil erosion, and improves crop yields.

Many cities and counties provide centralized composting, to help people keep compostables out of the municipal waste stream. You can also compost your own organic waste. All you need to do is to pile up lawn clippings, vegetable waste, fallen leaves, wood chips, or other organic matter in an out-of-the way place, keep it moist, and turn it over every week or so (fig. 14.13). Within a few months, naturally occurring microorganisms will decompose the organic material into a rich, pleasant-smelling compost that you can use to enrich your yard or garden.

As noted in the opening case study, some composting systems produce methane fuel. Worldwide, at least one-fifth of municipal waste is organic kitchen and garden refuse. In developing countries up to 85 percent of the waste stream is food, textiles, vegetable matter, and other biodegradable materials.

Methane is captured from this material at many landfills, but it's much more efficient to convert organic waste to methane in a contained, anaerobic digester. Germany and Switzerland now have at least 30 municipal-scale waste-to-methane plants. Anaerobic digestion also can be done on a small scale. Millions of household methane generators provide fuel for cooking and lighting for homes in China and India (chapter 13). In the United States some farmers produce all the fuel they need to run their farms—both for heating and for running trucks and tractors—by generating methane from animal manure.

Reuse is even better than recycling

Even better than recycling or composting is cleaning and reusing materials in their present form, thus saving the cost and energy of remaking them into something else. We do this already with some specialized items. Auto parts are regularly sold from junk yards, especially for older car models. In some areas stained-glass

▼ **FIGURE 14.12** Keeping material out of landfills has multiple benefits. Consumers can help by buying recycled products.

▼ **FIGURE 14.13**
Composting keeps tremendous volumes of material out of the waste stream, as well as providing rich nutrients for gardens.

Garbage: Liability or resource?

Municipal solid waste includes all our mixed refuse. Most of us don't spend much time thinking about where our waste ends up, but as you know from the principle of conservation of matter (chapter 2), materials are never destroyed or created, they're just transformed from one shape to another. Elements in our waste such as aluminum, lead, carbon, or nitrogen don't disappear. They may sit in a landfill for centuries, or they may be incinerated and emitted into the atmosphere, or they may be recycled and transformed into another useful object. The question is which of these is the most efficient use of our resources and environment.

Materials in our municipal waste stream have been hard to extract and reuse because they're usually all mixed together, although new sorting and recycling systems can separate mixed waste. Waste can be a liability or a resource. It all depends on how much we produce, how much we landfill and incinerate, and how much we recycle.

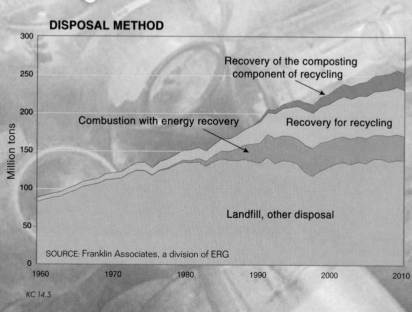

KC 14.1

What are the major types of waste we produce, and how much do we recycle?

We produce far more waste than our grandparents did.

The EPA has tracked overall production of municipal solid waste in the United States since 1960. Disposable paper and plastic products have grown most dramatically in the past 50 years. Recovery rates are worst for plastics (because mixed waste contaminates plastics and new plastic is inexpensive) and food products (because these are relatively hard to store and transport to central recycling facilities). Metals, glass, and yard compost have relatively high recycling rates.

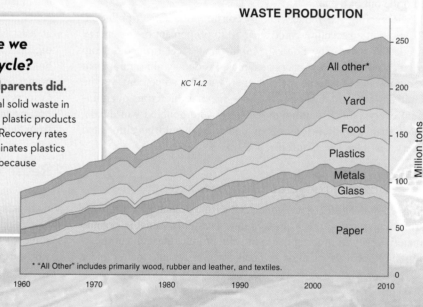

WASTE PRODUCTION

KC 14.2

All other*
Yard
Food
Plastics
Metals
Glass
Paper

Million tons
250 — 200 — 150 — 100 — 50 — 0

* "All Other" includes primarily wood, rubber and leather, and textiles.

1960   1970   1980   1990   2000   2010

Where does it all go?

As the available landfills decline, we are recycling and incinerating more municipal waste.

KC 14.4

DISPOSAL METHOD

Million tons
300 — 250 — 200 — 150 — 100 — 50 — 0

Recovery of the composting component of recycling

Combustion with energy recovery

Recovery for recycling

Landfill, other disposal

SOURCE: Franklin Associates, a division of ERG

1960   1970   1980   1990   2000   2010

KC 14.3

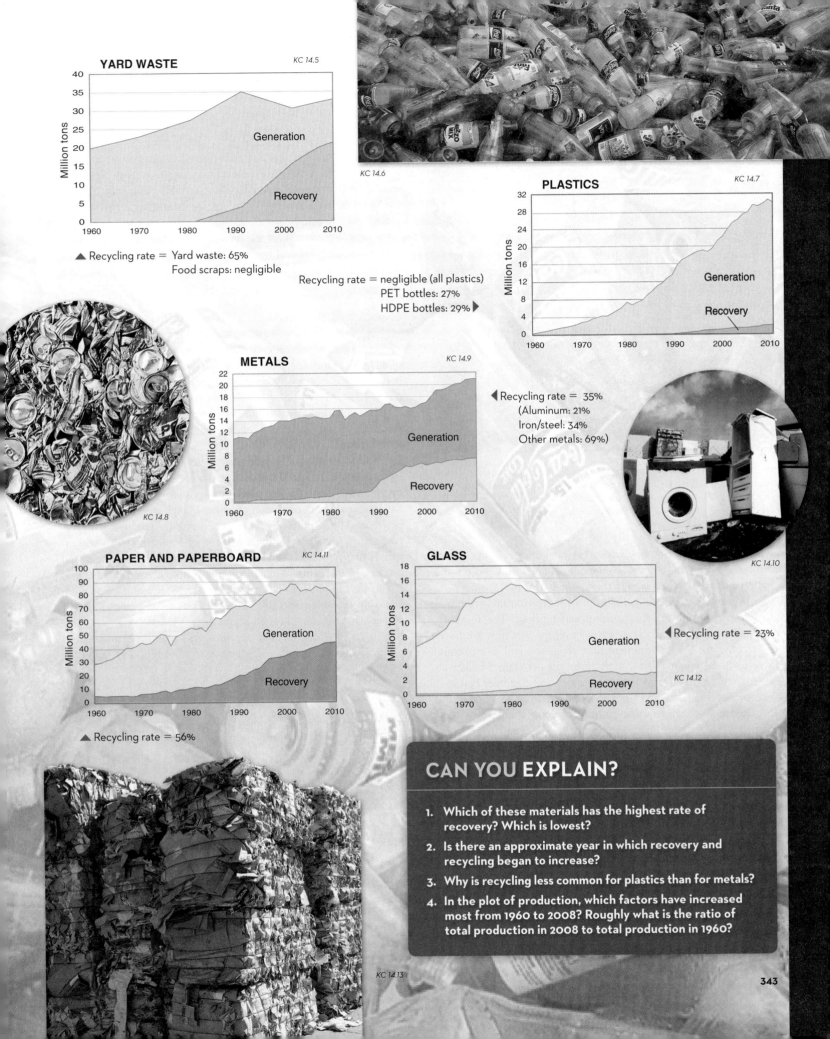

YARD WASTE

KC 14.5

Million tons — Generation, Recovery
(1960–2010, 0–40)

▲ Recycling rate = Yard waste: 65%
Food scraps: negligible

Recycling rate = negligible (all plastics)
PET bottles: 27%
HDPE bottles: 29% ▶

KC 14.6

PLASTICS

KC 14.7

Million tons — Generation, Recovery
(1960–2010, 0–32)

METALS

KC 14.9

Million tons — Generation, Recovery
(1960–2010, 0–22)

◀ Recycling rate = 35%
(Aluminum: 21%
Iron/steel: 34%
Other metals: 69%)

KC 14.8

KC 14.10

PAPER AND PAPERBOARD

KC 14.11

Million tons — Generation, Recovery
(1960–2010, 0–100)

▲ Recycling rate = 56%

GLASS

Million tons — Generation, Recovery
(1960–2010, 0–18)

◀ Recycling rate = 23%

KC 14.12

CAN YOU EXPLAIN?

1. Which of these materials has the highest rate of recovery? Which is lowest?

2. Is there an approximate year in which recovery and recycling began to increase?

3. Why is recycling less common for plastics than for metals?

4. In the plot of production, which factors have increased most from 1960 to 2008? Roughly what is the ratio of total production in 2008 to total production in 1960?

KC 14.13

▲ **FIGURE 14.14** Reusing discarded products is a creative and efficient way to reduce waste. This recycling center in Berkeley, California, provides building supplies and saves money for the community.

windows, brass fittings, fine woodwork, and bricks salvaged from old houses bring high prices. Some communities sort and reuse a variety of materials received in their dumps (fig. 14.14).

In many cities, glass and plastic bottles are routinely returned to beverage producers for washing and refilling. The reusable, refillable bottle is the most efficient beverage container we have. It is better for the environment than remelting and more profitable for local communities. A reusable glass container makes an average of 15 round-trips between factory and customer before it becomes so scratched and chipped that it has to be recycled.

Reusable containers also favor local bottling companies and help preserve regional businesses. Since the advent of cheap, lightweight, disposable food and beverage containers, many small, local breweries, canneries, and bottling companies have been forced out of business by national or multinational conglomerates. Big companies can afford to ship food and beverages thousands of miles, as long as it is a one-way trip. National companies favor recycling, rather than collection and reuse, because they don't want the responsibility for collecting and reusing containers.

In many less-affluent nations, reusing manufactured goods is an established tradition. If manufactured products are expensive and labor is cheap, it pays to salvage, clean, and repair products. Cairo, Manila, Mexico City, and many other cities have large populations of poor people who make a living by scavenging. Entire ethnic populations may survive on scavenging, sorting, and reprocessing

scraps from city dumps. These people provide essential but unpaid services to society by helping to reduce and reuse the mountains of waste products that governments can't afford to manage.

Reducing waste is the cheapest option

Most of our attention in waste management focuses on recycling, but slowing the production of throw-away products is by far the most effective way to save energy, materials, and money. Among the "three Rs"—reduce, reuse, recycle—the most important strategy is the first. Industries are increasingly finding that reducing saves money. Soft-drink makers use less aluminum per can than they did 20 years ago, and plastic bottles use less plastic. 3M has saved over $500 million in the past 30 years by reducing its use of raw materials, reusing waste products, and increasing efficiency. Rethinking consumption habits can be done at any scale, from the nation, city, or corporation to the individual (see What Can You Do?, p. 345).

In recent decades, we have greatly increased our waste production rather than reducing it. As consumer goods have multiplied and as global economies have grown, all of us have done our part for the economy by consuming, and discarding, more things (fig. 14.15). Moreover, as developing countries become wealthier they are catching up with the high levels of waste production in wealthier countries and the conundrum of disposal.

Excessive packaging of food and consumer products is one of our greatest sources of unnecessary waste. Paper, plastic, glass, and metal packaging materials make up 50 percent of our domestic trash by volume. Much of that packaging is primarily for marketing and has little to do with product protection. Manufacturers and retailers can reduce these practices if consumers ask for products with less packaging. Canada's National Packaging Protocol (NPP) recommends that packaging minimize depletion of virgin resources and production of toxins in manufacturing. The preferred hierarchy is (1) no packaging, (2) minimal packaging, (3) reusable packaging, and (4) recyclable packaging. This plan sets an ambitious target of 50 percent reduction in excess packaging.

In 2008, China banned ultrathin (less than 0.025 mm thick) plastic bags and called for a return to reusable cloth bags for

▲ **FIGURE 14.15** How much more stuff do we need?

▲ **FIGURE 14.16** Hazardous waste is dangerous even with small exposure. Here a worker tests for radioactive soil.

shopping. This could eliminate up to 3 billion plastic bags used every day in China. Japan, Ireland, South Africa, and Taiwan also have discouraged single-use plastic bags through taxes or prohibitions. In 2007, San Francisco became the first American city to outlaw petroleum-based plastic grocery bags.

Where disposable packaging is necessary, we still can reduce the volume of waste in our landfills by using materials that are compostable or degradable. **Photodegradable plastics** break down when exposed to ultraviolet radiation. **Biodegradable plastics** incorporate such materials as cornstarch that microorganisms can decompose. Several states have introduced legislation requiring biodegradable or photodegradable six-pack beverage yokes, fast-food packaging, and disposable diapers. These degradable plastics often don't decompose completely, however; many kinds only break down to small particles that remain in the environment.

14.4 HAZARDOUS AND TOXIC WASTES

The most dangerous aspect of the waste stream is that it often contains highly toxic and hazardous materials that are injurious to both human health and environmental quality (fig. 14.16). We now produce and use a vast array of flammable, explosive, caustic, acidic, and highly toxic chemical substances for industrial, agricultural, and domestic purposes. According to the EPA, U.S. industries generate about 900 million metric tons of officially classified hazardous wastes each year, about 3 metric tons for each person in the country. In addition, considerably more toxic and hazardous waste material is generated by industries or processes

not regulated by the EPA. Shockingly, at least 40 million metric tons (22 billion lb) of toxic and hazardous wastes are released into the air, water, and land in the United States each year. The biggest sources of these toxins are the chemical and petroleum industries (fig. 14.17).

Hazardous waste includes many dangerous substances

Legally, a **hazardous waste** is any discarded material, liquid or solid, that contains substances known to be (1) fatal to humans or laboratory animals in low doses; (2) toxic, carcinogenic, mutagenic, or teratogenic to humans or other life-forms; (3) ignitable with a flash point less than 60°C; (4) corrosive; or (5) explosive or highly reactive (undergoes violent chemical reactions either by itself or when mixed with other materials). Notice that this definition includes both toxic and hazardous materials, as defined in chapter 8. Certain compounds are exempt from regulation as hazardous waste if they are accumulated in less than 1 kg (2.2 lb) of

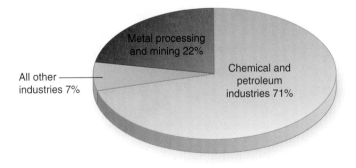

▲ **FIGURE 14.17** Producers of hazardous wastes in the United States.
SOURCE: U.S. EPA.

ment represents a serious environmental problem. And orphan wastes left behind by abandoned industries remain a serious threat to both environmental quality and human health. For years little attention was paid to this material. Wastes stored on private property, buried, or allowed to soak into the ground were considered of little concern to the public. An estimated 5 billion metric tons of highly poisonous chemicals were improperly disposed of in the United States between 1950 and 1975 before regulatory controls became more stringent.

Federal legislation regulates hazardous waste

Two important federal laws regulate hazardous waste management and disposal in the United States. The Resource Conservation and Recovery Act (RCRA, pronounced "rickra") of 1976 is a comprehensive program that requires rigorous testing and management of toxic and hazardous substances. At every step, generators, shippers, users, and disposers of these materials must keep an account of everything they handle and what happens to it from generation (cradle) to ultimate disposal (grave) (fig. 14.18).

The Comprehensive Environmental Response, Compensation, and Liability Act (CERCLA or Superfund Act), passed in 1980, is aimed at rapid containment, cleanup, or remediation of abandoned toxic waste sites. The act establishes a National Priority List (NPL) of sites most in need of remediation. In 2010 there were approximately 1,280 sites on the list, with 60 waiting for a

commercial chemicals or 100 kg of contaminated soil, water, or debris. Even larger amounts (up to 1,000 kg) are exempt when stored at an approved waste treatment facility for the purpose of being beneficially used, recycled, reclaimed, detoxified, or destroyed.

Most hazardous waste is recycled, converted to nonhazardous forms, stored, or otherwise disposed of on-site by the generators—chemical companies, petroleum refiners, and other large industrial facilities—so that it doesn't become a public problem. Still, the hazardous waste that does enter the waste stream or the environ-

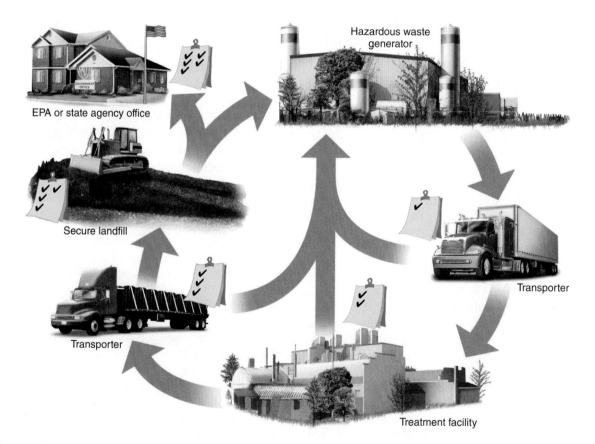

EPA or state agency office

Secure landfill

Transporter

Hazardous waste generator

Transporter

Treatment facility

▲ **FIGURE 14.18** Toxic and hazardous wastes must be tracked from "cradle to grave" by detailed shipping manifests (records).

decision and over 1,000 sites designated as "completed." CER-CLA authorizes the EPA to take emergency actions when there is a threat that toxic material could leak into the environment. The EPA also is empowered to sue responsible parties for the recovery of treatment costs.

About 30 percent of sites on the NPL are "orphan" sites, whose owners have disappeared or gone out of business. For these, CERCLA established a "Superfund," a pool of money to cover remediation costs until responsible parties can be located.

The government does not have to prove that anyone violated a law or what role he or she played in a Superfund site. Rather, liability under CERCLA is "strict, joint, and several," meaning that anyone associated with a site can be held responsible for the entire cost of cleaning it up, no matter how much of the mess they made. In some cases property owners have been assessed millions of dollars for removal of wastes left there years earlier by previous owners. This strict liability has been a headache for the real estate and insurance businesses.

CERCLA was amended in 1995 to make some of its provisions less onerous. In cases where treatment is unavailable or too costly and it is likely that a less costly remedy will become available within a reasonable time, interim containment is now allowed. The EPA also now has the discretion to set site-specific cleanup levels, rather than adhere to rigid national standards.

CERCLA was modified in 1984 by the Superfund Amendments and Reauthorization Act (SARA). SARA also established a community "right to know," the notion that the public had a right to know about production, use, or transportation of toxic materials where they live. A key part of public information and emergency planning is the **Toxic Release Inventory**, a listing of addresses where regulated materials are handled. This inventory requires 20,000 manufacturing facilities to report annually on releases of more than 300 toxic materials. The EPA publishes this list, and from it you can find specific information in the inventory about what is in your neighborhood.

Superfund sites are listed for federally funded cleanup

The EPA estimates that there are at least 36,000 seriously contaminated sites in the United States. The General Accounting Office (GAO) places the number much higher, perhaps more than 400,000 when all are identified. Originally, about 1,671 sites were placed on the National Priority List for cleanup with financing from the federal Superfund program. The **Superfund** is a revolving pool designed to (1) provide an immediate response to emergency situations that pose imminent hazards, and (2) to clean up or remediate abandoned or inactive sites. Without this fund, sites would languish for years or decades while the courts decided who was responsible for paying for the cleanup (fig. 14.19). Originally a $1.6 billion pool, the fund peaked at $3.6 billion. From its inception, the fund was financed by taxes on producers of toxic and hazardous wastes. Industries opposed this "polluter pays" tax, because current manufacturers are often not the ones responsible for the original contamination. In 1995 Congress agreed to let the tax expire. Since then the Superfund

▲ **FIGURE 14.19** Toxic and hazardous waste is a messy business, but disposal must be secure and permanent.

has dwindled, and the public has picked up an increasing share of the bill. In the 1980s the public covered less than 20 percent of the Superfund. Since 2004, however, general revenues (public tax dollars) have paid the entire cost off a greatly reduced program, and the industry share has been zero. President Barack Obama has proposed reducing the public tax burden by reinstating the "polluter pays" tax.

Total costs for hazardous waste cleanup in the United States are estimated to be between $370 billion and $1.7 trillion, depending on how clean sites must be and what methods are used. For years, Superfund money was spent mostly on lawyers and consultants, and cleanup efforts were often bogged down in disputes over liability and best cleanup methods. During the 1990s, however, progress improved substantially, with a combination of rule adjustments and administrative commitment to cleanup. By 2004, more than half (over 1,000) of the original NPL sites were listed as completed in cleanup or containment.

What qualifies a site for the NPL? These sites are considered to be especially hazardous to human health and environmental quality because they are known to be leaking or have a potential for leaking supertoxic, carcinogenic, teratogenic, or mutagenic materials (chapter 8). The ten substances of greatest concern or most commonly detected at Superfund sites are lead, trichloroethylene, toluene, benzene, PCBs, chloroform, phenol, arsenic, cadmium, and chromium. These and other hazardous or toxic materials are known to have contaminated groundwater at 75 percent of the sites now on the NPL. In addition, 56 percent of these sites have contaminated surface waters, and airborne materials are found at 20 percent of the sites. Seventy million Americans, including 10 million children, live within 6 km of a Superfund site.

Where are these thousands of hazardous waste sites, and how did they get contaminated? Old industrial facilities, such as smelters, mills, petroleum refineries, and chemical manufacturing plants, are highly likely to have been sources of toxic wastes. Regions of the country with high concentrations of aging factories, such as the "rust belt" around the Great Lakes or the Gulf Coast petrochemical centers, have large numbers of Superfund sites (fig. 14.20). Mining districts also are prime sources of toxic and hazardous waste. Within cities, factories and places such as railroad

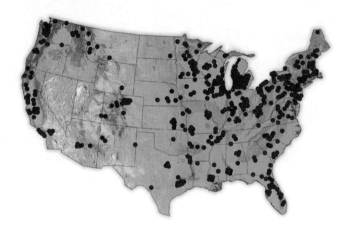

● Hazardous waste site

▬ Aquifers

▲ **FIGURE 14.20** Hazardous waste sites are often located on aquifer recharge zones, making groundwater contamination a common risk. SOURCE: U.S. EPA.

yards, bus repair barns, and filling stations, where solvents, gasoline, oil, and other petrochemicals were spilled or dumped on the ground, often are highly contaminated.

Some of the most infamous toxic waste sites were old dumps where many different materials were mixed together indiscriminately. For instance, Love Canal in Niagara Falls, New York, was an open dump that both the city and nearby chemical factories used as a disposal site. More than 20,000 tons of toxic chemical waste were buried under what later became a housing development. Another infamous example occurred in Hardeman County, Tennessee, where about a quarter of a million barrels of chemical waste were buried in shallow pits that subsequently leaked toxins into the groundwater.

Brownfields present both liability and opportunity

Among the biggest problems in cleaning up hazardous waste sites are questions of liability and the degree of purity required. In many cities, these problems have created large areas of contaminated properties, known as **brownfields**, that have been abandoned or are not being used to their potential because of real or suspected pollution. Up to one-third of all commercial and industrial sites in the urban core of many big cities fall in this category. In heavy industrial corridors the percentage typically is higher.

For years no one was interested in redeveloping brownfields because of liability risks. Who would buy a property, knowing that they might be forced to spend years in litigation and negotiations and be forced to pay millions of dollars for pollution they didn't create? Even if a site has been cleaned to current standards, there is a worry that additional pollution might be found in the future or that more stringent standards might be applied.

In many cases, property owners complain that unreasonably high levels of purity are demanded in remediation programs. Consider the case of Columbia, Mississippi. For many years a 35-ha (81-acre) site

in Columbia was used for turpentine and pine tar manufacturing. Soil tests showed concentrations of phenols and other toxic organic compounds exceeding federal safety standards. The site was added to the Superfund NPL, and remediation was ordered. Some experts recommended that the best solution was to simply cover the surface with clean soil and enclose the property with a fence to keep people out. The total costs would have been about $1 million. Instead, the EPA ordered Reichhold Chemical, the last known property owner, to excavate more than 12,500 tons of soil and haul it to a commercial hazardous waste dump in Louisiana at a cost of some $4 million. The intention is to make the site safe enough to be used for any purpose, including housing—even though no one has proposed building anything there. According to the EPA, the dirt must be clean enough for children to play in—even eat—without risk.

Similarly, in places where contaminants have seeped into groundwater, the EPA generally demands that cleanup be carried to drinking-water standards. Many critics believe that these pristine standards are unreasonable. Former congressman Jim Florio, a principal author of the original Superfund Act, says, "It doesn't make any sense to clean up a rail yard in downtown Newark so it can be used as a drinking water reservoir." Depending on where the site is, what else is around it, and what its intended uses are, much less stringent standards may be perfectly acceptable.

Brownfield redevelopment is increasingly seen as an opportunity for rebuilding cities, creating jobs, increasing the tax base, and preventing needless destruction of open space at urban margins. In 2002 the EPA established a new brownfields revitalization fund designed to encourage restoration of more sites, as well as more kinds of sites. In some communities former brownfields are being turned into "eco-industrial parks" that feature environmentally friendly businesses and bring in much-needed jobs to inner-city neighborhoods (chapter 15).

Hazardous waste must be processed or stored permanently

What shall we do with toxic and hazardous wastes? In our homes, we can reduce waste generation and choose less toxic materials. Buy only what you need for the job at hand. Use up the last little bit, or share leftovers with a friend or neighbor. Many common materials that you probably already have make excellent alternatives to commercial products.

Produce Less Waste As with other wastes, the safest and least expensive way to avoid hazardous waste problems is to avoid creating the wastes in the first place. Manufacturing processes can be modified to reduce or eliminate waste production. In Minnesota, the 3M Company reformulated products and redesigned manufacturing processes to eliminate more than 140,000 metric tons of solid and hazardous wastes, 4 billion liters (1 billion gal) of wastewater, and 80,000 metric tons of air pollution each year. It frequently found that these new processes not only spared the environment but also saved money by using less energy and fewer raw materials.

Recycling and reusing materials also eliminates hazardous wastes and pollution. Many waste products of one process or industry are valuable commodities in another. Already, about 10 percent of the wastes that would otherwise enter the waste stream in the United States are sent to surplus material exchanges, where they are sold as raw materials for use by other industries. This figure could probably be raised substantially with better waste management. In Europe at least one-third of all industrial wastes are exchanged through clearinghouses, where beneficial uses are found. This represents a double savings: the generator doesn't have to pay for disposal, and the recipient pays little, if anything, for raw materials.

Convert to Less Hazardous Substances Several processes are available to make hazardous materials less toxic. *Physical treatments* tie up or isolate substances. Charcoal or resin filters absorb toxins. Distillation separates hazardous components from aqueous solutions. Precipitation and immobilization in ceramics, glass, or cement isolate toxins from the environment, so that they become essentially nonhazardous. One of the few ways to dispose of metals and radioactive substances is to fuse them in silica at high temperatures to make a stable, impermeable glass that is suitable for long-term storage. Plants, bacteria, and fungi can also concentrate or detoxify contaminants (see Exploring Science, p. 350).

Incineration is applicable to mixtures of wastes. A permanent solution to many problems, it is quick and relatively easy, but not necessarily cheap—nor always clean—unless done correctly. Wastes must be heated to over 1,000°C (2,000°F) for a sufficient period of time to complete destruction. The ash resulting from thorough incineration is reduced in volume up to 90 percent and often is safer to store in a landfill or another disposal site than the original wastes.

Chemical processing can transform materials to make them nontoxic. Included in this category are neutralization, removal of metals or halogens (chlorine, bromine, etc.), and oxidation. The Sunohio Corporation of Canton, Ohio, for instance, has developed a process called PCBx, in which chlorine in such molecules as PCBs is replaced with other ions that render the compounds less toxic. A portable unit can be moved to the location of the hazardous wastes, eliminating the need for shipping them.

Store Permanently Inevitably, there are some materials we can't destroy, make into something else, or otherwise eliminate. We will have to store them out of harm's way (fig. 14.21).

Permanent retrievable storage involves placing waste storage containers in a secure place such as a salt mine or bedrock cavern, where they can be inspected periodically and retrieved if necessary. This approach is expensive because it requires monitoring, but it has the advantage that we don't completely lose control of highly toxic substances that could eventually leak into groundwater if they were buried in a landfill. If we learn someday that our disposal methods were bad, we can retrieve waste and treat it more effectively. Retrieving waste from storage in a mine is much cheaper and more effective than digging up and remediating buried pollutants from a landfill.

Secure landfills are the most popular solutions for hazardous waste disposal, however. Although many landfills have been environmental disasters, newer techniques make it possible to create safe, secure modern landfills that can contain many hazardous wastes. As with a modern solid waste landfill, the first line of defense in

▲ **FIGURE 14.21** Hazardous substances we can't decontaminate must be catalogued, contained, and stored permanently. Here workers retrieve improperly buried waste from the Hanford nuclear site in Washington State.

a secure landfill is a thick bottom cushion of compacted clay that surrounds the pit like a bathtub (fig. 14.22). Moist clay is flexible and resists cracking if the ground shifts. It is impermeable to groundwater and will safely contain wastes. A layer of gravel is spread over the clay liner, and perforated drainpipes are laid in a grid to collect any seepage that escapes from the stored material. A thick polyethylene liner, protected from punctures by soft padding materials, covers the gravel bed. A layer of soil or absorbent sand cushions the inner liner, and the wastes are packed in drums, which then are placed into the pit, separated into small units by thick berms of soil or packing material.

When the landfill has reached its maximum capacity, a cover much like the bottom sandwich of clay, plastic, and soil—in that order—caps the site. Vegetation stabilizes the surface and improves its appearance. Sump pumps collect any liquids that filter through the landfill, either from rainwater or leaking drums. This

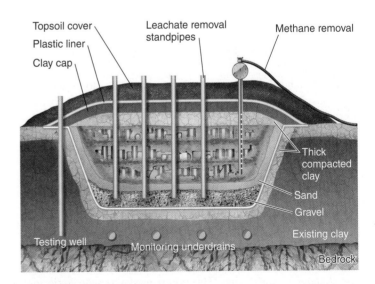

▲ **FIGURE 14.22** A secure landfill has a thick plastic liner, and two or more layers of compacted clay and a gravel bed, from which drains collect material leaching from the landfill. Testing wells allow monitoring for escaping contaminants or combustible methane.

Cleaning up the thousands of hazardous waste sites at factories, farms, and gas stations is an expensive project. In the United States alone, waste cleanup is projected to cost at least $700 billion. Usually hazardous waste remediation (cleanup) involves digging up soil and incinerating it, potentially releasing toxins into the air, or trucking it to a secure landfill. Contaminated groundwater is frequently pumped out of the ground; hopefully, contaminants are retrieved at the same time.

How do plants, bacteria, and fungi do all this? Many of the biophysical details are poorly understood, but in general, plant roots are designed to efficiently extract nutrients, water, and trace minerals from soil and groundwater. The mechanisms involved may aid extraction of metallic and organic contaminants. Some plants also use toxic elements as a defense against herbivores: locoweed, for example, selectively absorbs elements such as selenium, concentrating toxic levels in its leaves. Absorption can be extremely effective. Bracken fern growing in Florida has been found to contain arsenic at concentrations more than 200 times higher than in the soil in which it was growing.

Genetically modified plants are also being developed to process toxins. Poplars have been developed to process toxins, using a gene borrowed from bacteria that transforms a toxic compound of mercury into a safer form. In another experiment, a gene for producing mammalian liver enzymes, which specialize in breaking down toxic organic compounds, was inserted into tobacco plants. The plants succeeded in producing the liver enzymes and breaking down toxins absorbed through their roots.

These remediation methods are not without risks. Insects could consume leaves containing concentrated substances, allowing contaminants to enter the food chain. Some absorbed contaminants are volatilized, or emitted in gaseous form, through pores in plant leaves. Once contaminants are absorbed into plants, the plants themselves are usually toxic and must be landfilled. But the cost of phytoremediation can be less than half the cost of landfilling or treating toxic soil, and the volume of plant material requiring secure storage is a fraction of the volume of the contaminated dirt.

Cleaning up hazardous and toxic waste sites will be a big business for the foreseeable future, in North America and around the world. Innovations such as bioremediation offer promising prospects for business development, as well as for environmental health and saving taxpayer money.

A promising alternative to these methods involves **bioremediation**, or biological waste treatment. Microscopic bacteria and fungi can absorb, accumulate, and detoxify a remarkable variety of toxic compounds. They can also accumulate heavy metals, and some have been developed that can metabolize (break down) PCBs. Aquatic plants such as water hyacinths and cattails can also be used to purify contaminated effluent.

An increasing variety of plants have been used in phytoremediation (cleanup using plants). Some types of mustard can extract lead, arsenic, zinc, and other metals from contaminated soil. Radioactive strontium and cesium were extracted from soil near the Chernobyl nuclear power plant using common sunflowers. Poplar trees can absorb and break down toxic organic chemicals. Natural bacteria in groundwater, when provided with plenty of oxygen, can neutralize contaminants in aquifers. Experiments have shown that pumping air *into* groundwater can be a more effective cleanup method than pumping water *out*.

leachate is treated and purified before being released. Monitoring wells check groundwater around the site to ensure that no toxins have escaped.

Most landfills are buried below ground level to be less conspicuous; however, in areas where the groundwater table is close to the surface, it is safer to build aboveground storage. The same protective construction techniques are used as in a buried pit. An advantage to such a facility is that leakage is easier to monitor because the bottom is at ground level.

Transportation of hazardous wastes to disposal sites is of concern because of the risk of accidents. Emergency-preparedness officials conclude that the greatest risk in most urban areas is not nuclear war or natural disaster but crashes involving trucks or trains carrying hazardous chemicals through densely packed urban corridors. Another worry is who will bear financial responsibility for abandoned waste sites. Hazardous wastes remain toxic long after the businesses that created them are gone. As is the case with nuclear wastes (chapter 13), we may need new institutions for perpetual care of these wastes.

CONCLUSION

Modern society produces a prodigious amount of waste. Each of us today produces far more waste, and more toxic substances, than our grandparents did. Growing amounts of packaging, including plastics, are among our largest-volume problems; e-waste, with its mixed materials, and hazardous waste are growing and expensive challenges in waste management. Government policies and economies of scale make it cheaper and more convenient to extract virgin raw materials to make new consumer products than to reuse or recycle used materials. But we all bear the cost of disposing of this constantly increasing stream of waste products.

The rising cost and declining availability of landfills have led to new and creative strategies to reduce, reuse, and recycle waste. Biogas, recycling programs, and incentives to reduce packaging materials, as demonstrated in Kristianstad and many other European cities, are among the approaches that have proven both successful and economically beneficial. Incinerators, which have a poor economic and public health record in the United States, have been improved in other countries and are contributing to waste management solutions.

Controlling the production and disposal of hazardous waste requires careful oversight by government agencies and waste-management regulations. These rules and agencies are put in place to defend public health. Without them, disastrous cases like Love Canal, with cleanup paid for by the public rather than by polluters, develop all too easily.

A first step toward reducing our waste production is to understand how much we produce and what we produce. Another key step is to make our waste disposal more visible. Paying attention to recycling, reusing, and reducing household and hazardous waste can greatly improve our awareness of our environmental liabilities.

PRACTICE QUIZ

1. Explain the basic process of producing methane biogas from organic waste.
2. What are *solid wastes* and *hazardous wastes*? What is the difference between them?
3. Describe the difference between an open dump, a sanitary landfill, and a modern, secure, hazardous waste disposal site.
4. Describe some concerns about waste incineration.
5. List some benefits and drawbacks of recycling wastes. What are the major types of materials recycled from municipal waste, and how are they used?
6. What is *e-waste*? How is most of it disposed of, and what are some strategies for improving recycling rates?
7. What is *composting*, and how does it fit into solid waste disposal?
8. What materials are most recycled in the United States?
9. What are *brownfields*, and why do cities want to redevelop them?
10. What are *bioremediation* and *phytoremediation*? What are some advantages to these methods?

CRITICAL THINKING AND DISCUSSION

Apply the principles you have learned in this chapter to discuss these questions with other students.

1. A toxic waste disposal site has been proposed for the Pine Ridge Indian Reservation in South Dakota. Many tribal members oppose this plan, but some favor it because of the jobs and income it will bring to an area with 70 percent unemployment. If local people choose immediate survival over long-term health, should we object or intervene?
2. Should industry officials be held responsible for dumping chemicals that were legal when they did it but are now known to be extremely dangerous? At what point can we argue that they should have known about the hazards involved?
3. Suppose that your brother or sister has decided to buy a house next to a toxic waste dump because it costs $20,000 less than a comparable house elsewhere. What do you say to him or her?
4. Is there a fundamental difference between incinerating municipal, medical, or toxic industrial waste? Would you oppose an incinerator in your neighborhood for one type of waste but not others? Why or why not?
5. Some scientists argue that permanent retrievable storage of toxic and hazardous wastes is preferable to burial. How can we be sure that material that will be dangerous for thousands of years will remain secure? If you were designing such a repository, how would you address this question?

DATA ANALYSIS   How Much Waste Do You Produce, and How Much Do You Know How to Manage?

As people become aware of waste disposal problems in their communities, more people are recycling more materials. Some things are easy to recycle, such as newsprint, office paper, or aluminum drink cans. Other things are harder to classify. Most of us give up pretty quickly and throw things in the trash if we have to think too hard about how to recycle them. This exercise asks you to collect your own data by surveying your class about recycling know-how. Go to Connect for details and to assess your data.

TO ACCESS ADDITIONAL RESOURCES FOR THIS CHAPTER, PLEASE VISIT CONNECT AT www.mcgrawhillconnect.com.

You will find LearnSmart, an adaptive learning system, Google Earth™ exercises, additional Case Studies, Data Analysis exercises, and an interactive ebook.

CHAPTER

15 Economics and Urbanization

Car-free roads provide a cleaner, safer, healthier environment for residents of Vauban, Germany.

LEARNING OUTCOMES

After studying this chapter, you should be able to answer the following questions:

▶ How have the size and location of the world's largest cities changed over the past century?

▶ Define *slum* and *shantytown*, and describe the conditions you might find in them.

▶ What is urban sprawl? How have automobiles contributed to sprawl?

▶ What are some principles of smart growth and new urbanism?

▶ Describe sustainable development and why it's important.

▶ What value do we get from free ecological services?

▶ What's the difference between GNP and GPI?

▶ What do we mean by internalizing external costs?

Vauban: A Car-free Suburb

What would it be like to live in a city without automobiles? Residents of Vauban, a suburb of Freiburg, Germany, have a lifestyle that suggests it might be both enjoyable and economical. In Vauban, it's so easy to get around by tram, bicycle, and on foot that there is little need to depend on cars (fig. 15.1). The community is designed using "smart growth" principles with stores, banks, schools, and restaurants within easy walking distance of homes. Jobs and office space are available nearby, and trams to the city center run every few minutes through the center and around the edges of Vauban. Residential streets are narrow and vehicle-free, making a great place for bicycles and playing children. Cars must be parked in a large municipal ramp at the edge of town, and buying a space there costs $40,000. Consequently, nearly three-quarters of Vauban's families don't own a car, and more than half sold their car to move there.

Fewer vehicles means less air pollution and greater safety for pedestrians, but most families moved to Vauban not for environmental reasons but because they believe a car-free lifestyle is healthier for children. Schools, child-care services, playgrounds, and sports facilities are a short bike ride from all houses. Children can play outside and can walk or bike to school without having to cross busy streets.

In American cities, up to one-third of the land area is dedicated to cars, mainly for parking and roads. In Vauban, reducing car dependence has saved so much space that neighborhoods have abundant green space, gardens, and play areas while still being small enough for easy walking.

A highly successful and growing car-sharing program makes it still easier to live without cars. Starting in about 1992, the city's car-sharing program has grown to some 2,500 members, who save money and parking space by using shared cars. The cars are available all around town, and they can be reserved online or by mobile phone. In addition, a single monthly bus ticket covers all regional trains and buses, making it especially easy to get around by public transportation.

A car-free lifestyle makes economic sense. Owning and operating a vehicle in Germany is even more expensive than it is in the United States, where the average car costs about $9,000 per year.

▲ **FIGURE 15.1** Narrow streets in Vauban are designed for children and bicycles first, with limited car use.

Residents of areas like Vauban can put that money to other uses.

Vauban's comfortable row houses, with balconies and private gardens, are designed to conserve energy but maximize quality of life. Clever use of space, beautiful woodwork, large balconies, and large, superinsulated windows make the homes feel spacious while maintaining a small footprint. Just having shared walls minimizes energy losses. Many houses are so efficient that they don't need a heating system at all. In addition, a highly efficient wood-burning cogeneration plant provides much of the space heating and electricity for the district, and rooftop solar collectors and photovoltaic panels provide hot water and power for individual homes. Though not entirely carbon neutral, Vauban is highly sustainable. Many of the houses produce more energy than they consume.

Similar projects are being built across Europe and even in some developing countries, such as China. On Dongtan Island in the mouth of the Yangtze River near Shanghai, the Chinese government is planning an eco-city for 50,000 people that is expected to be energy, water, and food self-sufficient. In the United States, the Environmental Protection Agency is promoting "car reduced" communities. In California, for example, developers are planning a Vauban-like community called Quarry Village on the outskirts of Oakland, accessible to the Bay Area Rapid Transit system and to the California State University campus in Hayward.

Decades of advertisements and government policies in the United States have persuaded most people that the dream home is a single-family residence on a spacious lot in the suburbs, where a car—regardless of the costs in energy use, insurance, accidents, or land consumption—is essential for every trip, no matter how short the distance. Whether we can break those patterns remains to be seen.

Vauban illustrates a number of ways urban design can help us live sustainably with our environment and our neighbors. In this chapter we'll look at other aspects of city planning and urban environments as well as some principles of ecological economics that help us understand the nature of resources and the choices we face both as individuals and as communities. ■

15.1 CITIES ARE PLACES OF CRISIS AND OPPORTUNITY

More than half of humans now live in cities, and in the next quarter century that proportion will approach three-quarters of all humans. This is a dramatic change from our previous history, in which most people lived by hunting and gathering, farming, or fishing. Since the beginning of the Industrial Revolution about 300 years ago, cities have grown rapidly in both size and power (fig. 15.2). In 1950, 38 percent of the world's population lived in cities; by 2030 that proportion will nearly double (table 15.1).

The vast majority of urban growth will occur in less-developed countries (fig. 15.3). Populations in these cities are expanding far faster than infrastructure, such as roads, transportation, housing,

▼ **FIGURE 15.2** In less than 20 years, Shanghai, China, has built Pudong, a new city of 1.5 million residents and 500 skyscrapers on former marshy farmland across the Huang Pu River from the historic city center. This kind of rapid urban growth is occurring in many developing countries.

TABLE 15.1 | Urban Share of Total Population (Percentage)

	1950	2000	2030*
Africa	18.4	40.6	57.0
Asia	19.3	43.8	59.3
Europe	56.0	75.0	81.5
Latin America	40.0	70.3	79.7
North America	63.9	77.4	84.5
Oceania	32.0	49.5	60.7
World	38.3	59.4	70.5

*Projected.

SOURCE: Data from United Nations Population Division, 2003.

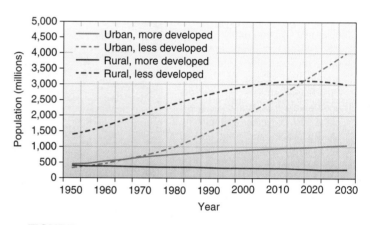

▲ **FIGURE 15.3** Growth of urban and rural populations in more-developed regions and in less-developed regions.   SOURCE: United Nations Population Division, World Urbanization Prospects, 2004.

water supplies, sewage treatment, and schools, can adapt. Building new infrastructure is especially hard in poor countries, where incomes are low and tax collection is insufficient to support public services. Despite these challenges, cities are also places where innovation occurs. Ideas mix and experimentation happens in urban areas. In cities there arise diverse employment opportunities and new economies, as well as concentrations of poverty. Huge **urban agglomerations** (merging of multiple municipalities) are forming throughout the world. Some have become **megacities** (with populations of over 10 million people). Though these cities pave over vast landscapes and consume large amounts of resources, their per capita resource use is relatively efficient. Environmental degradation would probably be much worse if that many people were spread across the countryside. Cities, for all their ills, are one of the places where we can learn new ways to live sustainably. New York City, one of the largest in the world, has established new codes for "green" building, water conservation, and recycling. More New Yorkers use public transportation and walk to get around than in any other major American city.

Cities can be engines of economic progress and social reform. Some of the greatest promise for innovation comes from cities like Vauban, where innovative leaders can focus knowledge and resources on common problems. Cities can be efficient places to live, where mass transportation can move people around and goods and services are more readily available than in the country. Concentrating people in urban areas leaves open space available for farming and biodiversity. But cities can also be dumping grounds for poverty, pollution, and unwanted members of society. Providing food, housing, transportation, jobs, clean water, and sanitation to the 2 or 3 billion new urban residents expected to crowd into cities—especially those in the developing world—may be one of the preeminent challenges of this century.

Large cities are expanding rapidly

You can already see the dramatic shift in size and location of big cities. In 1900 only 13 cities in the world had populations over 1 million (table 15.2). All of those cities except Tokyo and Peking (Beijing) were in Europe or North America. By 2012 there were at least 400 cities—100 of them in China alone—with more than 1 million residents. Of the 13 largest of these metropolitan areas, none are in Europe. Only New York City and Los Angeles are in a developed country. It's expected that by 2025 at least 93 cities will have populations over 5 million, and three-fourths of those cities will be in developing countries (fig. 15.4).

China represents the largest demographic shift in human history. Since the end of Chinese collectivized farming and factory work in 1986, around 250 million people have moved from rural areas to cities. And in the next 25 years an equal number are expected to join this vast exodus. In addition to expanding existing cities, over the next 20 years China plans to build 400 new urban centers with populations of at least 500,000. Already at least

TABLE 15.2	The World's Largest Urban Areas (Populations in Millions)		
1900		**2011**	
London, England	6.6	Tokyo, Japan	34.3
New York, USA	4.2	Guangzhou, China	25.2
Paris, France	3.3	Seoul, Korea	25.1
Berlin, Germany	2.4	Shanghai, China	24.8
Chicago, USA	1.7	Delhi, India	23.3
Vienna, Austria	1.6	Mumbai, India	23.0
Tokyo, Japan	1.5	Mexico City, Mexico	22.9
St. Petersburg, Russia	1.4	New York, USA	22.0
Philadelphia, USA	1.4	São Paulo, Brazil	20.9
Manchester, England	1.3	Manila, Philippines	20.3
Birmingham, England	1.2	Jakarta, Indonesia	18.9
Moscow, Russia	1.1	Los Angeles, USA	18.1
Peking, China*	1.1	Karachi, Pakistan	17.0

*Now spelled Beijing.

SOURCE: Data from T. Chandler, *Three Thousand Years of Urban Growth*, 1974, Academic Press; and *Th. Brinkhoff: The Principal Agglomerations of the World*, 2011.

half of the concrete and one-third of the steel used in construction around the world each year is consumed in China.

Consider Shanghai, China, for example. In 1985 the city had a population of about 10 million. It's now nearly 25 million in the city itself and the greater Shanghai metropolitan area may be over 100 million. In the past decade, Shanghai built 4,000 skyscrapers (buildings with more than 25 floors). The city already has

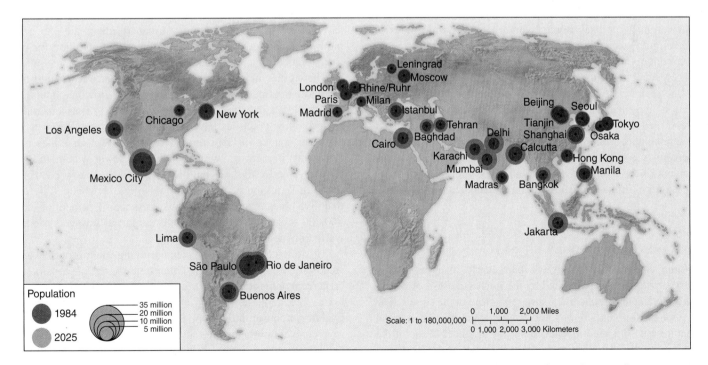

▲ **FIGURE 15.4** By 2025, at least 400 cities will have populations of 1 million or more, and 93 supercities will have populations above 5 million. Three-fourths of the world's largest cities will be developing countries that already have trouble housing, feeding, and employing their people.

twice as many tall buildings as Manhattan, and proposals have been made for 1,000 more. Most of this growth has taken place in a swampy area called Pudong, across the Huang Pu River from the historic city center (see fig. 15.2). Pudong is now sinking about 1.5 cm per year due to groundwater drainage and the weight of so many buildings.

Other Chinese cities also have plans for massive building projects. Harbin, a city of about 9 million people and the capital of Heilongjiang Province, for example, recently announced plans to relocate across the Songhua River on 740 km^2 (285 mi^2, or roughly the size of New York City) of former farmland. Residents hope these new towns will be both more livable for their residents and more ecologically sustainable than the old cities they're replacing. The Chinese government has signed several long-term contracts with international engineering firms to build at least five "eco-cities," each the size of a large Western capital. Plans call for these cities to be self-sufficient in energy, water, and most food products, with the aim of zero emissions of greenhouse gases from transportation.

Immigration is driven by push and pull factors

People migrate to cities for many reasons. In China over the past 20 years—or in America during the twentieth century—mechanization eliminated jobs and drove people off the land. Where cropland is owned by a minority of wealthy landlords, as is the case in many developing countries, subsistence farmers are often ejected when new cash crops or cattle grazing become economically viable. Many people also move to the city for the opportunities and independence offered there. Cities offer jobs, better housing, entertainment, and freedom from the constraints of village traditions. Possibilities exist in the city for upward social mobility, prestige, and power not ordinarily available in the country. Cities support specialization in arts, crafts, and professions for which markets don't exist elsewhere.

Government policies often favor urban over rural areas in ways that both push and pull people into cities. Developing countries commonly spend most of their budgets on improving urban areas (especially around the capital city, where leaders live). This gives the major cities a virtual monopoly on new jobs, housing, education, and finance, all of which bring in rural people searching for a better life. Lima, for example, has only 20 percent of Peru's population, but has 50 percent of the national wealth, 60 percent of the manufacturing, 65 percent of the retail trade, 73 percent of the industrial wages, and 90 percent of all banking in the country. Similar statistics pertain to many national capitals.

Congestion, pollution, and water shortages plague many cities

First-time visitors to a supercity—particularly in a developing country—often are overwhelmed by the immense crush of pedestrians and vehicles of all sorts jostling for space in the streets. The noise, congestion, and confusion of traffic make it seem suicidal to venture onto the street. Delhi, India, for instance, is one of the most densely populated cities in the world (fig. 15.5). Traffic is chaotic almost all the time. People often spend three or four hours each way commuting to work from outlying areas.

▲ **FIGURE 15.5** Traffic in developing countries, such as New Delhi, India, has grown far faster than the road network. Monumental traffic jams occur at almost any time of day.

Pollution from burgeoning traffic and from unregulated factories degrades air quality in many urban areas. China's spectacular economic growth has resulted in a flood of private automobiles mainly in cities. In Beijing, for instance, the number of cars on the streets has more than doubled in just the past 5 years to 5 million. China is now the world's largest producer of greenhouse gases, and the World Bank warns that it is home to 16 of the world's 20 cities with the worst air pollution. Chinese health authorities say that a third of the country's urban residents are exposed to harmful air pollution levels. They blame this pollution for more than 400,000 premature deaths each year.

Few cities in developing countries can afford to build modern waste treatment systems for their rapidly growing populations. The World Bank estimates that only one-third of urban residents in developing countries have satisfactory sanitation services. The 2010 earthquake in Haiti reminded us that Port au Prince is the largest city in the world with no municipal sewer system. In Egypt, Cairo's sewer system was built about 50 years ago to serve a population of 2 million people. It's now being overwhelmed by more than five times that many residents. Less than 1 percent of India's 500,000 towns and villages have even partial sewage systems or water treatment facilities.

It's often difficult to find clean drinking water for urban areas. According to Qiu Baoxing, Chinese minister of construction, 70 percent of his country's surface water is so polluted by industrial toxins, human waste, and agricultural chemicals that it is unsuited for human consumption. One hundred of China's 660 cities face severe water shortages, he reported. Worldwide, according to the United Nations, at least 1 billion people don't have safe drinking water, and twice that many don't have adequate sanitation. Scarcity of clean water is one of our greatest environmental health crises.

People for Community Recovery

The Lake Calumet Industrial District on Chicago's far South Side is an environmental disaster area. A heavily industrialized center of steel mills, oil refineries, railroad yards, coke ovens, factories, and waste disposal facilities, much of the site is now a marshy wasteland of landfills, toxic waste lagoons, and slag dumps, around a system of artificial ship channels.

At the southwest corner of this degraded district sits Altgeld Gardens, a low-income public housing project built in the late 1940s by the Chicago Housing Authority. The 2,000 units of "The Gardens" or "The Projects," as they are called by the largely minority residents, are low-rise row houses, many of which are vacant or in poor repair. But residents of Altgeld Gardens are doing something about their neighborhood. People for Community Recovery (PCR) is a grassroots citizen's group organized to work for a clean environment, better schools, decent housing, and job opportunities for the Lake Calumet neighborhood.

PCR was founded in 1982 by Mrs. Hazel Johnson, an Altgeld Gardens resident whose husband died from cancer that may have been pollution-related. PCR has worked to clean up more than two dozen waste sites and contaminated properties in their immediate vicinity. Often this means challenging authorities to follow established rules and enforce existing statutes. Public protests, leafleting, and community meetings have been effective in public education about the dangers of toxic wastes and have helped gain public support for cleanup projects. PCR's efforts successfully blocked construction of new garbage and hazardous waste landfills, transfer stations, and incinerators in the Lake Calumet district. Pollution prevention programs have been established at plants still in operation. And PCR helped set up a community monitoring program to stop illegal dumping and to review toxic inventory data from local companies.

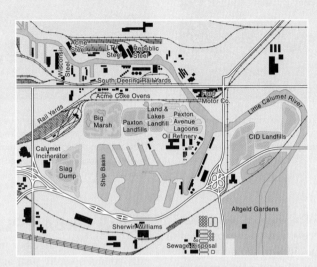

▲ The Calumet industrial district in South Chicago.

Education is an important priority for PCR. An environmental education center administered by community members organizes workshops, seminars, fact sheets, and outreach for citizens and local businesses. A public health education and screening program has been set up to improve community health. Partnerships have been established with nearby Chicago State University to provide technical assistance and training in environmental issues.

PCR also works on economic development. Environmentally responsible products and services are now available to residents. Jobs are being created as green businesses are brought into the community. Wherever possible, local people and minority contractors from the area are hired to clean up waste sites and restore abandoned buildings. Job training for youth and adults as well as retraining for displaced workers is a high priority.

In the 1980s a young community organizer named Barack Obama worked with PCR on jobs creation, housing issues, and education. He credits the lessons he learned there for much of his subsequent political successes. In his best-selling memoir, *Dreams from My Father,* Obama devotes more than 100 pages to his formative experiences at Altgeld Gardens and other nearby neighborhoods.

PCR and Mrs. Johnson have received many awards for their fight against environmental racism and despair. In 1992, PCR was the recipient of the President's Environmental and Conservation Challenge Award. PCR is the only African-American grassroots organization in the country to receive this prestigious award.

Although Altgeld Gardens is far from clean, much progress has been made. Perhaps the most important accomplishment is community education and empowerment. Residents have learned how and why they need to work together to improve their living conditions. Could these same lessons be useful in your city or community? What could you do to help improve urban environments where you live?

Pollution isn't limited to the developing world, however. Many cities in industrialized countries have a tragic heritage of pollution and abandoned factories (see What Do You Think? on this page). But community organizing can make a difference.

Many cities lack sufficient housing

The United Nations estimates that at least 1 billion people live in crowded, unsanitary slums of the central cities or in the vast shantytowns and squatter settlements that ring the outskirts of most major cities in the developing world. Around 100 million people have no home at all. In Mumbai (formerly Bombay), India, for example, it's estimated that half a million people sleep on the streets, sidewalks, and traffic circles because they can find no other place to live.

Slums are generally legal but inadequate multifamily tenements or rooming houses, often converted from some other use. Families live crowded in small rooms with inadequate ventilation and sanitation. Often these structures are rickety and unsafe. In 1999, for example, a 7.4 magnitude earthquake hit eastern Turkey, killing more than 14,000 people when shoddy, poorly built apartments collapsed.

▲ **FIGURE 15.6** Homeless people have built shacks along this busy railroad track in Jakarta. It's a dangerous place to live, with many trains per day using the tracks, but for the urban poor there are few other choices.

▲ **FIGURE 15.7** Huge houses on sprawling lots consume land, alienate us from our neighbors, and make us ever more dependent on automobiles. They also require a lot of lawn mowing! SOURCE: ©2003 Regents of the University of Minnesota. All rights reserved. Used with permission of the Design Center for American Urban Landscape.

Shantytowns, with shacks built of corrugated metal, discarded packing crates, brush, plastic sheets, and other scavenged materials, grow on vacant land in many cities in the developing world (fig. 15.6). They can house millions of people but generally lack clean water, sanitation, or safe electrical power. Shantytowns are usually illegal, but they quickly fill in the empty space in towns where squatters can build shelters close to jobs. With little or no public services, shantytowns often fill with trash and debris. Many governments try to clean out illegal settlements by torching or bulldozing the huts and sending riot police to drive out residents, but people either move back in or relocate to another shantytown. In 2005 the government of Zimbabwe destroyed the homes of some 700,000 people in shantytowns around the capital of Harare. Families were evicted in the middle of the night during the coldest weather of the year, often with only minutes to gather their belongings. President Robert Mugabe justified this blitzkrieg as necessary to control crime, but critics claimed it was mainly to remove political opponents.

Two-thirds of the population of Kolkata are thought to live in unplanned squatter settlements, and nearly half the 25 million residents of Mexico City occupy the unauthorized *colonias* around the city. Many shantytowns occupy the most polluted, dangerous parts of cities where no one else wants to live. In Bhopal, India, and Mexico City, for example, squatter settlements were built next to deadly industrial sites. In Brazil, shantytowns called *favelas* perch on steep hillsides unwanted for other building. As desperate and inhumane as conditions are in these slums and shantytowns, many people do more than merely survive there. They work hard, raise families, educate their children, and often improve their living standard little by little as they make some money.

Many countries are recognizing that the only way they can house all their citizens is to cooperate with shantytown dwellers. Recognizing land rights, providing financing for home improvements, and supporting community efforts to provide water, sewers, and power can greatly improve living conditions for many poor people.

15.2 URBAN PLANNING

How can we live together in cities in ways that are environmentally sound, socially just, and economically sustainable? Starting with Greek cities thousands of years ago, planners have debated the best ways for us to organize ourselves.

Transportation is crucial in city development

Getting people around within a large urban area has become one of the most difficult problems that many city officials face. Freeway construction, which began in America in the 1950s, allowed people to move out into the countryside. Cities that were once compact began to spread over the landscape, consuming space and wasting resources. This pattern of development is known as **sprawl**. While there is no universally accepted definition of the term, sprawl generally includes the characteristics outlined in table 15.3.

In most American metropolitan areas, the bulk of new housing is in large, tract developments that leapfrog out beyond the city

TABLE 15.3 | Characteristics of Urban Sprawl

1. Unlimited outward extension
2. Low-density residential and commercial development
3. Leapfrog development that consumes farmland and natural areas
4. Fragmentation of power among many small units of government
5. Dominance of freeways and private automobiles
6. No centralized planning or control of land uses
7. Widespread strip-malls and "big-box" shopping centers
8. Great fiscal disparities among localities
9. Reliance on deteriorating older neighborhoods for low-income housing
10. Decaying city centers as new development occurs in previously rural areas

SOURCE: Data from PlannersWeb, Burlington, Vermont, 2001.

edge in a search for inexpensive rural land with few restrictions on land use or building practices (fig. 15.7). Housing experts claim that overbuilding during the past decade's housing bubble left us with 40 million McMansions in drive-only suburbs that no one now wants.

The U.S. Department of Housing and Urban Development calculates that urban sprawl consumes some 200,000 ha (roughly 500,000 acres) of farmland and open space every year. Although the price of raw farmland generally is less than comparable urban property, there are external costs in the form of new roads, sewers, water mains, power lines, schools, shopping centers, and other infrastructure required by this low-density development.

Because many Americans live far from where they work, shop, or recreate, they consider it essential to own a private automobile. The average U.S. driver spends 443 hours per year behind a steering wheel, or the equivalent of one full 8-hour day per week in an automobile. The freeway system was designed to allow drivers to travel at high speeds from source to destination without ever having to stop (fig. 15.8). As more and more vehicles clog highways, however, the reality is far different.

Altogether, the average driver spends about 100 hours per year (2.5 weeks of work) in bumper-to-bumper traffic, and congestion costs the United States $78 billion per year in wasted time and fuel. Some people argue that the existence of traffic jams in cities shows that more highways are needed. Often, however, building more traffic lanes simply encourages more people to drive farther and put more cars on the road. Meanwhile, about one-third of Americans are too young, too old, or too poor to drive. For these people, car-oriented development causes isolation and makes daily tasks like grocery shopping difficult. Parents spend long hours transporting young children. Teenagers and aging grandparents are forced to drive, often presenting a hazard on public roads.

As the opening case study for this chapter shows, it's possible to build cities without private autos. Most European urban areas have good mass transit systems that have allowed them to preserve historic city centers and remain relatively compact while avoiding the sprawl engendered by an American-style freeway system. Many American cities are now rebuilding the public transportation systems that were abandoned in the 1950s (fig. 15.9). Consider how differ-

▲ **FIGURE 15.9** Many American cities are now restoring light-rail systems that were abandoned in the 1950s when freeways were built. Light rail is energy efficient and popular, but it can cost up to $100 million per mile ($60 million per kilometer).

ent your life might be if you lived an automobile-free life in a city with good mass transit.

A famous example of successful mass transit is found in Curitiba, Brazil. High-speed, bi-articulated buses, each of which can carry 270 passengers, travel on dedicated roadways closed to all other vehicles (see Key Concepts, p. 360). This bus-rapid-transit system is linked to 340 feeder routes extending throughout the city. Everyone in the city is within walking distance of a bus stop that has frequent, convenient, affordable service. Curitiba's buses carry some 1.9 million passengers per day, or about three-quarters of all personal trips within the city. Working with existing roadways for the most part, the city was able to construct this system for one-tenth the cost of a light-rail system or freeway system, and one-hundredth the cost of a subway.

Rebuilding cities

Vauban, Germany, featured in the opening case study for this chapter, wasn't a "greenfield" project built on farmland outside Freiburg. Instead, it was a former army base inside the city limits. The reason its streets are too narrow for automobile traffic is that they were intended for marching between barracks. It's a clever recycling of formerly occupied space.

There's an opportunity—some would say a necessity—to rebuild and redesign many aging cities in America's industrial "rust belt." A declining manufacturing base coupled with middle-class flight to the suburbs has hollowed-out many cities. Detroit's population, for example, has declined from 1.85 million in the 1950s to about 700,000 today. At least 100,000 abandoned, homes and businesses have

▼ **FIGURE 15.8** Freeways give us the illusion of speed and privacy, but they consume land, encourage sprawl, and create congestion as people move farther from the city to get away from traffic and then have to drive to get anywhere.

KC 15.1

What makes a city green?

Efficiency. Over half of humans now live in cities. Environmental scientists have often criticized cities for expanding into farmland and for their tremendous consumption of energy, water, food, concrete, and land. But the environmental cost per person is usually lower for urban living than for suburban or rural living, especially in wealthy countries. Because they are compact, cities require fewer miles of roads, water and sewer lines, less heating, and fewer private cars per household. Because distances are shorter, roads and utility infrastructure are shared, apartments or row houses share heat, and public transportation reduces the need for driving to work.

Polluted cities can be unhealthy, but well-organized cities can provide cultural resources and preserve environmental resources in many beneficial ways.

Here are 10 features that make cities healthy for people and the environment.

KC 15.2

1. Public transit
High-density areas can afford to support reliable, efficient transit systems, where many riders share the cost of getting around, such as this bus-rapid transit system in Curitiba, Brazil. Public transit uses far less space, energy, and materials than does private travel.

KC 15.3a

2. Safe walking and bike routes
Freedom from dependence on cars increases mobility for young people, old people, and others without cars. Cities with separated walk-ways and bikeways are friendly for children and families; they also provide exercise and save money.

KC 15.3b

KC 15.4

3. Compact building
A compact urban design greatly increases efficiency of land use, reduces transit distances, and increases heating or cooling efficiency, as buildings share walls. Reduced dependence on cars, and car sharing, can help control the problem of parking shortages.

◄ Amsterdam's row houses give the city its historic identity as well as efficiency.

KC 15.6

4. Mixed-use planning
Integrating housing with shopping, entertainment, and office space provides jobs and services where people live. These neighborhoods can encourage walking and build community, as people spend less time in travel to shopping and work.

A used bookstore and cafe share space with housing in the historic city center of Trondheim, Norway. ▶

KC 15.5

10. Farmland conservation
Sprawling suburbs gobble up farmland, woodlands, and wetlands. This is the fastest type of land-use change in most developing countries. Compact cities minimize destruction of farmland, habitat, recreational space, and watersheds.

KC 15.8

KC 15.7

9. Local food
Local farm economies are more viable if farmers can sell direct to consumers—something that is much easier where there are lots of buyers in one place. Cities have become an essential income source for many produce farmers.

▲ The St. Johnsbury, Vermont farmers' market provides fresh, locally grown food for urban residents.

KC 15.9

8. Energy efficiency
Alternative energy is easier to use right at the source. Rooftop solar energy, district heating, and other strategies aid efficiency.

KC 15.10

7. Green infrastructure
New techniques moderate the impact of impervious surfaces, including permeable pavement, green roofs, and better building design.

▲ This biomass-burning plant in Copenhagen, and others like it, provide nearly all heating for Denmark's major cities.

▲ A "green" parking lot in Chengdu, China supports both traffic and vegetation, allowing rainfall to percolate into the ground.

6. Recycling programs
Recycling collection is easiest where transportation is minimal and where recyclable materials are abundant.

◀ These bins in Kuala Lumpur, Malaysia accept all kinds of recyclables.

KC 15.11

CAN YOU EXPLAIN?

1. **What factors can make per capita energy use low in urban areas?**

2. **Which of the green factors listed would be easiest to enhance where you live? Why?**

3. **Which do you find most and least appealing? Why?**

5. Green space
Recreational space has physical and emotional benefits for urban residents. Living vegetation and soils cool the local microclimate, store nutrients and moisture, and provide habitat for birds and other wildlife.

◀ Here a visitor watches skaters in New York's Central Park.

▲ **FIGURE 15.10** Many old, industrial cities in America's "rust belt" are tearing down tens of thousands of abandoned, derelict houses. This opens up an opportunity for redevelopment with gardens, green belts, diverse flexible housing, walkable neighborhoods, and other features of smart growth.

been demolished (fig. 15.10). Whole blocks are now uninhabited. It costs the city at least $3 million per year just to mow the weeds on empty lots. At least 40 mi² (104 km²) of land inside the city limits are vacant. Urban pioneers are starting farms and establishing a subsistence, barter-based economy. And the city plans to tear down at least 10,000 more homes and resettle residents into compact, more easily served neighborhoods.

The question is, how will those new urban spaces be designed? Will they remain auto-centered, or will they have a more energy-efficient, pedestrian-friendly, flexible design intended for building community rather than being simply bedroom districts? And could some of the vacant land be used for commercial agriculture? A wealthy money manager named John Hantz recently announced plans to spend up to $30 million of his personal fortune to turn about 50 ha (124 acre) on Detroit's east side into the world's largest urban farm. He hopes to provide hundreds of jobs, supply local markets and restaurants with fresh produce, eliminate a big chunk of urban blight, and stimulate city development.

Other cities are reinvigorating decaying downtowns and creating sustainable eco-districts following principles of smart growth. Denver, for example has turned two square blocks in Lower Downtown into a human-scale neighborhood with walking streets, bike paths, rooftop gardens, solar panels, and energy retrofits that have cut energy consumption in half. The area has become a night-life epicenter drawing crowds of shoppers, diners, and partygoers. Another successful revitalization project is on Manhattan's West Side in New York City. An abandoned elevated rail line running along 10th Avenue from 14th Street to 34th Street has been converted into a linear park called the High Line. Landscaping, public gathering spaces, food stands, performance areas, and a two-mile (3.2 km) promenade turn this into welcome open space in the crowded city.

We can make our cities more livable

Are there alternatives to unplanned sprawl and wasteful resource use? One option proposed by many urban planners is **smart growth**, which makes effective use of land resources and existing infrastructure by encouraging in-fill development that avoids costly

duplication of services and inefficient land use (table 15.4). Smart growth aims to provide a mix of land uses to create a variety of affordable housing choices and opportunities. It also attempts to provide a variety of transportation choices, including pedestrian-friendly neighborhoods. This approach to planning also seeks to maintain a unique sense of place by respecting local cultural and natural features.

By making land-use planning open and democratic, smart growth makes urban expansion fair, predictable, and cost-effective. All stakeholders are encouraged to participate in creating a vision for the city and to collaborate with rather than confront each other. Goals are established for staged and managed growth in urban transition areas with compact development patterns. This approach is not opposed to growth. It recognizes that the goal is not to block growth but to channel it to areas where it can be sustained over the long term. Smart growth strives to enhance access to equitable public and private resources for everyone and to promote the safety, livability, and revitalization of existing urban and rural communities.

Smart growth protects environmental quality. It tries to reduce traffic and to conserve farmlands, wetlands, and open space. As cities grow and transportation and communications enable more community interaction, the need for regional planning becomes greater and more pressing. Community and business leaders must make decisions based on a clear understanding of regional growth needs and how infrastructure can be built most efficiently and for the greatest good.

New urbanism incorporates smart growth

Rather than abandon the cultural history and infrastructure investment in existing cities, a group of architects and urban planners is attempting to redesign metropolitan areas to make them more appealing, efficient, and livable. Vauban, Germany, follows many of these principles of green design and smart growth. Other European cities—such as Stockholm, Sweden; Helsinki, Finland; Leicester, England; and Neerlands, the Netherlands—have a long history of innovative urban planning. In the United States, Andres Duany, Elizabeth Plater-Zyberk, Peter Calthorpe, and Sym Van Der Ryn have been leaders in this movement.

Using what is sometimes called a neotraditionalist approach, these designers attempt to recapture some of the best features of small towns and livable cities of the past. They are designing urban neighborhoods that integrate houses, offices, shops, and civic buildings. Ideally, no house should be more than a five-minute

TABLE 15.4 | Goals for Smart Growth

1. Create a positive self-image for the community.
2. Make the downtown vital and livable.
3. Alleviate substandard housing.
4. Solve problems with air, water, toxic waste, and noise pollution.
5. Improve communication between groups.
6. Improve community member access to the arts.

SOURCE: Data from Vision 2000, Chattanooga, Tennessee.

walk from a neighborhood center with a convenience store, a coffee shop, a bus stop, and other amenities. A mix of apartments, townhouses, and detached houses in a variety of price ranges ensures that neighborhoods will include a diversity of ages and income levels. Some design principles of this movement include:

- Limit city size or organize cities in modules of 30,000 to 50,000 people—large enough to be a complete city but small enough to be a community.

- Maintain greenbelts in and around cities. These provide recreational space and promote efficient land use, as well as help improve air quality, moderate temperature, and reduce water pollution.

- Determine in advance where development will take place. This protects property values and prevents chaotic development. Planning can also protect historical sites, agricultural resources, and ecological services of wetlands, clean rivers, and groundwater replenishment.

- Locate everyday shopping and services so people can meet daily needs with greater convenience, less stress, less automobile dependency, and less use of time and energy (fig. 15.11). This might be accomplished by encouraging small-scale commercial development in or close to residential areas.

- Encourage walking or the use of small, low-speed, energy-efficient vehicles (microcars, golf carts, bicycles, etc.) for many local trips now performed in full-size automobiles. Creating special traffic lanes, reducing the number or size of parking spaces, and closing shopping streets to big cars might encourage such alternatives.

- Promote more diverse, flexible housing as an alternative to conventional detached, single-family houses. In-fill building between existing houses saves energy, reduces land costs, and might help provide a variety of living arrangements. Allowing single-parent families or groups of unrelated adults to share housing and to use facilities cooperatively also provides alternatives to those not living in a traditional nuclear family.

- Make cities more self-sustainable by growing food locally, recycling wastes and water, using renewable energy sources, reducing noise and pollution, and creating a cleaner, safer environment. Encourage community gardening (fig. 15.12). Reclaimed inner-city space or a greenbelt of agricultural and forestland around the city provides food and open space, and also contributes valuable ecological services, such as purifying air, supplying clean water, and protecting wildlife habitat and recreation land.

▲ **FIGURE 15.12** Many cities have large amounts of unused open space that could be used to grow food. Residents often need help decontaminating soil and gaining access to the land.

▲ **FIGURE 15.13** This award-winning green roof on the Chicago City Hall is functional as well as beautiful. It reduces rain runoff by about 50 percent, and keeps the surface as much as 30°F cooler than a conventional roof on hot summer days.

▶ **FIGURE 15.11** This walking street in Queenstown, New Zealand, provides opportunities for shopping, dining, and socializing in a pleasant outdoor setting.

- Equip buildings with "green roofs" or rooftop gardens that improve air quality, conserve energy, reduce stormwater runoff, reduce noise, and help reduce urban heat island effects. Intensive gardens can include large trees, shrubs, and flowers, and may require regular maintenance (fig. 15.13). Extensive gardens require less soil, add less weight to the building, and usually have simple plantings of prairie plants or drought-resistant species, such as sedum, that require minimum care. They can last twice as long as conventional roofs. In Europe more than 1 million m^2 of green roofs are installed every year. Urban roofs are also a good place for solar collectors or wind turbines.

- Plan cluster housing, or open-space zoning, which preserves at least half of a subdivision as natural areas, farmland, or other forms of open space. Studies have shown that people who move to the country don't necessarily want to live miles from the nearest neighbor; what they most desire is long views across an interesting landscape and an opportunity to see wildlife. By carefully clustering houses on smaller lots, a conservation subdivision can provide the same number of buildable lots as a conventional subdivision and still preserve 50 to 70 percent of the land as open space (fig. 15.14). This not only reduces development costs (less distance to build roads, lay telephone lines, sewers, power cables, and so on), but also helps foster a greater sense of community among new residents.

- Preserve urban habitat. It can make a significant contribution toward saving biodiversity as well as improving mental health and giving us access to nature.

These planning principles aren't just a matter of aesthetics. Dr. Richard Jackson, former director of the National Center for Environmental Health in Atlanta, points out a strong association between urban design and our mental and physical health. As our cities have become ever more spread out and impersonal, we have fewer opportunities for healthful exercise and socializing. Chronic diseases, such as cardiovascular diseases, asthma, diabetes, obesity, and depression, are becoming the predominant health concerns in the United States.

"Despite common knowledge that exercise is healthful," Dr. Jackson says, "fewer than 40% of adults are regularly active, and 25% do no physical activity at all. The way we design our communities makes us increasingly dependent on automobiles for the shortest trip, and recreation has become not physical but observational." Long commutes and a lack of reliable mass transit and

▲ **FIGURE 15.14** Jackson Meadows, an award-winning cluster development near Stillwater, Minnesota, groups houses at sociable distances and preserves surrounding open space for walking, gardening, and scenic views from all houses.

walkable neighborhoods mean that we spend more and more time in stressful road congestion. "Road rage" isn't imaginary. Every commuter can describe unpleasant encounters with rude drivers. Urban design that offers the benefits of more walking, more social contact, and surroundings that include water and vegetation can provide healthful physical exercise and psychic respite.

15.3 ECONOMICS AND SUSTAINABLE DEVELOPMENT

Like many of our environmental issues, improving urban conditions will ultimately be decided by economics and policy decisions. We'll discuss policy in chapter 16. In the next part of this chapter, we'll review some of the principles of environmental economics.

Can development be sustainable?

By now it is clear that security and living standards for the world's poorest people are inextricably linked to environmental protection. One of the most important questions in environmental science is how we can continue improvements in human welfare within the limits of the earth's natural resources. *Development* means improving people's lives. *Sustainability* means living on the earth's renewable resources without damaging the ecological processes that support us all. **Sustainable development** is an effort to marry these two ideas. A popular definition describes this goal as "meeting the needs of the present without compromising the ability of future generations to meet their own needs."

But is this possible? As you've learned elsewhere in this book, many people argue that our present population and economic levels are exhausting the world's resources. There's no way, they insist, that more people can live at a higher standard without irreversibly degrading our environment. Others claim that there's enough for everyone if we just share equitably and live modestly. Let's look a little deeper into this important debate.

Our definitions of resources shape how we use them

To understand the problems and promise of sustainability, you need to understand the different kinds of resources we use. The way we treat resources depends largely on how we view and define them. **Classical economics**, developed in the 1700s by philosophers such as Adam Smith (1723–1790) and Thomas Malthus (1766–1834), assumes that natural resources are finite—that resources such as oil, gold, water, and land exist in fixed amounts. According to this view, as populations grow, scarcity of these resources reduces quality of life, increases competition, and ultimately causes populations to fall again. In a free market, where buyers and sellers make free, independent decisions to buy and sell, the price of a commodity depends on the available supply (it's cheap when plenty is available) and the demand for it (buyers pay more when they must compete to get the resource). Perhaps the purest expression of this system is a farmers' market, where the price of goods is determined primarily by supply and demand (fig. 15.15).

▲ **FIGURE 15.15** A farmers' market is a good example of classical economics. When supplies are abundant, prices fall, and farmers don't plant those crops. When there's a shortage of a particular commodity, prices rise, and farmers will work to bring more of that crop to market.

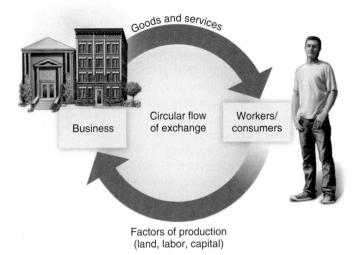

▲ **FIGURE 15.16** The neoclassical model of the economy focuses on the flow of goods, services, and factors of production (land, labor, capital) between businesses and individual workers and consumers. The social and environmental consequences of these relationships are irrelevant in this view.

The nineteenth-century economist John Stuart Mill agreed that most resources are finite, but he developed the idea of a **steady-state economy**. Rather than boom-and-bust cycles of population and resource use, as envisioned by Malthus, Mill proposed that economies can achieve an equilibrium of resource use and production. Intellectual and moral development continues, he argued, once this stable, secure state is achieved.

Neoclassical economics, developed in the nineteenth century, expanded the idea of resources to include labor, knowledge, and capital. Labor and knowledge are resources because they are necessary to create goods and services; they are not finite because every new person can add more labor and energy to an economy. **Capital** is any form of wealth that contributes to the production of more wealth. Money can be invested to produce more money. Mineral resources can be developed and manufactured into goods that return more money. Economists distinguish several kinds of capital:

1. Natural capital: goods and services provided by nature

2. Human capital: knowledge, experience, human enterprise

3. Manufactured (built) capital: tools, buildings, roads, technology

To this list some theorists would add social capital, the shared values, trust, cooperation, and organization that can develop in a group of people but cannot exist in one individual alone.

Because the point of capital is the production of more capital (that is, wealth), neoclassical economics emphasizes the idea of growth. Growth results from the flow of resources, goods, and services (fig. 15.16). Continued growth is always necessary for continued prosperity, according to this view. Natural resources contribute to production and growth, but they are not critical supplies that limit growth. They are not limiting because resources are considered to be interchangeable and substitutable. Neoclassical economics predicts that as one resource becomes scarce, a substitute will be found.

Because production of wealth is central to neoclassical economics, an important measure of growth and wealth is consumption. If a society consumes more oil, minerals, and food, it is presumably becoming wealthier. This idea has extended to the idea of *throughput*, the amount of resources a society uses and discards. More throughput is a measure of greater consumption and greater wealth, according to this view. Throughput is commonly measured in terms of **gross national product (GNP)**, the sum of all products and services bought and sold in an economy. Because GNP includes activities of offshore companies, economists sometimes prefer **gross domestic product (GDP)**, which more accurately reflects the local economy by accounting for only those goods and services bought and sold locally.

Natural resource economics extends the neoclassical viewpoint to treat natural resources as important waste sinks (absorbers), as well as sources of raw materials. Natural capital (resources) is considered more abundant, and therefore cheaper, than built or human-made capital.

Ecological economics incorporates principles of ecology

Ecological economics applies ecological ideas of system functions and recycling to the definition of resources. This school of thought also recognizes efficiency in nature, and it acknowledges the importance of ecosystem functions for the continuation of human economies and cultures. In nature, one species' waste is another's food, so that nothing is wasted. We need an economy that recycles materials and uses energy efficiently, much as a biological community does. Ecological economics also treats the natural environment as part of our economy, so that natural capital becomes a key consideration in economic calculations. Ecological functions, such as absorbing and purifying wastewater, processing air pollution, providing clean water, carrying out photosynthesis, and creating soil, are known as **ecological services** (table 15.5). These services are free: we don't pay for them directly (although we often pay indirectly when we suffer from their absence). Therefore, they are often excluded from conventional economic accounting, a situation that ecological economists attempt to rectify (fig. 15.17).

Many ecological economists also promote the idea of a steady-state economy. As with John Stuart Mill's original conception of steady states, these economists argue that economic health can be maintained without constantly growing consumption and throughput. Instead, efficiency and recycling of resources can allow steady prosperity where there is little or no population growth. Low birth rates and death rates (like *K*-adapted species; see chapter 3), political and social stability, and reliance on renewable energy would characterize such

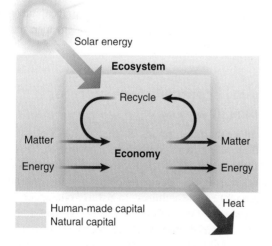

◀ **FIGURE 15.18** Living organisms are unique resources in that they reproduce themselves indefinitely, and yet, once lost through overexploitation or habitat destruction, they will probably never be replaced.

a steady-state economy. Like Mill, these economists argue that human and social capital—knowledge, happiness, art, life expectancies, and cooperation—can continue to grow even without constant expansion of resource use.

Both ecological economics and neoclassical economics distinguish between renewable and nonrenewable resources. **Nonrenewable resources** exist in finite amounts: minerals, fossil fuels, and also groundwater that recharges extremely slowly are all fixed, at least on a human time scale. **Renewable resources** are naturally replenished and recycled at a fairly steady rate. Fresh water, living organisms, air, and food resources are all renewable (fig. 15.18).

These categories are important, but they are not as deterministic as you might think. Nonrenewable resources, such as iron and gold, can be extended through more efficient use: cars now use less steel than they once did, and gold is mixed in alloys to extend its use. Substitution also reduces demand for these resources: car parts once made of iron are now made of plastic and ceramics; copper wire, once stockpiled to provide phone lines, is now being replaced with cheap, lightweight fiber-optic cables made from silica (sand). Recycling also extends supplies of nonrenewable resources. Aluminum, platinum, gold, silver, and many other valuable metals are routinely recycled now, further reducing the demand for extracting new sources.

The only limit to recycling is usually the relative costs of extracting new resources compared with collecting used materials. Recoverable sources of nonrenewable resources are also expanded by technological improvements. New methods make it possible to mine very dilute metal ores, for example. Gold ore of extremely low concentrations is now economically recoverable—that is, you can make money on it—even though this greater efficiency, more discoveries, and resource substitution can make the price of such metals fall. Scarcity of resources, seen by classical economists as the trigger for conflict and suffering, can actually provide the impetus for much of the innovation that leads to substitution, recycling, and efficiency.

On the other hand, renewable resources can become exhausted if they are managed badly. This is especially apparent in biological resources, such as the passenger pigeon, American bison, and Atlantic cod. All these species once existed in extraordinary numbers, but within a few years, each was brought to the brink of extinction (or eliminated entirely) by overharvesting.

TABLE 15.5 | Important Ecological Services

We depend on our environment to continually provide:

1. A regulated global energy balance and climate; chemical composition of the atmosphere and oceans; water catchment and groundwater recharge; production and recycling of organic and inorganic materials; maintenance of biological diversity.

2. Space and suitable substrates for human habitation, crop cultivation, energy conversion, recreation, and nature protection.

3. Oxygen, fresh water, food, medicine, fuel, fodder, fertilizer, building materials, and industrial inputs.

4. Aesthetic, spiritual, historic, cultural, artistic, scientific, and educational opportunities and information.

SOURCE: Data from R. S. de Groot, *Investing in Natural Capital*, 1994.

Solar energy

Ecosystem

Recycle

Matter   Matter

Economy

Energy   Energy

Heat

☐ Human-made capital
☐ Natural capital

▲ **FIGURE 15.17** Ecological economics includes services such as recycling and resource provision and economic accounting. An effort is made to internalize, rather than externalize, natural resources and services. SOURCE: Herman Daly in A. M. Jansson et al., *Investing in Natural Capital.* ISEE.

Scarcity can lead to innovation

Are we about to run out of essential natural resources? In the 1970s a team of scientists from MIT, headed by Donnela Meadows, created a complex computer model of the world economy. They examined different rates of resource depletion, growing population, pollution, and industrial output. Most of their models predicted that our population will boom and bust as we run out of natural resources and drown in pollution, more or less like the population dynamics models we discussed in chapter 3. A typical run of their computer program from *The Limits to Growth* (1972) is shown in figure 15.19.

The prospect that unchecked population growth and overconsumption would lead to a crisis has had a powerful impact on environmentalists for many years. However, an underlying assumption of this view is the idea that the natural resources and ecological services on which we depend are irreplaceable. Ultimately, of course, many of them are irreplaceable, but that ultimate limit may be far off.

Many economists criticized this model because it underestimated technical progress and innovation that could mitigate the effects of scarcity. They point out that scarcity can stimulate research and development that result in new processes and materials that extend our resources significantly.

As we discussed in chapter 4, Thomas Malthus warned two centuries ago that food production couldn't continue to keep up with population growth, and that starvation, poverty, crime, and war were inevitable unless we slowed birth rates drastically. Malthus didn't foresee agricultural progress that would enable us to produce more food per person for today's world population of seven billion than was available for the one billion in his day. Similarly, one of the greatest worries in recent years is that we are approaching (or may have already passed) peak oil. How will a society so dependent on oil manage if our supply dries up? But as

▲ **FIGURE 15.20** Scarcity can stimulate innovation. As fossil fuel supplies are depleted, alternative, renewable energy sources, such as those being demonstrated here at a solar fair, become more feasible.

you know from chapter 13, there's more than enough renewable energy available to meet all our current needs. The fact that we're using up the cheap, easily extracted fossil fuel supplies is now making solar, wind, and other renewable energy technologies competitive (fig. 15.20). As Sheik Yamani, the former Saudi oil minister said, "The stone age didn't end because we ran out of stones, and the oil age won't end because we have run out of oil."

Communal property resources are a classic problem in economics

One of the difficulties of economics and resource management is that there are many resources we all share but nobody owns. Clean air, fish in the ocean, clean water, wildlife, and open space are all natural amenities that we exploit but that nobody clearly controls.

In 1968, biologist Garret Hardin wrote "*The Tragedy of the Commons*," an article describing how commonly held resources are degraded and destroyed by self-interest. Using the metaphor of the "commons," or community pastures in colonial New England villages, Hardin theorized that it behooves each villager to put more cows on the pasture. Each cow brings more wealth to the individual farmer, but the costs of overgrazing are shared by all. The individual farmer, then, suffers only part of the cost, but gets to keep all the profits from the extra cows she or he put on the pasture. Consequently, the commons becomes overgrazed, exhausted, and depleted. This dilemma is also known as the "free rider problem." The best solution, Hardin argued, is to either give coercive power to the government or privatize resources so that a single owner controls resource use.

This metaphor has been applied to many resources, especially to human population growth. According to this view, it benefits poor villagers to produce a few more children, but collectively these children consume all the resources available, making us all poorer in the end. The same argument has been applied to many resource overuse problems, such as depletion of ocean fisheries, pollution, African famines, and urban crime.

Recent critics have pointed out that what Hardin was really describing was not a commons, or collectively owned and managed resource, but an **open access system**, in which there are no rules to

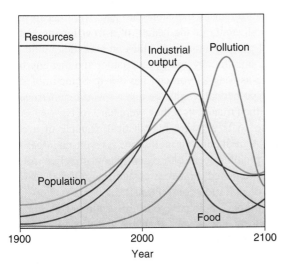

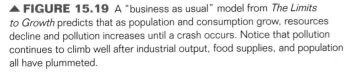

▲ **FIGURE 15.19** A "business as usual" model from *The Limits to Growth* predicts that as population and consumption grow, resources decline and pollution increases until a crash occurs. Notice that pollution continues to climb well after industrial output, food supplies, and population all have plummeted.

manage resource use. The work of Nobel laureate Elinor Ostrom and others shows that many common resources have been managed successfully for centuries by cooperative agreements among users. Native American management of wild rice beds, Swiss village-owned mountain forests and pastures, Maine lobster fisheries, and communal irrigation systems in Spain, Bali, Laos, and many other countries have all remained viable for centuries under communal management (fig. 15.21).

Each of these "commons," or **communal resource management systems**, shares a number of features: (1) community members have lived on the land or used the resource for a long time and anticipate that their children and grandchildren will as well, thus giving them a strong interest in sustaining the resource and maintaining bonds with their neighbors; (2) the resource has clearly defined boundaries; (3) the community group size is known and enforced; (4) the resource is relatively scarce and highly variable, so that the community is forced to be interdependent; (5) the management strategies appropriate for local conditions have evolved over time and are collectively enforced—that is, those affected by the rules have a say in them; (6) the resource and its use are actively monitored, discouraging anyone from cheating or taking too much; (7) conflict-resolution mechanisms reduce discord; and (8) incentives encourage compliance with rules, while sanctions for noncompliance keep community members in line.

Rather than being the only workable solution to problems in common-pool resources, privatization and increasing external controls often prove to be disastrous. Where villages have owned and operated local jointly held forests and fishing grounds for generations, nationalization and commodification of resources generally have led to rapid destruction of both society and ecosystems. Where communal systems once enforced restraint over harvesting, privatization encouraged narrow self-interest and allowed outsiders to take advantage of the weakest members of the community.

▲ **FIGURE 15.21** Communal irrigation systems on the island of Bali have been managed for centuries by village cooperatives called Subaks. A complex system coordinated by Hindu temple priests regulates water delivery so everyone shares.

15.4 NATURAL RESOURCE ACCOUNTING

Decision making about sustainable resource use often entails **cost-benefit analysis (CBA)**, the process of accounting and comparing the costs of a project and its benefits. Ideally this process assigns values to social and environmental effects of a given undertaking, as well as the value of the resources consumed or produced. However, the results of CBA often depend on how resources are accounted for and measured in the first place. CBA is one of the main conceptual frameworks of resource economics, and it is used by decision makers around the world as a way of justifying the building of dams, roads, and airports, as well as in considering what to do about biodiversity loss, air pollution, and global climate change. CBA is a useful way of rational decision making about these projects. It is also widely disputed because it tends to discount the value of natural resources, ecological services, and human communities, and it is used to justify projects that jeopardize all these resources.

In CBA the monetary value of all benefits of a project are counted up and compared with the monetary costs of the project. Usually the direct expenses of a project are easy to ascertain: how much will you have to pay for land, materials, and labor? The monetary worth of lost opportunities—to swim or fish in a river or to see birds in a forest—on the other hand, is much harder to appraise, as are inherent values of the existence of wild species or wild rivers. What is a bug or a bird worth, for instance, or the opportunity for solitude or inspiration? Eventually the decision maker compares all the costs and benefits to see whether the project is justified or whether an alternative action might bring greater benefit at less cost.

Critics of CBA point out its absence of standards, inadequate attention to alternatives, and the placing of monetary values on intangible and diffuse or distant costs and benefits. Who judges how costs and benefits will be estimated? How can we compare things as different as the economic gain from cheap power with loss of biodiversity or the beauty of a free-flowing river? Critics claim that placing monetary values on everything could lead to a belief that only money and profits matter and that any behavior is acceptable as long as you can pay for it. Sometimes speculative or even hypothetical results are given specific numerical values in CBA and then treated as if they were hard facts.

Figure 15.22 shows a hypothetical example of a cost-benefit analysis for reducing air pollution. As you can see, the initial efforts in pollution control are highly cost-effective. As more pollutants are removed, however, costs began to rise. At some point the costs of pollution control equal the benefits. Economists call this the optimum point. Beyond this optimum, the costs of pollution removal are greater than the benefits. As we've already noted, however, benefits can be intangible and widely dispersed, so they're often discounted by those who bear the costs.

Values such as wildlife, nonhuman ecological systems, and ecological services can be incorporated with natural resource accounting. In theory this accounting contributes to sustainable resource use because it can put a value on long-term or intangible goods that are necessary but often disregarded in economic decision making.

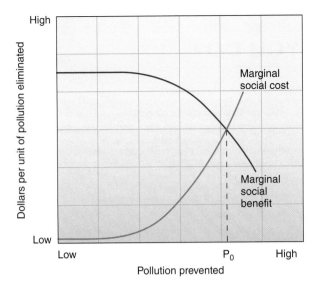

▲ FIGURE 15.22 To achieve maximum economic efficiency, regulations should require pollution prevention up to the optimum point (P_0) at which the costs of eliminating pollution just equal the social benefits of doing so.

Active LEARNING

Costs and Benefits

Figure 15.22 shows a hypothetical example of the costs and benefits of reducing air pollution.

1. For the first 25 percent (or so) of reduction (on the X axis), the benefits are uniformly high. What are some of the economic benefits that might be represented by this part of the graph?

2. What might be some of the low-cost pollution reduction steps represented by the "costs" curve in the first part of the graph?

3. Why is the P_0 considered the optimum point?

ANSWERS: 1. Benefits include improved health; less damage to buildings, crops, and materials; and improved quality of life. 2. Conservation. Improved planning and efficiency in transportation, buildings, and energy production; replacing old, inefficient industrial plants and vehicles. 3. Because it's the point at which the costs just equal the benefits.

One important part of natural resource accounting is assigning a value to ecological services (table 15.6). The total value of nature's services of $33.3 trillion per year is about half the current annual world GDP. Another is using alternative measures of wealth and development. As mentioned earlier, GDP is a widely used measure of wealth that is based on rates of consumption and throughput. It doesn't account, however, for natural resource depletion or ecosystem damage.

The World Resources Institute, for example, estimates that soil erosion in Indonesia reduces the value of crop production there

about 40 percent per year. If natural capital were taken into account, Indonesian GDP would be reduced by at least 20 percent annually. Similarly, Costa Rica experienced impressive increases in timber, beef, and banana production between 1970 and 1990. But decreased natural capital during this period, represented by soil erosion, forest destruction, biodiversity losses, and accelerated water runoff, added up to at least $4 billion, or about 25 percent of annual GDP. A number of countries, including Canada and China, now use a "green GDP" that measures environmental costs as part of economic accounting.

Internalizing external costs

One of the factors that can make resource-exploiting enterprises look good in cost-benefit analysis is externalizing costs. **Externalizing costs** is the act of disregarding or discounting resources or goods that contribute to producing something but for which the producer does not actually pay. Usually external costs are diffuse and difficult to quantify. Generally they belong to society at large, not to the individual user. When a farmer harvests a crop in the fall, for example, the values of seeds, fertilizer, and the sale of the crop are tabulated; the values of soil lost to erosion, water quality lost to nonpoint-source pollution, and depleted fish populations are not accounted for. These are most often costs shared by the whole society, rather than borne by the resource user. They are external to the accounting system, and they are generally ignored in cost-benefit analysis—or when the farmer evaluates whether the year was profitable. Larger enterprises, such as dam building, logging, and road building, generally externalize the cost of ecological services lost along the way.

One way to use the market system to optimize resource use is to make sure that those who reap the benefits of resource use also bear all the external costs. This is referred to as **internalizing costs**. Calculating the value of ecological services or diffuse pollution is not easy, but it is an important step in sustainable resource accounting.

ECOSYSTEM SERVICES	VALUE (TRILLION $ U.S.)
Soil formation	17.1
Recreation	3.0
Nutrient cycling	2.3
Water regulation and supply	2.3
Climate regulation (temperature and precipitation)	1.8
Habitat	1.4
Flood and storm protection	1.1
Food and raw materials production	0.8
Genetic resources	0.8
Atmospheric gas balance	0.7
Pollination	0.4
All other services	1.6
Total value of ecosystem services	33.3

TABLE 15.6 Estimated Annual Value of Ecological Services

SOURCE: Adapted from R. Costanza, et al., "The Value of the World's Ecosystem Services and Natural Capital," in *Nature*, vol. 387, 1997.

New approaches measure real progress

A number of systems have been proposed as alternatives to GNP that reflect genuine progress and social welfare. The economist Herman Daly and philosopher John Cobb proposed a **genuine progress index (GPI)**, which takes into account real per capita income, quality of life, distributional equity, natural resource depletion, environmental damage, and the value of unpaid labor. They point out that, while per capita GDP in the United States nearly doubled between 1970 and 2005, per capita GPI increased only 4 percent (fig. 15.23). Some social service organizations would add to this index the costs of social breakdown and crime, which would decrease real progress even further over this time span. Bhutan measures Gross Domestic Happiness as their indicator of progress.

The United Nations Development Programme (UNDP) uses a benchmark called the **human development index (HDI)** to track social progress. HDI incorporates life expectancy, educational attainment, and standard of living as critical measures of development. Gender issues are accounted for in the gender development index (GDI), which is simply HDI adjusted or discounted for inequality or achievement between men and women.

In its annual Human Development Report, the UNDP compares country-by-country progress. As you might expect, the highest development levels are generally found in North America, Europe, and Japan. In 2012 Norway ranked first in the world in both HDI and GDI. The United States was fourth in HDI but it was 49th in a measure of environmental

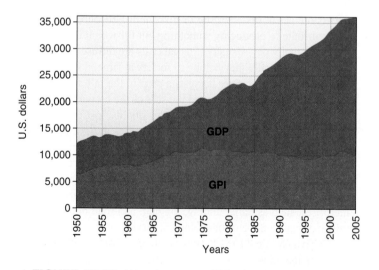

▲ **FIGURE 15.23** Although per capita GDP in the United States nearly doubled between 1970 and 2000 in inflation-adjusted dollars, a genuine progress index that takes into account natural resource depletion, environmental damage, and options for future generations hardly increased at all.

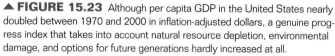

What Can YOU DO?

Personally Responsible Consumerism

Each of us can do many things to lower our ecological impacts and support "green" businesses through responsible consumerism and ecological economics.

- Practice living simply. Ask yourself if you really need more material goods to make your life happy and fulfilled.

- Rent, borrow, or barter when you can. Can you reduce the amount of stuff you consume by renting, instead of buying, machines and equipment you actually use only rarely?

- Recycle or reuse building materials: doors, windows, cabinets, appliances. Shop at salvage yards, thrift stores, yard sales, or other sources of used clothes, dishes, appliances, etc.

- Consult the *National Green Pages* from Co-Op America for a list of eco-friendly businesses. Write to companies from which you buy goods or services and ask them what they are doing about environmental protection and human rights.

- Buy "green" products. Look for efficient, high-quality materials that will last and that are produced in the most environmentally friendly manner possible. Subscribe to clean-energy programs if they are available in your area. Contact your local utility and ask that it provide this option if it doesn't now.

- Buy locally grown or locally made products made under humane conditions by workers who receive a fair wage.

- Think about the total life-cycle costs of the things you buy, especially big purchases, such as cars. Try to account for the environmental impacts, energy use, and disposal costs, as well as initial purchase price.

- Stop junk mail. Demand that your name be removed from mass-mailing lists.

- Invest in socially and environmentally responsible mutual funds or "green" businesses when you have money for investment.

protection and governance. The 29 countries with the lowest HDI in 2012 were all in Africa. Haiti was the lowest outside Africa.

Although poverty remains widespread in many places, encouraging news also can be found in development statistics. Poverty has fallen more in the past 50 years, the UNDP reports, than in the previous 500 years. Child death rates in developing countries as a whole have been more than halved. Average life expectancy has increased by 30 percent, while malnutrition rates have declined by almost a third. The proportion of children who lack primary school has fallen from more than half to less than a quarter. And the share of rural families without access to safe water has fallen from nine-tenths to about one-quarter.

Some of the greatest progress has been made in Asia. China and a dozen other countries with populations that add up to more than 1.6 billion have decreased the proportion of their people living below the poverty line by half. Overall, according to the World Bank, the number of people living in extreme poverty (less than US$1.25/day) has fallen from 1.8 billion in 1990 to about 900 million in 2012. Even in sub-Saharan Africa, the number of people living in poverty has dropped below 50 percent of the population for the first time ever.

15.5 TRADE, DEVELOPMENT, AND JOBS

Trade can be a powerful tool in expanding our resources and raising standards of living. Think of the things you now enjoy that might not be available if you had to live exclusively on what's available in your immediate neighborhood. Too often, however, the poorest, least powerful people suffer in this global marketplace. The banking and trading systems that regulate credit, currency exchange, shipping rates, and commodity prices were set up by the richer and more powerful nations in their own self-interest. The General Agreement on Tariffs and Trade (GATT) and World Trade Organization (WTO) agreements, for example, negotiated primarily between the largest industrial nations, regulate 90 percent of all international trade.

These systems often keep the less-developed countries in a perpetual role of resource suppliers to the more-developed countries. The producers of raw materials, such as mineral ores or agricultural products, get very little of the income generated from international trade. As a prerequisite for international development loans, the IMF frequently requires debtor nations to adopt harsh "structural adjustment" plans that slash welfare programs and impose cruel hardships on poor people. The WTO has issued numerous rulings that favor international trade over pollution prevention or protection of endangered species.

Microlending helps the poorest of the poor

No single institution has more influence on financing and policies of developing countries than the World Bank. Of some $25 billion loaned each year for development projects by international agencies, about two-thirds comes from the World Bank. Founded in 1945 to fund reconstruction of Europe and Japan, the World Bank shifted its emphasis to aid developing countries in the 1950s. Many of its projects have had adverse environmental and social effects, however. Its loans often go to corrupt governments and fund ventures such as nuclear power plants, huge dams, and giant water diversion schemes. Former U.S. treasury secretary Paul O'Neill said that these loans have driven poor countries "into a ditch" by loading them with unpayable debt. He said that funds should not be loans, but rather grants to fight poverty.

Global aid from the WTO usually aids banks and industries more than it helps the impoverished populations who most need assistance. Often structural adjustment leads the poorest to pay back loans negotiated by their governments and industries. These concerns led Dr. Muhammad Yunus of Bangladesh to initiate the microlending plan of the Grameen Bank (see What Do You Think?, p. 372).

Microlending programs have assisted billions of people—most of them low-status women who have no other way to borrow money at reasonable interest rates. This model is now being used by hundreds of other development agencies around the world. Even in the United States, organizations assist micro-enterprises with loans, grants, and training. The Women's Self-Employment Project in Chicago, for instance, teaches job skills to single mothers in housing projects. Similarly, "tribal circle" banks on Native American reservations successfully finance microscale economic development ventures.

One of the most important innovations of the Grameen Bank is that borrowers take out loans in small groups. Everyone in the group is responsible for the others' performance. The group not only guarantees loan repayment, it helps businesses succeed by offering support, encouragement, and advice. Where banks depend on the threat of foreclosure and a low credit rating to ensure debt repayment, the Grameen Bank has something at least as powerful for poor villagers—the threat of letting down your neighbors and relatives. Becoming a member of a Grameen group also requires participation in a savings program that fosters self-reliance and fiscal management.

The process of running a successful business and repaying the loan transforms many individuals. Women who previously had little economic power, influence, or self-esteem are empowered with a sense of pride and accomplishment. Dr. Yunus also discovered that money going to families through women helped the families much more than the same amount of money in men's salaries. Women were more likely to spend money on children's food or education, producing generational benefits with the increased income.

The most recent venture for the Grameen Bank is providing mobile phone service to rural villages. Supplying mobile phones to poor women not only allows them to communicate, it provides another business opportunity. They rent out their phones to neighbors, giving the owner additional income and linking the whole village to the outside world. Suddenly people who had no access to communication can talk with their relatives, order supplies from the city, check on prices at the regional market, and decide when and where to sell their goods and services. This is a great example of "bottom-up development." Founded in 1996, Grameen Phone now has 32 million subscribers and is Bangladesh's largest mobile phone company.

Active LEARNING

Try Your Hand at Microlending

The best way to observe microlending at work is to try it out. Collect donations of $1–$5 from people in your class, until the total is $25. Go to www.kiva.org, a microlending organization that pools small loans, and select a business to support. You can use PayPal to send the money, and for the next year, you will receive periodic reports on how the business is going. To evaluate whether you're getting a good rate of return, consider that for most stock market investments, about 5–10 percent annual return is reasonable.

1. What is 5–10 percent of $25?

Also poll the class to get these averages:

2. What percentage interest do you earn from your bank accounts?

3. What percentage do you pay to credit card companies?

ANSWERS: 1. $1.25–$2.50. 2., 3. Answers will vary, but probably <2% and >15%.

Loans That Change Lives

Ni Made is a young mother of two children who lives in a small Indonesian village. Her husband is a day laborer who makes only a few dollars per day—when he can find work. To supplement their income, Made goes to the village market every morning to sell a drink she makes out of boiled pandamus leaves, coconut milk, and pink tapioca. A small loan would allow her to rent a covered stall during the rainy season and to offer other foods for sale. The extra money she could make could change her life, but traditional banks consider Made too risky to lend to, and the amounts she needs too small to bother with.

Around the world, billions of poor people find themselves in the same position as Made; they're eager to work to build a better life for themselves and their families, but lack resources to succeed. Now, however, a financial revolution is sweeping around the world. Small loans are becoming available to the poorest of the poor. This new approach was invented by Dr. Muhammad Yunus, professor of rural economics at Chittagong University in Bangladesh. Talking to a woman who wove bamboo mats in a village near his university, Dr. Yunus learned that she had to borrow the few taka she needed each day to buy bamboo and twine. The exorbitant interest rate charged by the village moneylenders consumed nearly all her profits. Always living on the edge, this woman, and many others like her, couldn't climb out of poverty.

To break this predatory cycle, Dr. Yunus gave the woman and several of her neighbors small loans totaling about 1,000 taka

▼ A small amount of seed money would allow this young mother to expand her business and help provide for her family.

(about $20). To his surprise, the money was paid back quickly and in full. So he offered similar amounts to other villagers with similar results. In 1983, Dr. Yunus started the Grameen (village) Bank to show that "given the support of financial capital, however small, the poor are fully capable of improving their lives." His experiment has been tremendously successful. In 2006, Dr. Yunus won the Nobel Peace Prize for his work. By 2012 the Grameen Bank had more than 8.5 million customers, 96 percent of them women. It had loaned more than US$11.7 billion with 97 percent repayment, nearly twice the collection rate of commercial Bangladesh banks.

The Grameen Bank provides credit to poor people in rural Bangladesh without the need for collateral. It depends, instead, on mutual trust, accountability, participation, and creativity of the borrowers themselves. Microcredit is now being offered by hundreds of organizations in 43 other countries—including the United States. Institutions from the World Bank to religious charities make small loans to worthy entrepreneurs. Wouldn't you like to be part of this movement? Well, now you can. You don't have to own a bank to help someone in need.

A brilliant way to connect entrepreneurs in developing countries with lenders in wealthy countries is offered by Kiva, a San Francisco-based technology start-up. The idea for Kiva, which means "unity" or "cooperation" in Swahili, came from Matt and Jessica Flannery. Jessica had worked in East Africa with a California nonprofit that provides training, capital, and mentoring to small businesses in developing countries. Jessica and Matt wanted to help some of the people she had met, but they weren't wealthy enough to get into microfinancing on their own. Joining with four other young people with technology experience, they created Kiva, which uses the power of the Internet to help the poor.

Kiva partners with about 150 development nonprofits with staff in 60 developing countries. The partners identify hardworking entrepreneurs who deserve help. They then post a photo and brief introduction to each one on the Kiva webpage. You can browse the collection to find someone whose story touches you. The minimum loan is generally $25. Your loan is bundled with that of others until it reaches the amount needed by the borrower. You make your loan using your credit card (through PayPal, so it's safe and easy). Your loan is generally repaid within 12 to 18 months (although without interest). At that point, you can either withdraw the money, or use it to make another loan.

The in-country staff keeps track of the people you're supporting and monitors their progress, so you can be confident that your money will be well used. Loan requests often are on their webpage for only a few minutes before being filled. In just seven years, Kiva raised more than $280 million from 685,000 lenders to help 720,000 entrepreneurs around the world fulfill their dreams. Wouldn't you like to take part in this innovative person-to-person human development project? Check out Kiva.org.

Market mechanisms can reduce pollution

As you've read by now, global climate change is probably the most serious environmental problem we face. In 2006 the British economist Sir Nicolas Stern warned us that the damage caused by climate change 50 years from now could be equivalent to losing as much as 20 percent of the global GDP every year. Or to put it another way, every dollar we invest now to reduce greenhouse gas emissions will save $20 in the future.

But how can we bring about a transition to a low-carbon economy? Many environmentalists would like a prescriptive approach: "Just stop burning fossil fuels." But as we've seen in many areas, telling people not to do something just encourages them to do it.

Many economists believe market forces can reduce pollution more efficiently than rules and regulations. Assessing a tax on each ton of carbon emitted, for example, encourages businesses to reduce greenhouse gases, but still allows them to search for the most cost-effective ways to achieve this goal. And the incentives promote a continuing search for even better ways to reduce emissions. The more you reduce your discharges, the more you save.

Another approach is to set up a **cap-and-trade** system. The first step is to mandate upper limits (the cap), on how much each country, sector, or specific industry is allowed to emit. Companies that can cut pollution by more than they're required to can sell the credit to other companies that have more difficulty meeting their mandated levels. This sets up the same kind of incentives as a tax, but doesn't require as much government oversight.

We have a good model for cap-and-trade in sulfur emission. A market created in 1990 has been remarkably effective in reducing sulfur pollution. Most observers agree that this market has been much more cost-effective than rigid rules and directives would have been.

15.6 GREEN BUSINESS AND GREEN DESIGN

Businesspeople and consumers are increasingly aware of the unsustainability of producing the goods we use every day. Recently, business innovators have tried to develop green enterprises that produce environmentally and socially sound products. "Green" companies, such as the Body Shop, Patagonia, Aveda, Malden Mills, and Johnson and Johnson, have shown that operating according to the principles of sustainable development and environmental protection can be good for public relations, employee morale, and sales (table 15.7).

Green business works because consumers are becoming aware of the ecological consequences of their purchases. Increasing interest in environmental and social sustainability has caused an explosive growth of green products. The *National Green Pages* published by Co-Op America currently lists more than 2,000 green companies. You can find eco-travel agencies, telephone companies that donate profits to environmental groups, entrepreneurs selling organic foods, shade-grown coffee, straw-bale houses, paint thinner made from orange peels, sandals made from recycled auto tires, and a plethora of hemp products, including burgers, ale, clothing, shoes, rugs, and shampoo.

TABLE 15.7 | Goals for an Eco-Efficient Economy

- Introduce no hazardous materials into the air, water, or soil.
- Measure prosperity by how much natural capital we can accrue in productive ways.
- Measure productivity by how many people are gainfully and meaningfully employed.
- Measure progress by how many buildings have no smokestacks or dangerous effluents.
- Make the thousands of complex governmental rules that now regulate toxic or hazardous materials unnecessary.
- Produce nothing that will require constant vigilance from future generations.
- Celebrate the abundance of biological and cultural diversity.
- Live on renewable solar income rather than fossil fuels.

Although these eco-entrepreneurs represent a tiny sliver of the $7 trillion per year U.S. economy, they often are pioneers in developing new technologies and offering innovative services. Markets also grow over time: organic food marketing has grown from a few funky local co-ops to a $27 billion market segment. Most supermarket chains now carry some organic food choices. Similarly, natural-care health and beauty products reached $8.5 billion in sales in 2011 out of a $55 billion industry. By supporting these products, you can ensure that they will continue to be available and, perhaps, even help expand their penetration into the market.

Corporations committed to eco-efficiency and clean production include such big names as Monsanto, 3M, DuPont, and Duracell. Applying the famous three *R*s—reduce, reuse, recycle—these firms have saved money and gotten welcome publicity. Savings can be substantial. Pollution-prevention programs at 3M, for example, have saved $857 million over the past 25 years. In a major public relations achievement, DuPont has cut its emissions of airborne cancer-causing chemicals almost 75 percent since 1987. Small operations can benefit as well.

Green design is good for business and the environment

Architects are starting to get on board the green bandwagon, too. Acknowledging that heating, cooling, lighting, and operating buildings is one of our biggest uses of energy and resources, architects such as William McDonough are designing "green office" projects. Among McDonough's projects are the Environmental Defense Fund headquarters in New York City; the Environmental Studies Center at Oberlin College in Ohio; the European headquarters for Nike in Hilversum, the Netherlands; and the Gap corporate offices in San Bruno, California (fig. 15.24). Each uses a combination of energy-efficient designs and technologies, including natural lighting and efficient water systems (table 15.8).

The Gap office building, for example, is intended to promote employee well-being and productivity, as well as efficiency. It has high ceilings, abundant skylights, windows that open, a full-service

TABLE 15.8 | McDonough Design Principles

Inspired by the way living systems actually work, Bill McDonough offers three simple principles for redesigning processes and products:

- *Waste equals food.* This principle encourages elimination of the concept of waste in industrial design. Every process should be designed so that the products themselves, as well as leftover chemicals, materials, and effluents, can become "food" for other processes.

- *Rely on current solar income.* This principle has two benefits: First, it diminishes, and may eventually eliminate, our reliance on hydrocarbon fuels. Second, it means designing systems that sip energy rather than gulping it down.

- *Respect diversity.* Evaluate every design for its impact on plant, animal, and human life. What effects do products and processes have on identity, independence, and integrity of humans and natural systems? Every project should respect the regional, cultural, and material uniqueness of its particular place.

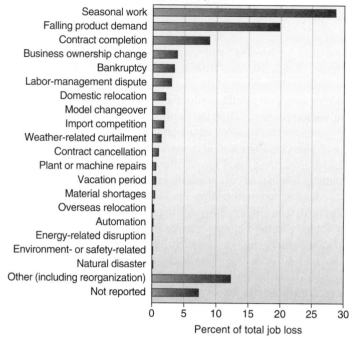

Reason for Layoffs

▲ **FIGURE 15.25** Although opponents of environmental regulation often claim that protecting the environment costs jobs, studies by economist E. S. Goodstein show that only 0.1 percent of all large-scale layoffs in the United States were the result of environmental laws. SOURCE: Data from E. S. Goodstein, Economic Policy Institute, Washington, D.C.

fitness center (including pool), and a landscaped atrium for each office bay that brings the outside in. The roofs are covered with native grasses. Warm interior tones and natural wood surfaces (all wood used in the building was harvested by certified sustainable methods) give a friendly feeling. Paints, adhesives, and floor coverings are low-toxicity, and the building is one-third more energy-efficient than strict California laws require. The pleasant environment helps improve employee effectiveness and retention. Gap, Inc., estimates that the increased energy and operational efficiency will have a four- to eight-year payback.

Environmental protection creates jobs

For years business leaders and politicians have portrayed environmental protection and jobs as mutually exclusive. They claim that pollution control, protection of natural areas and endangered species, and limits on use of nonrenewable resources will strangle the economy and throw people out of work. Ecological economists dispute this claim, however. Their studies show that only 0.1 percent of all large-scale layoffs in the United States in recent years were due to government regulations (fig. 15.25). Environmental protection, they argue, is not only necessary for a healthy economic system; it actually creates jobs and stimulates business.

Green businesses often create far more jobs and stimulate local economies far more than environmentally destructive ones. Wind energy, for example, provides about five times as many jobs per kilowatt-hour of electricity than does coal-fired power (chapter 13).

As chapter 13 shows, China has emerged as the world leader in sustainable energy. Recognizing the multibillion-dollar economic potential of "green" business, China is investing at least $8 billion per year on research and development, and now it is selling about $12 billion worth of equipment and services per year worldwide. Japan, also, is marketing advanced waste incinerators, pollution-control equipment, alternative energy sources, and water treatment systems. Superefficient "hybrid" gas-electric cars are helping Japanese automakers flourish, while U.S. corporations that once dominated the world slide into bankruptcy. Unfortunately, the United States has been resisting international pollution-control conventions, rather than recognizing the potential for economic growth and environmental protection in the field of "green" business.

CONCLUSION

More than half of us now live in cities, and in a generation three-quarters of us probably will be urban dwellers. Most of this

▼ **FIGURE 15.24** The award-winning Gap, Inc. corporate offices in San Bruno, California, demonstrate some of the best features of environmental design. A roof covered with native grasses provides insulation and reduces runoff. Natural lighting, an open design, and careful relation to its surroundings all make this a pleasant place to work.

growth will occur in the large cities of developing countries, where resources are already strained. Cities draw immigrants from the countryside by offering jobs, social mobility, education, and other opportunities unavailable in rural areas. Weak rural economies and a lack of access to land also push people to cities. Shortages of water and housing are especially urgent problems in fast-growing cities of the developing world. Illegal shantytowns often develop as people seek a place to live on the outskirts of cities. Crime and pollution plague residents of these developments, who often have nowhere else to go.

Urban planning tries to minimize the strains of urbanization. Good planning saves money, because providing roads and services to sprawling suburbs is very expensive. Transportation, which is key to our economy, is a critical part of planning because it determines how far-flung urban development will be. Smart growth, cluster development, and improved standards for environmentally conscious building are also important in urban planning.

Economic policies are often at the root of the success or failure of cities. These policies build from some basic sets of assumptions about the nature of resources. Classical economics assumes that resources are finite, so that we compete to control them. Neoclassical economics assumes that resources are based on capital, which can include knowledge and social capital as well as resources, and that constant growth is both possible and essential. Natural resource economics extends neoclassical ideas to internalize the value of ecological services in economic accounting.

The "tragedy of the commons" is a classic description of our inability to take care of public resources. Subsequent explanations have pointed that collective rules of ownership are necessary for the survival of shared resources. Our ability to agree on these rules appears to depend on a number of factors, including the scarcity of the resource and our ability to monitor its use.

Because classic productivity indices count many social and environmental ills as positive growth, alternatives have been proposed, including the genuine progress index (GPI). Measures like the GPI can be used to help ensure fair and responsible growth in developing areas. Microlending is another innovative strategy that promotes equity in economic growth. Green business and green design are fast-growing parts of many economies. These approaches save money by minimizing consumption and waste.

PRACTICE QUIZ

1. How many people now live in urban areas?
2. How many cities were over 1 million in 1900? How many are now?
3. Why do people move to urban areas?
4. What is the difference between a shantytown and a slum?
5. Define *sprawl*.
6. In what ways are cities ecosystems?
7. Define *smart growth*.
8. Describe a "green" roof.
9. Describe a few ways in which Vauban is self-sufficient and sustainable.
10. Define *sustainable development*.
11. Briefly summarize the differences in how neoclassical and ecological economics view natural resources.
12. What is the estimated economic value of all the world's ecological services?
13. How is it that nonrenewable resources can be extended indefinitely, while renewable resources are exhaustible?
14. In your own words, describe what is shown in figure 15.22.
15. What's the difference between open access and communal resource management?
16. Describe the genuine progress index (GPI).
17. What is *microlending*?

CRITICAL THINKING AND DISCUSSION

Apply the principles you have learned in this chapter to discuss these questions with other students.

1. Some people—especially automakers—claim that Americans will never give up their cars. Do you agree? What might persuade you to change to a car-free lifestyle?
2. This chapter presents a number of proposals for suburban redesign. Which of them would be appropriate or useful for your community? Try drawing up a plan for the ideal design of your neighborhood.
3. A city could be considered an ecosystem. Using what you learned in chapters 2 and 3, describe the structure and function of a city in ecological terms.
4. If you were doing a cost-benefit study, how would you assign a value to the opportunity for good health or the existence of rare and endangered species in faraway places? Is there a danger or cost in simply saying some things are immeasurable and priceless and therefore off limits to discussion?
5. What would be the effect on the developing countries of the world if we were to change to a steady-state economic system? How could we achieve a just distribution of resource benefits while still protecting environmental quality and future resource use?
6. When an ecologist warns that we are using up irreplaceable natural resources and an economist rejoins that ingenuity and enterprise will find substitutes for most resources, what underlying premises and definitions shape their arguments?

Urbanization and economic growth are two closely related changes going on in societies today. How are these two processes related? How do they compare in different parts of the world? Is the United States more urbanized or less so than other regions? How do economic growth rates compare in different regions?

Gapminder.org is a rich source of data on global population, health, and development that we examined in the Data Analysis exercise for chapter 4. The site includes animated graphs showing change over time. Go to Connect to find a link to Gapminder graphs, and answer questions about what they tell you.

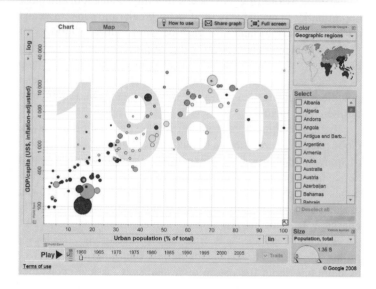

TO ACCESS ADDITIONAL RESOURCES FOR THIS CHAPTER, PLEASE VISIT CONNECT AT www.mcgrawhillconnect.com.

You will find LearnSmart, an adaptive learning system, Google Earth™ exercises, additional Case Studies, Data Analysis exercises, and an interactive ebook.

Hundreds of Indian students created an elephant threatened by climate change. Intended to be visible from space, these EARTH art projects are grassroots efforts to create a new sense of urgency and of possibility for our planet.

LEARNING OUTCOMES

After studying this chapter, you should be able to answer the following questions:

▶ What is environmental policy, and how is it formed?

▶ What is NEPA, and what does it do?

▶ Describe some of the important U.S. environmental laws.

▶ Describe several international environmental laws and conventions.

▶ How do the different branches of government influence environmental policy?

▶ Explain some of the ways students can contribute to environmental protection.

▶ What are the Millennium Development Goals?

350.org: Making a Change

Faced with environmental challenges on all sides, many students want to stand up and do something useful. It's hard to know how to start, though. Is it possible for students to implement real change on issues they care about? This is the question that faced a small group of students at Middlebury College in Vermont several years ago. The group was concerned about climate change, the most important environmental issue today. Was there anything they could do to mobilize action and make a difference? With a combination of creativity, energy, tenacity, and a sense of humor, this small group started a movement that has promoted education, understanding, and community building worldwide.

It all started when author Bill McKibben, writer in residence at Middlebury College, organized a group of students to take on the problem of global climate change. Previous popular movements, such as the civil rights movement, had gained momentum with national rallies, concerts, and media events. But the students didn't have the money or resources to put together a national rally. Furthermore, it didn't seem right to encourage thousands of people to fly or drive long distances to protest excessive burning of fossil fuels.

They decided instead to use the power of the Internet and social networks. They could mobilize activists to hold meaningful and newsworthy events in their own neighborhoods. By acting together, all on the same day, and by publishing photos and news releases to show the interconnections between actions, they could create a meta-event, one with greater power to influence local citizens and decision makers than a single event in one place might have.

The event they created was called Step It Up. In 2007, with a minuscule budget and little previous experience, they inspired tens of thousands of citizens to participate in more than 1,400 events in iconic places in all 50 of the United States. These creative actions—from skiing down a melting glacier to protest global warming, to planting endangered chestnut trees to absorb CO_2, to flying thousands of handmade kites with environmental messages—were designed to both attract attention and educate the public about the need to cut carbon emissions 80 percent by 2050.

This first event was so successful that the group broadened the campaign to the international stage. They adopted a new name, 350.org, referring to 350 parts per million, the concentration of atmospheric CO_2 that climate scientists say is the upper limit if we are to avoid catastrophic climate change in the coming decades. (The name also makes it easy to find the group online.) Since we're already approaching 400 ppm, and politicians are debating whether we might hold emissions to 450 ppm, the students have chosen an ambitious goal in trying to get us back to 350 ppm. But why not dream big?

On October 24, 2009, the 350 team mobilized 5,200 events in 181 countries. CNN called it "the most widespread day of political action in the planet's history." On October 10, 2010 (10/10/10),

they organized a "global work party" with more than 7,000 projects in 187 countries. People put up solar panels, dug community gardens, planted trees, and did other positive actions to help reduce carbon emissions. Working together on pragmatic local projects empowers people, gives them hope and purpose, and helps build grassroots networks.

One of the smart strategies of 350.org has been to focus on young people. Unlike many in older generations, young people have energy, creativity, and an obvious motivation to care about the future.

One of the most creative projects the team has launched is 350 EARTH ART. So far, more than 15 major art installations have been created involving thousands of people. Each is designed to be visible from space (fig. 16.1). Not only did the art pieces turn out beautifully, they captured media attention and demonstrated to political leaders a widespread desire for environmental protection. These art projects also unleash creativity, get people motivated, and offer a hopeful way to express opinions about the future of our world. You can get involved, too. Talk to your friends, and go to 350.org for ideas about planning an event, communicating with media, inviting elected officials, and other useful resources.

In this chapter we'll look at a few of the countless ways individuals and groups are working to influence environmental policy, protect the earth, and build a sustainable future. ∎

▼ **FIGURE 16.1** On 10/10/10, over a thousand New Mexicans of all ages flooded the Santa Fe River's dry riverbed with blue-painted cardboard and other blue materials to show where the River should be flowing.

16.1 ENVIRONMENTAL POLICY AND LAW

People who engage in public actions like those of 350.org seek to influence public policy. That is, they seek to shape the rules or decisions that influence how we act, as individuals and as a society. Policy acts at all scales: you might have a policy always to get your homework done on time. On a national level, we have policies to protect property, individual rights, public health, and other priorities. International policies and agreements regulate international trade in hazardous chemicals or in endangered species. Recently there have been efforts to enact policies aimed at slowing climate change—such as EPA rules on carbon dioxide emissions and incremental increases in funding for alternative energy development. Do those policies do enough? Do they go too far? Those are questions that we debate in a democratic society. The more you know about policy formation, the more you will understand about the rules that shape our use of environmental resources.

Environmental policy involves rules designed to protect natural resources and public health. The safety of our air, drinking water, and food are all protected by laws developed by past generations of voters and policy makers. Access to public lands and public waterways is also protected by laws. Most of the time we forget about these rules, and we take these protections for granted.

Theoretically, in a democratic system, these policies are established through negotiation and compromise. Open debate allows all voices to be heard, and policy decisions promote collective well-being. Elected representatives defend policies they think will benefit their constituents. In fact, the rules that result from political wrangling are imperfect, and usually nobody is completely satisfied. But ideally rules need to be reasonably palatable to a majority of voters in order to pass a vote. The best way to ensure that policies serve the general public interest is to ensure that the public is active, well informed, and involved in policy making.

What drives policy making?

Power and influence inevitably control much of our policy making. Economic interest groups, industry associations, and powerful individuals often have disproportionate access to lawmakers. Public interest groups work to gain similar access by developing broad support, by bringing citizens together to write letters or meet with legislators, and by drawing attention to a cause or an issue, in hopes that voters contact their legislators (fig. 16.2).

Even though self-interest is always present in politics, public citizenship is also a powerful force. Altruistic and community-oriented impulses are widespread, as shown by actions of groups like 350.org. These motivations are evident in many of our current public policies, such as the Clean Water Act, the Clean Air Act, the

▲ **FIGURE 16.2** This Moving Planet march was one of 2,000 events organized by 350.org in 175 countries in 2011. Actions, such as this, can be an effective way to get a message out to influence decision makers.

Voting Rights Act, and many other laws that defend public interests and collective well-being.

Citizen movements for environmental quality often have had far-reaching effects. For example, environmental protests helped precipitate the collapse of the Soviet Union in 1992, when political dissent wasn't allowed but environmental demonstrations were acceptable. Similarly, China's government has recently modified many policies in response to environmental complaints. In 2011 China had 70,000 public events to protest pollution, environmental health issues, and other abuses. A campaign led by artists, students, and writers forced the government to cancel plans for a series of 13 large dams on the Nu River in mountainous southwestern China, in an area of extraordinary biological and cultural diversity. There are now more than 2,000 social and environmental organizations in China, and for the first time they have officially recognized status.

Globally, support for environmental protection is widespread. A 2007 BBC poll of 22,000 residents of 21 countries found that 70 percent of respondents said they were personally ready to make sacrifices to protect the environment (fig. 16.3). Overall,

% Yes (combination of "definitely" and "probably" necessary)
% No (combination of "definitely not" and "probably not" necessary)

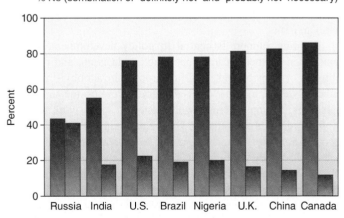

▲ **FIGURE 16.3** About 70 percent of the 22,000 people in 21 countries polled by the BBC in 2007 agreed with the statement, "I am ready to make significant changes to the way I live to help prevent global warming or climate change.

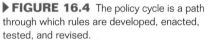

▶ FIGURE 16.4 The policy cycle is a path through which rules are developed, enacted, tested, and revised.

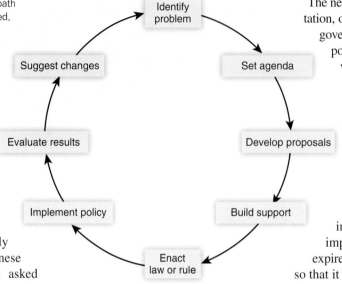

The next step in a policy cycle is implementation, or carrying out the new rules. Ideally, government agencies faithfully carry out policy directives as they provide services and enforce rules and regulations. Continued public attention is needed to make sure the government enforces its own rules. Many of our worst problems in air pollution, and related respiratory ailments, for example, could be controlled if we enforced existing air quality rules more effectively.

Once a rule is enacted, it almost invariably requires reevaluation and improvement after time. Often laws expire after a designated number of years, so that it is necessary to "reauthorize" a law, or vote again to continue it.

83 percent agreed that individuals would definitely or probably have to make lifestyle changes to reduce the amount of climate-changing gases they produce. Concern about environmental quality varied by country, however: just over 40 percent of Russians polled were willing to change their lifestyle to prevent global warming, compared to nearly 90 percent of Canadians. The Chinese were the most enthusiastic when asked about energy taxes to prevent climate change. Eighty-five percent of the Chinese polled agreed that such taxes are necessary.

Policy creation is ongoing and cyclic

How do policy issues and options make their way onto the stage of public debate? We often describe policy development as a cycle (fig. 16.4): A problem is identified. Plans to resolve the problem are developed through discussion, and new rules are proposed. Popular support is built to gather votes for the new rule. If the rule is agreed to, then it is implemented. Evaluation then leads to identification of flaws in the rule, and the cycle starts again.

Problem identification might be done by an individual, a group, or elected officials. A citizens group concerned about climate change or about polluted water might contact representatives to ask for pollution controls. Hunting and fishing groups might be concerned about the loss of wetland habitat, and they might propose new protections for wetlands and habitat. Often industry groups propose or write new rules that reduce the costs of doing business. Seizing the initiative in issue identification can allow a group to define terms, set the agenda, and legitimate (or delegitimate) an issue.

Building support is central to policy development. Proponents often do media campaigns, public education, and personal lobbying of decision makers. Often groups hire a lobbyist who can dedicate weeks, months, or even years to develop the support of legislators. Special interests with money to spend work to sway public opinion through media. They may spend millions of dollars on campaigns to convince the public to support a particular view on a public policy issue.

A common strategy is to finance commentators on radio and television who, while in an apparent position of neutrality, can promote a particular point of view. When watching the media, it's always a good idea to "follow the money" if you want to learn more about the views expressed. The tremendous amount of money spent on news media and advertising indicates how important it is to enlist public support in order to pass (or block) new policies.

Are we better safe than sorry?

A fundamental concept in policy is the **precautionary principle**, the idea that if an activity threatens to harm health or the environment, we should fully understand risks before initiating that activity. According to this principle, for example, we shouldn't mass-market new chemicals, new cars, or new children's toys until we're sure they are safe. These are four widely accepted tenets of this principle:

- People have a duty to take steps to prevent harm. If you suspect something bad might happen, you have an obligation to try to stop it.

- The burden of proof for a new technology, process, activity, or chemical lies with its developers, not with the general public.

- Before using a new technology, process, or chemical, or starting a new activity, there is an obligation to examine a full range of alternatives, including the alternative of not using it.

- Decisions using the precautionary principle must be open and democratic and must include the affected parties.

The European Union has adopted this precautionary principle as the basis of its environmental policy. In the United States, opponents of this approach claim that it threatens productivity and innovation. However, many American firms that do business in Europe—including virtually all of the largest corporations—are having to change their manufacturing processes to adapt to more careful EU standards. For example, lead, mercury, and other hazardous materials must be eliminated from electronics, toys, cosmetics, clothing, and a variety of other consumer products. A proposal being debated by the EU would require testing of thousands of chemicals, cost industry billions of dollars, and lead to many more products and compounds being banned as they are shown to be unsafe to the public. What would you do about this? Is this proposal just common sense, or does it interfere with reasonable trade if people want to buy those products?

▲ **FIGURE 16.5** Every major federal project in the United States must be preceded by an Environmental Impact Statement.

16.2 MAJOR ENVIRONMENTAL LAWS

We depend on countless laws to protect the water we drink, the air we breathe, the food we eat, and the biodiversity that surrounds us. Most of these laws work to negotiate a balance between the differing interests and needs of various groups, or between private interests and the public interest. In this section we'll examine some of the most important environmental policies in the United States. Many other countries and regions have comparable laws, designed to prevent damage to health, property, and the natural resources on which we depend.

NEPA (1969) establishes public oversight

The cornerstone of U.S. environmental policy is the **National Environmental Policy Act (NEPA).** Signed into law by President Nixon in 1970, NEPA is a model for many other countries.

NEPA does three important things: (1) it establishes the Council on Environmental Quality (CEQ), the oversight board for general environmental conditions, (2) it directs federal agencies to take environmental consequences into account in decision making, and (3) it requires that an **environmental impact statement (EIS)** be published for every major federal project that is likely to have an important effect on environmental quality. NEPA doesn't forbid environmentally destructive activities if they otherwise comply with relevant laws, but it requires that agencies admit publicly what they plan to do. If embarrassing information is revealed publicly, it becomes harder for agencies to ignore public opinion. An EIS can provide valuable information about government actions to public interest groups that wouldn't otherwise have access to these resources.

What kinds of projects require an EIS? The activity must be federal and it must be major, with a significant environmental impact (fig. 16.5). Whether specific activities meet these characteristics is often a subjective decision. Each case is unique and

depends on context, geography, the balance of beneficial versus harmful effects, and whether any areas of special cultural, scientific, or historical importance might be affected. A complete EIS for a project is usually time-consuming and costly. The final document is often hundreds of pages long and generally takes six to nine months to prepare. Sometimes just requesting an EIS is enough to sideline a questionable project. In other cases, the EIS process gives adversaries time to rally public opposition and information with which to criticize what's being proposed. If agencies don't agree to prepare an EIS voluntarily, citizens can petition the courts to force them to do so.

Every EIS must contain the following elements: (1) purpose and need for the project, (2) alternatives to the proposed action (including taking no action), and (3) a statement of positive and negative environmental impacts of the proposed activities. In addition, an EIS should make clear the relationship between short-term resource effects and long-term productivity, as well as any irreversible impacts on resources resulting from the project.

Areas in which lawmakers have tried recently to ignore or limit NEPA include forest policy, energy exploration, and marine wildlife protection. The "Healthy Forest Initiative," for example, called for bypassing EIS reviews for logging or thinning projects, and prohibited citizen appeals of forest management plans. Similarly, when the Bureau of Land Management proposed 77,000 coal-bed methane wells in Wyoming and Montana, promoters claimed that water pollution and aquifer depletion associated with this technology didn't require environmental review (chapter 13).

The Clean Air Act (1970) regulates air emissions

The first major environmental legislation to follow NEPA was the Clean Air Act (CAA) of 1970. Air quality has been a public concern since the beginning of the Industrial Revolution, when coal smoke, airborne sulfuric acid, and airborne metals such as mercury became common in urban and industrial areas around the world (fig. 16.6). Sometimes these conditions produced public health crises: One infamous event was the 1952 Great Smog of London, when several days of cold, still weather trapped coal smoke in the city and killed some 4,000 people from infections and asphyxiation. Another 8,000 died from respiratory illnesses in the months that followed (chapter 10).

▲ FIGURE 16.6 The Clean Air Act has greatly reduced the health and economic losses associated with air pollution.

Although crises of this magnitude have been rare, chronic exposure to bad air has long been a leading cause of illness in many areas. The Clean Air Act provided the first nationally standardized rules in the United States to identify, monitor, and reduce air contaminants. The core of the act is an identification and regulation of seven major "criteria pollutants," also known as "conventional pollutants": sulfur oxides, lead, carbon monoxide, nitrogen oxides (NO_x), particulates (dust), volatile organic compounds, and metals and halogens (such as mercury and bromine compounds).

Most of these pollutants have declined dramatically since 1970. An exception is NO_x, which derives from internal combustion engines such as those in our cars. Further details on these pollutants are covered in chapter 10.

The Clean Water Act (1972) protects surface water

Water protection has been a goal with wide public support, in part because clean water is both healthy and an aesthetic amenity. The act aimed to make the nation's waters "fishable and swimmable," that is, healthy enough to support propagation of fish and shellfish that could be consumed by humans, and low in contaminants so that they were safe for swimming and recreation.

The first goal of the Clean Water Act (CWA) was to identify and control point-source pollutants—end-of-the-pipe discharges from factories, municipal sewage treatment plants, and other sources. Discharges are not eliminated, but effluent at pipe outfalls must be tested, and permits are issued that allow moderate discharges of low-risk contaminants such as nutrients or salts. Metals, solvents, oil, high counts of fecal bacteria, and other more serious contaminants must be captured before water is discharged from a plant.

By the late 1980s, point sources were increasingly under control, and the CWA was used to address nonpoint sources, such as runoff from urban storm sewers. The act has also been used to promote watershed-based planning, in which communities and agencies collaborate to reduce contaminants in their surface waters. As with the CAA, the CWA provides funding to aid pollution-control projects. Those funds have declined in recent years, however, leaving many municipalities struggling to pay for aging and deteriorating sewage treatment facilities. For more detail on the CWA and water pollution control, see chapter 11 and Key Concepts (p. 384).

The Endangered Species Act (1973) protects wildlife

The Endangered Species Act (ESA) provides a structure for identifying and listing species that are vulnerable, threatened, or endangered. Once a species is listed as endangered, this act provides rules for protecting it and its habitat, ideally in order to help make recovery possible (fig. 16.7). Listing of a species has become a controversial process, because habitat conservation tends to get in the way of land development. For example, many ESA controversies arise when developers want to put new housing developments in scenic areas where the last remnants of a species occur. To reduce disputes, the ESA provides habitat and land-use planning assistance and grants, as well as guaranteeing landowner rights when an effective habitat conservation plan has been developed. Landowners can also get a tax reduction in exchange for habitat conservation. These strategies increasingly allow for both development and species protection.

The ESA maintains a worldwide list of endangered species. In 2010 the list included 1,969 threatened and endangered species, 753 of them plants. The responsibility for studying and attempting to restore threatened and endangered species lies mainly with the Fish and Wildlife Service and the National Oceanic and Atmospheric Administration. You can read more about endangered species, biodiversity, and the ESA in chapter 5.

The Superfund Act (1980) addresses hazardous sites

Most people know this law as the Superfund Act because it created a giant fund to help remediate abandoned toxic sites. The proper name of this law is informative, though: the Comprehensive

▶ FIGURE 16.7

The Endangered Species Act is charged with protecting species and their habitat. The black-footed ferret was declared extinct in 1979, but a remnant population was discovered in 1981, and captive breeding programs have restored the species to more than 1,000 mature, wild-born animals in eight states.

Environmental Response, Compensation, and Liability Act (CERCLA). The act aims to be comprehensive, addressing abandoned sites, emergency spills, or uncontrolled contamination, and it allows the EPA to try to establish liability, so that polluters help to pay for cleanup. Because it's much cheaper to make toxic waste than to clean it up, we have thousands of chemical plants, gas stations, and other sites that have been abandoned because they were too expensive to clean properly. The EPA is responsible for finding a private party to do cleanup, and the Superfund was established to cover the costs, which can be in the billions of dollars.

Until recently, the fund was supplied mainly by contributions from industrial producers of hazardous wastes. In 1995, however, Congress voted to end that source, and the Superfund was allowed to dwindle to negligible levels. Since then, cleanup has been paid for by public taxpayers. According to the EPA, one in four Americans lives within 3 miles of a hazardous waste site. The Superfund program has identified more than 47,000 sites that may require cleanup. The most serious of these (or the most serious for which proponents have been sufficiently vigorous) have been put on a National Priorities List. About 1,600 sites have been put on this list, and over 1,000 cleanups have been completed. The total cost of remediation is expected to be somewhere between $370 billion and $1.7 trillion.

16.3 HOW ARE POLICIES IMPLEMENTED?

The laws discussed below are among our most important environmental protections (table 16.1). Each of these laws resulted from action at many local and national levels. Individual citizens lobbied for change. State and federal representatives negotiated policies. Courts tested the legitimacy of the law. Local, state, and federal agencies worked to implement the law.

We examine here the ways that laws are enacted in the United States at the federal level. Environmental law can be established or modified in each of the three branches of government—legislative, judicial, and executive. Many of the 50 states have bodies with similar structure and function.

The legislative branch establishes statutes (laws)

Federal laws, also called **statutes**, are enacted by Congress and signed by the president. Thousands of proposed laws, or bills, are introduced every year in Congress. Some are very narrow, providing funds to build a specific section of road or to help a particular person, for instance. Others are extremely broad, perhaps overhauling the Social Security system or changing the tax code. Sometimes a bill will have 100 or more coauthors if it is a prominent issue.

TABLE 16.1 | Major U.S. Environmental Laws

LEGISLATION	PROVISIONS
Wilderness Act of 1964	Established the national wilderness preservation system.
National Environmental Policy Act of 1969	Declared national environmental policy, required environmental impact statements, created Council on Environmental Quality.
Clean Air Act of 1970	Established national primary and secondary air quality standards. Required states to develop implementation plans. Major amendments in 1977 and 1990.
Clean Water Act of 1972	Set national water quality goals and created pollutant discharge permits. Major amendments in 1977 and 1996.
Federal Pesticides Control Act of 1972	Required registration of all pesticides in U.S. commerce. Major modifications in 1996.
Marine Protection Act of 1972	Regulated dumping of waste into oceans and coastal waters.
Coastal Zone Management Act of 1972	Provided funds for state planning and management of coastal areas.
Endangered Species Act of 1973	Protected threatened and endangered species. Directed FWS to prepare recovery plans.
Safe Drinking-Water Act of 1974	Set standards for safety of public drinking-water supplies and to safeguard groundwater. Major changes made in 1986 and 1996.
Toxic Substances Control Act of 1976	Authorized EPA to ban or regulate chemicals deemed a risk to health or the environment.
Federal Land Policy and Management Act of 1976	Charged the BLM with long-term management of public lands. Ended homesteading and most sales of public lands.
Resource Conservation and Recovery Act of 1976	Regulated hazardous-waste storage, treatment, transportation, and disposal. Major amendments in 1984.
National Forest Management Act of 1976	Gave statutory permanence to national forests. Directed USFS to manage forests for "multiple use."
Surface Mining Control and Reclamation Act of 1977	Limited strip-mining on farmland and steep slopes. Required restoration of land to original contours.
Alaska National Interest Lands Act of 1980	Protected 40 million ha (100 million acres) of parks, wilderness, and wildlife refuges.
Comprehensive Environmental Response, Compensation and Liability Act of 1980	Created $1.6 billion "Superfund" for emergency response, spill prevention, and site remediation for toxic wastes. Established liability for cleanup costs.
Superfund Amendments and Reauthorization Act of 1994	Increased Superfund to $8.5 billion. Shared responsibility for cleanup among potentially responsible parties. Emphasized remediation and public "right to know."

SOURCE: Data from N. Vig and M. Kraft, *Environmental Policy in the 1990s*, 3rd Congressional Quarterly Press.

How does the Clean Water Act benefit you?

Environmental policies are rules we establish to protect public health and resources. The 1972 Clean Water Act is one of the most important and effective environmental laws in America. Dramatic improvements in water quality have resulted from the billions of dollars in public and private investment in pollution control. While we still haven't met all the goals of this momentous act, the water in your neighborhood is almost certainly cleaner than it was 40 years ago.

To learn the details for yourself, find further explanation, and the text of the law, at http://water.epa.gov/lawsregs/lawsguidance

What does the Clean Water Act do?
- Establishes rules for regulating discharges of pollutants into water;
- Charges the EPA with establishing and regulating standards for water quality;
- Makes it illegal to pollute navigable waters from a point source (such as a discharge pipe) without a permit.

◀ One of the first steps in cleaning up surface waters was to stop the dumping of untreated sewage and industrial effluents. Over the past 40 years, the United States has spent at least $2 trillion in public and private funds on point-source pollution control.

KC 16.1

◀ Nonpoint-source pollution, such as runoff from city streets and random dumping, is harder to control, but there has been dramatic improvement in this area as well.

KC 16.2

◀ A major goal of the Clean Water Act was to make all U.S. surface waters "fishable and swimmable." This aim has been partly a success and partly a failure. The EPA has reported that over 90 percent of all monitored river miles and 87 percent of assessed lake acres meet this goal. However, many waters remain impaired. You can't necessarily eat all the fish you catch, and you often should avoid drinking the water in which you swim.

KC 16.3

Regular monitoring is an important part of protecting and improving water quality ▶. The EPA monitors 4,000 watersheds for bacteria, nutrients, metals, clarity, and other standard measures. States are required to develop total maximum daily loads for each expected pollutant and each listed water body. Biological organisms often make good indicators of water quality. You can learn about water quality in your area from www.epa.gov/owow/tmdl/.

KC 16.4

Agricultural pollution remains a serious problem. A majority of remaining water pollution in the United States is thought to come from soil erosion, nutrient runoff from farm fields, feedlots, and other farm operations. Hundreds of millions of tons of fertilizers, pesticides, and manure wash into rivers and lakes. Excess fertilizer causes vast dead zones or explosive growth of harmful algae in the estuaries of major rivers. ▶

Novel pollutants—chemicals the EPA doesn't yet test for—are also a growing concern.

KC 16.5

We may never return to the days when you could safely drink surface waters almost anywhere in the country, but the Clean Water Act has resulted in remarkable progress in restoring water quality in most places. ▶

Like other environmental laws, this one is imperfect and incomplete, but it provides safeguards we depend on, usually without even knowing they're there.

KC 16.6

CAN YOU **EXPLAIN?**

1. Ask your parents or grandparents what water quality was like in your area 40 years ago. What progress has been made?

2. What water quality problems do you think remain most troublesome?

3. What do you regard as the most valuable provisions of the Clean Water Act?

385

All bills and all public laws (bills passed by Congress) are available for you to examine, as part of the public record. You can find the details of any national law by looking at www.thomas.gov. The EPA also provides access to the text of environmental legislation at www.epa.gov/lawsregs/.

Citizens can be involved in the legislative process by writing or calling their elected representatives and by appearing at public hearings. A personal letter or statement is always more persuasive than simply signing a petition. Some politicians treat all the copies of a similar form letter or e-mail as a single statement, because they all contain the same information. Still, a petition with a million signatures will probably catch the attention of a legislator—especially if they are all potential voters.

Being involved in local election campaigns can greatly increase your access to legislators. Media attention can also sway the opinions of decision makers. Drawing public attention to an issue or campaign can also help influence public opinion and enthusiasm for an issue (fig. 16.8). It's hard for a single individual to have much impact, but if you join a group, you can have a collective impact on public policy.

The judicial branch resolves legal disputes

The judicial branch of government decides (1) what the precise meaning of a law is, (2) whether or not laws have been broken, and (3) whether a law violates the Constitution. The cumulative body of legal opinions from court cases is known as case law. Often this law involves interpreting what a law really means. Interpretation is needed because legislation is frequently written in vague and general terms so as to make it widely enough accepted to gain passage. When trying to interpret a law, the courts depend on the legislative record from hearings, such as what was said by whom, to determine congressional intent.

When a law may have been broken, it becomes a matter of **criminal law**. Serious crimes such as murder or theft, as well as criminal violations of environmental laws, are matters of criminal law. Charges are usually brought by the state, rather than by individuals, and usually proof of intent or willful neglect of responsibilities is needed for a criminal conviction. **Civil law**, on the other hand, aims to resolve disputes between individuals and corporations. Civil disputes can involve issues such as property rights, injury, or personal freedom, and proof of criminal intent is not necessarily required. Fines can result from either civil disputes or criminal suits, but jail time results only from criminal cases. Civil suits can also be used to stop activities, such as endangering water resources, endangered species, or public health.

Damages can be difficult to collect after a successful suit. For example, when Texaco stopped oil drilling in Ecuador in 1992, after nearly 30 years of operations, it left behind what critics describe as a toxic dump of 6.8 million liters (1.8 million gallons) of spilled crude oil and drilling wastes—almost twice as much as Alaska's *Exxon Valdez* spill. In 2003, lawyers representing 30,000 Ecuadoran natives filed a lawsuit against Texaco claiming environmental damage and adverse health effects from more than 350 open toxic waste pits. The case dragged on until 2011, when an Ecuadoran court ordered Chevron, which had bought Texaco (and its liabilities) in 2001, to pay $18 billion in damages. A district judge in Manhattan immediately issued an injunction to block payment, but a U.S. Appeals Court threw out that injunction. Still, in 2012, some 50 years after the drilling started, no payments had been made to those who suffered damage (or their heirs).

If a lawsuit questions the legitimacy of a law itself, the suit may be decided by the U.S. Supreme Court (fig. 16.9). The nine justices of the federal Supreme Court decide whether a law is consistent with the Constitution of the United States. States also have supreme courts, which determine whether a law is consistent with the state's constitution.

Because the justices of the court interpret the Constitution and its meaning, they have far-reaching influence over our laws, policies, and practices. For example, in 2007 in the case *Massachusetts v. EPA*, the court determined that the EPA had a responsibility to regulate greenhouse gases, because the climate-warming impact of these gases threaten to "endanger public health and to endanger public welfare." This Court ruling forced the EPA to issue new policies in 2009, increasing monitoring nationwide and reducing allowable emissions of these gases.

▲ **FIGURE 16.8** Global awareness of environmental issues can push countries to comply with treaties. Here a youth group from the Maldives, an island nation threatened by rising sea levels, stages a protest as part of the global 350.org movement for controlling climate change. SOURCE: 350.org.

▲ **FIGURE 16.9** The Supreme Court decides pivotal cases, many of them bearing on resources or environmental health.

Perhaps the most widely debated Supreme Court case in recent years was the 2010 decision *Citizens United v. Federal Election Commission*. This decision determined that the government may not limit political spending by corporations. Reversing 70 years of precedent and multiple previous court rulings, the *Citizens United* case was decided on a 5-to-4 vote, and it unleashed a torrent of political spending in the 2012 elections. That election quickly became the most expensive presidential election in U.S. history. Political action committees, which by law must be unaffiliated with candidates, but are often staffed by former top aids and friends of candidates, spent hundreds of millions of dollars to attack opponents. A handful of billionaires personally selected and supported their favorite candidates. Victory isn't necessarily guaranteed to these financers, but historically the better-funded candidate usually wins. Many regard this Supreme Court decision as a great threat to democracy.

The executive branch oversees administrative rules

Within the executive branch—headed by the president, or by the governor of a state—various agencies, boards, and commissions oversee environmental policies. These bodies set rules, decide disputes, and investigate misconduct. At the federal level, executive rules are made that can have far-reaching environmental consequences. Executive rules can be made quickly and with little interference from Congress. They are often a strategy to change public policy with little public oversight or discussion. For example, President Clinton's administration established rules protecting 58 million acres of roadless wilderness areas from logging and development. President Bush quickly reversed those rules after his election, opening roadless acres to development. The Obama administration then overturned the Bush rules, returning protections to roadless wilderness areas. For better and for worse, executive rules can make rapid and profound changes in environmental policy.

The executive branch includes administrative agencies, such as the Environmental Protection Agency (EPA), which oversee and enforce public laws. The EPA is the primary agency with responsibility for protecting environmental quality in the United States. The EPA was created in 1970 at the same time as NEPA.

The EPA and other administrative agencies are headed by persons appointed by the president. The heads of these agencies are thus responsible to the political interests of the president and to the legal responsibilities of the agency. Other agencies that have profound environmental impacts are the Department

◄ FIGURE 16.10 Smokey Bear is the public face of an executive agency, the U.S. Forest Service.

of the Interior, which oversees public lands and national parks, and the Department of Agriculture, which oversees the nation's forests and grasslands, as well as agricultural issues. The Department of the Interior is home to the U.S. Fish and Wildlife Service, which operates more than 500 national wildlife refuges and administers endangered species protection. The Department of Agriculture is home to the U.S. Forest Service, which manages about 175 national forests and grasslands, totaling some 78 million ha (193 million acres). With 39,000 employees, the Forest Service is nearly twice as large as the EPA (fig. 16.10).

How much government do we want?

In his 1981 inaugural address, President Ronald Reagan famously said, "Government is not the solution to our problem; government is the problem." In this, he invoked a perennial debate in American politics: Is government a power that undermines personal liberties? Or is government a defender of liberties and rights, in a world where bullies too easily have their way? Do governments defend their own interests? Or are they a mechanism for defending the public interest and the common good?

The answer sometimes depends on when you ask. During political campaigns, many people decry the size and cost of government agencies. During times of crisis, most of us assume the government will step in to help (fig. 16.11). Strangely, in some surveys, more than half the people who receive food stamps, Social Security payments, public schooling, and police and fire

▲ FIGURE 16.11 When tornadoes, floods, hurricanes or other natural disasters afflict us, we expect government agencies to help us.

protection, or who drive on public highways, claim they have never benefited from a government program. Similarly, corporations eager to avoid taxes nonetheless rely on the state to provide educated workers, roads, communication infrastructure, and professional police forces to maintain order.

Debates about the proper size and role of government are always present. We value self-reliance and rugged individualism. Yet we also want someone to protect us from contaminated food and drugs, to educate our children, and to provide roads, bridges, and safe drinking water. President Reagan was among those who favor "free market" capitalism, with businesses unfettered by rules such as the Clean Air Act or the Clean Water Act. Grover Norquist, president of the Americans for Tax Reform, famously said he'd like to "shrink the government down to the size where we can drown it in the bathtub." Other observers note that antigovernment language and imagery focuses on freedom for individuals and small business owners, but the elimination of public health and safety regulations undermines the interests of those it claims to protect. Advantages are often enjoyed by the biggest players in a rule-less game.

Part of the reason these disputes persist may be that both views are partially correct. Regulations, such those imposing expensive pollution abatement technologies for polluters, require private businesses to bear the immediate cost of protecting public resources. Businesses are squeezed between shareholders' demands for ever-higher profits and agency demands for safer, sometimes costly, operating standards. Viewed another way, regulations require businesses to clean up their own messes. Those costs can be passed on to consumers, who then are asked to bear the costs of the benefits they enjoy. Opponents of agency regulations point out that corporations produce the economic vitality on which our prosperity depends. Proponents of regulations point out that corporations couldn't prosper without subsidies, tax breaks, transportation infrastructure, and a healthy and educated workforce. These costs are necessary to doing business but are normally external to the accounting of business costs and profits.

Since about 1981 the small-government philosophy has dismantled much of the regulatory structure set up during President Nixon's term in office. Often this dismantling has been done by agency heads, who have been appointed despite openly opposing the existence of those agencies and their laws. For example, President George W. Bush appointed Christopher Cox, a proponent of bank deregulation, to chair the Securities and Exchange Commission, which oversees Wall Street trading. Subsequent dismantling of trading rules led to risky behavior by banks, which culminated in the Wall Street collapse in 2007–2008. Business failures and high unemployment spread nationwide and have lasted for years. President Bush also oversaw dramatic reductions in USDA food safety inspections, on the grounds that they represented unnecessary interference in the private business of the food industry. Increasingly frequent food contamination scares have made many Americans rethink the importance of government inspectors in the food system.

These debates probably will always be with us. Which view is most correct depends on many factors: your interest group, life experience, philosophical perspective, and economic position, the time frame you analyze, and other priorities. What factors influence your view on these issues? Do you think there is room for compromise? If not, why not? If yes, where?

16.4 INTERNATIONAL POLICIES

With growing recognition of the interconnections in our global environment, nations have become increasingly interested in signing on to international agreements (often called "treaties" or "conventions") for environmental protection (table 16.2). A principal motivation in these treaties is the recognition that countries can no longer act alone to protect their own resources and interests. Water resources, the atmosphere, trade in endangered species, and many other concerns cross international boundaries. Over time, the number of parties taking part in negotiations has grown, and the speed at which agreements take force has increased. The Convention on International Trade in Endangered Species (CITES), for example, was not enforced until 14 years after its ratification in 1973, but the Convention on Biological Diversity (1991) was enforceable after just one year and had 160 signatories only four years after introduction.

Over the past 25 years, more than 170 treaties and conventions have been negotiated to protect our global environment. Designed to regulate activities ranging from intercontinental shipping of hazardous waste, to deforestation, overfishing, trade in endangered species, global warming, and wetland protection, these agreements theoretically cover almost every aspect of human impacts on the environment.

Many of these policies have emerged from major international meetings. One of the first of these was the 1972 UN Conference on the Human Environment in Stockholm, which set an agenda for subsequent meetings. This conference gathered representatives of 113 countries and several dozen nongovernmental organizations. A much larger gathering was the "Earth Summit" 20 years later.

TABLE 16.2 | Some Important International Treaties

CBD: Convention on Biological Diversity 1992 (1993)*

CITES: Convention on International Trade on Endangered Species of Wild Fauna and Flora 1973 (1987)

CMS: Convention on the Conservation of Migratory Species of Wild Animals 1979 (1983)

Basel: Basel Convention on the Transboundary Movements of Hazardous Wastes and their Disposal 1989 (1992)

Ozone: Vienna Convention for the Protection of the Ozone Layer and Montreal Protocol on Substances that Deplete the Ozone Layer 1985 (1988)

UNFCCC: United Nations Framework Convention on Climate Change 1992 (1994)

CCD: United Nations Convention to Combat Desertification in those Countries Experiencing Serious Drought and/or Desertification, Particularly in Africa 1994 (1996)

Ramsar: Convention on Wetlands of International Importance especially as Waterfowl Habitat 1971 (1975)

Heritage: Convention Concerning the Protection of the World Cultural and Natural Heritage 1972 (1975)

*The first date listed is when the treaty was enacted. The second date is when it went into force.

Officially called the UN Conference on Environment and Development (UNCED), this meeting was held in Rio de Janeiro in 1992. This time over 110 nations participated, as well as 2,400 nongovernmental organizations. The Rio meeting was such a watershed event that it was repeated in 2012, on its 20th anniversary. The Agenda 21 document produced from the first Rio meeting has laid out principles of sustainability and equity that have guided much policy making since 1992.

Major international agreements

International accords and conventions have emerged slowly but fairly steadily from meetings such as those in Stockholm and Rio (table 16.2). A few of the important benchmark agreements are discussed here.

The **Convention on International Trade in Endangered Species (CITES, 1973)** declared that wild flora and fauna are valuable, irreplaceable, and threatened by human activities. To protect disappearing species, CITES maintains a list of threatened and endangered species that may be affected by trade. As with most international agreements, this one takes no position on movement or loss of species within national boundaries, but it establishes rules to restrict unauthorized or illegal trade across boundaries. In particular, an export permit is required specifying that a state expert declares an export is legal, that it is not cruel, and that it will not threaten a wild population.

The **Montreal Protocol (1987)** protects stratospheric ozone. This treaty committed signatories to phase out the production and use of several chemicals that break down ozone in the atmosphere. The ozone "hole," a declining concentration of ozone (O_3) molecules over the south pole, threatened living things: ozone high in the atmosphere blocks cancer-causing ultraviolet radiation, keeping it from reaching the earth's surface. The stable chlorine- and fluorine-based chemicals at fault for reducing ozone are used mainly as refrigerants. Alternative refrigerants have since been developed, and the use of chlorofluorocarbons (CFCs) and related molecules has plummeted. Although the ozone "hole" has not disappeared, it has declined as predicted by atmospheric scientists since the phase-out of CFCs. The Montreal Protocol is often held up as an example of a highly successful and effective international environmental agreement.

The Montreal Protocol was effective because it bound signatory nations not to purchase CFCs or products made using them from countries that refused to ratify the treaty. This trade restriction put substantial pressure on producing countries. Initially the protocol called for only a 50 percent reduction in CFC production, but subsequent research showed that ozone was being depleted faster than previously thought (chapter 10). The protocol was strengthened to an outright ban on CFC production, in spite of the objection of a few countries.

The **Basel Convention (1992)** restricts shipment of hazardous waste across boundaries. The aim of this convention, which has 172 signatories, is to protect health and the environment, especially in developing areas, by stating that hazardous substances should be disposed of in the states that generated them. Signatories are required to prohibit the export of hazardous wastes unless the receiving state gives prior informed consent, in writing, that a shipment is allowable. Parties are also required to minimize production of hazardous materials and to ensure that there are safe disposal facilities within their own boundaries. This convention establishes that it is the responsibility of states to make sure that their own corporations comply with international laws. The Basel Convention was enhanced by the Rotterdam Convention (1997), which places similar restrictions on unauthorized transboundary shipment of industrial chemicals and pesticides.

The **UN Framework Convention on Climate Change (1994)** directs governments to share data on climate change, to develop national plans for controlling greenhouse gases, and to cooperate in planning for adaptation to climate change. Where the UNFCC encouraged reduction in greenhouse gas (GHG) emissions, the **Kyoto Protocol (1997)** set binding targets for signatories to reduce greenhouse gas emissions to less than 1990 levels by 2012. While the idea of binding targets is strong, and some countries (such as Sweden) are likely to achieve their goals, most countries are still falling short of their target.

The protocol has been controversial because it sets tighter restrictions on industrialized countries, which are responsible for roughly 90 percent of GHG emissions up to the present, than for developing countries. Signatories are required to report their GHG emissions in order to document changes in their production. The protocol went into force in 2005, when 198 states and the European Union had signed the agreement. These signatories contribute almost 64 percent of global GHG emissions. The United States remains the only signatory that hasn't ratified the Kyoto framework. Canada ratified it in 2002, but then the conservative government of Stephen Harper withdrew its consent.

Since the Kyoto Protocol was approved in 1997, a series of meetings called Conferences of Parties (COP) have convened to try to reach agreement on how to reduce greenhouse gases. The main disagreement is over how much burden developing nations should bear in this crisis. They argue that the rich nations have created most of the problem and should, therefore, take most of the responsibility for solving it. Industrialized countries, on the other hand, argue that energy consumption is rising so rapidly in countries such as China and India that if they don't do something to curb future emissions, nothing the richer countries can do now will do much good (fig. 16.12). Both sides fear their economies will be damaged by restrictions on fossil fuel use. The question is, who moves first?

A sequence of meetings has repeatedly failed to reach substantive solutions, although incremental agreements have been made. Recent meetings, for example, have produced agreements in principle that wealthy countries should aid poorer countries in developing alternative energy strategies and in reducing emissions from deforestation and forest degradation (REDD—see chapter 6).

Enforcement often relies on national pride

Enforcement of international agreements frequently depends on how much countries care about their international reputation. Except in extreme cases such as genocide, the global community is unwilling to send an external police force into a country, because

▲ **FIGURE 16.12** President Barack Obama meets with Chinese Premier Wen Jiabao at the United Nations Climate Change Conference in Denmark in 2009. SOURCE: Office White House Photo by Pete Souza.

▲ **FIGURE 16.13** CITES listing was proposed for bluefin tuna, shown here at Tokyo's Tsukiji market. In 2010, Japan almost single-handedly blocked the listing.

states are wary of interfering with the internal sovereignty of other states. However, most countries also are reluctant to appear irresponsible or immoral in the eyes of the international community, so moral persuasion and public embarrassment can be effective enforcement strategies. Shining a spotlight on transgressions will often push a country to comply with international agreements.

But national pride can also stand in the way of species protection. In 2010, Japan almost single-handedly derailed CITES protection for bluefin tuna, despite the fact that these beautiful, highly evolved, long-lived top predators are now at less than 15 percent of historic levels. They may already be headed for extinction. Part of Japan's objections to fishing limits is that they don't like to be told what they can or can't eat. Economics also plays a role. In 2012 a single 593-pound (263 kg) bluefin tuna caught off northeastern Japan fetched a record $736,000 (56.49 million yen), or about $1,238 per pound, at the first auction of the year at Tokyo's Tsukiji market (fig. 16.13).

Often international negotiators aim for unanimous agreement to ensure strong acceptance of international policies. While this approach makes for a strong agreement, a single recalcitrant nation can have veto power over the wishes of the vast majority. For instance, more than 100 countries at the UN Conference on Environment and Development (UNCED), held in Rio de Janeiro in 1992, agreed to restrictions on the release of greenhouse gases. At the insistence of U.S. negotiators, however, the climate convention was reworded so that it only urged—but did not require—nations to stabilize their emissions.

When a consensus cannot be reached, negotiators may seek an agreement acceptable to a majority of countries. This approach was used in negotiating the Kyoto Protocol on climate change, which sought, and eventually achieved, agreement from a majority of countries. Only signing countries are bound by such a treaty, but nonsigning countries may comply anyway, to avoid international embarrassment.

When strong accords with meaningful sanctions cannot be passed, sometimes the pressure of world opinion generated by revealing the sources of pollution can be effective. Activists can use this information to expose violators. For example, the environmental group Greenpeace discovered monitoring data in 1990 showing that Britain was disposing of coal ash in the North Sea. Although not explicitly forbidden by the Oslo Convention on ocean dumping, this evidence proved to be an embarrassment, and the practice was halted.

Trade sanctions can be an effective tool to compel compliance with international treaties. The Montreal Protocol used the threat of trade sanctions very effectively to cut CFC production dramatically (see chapter 9). On the other hand, trade agreements also can work against environmental protection. The World Trade Organization (WTO) was established in 1995 to promote free international trade and to encourage economic development. The WTO's emphasis on unfettered trade, however, has led to weakening of local environmental rules.

In 1990 the United States banned the import of tuna caught using methods that kill thousands of dolphins each year. Shrimp caught with nets that kill endangered sea turtles were also banned. Mexico filed a complaint with the WTO, contending that dolphin-safe tuna laws were an illegal barrier to trade. Thailand, Malaysia, India, and Pakistan filed a similar suit against turtle-friendly shrimp laws. The WTO ordered the United States to allow the import of both tuna and shrimp from countries that allow fisheries to kill dolphins and turtles. Environmentalists point out that the WTO has never ruled against a corporation because it is composed of industry leaders. As such, the WTO mainly defends the interests of the business community, rather than the broader public interest.

16.5 WHAT CAN INDIVIDUALS DO?

Global, federal, and state lawmakers are clearly important in environmental policy, but these entities are all composed of individuals who have decided to commit their energy, education, or careers to causes they find important. Whatever your skills and interests, you can participate in policy formation and help in protecting our common environment. If you enjoy science, there are many disciplines that contribute to environmental science. As you know by now, biology, chemistry, geology, ecology, climatology, geography, hydrology, and other sciences all provide essential ideas and data to environmental science.

Skills in art, writing, communication, working with children, history, politics, economics, and many other areas are also critical for developing ideas and engaging the public. Communicating the ideas of environmental science requires educators, policy makers, artists, and writers. Lawyers and other specialists are needed to develop and improve environmental laws and regulations. Engineers are needed to develop technologies and products to clean up pollution and to prevent its production in the first place. Economists and social scientists are needed to evaluate the costs of pollution and resource depletion and to develop equitable and appropriate solutions for different parts of the world. In addition, businesses will be looking for a new class of environmentally literate and responsible leaders who can help improve the green credentials of their products and services.

Environmental educators are also needed to help train an environmentally literate populace. We urgently need more teachers at every level who are trained in environmental education, and who are familiar with the outdoors. You can even contribute to environmental awareness just by enjoying your surroundings (fig. 16.14). As author Edward Abbey wrote,

> It is not enough to fight for the land; it is even more important to enjoy it. While it is still there. So get out there and mess around with your friends, ramble out yonder and explore the forests, encounter the grizz, climb the mountains. Run the rivers, breathe deep of that yet sweet and lucid air, sit quietly for a while and contemplate the precious stillness, that lovely mysterious and awesome space.

Environmental literacy is a policy aim

The National Environmental Education Act, passed by Congress in 1990, identified environmental education as a national priority. The act established two broad goals: (1) to improve public understanding of our environment, and (2) to encourage postsecondary students to pursue careers related to the environment. Learning objectives (table 16.3) include awareness and appreciation of our environment, knowledge of basic ecological concepts, acquaintance with a broad range of current environmental issues, and experience in using investigative, critical-thinking, and problem-solving skills in solving environmental problems (fig. 16.15). **Environmental literacy** is a term for a working knowledge of our

TABLE 16.3 | Outcomes from Environmental Education

The natural context: An environmentally educated person understands the scientific concepts and facts that underlie environmental issues and the interrelationships that shape nature.

The social context: An environmentally educated person understands how human society is influencing the environment, as well as the economic, legal, and political mechanisms that provide avenues for addressing issues and situations.

The valuing context: An environmentally educated person explores his or her values in relation to environmental issues; from an understanding of the natural and social contexts, the person decides whether to keep or change those values.

The action context: An environmentally educated person becomes involved in activities to improve, maintain, or restore natural resources and environmental quality for all.

SOURCE: Data from *A Greenprint for Minnesota*, Minnesota Office of Environmental Education, 1993.

▼ **FIGURE 16.14** Enjoying and learning about our natural world may be one of the most important things you can do to aid in environmental protection.

▲ **FIGURE 16.15** Environmental education helps develop awareness and appreciation of ecological systems and how they work.

environment and its systems. Former EPA administrator William K. Reilly said that environmental literacy is essential both to stewardship of our earth and to civic participation in general. Reilly argued that everyone needs some environmental literacy. It isn't enough, he said, "for a few specialists to know what is going on while the rest of us wander about in ignorance."

You have made a great start toward learning about your environment by reading this book and taking a class in environmental science. Happily, pursuing environmental literacy is enjoyable as well as important. There are thousands of excellent books you can read. Some of them have been persistently important for many readers over time (table 16.4).

Citizen science lets everyone participate

Many students are discovering they can make authentic contributions to scientific knowledge through active learning and undergraduate research programs. Internships in agencies or environmental organizations are one way of doing this. Another is to get involved in organized **citizen science** projects in which ordinary people join with established scientists to answer real scientific questions. Community-based research was pioneered in the Netherlands, where several dozen research centers now study environmental issues ranging from water quality in the Rhine River, cancer rates by geographic area, and substitutes for harmful organic solvents. In each project, students and neighborhood groups team with scientists and university personnel to collect data. Their results have been incorporated into official government policies.

Similar opportunities exist in many places. The Audubon Christmas Bird Count (see Exploring Science, p. 393) is a good example. Earthwatch offers a much smaller but more intense opportunity to take part in research. Every year hundreds of Earthwatch projects each field a team of a dozen or so volunteers who spend a week or two working on issues ranging from loon nesting behavior to archaeological digs. The American River Watch

TABLE 16.4 | Most Influential Books on the Environment

What are some of the most influential readings that have informed environmental scientists? Surveys[1] of environmental leaders frequently put the same books at the top of the list. Here are votes for the top ten.

A Sand County Almanac by Aldo Leopold (100)[2]

Silent Spring by Rachel Carson (81)

State of the World by Lester Brown and the Worldwatch Institute (31)

The Population Bomb by Paul Ehrlich (28)

Walden by Henry David Thoreau (28)

Wilderness and the American Mind by Roderick Nash (21)

Small Is Beautiful: Economics as If People Mattered by E. F. Schumacher (21)

Desert Solitaire: A Season in the Wilderness by Edward Abbey (20)

The Closing Circle: Nature, Man, and Technology by Barry Commoner (18)

The Limits to Growth: A Report for the Club of Rome's Project on the Predicament of Mankind by Donella H. Meadows, et al. (17)

[1]Robert Merideth, 1992, G. K. Hall/Macmillan, Inc.
[2]Indicates number of votes for each book. Because the preponderance of respondents were from the United States (82 percent), American books are probably overrepresented.

organizes teams of students to measure water quality. You might be able to get academic credit as well as helpful practical experience in one of these research projects.

How much is enough?

Technology has made consumer goods and services so cheap and readily available that it's hard to grasp the impacts of our consumption patterns. But we do know that we in the industrialized world use vastly disproportionate amounts of resources, and we know that everyone in the world cannot consume at our level. Can we consume less and still be happy? This is partly a philosophical question. Some people say no. Others say we can become *happier* by consuming less—and by reducing the credit card debt we use to pay for the things we buy.

A century ago, economist and social critic Thorstein Veblen wrote *The Theory of the Leisure Class*, in which he coined the term **conspicuous consumption**. He used the term to describe things we buy just to impress others, things we don't really want or need. Veblen's ideas are more relevant today than ever. The average American consumes twice as many goods and services as in 1950. The average house is now more than twice as big as it was 50 years ago, even though the typical family is half as large. Shopping shapes our identity and consumes our time. We even support commerce when we watch TV or network on Facebook, both funded by commercial advertisers. As more of our daily lives and schedules become commodified, we have less time to invest in family, community, religion, and civic engagement. Social observers frequently point out that the things we buy don't really make us young, beautiful, smart, and interesting as promised. By giving so much attention to earning and spending money, we lose the time to have real friends, to cook real food, to have creative hobbies, or to do work that makes us feel we have accomplished something with our lives.

Some social critics call this accelerated consumerism "**affluenza**." A growing number of people find themselves stuck in a vicious circle: They work frantically at an unfulfilling job, to buy things they don't need, so they can save time to work even longer hours (fig. 16.16). Seeking a measure of balance in their lives, some opt out of the rat race and adopt simpler, less-consumptive lifestyles. As Thoreau wrote in *Walden*, "Our life is frittered away by detail . . . simplify, simplify."

Choosing to consume less can be an easy way to reduce your global environmental footprint and save money. Cook simple foods with friends instead of eating prepared foods. Grow a garden. Spend less time shopping and more time talking and having fun with family and friends. Although individual choices may make a small impact, collectively they have global effects (see What Can You Do?, p. 395).

16.6 CAMPUS GREENING

Colleges and universities are powerful catalysts for change. Because a fundamental purpose of universities is to collaborate in exploring new ideas, students and faculty in schools worldwide are working to develop strategies for more sustainable living and for restoring environmental quality.

Informed citizens are essential to good policy making. Knowing your area, your neighbors, local environmental issues, and the ways local concerns relate to national and global issues is a first step toward becoming an informed voter and environmental citizen. In recent years, opportunities to become educated in environmental science have increased dramatically, as "citizen science" projects have become more and more widely practiced. Part fun and part basic research support, citizen science projects provide basic data that researchers could never get without a wide network of interested volunteers.

The Christmas Bird Count is probably the most widespread and longest-lasting citizen science project in the world. Every Christmas since 1900, groups of volunteers have gone out to count birds in designated areas. For the 112th count, over 2,200 teams, with more than 60,000 participants, counted over 61 million birds belonging to 2,309 species. Most counts are in the United States or Canada, but teams from the Caribbean, Pacific Islands, and Central and South America also participate. Participants enter their bird counts on standardized online forms, and compiled data can be viewed and investigated online, almost as soon as they are submitted.

Frank Chapman, the editor of *Bird-Lore* magazine and an officer in the newly formed Audubon Society in 1900, started the Christmas Bird Count as an alternative to the tradition of Christmas bird hunts. Competition among teams to outshoot each other was taking a devastating toll on birds and mammals. Chapman suggested an alternative contest: to see which team could identify the most birds, and the most species, in a day. The competition has grown and spread. In the 100th annual count, the winning team was in Monte Verde, Costa Rica, with an amazing 343 species tallied in a single day.

What is the larger purpose to all this counting? The tens of thousands of bird-watchers participating in the count gather vastly more information about the abundance and distribution of birds than biologists could ever achieve alone. A century of accumulated data allows ecologists and ornithologists observe changes in bird migrations, populations, and habitat availability. Geographic range shifts and responses to climate change or El Niño cycles become evident. Ultimately, the data informs basic policy

▲ Participating in citizen-science projects is a good way to learn about science and your local environment.

questions regarding biodiversity, climate regulation, and conservation. You can explore the data yourself on the BirdSource website (www.birdsource.org).

The success of the Christmas Bird Count has inspired other citizen science projects. Project Feeder Watch, which began in the 1970s, has more than 15,000 participants, from schoolchildren to dedicated birders. The Great Backyard Bird Count collects records of resident summer birds. Beyond birds, schoolchildren and adults have been enlisted to observe fish, frogs, and butterflies. Farmers monitor pasture and stream health. Volunteers monitor water quality in local streams and rivers; and nature reserves solicit volunteers to help gather ecological data. You can learn more about your local environment, and contribute to scientific research, by participating in a citizen science project. Contact your local Audubon chapter or your state's Department of Natural Resources to find out what you can do.

At first glance, a sociable day of counting birds or butterflies might seem unrelated to either environmental science or environmental policy. But in a society where citizens have a voice, or where they wish to have a voice, these projects can inform, educate, inspire, and empower people to act on matters of social concern and of environmental policy. It's also a way to discover that data collection is really fun, as well as useful, when you do it together with friends.

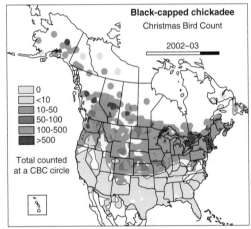

Black-capped chickadee
Christmas Bird Count

2002–03

	0
	<10
	10-50
	50-100
	100-500
	>500

Total counted at a CBC circle

▲ Volunteer data collection can produce a huge, valuable data set. Christmas Bird Count data, such as this map, are available online.   Data from Audubon Society.

Student organizations often are among our most active and effective groups on environmental policy questions. Like 350.org, countless student-led efforts have pushed for better environmental

policies. Another large and successful group is the Student Environmental Action Coalition (SEAC), formed in 1988 by students at the University of North Carolina at Chapel Hill. This umbrella

▲ **FIGURE 16.16** Is this our highest purpose? SOURCE: © 1990 Bruce von Alten.

organization and campus network has grown to include more than 30,000 members in some 500 campus environmental groups. SEAC also functions as an information clearinghouse and a training center for student leaders. Member groups undertake a diverse spectrum of activities ranging from recycling programs to protesting environmentally destructive public policies and projects. National conferences bring together thousands of activists who share tactics and inspiration while also having fun.

A network of Public Interest Research Groups (PIRGs) do policy research on campuses across the United States. The PIRGs give students opportunities to explore and promote environmental and social issues. Instead of doing research because it's assigned, participants can research problems they really care about.

If you are interested in a career in environmental policy, law, science, journalism, or other related fields, getting involved in student groups is a good way to build knowledge and experience. Many of today's environmental and political leaders gained experience in policy, leadership, and organizing by working with student groups (fig. 16.17). These activities can be as informative and life-changing as the courses you take, because they give you the experience of being in charge and making decisions.

Electronic media are changing the world

One of the lessons of 350.org is that new media provide new opportunities to organize action. The power of social media to organize mass demonstrations and change political systems—sometimes practically overnight—was demonstrated across North Africa and the Middle East in 2011. Youth with few resources other than cell phones, blogs, and a social network can have a powerful impact.

The Occupy Wall Street movement was another media-rich project. Events, updates, images, and tweets were broadcast worldwide. International audiences watched protests and commented on

blogs. Supporters in Italy and Germany sent online orders for pizzas in New York to feed protesters. With global visibility, these protests opened up public policy debates that had languished for decades.

You may not want to overthrow a government, but you can influence public opinion and political decision making in ways never possible a generation ago. The linkages of social media, online videos, and blogs create a virtual media network that competes with traditional media networks. Where media was once controlled by a handful of corporations and publishing moguls, now all of us have the power to communicate widely. The question is how you will choose to use these new tools.

Schools are embracing green building

Colleges and universities often serve as incubators for new ideas. Students and faculty have the knowledge, energy, enthusiasm, and opportunity to do basic research and innovation. But school administrators also are motivated to promote more sustainable strategies—especially when those strategies involve energy savings that cut costs in the long term.

Green building is one of the areas in which colleges and universities often show leadership. More than 110 colleges have buildings certified to LEED (Leadership in Energy and Environmental Design) standards. Prize-winning sustainable designs can be found at large institutions, such as Stanford University, and small colleges, such as Oberlin College in Ohio or Middlebury in Vermont. Oberlin's Center for Environmental Studies, designed by architect Bill McDonough, features 370 m² of photovoltaic panels on its roof, a geothermal well to help heat and cool the building, large south-facing windows for passive solar gain, and a "living machine" for water treatment, including plant-filled tanks in an indoor solarium and a constructed wetland outside (fig. 16.18).

▲ **FIGURE 16.17** Working together with others can give you energy, inspiration, and a sense of accomplishment.

What Can YOU DO?

Reducing Your Impact

Each of us can conserve much of the water we use and avoid water pollution in many simple ways.

Purchase Less

Ask yourself whether you really need more stuff. Avoid buying things you don't need or won't use.

Use items as long as possible (and don't replace them just because a new product becomes available).

Use the library instead of purchasing books you read.

Give nonmaterial gifts such as music or tickets to a play.

Reduce Excess Packaging

Carry reusable bags when shopping and refuse bags for small purchases.

Buy items in bulk or with minimal packaging; avoid single-serving foods. Choose packaging that can be recycled or reused.

Avoid Disposable Items

Use cloth napkins, handkerchiefs, and towels.

Bring a washable cup to meetings; use washable plates and utensils rather than single-use items.

Buy pens, razors, flashlights, and cameras with replaceable parts.

Choose items built to last and have them repaired; you will save materials and energy while providing jobs in your community.

Conserve Energy

Plan to live in a place where you can walk, bicycle, or use public transportation to get to work or school.

Carpool and combine trips to reduce car mileage.

Dry your clothes on a clotheslines to avoid using a clothes dryer.

Turn off lights, water, heat, and air conditioning when not in use; close windows when you leave a room; use fans instead of air conditioning.

Save Water

Water lawns and gardens at night, to minimize evaporation.

Use water-saving devices and fewer flushes with toilets.

Don't leave water running when washing hands, food, dishes, and teeth.

Based on material by Karen Oberhauser, Bell Museum Imprint, University of Minnesota, 1992. Used by permission.

▲ **FIGURE 16.18** The Adam Joseph Lewis Center for Environmental Studies at Oberlin College is designed to be self-sustaining even in northern Ohio's cool, cloudy climate. Large, south-facing windows let in sunlight, while 370 m² of solar panels on the roof generate electricity. Natural wastewater treatment, including a constructed wetland, purifies wastewater.

and compares it to long-term averages. Classrooms within the dorm offer environmental science classes in which students can see sustainability in action. Green dorms are also popular with students. Natural lighting, clean air, allergen-free materials, and other features make buildings more comfortable and less stressful for residents. One of the largest green dorms in the country is at the University of South Carolina, where there are long waiting lists for a room in the new, well-designed building.

Audits help reduce energy consumption

Campus audits are an area where students, faculty, and administrators collaborate to examine water and energy use, waste production and disposal, paper consumption, recycling, buying locally produced food, and other strategies to cut resource consumption. There are several different auditing systems that help guide and organize data collection on campus consumption of energy and materials. Measures can include policies on green building, food and recycling programs, climate change impacts, and energy consumption. Courses and facilities aimed at teaching more efficient and sustainable practices can also be part of a campus audit.

Audits are useful because, as in all of environmental science, we need meaningful data to evaluate how a system is working. Audits show where a campus is performing well and where to focus efforts. One of the leading campus audit programs is run by the Association for Advancement of Sustainability in Higher

These facilities can become important educational opportunities. In conjunction with a new, energy-efficient field station, Stanford's students worked with the administration to develop Guidelines for Sustainable Buildings, covering everything from energy-efficient lighting to native landscaping. Carnegie Mellon University students helped design a green roof for a new dorm on their campus in Pittsburgh. Students can now monitor how the living roof is reducing storm water drainage and improving water quality. A kiosk inside the dorm shows daily energy use

Education (AASHE). Institutions that do an AASHE-based audit can compare their performance to that of some 870 other educational institutions, and they can plot progress over time.

A recent study by the Sustainable Endowments Institute evaluated more than 100 of the leading colleges and universities in the United States. The report card ranked Dartmouth, Harvard, Stanford, and Williams as the top of the "A list" of 23 greenest campuses. Berea College in Kentucky got special commendation as a small school with a strong commitment to sustainability. Its "ecovillage" has a student-designed house that produces its own electricity and treats waste water in a living system. The college has a full-time sustainability coordinator to provide support to campus programs, community outreach, and teaching.

The Campus Climate Challenge, recently launched by a coalition of nonprofit groups, seeks to engage students, faculty, and staff at 500 college campuses in the United States and Canada in a long-term campaign to eliminate global-warming pollution (www.energyaction.net). This challenge has pushed campuses to invest in clean energy, set strict green building standards for new construction, purchase fuel-efficient vehicles, and adopt other policies to save energy and reduce their greenhouse gas emissions. Concordia University in Austin, Texas, was the first college or university in the country to purchase all of its energy from renewable sources. The 5.5 million kilowatt-hours of "green power" it uses each year will eliminate about 8 million pounds of CO_2 emissions annually, the equivalent of planting 1,000 acres of trees or taking 700 cars off the roads.

A growing number of colleges are cutting reliance on fossil fuels by installing their own wind turbines, geothermal heating and cooling, and solar energy systems (fig. 16.19). The University of Minnesota at Morris has two turbines that are expected to provide all the school's electrical needs as well as a surplus that can be sold to the local utility. The Morris campus also has a cellulose gasifier to burn corn stalks and cobs and heat and cool the campus.

▼ **FIGURE 16.19** Many schools have installed their own wind turbines. Although a large turbine may cost about $2 million, the payback can be as short as ten years.

At many schools, students have persuaded the administration to buy locally produced food and to provide organic, vegetarian, and fair-trade options in campus cafeterias. Administrators often support green plans because they garner public goodwill and serve as recruiting tools for prospective students. These plans also provide a powerful teaching tool and everyday reminder that individual actions matter.

As a final gesture with potentially powerful effects, graduating students at more than 100 universities and colleges across America have taken this pledge at graduation time:

> "I pledge to explore and take into account the social and environmental consequences of any job I consider and will try to improve these aspects of any organization for which I work."

16.7 THE CHALLENGE OF SUSTAINABLE DEVELOPMENT

The idea of sustainability is that if we hope to be here for the long term, we can't deplete the natural systems we depend on for food, water, energy, fiber, waste disposal, and other life-support services. The idea of **sustainable development** is to share opportunity by improving people's lives in impoverished regions, and to extend human well-being over many generations. To be truly enduring, these benefits must be available to all humans, not just to members of a privileged group.

Often we might question whether these ideals are even possible. Absolute sustainability may be an unachievable goal. Improving livelihoods requires increased consumption, if we follow models developed in history. Yet sustainability may also be a goal worth aiming for, like world peace or universal access to potable water and safe food. Critics often point out that the idea of sustainability is so broad that nobody knows exactly what it means. On the other hand, the idea has been a persistently useful organizing idea for efforts to conserve life-support systems for our grandchildren and great-grandchildren.

These concerns become increasingly evident as the developing countries of the world become more affluent, and as they adopt the extravagant consumption patterns of richer countries (fig. 16.20). Automobile production in China, for example, has increased from only 5,200 in 1985 to 18 million in 2011. There are now more than 100 million motor vehicles on the road in China, and the total is expected to pass 200 million by 2020. What will be the effect on air quality, world fossil fuel supplies, and global climate if that growth rate continues? Already, two-thirds of the children in Shenzhen, China's wealthiest province, suffer from lead poisoning, probably caused by use of leaded gasoline. And as discussed in chapter 8, diseases associated with affluent lifestyles—obesity, diabetes, heart attacks, depression, and traffic accidents—are becoming the leading causes of illness and mortality worldwide.

We would all benefit by helping developing countries access more efficient, less-polluting technologies. Education, democracy, and access to ideas and information are essential for sustainability. Many scholars and social activists believe that poverty is at the core of many of the world's most serious human problems: hunger,

▲ **FIGURE 16.20** Rising standards of living in developing countries, such as China, have brought better livelihoods but more consumption. What is the best way to plan for the future?

child deaths, migrations, insurrections, and environmental degradation. One way to alleviate poverty is to foster economic growth so there can be a bigger share for everyone.

Strong economic growth already is occurring in many places. The World Bank projects that if current trends continue, economic output in developing countries will rise by 4 to 5 percent per year in the next 40 years. Economies of industrialized countries are expected to grow more slowly but could still triple over that period. Altogether, the total world output could be quadruple what it is today. That growth could draw down resources, or it could provide funds to clean up environmental damage caused by earlier, wasteful technologies and misguided environmental policies. It is expected to cost $350 billion per year to develop renewable energy sources, stop soil erosion, protect ecosystems, control population growth, and provide a decent standard of living for the world's poor. This is a great deal of money, but it is small compared to over $1 trillion or more per year spent on wars and military equipment.

Economic growth usually implies an increase in average welfare or well-being. Sustainable development implies that growth can be based on nonconsumptive activities, such as education or arts, as well as on carefully managed renewable resources, such as soils, forests, and fisheries (fig. 16.21). The World Commission on Environment and Development defined sustainable development in *Our Common Future* as "meeting the needs of the present without compromising the ability of future generations to meet their own needs". Some goals of sustainable development include these points:

- A demographic transition to a stable world population of low birth and death rates.
- An energy transition to high efficiency in production and use, coupled with increasing reliance on renewable resources.
- A resource transition to reliance on nature's "income" without depleting its "capital."

- An economic transition to sustainable development and a broader sharing of its benefits.
- A political transition to global negotiation grounded in complementary interests between North and South, East and West.
- An ethical or spiritual transition to attitudes that do not separate us from nature or each other.

In 2000, UN secretary-general Kofi Annan called for a millennium assessment of the consequences of ecosystem change on human well-being as well as the scientific basis for actions to enhance the conservation and sustainable use of those systems. More than 1,360 experts from around the world worked on technical reports about the conditions and trends of ecosystems, scenarios for the future, and possible responses.

The primary aims of the Millennium Development Goals are to measurably improve health, education, and welfare by 2015. Eight specific concerns are ending hunger and poverty, improving access to education, equity for women (who have primary responsibility for children), improving infant and maternal health, combating major diseases, and supporting environmental quality (table 16.5). Together these goals seek to improve opportunities for future generations around the world (fig. 16.22).

Within these goals are discussions of many of the topics you have read about in this book: water resources, food and arable land, sustainable energy, and many other topics. Are these aims achievable? What can we do—individually and collectively—to work toward these objectives?

▲ **FIGURE 16.21** A model for integrating ecosystem health, human needs, and sustainable economic growth.

TABLE 16.5 | Millennium Development Goals

GOALS (AND SPECIFIC OBJECTIVES)

1. Eradicate extreme poverty and hunger.
   a. Cut by half the proportion of people living on less than $1/day.
   b. Reduce by half the proportion of hungry people.

2. Achieve universal primary education.
   a. Ensure that all boys and girls complete primary school.

3. Promote gender equality and empower women.
   a. Eliminate gender disparity in primary and secondary education by 2015.

4. Reduce child mortality.
   a. Reduce by two-thirds the mortality rate among children under five.

5. Improve maternal health.
   a. Reduce by three-quarters the maternal mortality ratio.

6. Combat HIV/AIDS, malaria, and other diseases.
   a. Halt and begin to reverse the spread of HIV/AIDS.
   b. Halt and begin to reverse the spread of malaria and other major diseases.

7. Ensure environmental sustainability.
   a. Integrate the principles of sustainable development into policies and programs; reverse the loss of environmental resources.
   b. Reduce by half the proportion of people without sustainable access to safe drinking water.
   c. Improve the lives of 100 million slum dwellers by 2020.

8. Develop a global partnership for development.
   a. Develop further an open trading and financial system that is rule-based, predictable, and nondiscriminatory, including good governance, development, and poverty reduction.
   b. Address the least-developed countries' special needs. Develop tariff-free and quota-free access for their exports; enhanced debt relief for heavily indebted poor countries.

▲ **FIGURE 16.22** Sustainable development aims to meet the needs of the present without compromising the ability of future generations to meet their own needs.

CONCLUSION

The policy choices we make now, at scales from national policy to individual consumption patterns, will influence environmental quality and natural resources for future generations. In this chapter we have examined some of some basic environmental policies, and the ways policies are formed. There are many players in the formation of environmental policy: legislative, administrative, and judicial branches of governments, at local, state and federal scales, all have power to shape policy. Individuals also have opportunities to influence policy, especially when they act together.

Student activism has long been essential both for shaping policy and for building leadership skills and experience in the leaders of tomorrow. Working with local groups on local issues is an essential opportunity to gain skills that can be useful in your career and life. You can apply these skills in a variety of fields, at national, local, and international scales.

Sustainable development is one of our greatest challenges. People everywhere wish to have the same comfort and opportunities enjoyed by those in wealthier regions. Finding ways to share opportunity without destroying resources requires creativity and commitment (fig. 16.22). As summarized in the millennium assessment, these are some of the challenges we must address:

- All of us depend on nature and ecosystem services to provide the conditions for a decent, healthy, and secure life.

- Growing demands for food, fresh water, and energy have made unprecedented demands on ecosystems and resources. These changes have improved the lives of billions but also weakened nature's ability to deliver other key services, such as purification of air and water or production of foods, fish, and climate regulation. Dramatic acceleration of species extinctions, pressure on ecosystems, and consumption of resources threatens our own well-being.

- Currently available technology and knowledge can greatly reduce our impacts on ecosystems. Better accounting for the value of services and the cost of lost resources is important for helping us rethink our use of these systems. A central remaining question is whether we can find ways to collaborate across sectors of governments, businesses, international institutions, and communities in order to find creative solutions for the future.

PRACTICE QUIZ

1. What is a *policy*? How are policies formed?
2. Describe three important provisions of NEPA.
3. List four important U.S. environmental laws (besides NEPA), and briefly describe what each does.
4. Why are international environmental conventions and treaties often ineffective? What can make them more successful?
5. Why is the U.S. Supreme Court case *Citizens United v. Federal Election Commission* controversial?
6. List two broad goals of environmental education identified by the National Environmental Education Act.
7. What is *citizen science*, and what are some of its benefits? Describe one citizen science project.
8. List five things each of us could do to help preserve our common environment.
9. Describe some things schools and students have done to promote sustainable living.
10. Define *sustainable development* and describe some of its principal tenets.

CRITICAL THINKING AND DISCUSSION

Apply the principles you have learned in this chapter to discuss these questions with other students.

1. In your opinion, how much environmental protection is too much? Think of a practical example in which some stakeholders may feel oppressed by governmental regulations. How would you justify or criticize these regulations?
2. Which is a better choice at the grocery store, paper or plastic? How would you evaluate the trade-offs between packaging choices? What evidence would you look for to make this decision in your life?
3. Do you agree with Margaret Mead that a small group of committed individuals is the only thing that can change the world? What do you think she meant? Think of some examples of groups of individuals who have changed the world. How did they do it?
4. Suppose that you were going to make a presentation on sustainability to your school administrators. What suggestions would you make for changes on your campus? What information would you need to support these proposals?
5. Debate this question: Is sustainability an achievable goal? What do you think are the main barriers to attaining this objective? What ideas in this book give you reason to believe sustainability is possible? What are the alternatives?

DATA ANALYSIS Campus Environmental Audit

If you want to understand how sustainable your campus is, the first step is to gather data. A campus environmental audit can be a huge, professionally executed task, but you can make a reasonable approximation if you work with your class to gather some basic information. Among the many places to find established listings of factors to consider, one widely used audit is that of the Association for the Advancement of Sustainability in Higher Education. You can find details of their full audit system online, but you can do a simplified version yourself or in collaboration with others in your class. Factors involved in an audit include building efficiency, transportation alternatives, waste management, dining facilities, administrative actions, and other factors. How sustainable is your school? Go to Connect to find a basic campus environmental audit. Then you can collect and evaluate your own data to find out.

Appendixes

Vegetation

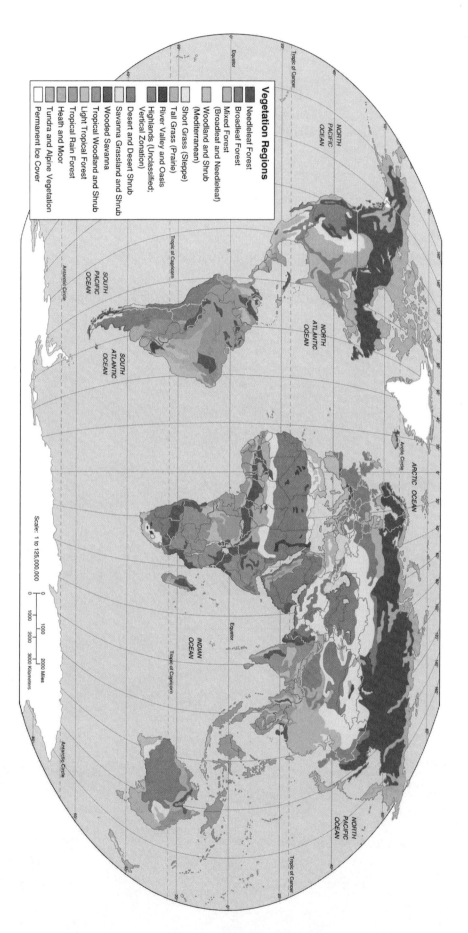

Vegetation Regions

- Needleleaf Forest
- Broadleaf Forest
- Mixed Forest
  (Broadleaf and Needleleaf)
- Woodland and Shrub
  (Mediterranean)
- Short Grass (Steppe)
- Tall Grass (Prairie)
- River Valley and Oasis
- Highlands (Unclassified;
  Vertical Zonation)
- Desert and Desert Shrub
- Savanna Grassland and Shrub
- Wooded Savanna
- Tropical Woodland and Shrub
- Light Tropical Forest
- Tropical Rain Forest
- Heath and Moor
- Tundra and Alpine Vegetation
- Permanent Ice Cover

Scale: 1 to 125,000,000

0   1000   2000 Miles
0   1000   2000   3000 Kilometers

Vegetation is the most visible consequence of the distribution of temperature and precipitation. The global distribution of vegetation types and the global distribution of climate are closely related, but not all vegetation types are the consequence of temperature and precipitation or other climatic variables. Many types of vegetation, in many areas of the world, are the consequence of human activities, particularly the grazing of domesticated livestock, burning, and forest clearance.

World Population Density

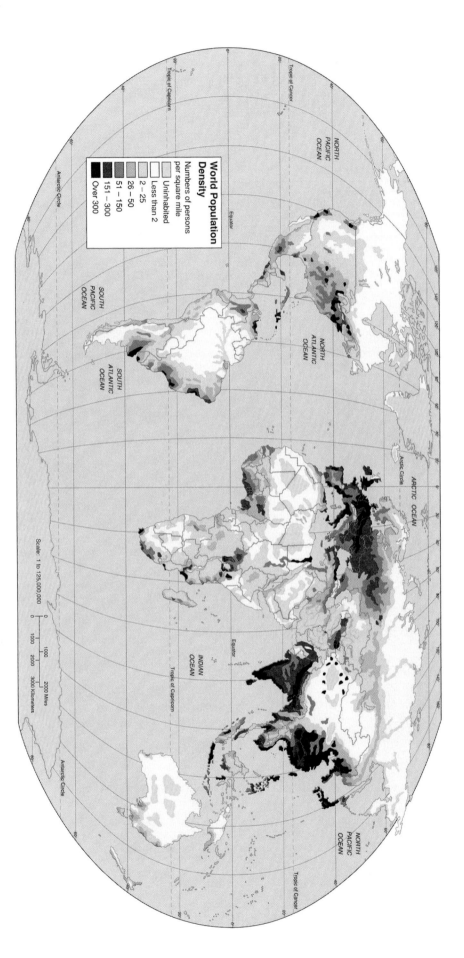

World Population Density

Numbers of persons per square mile

- Uninhabited
- Less than 2
- 2 – 25
- 26 – 50
- 51 – 150
- 151 – 300
- Over 300

Scale: 1 to 125,000,000

No feature of human activity is more reflective of environmental conditions than where people live. In the areas of densest population, a mixture of natural and human factors have combined to allow maximum food production, maximum urbanization, and especially concentrated economic activity. Three such great concentrations appear on the map—East Asia, South Asia, and Europe—with a fourth lesser concentration in eastern North America (the "Megalopolis" region of the United States and Canada). The areas of future high density (in addition to those already existing) are likely to be in Middle and South America and Africa, where population growth rates are well above the world average. Population that is extremely dense or growing at an excessive rate when measured against a region's habitability is one of the greatest indicators of environmental deterioration.

Temperature Regions and Ocean Currents

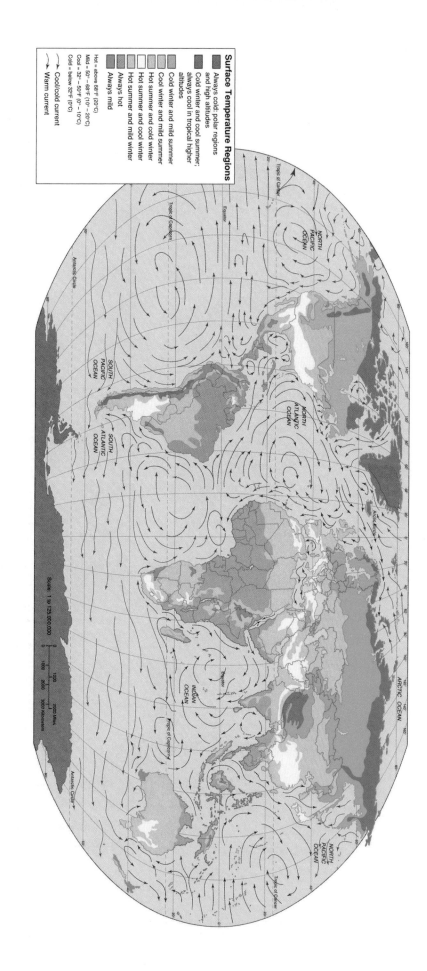

Surface Temperature Regions

- Always cold: polar regions and high altitudes
- Cold winter and cool summer; always cool in tropical higher altitudes
- Cold winter and mild summer
- Cool winter and mild summer
- Cool winter and cold winter
- Hot summer and cold winter
- Hot summer and cool winter
- Always hot
- Always mild

Hot = above 68°F (20°C)
Mid = 50° – 68°F (10° – 20°C)
Cool = 32° – 50°F (0° – 10°C)
Cold = below 32°F (0°C)

→ Cool/cold current
→ Warm current

Scale: 1 to 125,000,000

Along with precipitation, temperature is one of the two most important environmental variables, defining the climatic conditions so essential for the distribution of human activities and the human population. Ocean currents exert a significant influence over the climate of adjacent continents and are the most important mechanism for redistributing surplus heat from the equatorial region into middle and high latitudes.

Glossary

A

abundance The number of individuals of a species in an area.

acid precipitation Acidic rain, snow, or dry particles deposited from the air due to increased acids released by anthropogenic or natural resources.

acids Substances that release hydrogen atoms in water.

active solar systems Mechanical systems that use moving substances to collect and transfer solar energy.

acute effects A sudden onset of symptoms or effects of exposure to some factor.

acute poverty Insufficient income or access to resources needed to provide the basic necessities for life, such as food, shelter, sanitation, clean water, medical care, and education.

adaptation Physical changes that allow organisms to survive in a given environment.

adaptive management A management plan designed from the outset to "learn by doing" and to actively test hypotheses and adjust treatments as new information becomes available.

administrative law Executive orders, administrative rules and regulations, and enforcement decisions by administrative agencies and special administrative courts.

aerosols Minute particles or liquid droplets suspended in the air.

affluenza An addiction to spending and consuming beyond one's needs.

albedo A description of a surface's reflective properties.

allergens Substances that activate the immune system and cause an allergic response; may not be directly antigenic themselves but may make other materials antigenic.

allopatric speciation Species that arise from a common ancestor through geographic isolation or some other barrier to reproduction.

ambient air The air immediately around us.

amorphous silicon collectors Photovoltaic cells made from randomly assembled silicon molecules rather than silicon crystals. Amorphous collectors are less efficient but far cheaper than crystalline collectors.

analytical thinking A way of systematic analysis that asks, "How can I break this problem down into its constituent parts?"

anemia Low levels of hemoglobin due to iron deficiency or lack of red blood cells.

anthropocentric Believing that humans hold a special place in nature; being centered primarily on humans and human affairs.

antigens Substances that stimulate the production of, and react with, specific antibodies.

aquifers Porous, water-bearing layers of sand, gravel, and rock below the earth's surface; reservoirs for groundwater.

arithmetic scale A pattern of growth that increases at a constant amount per unit time, such as 1, 2, 3, 4 or 1, 3, 5, 7.

atmospheric deposition Sedimentation of solids, liquids, or gaseous materials from the air.

atom The smallest particle that exhibits the characteristics of an element.

atomic number The characteristic number of protons per atom of an element.

autotroph An organism that synthesizes food molecules from inorganic molecules by using an external energy source, such as light energy.

B

barrier islands Low, narrow, sandy islands that form offshore from a coastline.

Basel Convention Restricts shipment of hazardous waste across international boundaries.

bases Substances that readily bond with hydrogen ions in an aqueous solution.

Batesian mimicry Evolution by one species to resemble another species that is protected from predators by a venomous stinger, bad taste, or some other defensive adaptation.

benthic The bottom of a sea or lake.

binomials Scientific or Latin names that combine the genus and species, e.g., *Zea mays*.

bioaccumulation The selective absorption and concentration of molecules by cells.

biocentrism The belief that all creatures have rights and values; being centered on nature rather than humans.

biochemical oxygen demand (BOD) A standard test for measuring the amount of dissolved oxygen utilized by aquatic microorganisms.

biodegradable plastics Plastics that can be decomposed by microorganisms.

biodiversity The genetic, species, and ecological diversity of the organisms in a given area.

biofuel Fuel made from biomass.

biogeochemical cycles Movement of matter within or between ecosystems; caused by living organisms, geologic forces, or chemical reactions. The cycling of nitrogen, carbon, sulfur, oxygen, phosphorus, and water are examples.

biological community The populations of plants, animals, and microorganisms living and interacting in a certain area at a given time.

biological controls Use of natural predators, pathogens, or competitors to regulate pest populations.

biomagnification Increase in concentration of certain stable chemicals (for example, heavy metals or fat-soluble pesticides) in successively higher trophic levels of a food chain or web.

biomass The accumulated biological material produced by living organisms.

biomass fuel Organic material produced by plants, animals, or microorganisms that can be burned directly as a heat source or converted into a gaseous or liquid fuel.

biomass pyramid A metaphor or diagram that explains the relationship between the amounts of biomass at different trophic levels.

biomes Broad, regional types of ecosystems characterized by distinctive climate and soil conditions and distinctive kinds of biological community adapted to those conditions.

bioremediation Use of biological organisms to remove pollution or restore environmental quality.

biosphere The zone of air, land, and water at the surface of the earth that is occupied by organisms.

biosphere reserves World heritage sites identified by the IUCN as worthy for national park or wildlife refuge status because of high biological diversity or unique ecological features.

biotic potential The maximum reproductive rate of an organism, given unlimited resources and ideal environmental conditions.

birth control Any method used to reduce births, including celibacy, delayed marriage, contraception; devices or medications that prevent implantation of fertilized zygotes and induced abortions.

blind experiments A design in which researchers don't know which subjects were given experimental treatment until after data have been gathered and analyzed.

bogs Areas of waterlogged soil that tend to be peaty; fed mainly by precipitation; low productivity; some bogs are acidic.

boreal forest A broad band of mixed coniferous and deciduous trees that stretches across northern North America (and Europe and Asia); its northernmost edge, the taiga, intergrades with the arctic tundra.

brownfields Abandoned or underused urban areas in which redevelopment is blocked by liability or financing issues related to toxic contamination.

C

cancer Invasive, out-of-control cell growth that results in malignant tumors.

cap-and-trade agreement A policy to set pollution limits, then allow companies to buy and sell their allotted rights to emit pollutants.

capital Any form of wealth, resources, or knowledge available for use in the production of more wealth.

carbohydrate An organic compound consisting of a ring or chain of carbon atoms with hydrogen and oxygen attached; examples are sugars, starches, cellulose, and glycogen.

carbon cycle The circulation and reutilization of carbon atoms, especially via the processes of photosynthesis and respiration.

carbon management Projects to reduce carbon dioxide emissions from fossil fuel or to ameliorate their effects.

carbon monoxide Colorless, odorless, nonirritating but highly toxic gas produced by incomplete combustion of fuel, incineration of biomass or solid waste, or partially anaerobic decomposition of organic material.

carbon neutral Producing no net carbon dioxide emissions.

carbon sink Places of carbon accumulation, such as in large forests (organic compounds) or ocean sediments (calcium carbonate).

carcinogens Substances that cause cancer.

carnivores Organisms that mainly prey upon animals.

carrying capacity The maximum number of individuals of any species that can be supported by a particular ecosystem on a long-term basis.

case law Precedents from both civil and criminal court cases.

Catalytic converter The device on an automobile that uses platinum-palladium and rhodium catalysts to remove NO_x, hydrocarbons, and carbon monoxide from the exhaust.

cell Minute compartments surrounded by semipermeable membranes within which the processes of life are carried out by all living organisms.

cellular respiration The process in which a cell breaks down sugar or other organic compounds to release energy used for cellular work; may be anaerobic or aerobic, depending on the availability of oxygen.

chain reaction A self-sustaining reaction in which the fission of nuclei produces subatomic particles that cause the fission of other nuclei.

chaparral A biological community characterized by thick growth of thorny, evergreen shrubs typical of a Mediterranean climate.

chemical bond The force that holds molecules together.

chemical energy Potential energy stored in chemical bonds of molecules.

chemosynthesis Extracting energy for life from inorganic chemicals, such as hydrogen sulfide, rather than from sunlight.

chlorinated hydrocarbons Hydrocarbon molecules to which chlorine atoms are attached. Often used as pesticides and are both highly toxic and long-lasting in the environment.

chlorofluorocarbons Chemical compounds with a carbon skeleton and one or more attached chlorine and fluorine atoms. Commonly used as refrigerants, solvents, fire retardants, and blowing agents.

chloroplasts Chlorophyll-containing organelles in eukaryotic organisms; sites of photosynthesis.

chronic effects Long-lasting results of exposure to a toxin; can be a permanent change caused by a single, acute exposure or a continuous, low-level exposure.

citizen science Projects in which trained volunteers work with scientific researchers to answer real-world questions.

civil law A body of laws regulating relations between individuals or between individuals and corporations concerning property rights, personal dignity and freedom, and personal injury.

classical economics Modern, Western economic theories of the effects of resource scarcity, monetary policy, and competition on supply and demand of goods and services in the marketplace. This is the basis for the capitalist market system.

clear-cutting Cutting every tree in a given area, regardless of species or size; an appropriate harvest method for some species; can be destructive if not carefully controlled.

climate A description of the long-term pattern of weather in a particular area.

climax community A long-lasting, self-sustaining community resulting from ecological succession that is resistant to disturbance.

closed-canopy A forest where tree crowns spread over 20 percent of the ground; has the potential for commercial timber harvests.

closed system A system in which there is no exchange of energy or matter with its surroundings.

cloud forests High mountain forests where temperatures are uniformly cool and fog or mist keeps vegetation wet all the time.

coevolution The process in which species exert selective pressure on each other and gradually evolve new features or behaviors as a result of those pressures.

cogeneration The simultaneous production of electricity and steam or hot water in the same plant.

coliform bacteria Bacteria that live in the intestines (including the colon) of humans and other animals; used as a measure of the presence of feces in water or soil.

commensalism A symbiotic relationship in which one member is benefited and the second is neither harmed nor benefited.

communal resource management systems Resources managed by a community for long-term sustainability.

community (ecological) structure The patterns of spatial distribution of individuals, species, and communities.

competitive exclusion A theory that no two populations of different species will occupy the same niche and compete for exactly the same resources in the same habitat for very long.

complexity The number of species at each trophic level and the number of trophic levels in a community.

composting The biological degradation of organic material under aerobic (oxygen-rich) conditions to produce compost, a nutrient-rich soil amendment and conditioner.

compound Substances composed of different kinds of atoms.

confidence limits Upper and lower values in which the true value (such as a mean) is likely to fall.

confined animal-feeding operation Feeding large numbers of livestock at a high density in pens or barns.

conifer A needle-bearing tree that produces seeds in cones.

coniferous See *conifer*.

conservation medicine Attempts to understand how changes we make in our environment threaten our health as well as that of natural communities on which we depend.

conservation of matter In any chemical reaction, matter changes form; it is neither created nor destroyed.

conspicuous consumption A term coined by economist and social critic Thorstein Veblen to describe buying things we don't want or need to impress others.

constructed wetlands Artificially constructed wetlands.

consumers Organisms that obtain energy and nutrients by feeding on other organisms or their remains. See also *heterotroph*.

consumption The fraction of withdrawn water that is lost in transmission or that is evaporated, absorbed, chemically transformed, or otherwise made unavailable for other purposes as a result of human use.

contour plowing Plowing along hill contours; reduces erosion.

control rods Neutron-absorbing material inserted into spaces between fuel assemblies in nuclear reactors to regulate fission reaction.

controlled studies Comparisons made between two populations that are identical (as far as possible) in every factor except the one being studied.

convection currents Rising or sinking air currents that stir the atmosphere and transport heat from one area to another. Convection currents also occur in water.

Convention on International Trade in Endangered Species (CITIES) An international convention to protect endangered species.

conventional (criteria) pollutants The seven substances (sulfur dioxide, carbon monoxide, particulates, hydrocarbons, nitrogen oxides, photochemical oxidants, and lead) identified by the Clean Air Act as the most serious threat of all pollutants to human health and welfare.

convergent evolution Species evolve from different origins but under similar environmental conditions to have similar traits.

coral bleaching Whitening of corals when stressors, such as high temperatures, induce corals to expel their colorful single-celled protozoa, known as zooxanthellae, or when zooxanthellae die. Death of the coral reef may result.

coral reefs Prominent oceanic features composed of hard, limy skeletons produced by coral animals; usually formed along edges of shallow, submerged ocean banks or along shelves in warm, shallow, tropical seas.

core The dense, intensely hot mass of molten metal, mostly iron and nickel, thousands of kilometers in diameter at the earth's center.

core habitat A habitat patch large enough and with ecological characteristics suitable to support a critical mass of the species that make up a particular community.

Coriolis effect The tendency for air above the earth to appear to be deflected to the right (in the Northern Hemisphere) or the left (in the South) because of the earth's rotation.

corridors Strips of natural habitat that connect two adjacent nature preserves to allow migration of organisms from one place to another.

cost-benefit analysis (CBA) An evaluation of large-scale public projects by comparing the costs and benefits that accrue from them.

cover crops Plants, such as rye, alfalfa, or clover, that can be planted immediately after harvest to hold and protect the soil.

creative thinking Original, independent thinking that asks, "How might I approach this problem in new and inventive ways?"

criminal law A body of court decisions based on federal and state statutes concerning wrongs against persons or society.

criteria pollutants See *conventional pollutants*.

critical factor The single environmental factor closest to a tolerance limit for a given species at a given time.

critical thinking An ability to evaluate information and opinions in a systematic, purposeful, efficient manner.

crude birth rate The number of births per thousand persons in a given year (using the midyear population).

crude death rate The number of deaths per thousand persons in a given year; also called crude mortality rate.

crust The cool, lightweight, outermost layer of the earth's surface that floats on the soft, pliable underlying layers; similar to the "skin" on a bowl of warm pudding.

cultural eutrophication An increase in biological productivity and ecosystem succession caused by human activities.

D

debt-for-nature swaps Forgiveness of international debt in exchange for nature protection in developing countries.

deciduous Trees and shrubs that shed their leaves at the end of the growing season.

decomposer Fungus or bacterium that breaks complex organic material into smaller molecules.

deductive reasoning "Top down" reasoning in which we start with a general principle and derive a testable prediction about a specific case.

deforestation Removing trees from a forest.

demographic transition A pattern of falling death rates and birth rates in response to improved living conditions; typically leads to rapid then stabilizing population growth.

demography The statistical study of human populations relating to growth rate, age structure, geographic distribution, etc., and their effects on social, economic, and environmental conditions.

density-dependent factors Either internal or external factors that affect growth rates of a population depending on the density of the organisms in the population.

density-independent A population is affected no matter what its size.

deoxyribonucleic acid (DNA) The double helix of genetic material that determines heredity.

dependency ratio The number of nonworking members compared with working members for a given population.

dependent variable Also known as the response variable; is one affected by other variables.

desalinization (or desalination) Removal of salt from water by distillation, freezing, or ultrafiltration.

desertification Denuding and degrading a once fertile land, initiating a desert-producing cycle that feeds on itself and causes long-term changes in soil, climate, and biota of an area.

deserts Biomes characterized by low moisture levels and infrequent and unpredictable precipitation. Daily and seasonal temperatures fluctuate widely.

detritivore Organisms that consume organic litter, debris, and dung.

dieback A sudden population decline; also called a population crash.

disability-adjusted life years (DALYs) A health measure that assesses the total burden of disease by combining premature deaths and loss of a healthy life that result from illness or disability.

discharge The amount of water that passes a fixed point in a given amount of time; usually expressed as liters or cubic feet of water per second.

discount rate The amount we discount or reduce the value of a future payment. When you borrow money from the bank at 10 percent annual interest, you are in effect saying that having the money now is worth 10 percent more to you than having the same amount one year from now.

disease A deleterious change in the body's condition in response to destabilizing factors, such as nutrition, chemicals, or biological agents.

dissolved oxygen (DO) content Amount of oxygen dissolved in a given volume of water at a given temperature and atmospheric pressure; usually expressed in parts per million (ppm).

disturbance Any force that disrupts the established patterns and processes, such as species diversity and abundance, community structure, community properties, or species relationships.

disturbance-adapted species Species that depend on repeated disturbance for their survival and propagation.

diversity The number of species present in a community (species richness), as well as the relative abundance of each species.

DNA Deoxyribonucleic acid; the long, double-helix molecule in the nucleus of cells that contains the genetic code and directs the development and functioning of all cells.

double-blind experiment Neither the subject (participant) nor the experimenter knows which participants are receiving the experimental or the control treatments until after data have been gathered and analyzed.

dust domes High concentrations of dust and aerosols in the air over cities.

E

earthquakes Sudden, violent movement of the earth's crust.

ecological diseases Sudden, wide-spread epidemics among livestock and wild species.

ecological economics Application of ecological insights to economic analysis; incorporating ecological principles and priorities into economic accounting systems.

ecological footprint An estimate of our individual and collective environmental impacts. It is usually calculated and expressed as the area of bioproductive land required to support a particular lifestyle.

ecological niche The functional role and position of a species in its ecosystem, including what resources it uses, how and when it uses the resources, and how it interacts with other species.

ecological services Processes or materials, such as clean water, energy, climate regulation, and nutrient cycling, provided by ecosystems.

ecological succession The process by which organisms gradually occupy a site, alter its ecological conditions, and are eventually replaced by other organisms.

ecosystem A specific biological community and its physical environment interacting in an exchange of matter and energy.

ecosystem management An integration of ecological, economic, and social goals in a unified systems approach to resource management.

ecosystem restoration To reinstate an entire community of organisms to as near its natural condition as possible.

ecosystem services Resources or services provided by environmental systems.

ecotones Boundaries between two types of ecological communities.

ecotourism A combination of adventure travel, cultural exploration, and nature appreciation in wild settings.

edge effects A change in species composition, physical conditions, or other ecological factors at the boundary between two ecosystems.

electron A negatively charged subatomic particle that orbits around the nucleus of an atom.

element A substance that cannot be broken into simpler units by chemical means.

El Niño A climatic change marked by shifting of a large warm water pool from the western Pacific Ocean toward the east. Wind direction and precipitation patterns are changed over much of the Pacific and perhaps around the world.

emergent disease A new disease or one that has been absent for at least 20 years.

emergent properties Properties that make a system more than the sum of its parts.

emigration The movement of members from a population.

emission standards Regulations for restricting the amounts of air pollutants that can be released from specific point sources.

endangered species A species considered to be in imminent danger of extinction.

endemic species A species that is restricted to a single region, country, or other area.

endocrine hormone disrupters Chemicals that interfere with the function of endocrine hormones such as estrogen, testosterone, thyroxine, adrenaline, or cortisone.

energy The capacity to do work, such as moving matter over a distance.

energy intensity The amount of energy needed to provide the goods and services consumed in an economy.

energy recovery The incineration of solid waste to produce useful energy.

entropy A measure of disorder and usefulness of energy in a system.

environment The circumstances or conditions that surround an organism or a group of organisms as well as the complex of social or cultural conditions that affect an individual or a community.

environmental health The science of external factors that cause disease, including elements of the natural, social, cultural, and technological worlds in which we live.

environmental impact statement (EIS)   An analysis of the effects of any major program or project planned by a federal agency; required by provisions in the National Environmental Policy Act of 1970.

environmental law   Legal rules, decisions, and actions concerning environmental quality, natural resources, and ecological sustainability.

environmental literacy   A basic understanding of ecological principles and the ways society affects, or responds to, environmental conditions.

environmental policy   The official rules or regulations concerning the environment adopted, implemented, and enforced by some government agency.

environmental science   The systematic, scientific study of our environment as well as our role in it.

epigenetics   Effects (both positive and negative) expressed in future generations that are not caused by nuclear mutations and are not inherited by normal Mendelian genetics.

epigenome   DNA and its associated proteins and other small molecules that regulate gene function in ways that can affect multiple generations.

epiphyte   A plant that grows on a substrate other than the soil, such as the surface of another organism.

equilibrium   A system in a stable balance.

estuaries   Bays or drowned valleys where a river empties into the sea.

eutrophic   Rivers and lakes rich in organic material (*eu* = well; *trophic* = nourished).

evolution   A theory that explains how random changes in genetic material and competition for scarce resources cause species to change gradually.

evolutionary species concept   A definition of species that depends on evolutionary relationships.

e-waste   Discarded electronic equipment, including TVs, cell phones, computers, etc.

exotic organisms   Alien species introduced by human agency into biological communities where they would not naturally occur.

explanatory variables   Independent variables that help explain differences in the dependent variable.

exponential growth   Growth at a constant rate of increase per unit of time; can be expressed as a constant fraction or exponent. See also *geometric growth*.

externalizing costs   Shifting expenses, monetary or otherwise, to someone other than the individuals or groups who use a resource.

extinction   The irrevocable elimination of species; can be a normal process of the natural world as species outcompete or kill off others or as environmental conditions change.

F

family planning   Controlling reproduction; planning the timing of birth and having only as many babies as are wanted and can be supported.

famines   Acute food shortages characterized by large-scale loss of life, social disruption, and economic chaos.

fauna   All of the animals present in a given region.

fecundity   The physical ability to reproduce.

federal laws (statutes)   Laws passed by the federal legislature and signed by the chief executive.

fens   Wetlands fed mainly by groundwater.

feral   A domestic animal that has taken up a wild existence.

fetal alcohol syndrome   A tragic set of permanent physical, mental, and behavioral birth defects that result when mothers drink alcohol during pregnancy.

first law of thermodynamics   States that energy is conserved; that is, it is neither created nor destroyed under normal conditions.

flood   An overflow of water onto land that normally is dry.

floodplains   Low lands along riverbanks, lakes, and coastlines subjected to periodic inundation.

food security   The ability of individuals to obtain sufficient food on a day-to-day basis.

food web   A complex, interlocking series of individual food chains in an ecosystem.

fossil fuels   Petroleum, natural gas, and coal created by geologic forces from organic wastes and dead bodies of formerly living biological organisms.

fragmentation   Disruption of habitat into small, isolated fragments.

fuel assembly   A bundle of hollow metal rods containing uranium oxide pellets; used to fuel a nuclear reactor.

fuel cells   Mechanical devices that use hydrogen or hydrogen-containing fuel, such as methane, to produce an electric current. Fuel cells are clean, quiet, and highly efficient sources of electricity.

fugitive emissions   Substances that enter the air without going through a smokestack, such as dust from soil erosion, strip mining, rock crushing, construction, and building demolition.

fungi   Nonphotosynthetic, eukaryotic organisms with cell walls, filamentous bodies, and absorptive nutrition.

fungicide   A chemical that kills fungi.

G

gap analysis   A biogeographical technique of mapping biological diversity and endemic species to find gaps between protected areas that leave endangered habitats vulnerable to disruption.

gene   A unit of heredity; a segment of DNA nucleus of the cell that contains information for the synthesis of a specific protein, such as an enzyme.

generalists   Species that tolerate a wide range of conditions or exploit a wide range of resources.

genetic engineering   Laboratory manipulation of genetic material using molecular biology.

genetically modified organisms (GMOs) Organisms created by combining natural or synthetic genes using the techniques of molecular biology.

genuine progress index (GPI)   An alternative to GNP or GDP for economic accounting that measures real progress in quality of life and sustainability.

geographic isolation   Geographical changes that isolate populations of a species and prevent reproduction or gene exchange for a long enough time so that genetic drift changes the populations into distinct species.

geometric growth   Growth that follows a geometric pattern of increase, such as 2, 4, 8, 16, etc. See also *exponential growth*.

geothermal energy   Energy drawn from the internal heat of the earth, either through geysers, fumaroles, hot springs, or other natural geothermal features or through deep wells that pump heated groundwater.

GIS   Geographical information systems that use computers to combine and analyze geographical data.

global environmentalism   The extension of modern environmental concerns to global issues.

grasslands   Biomes dominated by grasses and associated herbaceous plants.

Great Pacific Garbage Patch   A vast area of the Pacific Ocean containing plastic debris concentrated by global ocean circulation currents. One of several oceanic garbage gyres.

green pricing   Plans in which consumers can voluntarily pay premium prices for renewable energy.

green revolution   Dramatically increased agricultural production brought about by "miracle" strains of grain; usually requires high inputs of water, plant nutrients, and pesticides.

greenhouse effect   Trapping of heat by the earth's atmosphere, which is transparent to incoming visible light waves but absorbs outgoing longwave infrared radiation.

greenhouse gas   A gas that traps heat in the atmosphere.

gross domestic product (GDP)   The total economic activity within national boundaries.

gross national product (GNP)   The sum total of all goods and services produced in a national economy. Gross domestic product (GDP) is used to distinguish economic activity within a country from that of offshore corporations.

gully erosion   Removal of layers of soil, creating channels or ravines too large to be removed by normal tillage operations.

H

habitat   The place or set of environmental conditions in which a particular organism lives.

half-life   The time required for one-half of a sample to decay or change into some other form.

hazardous air pollutants (HAPs)   A special category of toxins that cause cancer, nerve damage, disrupt hormone function, and fetal development. These persistent substances remain in ecosystems for long periods of time, and accumulate in animal and human tissues.

hazardous waste   Any discarded material containing substances known to be toxic, mutagenic, carcinogenic, or teratogenic to humans or other life-forms; ignitable, corrosive, explosive, or highly reactive alone or with other materials.

health   A state of physical and emotional well-being; the absence of disease or ailment.

heap-leach extraction   A technique for separating gold from extremely low-grade ores. Crushed ore is piled in huge heaps and sprayed with a dilute alkaline-cyanide solution, which percolates through the pile to extract the gold.

heat   Total kinetic energy of atoms or molecules in a substance not associated with the bulk motion of the substance.

heat islands   Areas of higher temperatures around cities.

herbicide A chemical that kills plants.

herbivores Organisms that eat only plants.

heterotroph An organism that is incapable of synthesizing its own food and, therefore, must feed upon organic compounds produced by other organisms.

high-level waste repository A place where intensely radioactive wastes can be buried and remain unexposed to groundwater and earthquakes for tens of thousands of years.

high-quality energy Intense, concentrated, and high-temperature energy that is considered high-quality because of its usefulness in carrying out work.

HIPPO Habitat destruction, Invasive species, Pollution, Population (human), and Overharvesting, the leading causes of extinction.

histogram A bar graph, generally with upright bars.

holistic science The study of entire, integrated systems rather than isolated parts. Often takes a descriptive or an interpretive approach.

homeostasis A dynamic, steady state in a living system maintained through opposing, compensating adjustments.

hormesis Nonlinear effects of toxic materials.

human development index (HDI) A measure of quality of life using data for life expectancy, child survival, adult literacy, education, gender equity, access to clean water and sanitation, and income.

hydrologic cycle The natural process by which water is purified and made fresh through evaporation and precipitation. This cycle provides all the freshwater available for biological life.

hypothesis A conditional explanation that can be verified or falsified by observation or experimentation.

I

I = PAT Our environmental impacts (I) are the product of our population size (P) times affluence (A) and the technology (T) used to produce the goods and services we consume.

igneous rocks Crystalline minerals solidified from molten magma from deep in the earth's interior; basalt, rhyolite, andesite, lava, and granite are examples.

independent variable One that does not respond to other variables in a particular test.

indicator species Species that tell us something about the health or condition of a biological community.

indicators Species that have very specific environmental requirements and tolerance levels that make them good indicators of pollution or other environmental conditions.

indigenous people Natives or original inhabitants of an area, those who have lived in a particular place for a very long time.

inductive reasoning "Bottom-up" reasoning in which we study specific examples and try to discover patterns and derive general explanations from collected observations.

insecticide A chemical that kills insects.

insolation Incoming solar radiation.

integrated gasification combined cycle (IGCC) A process in which a fuel (coal or biomass) is heated in the presence of high oxygen levels to produce a variety of gases, mostly hydrogen and carbon dioxide. Impurities, including CO_2, can easily be removed and the synthetic hydrogen gas, or syngas, is burned in a turbine to produce electricity. Superheated gas from the turbine is used to generate steam that produces more electricity, raising the efficiency of the system.

integrated pest management (IPM) An ecologically based pest-control strategy that relies on natural mortality factors, such as natural enemies, weather, cultural control methods, and carefully applied doses of pesticides.

Intergovernmental Panel on Climate Change (IPCC) A large group of scientists from many nations and a wide variety of fields assembled by the United Nations Environment Program and World Meteorological Organization to assess the current state of knowledge about climate change.

internalizing costs Planning so that those who reap the benefits of resource use also bear all the external costs.

international treaties and conventions Agreements between nations on important issues.

interspecific competition In a community, competition for resources between members of different species.

intraspecific competition In a community, competition for resources among members of the same species.

invasive species Organisms that thrive in new territory where they are free of predators, diseases, or resource limitations that may have controlled their population in their native habitat.

ionosphere The lower part of the thermosphere.

ions Electrically charged atoms that have gained or lost electrons.

island biogeography The study of rates of colonization and extinction of species on islands or other isolated areas based on size, shape, and distance from other inhabited regions.

isotopes Forms of a single element that differ in atomic mass due to a different number of neutrons in the nucleus.

J

J curve A growth curve that depicts exponential growth; called a J curve because of its shape.

joule A unit of energy. One joule is the energy expended in 1 second by a current of 1 amp flowing through a resistance of 1 ohm.

K

K-selected species Organisms whose population growth is regulated by internal (or intrinsic) as well as external factors. Large animals, such as whales and elephants, as well as top predators, generally fall in this category. They have relatively few offspring and often stabilize their population size near the carrying capacity of their environment.

keystone species A species whose impacts on its community or ecosystem are much larger and more influential than would be expected from mere abundance. This could be a top predator, a plant that

shelters or feeds other organisms, or an organism that plays a critical ecological role.

kinetic energy Energy contained in moving objects, such as a rock rolling down a hill, the wind blowing through the trees, or water flowing over a dam.

Kyoto Protocol An international treaty adopted in Kyoto, Japan, in 1997, in which 160 nations agreed to roll back CO_2, methane, and nitrous oxide emissions to reduce the threat of global climate change.

L

landscape ecology The study of the reciprocal effects of spatial pattern on ecological processes.

landslides Mass wasting or mass movement of rock or soil downhill. Often triggered by seismic events or heavy rainfall.

La Niña The opposite of El Niño.

latent heat Stored energy in a form that is not sensible (detectable by ordinary senses).

LD50 A chemical dose lethal to 50 percent of a test population.

life expectancy The average age that a newborn infant can expect to attain in a particular time and place.

limiting factors Chemical or physical factors that limit the existence, growth, abundance, or distribution of an organism.

limits to growth A belief that the world has a fixed carrying capacity for humans.

Living Machine® A wastewater treatment system composed of tanks or beds or constructed wetlands in which living organisms remove contaminants, nutrients, and pathogens from water.

logarithmic scale One that uses logarithms as units in a sequence that progresses by a factor of 10 in each step.

logical thinking A rational way of thought that asks, "How can orderly, deductive reasoning help me think clearly?"

logistic growth Growth rates regulated by internal and external factors that establish an equilibrium with environmental resources. See also *S curve*.

LULUs Locally Unwanted Land Uses, such as toxic waste dumps, incinerators, smelters, airports, freeways, and other sources of environmental, economic, or social degradation.

M

magma Molten rock from deep in the earth's interior; called lava when it spews from volcanic vents.

malnourishment A nutritional imbalance caused by lack of specific dietary components or inability to absorb or utilize essential nutrients.

Malthusian growth A population explosion followed by a population crash; also called irruptive growth.

Man and Biosphere (MAB) program A design for nature preserves that divides protected areas into zones with different purposes. A highly protected core is surrounded by a buffer zone and peripheral regions in which multiple-use resource harvesting is permitted.

mangrove forests Diverse groups of salt-tolerant trees and other plants that grow in intertidal zones of tropical coastlines.

manipulative experiment   Altering a particular factor for a test or experiment while holding all others (as much as possible) constant.

mantle   A hot, pliable layer of rock that surrounds the earth's core and underlies the cool outer crust.

marasmus   A widespread human protein deficiency disease caused by a diet low in calories and protein or imbalanced in essential amino acids.

marginal costs   The cost to produce one additional unit of a good or service.

marshes   Wetlands without trees; in North America, this type of land is characterized by cattails and rushes.

mass burn   The incineration of unsorted solid waste.

matter   Anything that takes up space and has mass.

mean   Average.

megacities   See *megalopolis*.

megalopolis   Also known as a megacity or supercity; megalopolis indicates an urban area with more than 10 million inhabitants.

mesosphere   The atmospheric layer above the stratosphere and below the thermosphere; the middle layer; temperatures are usually very low.

metamorphic rocks   Igneous and sedimentary rocks modified by heat, pressure, and chemical reactions.

methane hydrate   Small bubbles or individual molecules of methane (natural gas) trapped in a crystalline matrix of frozen water.

microlending   Small loans made to poor people who otherwise don't have access to capital.

midocean ridges   Mountain ranges on the ocean floor where magma wells up through cracks and creates new crust.

Milankovitch cycles   Periodic variations in tilt, eccentricity, and wobble in the earth's orbit; Milutin Milankovitch suggested these are responsible for cyclic weather changes.

millennium assessment   A set of ambitious environmental and human development goals established by the United Nations in 2000.

mineral   A naturally occurring, inorganic, crystalline solid with definite chemical composition, a specific internal crystal structure, and characteristic physical properties.

minimills   Mills that use scrap metal as their starting material.

minimum viable population   The number of individuals needed for long-term survival of rare and endangered species.

modern environmentalism   A fusion of conservation of natural resources and preservation of nature with concerns about pollution, environmental health, and social justice.

molecules   Combinations of two or more atoms.

monoculture forestry   Intensive planting of a single species; an efficient wood production approach, but one that encourages pests and disease infestations and conflicts with wildlife habitat or recreation uses.

Montreal Protocol   An international treaty to eliminate chloroflurocarbons that destroy stratospheric ozone.

morbidity   Illness or disease.

mortality   Death rate in a population, such as number of deaths per thousand people per year.

Müllerian (or Muellerian) mimicry   Evolution of two species, both of which are unpalatable and have poisonous stingers or some other defense mechanism, to resemble each other.

municipal solid waste   The mixed refuse produced by households and businesses.

mutagens   Agents, such as chemicals or radiation, that damage or alter genetic material (DNA) in cells.

mutation   A change, either spontaneous or by external factors, in the genetic material of a cell; mutations in the gametes (sex cells) can be inherited by future generations of organisms.

mutualism   A symbiotic relationship between individuals of two different species in which both species benefit from the association.

N

National Environmental Policy Act (NEPA)   The law that established the Council on Environmental Quality and that requires environmental impact statements for all federal projects with significant environmental impacts.

natural experiment   Observation of natural events to deduce causal relationships.

natural increase   Crude death rate subtracted from crude birth rate.

natural resource economics   Economics that takes natural resources into account as valuable assets.

natural resources   Goods and services supplied by the environment.

natural selection   The mechanism for evolutionary change in which environmental pressures cause certain genetic combinations in a population to become more abundant; genetic combinations best adapted for present environmental conditions tend to become predominant.

negative feedback loop   A signal or factor that tends to decrease a process or component.

negative feedbacks   Factors that result from a process and, in turn, reduce that same process.

neoclassical economics   The branch of economics that attempts to apply the principles of modern science to economic analysis in a mathematically rigorous, noncontextual, abstract, predictive manner.

net primary productivity   The amount of biomass produced by photosynthesis and stored in a community after respiration, emigration, and other factors that reduce biomass.

neurotoxins   Toxic substances, such as lead or mercury, that specifically poison nerve cells.

neutron   A subatomic particle, found in the nucleus of the atom, that has no electromagnetic charge.

new source review   A permitting process required by 1977 amendments to the Clean Air Act, required when industries expand or modify facilities. The rule is contentious because vague language in the law allows industries to avoid oversight.

NIMBY   Not-In-My-Back-Yard: the position of those opposed to LULUs.

nitrogen cycle   The circulation and reutilization of nitrogen in both inorganic and organic phases.

nitrogen-fixing bacteria   Bacteria that convert nitrogen from the atmosphere or soil solution into ammonia that can then be converted to plant nutrients by nitrite- and nitrate-forming bacteria.

nitrogen oxides (NO_x)   Highly reactive gases formed when nitrogen in fuel or combustion air is heated to over 650°C (1,200°F) in the presence of oxygen or when bacteria in soil or water oxidize nitrogen-containing compounds.

noncriteria pollutants   See *unconventional pollutants*.

nongovernmental organizations (NGOs)   Pressure and research groups, advisory agencies, political parties, professional societies, and other groups concerned about environmental quality, resource use, and many other issues.

nonpoint sources   Scattered, diffuse sources of pollutants, such as runoff from farm fields, golf courses, and construction sites.

nonrenewable resources   Minerals, fossil fuels, and other materials present in essentially fixed amounts (within human time scales) in our environment.

normal distribution   A bell-shaped curve or a Gaussian distribution.

nuclear fission   The radioactive decay process in which isotopes split apart to create two smaller atoms.

nuclear fusion   A process in which two smaller atomic nuclei fuse into one larger nucleus and release energy; the source of power in a hydrogen bomb.

nucleic acids   Large organic molecules made of nucleotides that function in the transmission of hereditary traits, in protein synthesis, and in control of cellular activities.

nucleus   The center of the atom; occupied by protons and neutrons. In cells, the organelle that contains the chromosomes (DNA).

O

obese   Pathologically overweight, having a body mass greater than 30 kg/m², or roughly 30 pounds above normal for an average person.

oil shales   Fine-grained sedimentary rock rich in solid organic material called kerogen. When heated, the kerogen liquefies to produce a fluid petroleum fuel.

old-growth forests   Forests free from disturbance for long enough (generally 150 to 200 years) to have mature trees, physical conditions, species diversity, and other characteristics of equilibrium ecosystems.

oligotrophic   Condition of rivers and lakes that have clear water and low biological productivity (*oligo* = little; *trophic* = nourished); are usually clear, cold, infertile headwater lakes and streams.

omnivores   Organisms that eat both plants and animals.

open access system   A commonly held resource for which there are no management rules.

open canopy   A forest where tree crowns cover less than 20 percent of the ground; also called woodland.

open system   A system that exchanges energy and matter with its environment.

organic compounds   Complex molecules organized around skeletons of carbon atoms arranged in rings or chains; includes biomolecules, molecules synthesized by living organisms.

organophosphates   Organic molecules to which a phosphate group is attached. A group of highly toxic pesticides that are primarily neurotoxins.

overgrazing   Allowing domestic livestock to eat so much plant material that it degrades the biological community.

overharvesting   Harvesting so much of a resource that it threatens its existence.

overnutrition   Receiving too many calories.

oxygen sag   Oxygen decline downstream from a pollution source that introduces materials with high biological oxygen demands.

ozone   A highly reactive molecule containing three oxygen atoms; a dangerous pollutant in ambient air. In the stratosphere, however, ozone forms an ultraviolet absorbing shield that protects us from mutagenic radiation.

P

paradigms   Overarching models of the world that shape our worldviews and guide our interpretation of how things are.

parasite   An organism that lives in or on another organism, deriving nourishment at the expense of its host, usually without killing it.

parasitism   A relationship in which one organism feeds on another without immediately killing it.

particulate material   Atmospheric aerosols, such as dust, ash, soot, lint, smoke, pollen, spores, algal cells, and other suspended materials; originally applied only to solid particles but now extended to droplets of liquid.

passive solar absorption   The use of natural materials or absorptive structures without moving parts to gather and hold heat; the simplest and oldest use of solar energy.

pastoralists   People who live by herding domestic animals.

pathogens   Organisms that produce disease in host organisms, disease being an alteration of one or more metabolic functions in response to the presence of the organisms.

peat   Deposits of moist, acidic, semidecayed organic matter.

pelagic   Zones in the vertical water column of a water body.

permafrost   A permanently frozen layer of soil that underlies the arctic tundra.

permanent retrievable storage   Placing waste storage containers in a secure location where they can be inspected periodically and retrieved, if necessary, for repacking or for transfer if a better means of disposal or reuse is developed.

persistent organic pollutants (POPs)   Chemical compounds that persist in the environment and retain biological activity for a long time.

pest   Any organism that reduces the availability, quality, or value of a useful resource.

pesticide   Any chemical that kills, controls, drives away, or modifies the behavior of a pest.

pesticide treadmill   A situation in which farmers must use increasingly complex and expensive cocktails of pesticides to combat pests: similar to addiction.

pH   A value that indicates the acidity or alkalinity of a solution on a scale of 0 to 14, based on the proportion of H^+ ions present.

phosphorus cycle   The movement of phosphorus atoms from rocks through the biosphere and hydrosphere and back to rocks.

photochemical oxidants   Products of secondary atmospheric reactions. See also *smog*.

photodegradable plastics   Plastics that break down when exposed to sunlight or to a specific wavelength of light.

photosynthesis   The biochemical process by which green plants and some bacteria capture light energy and use it to produce chemical bonds. Carbon dioxide and water are consumed while oxygen and simple sugars are produced.

photovoltaic cell   An energy-conversion device that captures solar energy and directly converts it to electrical current.

phylogenetic species concept   A definition of species that depends on genetic similarities (or differences).

phytoplankton   Microscopic, free-floating, autotrophic organisms that function as producers in aquatic ecosystems.

pioneer species   In primary succession on a terrestrial site, the plants, lichens, and microbes that first colonize the site.

plankton   Primarily microscopic organisms that occupy the upper water layers in both freshwater and marine ecosystems.

point sources   Specific locations of highly concentrated pollution discharge, such as factories, power plants, sewage treatment plants, underground coal mines, and oil wells.

policy   A societal plan or statement of intentions intended to accomplish some social or economic goal.

policy cycle   The process by which problems are identified and acted upon in the public arena.

pollution   To make foul, unclean, dirty; any physical, chemical, or biological change that adversely affects the health, survival, or activities of living organisms or that alters the environment in undesirable ways.

pollution charges   Fees assessed per unit of pollution based on the "polluter pays" principle.

population   All members of a species that live in the same area at the same time.

population crash   A sudden population decline caused by predation, waste accumulation, or resource depletion; also called a dieback.

population explosion   Growth of a population at exponential rates to a size that exceeds environmental carrying capacity; usually followed by a population crash.

population momentum   A potential for increased population growth as young members reach reproductive age.

positive feedback loop   A signal or factor that tends to increase a process or component.

positive feedbacks   Factors that result from a process and, in turn, increase that same process.

potential energy   Stored energy that is latent but available for use. A rock poised at the top of a hill or water stored behind a dam are examples of potential energy.

power   The rate of energy delivery; measured in horsepower or watts.

precautionary principle   The rule that we should leave a margin of safety for unexpected developments. This principle implies that we should strive to prevent harm to human health and the environment even if risks are not fully understood.

predator   An organism that feeds directly on other organisms in order to survive; live-feeders, such as herbivores and carnivores.

predator-mediated competition   A situation in which the effects of a predator dominate population dynamics.

preservation   A philosophy that emphasizes the fundamental right of living organisms to exist and to pursue their own ends.

primary pollutants   Chemicals released directly into the air in a harmful form.

primary producers   Photosynthesizing organisms.

primary productivity   Synthesis of organic materials (biomass) by green plants using the energy captured in photosynthesis.

primary standards   Regulations of the 1970 Clean Air Act; intended to protect human health.

primary succession   Ecological succession that begins in an area where no biotic community previously existed.

primary treatment   A process that removes solids from sewage before it is discharged or treated further.

principle of competitive exclusion   A result of natural selection whereby two similar species in a community occupy different ecological niches, thereby reducing competition for food.

probability   The likelihood that a situation, a condition, or an event will occur.

producer   An organism that synthesizes food molecules from inorganic compounds by using an external energy source; most producers are photosynthetic.

productivity   The amount of biomass (biological material) produced in a given area during a given period of time.

prokaryotic   Cells that do not have a membrane-bounded nucleus or membrane-bounded organelles.

pronatalist pressures   Influences that encourage people to have children.

prospective study   A study in which experimental and control groups are identified before exposure to some factor. The groups are then monitored and compared for a specific time after the exposure to determine any effects the factor may have.

proteins   Chains of amino acids linked by peptide bonds.

proton   A positively charged subatomic particle found in the nucleus of an atom.

R

r-selected species   Organisms whose population growth is regulated mainly by external factors. They tend to have rapid reproduction and high mortality of offspring. Given optimum environmental conditions, they can grow exponentially. Many "weedy" or pioneer species fit in this category.

radioactive decay   A change in the nuclei of radioactive isotopes that spontaneously emit high-energy electromagnetic radiation and/or subatomic particles while gradually changing into another isotope or different element.

random sample   A subset of a collection of items or observations chosen at random.

rational choice   Public decision making based on reason, logic, and science-based management.

recharge zones   Areas where water infiltrates into an aquifer.

reclamation Chemical, biological, or physical cleanup and reconstruction of severely contaminated or degraded sites to return them to something like their original topography and vegetation.

recycling Reprocessing of discarded materials into new, useful products; not the same as reuse of materials for their original purpose, but the terms are often used interchangeably.

reduced tillage systems Farming methods that preserve soil and save energy and water through reduced cultivation; includes minimum till, conserve-till, and no-till systems.

reflective thinking A thoughtful, contemplative analysis that asks, "What does this all mean?"

reformer A device that strips hydrogen from fuels such as natural gas, methanol, ammonia, gasoline, or vegetable oil so they can be used in a fuel cell.

refuse-derived fuel Processing of solid waste to remove metal, glass, and other unburnable materials; organic residue is shredded, formed into pellets, and dried to make fuel for power plants.

regenerative farming Farming techniques and land stewardship that restore the health and productivity of the soil by rotating crops, planting ground cover, protecting the surface with crop residue, and reducing synthetic chemical inputs and mechanical compaction.

relative humidity At any given temperature, a comparison of the actual water content of the air with the amount of water that could be held at saturation.

remediation Cleaning up chemical contaminants from a polluted area.

renewable resources Resources normally replaced or replenished by natural processes; resources not depleted by moderate use; examples include solar energy, biological resources such as forests and fisheries, biological organisms, and some biogeochemical cycles.

renewable water supplies Annual freshwater surface runoff plus annual infiltration into underground freshwater aquifers that are accessible for human use.

replacement rate The number of children per couple needed to maintain a stable population. Because of early deaths, infertility, and nonreproducing individuals, this is usually about 2.1 children per couple.

replication Repeating studies or tests.

reproducibility Making an observation or obtaining a particular result consistently.

residence time The length of time a component, such as an individual water molecule, spends in a particular compartment or location before it moves on through a particular process or cycle.

resilience The ability of a community or ecosystem to recover from disturbances.

resource partitioning In a biological community, various populations sharing environmental resources through specialization, thereby reducing direct competition. See also *ecological niche*.

resources In economic terms, anything with potential use in creating wealth or giving satisfaction.

restoration ecology Seeks to repair or reconstruct ecosystems damaged by human actions.

retrospective study A study that looks back in history at a group of people (or other organisms) who suffer from some condition to try to identify something in their past life that the whole group shares but that is not found in the histories of a control group

as near as possible to those being studied but who do not suffer from the same condition.

riders Amendments attached to bills in conference committee, often completely unrelated to the bill to which they are added.

rill erosion The removing of thin layers of soil as little rivulets of running water gather and cut small channels in the soil.

risk The probability that something undesirable will happen as a consequence of exposure to a hazard.

risk assessment Evaluation of the short-term and long-term risks associated with a particular activity or hazard; usually compared with benefits in a cost-benefit analysis.

rock A solid, cohesive aggregate of one or more crystalline minerals.

rock cycle The process whereby rocks are broken down by chemical and physical forces; sediments are moved by wind, water, and gravity; sedimented and reformed into rock; and then crushed, folded, melted, and recrystallized into new forms.

rotational grazing Confining grazing animals in a small area for a short time to force them to eat weedy species as well as the more desirable grasses and forbes.

runoff The excess of precipitation over evaporation; the main source of surface water and, in broad terms, the water available for human use.

S

S curve A curve that depicts logistic growth; called an S curve because of its shape.

salinity The amount of dissolved salts (especially sodium chloride) in a given volume of water.

salinization A process in which mineral salts accumulate in the soil, killing plants; occurs when soils in dry climates are irrigated profusely.

salt marsh A wetland with salt water and salt tolerant plants, usually coastal.

saltwater intrusion The movement of saltwater into freshwater aquifers in coastal areas where groundwater is withdrawn faster than it is replenished.

sample To analyze a small but representative portion of a population to estimate the characteristics of the entire class.

sanitary landfills Landfills in which garbage and municipal waste are buried every day under enough soil or fill to eliminate odors, vermin, and litter.

savannas An open prairie or grassland with scattered groves of trees.

scavengers In biology, organisms that consume carrion, or organisms not killed by the scavenger.

science The orderly pursuit of knowledge, relying on observations that test hypotheses in order to answer questions.

scientific consensus A general agreement among informed scholars.

scientific method A systematic, precise, objective study of a problem. Generally this requires observation, hypothesis development and testing, data gathering, and interpretation.

scientific theory An explanation or idea accepted by a substantial number of scientists.

sea-grass beds Large expanses of rooted, submerged, or emergent aquatic vegetation, such as eel grass or salt grass.

second law of thermodynamics States that, with each successive energy transfer or transformation in a system, less energy is available to do work.

secondary pollutants Chemicals modified to a hazardous form after entering the air or that are formed by chemical reactions as components of the air mix and interact.

secondary succession Succession on a site where an existing community has been disrupted.

secondary treatment Bacterial decomposition of suspended particulates and dissolved organic compounds that remain after primary sewage treatment.

secure landfills Solid waste disposal sites lined and capped with an impermeable barrier to prevent leakage or leaching.

sedimentary rocks Rocks composed of accumulated, compacted mineral fragments, such as sand or clay; examples include shale, sandstone, breccia, and conglomerates.

sedimentation The deposition of organic materials or minerals by chemical, physical, or biological processes.

selection pressure Limited resources or adverse environmental conditions that tend to favor certain adaptations in a population. Over many generations, this can lead to genetic change, or evolution.

selective cutting Harvesting only mature trees of certain species and size; usually more expensive than clear-cutting but less disruptive for wildlife and often better for forest regeneration.

shade-grown coffee and cocoa Plants grown under a canopy of taller trees, which provides habitat for birds and other wildlife.

shantytowns Groups of shacks built of inexpensive materials on empty land.

sheet erosion Peeling off thin layers of soil from the land surface; accomplished primarily by wind and water.

shelterwood harvesting Mature trees are removed from the forest in a series of two or more cuts, leaving young trees and some mature trees as a seed source for future regeneration.

sick building syndrome A cluster of allergies and other illnesses caused by sensitivity to molds, synthetic chemicals, or other harmful compounds trapped in insufficiently ventilated buildings.

sinkholes A large surface crater caused by the collapse of an underground channel or cavern; often triggered by groundwater withdrawal.

sludge A semisolid mixture of organic and inorganic materials that settles out of wastewater at a sewage treatment plant.

slums Legal but inadequate multifamily tenements or rooming houses.

smart growth The efficient use of land resources and existing urban infrastructure that encourages in-fill development, provides a variety of affordable housing and transportation choices, and seeks to maintain a unique sense of place by respecting local cultural and natural features.

smart metering A system of meters that give information about the source and price of electricity used by individual appliances and that can time usage to take advantage of the lowest cost power.

smelting Roasting ore to release metals from mineral compounds.

smog The combination of smoke and fog in the stagnant air of London; now often applied to photochemical pollution.

social justice Equitable access to resources and the benefits derived from them; a system that recognizes inalienable rights and adheres to what is fair, honest, and moral.

soil creep The slow, downhill movement of soil due to erosion.

sound science Although definitions differ, this generally means valid science according to basic scientific principles.

Southern Oscillation The combination of El Niño and La Niña cycles.

specialists Species that require a narrow range of conditions or exploit a very specific set of resources.

speciation Evolution of new species.

species All the organisms genetically similar enough to breed and produce live, fertile offspring in nature.

species diversity The number and relative abundance of species present in a community.

specific heat The amount of heat energy needed to change the temperature of a body. Water has a specific heat of 1, which is higher than most substances.

sprawl Unlimited, unplanned growth of urban areas that consumes open space and wastes resources.

stability In ecological terms, a dynamic equilibrium among the physical and biological factors in an ecosystem or a community; relative homeostasis.

state shift An abrupt response to a disturbance that causes a persistent change in a system to a new set of conditions and relationships.

statutory law Rules passed by a state or national legislature.

steady-state economy Characterized by low birth and death rates, use of renewable energy sources, recycling of materials, and emphasis on durability, efficiency, and stability.

stratosphere The zone in the atmosphere extending from the tropopause to about 50 km (30 mi) above the earth's surface; temperatures are stable or rise slightly with altitude; has very little water vapor but is rich in ozone.

streambank erosion Erosion along the edges of a stream.

stress Physical, chemical, or emotional factors that place a strain on an animal. Plants also experience physiological stress under adverse environmental conditions.

strip-cutting Harvesting trees in strips narrow enough to minimize edge effects and to allow natural regeneration of the forest.

strip-farming Planting different kinds of crops in alternating strips along land contours; when one crop is harvested, the other crop remains to protect the soil and prevent water from running straight down a hill.

strip-mining Extracting shallow mineral deposits (especially coal) by scraping off surface layers with giant earth-moving equipment; creates a huge open pit; an alternative to underground or deep open-pit mines.

Student Environment Action Coalition (SEAC) A grassroots coalition of student and youth environmental groups, working together to protect our planet and our future.

subduction (subducted) Where the edge of one tectonic plate dives beneath the edge of another.

subsidence Settling of the ground surface caused by the collapse of porous formations that result from withdrawal of large amounts of groundwater, oil, or other underground materials.

subsoil A layer of soil beneath the topsoil that has lower organic content and higher concentrations of fine mineral particles; often contains soluble compounds and clay particles carried down by percolating water.

sulfur cycle The chemical and physical reactions by which sulfur moves into or out of storage and through the environment.

sulfur dioxide (SO_2) A colorless, corrosive gas directly damaging to both plants and animals.

Superfund A fund established by Congress to pay for containment, cleanup, or remediation of abandoned toxic waste sites. The fund is financed by fees paid by toxic waste generators and by cost recovery from cleanup projects.

surface mining Some minerals are also mined from surface pits. See also *strip-mining*.

surface soil The A horizon in a soil profile; the soil just below the litter layer.

surface tension The tendency for a surface of water molecules to hold together, producing a surface that resists breaking.

sustainability Ecological, social, and economic systems that can last over the long term.

sustainable agriculture (regenerative farming) Ecologically sound, economically viable, socially just agricultural system. Stewardship, soil conservation, and integrated pest management are essential for sustainability.

sustainable development A real increase in well-being and standard of life for the average person that can be maintained over the long term without degrading the environment or compromising the ability of future generations to meet their own needs.

sustained yield Utilization of a renewable resource at a rate that does not impair or damage its ability to be fully renewed on a long-term basis.

swamps Wetlands with trees, such as the extensive swamp forests of the southern United States.

symbiosis The intimate living together of members of two species; includes mutualism, commensalism, and, in some classifications, parasitism.

sympatric speciation A gradual change (generally through genetic drift) so that offspring are genetically distinct from their ancestors even though they live in the same place.

synergism When an injury caused by exposure to two environmental factors together is greater than the sum of exposure to each factor individually.

synergistic effects The combination of several processes or factors is greater than the sum of their individual effects.

systems Networks of interdependent components and processes.

T

taiga The northernmost edge of the boreal forest, including species-poor woodland and peat deposits; intergrading with the arctic tundra.

tailings Mining waste left after mechanical or chemical separation of minerals from crushed ore.

taking The unconstitutional confiscation of private property.

tar sands Geologic deposits composed of sand and shale particles coated with bitumen, a viscous mixture of long-chain hydrocarbons.

tectonic plates Huge blocks of the earth's crust that slide around slowly, pulling apart to open new ocean basins or crashing ponderously into each other to create new, larger landmasses.

telemetry Locating or studying organisms at a distance using radio signals or other electronic media.

temperate rainforest The cool, dense, rainy forest of the northern Pacific coast; enshrouded in fog much of the time; dominated by large conifers.

temperature A measure of the speed of motion of a typical atom or molecule in a substance.

temperature inversions Atmospheric conditions in which a layer of warm air lies on top of cooler air and blocks normal convection currents. This can trap pollutants and degrade air quality.

teratogens Chemicals or other factors that specifically cause abnormalities during embryonic growth and development.

terracing Shaping the land to create level shelves of earth to hold water and soil; requires extensive hand labor or expensive machinery, but it enables farmers to farm very steep hillsides.

tertiary treatment The removal of inorganic minerals and plant nutrients after primary and secondary treatment of sewage.

thermal pollution Artificially raising or lowering of the temperature of a water body in a way that adversely affects the biota or water quality.

thermocline In water, a distinctive temperature transition zone that separates an upper layer that is mixed by the wind (the epilimnion) and a colder deep layer that is not mixed (the hypolimnion).

thermodynamics The branch of physics that deals with transfers and conversions of energy.

thermohaline circulation A large-scale oceanic circulation system in which warm water flows from equatorial zones to higher latitudes where it cools, evaporates, and becomes saltier and more dense, which causes it to sink and flow back toward the equator in deep ocean currents.

threatened species While still abundant in parts of its territorial range, this species has declined significantly in total numbers and may be on the verge of extinction in certain regions or localities.

thresholds Conditions where sudden change can occur in a system.

throughput The flow of energy and/or matter into and out of a system.

tide pools Small pools of water left behind by falling tides.

tolerance limits See *limiting factors.*

topsoil A layer of mixed organic and mineral soil material, also called the A horizon.

total fertility rate The number of children born to an average woman in a population during her entire reproductive life.

total growth rate The net rate of population growth resulting from births, deaths, immigration, and emigration.

total maximum daily loads (TMDL) The amount of particular pollutant that a water body can receive from both point and nonpoint sources and still meet water quality standards.

Toxic Release Inventory A program created by the Superfund Amendments and Reauthorization Act of 1984 that requires manufacturing facilities and waste handling and disposal sites to report annually on releases of more than 300 toxic materials. You can find out from the EPA whether any of these sites are in your neighborhood and what toxins they release.

toxins Poisonous chemicals that react with specific cellular components to kill cells or to alter growth or development in undesirable ways; often harmful, even in dilute concentrations.

tradable permits Pollution quotas or variances that can be bought or sold.

"Tragedy of the Commons" An inexorable process of degradation of communal resources due to selfish self-interest of "free riders" who use or destroy more than their fair share of common property. See *open access system.*

transpiration The evaporation of water from plant surfaces, especially through stomates.

trophic level Step in the movement of energy through an ecosystem; an organism's feeding status in an ecosystem.

tropical rainforests Forests near the equator in which rainfall is abundant—more than 200 cm (80 in.) per year—and temperatures are warm to hot year-round.

tropical seasonal forests Semi-evergreen or partly deciduous forests tending toward open woodlands and grassy savannas dotted with scattered, drought-resistant trees.

tropopause The boundary between the troposphere and the stratosphere.

troposphere The layer of air nearest to the earth's surface; both temperature and pressure usually decrease with increasing altitude.

tsunami Far-reaching waves caused by earthquakes or undersea landslides.

tundra Treeless arctic or alpine biome characterized by cold, dark winters; a short growing season; and potential for frost any month of the year; vegetation includes low-growing perennial plants, mosses, and lichens.

U

UN Framework Convention on Climate Change Directs governments to share data on climate change, to develop national plans for controlling greenhouse gases, and to cooperate in planning for adaptation to climate change.

unconventional pollutants Toxic or hazardous substances, such as asbestos, benzene, beryllium, mercury, polychlorinated biphenyls, and vinyl chloride, not listed in the original Clean Air Act because they were not released in large quantities; also called noncriteria pollutants.

urban agglomerations Urban areas where several cities or towns have coalesced.

utilitarian conservation The philosophy that resources should be used for the greatest good for the greatest number for the longest time.

V

vertical stratification The vertical distribution of specific subcommunities within a community.

vertical zonation Vegetation zones determined by climate changes brought about by altitude changes.

volatile organic compounds Organic chemicals that evaporate readily and exist as gases in the air.

volcanoes Vents in the earth's surface through which molten lava (magma), gases, and ash escape to create mountains.

vulnerable species Naturally rare organisms or species whose numbers have been so reduced by human activities that they are susceptible to actions that could push them into threatened or endangered status.

W

warm front A long, wedge-shaped boundary caused when a warmer advancing air mass slides over neighboring cooler air parcels.

waste stream The steady flow of varied wastes, from domestic garbage and yard wastes to industrial, commercial, and construction refuse.

waterlogging Water saturation of soil that fills all air spaces and causes plant roots to die from lack of oxygen; a result of overirrigation.

water scarcity Having less than 1,000 m³ (264,000 gal) of clean fresh water available per person per year.

watershed The land surface and groundwater aquifers drained by a particular river system.

water stress Countries that consume more than 10 percent of renewable water supplies.

water table The top layer of the zone of saturation; undulates according to the surface topography and subsurface structure.

watt One joule per second.

weather The physical conditions of the atmosphere (moisture, temperature, pressure, and wind).

weathering Changes in rocks brought about by exposure to air, water, changing temperatures, and reactive chemical agents.

wetlands Ecosystems of several types in which rooted vegetation is surrounded by standing water during part of the year. See also *swamps, marshes, bogs, fens.*

withdrawal A description of the total amount of water taken from a lake, a river, or an aquifer.

work The application of force through a distance; requires energy input.

world conservation strategy A proposal for maintaining essential ecological processes, preserving genetic diversity, and ensuring that utilization of species and ecosystems is sustainable.

Z

zero population growth (ZPG) A condition in which births and immigration in a population just balance deaths and emigration.

zone of aeration Upper soil layers that hold both air and water.

zone of saturation Lower soil layers where all spaces are filled with water.

Credits

PHOTOGRAPHS

Design Elements

Active Learning box (toad icon): © iStockphoto .com/ Alasdair James; What Can You Do? (hand/ earth image): © Istockphoto.com/urbancow; Texture for front/end matter: © iStockphoto.com/ mkphoto; Appendices opener (Landscape): © Brand X Pictures/PunchStock RF; Brief Contents (Earth): © StockTrek/Getty RF.

Chapter 1

Opener: © PunchStock RF; 1.1: © image 100/ PunchStock RF; 1.3: NASA/NOAA/GSFC/ SUOMI NPP/VLLRS/Norman Kuring; 1.6a NOAA Geophysical Fluid Dynamics Laboratory; 1.6b © Norbert Schiller/The Image Works; 1.6c: © Getty RF; 1.7a: © Dimas Ardian/Getty; 1.7b: © Christopher S. Collins, Pepperdine University; 1.7c: © William P. Cunningham; 1.9: © Vic Engelbert/Photo Researchers; 1.11: © Tetra images/PunchStock RF; 1.12: © The McGraw-Hill Companies, Inc. Barry Barker, photographer; KC 1.1 © Digital Vision/PunchStock RF; KC 1.2: © Cynthia Shaw; KC 1.3: © Dimas Ardian/Getty; KC 1.4: © William P. Cunningham; KC 1.5 © Getty RF; KC 1.6 Food and Agriculture Orga- nization photo/R. Faidutti; KC 1.7: © PhotoDisc/ Getty RF; KC 1.8: © Designpics.com/PunchStock RF; 1.13: David L. Hansen, University of Minnesota Agricultural Experiment Station; 1.15: © Mary Ann Cunningham; 1.16: © The McGraw-Hill Companies, Inc. John A. Karachewski, photographer; 1.17: © Mary Ann Cunningham; 1.18a: Courtesy of the Bancroft Library at the University of California, Berkeley #Muir, John – POR 65; 1.18b: Courtesy of Grey Towers National Historic Landmark; 1.18c: © Bettmann/ Corbis; 1.18d: © AP/Wide World Photos; 1.19: © Joe Klune/iStockphoto; 1.20a AP/Wide World Photos; 1.20b: © Tom Turner/The Brower Fund/ Earth Island Institute; 1.20c: Columbia University Archives; Columbia University in the City of New York; 1.20d: © AP/Wide World Photos; 1.21: © Dr. Parvinder Sethi.

Chapter 2

Opener: © iStockphoto; 2.4: © William P. Cunningham; p. 35: © G.I. Bernard/Animals Animals; 2.11: © William P. Cunningham; 2.12: NOAA; p. 40 Fig. 2: The SeaWIFS Project, NASA/Goddard Space Flight Center and ORBIMAGE; KC 2.3: © Creatas/PunchStock RF; 2.20: © Nigel Cattlin/Visuals Unlimited.

Chapter 3

Opener: © Galen Rowell/Corbis; 3.2: © Vol. 6/ Corbis RF; 3.3: © William P. Cunningham; 3.6: © Mary Ann Cunningham; KC 3.1a: © OS50/

Getty RF; KC 3.1b: © Cynthia Shaw; KC 3.1c: © David Shaw; KC 3.1d: © Tom Cooper; KC 3.1e: © PhotoDisc/Getty RF; KC 3.1f: © Ingram Publishing/Alamy RF; KC 3.1g: © David Shaw; KC 3.4: © Mary Ann Cunningham; KC 3.5: © Alamy RF; KC 3.6: © Creatas/PunchStock RF; KC 3.7: © David Zurick; KC 3.9 © Tom Cooper; KC 3.10 © The McGraw-Hill Companies, Inc. Jill Braaten, photographer; 3.11: © William P. Cunningham; 3.12: © Corbis RF; 3.13: © D.P. Wilson/Photo Researchers; 3.14: © Creatas/ PunchStock RF; 3.16a&b: © Edward Ross; 3.15: Courtesy Tom Finkle; 3.18a: © William P. Cunningham; 3.18b: © PhotoDisc RF; 3.18c: © William P. Cunningham; 3.23a: © Digital Vision/ Getty RF; 3.19a: © Gregory Ochocki/Photo Researchers; 3.23b: © Stockbyte RF; 3.19b: © PhotoLink/PhotoDisc/Getty RF; 3.19c: © Medioimages/PunchStock RF; 3.23c: © Visual Language Illustration/Veer RF; 3.21: © Tom & Pat Leeson/Photo Researchers; 3.24a: © Corbis RF; 3.24b: © Eric and David Hosking/Corbis; 3.24c: © image100/PunchStock RF; 3.26: © Vol. 262/ Corbis RF; p. 70: Courtesy of David Tilman; 3.30 & 3.31: © William P. Cunningham.

Chapter 4

Opener: © iStockphoto; 4.2: © Vol. EP39/Photo- Disc/Getty RF; 4.4: © William P. Cunningham; 4.6: © Frans Lemmens/Getty; KC 4.4: © Nigel Hicks/Alamy RF; KC 4.5: © Corbis RF; KC 4.6: © Getty RF; KC 4.7: © Goodshoot/PunchStock RF; KC 4.8: © Cynthia Shaw; KC 4.9: © William P. Cunningham; p. 86: © Alain Le Garsmeur/ Corbis; 4.12a: © William P. Cunningham; 4.112b: © Vol. 17/PhotoDisc/Getty RF; p. 89(1): Courtesy Library of Congress Prints and Photographs Division LC-DIG-nclc-01366; p. 89(2): Courtesy USDA Photograph Archives; p. 89(3): © Corbis RF; p. 89(4): © Stockbyte/Getty RF.

Chapter 5

Opener & 5.2: © Artur Stefanski; 5.7: © Vol. 60/ PhotoDisc RF; 5.8: © Vol. 6/Corbis RF; 5.9: © William P. Cunningham; 5.10: © Mary Ann Cunningham; 5.11: © William P. Cunningham; 5.12: © Vol. 90/Corbis RF; 5.13: © William P. Cunningham; 5.14: © Mary Ann Cunningham; 5.15: Courtesy of Sea WIFS/NASA; 5.17: NOAA; 5.18a: © Vol. 89/PhotoDisc RF; 5.18b: © Mary Ann Cunningham; 5.18c: © Andrew Martinez/ Photo Researchers; 5.18d: © Pat O'Hara/Corbis; 5.20a&b: © William P. Cunningham; 5.20c: © Mary Ann Cunningham; 5.21: © The McGraw-Hill Companies, Inc. Barry Barker, photographer; KC 5.1: © Cynthia Shaw; KC 5.5: © William P. Cunningham; KC 5.6: © IT Stock/age fotostock RF; KC 5.7: © Cynthia Shaw; p. 117 (top & bottom): © Paul Ojanen; 5.27: © Mary Ann

Cunningham; 5.28: Courtesy U.S. Fish and Wildlife Service, photographer Dave Menke; 5.29: Courtesy USGS; 5.30: © Mary Ann Cunningham; 5.31: © William P. Cunningham; 5.32: © Lynn Funkhouser/Peter Arnold/Photolibrary; 5.33: © 2009 J.T. Oris.

Chapter 6

Opener: © Antonio Scorza/AFP/Getty; 6.3: © Digital Vision/Getty RF; 6.5 & 6.6: © William P. Cunningham; 6.7a–c: Courtesy United Nations Environment Programme; 6.1: © Digital Vision/ PunchStock RF; 6.8: © William P. Cunningham; p. 134: © Amazon Conservation Team; 6.9: © William P. Cunningham; 6.10: © Gary Braasch/ Stone/Getty; KC 6.1: © Digital Vision/PunchStock RF; KC 6.2: NASA; KC 6.4: © Comstock/Alamy RF; KC 6.5a: © Creatas/PunchStock RF; KC 6.5b: © PhotoDisc/Getty RF; KC 6.6: © Mary Ann Cunningham; KC 6.7: Source: Data from United Nations Food and Agriculture Organization, 2002; KC 6.8 & 6.9: © Amazon Conservation Team; p. 138: Courtesy U.S. Fish & Wildlife Service/J & K Hollingworth; 6.11: Courtesy of John McColgan, Alaska Fires Service/Bureau of Land Management; 6.12: © William P. Cunningham; 6.14: Courtesy of Tom Finkle; 6.15, 6.16, 6.18: © William P. Cunningham; 6.20: © National Parks – Yellowstone – Wildlife Photo File, American Heritage Center, University of Wyoming #20475; 6.21: © William P. Cunningham; 6.22: © Digital Vision/Getty RF; 6.23: © PictureQuest RF; p. 147 (top & bottom): © The Jane Goodall Institute/Lilian Pintea www.janegoodall.org; 6.24: © William P. Cunningham; 6.27: Courtesy of R.O. Bierregaard.

Chapter 7

Opener: © Paulo Fridman/Corbis; 7.5: © Norbert Schiller/The Image Works; 7.6a: © Scott Daniel Peterson; 7.6b: © Lester Bergman/Corbis; 7.10: © William P. Cunningham; 7.12a: Photo by Jeff Vanuga, USDA Natural Resources Conservation Service; 7.13: © Novastock/PhotoEdit; 7.14: © FAO photo/R. Faidutti; 7.15: © William P. Cunningham; 7.17: © Soil & Land Resources Division, University of Idaho; 7.18a: © Ingram Publishing/SuperStock RF; 7.18b: © imagebroker/ Alamy RF; 7.20a: Photo by Lynn Betts, courtesy of USDA Natural Resources Conservation Service; 7.20b: Photo by Jeff Vanuga, courtesy of USDA Natural Resources Conservation Center; 7.20c: Courtesy Natural Resource Conservation Service; 7.21: © Corbis RF; KC 7.1 (background): National Agricultural Imagery Program, USDA; KC 7.5: © Golden Rice Humanitarian Board www.goldenrice.org; KC 7.6: © S. Meltzer/ PhotoLink/Getty RF; KC 7.7 & KC 7.8: © Corbis RF; 7.23: © Vol. 120/Corbis RF; 7.25: © William P. Cunningham; 7.27: © Michael Rosenfeld/

Index

Natural wastewater treatment
systems, 275, 278
Nature preserves. *See* Parks and
preserves
Negative feedback, 29, 30
Neoclassical economics, 365
Net primary productivity, 39, 69
Neurotoxins, 188–89.
See also Toxins
Neutrons, 31
New Delhi, 272
New Jersey, 87
New Madrid earthquake, 296
New source reviews, 244
New urbanism, 362–64
New York City, 335, 340, 354
Newsprint costs, 341
Niche evolution, 54–55
Nicotinamide adenine dinucleotide
phosphate (NADPH), 38
Niger, 81, 87, 88
Nigeria, 81
Nitrates, 43, 46, 167, 270, 271, 273
Nitrites, 43
Nitrogen
in Chesapeake Bay, 27
in fertilizers, 167
as key nutrient, 34
percentage in atmosphere, 209
uptake in plants, 43
Nitrogen cycle, 43, 46
Nitrogen-fixing bacteria, 43, 167
Nitrogen isotopes, 31
Nitrogen molecules, 31–32
Nitrogen oxides, 233, 306
Nitrous oxide, 46, 215, 216, 241,
244
Nixon, Richard, 381
No-fishing zones, 2
"No reasonable harm"
requirement, 198
No-take preserves, 146
No-till systems, 175, 177
Noah's flood, 18
Nobel Peace Prize, 23, 213, 232,
372
Nonmetal mineral resources, 288
Nonmetallic salts, 268
Nonpoint-source emissions
air pollution from, 232
water pollution from, 264, 270,
271, 274–75
Nonrenewable resources, 366
Normal distributions, 17
Norquist, Grover, 388
North Pacific gyre, 335
Northern spotted owl
conservationist lawsuit over,
138
habitat destruction, 68, 113, 123
image of, 138
as umbrella species, 123
Norway, partnership with
Indonesia, 128
Notestein, Frank, 89
Nu River dam proposal, 379

Nuclear fission, 311
Nuclear power generation
overview, 310–13
percentage of current energy
from, 304
to reduce carbon dioxide
emissions, 226
Nuclear Regulatory Commission,
310
Nuclei, atomic, 31
Nucleic acids, 32
Nucleotides, 33
Nuees ardentes, 296–97
Nutrients
control in Chesapeake Bay
ecosystem, 27
in fresh water, 266
human needs, 154, 157
important to plant growth, 167
in oceans, 105, 107
in tropical moist forests,
100–101
Nutrition, 154–58.
See also Food supplies;
Malnutrition

O

O horizon, 163
Obama, Barack, 139, 140, 312,
313, 315, 316, 322, 328,
347, 357, 387, 390
Oberlin College, Environmental
Studies Center, 394, 395
Obesity, 157–58
Observational experiments, 17
Observations, science's
dependence on, 14, 15
Occupy Wall Street movement,
9, 394
Ocean conveyor, 210
Ocean currents, 105, 210, 212–13
Ocean ecosystems, 40, 105–8
Ocean fisheries. *See* Fisheries
Ocean sediments, 43, 46–47
Oceans
buffering effects, 220
deep-sea organisms, 35, 105–8
as ecosystems, 105–8
as energy source, 325–26
low productivity, 71
ocean currents, A-4
percentage of earth's water
in, 253
pollution of, 273–74
preserves in, 146
remote sensing, 40
solid waste disposal in, 335
Ochre starfish, 60
Off-road vehicles, 145
Office of Technology Assessment,
(U.S.), 240
Ogallala Aquifer, 260
Oil
carbon in, 43
importance to modern
agriculture, 167

Oil drilling
exploration methods, 293
threat to oceans, 307–8
in wildlife preserves, 308
Oil palms, 316
Oil pollution, 273
Oil resource distribution, 307–9
Oil shales, 308, 309
Oil-soluble toxins, 192
Old-growth forests, 113, 124
disappearance, 113
forest management plan for,
134–35
largest remaining, 130
major features, 129–30
Oligotrophic waters, 265
Omnivores, 40, 60
On the Origin of Species
(Darwin), 53, 57
One-child-per-family policy
(China), 86
100–year floods, 297
O'Neill, Paul, 371
Open access systems, 367–68
Open dumps, 334–35
Open-pit mines, 293
Open systems, 29
Organ damage from toxins, 195
Organic agriculture, 169, 177
Organic chemicals, 268
Organic compounds, 32–33, 35,
235, 268
Organic farming, 177
Organic food marketing, 373
Organic waste, 337
Organismal theory of community,
73
Organisms, water pollution from,
264–66
Organization for Economic
Cooperation and
Development, 231
Organochlorines, 170
Organophosphates, 170, 189
Orphan wastes, 346
Orr, David, 283
Ostrom, Elinor, 368
Otters, 63
Our Common Future, 397
Overconsumption, 9, 392
Overeating, 157–58
Overfertilizing, 167
Overgrazing
degradation from, 102, 141–42
population crashes and, 64
Overharvesting, 120–21, 161, 366
Overweight, 157–58
Ovshinky, Stanford, 324
Owls, 138
See also Northern spotted owl
Oxidation, 32, 287
Oxygen
isotopes, 31
molecular, 31, 32
origins in atmosphere, 207
from photosynthesis, 38

removal from water, 265
Oxygen sag, 265, 266
Oysters, 186
Ozone
as air pollutant, 234–35
depletion from stratosphere,
239–40
UV absorption, 208, 239–40
Ozone hole, 239, 389

P

Pacala, Stephen, 224
Packaging, excessive, 344
Palau, 144
Panama, 10
Pandemics, 183, 184
Pantanal region, 143
Panther, Florida, 123
Paper parks, 143
Paper pulp, 130
Paper recycling, 341
Papua New Guinea, 130, 133
Parabolic mirrors, 323
Paracelsus, 195
Paradigm shifts, 18
Paramillo National Park, 144
Parasites, 39, 60
Parasitic diseases, 183, 184
Parasitism, 62
Parks and preserves
air pollution effects, 145, 243
current trends in, 7
enforcement challenges, 143,
144, 145–46
improving, 145–49
as island habitats, 116
national, 22, 142, 144–45
Parsimony, 14
Particulate material, 235, 243, 245
Partition, 57
Passenger pigeons, 120
Passive solar absorption, 322
Pastoralists, 141
Patchiness, 68
Pathogens. *See also* Diseases
defined, 183
as predators, 60
water pollution from, 264–65
Patterns of species, 68–69
PCBs, 238
PCBx, 349
Peacocks, 116
Peat, 109
Pegmatite, 286
Pelagic zones, 106
Pelamis wave converter, 326
Penguins, 216
Penicillin, 186
People for Community Recovery
(PCR), 357
Per capita energy use, 304–5
Per capita water supplies, 256
Perchlorate, 194
Peregrine falcons, 123
Perennial grasses, 316–17
Perfluorooctane sulfonate, 194